T0280357

Introduction to
Real Analysis

Textbooks in Mathematics

Series editors:

Al Boggess, Kenneth H. Rosen

Ordinary Differential Equations

An Introduction to the Fundamentals, Second Edition

Kenneth B. Howell

Differential Geometry of Manifolds, Second Edition

Stephen Lovett

The Shape of Space, Third Edition

Jeffrey R. Weeks

Differential Equations

A Modern Approach with Wavelets

Steven Krantz

Advanced Calculus

Theory and Practice, Second Edition

John Srdjan Petrovic

Advanced Problem Solving Using Maple

Applied Mathematics, Operations Research, Business Analytics, and Decision Analysis

William P Fox, William Bauldry

Nonlinear Optimization

Models and Applications

William P. Fox

Linear Algebra

James R. Kirkwood, Bessie H. Kirkwood

Train Your Brain

Challenging Yet Elementary Mathematics

Bogumil Kaminski, Pawel Pralat

Real Analysis

With Proof Strategies

Daniel W. Cunningham

Introduction to Real Analysis, 3rd Edition

Manfred Stoll

https://www.routledge.com/Textbooks-in-Mathematics/book-series/
CANDHTEXBOOMTH

Introduction to Real Analysis

Third edition

Manfred Stoll

CRC Press
Taylor & Francis Group
Boca Raton London New York

CRC Press is an imprint of the
Taylor & Francis Group, an **informa** business

A CHAPMAN & HALL BOOK

third edition published 2021
by CRC Press
6000 Broken Sound Parkway NW, Suite 300, Boca Raton, FL 33487-2742

and by CRC Press
2 Park Square, Milton Park, Abingdon, Oxon, OX14 4RN

Library of Congress Cataloging-in-Publication Data

Library of Congress Control Number: 2020943349

ISBN: 978-0-367-48688-4 hbk)
ISBN: 978-1-003-13735-1 (ebk)

Typeset in CMR10
by KnowledgeWorks Global Ltd.

Contents

Preface to the Third Edition

The major changes to the third edition involve the exposition of the topological concepts required in the study of analysis. In the first two editions metric spaces are scattered throughout the text. For example, the Notes section of Chapter 3 contained the definitions of a metric space, ϵ-neighborhoods, as well as open and compact sets. The Miscellaneous Exercises of this chapter also contained several relevant exercises. The concept of a metric and norm, as well as ϵ-neighborhoods are also included in the Text in Section 7.4.

In this third edition, I have decided to include the relevant topological concepts in Chapter 2, beginning with a discussion of metric spaces in Section 2.1. The usual concepts of open and closed sets, as well as limit points of a set are included in Section 2.2. While the emphasis is on the general, the examples emphasize these topological concepts on the real line. Section 2.3 contains a brief introduction to compact sets and their properties that require only the definition of compactness. Finally, the characterization of compact subsets of $\mathbb{R}$ is included in Section 2.4. Chapter 3 now contains all the topics previously included in Chapter 2, with the exception that the convergence of a sequence is defined in metric spaces. Likewise, limits of functions and continuity in Chapter 4 are also defined in metric spaces. Norms and normed linear spaces are still introduced in Chapter 7, with examples of the usual normed linear spaces in the remaining chapters.

Even with these changes, the emphasis is still on sequences of real numbers, the compact subsets of $\mathbb{R}$, as well as real-valued functions. The major theorems remain unchanged. The main advantages of the revision is that it unifies the subject matter, provides students with an introduction to metric spaces and abstract topological concepts, and provides a better preparation for advanced studies in analysis.

The third edition also includes additional exercises and expanded hints and solutions. I have also attempted to correct some of the errors that were present in the earlier editions. The Supplemental Readings sections have also been updated. The one item that has been deleted is the appendix on Logic and Proofs previously included in the second edition. In the author's opinion these are topics that are best covered in greater detail in a seperate course.

Manfred Stoll

Preface to the First Edition

The subject of real analysis is one of the fundamental areas of mathematics, and is the foundation for the study of many advanced topics, not only in mathematics, but also in engineering and the physical sciences. A thorough understanding of the concepts of real analysis has also become increasingly important for the study of advanced topics in economics and the social sciences. Topics such as Fourier series, measure theory and integration, are fundamental in mathematics and physics as well as engineering, economics, and many other areas.

Due to the increased importance of real analysis in many diverse subject areas, the typical first semester course on this subject has a varied student enrollment in terms of both ability and motivation. From my own experience, the audience typically includes mathematics majors, for whom this course represents the only rigorous treatment of analysis in their collegiate career, and students who plan to pursue graduate study in mathematics. In addition, there are mathematics education majors who need a strong background in analysis in preparation for teaching high school calculus. Occasionally, the enrollment includes graduate students in economics, engineering, physics, and other areas, who need a thorough treatment of analysis in preparation for additional graduate study either in mathematics or their own subject area. In an ideal situation it would be desirable to offer separate courses for each of these categories of students. Unfortunately, staffing and enrollment usually make such choices impossible.

In the preparation of the text there were several goals I had in mind. The first was to write a text suitable for a one-year sequence in real analysis at the junior or senior level, providing a rigorous and comprehensive treatment of the theoretical concepts of analysis. The topics chosen for inclusion are based on my experience in teaching graduate courses in mathematics, and reflect what I feel are minimal requirements for successful graduate study. I get to the least upper bound property as quickly as possible, and emphasize this important property in the text. For this reason, the algebraic properties of the rational and real number systems are treated very informally, and the construction of the real number system from the rational numbers is included only as a miscellaneous exercise. I have attempted to keep the proofs as concise as possible, and to let the subject matter progress in a natural manner. Topics or sections that are not specifically required in subsequent chapters are indicated by a footnote.

My second goal was to make the text understandable to the typical student enrolled in the course, taking into consideration the variations in abilities, background, and motivation. For this reason, chapters one through six have been written with the intent to be accessible to the average student, while at the same time challenging the more talented student through the exercises. The basic topological concepts of open, closed, and compact sets, as well as limits of sequences and functions are introduced for the real line only. However, the proofs of many of the theorems, especially those involving topological concepts, are presented in a manner that permit easy extensions to more abstract settings. These chapters also include a large number of examples and more routine and computational exercises. Chapters seven through ten assume that the students have achieved some level of expertise in the subject. In these chapters, function spaces are introduced and studied in greater detail. The theorems, examples, and exercises require greater sophistication and mathematical maturity for full understanding. From my own experiences, these are not unrealistic expectations.

The book contains most of the standard topics one would expect to find in an introductory text on real analysis—limits of sequences, limits of functions, continuity, differentiation, integration, series, sequences and series of functions, and power series. These topics are basic to the study of real analysis and are included in most texts at this level. In addition I have also included a number of topics that are not always included in comparable texts. For instance, Chapter 6 contains a section on the Riemann-Stieltjes integral, and a section on numerical methods. Chapter 7 also includes a section on square summable sequences and a brief introduction to normed linear spaces. Both of these concepts appear again in later chapters of the text.

In Chapter 8, to prove the Weierstrass approximation theorem, I use the method of approximate identities. This exposes the student to a very important technique in analysis that is used again in the chapter on Fourier series. The study of Fourier series, and the representation of functions in terms of series of orthogonal functions, has become increasingly important in many diverse areas. The inclusion of Fourier series in the text allows the student to gain some exposure to this important subject, without the necessity of taking a full semester course on partial differential equations. In the final chapter I have also included a detailed treatment of Lebesgue measure and the Lebesgue integral. The approach to measure theory follows the original method of Lebesgue, using inner and outer measure. This provides an intuitive and leisurely approach to this very important topic.

The exercises at the end of each section are intended to reinforce the concepts of the section and to help the students gain experience in developing their own proofs. Although the text contains some routine and computational problems, many of the exercises are designed to make the students think about the basic concepts of analysis and to challenge their creativity and logical thinking. Solutions and hints to selected exercises are included at the end of the text. These problems are marked by an asterisk (*).

At the end of each chapter I have also included a section of notes on the chapter, miscellaneous exercises, and a supplemental reading list. The notes in many cases provide historical comments on the development of the subject, or discuss topics not included in the chapter. The miscellaneous exercises are intended to extend the subject matter of the text or to cover topics that although important, are not covered in the chapter itself. The supplemental reading list provides references to topics that relate to the subject under discussion. Some of the references provide historical information; others provide alternate solutions of results or interesting related problems. Most of the articles appear in the American Mathematical Monthly or Mathematics Magazine, and should easily be accessible for students' reference.

To cover all the chapters in a one-year sequence is perhaps overly ambitious. However, from my own experience in teaching the course, with a judicious choice of topics it is possible to cover most of the text in two semesters. A one-semester course should at a minimum include all or most of the first five chapters, and part or all of Chapter 6 or Chapter 7. The latter chapter can be taught independently of Chapter 6; the only dependence on Chapter 6 is the integral test, and this can be covered without a theoretical treatment of Riemann integration. The remaining topics should be more than sufficient for a full second semester. The only formal prerequisites for reading the text is a standard three- or four-semester sequence in calculus. Even though an occasional talented student has completed one semester of this course during their sophomore year, some mathematical maturity is expected and the average student might be advised to take the course during their junior or senior year.

Manfred Stoll

To the Student

The difference between a course on calculus and a course on real analysis is analogous to the difference in the approach to the subject prior to the nineteenth century and since that time. Most of the topics in calculus were developed in the late seventeenth and eighteenth centuries by such prominent mathematicians as Newton, Leibniz, Bernoulli, Euler, and many others. Newton and Leibniz developed the differential and integral calculus; their successors extended and applied the theory to many problems in mathematics and the physical sciences. They had phenomenal insight into the problems, and were extremely proficient and ingenious in deriving complex formulas. What they lacked, however, were the tools to place the subject on a rigorous mathematical foundation. This did not occur until the nineteenth century with the contributions of Cauchy, Bolzano, Weierstrass, Cantor, and many others.

In calculus the emphasis is primarily on developing expertise in computational techniques and applications. In real analysis, you will be expected to understand the concepts and to develop the ability to prove results using the definitions and previous theorems. Understanding the concept of a limit, and proving results about limits, will be significantly more important than computing limits. To accomplish this, it is essential that all definitions and statements of theorems be learned precisely. Most of the proofs of the theorems and solutions of the problems are logical consequences of the definitions and previous results; some however do require ingenuity and creativity.

The text contains numerous examples and counter-examples to illustrate the particular topics under discussion. These are included to show why certain hypotheses are required, and to help develop a more thorough understanding of the subject. It is crucial that you not only learn what is true, but that you also have sufficient counter-examples at your disposal. I have included hints and answers to selected exercises at the end of the text; these are indicated by an asterisk (*). For some of the problems I have provided complete details; for others I have provided brief hints, leaving the details to you. As always, you are encouraged to first attempt the exercises and, to look at the hints or solutions only after repeated attempts have been unsuccessful.

At the end of each chapter I have included a supplemental reading list. The journal articles or books are all related to the topics in the chapter. Some provide historical information or extensions of the topics to more general settings; others provide alternate solutions of results in the text, or solutions of interesting related problems. All of the articles should be accessible in your library. They are included to encourage you to develop the habit of looking into the mathematical literature.

On reading the text you will inevitably encounter topics, formulas, or examples that may appear too technical and difficult to comprehend. Skip them for the moment; there will be plenty for you to understand in what follows. Upon later reading the section, you may be surprised that it is not nearly as difficult as previously imagined. Concepts that initially appear difficult become clearer once you develop a greater understanding of the subject. It is important to keep in mind that many of the examples and topics that appear difficult to you were most likely just as difficult to the mathematicians of the era in which they first appeared.

The material in the text is self-contained and independent of the calculus. I do not use any results from calculus in the definitions and development of the subject matter. Occasionally, however, in the examples and exercises, I do assume knowledge of the elementary functions and of notation and concepts that should have been encountered elsewhere. These concepts will be defined carefully at the appropriate place in the text.

Manfred Stoll

Acknowledgments

I would like to thank the students at the University of South Carolina who have learned this material from me, or my colleagues, from preliminary versions of this text. Your criticisms, comments, and suggestions were appreciated. I am also indebted to those colleagues, especially the late Jeong Yang, who agreed to use the manuscript in their courses.

Special thanks are also due to the many reviewers who examined the manuscripts for the first and second editions and provided constructive criticisms and suggestions for improvements. I would also like to thank the many readers who over the years have informed me of errors in the text. Hopefully all of the errors of the first and second edition have been corrected.

Finally, I would like to thank the staff at Addison-Wesley for their assistance in the publication of the first two editions. I would also like to thank Bob Ross of CRC Press for encouraging me to prepare a third edition, and his staff for their assistance in the preparation of the third edition.

Manfred Stoll

1

The Real Numbers

The key to understanding many of the fundamental concepts of calculus, such as limits, continuity, and the integral, is the least upper bound property of the real number system $\mathbb{R}$. As we all know, the rational number system contains gaps. For example, there does not exist a rational number r such that $r^2 = 2$, i.e., $\sqrt{2}$ is irrational. The fact that the rational numbers do contain gaps makes them inadequate for any meaningful discussion of the above concepts.

The standard argument used in proving that the equation $r^2 = 2$ does not have a solution in the rational numbers goes as follows: Suppose that there exists a rational number r such that $r^2 = 2$. Write $r = \frac{m}{n}$ where m, n are integers which are not both even. Thus $m^2 = 2n^2$. Therefore m^2 is even, and hence m itself must be even. But then m^2, and hence also $2n^2$ are both divisible by 4. Therefore n^2 is even, and as a consequence n is also even. This however contradicts our assumption that not both m and n are even. The method of proof used in this example is proof by contradiction; namely, we assume the negation of the conclusion and arrive at a logical contradiction.

The above argument shows that there does not exist a rational number r such that $r^2 = 2$. This argument was known to Pythagoras (around 500 B.C.), and even the Greek mathematicians of this era noted that the straight line contains many more points than the rational numbers. It was not until the nineteenth century, however, when mathematicians became concerned with putting calculus on a firm mathematical footing, that the development of the real number system was accomplished. The construction of the real number system is attributed to Richard Dedekind (1831–1916) and Georg Cantor (1845–1917), both of whom published their results independently in 1872. Dedekind's aim was the construction of a number system, with the same completeness as the real line, using only the basic postulates of the integers and the principles of set theory. Instead of constructing the real numbers, we will assume their existence and examine the least upper bound property. As we will see, this property is the key to many basic facts about the real numbers which are usually taken for granted in the study of calculus.

In Chapter 1 we will assume a basic understanding of the concept of a set and also of both the rational and real number systems. In Section 4 we will briefly review the algebraic and order properties of both the rational and real number systems and discuss the least upper bound property. By example we will show that this property fails for the rational numbers. In the subsequent two sections we will prove several elementary consequences of the least upper bound property. In Section 7 we define the notion of a countable set, and consider some of the basic properties of countable sets. Among the key results of this section are that the rational numbers are countable, whereas the real numbers are not.

1.1 Sets and Operations on Sets

Sets are constantly encountered in mathematics. One speaks of sets of points, collections of real numbers, and families of functions. A **set** is conceived simply as a collection of definable objects. The words **set**, **collection**, and **family** are all synonymous. The notation $x \in A$ means that x is an **element of** the set A; the notation $x \notin A$ means that x is not an element of the set A. The set containing no elements is called the **empty set** and will be denoted by $\emptyset$.

A set can be described by listing its elements, usually within braces $\{\ \}$. For example,

$$A = \{-1, 2, 5, 4\}$$

describes the set consisting of the numbers -1, 2, 4, and 5. More generally, a set A may be defined as the collection of all elements x in some larger collection satisfying a given property. Thus the notation

$$A = \{x : P(x)\}$$

defines A to be the set of all objects x having the property $P(x)$. This is usually read as "A equals the set of all elements x such that $P(x)$." For example, if x ranges over all real numbers, the set A defined by

$$A = \{x : 1 < x < 5\}$$

is the set of all real numbers which lie between 1 and 5. For this example, $3.75 \in A$ whereas $5 \notin A$. We will also use the notation $A = \{x \in X : P(x)\}$ to indicate that only those x which are elements of X are being considered.

Some basic sets that we will encounter throughout the text are the following:

$\mathbb{N}$ = the set of **natural numbers** or **positive integers** = $\{1, 2, 3, ...\}$

$\mathbb{Z}$ = the set of all **integers** = $\{..., -2, -1, 0, 1, 2, ...\}$,

$\mathbb{Q}$ = the set of **rational numbers** = $\{\,p/q \,:\, p, q \in \mathbb{Z},\ q \neq 0\}, and$

$\mathbb{R}$ = the set of **real numbers**.

In addition we will occasionally also encounter the set $\{0, 1, 2, 3, ...\}$ of **nonnegative integers**.

Real numbers which are not rational numbers are called **irrational** numbers. Since many fractions can represent the same rational number, two rational numbers $r_1 = p_1/q_1$ and $r_2 = p_2/q_2$ are equal if and only if $q_1 p_2 = p_1 q_2$.

Set-theoretically the rational numbers can be defined as sets of ordered pairs of integers (m, n), $n \neq 0$, where two ordered pairs (p_1, q_1) and (p_2, q_2) are said to be equivalent (represent the same rational number) if $p_1 q_2 = p_2 q_1$. We assume that the reader is familiar with the algebraic operations of addition and multiplication of rational numbers.

A set A is a **subset** of a set B or is **contained in** B, denoted $A \subset B$, if every element of A is an element of B. The set A is a **proper subset** of B, denoted $A \subsetneq B$ if A is a subset of B but there is an element of B which is not an element of A. Two sets A and B are **equal**, denoted $A = B$, if $A \subset B$ and $B \subset A$. By definition, the empty set $\emptyset$ is a subset of every set.

Set Operations

There are a number of elementary operations which may be performed on sets. If A and B are sets, the **union** of A and B, denoted $A \cup B$, is the set of all elements that belong either to A or to B or to both A and B. Symbolically,

$$A \cup B = \{\, x \,:\, x \in A \text{ or } x \in B \,\}.$$

The **intersection** of A and B, denoted $A \cap B$, is the set of elements that belong to both A and B; that is

$$A \cap B = \{x : x \in A \text{ and } x \in B\}.$$

Two sets A and B are **disjoint** if $A \cap B = \emptyset$. The **relative complement** $B \setminus A$, is the set of all elements which are in B but not in A. In set notation,

$$B \setminus A = \{x : x \in B \text{ and } x \notin A\}.$$

If the set A is a subset of some fixed set X, then $X \setminus A$ is usually referred to as the **complement of** A and is denoted by A^c. These basic set operations are illustrated in Figure 1.1 with the shaded areas representing $A \cup B$, $A \cap B$, and $B \setminus A$, respectively.

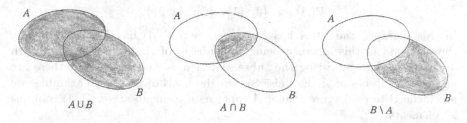

FIGURE 1.1
$A \cup B$, $A \cap B$, $B \setminus A$

There are several elementary set theoretic identities which will be encountered on numerous instances throughout the text. We state some of these in the following theorem; others are given in the exercises.

THEOREM 1.1.1 *If* A, B, *and* C *are sets, then*

    (a) $A \cap (B \cup C) = (A \cap B) \cup (A \cap C)$,

    (b) $A \cup (B \cap C) = (A \cup B) \cap (A \cup C)$,

    (c) $C \setminus (A \cup B) = (C \setminus A) \cap (C \setminus B)$,

    (d) $C \setminus (A \cap B) = (C \setminus A) \cup (C \setminus B)$.

The identities (a) and (b) are referred to as the **distributive laws**, whereas (c) and (d) are **De Morgan's laws**. If A and B are subsets of a set X, then De Morgan's laws can also be expressed as

$$(A \cup B)^c = A^c \cap B^c, \qquad (A \cap B)^c = A^c \cup B^c.$$

A more general version of both the distributive laws and De Morgan's laws will be stated in Theorems 1.7.12 and 1.7.13.

Proof. We will provide the proof of (a) to illustrate the method used in proving these results. The proofs of (b) – (d) are relegated to the exercises (Exercise 7). Suppose $x \in A \cap (B \cup C)$. Then $x \in A$ and $x \in B \cup C$. Since $x \in B \cup C$, $x \in B$ or $x \in C$. If $x \in B$, then $x \in A \cap B$ and therefore $x \in (A \cap B) \cup (A \cap C)$. Similarly, if $x \in C$ then $x \in A \cap C$ and thus again $x \in (A \cap B) \cup (A \cap C)$. This then proves that

$$A \cap (B \cup C) \subset (A \cap B) \cup (A \cap C).$$

To complete the proof, it still has to be shown that $(A \cap B) \cup (A \cap C) \subset A \cap (B \cup C)$, thereby proving equality. If $x \in (A \cap B) \cup (A \cap C)$, then by definition $x \in A \cap B$ or $x \in A \cap C$, or both. But if $x \in A \cap B$ then $x \in A$ and $x \in B$. Since $x \in B$ we also have $x \in B \cup C$. Therefore $x \in A \cap (B \cup C)$. Similarly, if $x \in A \cap C$ then $x \in A \cap (B \cup C)$. $\square$

If A is any set, the set of all subsets of A is denoted by $\mathcal{P}(A)$. The set $\mathcal{P}(A)$ is sometimes referred to as the **power set** of A. For example, if $A = \{1, 2\}$, then

$$\mathcal{P}(A) \;=\; \{\emptyset, \; \{1\}, \; \{2\}, \; \{1, 2\}\}.$$

In this example, the set A has 2 elements and $\mathcal{P}(A)$ has 4 or 2^2 elements, the elements in this instance being the subsets of A. If we take a set with 3 elements, then by listing the subsets of A it is easily seen that there are exactly 2^3 subsets of A (Exercise 8). On the basis of these two examples we are inclined to conjecture that if A contains n elements, then $\mathcal{P}(A)$ contains 2^n elements.

We now prove that this is indeed the case. We form subsets B of A by deciding for each element of A whether to include it in B, or to leave it out. Thus for each element of A there are exactly two possible choices. Since A has n elements, there are exactly 2^n possible decisions, each decision corresponding to a subset of A.

Finally, if A and B are two sets, the **Cartesian product** of A and B, denoted $A \times B$, is defined as the set of all **ordered pairs**[1] (a, b), where the first component a is from A and the second component b is from B, i.e.,

$$A \times B = \{(a, b) \ : \ a \in A, \ b \in B\}.$$

For example, if $A = \{1, 2\}$ and $B = \{-1, 2, 4\}$ then

$$A \times B = \{(1, -1), (1, 2), (1, 4), (2, -1), (2, 2), (2, 4)\}.$$

The Cartesian product of $\mathbb{R}$ with $\mathbb{R}$ is usually denoted by $\mathbb{R}^2$ and is referred to as the euclidean plane. If A and B are subsets of $\mathbb{R}$, then $A \times B$ is a subset of $\mathbb{R}^2$. The case where A and B are intervals is illustrated in Figure 1.2.

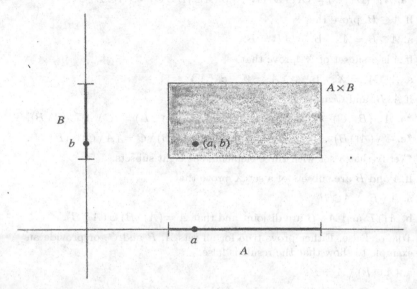

FIGURE 1.2
The Cartesian product $A \times B$

Exercises 1.1

1. Let $A = \{-1, 0, 1, 2\}$, $B = \{-2, 3\}$, and $C = \{-2, 0, 1, 5\}$.

    a. Find each of the following: $(A \cup B)$, $(B \cup C)$, $(A \cap B)$, $(B \cap C)$, $A \cap (B \cup C)$, $A \setminus B$, $C \setminus B$, $A \setminus (B \cup C)$.

    b. Find each of the following: $(A \times B)$, $(C \times B)$, $(A \times B) \cap (C \times D)$, $(A \cap C) \times B$.

    c. On the basis of your answer in (b), what might you conjecture about $(A \cap C) \times B$ for arbitrary sets A, B, C?

---

[1] A set theoretic definition of ordered pair can be given as follows: $(a, b) = \{\{a\}, \{a, b\}\}$. With this definition two ordered pairs (a, b) and (c, d) are **equal** if and only if $a = c$ and $b = d$ (Miscellaneous Exercise 1).

2. Let $A = \{x \in \mathbb{R} : -1 \leq x \leq 5\}$, $B = \{x \in \mathbb{R} : 0 \leq x \leq 3\}$,
    $C = \{x \in \mathbb{R} : 2 \leq x \leq 4\}$.

    *a. Find each of the following: $A \cap B$, $A \cap \mathbb{Z}$, $B \cap C$, $A \cup B$, $B \cup C$.

    *b. Find each of the following: $A \times B$, $A \times C$, $(A \times B) \cup (A \times C)$.

    c. Sketch the sets $(A \times B) \cup (A \times C)$ and $A \times (B \cup C)$.

3. If A, B, and C are sets, prove that

    a. $A \cap \emptyset = \emptyset$, $\quad A \cup \emptyset = A$, $\quad$ **b.** $A \cap A = A$, $\quad A \cup A = A$,

    c. $A \cap B = B \cap A$, $\quad A \cup B = B \cup A$.

4. Prove the following associate laws for the set operations $\cup$ and $\cap$:

    *a. $A \cap (B \cap C) = (A \cap B) \cap C$, $\quad$ **b.** $A \cup (B \cup C) = (A \cup B) \cup C$.

5. If $A \subset B$, prove that

    a. $A \cap B = A$, $\quad$ **b.** $A \cup B = B$.

6. If A is a subset of X, prove that

    a. $A \cup A^c = X$, $\quad$ **b.** $A \cap A^c = \emptyset$, $\quad$ **c.** $(A^c)^c = A$.

7. If A, B, and C are sets, prove that

    *a. $A \cup (B \cap C) = (A \cup B) \cap (A \cup C)$, $\quad$ **b.** $C \setminus (A \cup B) = (C \setminus A) \cap (C \setminus B)$,

    *c. $C \setminus (A \cap B) = (C \setminus A) \cup (C \setminus B)$, $\quad$ **d.** $(A \cup B) \setminus C = (A \setminus C) \cup (B \setminus C)$.

8. ***Verify that a set with three elements has eight subsets.

9. If A and B are subsets of a set X prove that

    a. $A \setminus B = A \cap B^c$,

    b. $A \cap B$ and $A \setminus B$ are disjoint and that $A = (A \cap B) \cup (A \setminus B)$.

10. True or False. Either prove true for all sets A, B and C, or provide an example to show that the result is false.

    a. $(A \cup B) \setminus A = B$.

    b. $(A \cup B) \setminus (A \cap B) = (A \setminus B) \cup (B \setminus A)$.

    c. $(A \cap B) \cup (B \cap C) \cup (A \cap C) = A \cap B \cap C$.

    d. $(A \cap B) \setminus C = A \cap (B \setminus C)$.

11. ***Prove that $A \times (B_1 \cup B_2) = (A \times B_1) \cup (A \times B_2)$.

12. Suppose A, C are subsets of X and B, D are subsets of Y. Prove that
    $(A \times B) \cap (C \times D) = (A \cap C) \times (B \cap D)$.

1.2   Functions

We begin this section with the fundamental concept of a function. In many texts a **function** or a mapping f from a set A to a set B is described as *a rule*

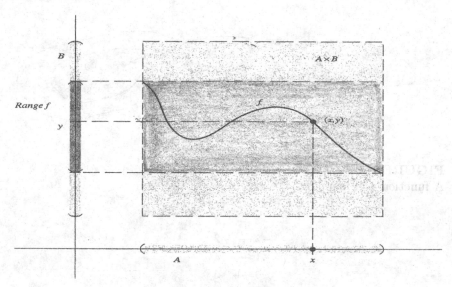

FIGURE 1.3
A function as a graph

that assigns to each element $x \in A$ a unique element $y \in B$. This is generally expressed by writing $y = f(x)$ to denote the value of the function f at x. The difficulty with this "definition" is that the terms "rule" and "assigns" are vague and difficult to define. Consequently we will define "function" strictly in terms of sets, using the notation and concepts introduced in the preceding section.

The motivation for the following definition is to think of the **graph** of a function; namely the set of ordered pairs (x, y) where y is given by the "rule" that defines the function.

DEFINITION 1.2.1 *Let A and B be any two sets. A **function** f from A into B is a subset of $A \times B$ with the property that each $x \in A$ is the first component of precisely one ordered pair $(x, y) \in f$; that is, for every $x \in A$ there exists $y \in B$ such that $(x, y) \in f$, and if (x, y) and (x, y') are elements of f, then $y = y'$. The set A is called the **domain** of f, denoted Dom f. The **range** of f, denoted Range f, is defined by*

$$\text{Range} \, f = \{y \in B : (x, y) \in f \text{ for some } x \in A\}.$$

If Range $f = B$, *then the function f is said to be **onto** B.* (See Figure 1.3)

If f is a function from A to B and $(x, y) \in f$, then the element y is called the **value of the function** f at x and we write

$$y = f(x) \quad \text{or} \quad f : x \to y.$$

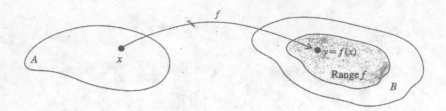

FIGURE 1.4
A function as a mapping

$$-3 \longrightarrow 2$$
$$-2 \longrightarrow -2$$
$$-1 \longrightarrow 4$$
$$0 \longrightarrow -6$$
$$1 \longrightarrow 4$$

FIGURE 1.5
The function of Example 1.2.2(a)

We also use the notation $f : A \to B$ to indicate that f is a function from (or *on*) A into (or *to*) B, and sometimes say that f **maps** A to B or is a **mapping** of A to B. If we think of a function $f : A \to B$ as mapping an element $x \in A$ to an element $y = f(x)$ in B, then this is often represented by a diagram as in Figure 1.4. If $f : A \to \mathbb{R}$, then f is said to be a **real-valued function** on A.

A function f from A to B is not just any subset of $A \times B$. The key phrase in Definition 1.2.1 is that each $x \in A$ is the first component of precisely one ordered pair $(x, y) \in f$. To better understand the notion of function we consider several examples.

EXAMPLES 1.2.2 (a) Let $A = \{-3, -2, -1, 0, 1\}$ and $B = \mathbb{Z}$. Consider the subset f of $A \times B$ given by

$$f = \{(-3, 2), (-2, -2), (-1, 4), (0, -6), (1, 4)\}.$$

Since each $x \in A$ belongs to precisely one ordered pair $(x, y) \in f$, f is a function from A into B with Range $f = \{-6, -2, 2, 4\}$. Figure 1.5 indicates what f does to each element of A. Even though the element 4 in B is the second component of the two distinct ordered pairs $(-1, 4)$ and $(1, 4)$, this does not contradict the definition of function.

(b) Let A and B be as in (a) and consider g defined by

$$g = \{(-3, 2), (-2, 4), (-2, 1), (-1, 4), (0, 5), (1, 1)\}.$$

Since both $(-2, 4)$ and $(-2, 1)$ are two elements of g with the same first component, g is not a function from A to B.

(c) In this example, we let $A = B = \mathbb{R}$, and let h be defined by

$$h = \{(x, y) \in \mathbb{R} \times \mathbb{R} : y = x^2 + 2\}.$$

This function is described by the equation $y = x^2 + 2$. The standard way of expressing this function is as

$$h(x) = x^2 + 2, \quad \text{Dom}\, h = \mathbb{R}.$$

This specifies both the equation defining the function and the domain of the function. For this example, Range $h = \{y \in \mathbb{R} : y \geq 2\}^2$.

(d) Let A be any nonempty set and let

$$i = \{(x, x) : x \in A\}.$$

Then i is a function from A onto A whose value at each $x \in A$ is x; i.e., $i(x) = x$. The function i is called the **identity function** on A.

(e) Let A and B be two nonempty sets and consider the **projection** function p from $A \times B$ to A defined by

$$p = \{((a, b), a) : (a, b) \in A \times B\}.$$

In this example, $\text{Dom}\, p = A \times B$. Since $p : (a, b) \to a$, we denote this simply by $p(a, b) = a$. For example, if $A = \{1, 2, 3\}$ and $B = \{-1, 1\}$, then $A \times B = \{(1, -1), (1, 1), (2, -1), (2, 1), (3, -1), (3, 1)\}$ and $p(1, -1) = 1$, $p(1, 1) = 1$, $p(2, -1) - 2$, etc. $\square$

As was indicated in (c) above, if our function h is given by an equation such as $y = x^2 + 2$ we will simply write $h(x) = x^2 + 2$, $\text{Dom}\, h = \mathbb{R}$, to denote the function h. It should be emphasized however that an equation such as $h(x) = x^2 + 2$ by itself does not define a function; the domain of h must also be specified. Thus

$$h(x) = x^2 + 2, \quad \text{Dom}\, h = \mathbb{R},$$

and

$$g(x) = x^2 + 2, \quad \text{Dom}\, y - \{u \in \mathbb{R} : -1 \leq x \leq 2\},$$

define two different functions.

---

[2]Since $x^2 \geq 0$ for all $x \in \mathbb{R}$, the range of h is a subset of $\{y \in \mathbb{R} : y \geq 2\}$. To obtain equality, we require that for every $y > 2$ there exists an $x \in \mathbb{R}$ such that $x^2 + 2 = y$. The existence of such a y will follow as a consequence of Example 1.4.6.

Image and Inverse Image

DEFINITION 1.2.3 *Let f be a function from A into B . If $E \subset A$, then $f(E)$, the* **image** *of E under f, is defined by*

$$f(E) = \{f(x) : x \in E\}.$$

If $H \subset B$, the **inverse image** *of H, denoted $f^{-1}(H)$, is defined by*

$$f^{-1}(H) = \{x \in A : f(x) \in H\}.$$

If H contains a single element of B, i.e., $H = \{y\}$, we will write $f^{-1}(y)$ instead of $f^{-1}(\{y\})$. Thus for $y \in B$,

$$f^{-1}(y) = \{x \in A : f(x) = y\}.$$

It is important to keep in mind that for $E \subset A$, $f(E)$ denotes a subset of B, while for $H \subset B$, $f^{-1}(H)$ describes a subset of the domain A. It should be clear that $f(A) = \text{Range } f$, and that f is onto B if and only if $f(A) = B$. To illustrate the notions of image and inverse image of a set we consider the following examples.

EXAMPLES 1.2.4 (a) As in Example 1.2.2 let $A = \{-3, -2, -1, 0, 1\}$, $B = \mathbb{Z}$, and $f : A \to \mathbb{Z}$ the function given by

$$f = \{(-3, 2), (-2, -2), (-1, 4), (0, -6), (1, 4)\}.$$

Consider the subset $E = \{-1, 0, 1\}$ of A. Then

$$f(E) = \{f(-1), f(0), f(1)\} = \{-6, 4\}.$$

If $H = \{0, 1, 2, 3, 4\}$, then

$$f^{-1}(H) = \{x \in A : f(x) \in H\} = \{-3, -1, 1\}.$$

Since both $f(-1)$ and $f(1)$ are equal to 4, $f^{-1}(4) = \{-1, 1\}$. On the other hand, since $(x, 0) \notin f$ for any $x \in A$, $f^{-1}(0) = \emptyset$.

(b) Consider the function $g : \mathbb{Z} \to \mathbb{Z}$ given by $g(x) = x^2$, and let $E = \{-1, -2, -3, \cdots\}$. Then

$$g(E) = \{(-n)^2 : n \in \mathbb{N}\} = \{1, 4, 9, \cdots\}.$$

On the other hand,
$$g^{-1}(g(E)) = \mathbb{Z} \setminus \{0\}.$$
For this example $E \subsetneq g^{-1}(g(E))$.

(c) Let h be the function defined by $h(x) = 2x + 3$, Dom $h = \mathbb{R}$. If $E = \{x \in \mathbb{R} : -1 \leq x \leq 2\}$, then

$$h(E) = \{2x + 3 : -1 \leq x \leq 2\} = \{y \in \mathbb{R} : 1 \leq y \leq 7\}.$$

For the set E we also have

$$h^{-1}(E) = \{x \in \mathbb{R} : 2x + 3 \in E\} = \{x : -2 \leq x \leq -\tfrac{1}{2}\}.$$

For each $y \in \mathbb{R}$, $x \in h^{-1}(y)$ if and only if $2x + 3 = y$, which upon solving for x gives $x = \tfrac{1}{2}(y - 3)$. Thus for each $y \in \mathbb{R}$, $h^{-1}(y) = \{\tfrac{1}{2}(y - 3)\}$. Since $h^{-1}(y) \neq \emptyset$ for each $y \in \mathbb{R}$, the function h maps $\mathbb{R}$ onto $\mathbb{R}$.   □

The operations of finding the image or inverse image of a set usually preserve the basic set operations of union and intersection. There is one important exception which is presented in part (b) of the next theorem.

THEOREM 1.2.5 *Let f be a function from A into B. If A_1 and A_2 are subsets of A, then*

    (a) $f(A_1 \cup A_2) = f(A_1) \cup f(A_2)$,

    (b) $f(A_1 \cap A_2) \subset f(A_1) \cap f(A_2)$.

Proof. To prove (a), let y be an element of $f(A_1 \cup A_2)$. Then $y = f(x)$ for some x in $A_1 \cup A_2$. Thus $x \in A_1$ or $x \in A_2$. Suppose $x \in A_1$. Then $y = f(x) \in f(A_1)$. Similarly, if $x \in A_2$, $y \in f(A_2)$. Therefore $y \in f(A_1) \cup f(A_2)$. Thus

$$f(A_1 \cup A_2) \subset f(A_1) \cup f(A_2).$$

Since it is clear that $f(A_1)$ and $f(A_2)$ are subsets of $f(A_1 \cup A_2)$, the reverse inclusion also holds, thereby proving equality.

Since $f(A_1 \cap A_2)$ is a subset of both $f(A_1)$ and $f(A_2)$, the relation stated in (b) is also true.   □

To see that equality need not hold in (b), consider the function $g(x) = x^2$, Dom $g = \mathbb{Z}$, of Example 1.2.4(b). If $A_1 = \{-1, -2, -3, \cdots\}$ and $A_2 = \{1, 2, 3, \cdots\}$, then $f(A_1) = f(A_2) = \{1, 4, 9, \cdots\}$, but $A_1 \cap A_2 = \emptyset$. Thus

$$f(A_1 \cap A_2) = f(\emptyset) = \emptyset \neq f(A_1) \cap f(A_2) = \{1, 4, 9, ...\}.$$

THEOREM 1.2.6 *Let f be a function from A to B. If B_1 and B_2 are subsets of B, then*

    (a) $f^{-1}(B_1 \cup B_2) = f^{-1}(B_1) \cup f^{-1}(B_2)$,

    (b) $f^{-1}(B_1 \cap B_2) = f^{-1}(B_1) \cap f^{-1}(B_2)$,

    (c) $f^{-1}(B \setminus B_1) = A \setminus f^{-1}(B_1)$.

Proof. The proof of Theorem 1.2.6 is left to the exercises (Exercise 8).   □

Inverse Function

DEFINITION 1.2.7 *A function f from A into B is said to be* **one-to-one** *if whenever $x_1 \neq x_2$, then $f(x_1) \neq f(x_2)$.*

Alternately, a function f is one-to-one if whenever (x_1, y) and (x_2, y) are elements of f then $x_1 = x_2$. From the definition it follows that f is one-to-one if and only if $f^{-1}(y)$ consists of at most one element of A for every $y \in B$. If f is onto B, then $f^{-1}(y) \neq \emptyset$ for every $y \in B$. Thus if f is one-to-one and onto B, then $f^{-1}(y)$ consists of exactly one element $x \in A$ and

$$g = \{(y, x) \in B \times A : f(x) = y\}$$

defines a function from B to A. This leads to the following definition.

DEFINITION 1.2.8 *If f is a one-to-one function from A onto B, let*

$$f^{-1} = \{(y, x) \in B \times A : f(x) = y\}.$$

The function f^{-1} from B onto A is called the **inverse function** *of f. Furthermore, for each $y \in B$,*

$$x = f^{-1}(y) \quad \text{if and only if} \quad f(x) = y.$$

There is a subtle point that needs to be clarified. If f is any function from A to B, then $f^{-1}(y)$ (technically $f^{-1}(\{y\})$) is defined for any $y \in B$ as the set of points x in A such that $f(x) = y$. However, if f is a one-to-one function of A onto B, then $f^{-1}(y)$ denotes the *value* of the inverse function f^{-1} at $y \in B$. Thus it makes sense to write $f^{-1}(y) = x$ whenever $(y, x) \in f^{-1}$. Also, if f is a one-to-one function of A into B, then f^{-1} defined by

$$f^{-1} = \{(y, x) : y \in \text{Range}\, f \text{ and } f(x) = y\}$$

is a function from Range f onto A.

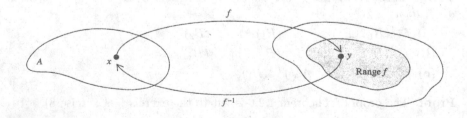

FIGURE 1.6
The inverse function

EXAMPLES 1.2.9 (a) Let h be the function of Example 1.2.4 (c); that is, $h(x) = 2x + 3$, Dom $h = \mathbb{R}$. The function h is clearly one-to-one and onto $\mathbb{R}$ with

$$x = h^{-1}(y) = \tfrac{1}{2}(y - 3), \quad \text{Dom } h^{-1} = \mathbb{R}.$$

(b) Consider the function f defined by the equation $y = x^2$. If we take for the domain of f all of $\mathbb{R}$, then f is not a one-to-one function. However, if we let

$$\text{Dom } f = A = \{x \in \mathbb{R} : x \geq 0\},$$

then f becomes a one-to-one mapping of A into A. To see that f is one-to-one, let $x_1, x_2 \in A$ with $x_1 \neq x_2$. Suppose $x_1 < x_2$. Then $x_1^2 < x_2^2$, that is, $f(x_1) \neq f(x_2)$. Therefore f is one-to-one. To show that f is onto A, we need to show that for each $y \in A$, $y > 0$, there exists a positive real number x such that

$$x^2 = y.$$

Intuitively we know that such an x exists; namely the square root of y. However, a rigorous proof of the existence of such an x will require the least upper bound property of the real numbers. In Example 1.4.6 we will prove that for each $y > 0$ there exists a unique positive real number x such that $x^2 = y$. The number x is called the **square root** of y and is denoted by $\sqrt{y}$. Thus the inverse function of f is given by

$$f^{-1}(y) = \sqrt{y}, \quad \text{Dom } f^{-1} = \{y \in \mathbb{R} : y \geq 0\}. \quad \square$$

Composition of Functions

Suppose f is a function from A to B and g is a function from B to C. If $x \in A$, then $f(x)$ is an element of B, the domain of g. Consequently we can apply the function g to $f(x)$ to obtain the element $g(f(x))$ in C. This process, illustrated in Figure 1.7, gives a new function h which maps $x \in A$ to $g(f(x))$ in C.

DEFINITION 1.2.10 *If f is a function from A to B and g is a function from B to C, then the function $g \circ f : A \to C$ defined by*

$$g \circ f = \{(x, z) \in A \times C : z = g(f(x))\}$$

is called the **composition** *of g with f.*

If f is a one-to-one function from A into B, then it can be shown that $(f^{-1} \circ f)(x) = x$ for all $x \in A$ and that $(f \circ f^{-1})(y) = y$ for all $y \in \text{Range } f$ (Exercise 10). This is illustrated in (b) of the following example.

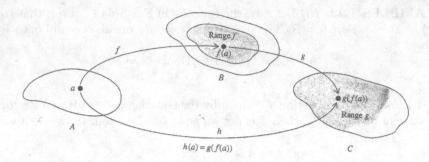

FIGURE 1.7
Composition of g with f

EXAMPLES 1.2.11 (a) If $f(x) = \sqrt{1+x}$ with $\mathrm{Dom}\, f = \{x \in \mathbb{R} : x \geq -1\}$ and $g(x) = x^2$, $\mathrm{Dom}\, g = \mathbb{R}$, then

$$(g \circ f)(x) = g(f(x)) = (\sqrt{1+x})^2 = 1 + x, \quad \mathrm{Dom}(g \circ f) = \{x \in \mathbb{R} : x \geq -1\}.$$

Even though the equation $(g \circ f)(x) = 1 + x$ is defined for all real numbers x, the domain of the composite function $g \circ f$ is still only the set $\{x \in \mathbb{R} : x \geq -1\}$. For this example, since $\mathrm{Range}\, g \subset \mathrm{Dom}\, f$, we can also find $f \circ g$; namely,

$$(f \circ g)(x) = f(g(x)) = \sqrt{1+x^2}, \quad \mathrm{Dom}\, f \circ g = \mathbb{R}.$$

(b) For the function f in (a), the inverse function f^{-1} is given by

$$f^{-1}(y) = y^2 - 1, \qquad \mathrm{Dom}\, f^{-1} = \mathrm{Range}\, f = \{y \in \mathbb{R} : y \geq 0\}.$$

Thus for $x \in \mathrm{Dom}\, f$,

$$(f^{-1} \circ f)(x) = f^{-1}(f(x)) = (f(x))^2 - 1 = (\sqrt{x+1})^2 - 1 = x,$$

and for $y \geq 0$,

$$(f \circ f^{-1})(y) = f(f^{-1}(y)) = \sqrt{(y^2 - 1) + 1} = y. \quad \square$$

Exercises 1.2

1. Let $A = \{-1, 0, 1, 2\}$ and $B = \mathbb{N}$. Which of the following subsets of $A \times B$ is a function from A into B.

    a. $f = \{(-1, 2), (0, 3), (2, 5)\}$

    *b. $g = \{(-1, 2), (0, 7), (1, -1), (1, 3), (2, 7)\}$

    c. $h = \{(-1, 2), (0, 2), (1, 2), (2, -1)\}$

    *d $k = \{((x, y) : y = 2x + 3, \, x \in A\}$

2. *a. Let $A = \{(x,y) \in \mathbb{R} \times \mathbb{R} : x^2 + y^2 = 1\}$. Is A a function? Explain your answer.

   b. Let $B = \{(x,y) \in \mathbb{R} \times \mathbb{R} : x^2 + y^2 = 1, y \geq 0\}$. If B a function? Explain your answer.

3. Let $f : \mathbb{N} \to \mathbb{N}$ be the function defined by $f(n) = 2n - 1$. Find $f(E)$ and $f^{-1}(E)$ for each of the following subsets E of $\mathbb{N}$.

   *a. $\{1,2,3,4\}$ b. $\{1,3,5,7\}$ c. $\mathbb{N}$

4. Let $f = \{(x,y) : x \in \mathbb{R}, y = x^3 + 1\}$.

   *a. Let $A = \{x : -1 \leq x \leq 2\}$. Find $f(A)$ and $f^{-1}(A)$.

   b. Show that f is a one-to-one function of $\mathbb{R}$ onto $\mathbb{R}$.

   *c. Find the inverse function f^{-1}.

5. Let f, g mapping $\mathbb{Z}$ into $\mathbb{Z}$ be given by $f(x) = x + 3$ and $g(x) = 2x$.

   a. Find $(f \circ g)(x)$ and $(g \circ f)(x)$.

   *b. Find $(f \circ g)(\mathbb{N})$ and $(g \circ f)(\mathbb{N})$.

6. For each of the following real-valued functions, find the range of the function f and determine whether f is one-to-one. If f is one-to-one, find the inverse function f^{-1} and specify the domain of f^{-1}.

   a. $f(n) = 5x + 4$,   Dom $f = \mathbb{R}$.

   *b. $f(x) = 3x - 2$,   Dom $f = \mathbb{R}$.

   c. $f(x) = \dfrac{x}{x-1}$,   Dom $f = \{x \in \mathbb{R} : 0 \leq x < 1\}$.

   d. $f(x) = \sin x$,   Dom $f = \{x \in \mathbb{R} : 0 \leq x \leq \pi\}$.

   *e. $f(x,y) = x$,   Dom $f = \mathbb{R}^2$.

   *f. $f(x) - \dfrac{1}{x^2+1}$,   Dom $f = \{x \in \mathbb{R} : -1 \leq x \leq 1\}$.

   g. $f = \{(x, x^2) : 0 \leq x \leq 1\}$.

7. Let $A = \{t \in \mathbb{R} : 0 \leq t < 2\pi\}$ and $B = \mathbb{R}^2$, and let $f : A \to B$ be defined by $f(t) = (\cos t, \sin t)$.

   *a. What is the range of f?

   *b. Find f^{-1} of each of the following points: $(1,0)$, $(0,-1)$, $(\frac{\sqrt{2}}{2}, \frac{\sqrt{2}}{2})$, $(0,1)$.

   c. Is the function f one-to-one?

8. Prove Theorem 1.2.6.

9. Let $f : A \to B$ and let $F \subset A$.

   a. Prove that $f(A) \setminus f(F) \subset f(A \setminus F)$.

   b. Give an example for which $f(A) \setminus f(F) \neq f(A \setminus F)$.

10. a. If f is a one-to-one function from A into B. Show that $(f^{-1} \circ f)(x) = x$ for all $x \in A$ and that $(f \circ f^{-1})(y) = y$ for all $y \in$ Range f.

    b. If g is a function from C into A and $h = f \circ g$, show that $g = f^{-1} \circ h$.

11. *Let $f : A \to B$ and $g : B \to A$ be functions satisfying $(g \circ f)(x) = x$ for all $x \in A$. Show that f is a one-to-one function. Must f be onto B?

12. If $f : A \to B$ and $g : B \to C$ are one-to-one functions, show that $(g \circ f)^{-1} = f^{-1} \circ g^{-1}$ on Range $(g \circ f)$.

1.3 Mathematical Induction

Throughout the text we will on occasion need to prove a statement, identity, or inequality involving the positive integer n. As an example, consider the following identity. For each $n \in \mathbb{N}$,

$$r + r^2 + \cdots r^n = \frac{r - r^{n+1}}{1 - r}, \qquad r \neq 1.$$

Mathematical induction is a very useful tool in establishing that such an identity is valid for all positive integers n.

THEOREM 1.3.1 (Principle of Mathematical Induction) *For each* $n \in \mathbb{N}$, *let* $P(n)$ *be a statement about the positive integer* n. *If*

(a) $P(1)$ *is true, and*

(b) $P(k+1)$ *is true whenever* $P(k)$ *is true,*

then $P(n)$ *is true for all* $n \in \mathbb{N}$.

The proof of this theorem depends on the fact that the positive integers are **well-ordered**; namely, every nonempty subset of $\mathbb{N}$ has a smallest element. This statement is usually taken as a postulate or axiom for the positive integers: we do so in this text. Since it will be used on several other occasions, we state it both for completeness and emphasis.

WELL-ORDERING PRINCIPLE Every nonempty subset of $\mathbb{N}$ has a smallest element.

The well-ordering principle can be restated as follows: If $A \subset \mathbb{N}$, $A \neq \emptyset$, then there exists $n \in A$ such that $n \leq k$ for all $k \in A$.

To prove Theorem 1.3.1 we will use the method of proof by contradiction. Most theorems involve showing that a statement P implies the statement Q; namely, if P is true, then Q is true. In a proof by contradiction one assumes that P is true and Q is false, and then shows that these two assumptions lead to a logical contradiction; namely show that some statement R is both true and false.

Proof of Theorem 1.3.1 Assume that the hypothesis of Theorem 1.3.1 are true, but that the conclusion is false; that is, there exists a positive integer n such that the statement $P(n)$ is false. Let

$$A = \{k \in \mathbb{N} : P(k) \text{ is false }\}.$$

By our assumption the set A is nonempty. Thus by the well-ordering principle A has a smallest element k_o. Since $P(1)$ is true, $k_o > 1$. Also, since k_o is the smallest element of A, $P(k_o - 1)$ is true. But then by hypothesis (b), $P(k_o)$ is also true, which is a contradiction. Consequently, $P(n)$ must be true for all $n \in \mathbb{N}$. $\square$

EXAMPLES 1.3.2 We now provide two examples to illustrate the method of proof by mathematical induction. The first example provides a proof of the identity in the introduction to the section. An alternate method of proof will be requested in the exercises (Exercise 7).

(a) To use mathematical induction, we let our statement $P(n)$, $n \in \mathbb{N}$, be as follows:

$$r + \cdots + r^n = \frac{r - r^{n+1}}{1 - r}, \qquad r \neq 1.$$

When $n = 1$ we have

$$r = \frac{r(1 - r)}{(1 - r)} = \frac{r - r^2}{1 - r} \qquad \text{provided} \quad r \neq 1.$$

Thus the identity is valid for $n = 1$. Assume $P(k)$ is true for $k \geq 1$, i.e.,

$$r + \cdots + r^k = \frac{r - r^{k+1}}{1 - r}, \qquad r \neq 1.$$

We must now show that the statement $P(k + 1)$ is true; that is

$$r + \cdots + r^{k+1} = \frac{r - r^{(k+1)+1}}{1 - r}, \qquad r \neq 1.$$

But

$$r + \cdots + r^{k+1} = r + \cdots + r^k + r^{k+1}$$

which by the induction hypothesis

$$= \frac{r - r^{k+1}}{1 - r} + r^{k+1} = \frac{r - r^{k+1} + (1 - r)r^{k+1}}{1 - r}$$

$$= \frac{r - r^{k+2}}{1 - r}, \qquad r \neq 1.$$

Thus the identity is valid for $k+1$, and hence by the principle of mathematical induction for all $n \in \mathbb{N}$.

(b) For our second example we use mathematical induction to prove **Bernoulli's inequality**. If $h > -1$, then

$$(1 + h)^n \geq 1 + nh \qquad \text{for all } n \in \mathbb{N}.$$

When $n = 1$, $(1 + h)^1 = 1 + h$. Thus since equality holds, the inequality is certainly valid. Assume that the inequality is true when $n = k$, $k \geq 1$. Then for $n = k + 1$,

$$(1 + h)^{k+1} = (1 + h)^k (1 + h),$$

which by the induction hypothesis and the fact that $(1 + h) > 0$

$$\geq (1 + k\,h)(1 + h) = 1 + (k + 1)h + k\,h^2$$
$$\geq 1 + (k + 1)h.$$

Therefore the inequality holds for $n = k + 1$, and thus by the principle of mathematical induction for all $n \in \mathbb{N}$. □

Although the statement of Theorem 1.3.1 starts with $n = 1$, the result is still true if we start with any integer $n_o \in \mathbb{Z}$. The **modified principle of mathematical induction** is as follows: If for each $n \in \mathbb{Z}$, $n \geq n_o$, $P(n)$ is a statement about the integer n satisfying

(a') $P(n_o)$ is true, and

(b') $P(k + 1)$ is true whenever $P(k)$ is true, $k \geq n_o$,

then $P(n)$ is true for all $n \in \mathbb{Z}$, $n \geq n_o$.

The proof of this follows from Theorem 1.3.1 by simply setting

$$Q(n) = P(n + n_o - 1), \qquad n \in \mathbb{N},$$

which is now a statement about the positive integer n.

Remark. In the principle of mathematical induction, the hypothesis that $P(1)$ be true is essential. For example, consider the statement $P(n)$:

$$n + 1 = n, \qquad n \in \mathbb{N}.$$

This is clearly false! However, if we assume that $P(k)$ is true, then we also obtain that $P(k+1)$ is true. Thus it is absolutely essential that $P(n_o)$ be true for at least one fixed value of n_o.

There is a second version of the principle of mathematical induction which is also quite useful.

THEOREM 1.3.3 (Second Principle of Mathematical Induction)
For each $n \in \mathbb{N}$, let $P(n)$ be a statement about the positive integer n. If

(a)   *$P(1)$ is true, and*

(b)   *for $k > 1$, $P(k)$ is true whenever $P(j)$ is true for all positive integers $j < k$,*

then $P(n)$ is true for all $n \in \mathbb{N}$.

Proof. (Exercise 3). □

Mathematical induction is also used in the **recursive** definition of functions defined for the positive integers. In this procedure, we give an initial value of the function f at $n = 1$, and then assuming that f has been defined for all integers $k = 1, ..., n$, the value of f at $n + 1$ is given in terms of the values of f at k, $k \leq n$. This is illustrated by the following examples.

EXAMPLES 1.3.4 (a) As an example, consider the function $f : \mathbb{N} \to \mathbb{N}$ defined by $f(1) = 1$ and $f(n + 1) = nf(n)$, $n \in \mathbb{N}$. The values of f for $n = 1, 2, 3, 4$ are given as follows:

$$f(1) = 1, \ f(2) = 1, \ f(3) = 2f(2) = 2 \cdot 1, \ f(4) = 3f(3) = 3 \cdot 2 \cdot 1.$$

Thus we conjecture that $f(n) = (n - 1)!$, where $0!$ is defined to be equal to one, and for $n \in \mathbb{N}$, $n!$ (read n **factorial**) is defined as

$$n! = n \cdot (n - 1) \cdots 2 \cdot 1.$$

This conjecture is certainly true when $n = 1$. Thus assume that it is true for $n = k$, $k \geq 1$, that is $f(k) = (k - 1)!$. Then for $n = k + 1$,

$$f(k + 1) = kf(k)$$

which by the induction hypothesis

$$= k \cdot (k - 1)! = k!.$$

Therefore the identity holds for $n = (k + 1)$, and thus by the principle of mathematical induction for all $n \in \mathbb{N}$.

(b) For our second example, consider the function $f : \mathbb{N} \to \mathbb{R}$ defined by $f(1) = 0$, $f(2) = \frac{1}{3}$, and for $n \geq 3$ by $f(n) = (\frac{n-1}{n+1})f(n - 2)$, Computing the values of f for $n = 3, 4, 5$, and 6, we have

$$f(3) = 0, \quad f(4) = \frac{1}{5}, \quad f(5) = 0, \quad f(6) = \frac{1}{7}.$$

From these values we conjecture that

$$f(n) = \begin{cases} 0, & \text{if } n \text{ is odd,} \\ \dfrac{1}{n+1}, & \text{if } n \text{ is even.} \end{cases}$$

To prove our conjecture we will use the second principle of mathematical induction. Our conjecture is certainly true for $n = 1, 2$. Suppose $n > 2$, and suppose our conjecture holds for all $k < n$. If n is odd, then so is $(n - 2)$, and thus by the induction hypothesis $f(n - 2) = 0$. Therefore $f(n) = 0$. On the other hand if n is even, so is $(n - 2)$. Thus by the induction hypothesis $f(n - 2) = 1/(n - 1)$. Therefore

$$f(n) = \left(\frac{n-1}{n+1}\right) f(n - 2) = \left(\frac{n-1}{n+1}\right) \frac{1}{n-1} = \frac{1}{n+1}. \quad □$$

Exercises 1.3

1. Use mathematical induction to prove that each of the following identities are valid for all $n \in \mathbb{N}$.

    a. $1 + 2 + 3 + \cdots + n = \dfrac{n(n+1)}{2}$.

    *b. $1 + 3 + 5 + \cdots + (2n - 1) = n^2$.

    c. $1^2 + 2^2 + \cdots + n^2 = \dfrac{n(n+1)(2n+1)}{6}$.

    *d. $1^3 + 2^3 + \cdots + n^3 = \left[\frac{1}{2} n(n+1)\right]^2$.

    e. $2 + 2^2 + 2^3 + \cdots + 2^n = 2(2^n - 1)$.

    *f. For $x, y \in \mathbb{R}$, $x^{n+1} - y^{n+1} = (x - y)(x^n + x^{n-1}y + \cdots + y^n)$.

    g. $\dfrac{1}{1(2)} + \dfrac{1}{2(3)} + \cdots + \dfrac{1}{n(n+1)} = \dfrac{n}{n+1}$.

2. Use mathematical induction to establish the following inequalities for $n \in \mathbb{N}$.

    *a. $2^n > n$ for all $n \in \mathbb{N}$ **b.** $2^n > n^2$ for all $n \geq 5$

    *c. $n! > 2^n$ for all $n \geq 4$ ***d** $1^3 + 2^3 + \cdots + n^3 < \frac{1}{2} n^4$ for all $n \geq 3$

3. Prove Theorem 1.3.3.

4. ***** Let $f : \mathbb{N} \to \mathbb{N}$ be defined by $f(1) = 5$, $f(2) = 13$, and for $n \geq 3$, $f(n) = 2f(n - 2) + f(n - 1)$. Prove that $f(n) = 3 \cdot 2^n + (-1)^n$ for all $n \in \mathbb{N}$.

5. For each of the following functions with domain $\mathbb{N}$, determine a formula for $f(n)$ and use mathematical induction to prove your conclusion.

    a. $f(1) = \frac{1}{2}$, and for $n > 1$, $f(n) = (n - 1)f(n - 1) - \dfrac{1}{n+1}$.

    *b. $f(1) = 1$, $f(2) = 4$, and for $n > 2$, $f(n) = 2f(n - 1) - f(n - 2) + 2$.

    c. $f(1) = 1$, and for $n > 1$, $f(n) = \dfrac{(n+1)}{3n} f(n - 1)$.

    *d. $f(1) = 1$, $f(2) = 0$, and for $n > 2$, $f(n) = -\dfrac{f(n - 2)}{n(n - 1)}$.

    e. For $a_1, a_2 \in \mathbb{R}$ arbitrary, let $f(1) = a_1$, $f(2) = a_2$, and for $n > 2$, $f(n) = -\dfrac{f(n - 2)}{n(n - 1)}$.

    *f. For $a_1, a_2 \in \mathbb{R}$ arbitrary, let $f(1) = a_1$, $f(2) = a_2$, and for $n > 2$, $f(n) = \left(\dfrac{n - 1}{n + 1}\right) f(n - 2)$.

6. Let $f : \mathbb{N} \to \mathbb{N}$ be defined by $f(1) = 1$, $f(2) = 2$, and $f(n + 2) = \dfrac{1}{2}(f(n + 1) + f(n))$.
   Use Theorem 1.3.3 to prove that $1 \leq f(n) \leq 2$ for all $n \in \mathbb{N}$.

7. ***** Prove that $r + r^2 + \cdots + r^n = \dfrac{r - r^{n+1}}{1 - r}$, $r \neq 1$, $n \in \mathbb{N}$, without using mathematical induction.

8. *Use mathematical induction to prove the **arithmetic-geometric mean inequality**: If $a_1, a_2, ..., a_n$, $n \in \mathbb{N}$, are nonnegative real numbers, then

$$a_1 a_2 \cdots a_n \leq \left(\frac{a_1 + a_2 + \cdots + a_n}{n} \right)^n,$$

with equality if and only if $a_1 = a_2 = \cdots = a_n$.

1.4 The Least Upper Bound Property

In this section, we will consider the concept of the least upper bound of a set and introduce the least upper bound or supremum property of the real numbers $\mathbb{R}$. Prior to introducing these new ideas we briefly review the algebraic and order properties of $\mathbb{Q}$ and $\mathbb{R}$.

Both the rational numbers $\mathbb{Q}$ and the real numbers $\mathbb{R}$ are algebraic systems known as *fields*. The key facts about a **field** which we need to know is that it is a set $\mathbb{F}$ with two operations, addition $(+)$ and multiplication $(\cdot)$, which satisfy the following axioms:

1. If $a, b \in \mathbb{F}$, then $a + b \in \mathbb{F}$ and $a \cdot b \in \mathbb{F}$.

2. The operations are **commutative**; that is, for all $a, b \in \mathbb{F}$

$$a + b = b + a \quad \text{and} \quad a \cdot b = b \cdot a.$$

3. The operations are **associative**; that is, for all $a, b, c \in \mathbb{F}$,

$$a + (b + c) = (a + b) + c \quad \text{and} \quad a \cdot (b \cdot c) = (a \cdot b) \cdot c.$$

4. There exists an element $0 \in \mathbb{F}$ such that $a + 0 = a$ for every $a \in \mathbb{F}$.

5. Every $a \in \mathbb{F}$ has an additive inverse; that is, there exists an element $-a$ in $\mathbb{F}$ such that

$$a + (-a) = 0.$$

6. There exists an element $1 \in \mathbb{F}$ with $1 \neq 0$ such that $a \cdot 1 = a$ for all $a \in \mathbb{F}$.

7. Every $a \in \mathbb{F}$ with $a \neq 0$ has a multiplicative inverse; that is, there exists an element a^{-1} in $\mathbb{F}$ such that

$$a \cdot a^{-1} = 1.$$

8. The operation of multiplication is **distributive** over addition; that is, for all $a, b, c \in \mathbb{F}$,

$$a \cdot (b + c) = a \cdot b + a \cdot c.$$

The element 0 is called the **zero** of $\mathbb{F}$ and the element 1 is called the **unit** of $\mathbb{F}$. For $a \neq 0$, the element a^{-1} is customarily written as $\frac{1}{a}$ or $1/a$. Similarly we write $a - b$ instead of $a + (-b)$, ab instead of $a \cdot b$, and a/b or $\frac{a}{b}$ instead of $a \cdot b^{-1}$.

The real numbers $\mathbb{R}$ contain a subset $\mathbb{P}$ known as the **positive real numbers** satisfying the following:

(O1) If $a, b \in \mathbb{P}$, then $a + b \in \mathbb{P}$ and $a \cdot b \in \mathbb{P}$.

(O2) If $a \in \mathbb{R}$ then one and only one of the following hold:

$$a \in \mathbb{P}, \quad -a \in \mathbb{P}, \quad a = 0.$$

Properties (O1) and (O2) are called the **order properties** of $\mathbb{R}$. Any field $\mathbb{F}$ with a nonempty subset satisfying (O1) and (O2) above is called an **ordered field**. For the real numbers we assume the existence of a positive set $\mathbb{P}$. For the rational numbers $\mathbb{Q}$, the set of **positive rational numbers** is given by $\mathbb{P} \cap \mathbb{Q}$ which can be proved to be equal to $\{ p/q : p, q \in \mathbb{Z}, q \neq 0, pq \in \mathbb{N} \}$.

Let a, b be elements of $\mathbb{R}$. If $a - b$ is positive, i.e., $a - b \in \mathbb{P}$, then we write $a > b$ or $b < a$. In particular, the notation $a > 0$ (or $0 < a$) means that a is a positive element. Also, $a \leq b$ (or $b \geq a$) if $a < b$ or $a = b$.

The following useful results are immediate consequences of the order properties and the axioms for addition and multiplication. Let a, b, c be elements of $\mathbb{R}$.

(a) If $a > b$, then $a + c > b + c$.

(b) If $a > b$ and $c > 0$, then $ac > bc$.

(c) If $a > b$ and $c < 0$, then $ac < bc$.

(d) If $a \neq 0$, then $a^2 > 0$.

(e) If $a > 0$, then $1/a > 0$; if $a < 0$, then $1/a < 0$.

To illustrate the method of proof, we provide the proof of (b). Suppose $a > b$; i.e, $a - b$ is positive. If c is positive, then by (O1) $(a - b)c$ is positive. By the distributive law,

$$(a - b)c = ac - bc.$$

Therefore $ac - bc$ is positive; that is, $ac > bc$. The proofs of the other results are left as exercises.

Upper Bound of a Set

We now turn our attention to the most important topic of this chapter; namely, the least upper bound or supremum property of $\mathbb{R}$. In Example 1.4.2(c) we will show that this property fails for the rational numbers $\mathbb{Q}$. First however, we define the concept of an upper bound of a set.

DEFINITION 1.4.1 *A subset E of $\mathbb{R}$ is* **bounded above** *if there exists $\beta \in \mathbb{R}$ such that $x \leq \beta$ for every $x \in E$. Such a β is called an* **upper bound** *of E.*

The concepts **bounded below** and **lower bound** are defined similarly. A set E is **bounded** if E is bounded both above and below. We now consider several examples to illustrate these concepts.

EXAMPLES 1.4.2 (a) Let $A = \{0, \frac{1}{2}, \frac{2}{3}, \frac{3}{4}, ...\} = \{1 - \frac{1}{n} : n = 1, 2, 3, ...\}$
Clearly A is bounded below by any real number $r \leq 0$ and also above by any real number $s \geq 1$.

(b) $\mathbb{N} = \{1, 2, 3, ...\}$. This set is bounded below; e.g., 1 is a lower bound. Our intuition tells us that $\mathbb{N}$ is not bounded above. It is obvious that there is no positive integer n such that $j \leq n$ for all $j \in \mathbb{N}$. However, what is not so obvious is that there is no real number β such that $j \leq \beta$ for all $j \in \mathbb{N}$. In fact, given $\beta \in \mathbb{R}$, the proof of the existence of a positive integer $n > \beta$ will require the least upper bound property of $\mathbb{R}$ (Theorem 1.5.1).

(c) $B = \{r \in \mathbb{Q} : r > 0 \text{ and } r^2 < 2\}$. Again it is clear that 0 is a lower bound for B, and that B is bounded above; e.g., 2 is an upper bound for B. What is not so obvious however is that B has no maximum. By the **maximum** or **largest element** of B we mean an element $\alpha \in B$ such that $p \leq \alpha$ for all $p \in B$. Suppose $p \in B$. Define the rational number q by

$$q = p + \left(\frac{2 - p^2}{p + 2}\right) = \frac{2p + 2}{p + 2}.$$

With q as defined, a simple computation gives

$$q^2 - 2 = \frac{2(p^2 - 2)}{(p + 2)^2}.$$

Since $p^2 < 2$, $q > p$ and $q^2 < 2$. Thus B has no largest element. Similarly, the set

$$\{p \in \mathbb{Q} : p > 0, \, p^2 > 2\}$$

has no minimum or smallest element. Intuitively, the largest element of B would satisfy $p^2 = 2$. However, as was shown in the introduction, there is no rational number p for which $p^2 = 2$. $\square$

Least Upper Bound of a Set

DEFINITION 1.4.3 *Let E be a nonempty subset of $\mathbb{R}$ that is bounded above. An element $\alpha \in \mathbb{R}$ is called the **least upper bound** or **supremum** of E if*
(i) *α is an upper bound of E, and*
(ii) *if $\beta \in \mathbb{R}$ satisfies $\beta < \alpha$, then β is not an upper bound of E.*

Condition (ii) is equivalent to $\alpha \leq \beta$ for all upper bounds β of E. Also by (ii), the least upper bound of a set is unique. If the set E has a least upper bound, we write

$$\alpha = \sup E$$

to denote that α is the supremum or least upper bound of E. The **greatest lower bound** or **infimum** of a nonempty set E is defined similarly, and if it exists, is denoted by $\inf E$.

There is one important fact about the supremum of a set which will be used repeatedly throughout the text. Due to its importance we state it as a theorem.

THEOREM 1.4.4 *Let A be a nonempty subset of $\mathbb{R}$ that is bounded above. An upper bound α of A is the supremum of A if and only if for every $\beta < \alpha$, there exists an element $x \in A$ such that*

$$\beta < x \le \alpha.$$

Proof. Suppose $\alpha = \sup A$. If $\beta < \alpha$, then β is not an upper bound of A. Thus there exists an element x in A such that $x > \beta$. On the other hand, since α is an upper bound of A, $x \le \alpha$.

Conversely, if α is an upper bound of A satisfying the stated condition, then every $\beta < \alpha$ is not an upper bound of A. Thus $\alpha = \sup A$. $\square$

EXAMPLES 1.4.5 In the following examples, let's consider again the three sets of the previous examples.

(a) As in Example 1.4.2(a), let $A = \{0, \frac{1}{2}, \frac{2}{3}, \frac{3}{4}, \cdots\}$. Since 0 is a lower bound of A and $0 \in A$, $\inf A = 0$. We now prove that $\sup A = 1$. Since $1 - \frac{1}{n} < 1$ for all $n = 1, 2, \ldots$, 1 is an upper bound. To show that $1 = \sup A$ we need to show that if $\beta \in \mathbb{R}$ with $\beta < 1$, then β is not an upper bound of A. Clearly if $\beta \le 0$, then β is not an upper bound of A. Suppose as in Figure 1.8, $0 < \beta < 1$. Then our intuition tells us that there exists an integer n_o such that

$$n_o > \frac{1}{1 - \beta}, \quad \text{or} \quad \beta < 1 - \frac{1}{n_o}.$$

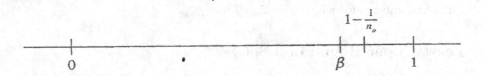

FIGURE 1.8
Proof that $\sup A = 1$ in Example 1.4.5(a)

But $1 - \frac{1}{n_o} \in A$, and thus β is not an upper bound. Therefore $\sup A = 1$. The existence of such an integer n_o will follow from Theorem 1.5.1. In this example, $\inf A \in A$ but $\sup A \notin A$.

(b) For the set $\mathbb{N}$, $\inf \mathbb{N} = 1$. Since $\mathbb{N}$ is not bounded above, $\mathbb{N}$ does not have an upper bound in $\mathbb{R}$.

(c) In this example, we prove that the supremum of the set

$$B = \{r \in \mathbb{Q} : r > 0 \quad \text{and} \quad r^2 < 2\},$$

if it exists, is not an element of $\mathbb{Q}$. Suppose $\alpha = \sup B$ exists and is in $\mathbb{Q}$. Since α is rational, $\alpha^2 \neq 2$. Thus $\alpha^2 < 2$ or $\alpha^2 > 2$. But if $\alpha \in B$, then since B contains no largest element, there exists $q \in B$ such that $q > \alpha$. This contradicts that α is an upper bound of B. Similarly, if $\alpha^2 > 2$, then there exists a $q < \alpha$ such that $q^2 > 2$. But then q is an upper bound of B, which is a contradiction of property (ii) of Definition 1.4.3. The least upper bound of B in $\mathbb{R}$ is $\sqrt{2}$ (Section 1.5, Exercise 9), which we know is not rational. $\square$

Least Upper Bound Property of $\mathbb{R}$

The following property, also referred to as the **completeness property** of $\mathbb{R}$, distinguishes the real numbers from the rational numbers and forms the foundation for many of the results in real analysis.

SUPREMUM OR LEAST UPPER BOUND PROPERTY OF $\mathbb{R}$
Every nonempty subset of $\mathbb{R}$ that is bounded above has a supremum in $\mathbb{R}$.

For our later convenience we restate the supremum property of $\mathbb{R}$ as the infimum property of $\mathbb{R}$.

INFIMUM OR GREATEST LOWER BOUND PROPERTY OF $\mathbb{R}$.
Every nonempty subset of $\mathbb{R}$ that is bounded below has an infimum in $\mathbb{R}$.

Although stated here as a property, which we will assume as a basic axiom about $\mathbb{R}$, the least upper bound property of $\mathbb{R}$ is really a *theorem* due to both Cantor and Dedekind, both of whom published their results independently in 1872. Dedekind, in the paper "Stetigkeit und irrationale Zahlen" (Continuity and irrational numbers), used algebraic techniques now known as the method of *Dedekind cuts* to construct the real number system $\mathbb{R}$ from the rational numbers $\mathbb{Q}$. He proved that the system $\mathbb{R}$ contained a natural subset of positive elements satisfying the order axioms (O1) and (O2), and furthermore, that $\mathbb{R}$ also satisfied the least upper bound property. The books by Burrill and by Spooner and Mentzger cited in the Supplemental Readings are devoted to number systems. Both texts contain Dedekind's construction of $\mathbb{R}$. Cantor on the other hand constructed $\mathbb{R}$ from $\mathbb{Q}$ using Cauchy sequences. In the miscellaneous exercises of Chapter 3 we will provide some of the key steps of this construction.

EXAMPLE 1.4.6 In this example, we show that for every positive real number $y > 0$, there exists a unique positive real number α such that $\alpha^2 = y$; i.e., $\alpha = \sqrt{y}$. The uniqueness of α was established in Example 1.2.9(b).

We only prove the result for $y > 1$, leaving the case $0 < y \leq 1$ to the exercises (Exercise 6). Let

$$C = \{x \in \mathbb{R} : x > 0 \text{ and } x^2 < y\}.$$

With $y = 2$, this set is similar to the set B of Example 1.4.5(c), except that here we consider all positive real numbers x for which $x^2 < y$. Since $y > 1, 1 \in C$ and thus C is nonempty. Also since $y > 1$, $y^2 > y$, and thus y is an upper bound of C. Hence by the least upper bound property, C has a supremum in $\mathbb{R}$. Let $\alpha = \sup C$. We now prove that $\alpha^2 = y$. To accomplish this we show that the assumptions $\alpha^2 < y$ and $\alpha^2 > y$ lead to contradictions. Thus $\alpha^2 = y$.

Define the real number β by

$$\beta = \alpha + \left(\frac{y - \alpha^2}{\alpha + y}\right) = \frac{y(\alpha + 1)}{\alpha + y}. \tag{1}$$

Then

$$\beta^2 - y = \frac{y(y - 1)(\alpha^2 - y)}{(\alpha + y)^2}. \tag{2}$$

If $\alpha^2 < y$, then by (1) $\beta > \alpha$, and by (2) $\beta^2 < y$. This contradicts that α is an upper bound for C. On the other hand, if $\alpha^2 > y$, then by (1) $\beta < \alpha$ and by (2), $\beta^2 > y$. Thus if $x \in \mathbb{R}$ with $x \geq \beta$, then $x^2 > y$. Therefore β is an upper bound of C. This contradicts that α is the least upper bound of C. Since β defined by (1) may not be rational, the same proof will not work for the set B of Example 1.4.5(c). However, using Theorem 1.5.2 of the following section, it is possible to also prove that $\sup B = \sqrt{2}$. $\square$

For convenience, we extend the definition of supremum and infimum of a subset E of $\mathbb{R}$ to include the case where E is not necessarily bounded above or below.

DEFINITION 1.4.7 *If E is a nonempty subset of $\mathbb{R}$, we set*

$$\sup E = \infty \qquad \text{if } E \text{ is not bounded above, and}$$
$$\inf E = -\infty \qquad \text{if } E \text{ is not bounded below.}$$

For the empty set $\emptyset$, every element of $\mathbb{R}$ is an upper bound of $\emptyset$. For this reason the supremum of the empty set $\emptyset$ is taken to be $-\infty$. Similarly, $\inf \emptyset = \infty$. Also, for the symbols $-\infty$ and ∞ we adopt the convention that $-\infty < x < \infty$ for every $x \in \mathbb{R}$.

Intervals

Using the order properties of $\mathbb{R}$, we can define certain subsets of $\mathbb{R}$ known as intervals.

DEFINITION 1.4.8 *For $a, b \in \mathbb{R}$, $a \leq b$, the* **open interval** (a, b) *is defined as*

$$(a, b) = \{x \in \mathbb{R} : a < x < b\},$$

whereas the **closed interval** $[a, b]$ *is defined as*

$$[a, b] = \{x \in \mathbb{R} : a \leq x \leq b\}.$$

In addition, we also have the **half-open (half-closed) intervals**

$$[a, b) = \{x \in \mathbb{R} : a \leq x < b\},$$
$$(a, b] = \{x \in \mathbb{R} : a < x \leq b\},$$

and the **infinite intervals**

$$(a, \infty) = \{x \in \mathbb{R} : a < x < \infty\},$$
$$[a, \infty) = \{x \in \mathbb{R} : a \leq x < \infty\},$$

with analogous definitions for $(-\infty, b)$ and $(-\infty, b]$. The intervals (a, ∞), $(-\infty, b)$ and $(-\infty, \infty) = \mathbb{R}$ are also referred to as open intervals, whereas the intervals $[a, \infty)$ and $(-\infty, b]$ are called closed intervals.

In the above, when $b = a$, $(a, a) = \emptyset$ and $[a, a] = \{a\}$. Although the empty set $\emptyset$ and the singleton $\{a\}$ do not fit our intuitive definition of an interval, we will include them as the degenerate case of open and closed intervals respectively. It should be noted that the intervals of the form (a, b), $(a, b]$, $[a, b)$, and $[a, b]$ with $a, b \in \mathbb{R}$, $a \leq b$, are all bounded subsets of $\mathbb{R}$.

An alternate way of defining intervals without use of the adjectives open and closed is as follows:

DEFINITION 1.4.9 *A subset J of $\mathbb{R}$ is an* **interval** *if whenever $x, y \in J$ with $x < y$, then every t satisfying $x < t < y$ is in J.*

This definition also allows the possibility that J is empty or a singleton. One can show that that every set J satisfying the above is one of the intervals defined in Definition 1.4.8 (Exercise 21).

Exercises 1.4

1. Use the axioms for addition and multiplication to prove the following: if $a \in \mathbb{R}$, then

   a. $a \cdot 0 = 0$.   **b.** $(-1) \cdot a = -a$.   **c.** $-(-a) = a$.

2. Let $a, b \in \mathbb{R}$. Prove the following:

   a. If $a \neq 0$, then $\frac{1}{a} \neq 0$.   **b.** If $a \cdot b = 0$, then either $a = 0$ or $b = 0$.

3. Let $a, b, c \in \mathbb{R}$. Prove the following:

   a. If $a > b$, then $a + c > b + c$.   **b.** If $a \neq 0$, then $a^2 > 0$.

   c. If $a > b$ and $c > 0$, then $ac > bc$.

   d. If $a > 0$ then $1/a > 0$, and if $a < 0$ then $1/a < 0$.

4. *If $a, b \in \mathbb{R}$, prove that $ab \leq \frac{1}{2}(a^2 + b^2)$.

5. Find the supremum and the infimum of each of the following sets:

   *a. $A = \{1, \frac{1}{2}, \frac{1}{4}, \frac{1}{8}, ...\} = \left\{ \dfrac{1}{2^{n-1}} : n \in \mathbb{N} \right\}$

   b. $B = \{\cos n\frac{\pi}{4} : n \in \mathbb{N}\}$

   *c. $C = \{(1 - (-1)^n)^n : n \in \mathbb{N}\}$

   d. $D = \{\sin n\frac{\pi}{2} ; n \in \mathbb{N}\}$

   e. $E = \{n \cos n\pi : n \in \mathbb{N}$

   *f. $F = \left\{ \dfrac{2 + n}{n} : n \in \mathbb{N} \right\}$

   g. $G = \left\{(-1)^n - \frac{1}{n} : n \in \mathbb{N} \right\}$

   *h. $H = \{x \in \mathbb{R} : x^2 < 4\}$.

   i. $I = \{x^2 : -2 < x < 2\}$

6. If $0 < y \leq 1$, prove that there exists a unique positive real number x such that $x^2 = y$.

7. Prove that there exists a positive real number x such that $x^2 = 2$.

8. *Let A be a nonempty subset of $\mathbb{R}$. If $\alpha = \sup A$ is finite, show that for each $\epsilon > 0$, there is an $a \in A$ such that $\alpha - \epsilon < a \leq \alpha$.

9. Let E be a nonempty subset of $\mathbb{R}$ that is bounded above, and set $U = \{\beta \in \mathbb{R} : \beta$ is an upper bound of $E\}$. Prove that $\sup E = \inf U$.

10. Let A be a nonempty subset of $\mathbb{R}$ and let $-A = \{-x : x \in A\}$. Prove that $\inf A = -\sup(-A)$.

11. If A and B are nonempty subsets of $\mathbb{R}$ with $A \subset B$, prove that $\inf B \leq \inf A \leq \sup A \leq \sup B$.

12. Suppose that A and B are bounded subsets of $\mathbb{R}$. Prove that $A \cup B$ is bounded and that $\sup(A \cup B) = \sup\{\sup A, \sup B\}$.

13. Use the least upper bound property of $\mathbb{R}$ to prove that every nonempty subset of $\mathbb{R}$ that is bounded below has an infimum.

14. For A, B subsets of $\mathbb{R}$, define
$$A + B = \{a + b : a \in A, b \in B\} \quad \text{and} \quad A \cdot B = \{ab : a \in A, b \in B\}.$$

    a. For $A = \{-1, 2, 4, 7\}$ and $B = \{-2, -1, 1\}$, find $A + B$ and $A \cdot B$.

    *b. If A and B are nonempty and bounded above, prove that $\sup(A + B) = \sup A + \sup B$.

    c. If A and B are nonempty subsets of the positive real numbers that are bounded above, prove that $\sup(A \cdot B) = (\sup A)(\sup B)$.

    d. Give an example of two nonempty bounded sets A and B for which $\sup(A \cdot B) \neq (\sup A)(\sup B)$.

15. Let f, g be real-valued functions defined on a nonempty set X satisfying Range f and Range g are bounded subsets of $\mathbb{R}$. Prove each of the following:

    *a. $\sup\{f(x) + g(x) : x \in X\} \leq \sup\{f(x) : x \in X\} + \sup\{g(x) : x \in X\}$.

    b. Provide an example for which equality does not hold in (a).

    c. $\inf\{f(x) : x \in X\} + \inf\{g(x) : x \in X\} \leq \inf\{f(x) + g(x) : x \in X\}$.

    *d. If $f(x) \leq g(x)$ for all $x \in X$, then $\sup\{f(x) : x \in X\} \leq \sup\{g(x) : x \in X\}$.

    e. For each $x \in X$, let $h(x) = \max\{f(x), g(x)\}$. Prove that $\sup\{h(x) : x \in X\} = \max\{\sup\{f(x) : x \in X\}, \sup\{g(x) : x \in X\}\}$.

16. Let $X = Y = [0, 1]$ and let $f : X \times Y \to \mathbb{R}$ be defined by $f(x, y) = 3x + 2y$.

    *a. For each $x \in X$, find $F(x) = \sup\{f(x, y) : y \in Y\}$; then find $\sup\{F(x) : x \in X\}$.

    b. For each $y \in Y$, find $G(y) = \sup\{f(x, y) : x \in X\}$; then find $\sup\{G(y) : y \in Y\}$.

    *c. Find $\sup\{f(x, y) : (x, y) \in X \times Y\}$. Compare your answer with your answer in parts (a) and (b).

17. Perform the computations of Exercise 16 with $X = [-1, 1]$, $Y = [0, 2]$, and $f(x, y) = 3x - 2y$.

18. Let X, Y be nonempty sets, and let f be a nonnegative real-valued function defined on $X \times Y$. For each $x \in X$ and $y \in Y$, define
$$F(x) = \sup\{f(x, y) : y \in Y\}, \qquad G(y) = \sup\{f(x, y) : x \in X\}.$$
Prove that
$$\sup\{F(x) : x \in X\} = \sup\{G(y) : y \in Y\} = \sup\{f(x, y) : (x, y) \in X \times Y\}.$$

19. Let X, Y be nonempty sets and let $f : X \times Y \to \mathbb{R}$ be a function with bounded range. Let
$$F(x) = \sup\{f(x, y) : y \in Y\} \text{ and } H(y) = \inf\{f(x, y) : x \in X\}.$$
Prove that
$$\sup\{H(y) : y \in Y\} \leq \inf\{F(x) : x \in X\}.$$

20. Let $X = Y = [0, 1]$. Perform the computations of Exercise 19 for each of the following functions $f(x, y)$.

    *a. $f(x, y) = 3x + 2y$ **b.** $f(x, y) = \begin{cases} 1, & x = y, \\ 0, & x \neq y. \end{cases}$

21. Let J be a subset of $\mathbb{R}$ that has the following property: if $x, y \in J$ with $x < y$, then $t \in J$ for every t satisfying $x < t < y$. Prove that J is an interval as defined in Definition 1.4.8.

1.5   Consequences of the Least Upper Bound Property

In this section, we look at a number of elementary properties of the real numbers which in more elementary courses are usually always taken for granted. As we will see however, these are all actually consequences of the least upper bound property of the real numbers.

THEOREM 1.5.1 (Archimedian Property) *If $x, y \in \mathbb{R}$ and $x > 0$, then there exists a positive integer n such that*

$$n x > y.$$

Proof. If $y \leq 0$, then the result is true for all n. Thus assume that $y > 0$. We will again use the method of proof by contradiction. Let

$$A = \{ nx : n \in \mathbb{N} \}.$$

If the result is false, that is, there does not exist an $n \in \mathbb{N}$ such that $nx > y$, then $nx \leq y$ for all $n \in \mathbb{N}$. Thus y is an upper bound for A. Thus since $A \neq \emptyset$, A has a least upper bound in $\mathbb{R}$. Let $\alpha = \sup A$. Since $x > 0$, $\alpha - x < \alpha$. Therefore $\alpha - x$ is not an upper bound and thus there exists an element of A, say mx such that

$$\alpha - x < mx.$$

But then $\alpha < (m + 1)x$, which contradicts the fact that α is an upper bound of A. Therefore, there exists a positive integer n such that $nx > y$. $\square$

Remark. One way in which the previous result is often used is as follows: given $\epsilon > 0$, there exists a positive integer n_o such that $n_o \epsilon > 1$. As a consequence,

$$\frac{1}{n} < \epsilon$$

for all integers n, $n \geq n_o$.

THEOREM 1.5.2 *If $x, y \in \mathbb{R}$ and $x < y$, then there exists $r \in \mathbb{Q}$ such that*

$$x < r < y.$$

Proof. Assume first that $x \geq 0$. Since $y - x > 0$, by Theorem 1.5.1 there exists an integer $n > 0$ so that

$$n(y - x) > 1 \quad \text{or} \quad ny > 1 + nx.$$

Again by Theorem 1.5.1, $\{k \in \mathbb{N} : k > nx\}$ is nonempty. Thus by the well ordering principle there exists $m \in \mathbb{N}$ such that

$$m - 1 \leq nx < m.$$

Therefore

$$nx < m \leq 1 + nx < ny,$$

or dividing by n,

$$x < \frac{m}{n} < y.$$

If $x < 0$ and $y > 0$, then the result is obvious. Finally, if $x < y < 0$, then by the above there exists $r \in \mathbb{Q}$ such that $-y < r < -x$, i.e., $x < -r < y$. $\square$

The conclusion of Theorem 1.5.2 is often expressed by the statement that the rational numbers are **dense** in the real numbers, that is, between any two real numbers there exists a rational number. A precise definition of "dense" is given in Definition 2.2.19.

Another consequence of the least upper bound property is the following theorem concerning the existence of nth roots.

THEOREM 1.5.3 *For every real number $x > 0$ and every positive integer n, there exists a unique positive real number y so that $y^n = x$*

The number y is written as $\sqrt[n]{x}$ or $x^{\frac{1}{n}}$ and is called the nth **root** of x. The uniqueness of y is obvious. Since the existence of y will be an immediate consequence of the intermediate value theorem (Theorem 4.2.11), we omit the details of the proof in the text. A sketch of the proof of Theorem 1.5.3 using the least upper bound property is included in the miscellaneous exercises. It should be emphasized that the proof of Theorem 4.2.11 also depends on the least upper bound property.

COROLLARY 1.5.4 *If a, b are positive real numbers and n is a positive integer, then*

$$(ab)^{\frac{1}{n}} = a^{\frac{1}{n}} b^{\frac{1}{n}}.$$

Proof. Set $\alpha = a^{1/n}$ and $\beta = b^{1/n}$. Then

$$ab = \alpha^n \beta^n = (\alpha\beta)^n.$$

Thus by uniqueness, $\alpha\beta = (ab)^{1/n}$. $\square$

Exercises 1.5

1. *If r and s are positive rational numbers, prove directly (without using the supremum property) that there exists an $n \in \mathbb{N}$ such that $nr > s$.

2. Given any $x \in \mathbb{R}$, prove that there exists a unique $n \in \mathbb{Z}$ such that $n - 1 \le x < n$.

3. If $r \ne 0$ is a rational number and x is an irrational number, prove that $r + x$ and rx are irrational.

4. *Prove directly (without using Theorem 1.5.2) that between any two rational numbers there exists a rational number.

5. If $x, y \in \mathbb{R}$ with $x < y$, show that $x < tx + (1-t)y < y$ for all t, $0 < t < 1$.

6. *a. Prove that between any two rational numbers there exists an irrational number.

    *b. Prove that between any two real numbers there exists an irrational number.

7. If $x > 0$, show that there exists $n \in \mathbb{N}$ such that $1/2^n < x$.

8. *Let $x, y \in \mathbb{R}$ with $x < y$. If $u \in \mathbb{R}$ with $u > 0$, show that there exists a rational number r such that $x < ru < y$.

9. Let $B = \{r \in \mathbb{Q} : r > 0 \text{ and } r^2 < 2\}$ and $\alpha = \sup B$. Prove that $\alpha^2 = 2$.

1.6 Binary and Ternary Expansions[3]

In our standard base 10 number system we use the integers $\{0, 1, ..., 9\}$ to represent real numbers. Base 10 however is not the only possible base. In base 3 (ternary) we use only the integers $\{0, 1, 2\}$, and in base 2 (binary) we use only $\{0, 1\}$. In this section, we will show how the least upper bound property may be used to prove the existence of the expansion of a real number x, $0 < x \le 1$, for a given base. For purposes of illustration, and for later use, we will use base 2 and 3, which are commonly referred to as the binary and ternary expansions respectively.

In base ten, given the decimal .1021, what we really mean is the real number given by

$$\frac{1}{10} + \frac{0}{10^2} + \frac{2}{10^3} + \frac{1}{10^4}.$$

However, in base 3, the expansion .1021 represents

$$\frac{1}{3} + \frac{0}{3^2} + \frac{2}{3^3} + \frac{1}{3^4},$$

---

[3]This section can be omitted on first reading. The ternary expansion of a real number is only required in Section 2.5 and can be covered at that point.

which is the ternary expansion of 34/81.

For a given x, $0 < x \leq 1$, the ternary or base 3 expansion of x is defined inductively as follows:

DEFINITION 1.6.1 *Let* $n_1 \in \{0, 1, 2\}$ *be the largest integer such that*

$$\frac{n_1}{3} < x.$$

Having chosen $n_1, ..., n_k$, *let* $n_{k+1} \in \{0, 1, 2\}$ *be the largest integer such that*

$$\frac{n_1}{3} + \frac{n_2}{3^2} + \cdots + \frac{n_k}{3^k} + \frac{n_{k+1}}{3^{k+1}} < x.$$

The expression $.n_1 n_2 n_3 .....$ *is called the* **ternary expansion** *of* x.

If we set

$$E = \left\{ \frac{n_1}{3} + \cdots + \frac{n_k}{3^k} : k = 1, 2, ... \right\},$$

then $E \neq \emptyset$ and E is bounded above by x. As we will shortly see, sup $E = x$. In terms of series, which will be covered in detail later, we have

$$x = \sum_{k=1}^{\infty} \frac{n_k}{3^k}.$$

The **binary expansion** of x, or the expansion of x to any other base, is defined similarly. For the binary expansion, the integer n_k at each step is chosen as the largest integer in $\{0, 1\}$.

EXAMPLES 1.6.2 (a) We now use the above definition to obtain the ternary expansion of $\frac{1}{3}$. At the first step, we must choose n_1 as the largest integer in $\{0, 1, 2\}$ such that

$$\frac{n_1}{3} < \frac{1}{3}.$$

This inequality fails for $n_1 = 1, 2$. Thus $n_1 = 0$. To find n_2, we choose the largest integer $n_2 \in \{0, 1, 2\}$ such that

$$\frac{0}{3} + \frac{n_2}{3^2} < \frac{1}{3},$$

which is satisfied by $n_2 = 2$. To find n_3 we must have

$$\frac{0}{3} + \frac{2}{3^2} + \frac{n_3}{3^3} < \frac{1}{3}.$$

It is left as an exercise to show that this is satisfied for $n_3 = 0, 1, 2$. Thus we take $n_3 = 2$. At this stage we conjecture that $n_k = 2$ for all $k \geq 2$, and that

$$\frac{1}{3} = .02222... \qquad \text{(base 3)}.$$

To see that this indeed is the case we use the fact that for the geometric series

$$\sum_{k=0}^{\infty} r^k = \frac{1}{(1-r)}, \qquad 0 < r < 1.$$

Thus

$$.0222... = \sum_{k=2}^{\infty} \frac{2}{3^k} = \frac{2}{9} \sum_{k=0}^{\infty} \left(\frac{1}{3}\right)^k = \frac{2}{9} \frac{1}{1 - \frac{1}{3}} = \frac{1}{3}.$$

In (c) we will illustrate how mathematical induction may also be used to prove such a result. The above ternary expansion is not unique. The number $\frac{1}{3}$ also has a finite expansion

$$\frac{1}{3} = .1000... \qquad \text{(base 3)}.$$

We will discuss this in more detail at the end of this section.

(b) The binary expansion of $\frac{1}{3}$ is given by

$$\frac{1}{3} = .010101... \qquad \text{(base 2)}.$$

This expansion can be obtained using the definition and induction (Exercise 4). Alternately, using the geometric series we have

$$.010101... = \sum_{k=1}^{\infty} \frac{1}{2^{2k}} = \frac{1}{4} \sum_{k=0}^{\infty} \left(\frac{1}{4}\right)^k = \frac{1}{3}.$$

(c) The ternary expansion of $\frac{1}{2}$ is given by

$$\frac{1}{2} = .1111..... \qquad \text{(base 3)}.$$

We now show in detail how this expansion is derived. Since $\frac{1}{3} < \frac{1}{2}$ and $\frac{2}{3} > \frac{1}{2}$, $n_1 = 1$. We will use the second principle of mathematical induction to prove that $n_k = 1$ for all $k \in \mathbb{N}$. By the above, the result is true for $k = 1$. Let $k > 1$ and assume that $n_j = 1$ for all $j < k$. By definition n_k is the largest integer in $\{0, 1, 2\}$ such that

$$\frac{1}{3} + \cdots + \frac{1}{3^{k-1}} + \frac{n_k}{3^k} < \frac{1}{2}. \tag{3}$$

Using the identity $r + r^2 + \cdots + r^n = \dfrac{r - r^{n+1}}{1 - r}, r \neq 1$, (Example 1.3.2(a)) with $n = k - 1$, we obtain

$$\frac{1}{3} + \cdots + \frac{1}{3^{k-1}} = \frac{\frac{1}{3} - \frac{1}{3^k}}{\frac{2}{3}} = \frac{1}{2}\left(1 - \frac{1}{3^{k-1}}\right).$$

Substituting into equation (3) and multiplying by 2 gives

$$1 - \frac{3}{3^k} + \frac{2n_k}{3^k} < 1.$$

This inequality is true if $n_k = 0$ or 1 and is false if $n_k = 2$. Since n_k is to be chosen as the largest integer of $\{0, 1, 2\}$ for which (3) holds, $n_k = 1$. Thus by Theorem 1.3.3 $n_k = 1$ for all $k \in \mathbb{N}$. $\square$

THEOREM 1.6.3 *Let $x \in \mathbb{R}$ with $0 < x \leq 1$, and with $\{n_k\}$ as defined in Definition 1.6.1 let*

$$E = \left\{ \frac{n_1}{3} + \cdots + \frac{n_k}{3^k} : k = 1, 2, .. \right\}.$$

Then $\sup E = x$.

Proof. Let $\alpha = \sup E$. Since x is an upper bound, $\alpha \leq x$. Suppose $\alpha < x$. Let k be the smallest positive integer such that

$$\frac{1}{3^k} < x - \alpha \qquad \text{or} \qquad \alpha + \frac{1}{3^k} < x.$$

Since $\alpha = \sup E$,

$$\frac{n_1}{3} + \cdots + \frac{n_k}{3^k} \leq \alpha.$$

But

$$\frac{n_1}{3} + \cdots + \frac{n_k}{3^k} + \frac{1}{3^k} = \frac{n_1}{3} + \cdots + \frac{n_{k-1}}{3^{k-1}} + \frac{n_k + 1}{3^k} < x.$$

If $n_k = 0$ or 1, then this contradicts the choice of n_k. If $n_k = 2$, then we have

$$\frac{n_1}{3} + \cdots + \frac{n_{k-1} + 1}{3^{k-1}} < x.$$

If any n_j, $1 \leq j \leq k - 1$ is 0 or 1, we have a contradiction to the choice of n_j. If all the $n_j = 2$, then we obtain

$$\frac{2}{3} + \frac{1}{3} < x \leq 1,$$

which is also a contradiction. $\square$

The expansion

$$x = .n_1 n_2 n_3 ....$$

is **finite** or **terminating** if there exists an integer $m \in \mathbb{N}$ such that $n_k = 0$ for all $k > m$. Otherwise, the expansion is **infinite** or **nonterminating**. The expansion of a real number in a given base is not always unique; when x has a finite expansion, it also has an infinite expansion as well. For the ternary expansion, when

$$x = \frac{a}{3^m}, \qquad a \in \mathbb{N} \quad \text{with} \quad 0 < a < 3^m,$$

and 3 does not divide a, then x has a finite expansion of the form

$$x = \frac{a_1}{3} + \cdots + \frac{a_m}{3^m}, \qquad a_m \in \{1, 2\},$$

or an infinite expansion of the form

$$x = \frac{a_1}{3} + \cdots + \frac{0}{3^m} + \sum_{k=m+1}^{\infty} \frac{2}{3^k} \quad \text{when } a_m = 1, \text{ or}$$

$$x = \frac{a_1}{3} + \cdots + \frac{1}{3^m} + \sum_{k=m+1}^{\infty} \frac{2}{3^k}, \quad \text{when } a_m = 2.$$

Exercises 1.6

1. Find the ternary expansion of each of the following.
   *a. $\frac{1}{4}$ b. $\frac{1}{13}$

2. Find the real number determined by each of the following finite or infinite binary expansions. In the case of an infinite expansion use the geometric series as in Example 1.6.2 (c) to determine your answer.
   a. .1010   b. .101010$\cdots$ *c. .0101
   *d. .010101$\cdots$ e. .001001   *f. .001001$\cdots$

3. Find the real number determined by each of the following finite or infinite ternary expansions.
   *a. .0022   b. .00222$\cdots$ c. .1010
   *d. .101010$\cdots$ e. .001001001$\cdots$ *f. .121212$\cdots$

4. *Find the binary expansion of $\frac{1}{3}$. Use induction to prove the result.

5. *Find both the finite and infinite binary expansions of $\frac{3}{16}$.

6. If $0 < x \leq 1$, prove that the infinite ternary expansion of x is unique.

7. If $x = \dfrac{a}{2^m}$, with $a \in \mathbb{N}$ odd, and $0 < a < 2^m$, show that x has a finite binary expansion of the form

   $$x = \frac{a_1}{2} + \cdots \frac{a_m}{2^m}, \qquad a_m = 1,$$

   and an infinite expansion

   $$x = \frac{a_1}{2} + \cdots + \frac{0}{2^m} + \sum_{k=m+1}^{\infty} \frac{1}{2^k}.$$

1.7   Countable and Uncountable Sets

In discussing sets, we all have an intuitive understanding of what it means for a set to be finite or infinite, and what it means for two finite sets to be of the same size; that is, to have the same number of elements. For example, the sets $A = \{2, 7, 11, 21\}$ and $B = \{7, 3, 19, 32\}$ both have the same number of elements; namely, four. We accomplish this by counting the number of elements in each of the two sets. Alternately, the same can be accomplished, without counting, by simply pairing up the elements; i.e.,

$$7 \leftrightarrow 2$$
$$3 \leftrightarrow 7$$
$$19 \leftrightarrow 11$$
$$32 \leftrightarrow 21$$

For infinite sets, the concept of two sets being of the same size or having the same number of elements is vague. For example, let S denote the squares of the positive integers; namely,

$$S = \{1^2, 2^2, 3^2, ...\}.$$

Then on one hand, S is a proper subset of the positive integers $\mathbb{N}$, yet as Galileo (1564–1642) observed, the sets $\mathbb{N}$ and S can be placed into a one-to-one correspondence as follows:

$$1 \longleftrightarrow 1^2$$
$$2 \longleftrightarrow 2^2$$
$$3 \longleftrightarrow 3^2$$
$$\vdots$$

This example caused Galileo, and many subsequent mathematicians, to conclude that the standard notion of "size of a set" did not apply to infinite sets. Cantor on the other hand realized that the concept of one-to-one correspondence raised many interesting questions about the theory of infinite sets. In this section, we take a closer look at infinite sets and what it means for an infinite set to be countable. We begin by defining the concept of equivalence of sets.

DEFINITION 1.7.1 *Two sets A and B are said to be* **equivalent** *(or to have the same* **cardinality***), denoted $A \sim B$, if there exists a one-to-one function of A onto B.*

The notion of equivalence of sets satisfies the following:
(i) $A \sim A$. (**reflexive**)
(ii) If $A \sim B$, then $B \sim A$. (**symmetric**)
(iii) If $A \sim B$ and $B \sim C$, then $A \sim C$. (**transitive**)

DEFINITION 1.7.2 *For each positive integer n, let $\mathbb{N}_n = \{1, 2, ..., n\}$. As in Section 1, $\mathbb{N}$ denotes the set of all positive integers. If A is a set, we say:*

(a) *A is **finite** if $A \sim \mathbb{N}_n$ for some n, or if $A = \emptyset$.*
(b) *A is **infinite** if A is not finite.*
(c) *A is **countable** if $A \sim \mathbb{N}$.*
(d) *A is **uncountable** if A is neither finite nor countable.*
(e) *A is **at most countable** if A is finite or countable.*

The reader might want to ponder Exercise 20 at this point. Countable sets are often called **denumerable** or **enumerable** sets. It should be pointed out that some textbooks, when using the term countable, include the possibility that the set is finite.

EXAMPLES 1.7.3 (a) As above, let $S = \{1^2, 2^2, 3^2, ...\}$. Then the function $g(n) = n^2$ is a one-to-one mapping of $\mathbb{N}$ onto S. Thus $S \sim \mathbb{N}$ and S is countable.

(b) For our second example, we show that $\mathbb{Z} \sim \mathbb{N}$. To see this, consider the function $f : \mathbb{N} \to \mathbb{Z}$ defined by

$$f(n) = \begin{cases} \dfrac{n}{2}, & (n \text{ even}), \\ -\dfrac{(n-1)}{2}, & (n \text{ odd}). \end{cases}$$

The following diagram illustrates what the mapping f does to the first few integers.

$$
\begin{array}{ccc}
1 & \longrightarrow & 0 \\
2 & \longrightarrow & 1 \\
3 & \longrightarrow & -1 \\
4 & \longrightarrow & 2 \\
5 & \longrightarrow & -2 \\
& \vdots &
\end{array}
$$

It is left as an exercise (Exercise 1) to show that this function is a one-to-one mapping of $\mathbb{N}$ onto $\mathbb{Z}$. Thus $\mathbb{Z} \sim \mathbb{N}$, and the set $\mathbb{Z}$ is also countable. $\square$

As another illustration of countable sets consider the following theorem.

THEOREM 1.7.4 $\mathbb{N} \times \mathbb{N}$ *is countable.*

Proof. For this example, it is easier to construct a one-to-one mapping of $\mathbb{N} \times \mathbb{N}$ onto $\mathbb{N}$. Such a function is given by

$$f(m, n) = 2^{m-1}(2n - 1).$$

It is left as an exercise (Exercise 3) to show that f as defined is a one-to-one mapping of $\mathbb{N} \times \mathbb{N}$ onto $\mathbb{N}$. $\square$

One of our goals in this section will be to show that the set $\mathbb{Q}$ of rational numbers is countable.

Sequences

DEFINITION 1.7.5 *If A is a set, by a* **sequence** *in A we mean a function f from $\mathbb{N}$ into A. For each $n \in \mathbb{N}$, let $x_n = f(n)$. Then x_n is called the nth* **term** *of the sequence f.*

For notational convenience, sequences are denoted by $\{x_n\}_{n=1}^{\infty}$ or just $\{x_n\}$, rather than the function f. Note however the distinction between $\{x_n\}_{n=1}^{\infty}$, which denotes the sequence, and $\{x_n : n = 1, 2, ...\}$ which denotes the range of the sequence. For example, $\{1 - (-1)^n\}$ denotes the sequence f where

$$f(n) = x_n = 1 - (-1)^n.$$

On the other hand, $\{x_n : n = 1, 2, ...\} = \{0, 2\}$.

By definition, if A is a countable set, then there exists a one-to-one function f from $\mathbb{N}$ onto A. Thus

$$A = \text{Range } f = \{x_n : n = 1, 2, ...\}.$$

The sequence f is called an **enumeration** of A, i.e., $A = \{x_n : n = 1, 2, ...\}$ with $x_n \neq x_m$ whenever $n \neq m$. This ability to enumerate elements of a countable set plays a key role in the proofs of some of the following results.

THEOREM 1.7.6 *Every infinite subset of a countable set is countable.*

Proof. Let A be a countable set and let $\{x_n : n = 1, 2, ...\}$ be an enumeration of A. Suppose E is an infinite subset of A. Then each $x \in E$ is of the form x_k for some $k \in \mathbb{N}$. We inductively construct a function $f : \mathbb{N} \to E$ as follows: let n_1 be the smallest positive integer such that $x_{n_1} \in E$. Such an integer exists by the well ordering principle. Having chosen $n_1, ..., n_{k-1}$, let n_k be the smallest integer greater than n_{k-1} such that $x_{n_k} \in E \setminus \{x_{n_1}, ..., x_{n_{k-1}}\}$. Set $f(k) = x_{n_k}$. Since E is infinite, f is defined on $\mathbb{N}$.

If $m > k$, then $n_m > n_k$ and thus $x_{n_m} \neq x_{n_k}$. Therefore f is one-to-one. The function f is onto since if $x \in E$, then $x = x_j$ for some j. By construction, $n_k = j$ for some k, and thus $f(k) = x$. $\square$

THEOREM 1.7.7 *If f maps $\mathbb{N}$ onto A, then A is at most countable.*

Proof. If A is finite, the result is certainly true. Suppose A is infinite. Since

f maps $\mathbb{N}$ onto A, each $a \in A$ is of the form $f(n)$ for some $n \in \mathbb{N}$. For each $a \in A$, by the well ordering principle

$$f^{-1}(\{a\}) = \{n \in \mathbb{N} : f(n) = a\}$$

has a smallest integer, which we denote by n_a. Consider the mapping $a \to n_a$ of A into $\mathbb{N}$. If $a \neq b$, then since f is a function, $n_a \neq n_b$. Also, since A is infinite, $\{n_a : a \in A\}$ is an infinite subset of $\mathbb{N}$. Thus the mapping $a \to n_a$ is a one-to-one mapping of A onto an infinite subset of $\mathbb{N}$. Therefore by Theorem 1.7.6 A is countable.   □

Indexed Families of Sets

In Section 1 we defined the union and intersection of two sets. We now extend these definitions to larger collections of sets. Recall that if X is a set, $\mathcal{P}(X)$ denotes the set of all subsets of X.

DEFINITION 1.7.8 *Let A and X be nonempty sets. An **indexed family** of subsets of X with **index set** A is a function from A into $\mathcal{P}(X)$.*

If $f : A \to \mathcal{P}(A)$, then for each $\alpha \in A$, we let $E_\alpha = f(\alpha)$. As for sequences, we denote this function by $\{E_\alpha\}_{\alpha \in A}$. If $A = \mathbb{N}$, then $\{E_n\}_{n \in \mathbb{N}}$ is called a **sequence of subsets** of X. In this instance, we adopt the more conventional notation $\{E_n\}_{n=1}^\infty$ to denote $\{E_n\}_{n \in \mathbb{N}}$.

EXAMPLES 1.7.9 The following are all examples of indexed families of sets.

(a) The sequence $\{\mathbb{N}_n\}_{n=1}^\infty$, where $\mathbb{N}_n = \{1, 2, ..., n\}$, is a sequence of subsets of $\mathbb{N}$.

(b) For each $n \in \mathbb{N}$, set $I_n = \{x \in \mathbb{R} : 0 < x < \frac{1}{n}\}$. Then $\{I_n\}_{n=1}^\infty$ is a sequence of subsets of $\mathbb{R}$.

(c) For each x, $0 < x < 1$, let

$$E_x = \{r \in \mathbb{Q} : 0 \leq r < x\}.$$

Then $\{E_x\}_{x \in (0,1)}$ is an indexed family of subsets of $\mathbb{Q}$. In this example, the open interval $(0, 1)$ is our index set.   □

DEFINITION 1.7.10 *Suppose $\{E_\alpha\}_{\alpha \in A}$ is an indexed family of subsets of X. The **union** of the family of sets $\{E_\alpha\}_{\alpha \in A}$ is defined to be*

$$\bigcup_{\alpha \in A} E_\alpha = \{x \in X : x \in E_\alpha \text{ for some } \alpha \in A\}.$$

*The **intersection** of the family of sets $\{E_\alpha\}_{\alpha \in A}$ is defined as*

$$\bigcap_{\alpha \in A} E_\alpha = \{x \in X : x \in E_\alpha \text{ for all } \alpha \in A\}.$$

If $A = \mathbb{N}$ we use the notation

$$\bigcup_{n=1}^{\infty} E_n \quad \text{and} \quad \bigcap_{n=1}^{\infty} E_n \quad \text{instead of} \quad \bigcup_{n \in \mathbb{N}} E_n \quad \text{and} \quad \bigcap_{n \in \mathbb{N}} E_n$$

respectively. Also, if $A = \mathbb{N}_k$, then

$$\bigcup_{n \in \mathbb{N}_k} E_n \quad \text{is denoted by} \quad \bigcup_{n=1}^{k} E_n,$$

with an analogous definition for the intersection. Occasionally, when the index set A is fixed in the discussion, we will use the shorthand notation $\bigcup_\alpha E_\alpha$ or $\bigcap_\alpha E_\alpha$ rather than $\bigcup_{\alpha \in A} E_\alpha$ or $\bigcap_{\alpha \in A} E_\alpha$.

EXAMPLES 1.7.11 We now consider the union and intersection of the families of sets given in the previous example.

(a) With $\mathbb{N}_n = \{1, 2, ..., n\}$, we have

$$\bigcap_{n=1}^{\infty} \mathbb{N}_n = \{1\}, \quad \text{and} \quad \bigcup_{n=1}^{\infty} \mathbb{N}_n = \mathbb{N}.$$

Since 1 is the only element which is in $\mathbb{N}_n$ for all n, $\bigcap_n \mathbb{N}_n = \{1\}$. For the union, since $\mathbb{N}_n \subset \mathbb{N}$ for all n, $\bigcup_n \mathbb{N}_n \subset \mathbb{N}$. On the other hand, if $n \in \mathbb{N}$, then $n \in \mathbb{N}_n$, and as a consequence, $\mathbb{N} \subset \bigcup_n \mathbb{N}_n$, which proves equality.

(b) As in the previous example, for $n \in \mathbb{N}$ let $I_n = \{x \in \mathbb{R} : 0 < x < \frac{1}{n}\}$. We first show that

$$\bigcap_{n=1}^{\infty} I_n = \emptyset.$$

Suppose not, then there exist $x \in \mathbb{R}$ such that $x \in I_n$ for all n, i.e.,

$$0 < x < \frac{1}{n}, \quad \text{for all} \quad n \in \mathbb{N}.$$

This however contradicts Theorem 1.5.1 which guarantees the existence of a positive integer n such that $nx > 1$. For the union, since $I_n \subset I_1$ for all $n \geq 1$,

$$\bigcup_{n=1}^{\infty} I_n = I_1 = \{x \in \mathbb{R} : 0 < x < 1\}$$

(c) We leave it as an exercise (Exercise 9) to show that if E_x is defined as in Example 1.7.9(c), i.e., $E_x = \{r \in \mathbb{Q} : 0 \leq r < x\}$, then

$$\bigcap_{x \in (0,1)} E_x = \{0\}, \quad \text{and} \quad \bigcup_{x \in (0,1)} E_x = \{r \in \mathbb{Q} : 0 \leq r < 1\}. \quad \square$$

As for a finite number of sets, we also have analogs of the distributive laws and De Morgan's laws for arbitrary unions and intersections.

THEOREM 1.7.12 (Distributive Laws) *If E_α, $\alpha \in A$, and E are subsets of a set X, then*

(a) $E \cap \left(\bigcup_{\alpha \in A} E_\alpha \right) = \bigcup_{\alpha \in A} (E \cap E_\alpha)$,

(b) $E \cup \left(\bigcap_{\alpha \in A} E_\alpha \right) = \bigcap_{\alpha \in A} (E \cup E_\alpha)$.

THEOREM 1.7.13 (De Morgan's Laws) *If $\{E_\alpha\}_{\alpha \in A}$ is a family of subsets of X, then*

(a) $\left(\bigcup_{\alpha \in A} E_\alpha \right)^c = \bigcap_{\alpha \in A} E_\alpha^c$,

(b) $\left(\bigcap_{\alpha \in A} E_\alpha \right)^c = \bigcup_{\alpha \in A} E_\alpha^c$.

The proofs of both of these theorems, as well as the following analogue of Theorems 1.2.5 and 1.2.6 are left to the exercises.

THEOREM 1.7.14 *Let f be a function from X into Y, and let A be a nonempty set.*

(a) *If $\{E_\alpha\}_{\alpha \in A}$ is a family of subsets of X, then*

$$f \left(\bigcup_{\alpha \in A} E_\alpha \right) = \bigcup_{\alpha \in A} f(E_\alpha),$$

$$f \left(\bigcap_{\alpha \in A} E_\alpha \right) \subset \bigcap_{\alpha \in A} f(E_\alpha).$$

(b) *If $\{B_\alpha\}_{\alpha \in A}$ is a family of subsets of Y, then*

$$f^{-1} \left(\bigcup_{\alpha \in A} B_\alpha \right) = \bigcup_{\alpha \in A} f^{-1}(B_\alpha),$$

$$f^{-1} \left(\bigcap_{\alpha \in A} B_\alpha \right) = \bigcap_{\alpha \in A} f^{-1}(B_\alpha).$$

The Countability of $\mathbb{Q}$

The countability of the set of rational numbers $\mathbb{Q}$ will follow as a corollary of the following theorem.

THEOREM 1.7.15 *If $\{E_n\}_{n=1}^{\infty}$ is a sequence of countable sets and*

$$S = \bigcup_{n=1}^{\infty} E_n,$$

then S is countable.

Proof. Since E_n is countable for each $n \in \mathbb{N}$, we can write

$$E_n = \{ x_{n,k} : k = 1, 2, \dots \}.$$

Since E_1 is an infinite subset of S, the set S itself is infinite. Consider the function $h : \mathbb{N} \times \mathbb{N} \to S$ by

$$h(n, k) = x_{n,k}.$$

The function h, although not necessarily one-to-one, is a mapping of $\mathbb{N} \times \mathbb{N}$ onto S. Thus since $\mathbb{N} \times \mathbb{N} \sim \mathbb{N}$, there exists a mapping of $\mathbb{N}$ onto S. Hence by Theorem 1.7.7 the set S is countable. $\square$

COROLLARY 1.7.16 $\mathbb{Q}$ *is countable.*

Proof. For each $m \in \mathbb{N}$, let

$$E_m = \left\{ \frac{n}{m} : n \in \mathbb{Z} \right\}.$$

Then E_m is countable, and since $\mathbb{Q} = \bigcup_{m=1}^{\infty} E_m$, by Theorem 1.7.15 the set $\mathbb{Q}$ is countable. $\square$

The Uncountability of $\mathbb{R}$

In November 1873, in a letter to Dedekind, Cantor asked whether the set $\mathbb{R}$ itself was countable. A month later he answered his own question by proving that $\mathbb{R}$ was not countable. We now prove, using Cantor's elegant "diagonal" argument, that the closed interval $[0, 1]$ is uncountable, and thus $\mathbb{R}$ itself is uncountable. For the proof we will use the fact that as in Section 6, every $x \in [0, 1]$ has a decimal expansion of the form $x = .n_1 n_2 \cdots$ with $n_i \in \{0, 1, 2, \dots, 9\}$. As for the binary and ternary expansions, the decimal expansion is not necessarily unique. Certain numbers such as $\frac{1}{10}$ have two expansions; namely

$$\frac{1}{10} = .100\cdots \qquad \text{and} \qquad \frac{1}{10} = .0999\cdots.$$

This however will not be crucial in the proof of the following theorem.

THEOREM 1.7.17 *The closed interval* $[0, 1]$ *is uncountable.*

Proof. Since there are infinitely many rational numbers in $[0, 1]$, the set is not finite. To prove that it is uncountable, we only need to show that it is not countable. To accomplish this, we will prove that every countable subset of $[0, 1]$ is a proper subset of $[0, 1]$. Thus $[0, 1]$ cannot be countable.

Let $E = \{x_n \; n = 1, 2, ...\}$ be a countable subset of $[0, 1]$. Then each x_n has a decimal expansion

$$x_n = .x_{n,1}x_{n,2}x_{n,3} \cdots$$

where for each $k \in \mathbb{N}$, $x_{n,k} \in \{0, 1, ..., 9\}$. We now define a new number

$$y = .y_1 y_2 y_3 \cdots$$

as follows: if $x_{n,n} \leq 5$, define $y_n = 6$; if $x_{n,n} \geq 6$, define $y_n = 3$. Then $y \in [0, 1]$, and since $y_n \neq 0$ or 9, y is not one of the real numbers with two decimal expansions. Also, since for each $n \in \mathbb{N}$ $y_n \neq x_{n,n}$, we have $y \neq x_n$ for any n. Therefore $y \notin E$; i.e., E is a proper subset of $[0, 1]$. $\square$

Another example of an uncountable set is given in the following theorem.

THEOREM 1.7.18 *If A is the set of all sequences whose elements are 0 or 1, then A is uncountable.*

Remark. The set A is the set of all functions f from $\mathbb{N}$ into $\{0, 1\}$. Thus a sequence $f \in A$ if and only if $f(n) = 0$ or 1 for all n.

Proof. As in the previous theorem, we will also prove that every countable subset of A is a proper subset of A, and thus A cannot be countable.

Let E be a countable subset of A and let $\{s_n : n = 1, 2, ...\}$ be an enumeration of the set E. For each n, s_n is a sequence of 0's and 1's. We construct a new sequence s as follows: for each $k \in \mathbb{N}$, let

$$s(k) = 1 - s_k(k).$$

Thus if $s_k(k) = 0$, $s(k) = 1$, and if $s_k(k) = 1$, $s(k) = 0$. Thus $s \in A$. Since for all $k \in \mathbb{N}$ $s(k) \neq s_k(k)$, we have $s \neq s_n$ for any $n \in \mathbb{N}$. Therefore $s \notin E$, i.e., $E \subsetneq A$. $\square$

In the previous two theorems we proved that the closed interval $[0, 1]$ and the set of all sequences of 0's and 1's are both uncountable sets. A natural question is are these two sets equivalent? Considering the fact that every real number $x \in [0, 1]$ has a binary expansion, which is really a sequence of 0's and 1's, one would expect that the answer is yes. This indeed is the case (Miscellaneous Exercise 5).

Exercises 1.7

1. **a.** Prove that the function f of Example 1.7.3(b) is a one-to-one function of $\mathbb{N}$ onto $\mathbb{Z}$.

    b. Find a one-to-one function of $\mathbb{Z}$ onto $\mathbb{N}$.

    *c. Find a one-to-one function from $\mathbb{N}$ onto $\mathbb{O}$, the set of all odd positive integers.

2. For each of the following determine whether the set is finite, infinite, countable, or uncountable.
    a. $\{1 - (-1)^n : n \in \mathbb{N}\}$ **b.** $\{n \cos n\pi : n \in \mathbb{N}\}$
    c. $\{2^n : n \in \mathbb{Z}\}$ **d.** $\left\{\frac{m}{2^n} : m, n \in \mathbb{N}, m \text{ odd}\right\}$
    e. $[0, 1] \setminus \mathbb{Q}$ **f.** $\{x \in [0, 1] : x \text{ has an infinite ternary expansion}\}$

3. Prove that the function f of Theorem 1.7.4 is a one-to-one function of $\mathbb{N} \times \mathbb{N}$ onto $\mathbb{N}$.

4. ***a.** If $a, b \in \mathbb{R}$ with $a < b$, prove that $(a, b) \sim (0, 1)$.

    b. Prove that $(0, 1) \sim (0, \infty)$.

5. Suppose X, Y, Z are sets. If $X \sim Y$ and $Y \sim Z$, prove that $X \sim Z$.

6. ***a** If $A \sim X$ and $B \sim Y$, prove that $(A \times B) \sim (X \times Y)$.

    b. If A and B are countable sets, prove that $A \times B$ is countable.

7. If $X \sim Y$, prove that $\mathcal{P}(X) \sim \mathcal{P}(Y)$.

8. Find $\displaystyle\bigcup_{n=1}^{\infty} A_n$ and $\displaystyle\bigcap_{n=1}^{\infty} A_n$ for each of the following sequences of sets $\{A_n\}$.
    *a. $A_n = \{x \in \mathbb{R} : -n < x < n\}$, $n \in \mathbb{N}$
    b. $A_n = \{x \in \mathbb{R} : -\frac{1}{n} < x < 1\}$, $n \in \mathbb{N}$
    *c. $A_n = \{x \in \mathbb{R} : -\frac{1}{n} < x < 1 + \frac{1}{n}\}$, $n \in \mathbb{N}$
    d. $A_n = \{x \in \mathbb{R} : 0 \le x \le 1 - \frac{1}{n}\}$ $n \in \mathbb{N}$
    *e. $A_n = \left[\frac{1}{n}, 1 - \frac{1}{n}\right]$, $n \in \mathbb{N}, n \ge 2$
    f. $A_n = \{x \in \mathbb{R} : n \le x < \infty\}$

9. For each $x \in (0, 1)$, let $E_x = \{r \in \mathbb{Q} : 0 \le r < x\}$. Prove that

$$\bigcap_{x \in (0,1)} E_x = \{0\} \quad \text{and} \quad \bigcup_{x \in (0,1)} E_x = \{r \in \mathbb{Q} : 0 \le r < x\}.$$

10. Prove Theorem 1.7.12.

11. Prove Theorem 1.7.13.

12. Let f be a function from X into Y.

    *a. If $\{E_\alpha\}_{\alpha \in A}$ is a family of subsets of X, prove that
    $f(\bigcup_\alpha F_\alpha) = \bigcup_\alpha f(E_\alpha)$.

    b. If $\{B_\alpha\}_{\alpha \in A}$ is a family of subsets of Y, prove that
    $f^{-1}(\bigcap_\alpha B_\alpha) = \bigcap_\alpha f^{-1}(B_\alpha)$.

13. **a.** If A is a countable subset of an uncountable set X, prove that $X \setminus A$ is uncountable.

    *b. Prove that the set of irrational numbers is uncountable.

14. Suppose f is a function from X into Y. If the range of f is uncountable, prove that X is uncountable.

15. *a. For each $n \in \mathbb{N}$, prove that the collection of all polynomials in x of degree less than or equal to n with rational coefficients is countable.

    *b. Prove that the set of all polynomials in x with rational coefficients is countable.

16. Prove that the set of all intervals with rational endpoints is countable.

17. Let A be a nonempty subset of $\mathbb{R}$ that is bounded above and let $\alpha = \sup A$. If $\alpha \notin A$, prove that for every $\epsilon > 0$, the interval $(\alpha - \epsilon, \alpha)$ contains infinitely many points of A

18. *a. Prove that $(0,1) \sim (0,1]$. (This problem is not easy!)

    b. Prove that $(0,1) \sim [0,1]$.

19. *A real number a is **algebraic** if there exists a polynomial $p(x)$ with integer coefficients such that $p(a) = 0$. Prove that the set of algebraic numbers is countable.

20. Prove that any infinite set contains a countable subset.

21. Prove that a set is infinite if and only if it is equivalent to a proper subset of itself.

22. *Prove that any function from a set A to the set $\mathcal{P}(A)$ of all subsets of A is not onto.

23. *Prove that $[0,1] \times [0,1] \sim [0,1]$.

Notes

The most important concept of this chapter is the least upper bound property of the real numbers. This property will be fundamental in the development of the underlying theory of calculus. In the present chapter we have already seen its application in proving the Archimedian property (Theorem 1.5.1). For the rational number system this property can be proved directly. For the real number system it was originally assumed as an axiom by Archimedes (287–212 B.C.). Cantor however proved that the Archimedian property was no axiom, but a proposition derivable from the least upper bound property.

In subsequent chapters the least upper bound property will occur in proofs of theorems, either directly or indirectly, with regular frequency. It will play a crucial role in the characterization of the compact subsets of $\mathbb{R}$ and in the study of sequences of real numbers. It will also be required in the proof of the intermediate value theorem for continuous functions. One of the corollaries of this theorem is Theorem 1.5.3 on the existence of nth roots of positive real numbers. Many other results in the text will depend on previous theorems which required the least upper bound property in their proofs.

The emphasis on the least upper bound property is not meant to overshadow the importance of the concepts of countable and uncountable sets. The fact that the rational numbers are countable, and thus can be enumerated, will be used on several occasions in the construction of examples. In all the examples and exercises, every infinite subset of $\mathbb{R}$ turns out to be either countable or equivalent to $[0, 1]$. Cantor also made this observation and it led him to ask whether this result was true for every infinite subset of $\mathbb{R}$. Cantor was never able to answer this question; nor has anyone else. The assertion that every infinite subset of $\mathbb{R}$ is either countable or equivalent to $[0, 1]$ is known as the **continuum hypothesis**. In 1938 Kurt Gödel proved that the continuum hypothesis is consistent with the standard axioms of set theory; that is, Gödel showed that continuum hypothesis cannot be disproved on the basis of the standard axioms of set theory. On the other hand, Paul Cohen in 1963 showed that the continuum hypothesis is undecidable on the basis of the current axioms of set theory.

Cantor's creation of the theory of infinite sets was motivated to a great extend by problems arising in the study of convergence of Fourier series. We will discuss some of these problems in greater detail in Chapters 9 and 10. Cantor's original work on the theory of infinite sets can be found in his monograph listed in the Supplemental Readings.

Miscellaneous Exercises

1. Let A and B be nonempty sets. For $a \in A$, $b \in B$, define the ordered pair (a, b) by
$$(a, b) = \{\{a\}, \{a, b\}\}.$$
Prove that two ordered pairs (a, b) and (c, d) are equal if and only if $a = c$ and $b = d$.

The following two exercises are detailed and lengthy. The first exercise is a sketch of the proof of Theorem 1.5.3. The second exercise shows how the least upper bound property may be used to define the exponential function b^x, $b > 1$.

2. Let
$$E = \{t \in \mathbb{R} : t > 0 \text{ and } t^n < x\}.$$
a. Show that $E \neq \emptyset$ by showing that $x/(x + 1) \in E$.

b. Show that $1 + x$ is an upper bound of E.

Let
$$y = \sup E.$$
The remaining parts of the exercise are to show that $y^n = x$. This will be accomplished by showing that $y^n < x$ and $y^n > x$ lead to contradictions, leaving $y^n = x$. To accomplish this, the following inequality will prove useful. Suppose $0 < a < b$, then
$$b^n - a^n = (b - a)(b^{n-1} + ab^{n-2} + \cdots + a^{n-1}) < n(b - a)b^{n-1}. \quad (*)$$

c. Show that the assumption $y^n < x$ contradicts $y = \sup E$ as follows: Choose $0 < h < 1$ such that

$$h < \frac{x - y^n}{n(y+1)^{n-1}}.$$

Use (*) to show that $y + h \in E$.

d. Show that the assumption $y^n > x$ also leads to a contradiction of the definition of y as follows: Set

$$k = \frac{y^n - x}{y^{n-1}}.$$

Show that if $t \geq y - k$, then $t \notin E$.

3. Fix $b > 1$.

   a. Suppose m, n, p, q are integers with $n > 0$ and $q > 0$. If $m/n = p/q$, prove that
   $$(b^m)^{1/n} = (b^p)^{1/q}.$$
   Thus if r is rational, b^r is well defined.

   b. If r, s are rational, prove that $b^{r+s} = b^r b^s$.

   c. If $x \in \mathbb{R}$, let $B(x) = \{b^t : t \in \mathbb{Q}, t \leq x\}$. Prove that $b^r = \sup B(r)$ when $r \in \mathbb{Q}$. Thus it now makes sense to define $b^x = \sup B(x)$ when $x \in \mathbb{R}$.

   d. Prove that $b^{x+y} = b^x b^y$ for all real numbers x, y.

The following result, known as the Schröder-Bernstein theorem, is non-trivial, but very important. It is included as an exercise to provide motivation for further thought and additional studies. A proof of the result can be found in the text by Halmos listed in the Supplemental Reading.

4. Let X and Y be infinite sets. If X is equivalent to a subset of Y, and Y is equivalent to a subset of X, prove that X is equivalent to Y.

5. As in Theorem 1.7.18, let A denote the set of all sequences of 0's and 1's. Use the previous result to prove that $A \sim [0,1]$.

6. **DEFINITION.** A **complex number** is an ordered pair (a, b) of real numbers. If $z = (a, b)$ and $w = (c, d)$, we write $z = w$ if and only if $a = c$ and $b = d$. For complex numbers z and w we define addition and multiplication as follows:

$$z + w = (a + c, b + d)$$
$$z \cdot w = (ac - bd, ad + bc).$$

The set of ordered pairs (a, b) of real numbers with the above operations of addition and multiplication is denoted by $\mathbb{C}$.

   a. Prove that $(\mathbb{C}, +, \cdot)$ with zero $\mathbf{0} = (0, 0)$ and unit $\mathbf{1} = (1, 0)$ is a field.

   b. Set $i = (0, 1)$. Show that $i^2 = -\mathbf{1}$.

   c. Prove that $\mathbb{C}$ is not an ordered field.

Supplemental Reading

Buck, R. C., "Mathematical induction and recursive definition," *Amer. Math. Monthly* **70** (1963), 128–135.

Burrill, Claude W., *Foundations of Real Numbers*, Mc Graw-Hill, Inc., New York, 1967.

Cantor, Georg, *Contributions to the Founding of the Theory of Transfinite Numbers*, (translated by Philip E.B. Jourdain), Open Court Publ. Co., Chicago and London, 1915.

Dauben, Joseph W., *Georg Cantor; his Mathematics and Philosophy of the Infinite*, Princeton University Press, Princeton, N.J., 1979.

Gascón, J., "Another proof that the real numbers are uncountable," *Amer. Math. Monthly* **122** (2015), 596–497.

Gödel, Kurt, "What is Cantor's continuum problem," *Amer. Math. Monthly* **54** (1947), 515–525.

Halmos, Paul, *Naive Set Theory*, Springer-Verlag, New York, Heidelberg, Berlin, 1974.

Nimbran, A. S., "One more proof of the irrationality of $\sqrt{2}$," *Amer. Math. Monthly* **121** (2014), 964.

Shrader-Frechette, M., "Complementary rational numbers," *Math. Mag.*, **51** (1978), 90–98.

Spooner, George and Mentzer, Richard, *Introduction to Number Systems*, Prentice-Hall, Inc., Englewood Cliffs, N.J., 1968

Tripathi, A., "An alternate method to compute the decimal expansion of rational numbers," *Math. Mag.* **90** (2017), 108–113.

2

Topology of the Real Line

In this chapter, we introduce some of the basic concepts fundamental to the study of limits and continuity, and study the structure of point sets in $\mathbb{R}$. The branch of mathematics concerned with the study of these topics–not only for the real numbers but also for more general sets–is known as topology. Modern point set topology dates back to the early part of the 20th century; its roots, however, date back to the 1850s and 1860s and the studies of Bolzano, Cantor, and Weierstrass on sets of real numbers. Many important mathematical concepts depend on the concept of a limit point of a set and the limit process, and one of the primary goals of topology is to provide an appropriate setting for the study of these concepts.

One of the first topics encountered in the study of calculus is the concept of limit, which requires the notion of closeness, or the distance between points becoming small. On the real line or in the euclidean plane, the distance between points is usually measured as the length of the straight line segment joining the points. However, in many instances in subsequent chapters our points will not be points on the line or in the plane, but rather functions defined on some set. For this reason we introduce the concept of distance on an arbitrary set and study metric spaces in general. In many instance, the proofs of the results are such that they are valid in any metric space, and will be stated as such. These more general results about arbitrary metric spaces will prove useful not only in subsequent chapter but also in other courses.

Even though we introduce abstract metric spaces, our primary emphasis in this chapter will be on the topology of the real line. A thorough understanding of the topics on the real line and the plane will prove invaluable when they are encountered again in more abstract settings. On first reading, the concepts introduced in this chapter may seem difficult and challenging. With perseverance, however, understanding will follow.

2.1 Metric Spaces

We begin our study of metric spaces with a review of distance between points in the real numbers or its geometric interpretation as the real line.

DEFINITION 2.1.1 *For a real number x the **absolute value** of x, denoted $|x|$, is defined by*

$$|x| = \begin{cases} x, & \text{if } x > 0, \\ -x, & \text{if } x \leq 0. \end{cases}$$

For example, $|4| = 4$ and $|-5| = 5$. From the definition, $|x| \geq 0$ for all $x \in \mathbb{R}$ and $|x| = 0$ if and only if $x = 0$. This last statement follows from the fact that if $x \neq 0$, then $-x \neq 0$ and thus $|x| > 0$. The following theorem, the proof of which is left to the exercises, summarizes several well known properties of absolute value.

THEOREM 2.1.2 (a) $|-x| = |x|$ *for all* $x \in \mathbb{R}$.
   (b) $|xy| = |x||y|$ *for all* $x, y \in \mathbb{R}$.
   (c) $|x| = \sqrt{x^2}$ *for all* $x \in \mathbb{R}$.
   (d) *If* $r > 0$, *then* $|x| < r$ *if and only if* $-r < x < r$.
   (e) $-|x| \leq x \leq |x|$ *for all* $x \in \mathbb{R}$.

The following inequality is very important and will be used frequently throughout the text.

THEOREM 2.1.3 (Triangle Inequality) *For all* $x, y \in \mathbb{R}$, *we have*

$$|x + y| \leq |x| + |y|.$$

Proof. The triangle inequality is easily proved as follows: For $x, y \in \mathbb{R}$,

$$0 \leq (x + y)^2 = x^2 + 2xy + y^2$$
$$\leq |x|^2 + 2|x||y| + |y|^2 = (|x| + |y|)^2.$$

Thus by Theorem 2.1.2(c),

$$|x + y| = \sqrt{(x + y)^2} \leq \sqrt{(|x| + |y|)^2} = |x| + |y|. \quad \square$$

As a consequence of the triangle inequality we obtain the following two useful inequalities, the proofs of which are left to the exercises.

COROLLARY 2.1.4 *For all* $x, y, z \in \mathbb{R}$ *we have*
   (a) $|x - y| \leq |x - z| + |z - y|$, *and*
   (b) $||x| - |y|| \leq |x - y|$.

In the following example we illustrate how properties of absolute value can be used to solve inequalities.

EXAMPLE 2.1.5 Determine the set of all real numbers x that satisfy the inequality $|2x + 4| < 8$. By Theorem 2.1.2(d), $|2x + 4| < 8$ if and only if $-8 < 2x + 4 < 8$, or equivalently, $-12 < 2x < 4$. Thus the given inequality is satisfied if and only if $-6 < x < 2$.

Geometrically, $|x|$ represents the distance from x to the origin 0. More generally, for $x, y \in \mathbb{R}$, the **euclidean distance** $d(x, y)$ is defined by

$$d(x, y) = |x - y|.$$

For example, $d(-1, 3) = |-1 - 3| = 4$ and $d(5, -2) = |5 - (-2)| = 7$. The distance d may be regarded as a function on $\mathbb{R} \times \mathbb{R}$ which satisfy the following properties: $d(x, y) \geq 0$, $d(x, y) = 0$ if and only if $x = y$, $d(x, y) = d(y, x)$, and

$$d(x, y) \leq d(x, z) + d(z, y)$$

for all $x, y, z \in \mathbb{R}$. We now extend the notion of distance to sets other than $\mathbb{R}$.

DEFINITION 2.1.6 *Let X be a nonempty set. A real valued function d defined on $X \times X$ satisfying*
    (1) $d(x, y) \geq 0$ *for all $x, y \in X$,*
    (2) $d(x, y) = 0$ *if and only if $x = y$,*
    (3) $d(x, y) = d(y, x)$,
    (4) $d(x, y) \leq d(x, z) + d(z, y)$ *for all $x, y, z \in \mathbb{R}$,*
is called a **metric** *or* **distance function** *on X. The set X with metric d is called a* **metric space,** *and is denoted by (X, d).*

Before we continue, let us reflect on properties (1) to (4) in the definition of a metric. The first two combined simply state that distance is nonnegative and that the distance between two points is zero if and only if they are the same. Property (3) is symmetry; namely the distance between x and y is the same as the distance between y and x. Property (4) is again called the **triangle inequality** for the metric d. This inequality simply states that the distance between x and y is less than or equal to the distance between x and z, and z and y, for any other point z in the space X. All of these properties are what we intuitively expect a distance function to satisfy.

EXAMPLES 2.1.7 We now provide several examples of metrics and metric spaces, including the standard metrics on the euclidean plane $\mathbb{R}^2$. For some of the examples we prove that the functions as defined are indeed metrics, the others are left as exercises.

    (a) The real numbers $\mathbb{R}$ with the euclidean metric $d(x, y) = |x - y|$ is certainly a metric space. The metric is referred to as the **usual metric** on $\mathbb{R}$.

    (b) Let X be any nonempty subset of $\mathbb{R}$. For $x, y \in X$ define

$$d(x, y) = |x - y|.$$

The distance between the points x and y is just the usual euclidean distance between x and y as points in $\mathbb{R}$. However, it is important to remember that our space is the set X, and not $\mathbb{R}$.

(c) Let X be any nonempty set with

$$d(p,q) = \begin{cases} 1, & \text{if } p \neq q, \\ 0, & \text{if } p = q. \end{cases}$$

This metric is usually referred to as the **trivial metric** on X.

(d) In this example, we consider euclidean space

$$\mathbb{R}^2 = \mathbb{R} \times \mathbb{R} = \{(x_1, x_2) : x_1, x_2 \in \mathbb{R}\}.$$

For convenience we denote a point $(p_1, p_2) \in \mathbb{R}^2$ by $\mathbf{p}$. For such a point $\mathbf{p}$ define

$$\|\mathbf{p}\|_2 = \sqrt{p_1^2 + p_2^2}.$$

The quantity $\|\mathbf{p}\|_2$ is called the **euclidean norm** of $\mathbf{p}$. Given two points $\mathbf{p} = (p_1, p_2)$ and $\mathbf{q} = (q_1, q_2)$ set

$$d_2(\mathbf{p}, \mathbf{q}) = \|\mathbf{p} - \mathbf{q}\|_2 = \sqrt{(p_1 - q_1)^2 + (p_2 - q_2)^2}.$$

By the Pythagorean theorem $d_2(\mathbf{p}, \mathbf{q})$ is the euclidean distance between the points $\mathbf{p}$ and $\mathbf{q}$.

Since the square root is nonnegative, $d_2(\mathbf{p}, \mathbf{q}) \geq 0$ for all $\mathbf{p}, \mathbf{q} \in \mathbb{R}^2$. Also, from the definition it follows immediately that $d_2(\mathbf{p}, \mathbf{q}) = 0$ if and only if $\mathbf{p} = \mathbf{q}$. Clearly, $d_2(\mathbf{p}, \mathbf{q}) = d_2(\mathbf{q}, \mathbf{p})$. Hence d_2 satisfies properties (1)–(3) of a metric. For the proof of property (4) we need that $\| \ \|_2$ satisfies the triangle inequality, namely

$$\|\mathbf{p} + \mathbf{q}\|_2 \leq \|\mathbf{p}\|_2 + \|\mathbf{q}_2\|.$$

Assuming that $\| \ \|_2$ satisfies the triangle inequality, given any three points $\mathbf{p}, \mathbf{q}, \mathbf{r} \in \mathbb{R}^2$, we have

$$\begin{aligned} d_2(\mathbf{p}, \mathbf{q}) = \|\mathbf{p} - \mathbf{q}\|_2 &= \|(\mathbf{p} - \mathbf{r}) + (\mathbf{r} - \mathbf{q})\|_2 \\ &\leq \|\mathbf{p} - \mathbf{r}\|_2 + \|\mathbf{r} - \mathbf{q}\|_2 = d_2(\mathbf{p}, \mathbf{r}) + d_2(\mathbf{r}, \mathbf{q}). \end{aligned}$$

A geometric proof of the triangle inequality follows from the simple fact that in a triangle, the length of any one side does not exceed the sum of the lengths of the other two sides. This inequality can also be proved algebrically (Exercise 9).

(e) Again we let $X = \mathbb{R}^2$. In the previous example the distance between points was measured as the length of the straight line segment joining the two points. This metric however is of little use if one is in a city, such as New York, where the streets are laid out in a rectangular pattern. In such a setting a more appropriate way to measure distance is along the actual path one needs to traverse to get from one point to another. Specifically, for $\mathbf{p}, \mathbf{q} \in \mathbb{R}^2$ set

$$d_1(\mathbf{p}, \mathbf{q}) = |p_1 - q_1| + |p_2 - q_2|.$$

Another metric often encountered in $\mathbb{R}^2$ is the metric d_∞ which for $\mathbf{p}, \mathbf{q} \in \mathbb{R}^2$ is defined as

$$d_\infty(\mathbf{p}, \mathbf{q}) = \max\{|p_1 - q_1|, |p_2 - q_2|\}.$$

That d_1 and d_∞ are indeed metrics on $\mathbb{R}^2$ is left to the exercises (Exercise 10).

(f) In the previous two examples we restricted our discussion to $\mathbb{R}^2$ primarily for purposes of illustration. It is easier to visualize these concepts in the plane as opposed to higher dimensional space. All three of these metrics however have natural extensions to $\mathbb{R}^n$, where for $n \in \mathbb{N}, n > 2$,

$$\mathbb{R}^n = \{(x_1, \ldots, x_n) : x_i \in \mathbb{R}, i = 1, \ldots, n\}.$$

Here $(x_1, \ldots, x_n)$ denotes the ordered n-tuple of real numbers $x_1, \ldots, x_n$. As for ordered pairs, two n-tuples $(x_1, \ldots, x_n)$ and $(y_1, \ldots, y_n)$ are equal if and only if $x_i = y_i$ for all $i = 1, \ldots, n$. The euclidean metric d_2 for $\mathbb{R}^n$ is defined as follows: if $\mathbf{p}, \mathbf{q} \in \mathbb{R}^n$ with $\mathbf{p} = (p_1, \ldots, p_n)$ and $\mathbf{q} = (q_1, \ldots, q_n)$, then

$$d_2(\mathbf{p}, \mathbf{q}) = \sqrt{(p_1 - q_1)^2 + \cdots + (p_n - q_n)^2}.$$

In the miscellaneous exercises we outline how to prove that d_2 is a metric on $\mathbb{R}^n$. An alternate proof will be provided in Chapter 7. The metrics d_1 and d_∞ also have analogous extensions to $\mathbb{R}^n$.

(g) For this example we let $X = \{\mathbf{p} \in \mathbb{R}^2 : \|\mathbf{p}\|_2 \le 1\}$. For $\mathbf{p}, \mathbf{q} \in X$, define

$$d(\mathbf{p}, \mathbf{q}) = \begin{cases} \|\mathbf{p} - \mathbf{q}\|_2, & \text{if } \mathbf{p}, \mathbf{q} \text{ are co-linear through } 0, \\ \|\mathbf{p}\|_2 + \|\mathbf{q}\|_2, & \text{otherwise.} \end{cases}$$

This metric space is sometimes referred to as the Washington D. C. space or the French Railway space. To see the connection one has to be familiar with how the main roads run in Washington or how the railroads run in France with regard to the city of Paris. In Washington (excluding the beltway) all the major roads run to the city center. Similarly in France, all the major rail lines run through Paris. Thus if one travels between points $\mathbf{p}$ and $\mathbf{q}$ on the same line, the distance is just the usual distance between the points. However, if $\mathbf{p}$ and $\mathbf{q}$ are on different lines, then the distance between $\mathbf{p}$ and $\mathbf{q}$ is the distance from $\mathbf{p}$ to Paris plus the distance from Paris to $\mathbf{q}$. By computing the distance between the points in Exercise 11 one can verify that this is how distance is measured in this metric. Plotting the points in the plane will further illustrate this metric.

(h) Let A be any nonempty set. A real-valued function f is **bounded** on A if there exists a positive constant M such that $|f(x)| \le M$ for all $x \in A$. Let X be the set of all bounded real-valued functions on A. For $f, g \in X$ define

$$d(f, g) = \sup\{|f(x) - g(x)| : x \in A\}.$$

Clearly, since f and g are bounded, $\{|f(x) - g(x)| : x \in A\}$ is bounded above, and thus $d(f, g) < \infty$.

Since this is our first example of a space where the elements are functions we proceed to show that d is a metric. Clearly d is nonnegative. Furthermore, since $|f(x) - g(x)| \leq d(f, g)$ for all $x \in A$, $d(f, g) = 0$ if and only if $f(x) = g(x)$ for all $x \in A$. That $d(f, g) = d(g, f)$ follows from the fact that $|f(x) - g(x)| = |g(x) - f(x)|$. It only remains to be shown that d satisfies the triangle inequality. Let f, g, h be bounded functions on A. Then for $x \in A$

$$|f(x) - g(x)| = |f(x) - h(x) + h(x) - g(x)|$$
$$\leq |f(x) - h(x)| + |h(x) - g(x)| \leq d(f, h) + d(h, g).$$

Therefore $d(f, h) + d(h, g)$ is an upper bound for the set $\{|f(x) - g(x)| : x \in A\}$, and as a consequence,

$$d(f, g) \leq d(f, h) + d(h, g).$$

This then proves that d is a metric on X. □

Exercises 2.1

1. Prove Theorem 2.1.2.

2. *Prove Corollary 2.1.4

3. Prove that for $x_1, \ldots, x_n \in \mathbb{R}$, $|x_1 + \cdots + x_n| \leq |x_1| + \cdots + |x_n|$.

4. If $a, b \in \mathbb{R}$, prove that $|ab| \leq \frac{1}{2}(a^2 + b^2)$.

5. Determine all $x \in \mathbb{R}$ that satisfy each of the following inequalities.
   *a. $|3x - 2| \leq 11$,   b. $|x^2 - 4| < 5$,
   *c. $|x| + |x - 1| < 3$,   d. $|x - 1| < |x + 1|$.

6. Determine and sketch the set of ordered pairs (x, y) in $\mathbb{R} \times \mathbb{R}$ that satisfy the following:
   a. $|x| = |y|$,   b. $|x| \leq |y|$,
   c. $|xy| \leq 2$,   d. $|x| + |y| \leq 1$.

7. Determine which of the following are metrics on $\mathbb{R}$.
   a. $d(x, y) = (x - y)^2$.   b. $d(x, y) = \sqrt[3]{|x - y|}$
   *c. $d(x, y) = \ln(1 + |x - y|)$.
   d. $d(x, y) = |3x - y|$.
   e. $d(x, y) = \sqrt{|x - y|^3}$.

8. If d is a metric on X, prove that $|d(x, z) - d(z, y)| \leq d(x, y)$.

9. For $\mathbf{p} = (p_1, p_2) \in \mathbb{R}^2$, let $\|\mathbf{p}\|_2$ be defined as in Example 2.1.7(d). Prove algebraically that
   $$\|\mathbf{p} + \mathbf{q}\|_2 \leq \|\mathbf{p}\|_2 + \|\mathbf{q}\|_2 \quad \text{for all } \mathbf{p}, \mathbf{q} \in \mathbb{R}^2.$$

10. Prove that d_1 and d_∞ as defined in Example 2.1.7(e) are metrics on $\mathbb{R}^2$.

11. Let d be the metric on $\mathbb{R}^2$ defined in Example 2.1.7(g).

    a. Compute the distance in this metric between each of the following pair of points: ***i.** $(\frac{1}{2}, \frac{1}{4}), (-\frac{1}{2}, -\frac{1}{4})$ ***ii.** $(\frac{1}{2}, \frac{1}{2}), (0, 1)$ **iii.** $(\frac{1}{2}, \frac{2}{3}), (\frac{1}{4}, \frac{1}{3})$

    b. Prove that d is a metric on $\mathbb{R}^2$.

12. Let $A = [0, 1]$ and let X denote the bounded functions on A. Let d be the metric on X defined in Example 2.1.7(h). Compute $d(f, g)$ for each of the following pairs of functions f and g.

    a. $f(x) = 1$, $g(x) = x$.   ***b.** $f(x) = x$, $g(x) = x^2$.

    c. $f(x) = x$, $g(x) = \begin{cases} 0, & 0 \le x < \frac{1}{2}, \\ \frac{1}{2}, & \frac{1}{2} \le x \le 1. \end{cases}$

13. Let ρ and σ be metrics on a set X. Show that each of the following are also metrics on X.

    i. 2ρ **ii.** $\rho + \sigma$ **iii.** $\max\{\rho, \sigma\}$.

14. ***Prove that d defined by**

    $$d(x, y) = \frac{|x - y|}{1 + |x - y|}, \quad x, y \in \mathbb{R}, \text{ is a metric on } \mathbb{R}.$$

15. Let $X = (0, \infty)$. For $x, y \in X$, set

    $$\rho(x, y) = \left| \frac{1}{x} - \frac{1}{y} \right|. \quad \text{Prove that } \rho \text{ is a metric on } X.$$

2.2   Open and Closed Sets

In Chapter 1 we have used the terms *open* and *closed* in describing intervals in $\mathbb{R}$. The purpose of this section is to give a precise meaning to the adjectives open and closed, not only for intervals, but also for arbitrary subsets of $\mathbb{R}$ or a metric space (X, d). Before defining what we mean by an open set we first define the concept of an interior point of a set.

DEFINITION 2.2.1 *Let (X, d) be a metric space and let $p \in X$. For $\epsilon > 0$, the set*

$$N_\epsilon(p) = \{x \in X : d(p, x) < \epsilon\}$$

*is called an ϵ-**neighborhood** of the point p.*

Whenever we use the term neighborhood, we will always mean an ϵ-neighborhood with $\epsilon > 0$.

EXAMPLES 2.2.2 In the following examples we consider the ϵ-neighborhoods of the metric spaces given in Examples 2.1.7

(a) Consider $\mathbb{R}$ with the usual metric $d(x, y) = |x - y|$. Then for fixed $p \in \mathbb{R}$ and $\epsilon > 0$,

$$N_\epsilon(p) = \{x : p - \epsilon < x < p + \epsilon\}$$

In this case, $N_\epsilon(p)$ is the open interval $(p - \epsilon, p + \epsilon)$ centered at p with radius ϵ.

(b) Let $X = [0, \infty)$ with $d(x, y) = |x - y|$. With $\epsilon = \frac{1}{2}$ and $p = \frac{1}{4}$,

$$N_{\frac{1}{2}}\left(\tfrac{1}{4}\right) = \{x \in [0, \infty) : |x - \tfrac{1}{4}| < \tfrac{1}{2}\}$$
$$= \{x \in [0, \infty) : -\tfrac{1}{4} < x < \tfrac{3}{4}\} = [0, \tfrac{3}{4}).$$

On the other hand, $N_{\frac{1}{4}}\left(\tfrac{1}{4}\right) = (0, \tfrac{1}{2})$. If $p = 0$, then for any $\epsilon > 0$,

$$N_\epsilon(0) = \{x \in [0, \infty) : |x| < \epsilon\} = [0, \epsilon).$$

Thus an ϵ-neighborhood of 0 is the half-open interval $[0, \epsilon)$. It is important to remember that in this example our space is the interval $[0, \infty)$ and not $\mathbb{R}$.

(c) Consider $\mathbb{R}^2$ with the metric d_2 of Example 2.1.7(d). For fixed $\mathbf{a} = (a_1, a_2)$ and $\epsilon > 0$,

$$N_\epsilon(\mathbf{a}) = \{x \in \mathbb{R}^2 : \|\mathbf{x} - \mathbf{a}\|_2 < \epsilon\}$$
$$= \{(x_1, x_2) : (x_1 - a_1)^2 + (x_2 - a_2)^2 < \epsilon^2\}.$$

This is easily recognized as the interior of a circle with center $\mathbf{a}$ and radius ϵ. (Figure 2.1) Although ϵ-neighborhoods in the plane $\mathbb{R}^2$ are typically drawn as circular regions, this is only the case for the metric d_2. For other metrics this need not be the case as is illustrated in Exercise 5 of this section.

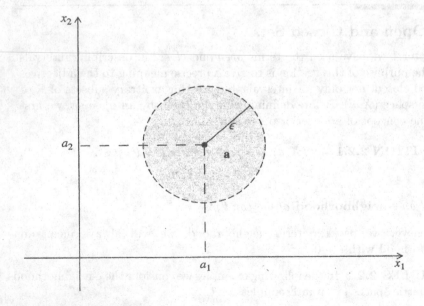

FIGURE 2.1
$N_\epsilon(\mathbf{a})$ for the metric d_2

(d) Let $A = [a, b]$ and X the set of real-valued functions on A with the metric as given in Example 2.1.7(h). For a fixed $f \in X$ and $\epsilon > 0$,

$$N_\epsilon(f) = \{g \in X : \sup |f(x) - g(x)| < \epsilon\}.\text{'}$$

Thus if $g \in N_\epsilon(f)$, then $|g(x) - f(x)| < \epsilon$ for all $x \in A$, or equivalently

$$f(x) - \epsilon < g(x) < f(x) + \epsilon. \quad \square$$

In the following definitions we will assume that X is a nonempty set with a metric d. If $X = \mathbb{R}$, unless otherwise specified d will denote the usual euclidean metric on $\mathbb{R}$.

DEFINITION 2.2.3 *Let E be a subset of X. A point $p \in E$ is called an* **interior point** *of E if there exists an $\epsilon > 0$ such that $N_\epsilon(p) \subset E$. The set of interior points of E is denoted by* $\mathrm{Int}(E)$, *and is called the* **interior** *of E.*

EXAMPLES 2.2.4 **(a)** Let $X = \mathbb{R}$ and let $E = (a, b]$ with $a < b$. Every p satisfying $a < p < b$ is an interior point of E. If ϵ is chosen such that

$$0 < \epsilon < \min\{|p - a|, |b - p|\},$$

then $N_\epsilon(p) \subset E$. The point b however is not an interior point. For every $\epsilon > 0$, $N_\epsilon(b) = (b - \epsilon, b + \epsilon)$ contains points which are not in E. Any x satisfying $b < x < b + \epsilon$ is not in E. This is illustrated in Figure 2.2. For this example, $\mathrm{Int}(E) = (a, b)$.

FIGURE 2.2
Epsilon neighborhoods of p and b in Example 2.2.4(a)

(b) Let E denote the set of irrational real numbers, i.e., $E = \mathbb{R} \setminus \mathbb{Q}$. If $p \in E$, then by Theorem 1.5.2, for every $\epsilon > 0$ there exists $r \in \mathbb{Q} \cap N_\epsilon(p)$. Thus $N_\epsilon(p)$ always contains a point of $\mathbb{R}$ not in E. Therefore no point of E is an interior point of E, i.e., $\mathrm{Int}(E) = \emptyset$. Using the fact that between any two real numbers there exists an irrational number, a similar argument also proves that $\mathrm{Int}(\mathbb{Q}) = \emptyset$.

(c) This example will illustrate that the space X itself is crucial in the definition of interior point. Let $E = [0, 1)$ with $d(x, y) = |x - y|$. We claim that every point of E is an interior point of E. Certainly every $p \in E$, $0 < p < 1$, is an interior point of E. The only point about which there may be some doubt is $p = 0$. However, if $0 < \epsilon < 1$, then as in Example 2.2.2(b)

$$N_\epsilon(0) = \{x \in [0, \infty) : |x| < \epsilon\} = [0, \epsilon),$$

which is a subset of E. Therefore 0 is an interior point of E. Even though this appears to contradict our intuition, it does not violate the definition of interior point.

(d) Let $X = \mathbb{R}^2$ with metric d_2, and let

$$E = \{(p_1, p_2) : 1 < p_1 < 3, 1 < p_2 < 2\}.$$

Given a point $\mathbf{p} = (p_1, p_2) \in E$, if we choose ϵ such that

$$0 < \epsilon < \min\{|p_1 - 1|, |p_1 - 3|, |p_2 - 1|, |p_2 - 2|\},$$

then $N_\epsilon(\mathbf{p}) \subset E$. Therefore every point of E is an interior point of E. □

Open and Closed Sets

Using the notion of an interior point we now define what we mean by an open set.

DEFINITION 2.2.5
(a) *A subset O of $\mathbb{R}$ is **open** if every point of O is an interior point of O.*
(b) *A subset F of $\mathbb{R}$ is **closed** if $F^c = \mathbb{R} \setminus F$ is open.*

Remark. From the definition of an interior point it should be clear that a set $O \subset \mathbb{R}$ is open if and only if for every $p \in O$ there exists an $\epsilon > 0$ (depending on p) so that $N_\epsilon(p) \subset O$. Both the definition of interior point and open depend on the metric of the given set. In situations where there is more than one metric defined on a given set, we will use the phrase *open with respect to d* to emphasize the metric.

EXAMPLES 2.2.6 (a) The entire set $\mathbb{R}$ is open. For any $p \in \mathbb{R}$ and $\epsilon > 0$, $N_\epsilon(p) \subset \mathbb{R}$. Since $\mathbb{R}$ is open, by definition the empty set $\emptyset$ is closed. However, the empty set is also open. Since $\emptyset$ contains no points at all, Definition 2.2.5(a) is vacuously satisfied. Consequently $\mathbb{R}$ is also closed.

(b) Every ϵ-neighborhood is open. Suppose $p \in \mathbb{R}$ and $\epsilon > 0$. If $q \in N_\epsilon(p)$, then $|p - q| < \epsilon$. Choose δ so that $0 < \delta \le \epsilon - |p - q|$. If $x \in N_\delta(q)$, then

$$|x - p| \le |p - q| + |x - q|$$
$$< |p - q| + \delta \le |p - q| + \epsilon - |p - q| = \epsilon.$$

Therefore $N_\delta(q) \subset N_\epsilon(p)$ (see Figure 2.3) Thus q is an interior point of $N_\epsilon(p)$. Since $q \in N_\epsilon(p)$ was arbitrary, $N_\epsilon(p)$ is open.

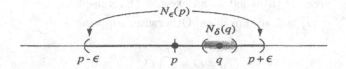

FIGURE 2.3
A delta neighborhood of q in Example 2.2.6(d)

(c) Let $E = (a, b]$, $a < b$, be as in Example 2.2.4(a). Since the point $b \in E$ is not an interior point of E, the set E is not open. The complement of E is given by

$$E^c = (-\infty, a] \cup (b, \infty).$$

An argument similar to the one given in Example 2.2.4(a) shows that a is not an interior point of E^c. Thus E^c is not open and hence by definition E is not closed. Hence E is neither open nor closed.

(d) Let $F = [a, b]$, $a < b$. Then

$$F^c = (-\infty, a) \cup (b, \infty),$$

and this set is open. This can be proved directly, but also follows as a consequence of Theorem 2.2.9(a) below.

(e) Consider the set $\mathbb{Q}$. Since no point of $\mathbb{Q}$ is an interior point of $\mathbb{Q}$ (Example 2.2.4(b)), the set $\mathbb{Q}$ is not open. Also, $\mathbb{Q}$ is not closed. □

The use of the adjective open in describing the intervals (a, b), (a, ∞), $(-\infty, b)$ and $(-\infty, \infty)$ is justified by the following theorem:

THEOREM 2.2.7 *Every open interval in $\mathbb{R}$ is an open subset of $\mathbb{R}$.*

Proof. Exercise 1. □

THEOREM 2.2.8 *Every ϵ-neighborhood is open.*

Proof. The proof is identical to the proof given in Example 2.2.6(b). □

THEOREM 2.2.9 *Let (X, d) be a metric space. Then*
(a) *for any collection $\{O_\alpha\}_{\alpha \in A}$ of open subsets of X, $\bigcup_{\alpha \in A} O_\alpha$ is open,* and
(b) *for any finite collection $\{O_1, ..., O_n\}$ of open subsets of X, $\bigcap_{j=1}^{n} O_j$ is open.*

Proof. The proof of (a) is left as an exercise (Exercise 2).

(b) If $\bigcap_{j=1}^{n} O_j = \emptyset$, we are done. Otherwise, suppose

$$p \in O = \bigcap_{j=1}^{n} O_j.$$

Then $p \in O_i$ for all $i = 1, ..., n$. Since O_i is open, there exists an $\epsilon_i > 0$ such that

$$N_{\epsilon_i}(p) \subset O_i.$$

Let $\epsilon = \min\{\epsilon_1, ..., \epsilon_n\}$. Then $\epsilon > 0$ and $N_\epsilon(p) \subset O_i$ for all i. Therefore $N_\epsilon(p) \subset O$, i.e., p is an interior point of O. Since $p \in O$ was arbitrary, O is open. $\square$

For closed subsets we have the following analogue of the previous result.

THEOREM 2.2.10 *Let (X, d) be a metric space. Then*

(a) *for any collection $\{F_\alpha\}_{\alpha \in A}$ of closed subsets of X, $\bigcap_{\alpha \in A} F_\alpha$ is closed,* and

(b) *for any finite collection $\{F_1, ..., F_n\}$ of closed subsets of X, $\bigcup_{j=1}^{n} F_j$ is closed.*

Proof. The proofs of (a) and (b) follow from the previous theorem and De-Morgan's laws:

$$\left(\bigcap_{\alpha \in A} F_\alpha\right)^c = \bigcup_{\alpha \in A} F_\alpha^c, \qquad \left(\bigcup_{j=1}^{n} F_j\right)^c = \bigcap_{j=1}^{n} F_j^c. \quad \square$$

Remark. The fact that the intersection of a finite number of open sets is open is due to the fact that the minimum of a finite number of positive numbers is positive. This guarantees the existence of an $\epsilon > 0$ such that the ϵ-neighborhood of p is contained in the intersection. For an infinite number of open sets, the choice of a positive ϵ may no longer be possible. This is illustrated by the following two examples.

EXAMPLES 2.2.11 We now provide two examples to show that part (b) of Theorem 2.2.9 is in general false for a countable collection of open sets. Likewise, part (b) of Theorem 2.2.10 is in general also false for an arbitrary union of closed sets (Exercise 15).

(a) For each $n = 1, 2, ...$, let $O_n = (-\frac{1}{n}, \frac{1}{n})$. Then each O_n is open, but

$$\bigcap_{n=1}^{\infty} O_n = \{0\},$$

which is not open.

(b) Alternately, if we let $G_n = (0, 1 + \frac{1}{n})$, $n = 1, 2, ...$, then again each G_n is open, but

$$\bigcap_{n=1}^{\infty} G_n = (0, 1],$$

which is neither open, nor closed. $\square$

Limit Points

DEFINITION 2.2.12 *Let E be a subset of a metric space X.*
   (a) *A point $p \in X$ is a* **limit point** *of E if every ϵ-neighborhood $N_\epsilon(p)$ of p contains a point $q \in E$ with $q \neq p$.*
   (b) *A point $p \in E$ that is not a limit point of E is called an* **isolated** *point of E.*

Remark. In the definition of limit point it is not required that p is a point of E. Also, a point $p \in E$ is an isolated point of E if there exists an $\epsilon > 0$ such that $N_\epsilon(p) \cap E = \{p\}$.

EXAMPLES 2.2.13 **(a)** $E = (a, b), a < b$. Every point $p, a < p < b$, is a limit point of E. This follows from the fact that for any $\epsilon > 0$ there exists a point $x \in (a, b)$ satisfying $p < x < p + \epsilon$. These however are not the only limit points. Both a and b are limit points of E, but they do not belong to E.
   (b) $E = \{\frac{1}{n} : n = 1, 2, ...\}$. Each $\frac{1}{n}$ is an isolated point of E. If ϵ is chosen so that

$$0 < \epsilon < \frac{1}{n(n+1)} = \frac{1}{n} - \frac{1}{n+1},$$

Then $N_\epsilon(\frac{1}{n}) = \{\frac{1}{n}\}$. Hence no point of E is a limit point of E. However, 0 is a limit point of E which does not belong to E. To see that 0 is a limit point, given $\epsilon > 0$ choose $n \in \mathbb{N}$ so that $1/n < \epsilon$. Such a choice of n is possible by Theorem 1.5.1. Then $1/n \in N_\epsilon(0) \cap E$, and thus 0 is a limit point of E.
   (c) Let $E = \mathbb{Q} \cap [0, 1]$. If $p \notin [0, 1]$ then p is not a limit point of E. For if $p > 1$, then for $\epsilon = \frac{1}{2}(p - 1)$ we have $N_\epsilon(p) \cap E = \emptyset$. Likewise when $p < 0$. On the other hand, every $p \in [0, 1]$ is a limit point of E. Let $\epsilon > 0$ be given. Suppose first that $0 \leq p < 1$. Then by Theorem 1.5.2 there exists $r \in \mathbb{Q}$ such that $p < r < \min\{p + \epsilon, 1\}$. When $p = 1$, Theorem 1.5.2 also guarantees the existence of an $r \in \mathbb{Q} \cap [0, 1]$ with $p - \epsilon < r < p$. Thus for every $\epsilon > 0$, $N_\epsilon(p)$ contains a point $r \in E$ with $r \neq p$. The same argument also proves that every point of $\mathbb{R}$ is a limit point of $\mathbb{Q}$. $\square$

The following theorem provides a characterization of the closed subsets of a metric space (X, d).

THEOREM 2.2.14 *A subset F of a metric space X is closed if and only if F contains all its limit points.*

Proof. Suppose F is closed. Then by definition F^c is open and thus for every $p \in F^c$ there exists $\epsilon > 0$ such that $N_\epsilon(p) \subset F^c$, that is, $N_\epsilon(p) \cap F = \emptyset$. Consequently no point of F^c is a limit point of F. Therefore F must contain all its limit points.

Conversely, let F be a subset of X that contains all its limit points. To show F is closed we must show F^c is open. Let $p \in F^c$. Since F contains all its limit points, p is not a limit point of F. Thus there exists an $\epsilon > 0$ such that $N_\epsilon(p) \cap F = \emptyset$. Hence $N_\epsilon(p) \subset F^c$ and p is an interior point of F^c. Since $p \in F^c$ was arbitrary, F^c is open and therefore F is closed. $\square$

THEOREM 2.2.15 *Let E be a subset of a metric space X. If p is a limit point of E, then every ϵ-neighborhood of p contains infinitely many points of E.*

Proof. Suppose there exists an ϵ-neighborhood of p that contains only finitely many points of E, say $q_1, \ldots, q_n$ with $q_i \neq p$. Let

$$\epsilon = \min\{d(q_i, p) : i = 1, \ldots, n\}.$$

Then $N_\epsilon(p)$ contains at most p. Thus p is not a limit point of E. $\square$

COROLLARY 2.2.16 *A finite set has no limit points.*

Closure of a Set

DEFINITION 2.2.17 *If E is a subset of a metric space X, let E' denote the set of limit points of E. The **closure** of E, denoted $\overline{E}$ is defined as*

$$\overline{E} = E \cup E'.$$

THEOREM 2.2.18 *If E is a subset of a metric space X, then*

(a) $\overline{E}$ *is closed.*

(b) $E = \overline{E}$ *if and only if E is closed.*

(c) $\overline{E} \subset F$ *for every closed set $F \subset X$ such that $E \subset F$.*

Proof. (a) To show that $\overline{E}$ is closed, we must show that $\overline{E}^c$ is open. Let $p \in \overline{E}^c$. Then $p \notin E$ and p is not a limit point of E. Thus there exists an $\epsilon > 0$ such that

$$N_\epsilon(p) \cap E = \emptyset.$$

We complete the proof by showing that $N_\epsilon(p) \cap E'$ is also empty and thus $N_\epsilon(p) \cap \overline{E} = \emptyset$. Therefore $N_\epsilon(p) \subset \overline{E}^c$, i.e., p is an interior point of $\overline{E}^c$.

Suppose $N_\epsilon(p) \cap E' \neq \emptyset$. Let $q \in N_\epsilon(p) \cap E'$, and choose $\delta > 0$ such that $N_\delta(q) \subset N_\epsilon(p)$. Since $q \in E'$, q is a limit point of E and thus $N_\delta(q) \cap E \neq \emptyset$. But this implies that $N_\epsilon(p) \cap E \neq \emptyset$, which is a contradiction. Therefore, $N_\epsilon(p) \cap E' = \emptyset$, which proves the result.

(b) If $E = \overline{E}$, then E is closed. Conversely, if E is closed, then $E' \subset E$ and thus $\overline{E} = E$.

(c) If $E \subset F$ and F is closed, then $E' \subset F$. Thus $\overline{E} \subset F$. □

DEFINITION 2.2.19 *A subset D of a metric space X is **dense** in X if* $\overline{D} = X$.

The rationals $\mathbb{Q}$ are dense in $\mathbb{R}$. By Example 2.2.13(c), every point of $\mathbb{R}$ is a limit point of $\mathbb{Q}$. Hence $\overline{\mathbb{Q}} = \mathbb{R}$. This explains the comment following Theorem 1.5.2. The rationals are not only dense; they are also countable. The existence of countable dense subsets play a very important role in analysis. They allow us to approximate arbitrary elements in a set by elements chosen from a countable subset of $\mathbb{R}$. Since the rationals are dense in $\mathbb{R}$, given any $p \in \mathbb{R}$ and $\epsilon > 0$, there exists $r \in \mathbb{Q}$ such that $|p - r| < \epsilon$. Additional examples of this will occur elsewhere in the text.

Characterization of the Open Subsets of $\mathbb{R}$[1]

If $\{I_n\}$ is any finite or countable collection of open intervals, then by Theorem 2.2.9, $U = \bigcup_n I_n$ is an open subset of $\mathbb{R}$. Conversely, every open subset of $\mathbb{R}$ can be expressed as a finite or countable union of open intervals (see Exercise 22). However, a much stronger result is true. We now prove that every open set can be expressed as a finite or countable union of pairwise disjoint open intervals. A collection $\{I_n\}$ of subsets of $\mathbb{R}$ is **pairwise disjoint** if $I_n \cap I_m = \emptyset$ whenever $n \neq m$.

THEOREM 2.2.20 *If U is an open subset of $\mathbb{R}$, then there exists a finite or countable collection $\{I_n\}$ of pairwise disjoint open intervals such that*

$$U = \bigcup_n I_n.$$

Proof. Let $x \in I$. Since U is open, there exists an $\epsilon > 0$ such that

$$(x - \epsilon, x + \epsilon) \subset U.$$

In particular $(s, x]$ and $[x, t)$ are subsets of U for some $s < x$ and some $t > x$. Define r_x and l_x as follows:

$$r_x = \sup\{t : t > x \text{ and } [x, t) \subset U\}, \qquad \text{and}$$
$$l_x = \inf\{s : s < x \text{ and } (s, x] \subset U\}.$$

Then $x < r_x \leq \infty$ and $-\infty \leq l_x < x$. For each $x \in U$, let $I_x = (l_x, r_x)$. Then
(a) $I_x \subset U$,

---

[1]This topic can be omitted upon first reading of the text. The structure of open sets will only be required in Chapter 10 in defining the measure of an open subset of $\mathbb{R}$.

(b) If x, $y \in U$, then either $I_x = I_y$ or $I_x \cap I_y = \emptyset$.

The proofs of (a) and (b) are left as exercises (Exercise 21).

To complete the proof, we let $\mathcal{I} = \{I_x : x \in U\}$. For each interval $I \in \mathcal{I}$, choose $r_I \in \mathbb{Q}$ such that $r_I \in I$. If I, $J \in \mathcal{I}$ are distinct intervals, then $r_I \neq r_J$. Therefore the mapping $I \to r_I$ is a one-to-one mapping of $\mathcal{I}$ into $\mathbb{Q}$. Thus the collection $\mathcal{I}$ is at most countable and therefore can be enumerated as $\{I_j\}_{j \in A}$, where A is either a finite subset of $\mathbb{N}$, or $A = \mathbb{N}$. Clearly

$$U = \bigcup_{j \in A} I_j,$$

and by (b), if $n \neq j$, then $I_n \cap I_j = \emptyset$. Thus the collection $\{I_j\}_{j \in A}$ is pairwise disjoint. $\square$

Relatively Open and Closed Sets

One of the reasons for studying topological concepts is to enable us to study properties of continuous functions. In most instances, the domain of a function is not all of $\mathbb{R}$ but rather a proper subset of $\mathbb{R}$ as is the case with $f(x) = \sqrt{x}$ for which Dom $f = [0, \infty)$. When discussing a particular function we will always restrict our attention to the domain of the function rather than all of $\mathbb{R}$. With this in mind we make the following definition.

DEFINITION 2.2.21 *Let Y be a subset of a metric space X.*

(a) *A subset U of Y is* **open in** *(or open relative to) Y if for every $p \in U$, there exists $\epsilon > 0$ such that $N_\epsilon(p) \cap Y \subset U$.*

(b) *A subset C of Y is* **closed in** *(or closed relative to) Y if $Y \setminus C$ is open in Y.*

EXAMPLE 2.2.22 Let $X = [0, \infty)$ and let $U = [0, 1)$. Then U is not open in $\mathbb{R}$ but is open in X. (Why?) $\square$

The following theorem, the proof of which is left as an exercise (Exercise 24), provides a simple characterization of what it means for a set to be open or closed in a subset of X.

THEOREM 2.2.23 *Let Y be a subset of a metric space X.*

(a) *A subset U of Y is open in Y if and only if $U = Y \cap O$ for some open subset O of X.*

(b) *A subset C of Y is closed in Y if and only if $C = Y \cap F$ for some closed subset F of X.*

Connected Sets[2]

Our final topic of this section involves the notion of a *connected set*. The idea of connectedness is just one more of the many mathematical concepts which have their roots in the studies of Cantor on the structure of subsets of $\mathbb{R}$. When we use the term *connected subset* of $\mathbb{R}$, intuitively we are inclined to think of an interval as opposed to sets such as the positive integers $\mathbb{N}$ or $(0,1) \cup \{2\}$. We make this precise with the following definition.

DEFINITION 2.2.24 *A subset A of a metric space X is* **connected** *if there* **do not** *exist two disjoint open sets U and V such that*

(a) $A \cap U \neq \emptyset$ and $A \cap V \neq \emptyset$, and

(b) $(A \cap U) \cup (A \cap V) = A$.

The definition for a connected set differs from most definitions in that it defines connectedness by negation; i.e., defining what it means for a set not to be connected. According to the definition, a set A is not connected if there exist disjoint open sets U and V satisfying both (a) and (b). As an example of a subset of $\mathbb{R}$ which is not connected, consider the set of positive integers $\mathbb{N}$. If we let $U = (\frac{1}{2}, \frac{3}{2})$ and $V = (\frac{3}{2}, \infty)$, then U and V are disjoint open subsets of $\mathbb{R}$ with

$$U \cap \mathbb{N} = \{1\} \quad \text{and} \quad V \cap \mathbb{N} = \{2, 3, ...\}$$

that also satisfy $(U \cap \mathbb{N}) \cup (V \cap \mathbb{N}) = \mathbb{N}$. That the interval (a, b) is connected is a consequence of the following theorem, the proof of which is left to the exercises (Exercise 27).

THEOREM 2.2.25 *A subset of $\mathbb{R}$ is connected if and only if it is an interval.*

Exercises 2.2

1. Prove Theorem 2.2.7.

2. *Prove Theorem 2.2.9(a).

3. *a Show that every finite subset of $\mathbb{R}$ is closed.

    b. Show that the intervals $(-\infty, a]$ and $[a, \infty)$ are closed subsets of $\mathbb{R}$.

4. Let X be the metric space of Example 2.1.7(g). Let $\mathbf{p} = (\frac{1}{2}, 0)$. Describe the ϵ-neigborhoods of $\mathbf{p}$ for each of the following values of ϵ: $\epsilon = \frac{1}{4}$, $\epsilon = \frac{1}{2}$, $\epsilon = \frac{3}{4}$.

---

[2]This concept, although important and used implicitly in several instances in the text, will not be required specifically in subsequent chapters except in a few exercises. Thus the topic of connectedness can be omitted upon first reading of the text.

5. For $j = 1, 2, \infty$, let d_j denote the metrics given in Examples 2.1.7 (d) and (e). For each j and $\epsilon > 0$, let $N_\epsilon^j(\mathbf{p})$ denote the ϵ-neighborhood of the point $\mathbf{p} \in \mathbb{R}^2$.

   a. For $\mathbf{p} = (0,0)$ and $\epsilon = 1$, sketch the ϵ-neighborhoods $N_\epsilon^j(\mathbf{p})$ for $j = 1, \infty$.

   *b. Prove that for $\epsilon > 0$,
   $$N_\epsilon^1(P) \subset N_\epsilon^2(P) \subset N_\epsilon^\infty(P) \subset N_{2\epsilon}^1(P).$$

   c. Using the results of (b), prove that a set U is open with respect to one of the metrics d_1, d_2, d_∞ if and only if it is open with respect to the other two.

6. If U and V are open subsets of $\mathbb{R}$, prove that $U \times V$ is an open subset of $\mathbb{R}^2$.

7. Consider the metric space (X, d) of Example 2.1.7(c). Prove that every subset of X is open.

8. For the following subsets E of $\mathbb{R}$, find each of the following: Int(E), E', Isolated points of E, if any, and $\overline{E}$. Determine whether the set E is open, closed, or neither.

   *a. $(0,1) \cup \{2\}$ **b.** (a,b) **c.** $(a,b]$ ***d.** $\{\frac{1}{n} : n \in \mathbb{N}\}$ **e.** $\mathbb{Q} \cap [0,1]$.

9. Let $E \subset \mathbb{R}$. A point $p \in \mathbb{R}$ is a **boundary point** of E if for every $\epsilon > 0$, $N_\epsilon(p)$ contains both points of E and points of E^c. Find the boundary points of each of the following sets,

   *a. (a,b) **b.** $E = \{\frac{1}{n} : n \in \mathbb{N}\}$ **c.** $\mathbb{N}$ **d.** $\mathbb{Q}$

10. **a.** Prove that a set $E \subset \mathbb{R}$ is open if and only if E does not contain any of its boundary points.

    b. Prove that a set $E \subset \mathbb{R}$ is closed if and only if E contains all its boundary points.

11. For each of the following subset E of $\mathbb{R}^2$, find Int(E) and $\overline{E}$.
    a. $E = (1,2) \times [-1,1]$ **b.** $E = \{(x,y) : -1 < x \leq 2, y \in \mathbb{R}\}$
    c. $E = \{(x,y) : y = x\}$ **d.** $E = \{(x,y) : y \leq x + 1\}$
    e. $E = \{\mathbf{p} \in \mathbb{R}^2 : 0 < d_2(\mathbf{p}, 0) < 1\}$

12. **a.** Construct a set with exactly two limit points.

    b. Find an infinite subset of $\mathbb{R}$ with no limit points.

    c. Construct a countable subset of $\mathbb{R}$ with countably many limit points.

    d. Find a countable subset of $\mathbb{R}$ with uncountably many limit points.

13. Let $X = (0, \infty)$. For each of the following subsets of X determine whether the given set is open in X, closed in X, or neither.

    *a. $(0,1]$ **b.** $(0,1)$ ***c.** $(0,1] \cup (2,3)$ **d.** $(0,1] \cup \{2\}$

14. For each of the following subsets of $\mathbb{Q}$, determine whether the set is open in $\mathbb{Q}$, closed in $\mathbb{Q}$, both open and closed in $\mathbb{Q}$, or neither.

    a. $A = \{p \in \mathbb{Q} : 1 < p < 2\}$.   **b.** $B = \{p \in \mathbb{Q} : 2 < p^2 < 3\}$.   **c.** $\mathbb{N}$.

15. **a.** Prove Theorem 2.2.10(a).

    b. Give an example of a countable collection $\{F_n\}_{n=1}^{\infty}$ of closed subsets of $\mathbb{R}$ such that $\bigcup_{n=1}^{\infty} F_n$ is not closed.

16. *Let A be a nonempty subset of $\mathbb{R}$ that is bounded above, and let $\alpha = \sup A$. If $\alpha \notin A$, prove that α is a limit point of A

17. Let (X, d) be a metric space, and $E \subset X$.

    *a. Prove that $\text{Int}(E)$ is open.

    b. Prove that E is open if and only if $E = \text{Int}(E)$.

    c. If $G \subset E$ and G is open, prove that $G \subset \text{Int}(E)$.

18. Let A, B be subsets of $\mathbb{R}$.

    a. If $A \subset B$, show that $\text{Int}(A) \subset \text{Int}(B)$.

    b. Show that $\text{Int}(A \cap B) = \text{Int}(A) \cap \text{Int}(B)$.

    c. Is $\text{Int}(A \cup B) = \text{Int}(A) \cup \text{Int}(B)$?

    d. Are the results of (a) and (b) still true if A and B are subsets of a metric space X ?

19. Let A, B be subsets of a metric space X.

    *a. Show that $\overline{(A \cup B)} = \overline{A} \cup \overline{B}$.

    b Show that $\overline{A \cap B} \subset \overline{A} \cap \overline{B}$.

    c. Give an example for which the containment in (b) is proper.

20. Let $D_0 = \{0, 1\}$, and for each $n \in \mathbb{N}$, let
$D_n = \{a/2^n : a \in \mathbb{N}, a \text{ is odd}, 0 < a < 2^n\}$. Let $D = \bigcup_{n=0}^{\infty} D_n$.

    Prove that D is a countable dense subset of $[0, 1]$.

21. Prove statements (a) and (b) of Theorem 2.2.20.

22. *Prove that there exists a countable collection $\mathcal{I}$ of open intervals such that if U is an open subset of $\mathbb{R}$, there exists a finite or countable collection $\{I_n\} \subset \mathcal{I}$ with $U = \bigcup I_n$.

23. *If U is an open subset of $\mathbb{R}$, prove that $E \subset U$ is open in U if and only if E is an open subset of $\mathbb{R}$.

24. Prove Theorem 2.2.23.

25. For each of the following, use the definition to prove that the given set is not connected.

    *a. $(0, 1) \cup \{2\}$ **b.** $\{\frac{1}{n} : n = 1, 2, ...\}$ **c.** $\{p \in \mathbb{Q} : p > 0 \text{ and } 1 < p^2 < 3\}$.

26. *If A is connected, prove that $\overline{A}$ is connected.

27. *Prove Theorem 2.2.25.

2.3   Compact Sets

In this section, we introduce the concept of a *compact set* in the setting of
metric spaces. A characterization of the compact subsets of $\mathbb{R}$ is provided in
Section 2.4. The notion of a compact set is very important in the study of
analysis, and many significant results in the text will depend on the fact that
every closed and bounded interval in $\mathbb{R}$ is compact. The modern definition of
a compact set given in 2.3.3 dates back to the second half of the nineteenth
century and the studies of Heine and Borel on compact subsets of $\mathbb{R}$.

DEFINITION 2.3.1 *Let E be a subset of a metric space (X, d). A collection
$\{O_\alpha\}_{\alpha \in A}$ of open subsets of X is an* **open cover** *of E if*

$$E \subset \bigcup_{\alpha \in A} O_\alpha.$$

An alternate definition is as follows: The collection $\{O_\alpha\}_{\alpha \in A}$ of open sets
is an open cover of E if for each $p \in E$, there exists an $\alpha \in A$ such that
$p \in O_\alpha$.

EXAMPLES 2.3.2 **(a)** Let $E = (0, 1)$ and $O_n = (0, 1 - \frac{1}{n})$, $n = 2, 3, \dots$.
Then $\{O_n\}_{n=2}^{\infty}$ is an open cover of E. To see this, suppose $x \in E$. Then since
$x < 1$, there exists an integer n such that $x < 1 - 1/n$. Thus $x \in O_n$, and as
a consequence

$$E \subset \bigcup_{n=2}^{\infty} O_n,$$

which proves the assertion. In fact, since $O_n \subset E$ for each n, we have $E = \bigcup_{n=2}^{\infty} O_n$.

   (b) Let $F = [0, \infty)$ and for each $n \in \mathbb{N}$ let $U_n = (-1, n)$. Then $\{U_n\}_{n \in \mathbb{N}}$
is an open cover of F. $\square$

DEFINITION 2.3.3 *A subset K of X is* **compact** *if every open cover of
K has a finite subcover of K; that is, if $\{O_\alpha\}_{\alpha \in A}$ is an open cover of K, then
there exists $\alpha_1, \dots, \alpha_n \in A$ such that*

$$K \subset \bigcup_{j=1}^{n} O_{\alpha_j}.$$

EXAMPLES 2.3.4 (a) Every finite set is compact. Suppose $E = \{p_1, ..., p_n\}$ is a finite subset of $\mathbb{R}$ and $\{O_\alpha\}_{\alpha \in A}$ is an open cover of E. Then for each j, $j = 1, ..., n$, there exists $\alpha_j \in A$ such that $p_j \in O_{\alpha_j}$. But then $\{O_{\alpha_j}\}_{j=1}^n$ is a finite sub-collection which covers E.

(b) The open interval $(0, 1)$ is not compact. For the open cover $\{O_n\}_{n=2}^\infty$ of $(0, 1)$ in Example 2.3.2(a), no finite sub-collection can cover $(0, 1)$. Suppose on the contrary that a finite number, say $O_{n_1}, ..., O_{n_k}$, cover $(0, 1)$. Let $N = \max\{n_1, ..., n_k\}$. Then

$$(0, 1) \subset \bigcup_{j=1}^k O_{n_j} \subset \bigcup_{n=2}^N O_n = (0, 1 - \tfrac{1}{N}),$$

which is a contradiction.

(c) The closed set $F = [0, \infty)$ is not compact. For the open cover $\mathcal{U} = \{(-1, n)\}_{n \in \mathbb{N}}$ of F, no finite sub-collection can cover F. If there exist a finite number of sets in $\mathcal{U}$ which cover F, then there exists $N \in \mathbb{N}$ such that $F \subset (-1, N)$. (Why?) This however is a contradiction. $\square$

Properties of Compact Sets

Before we provide a characterization of the compact subsets of $\mathbb{R}$, we first prove several properties of compact sets that require only the definition of compactness. As a consequence, all three of the following theorems are true in more general settings; e.g. in n–dimensional space $\mathbb{R}^n$ as well as in a general metric space (X, d).

THEOREM 2.3.5

(a) *Every compact subset of a metric space is closed.*

(b) *Every closed subset of a compact set is compact.*

Proof. (a) To show that K is closed, we need to show that K^c is open. Let $p \in K^c$ be arbitrary. For each $q \in K$, choose $\epsilon_q > 0$ such that

$$N_{\epsilon_q}(q) \cap N_{\epsilon_q}(p) = \emptyset.$$

Any ϵ_q satisfying $0 < \epsilon_q < \tfrac{1}{2}d(p, q)$ will work. Then $\{N_{\epsilon_q}(q)\}_{q \in K}$ is an open cover of K. Since K is compact, there exists $q_1, ..., q_n$ such that

$$K \subset \bigcup_{j=1}^n N_{\epsilon_{q_j}}(q_j).$$

Let $\epsilon = \min\{\epsilon_{q_j} : j = 1, ..., n\}$, which is positive. Then

$$N_\epsilon(p) \cap N_{\epsilon_{q_j}}(q_j) = \emptyset \qquad \text{for all } j = 1, ..., n.$$

Thus $N_\epsilon(p) \cap K = \emptyset$, i.e., $N_\epsilon(p) \subset K^c$. Therefore p is an interior point of K^c and thus K^c is open.

(b) Let F be a closed subset of the compact set K and let $\{O_\alpha\}_{\alpha \in A}$ be an open cover of F. Then

$$\{O_\alpha\}_{\alpha \in A} \cup \{F^c\}$$

is an open cover of K. Since K is compact, a finite number of these will cover K, and hence also F. $\square$

COROLLARY 2.3.6 *If F is closed and K is compact, then $F \cap K$ is compact.*

As a consequence of the previous theorem, the open interval $(0, 1)$ is not compact since it is not closed. However, being closed is not sufficient for a set to be compact. The half-open interval $[0, \infty)$ is closed in $\mathbb{R}$, but as shown in Example 2.3.4(c) is not compact. In Theorem 2.4.2 we will provide necessary and sufficient conditions for a subset of $\mathbb{R}$ to be compact.

Remark. In proving that the compact set K was closed, compactness allowed us to select a finite subcover from the constructed open cover of K. Finiteness then assured that the ϵ as defined was positive. This method of first constructing an open cover possessing certain properties and then using compactness to assure the existence of a finite subcover will be used on other occasions in the text.

THEOREM 2.3.7 *If E is an infinite subset of a compact set K, then E has a limit point in K.*

Proof. If no point of K is a limit point of E, then for each $q \in K$, there exists a neighborhood N_q of q so that N_q contains at most one point of E, namely q if $q \in E$. Since E is infinite, no finite sub-collection of $\{N_q\}_{q \in K}$ can cover E, and consequently no finite sub-collection of $\{N_q\}_{q \in K}$ can cover K, which is a contradiction. $\square$

Another useful consequence of compactness is the following analogue of what is known as the nested intervals property (see Exercise 3 of Section 2.4).

THEOREM 2.3.8 *If $\{K_n\}_{n=1}^\infty$ is a sequence of nonempty compact subsets of X with $K_n \supset K_{n+1}$ for all n, then*

$$K = \bigcap_{n=1}^\infty K_n$$

is nonempty and compact.

Proof. We first show that $\bigcap_{n=1}^{\infty} K_n \neq \emptyset$. Let $O_n = K_n^c$. By Theorem 2.3.5 K_n is closed and thus O_n is open. Furthermore,

$$\bigcap_{n=1}^{\infty} K_n = \emptyset \quad \text{if and only if} \quad \bigcup_{n=1}^{\infty} O_n = X.$$

Thus if $\bigcap_{n=1}^{\infty} K_n = \emptyset$, then $\{O_n\}_{n=1}^{\infty}$ is an open cover of X, and thus also of K_1. But K_1 is compact. Therefore there exists $n_1 < \cdots < n_k$ such that

$$K_1 \subset \bigcup_{j=1}^{k} O_{n_j}.$$

But then $K_1 \cap K_{n_1} \cap \cdots \cap K_{n_k} = \emptyset$. This however is a contradiction, since the intersection is equal to K_{n_k}, which by hypothesis is nonempty. Thus $K = \bigcap K_n \neq \emptyset$. By Theorem 2.2.10 K is closed, and hence by Theorem 2.3.5(b), K is compact. $\square$

Exercises 2.3

1. Let $A = \left\{ \frac{1}{n} : n = 1, 2, ... \right\}$.

   a. Show that the set A is not compact.

   *b. Prove directly (using the definition) that $K = A \cup \{0\}$ is compact.

2. Show that $(0, 1]$ is not compact by constructing an open cover of $(0, 1]$ that does not have a finite subcover.

3. Suppose A and B are compact subsets of a metric space X.

   *a. Prove (using only the definition) that $A \cup B$ is compact.

   b. Prove that $A \cap B$ is compact.

4. *Let K be a nonempty compact subset of $\mathbb{R}$. Prove that $\sup K$ and $\inf K$ exist and are in K.

5. Construct a compact subset K of $\mathbb{R}$ with an infinite number of isolated points. Justify that your set K is compact.

6. Suppose K is an infinite compact subset of a metric space (X, d). Prove that there exists a countable subset D of K such that $\overline{D} = K$.

2.4 Compact Subsets of $\mathbb{R}$

We now turn to our goal of providing a characterization of the compact subsets of the real line $\mathbb{R}$. The first of the two main results is attributed to Eduard Heine (1821–1881) and Emile Borel (1871–1956), whereas the second is due to Bernhard Bolzano (1781–1848) and Karl Weierstrass (1815–1897). The two theorems rank very high among the many important advances in the foundations of analysis during the nineteenth century. The importance of these results will become evident in later chapters. As is to be expected, the least upper bound property of $\mathbb{R}$ will play a crucial role in the proofs of these theorems.

THEOREM 2.4.1 (Heine-Borel) *Every closed and bounded interval* $[a, b]$ *is compact.*

Proof. Let $\mathcal{U} = \{U_\alpha\}_{\alpha \in A}$ be an open cover of $[a, b]$ and let

$$E = \{r \in [a, b] : [a, r] \text{ is covered by a finite number of the sets } U_\alpha\}.$$

The set E is bounded above by b, and since $a \in U_\alpha$ for some $\alpha \in A$, E is nonempty. Thus by the least upper bound property the supremum of E exists in $\mathbb{R}$. Let $\alpha = \sup E$. Since b is an upper bound of E, $\alpha \leq b$.

We first show that $\alpha \in E$, i.e., $[a, \alpha]$ is covered by a finite number of sets in $\mathcal{U}$. Since $\alpha \in [a, b]$, $\alpha \in U_\beta$ for some $\beta \in A$. Since U_β is open, there exists $\epsilon > 0$ such that $(\alpha - \epsilon, \alpha + \epsilon) \subset U_\beta$. Furthermore, since $\alpha - \epsilon$ is not an upper bound of E, there exists $r \in E$ such that $\alpha - \epsilon < r \leq \alpha$. But then $[a, r]$ is covered by a finite number, say $U_{\alpha_1}, ..., U_{\alpha_n}$, of sets in $\mathcal{U}$. But then the finite collection $\{U_{\alpha_1}, ..., U_{\alpha_n}, U_\beta\}$ covers $[a, \alpha]$. Therefore, $\alpha \in E$.

To conclude the proof we show that $\alpha = b$. Suppose $\alpha < b$. If we choose $s < b$ such that $\alpha < s < \alpha + \epsilon$, then the collection $\{U_{\alpha_1}, ..., U_{\alpha_n}, U_\beta\}$ also covers $[a, s]$. Thus $s \in E$ which contradicts that $\alpha = \sup E$. Hence α must equal b. $\square$

The statement of the Heine-Borel theorem was initially due to Heine, a student of Weierstrass, who used the result implicitly in the 1870s in his studies on continuous functions. The theorem was proved by Borel in 1894 for the case where the open cover was countable. For an arbitrary open cover the result was finally proved in 1904 by Henri Lebesgue (1875–1941). Using the Heine-Borel theorem we now prove the following characterization of the compact subsets of $\mathbb{R}$.

THEOREM 2.4.2 (Heine-Borel-Bolzano-Weierstrass) *Let K be a subset of $\mathbb{R}$. Then the following are equivalent:*

(a) *K is closed and bounded.*

(b) *K is compact.*

(c) *Every infinite subset of K has a limit point in K.*

Remark. A subset E of $\mathbb{R}$ is **bounded** if it is bounded both above and below, i.e., there exists a constant M such that $|x| \leq M$ for all x in E.

Proof. (a) $\Rightarrow$ (b). Since K is bounded, there exists a positive constant M so that $K \subset [-M, M]$. Since $[-M, M]$ is compact and K is closed, by Theorem 2.3.5(b) K is compact.

(b) $\Rightarrow$ (c). This is Theorem 2.3.7.

(c) $\Rightarrow$ (a). Suppose the set K is not bounded. Then for every $n \in \mathbb{N}$ there exists $p_n \in K$ with $p_n \neq p_m$ for $n \neq m$ such that $|p_n| > n$ for all n. Then $\{p_n : n = 1, 2, ...\}$ is an infinite subset of $\mathbb{R}$ with no limit point in $\mathbb{R}$, and hence none in K, which is a contradiction. Thus K is bounded.

Let p be a limit point of K. By definition of limit point, for each $n \in \mathbb{N}$ there exist $p_n \in K$ with $p_n \neq p$ such that

$$|p_n - p| < \frac{1}{n}.$$

Then $S = \{p_n : n = 1, 2, ...\}$ is an infinite subset of K, and p is a limit point of S. To complete the proof we must show that p is the only limit point of S, and hence by hypothesis must be in K, i.e., K is closed.

Suppose $q \in \mathbb{R}$ with $q \neq p$. Let $\epsilon = \frac{1}{2}|p - q|$ and choose $N \in \mathbb{N}$ such that $1/N < \epsilon$. Then for all $n \geq N$

$$|p - q| \leq |p_n - q| + |p_n - p| < |p_n - q| + \frac{1}{n} < |p_n - q| + \epsilon.$$

Therefore, for all $n \geq N$,

$$|p_n - q| > |p - q| - \epsilon = \tfrac{1}{2}|p - q|.$$

Thus $N_\epsilon(q)$ contains at most finitely many p_n, and as a consequence q cannot be a limit point of S. Therefore no $q \in \mathbb{R}$ with $q \neq p$ is a limit point of S. $\square$

Statement (c) of the previous theorem is basically what is referred to as the Bolzano-Weierstrass theorem, which we state for completeness. The proof of the result follows immediately from Theorems 2.3.7 and 2.4.2.

THEOREM 2.4.3 (Bolzano-Weierstrass) *Every bounded infinite subset of $\mathbb{R}$ has a limit point.*

The theorem was originally proved by Bolzano, and modified slighly in the 1860s by Weierstrass. This result can also be proved directly using the nested interval property (Exercise 7). Although we restricted ourselves to $\mathbb{R}$, the analogous statement of Theorem 2.4.2 is also true in $\mathbb{R}^n$. The proof of the following theorem, for $n = 2$, is left as an exercise. The proof for $n > 2$ is identical to the case $n = 2$.

THEOREM 2.4.4 *A subset E of $\mathbb{R}^n$ is compact if and only if E is closed and bounded.*

As in the Remark following Theorem 2.4.2, a subset E of $\mathbb{R}^n$ is **bounded** if and only if there exists a positive constant M such that $d_2(\mathbf{p}, 0) \leq M$ for all $\mathbf{p} \in E$. This is equivalent to

$$|p_i| \leq M, \qquad i = 1, \ldots, n$$

for all $\mathbf{p} = (p_1, \ldots, p_n) \in E$.

Exercises 2.4

1. * Find a countable collection $\{K_n\}_{n=1}^{\infty}$ of compact subsets of $\mathbb{R}$ such that $\bigcup_{n=1}^{\infty} K_n$ is not compact.

2. **a.** Suppose I and J are closed and bounded intervals in $\mathbb{R}$. Prove that $I \times J$ is a compact subset of $\mathbb{R}^2$.

    b. Prove that a subset E of $\mathbb{R}^2$ is compact if and only if it is closed and bounded.

3. *(Nested Intervals Property) Let $\{I_n\}$ be a countable family of nonempty closed and bounded intervals satisfying $I_n \supset I_{n+1}$ for all n. Prove that there exists $a \leq b$ in $\mathbb{R}$ such that

    $$\bigcap_{n=1}^{\infty} I_n = [a, b].$$

4. Let d denote the usual metric on $\mathbb{R}$ and let ρ be the metric on $\mathbb{R}$ given by

    $$\rho(x, y) = \frac{|x - y|}{1 + |x - y|}.$$

    a. Prove that a subset of $\mathbb{R}$ is open with respect to the metric d if and only if it is open with respect to ρ.

    b. Show that $[0, \infty)$ is closed and bounded in the metric ρ but that $[0, \infty)$ is not a compact subset of the metric space $(\mathbb{R}, \rho)$.

5. Let $X = \mathbb{Q}$ with metric $d(p, q) = |p - q|$. Let $E = \{p \in \mathbb{Q} : p \geq 0, p^2 < 2\}$. Show that E is closed and bounded in $\mathbb{Q}$, but not compact.

6. This exercise outlines an alternate proof of the Heine-Borel Theorem. Suppose $[a, b]$ is not compact. Then there exists an open cover $\mathcal{U} = \{U_\alpha\}_{\alpha \in A}$ of $[a, b]$ such that no finite sub-collection of $\mathcal{U}$ covers $[a, b]$. We now proceed to show that this leads to a contradiction. Divide $[a, b]$ into two closed subintervals $\left[a, \frac{a+b}{2}\right]$ and $\left[\frac{a+b}{2}, b\right]$ each of length $(b - a)/2$. At least one of these, call it I_1 cannot be covered by a finite number of the U_α. Repeating this process obtain a sequence $\{I_n\}$ of closed and bounded intervals satisfying (a) $[a, b] \supset I_1 \supset I_2 \supset \cdots \supset I_n \supset \cdots$, (b) length of $I_n = (b - a)/2^n$, and (c) for each n, I_n is not covered by a finite number of the U_α. Now use Exercise 3 to obtain a contradiction.

7. Prove the Bolzano-Weierstrass Theorem using the nested intervals property.

2.5   The Cantor Set

In this section, we will construct a compact subset of $[0, 1]$, known as the Cantor set, that has a number of interesting properties. This set is constructed by induction, the first two stages of which are illustrated in Figure 2.4.

Let $P_0 = [0, 1]$. From P_0 remove the middle third open interval $(\frac{1}{3}, \frac{2}{3})$. This leaves two disjoint closed intervals

$$J_{1,1} = \left[0, \tfrac{1}{3}\right], \quad J_{1,2} = \left[\tfrac{2}{3}, 1\right].$$

Set $P_1 = J_{1,1} \cup J_{1,2}$.

From each of $J_{1,1}$ and $J_{1,2}$ remove the middle third open intervals

$$\left(\tfrac{1}{3^2}, \tfrac{2}{3^2}\right) \quad \text{and} \quad \left(\tfrac{7}{3^2}, \tfrac{8}{3^2}\right)$$

of length $\frac{1}{3^2}$. This leaves 2^2 disjoint closed intervals $J_{2,1}, J_{2,2}, J_{2,3}, J_{2,4}$ of length $\frac{1}{3^2}$; namely

$$\left[0, \tfrac{1}{3^2}\right], \quad \left[\tfrac{2}{3^2}, \tfrac{3}{3^2}\right], \quad \left[\tfrac{6}{3^2}, \tfrac{7}{3^2}\right], \quad \left[\tfrac{8}{3^2}, \tfrac{9}{3^2}\right].$$

Set $P_2 = J_{2,1} \cup J_{2,2} \cup J_{2,3} \cup J_{2,4}$. In Figure 2.4, the shaded intervals indicate the open intervals that are removed at each stage of the construction.

FIGURE 2.4
Construction of the Cantor set

We continue this process inductively. At the nth step, each P_n is the union of 2^n disjoint closed intervals each of length $1/3^n$, i.e.,

$$P_n = \bigcup_{j=1}^{2^n} J_{n,j},$$

where for each j, $J_{n,j}$ is a closed interval of the form

$$J_{n,j} = \left[\frac{x_j}{3^n}, \frac{x_j + 1}{3^n}\right].$$

Since each P_n is a finite union of closed intervals, P_n is closed and bounded, hence compact. Furthermore, since $P_o \supset P_1 \supset P_2 \supset \cdots$, by Theorem 2.3.8,

$$P = \bigcap_{n=0}^{\infty} P_n$$

is a nonempty compact subset of $[0, 1]$. The set P is called the **Cantor ternary set**.

We now consider some of the properties of the set P.

Property 1 P is compact and nonempty.

Property 2 P contains all the endpoints of the closed intervals $\{J_{n,k}\}$, $n = 1, 2, ..., k = 1, 2, ..., 2^n$.

Property 3   Every point of P is a limit point of P.

Proof. Let $p \in P$ and let $\epsilon > 0$ be given. Choose $m \in \mathbb{N}$ such that $1/3^m < \epsilon$. Since $p \in P_m$, $p \in J_{m,k}$ for some k, $1 \le k \le 2^m$. But

$$J_{m,k} = \left[\frac{x_k}{3^m}, \frac{x_k + 1}{3^m} \right].$$

Since length of $J_{m,k} = 1/3^m < \epsilon$, $J_{m,k} \subset N_\epsilon(p)$. Thus both endpoints of $J_{m,k}$ are in $P \cap N_\epsilon(p)$, and at least one of these is distinct from p. $\square$

Property 4   The sum of the lengths of the intervals removed is 1.

Proof. At step 1, we removed one interval of length $1/3$. At the second step, we removed two intervals of length $1/3^2$. At the nth step, to obtain P_n, we removed 2^{n-1} intervals of length $1/3^n$. Thus we obtain that

$$\text{Sum of the lengths of the intervals removed } = \frac{1}{3} + 2\frac{1}{3^2} + \cdots + 2^{n-1}\frac{1}{3^n} + \cdots$$

$$= \sum_{n=1}^{\infty} \frac{2^{n-1}}{3^n} = \frac{1}{3} \sum_{n=0}^{\infty} \left(\frac{2}{3} \right)^n$$

$$= \frac{1}{3} \frac{1}{1 - \frac{2}{3}} = 1. \quad \square$$

As a consequence of Property 4,

Property 5 P contains no intervals.

For $x \in [0, 1]$ let $x = .n_1 n_2 n_3...$ be the ternary expansion. As we indicated in Section 1.6, this expansion is unique except when

$$x = \frac{a}{3^m}, \quad a \in \mathbb{N} \quad \text{with} \quad 0 < a < 3^m,$$

where 3 does not divide a. In this case x has two expansions: a finite expansion

$$x = \frac{a_1}{3} + \cdots + \frac{a_m}{3^m}, \qquad a_m \in \{1, 2\}$$

and an infinite expansion. If $a_m = 2$, we will use the finite expansion. If $a_m = 1$, we will use the infinite expansion

$$x = \frac{a_1}{3} + \cdots + \frac{0}{3^m} + \sum_{k=m+1}^{\infty} \frac{2}{3^k}.$$

With this convention we have

Property 6 If for each $x \in [0,1]$, $x = .n_1n_2n_3....$ is the ternary expansion of x, then

$$x \in P \qquad \text{if and only if} \qquad n_k \in \{0,2\}.$$

Proof. Exercise 2. □

As a consequence of Property 6 and Theorem 1.7.18,

Property 7 P is uncountable.

For each n, the set P_n has only a finite number of endpoints. As a consequence, the set of points of P which are endpoints of some open interval removed in the construction is countable. Since P is uncountable, P contains points other than endpoints. By Exercise 1 of Section 1.6, the ternary expansion of $\frac{1}{4}$ is

$$\frac{1}{4} = .020202.....$$

Thus $\frac{1}{4} \in P$, but $\frac{1}{4}$ is not an endpoint of P.

Remark. By Property 4, the sum of the lengths of the intervals removed is 1. This seems to imply that P is in some sense very "small." On the other hand, by Property 7 P is uncountable, which seems to imply that P is "large."

Exercises 2.5

1. Determine whether $\frac{1}{13}$ is in the Cantor set.

2. Prove Property 6 of the Cantor set.

3. Let $0 < \alpha < 1$. Construct a closed subset F of $[0,1]$ in a manner similar to the construction of the Cantor set such that the sum of the lengths of all the intervals removed is α.

4. *Prove that the Cantor set P is equivalent to $[0,1]$.

Notes

Without a doubt, the most important concept of this chapter is compactness. The fact that every open cover of a compact set has a finite subcover will be crucial in the study of continuous functions, especially uniform continuity. As we will see in many instances, the applications of compactness depend on the ability to choose a finite subcover from a particular open cover. A good example of this is the proof of Theorem 2.3.5. Other instances will occur later in the text.

Since compactness is the most important concept, Theorems 2.4.1 and 2.4.2 are the two most important results. In the Heine-Borel theorem we proved that every closed and bounded interval is compact, whereas in the Heine-Borel-Bolzano-Weierstrass theorem we characterized the compact subsets of $\mathbb{R}$. In Theorem 2.3.7 we proved that if K is a compact set then every infinite subset of K has a limit point in K. The converse of this result is also true, not only in $\mathbb{R}$ (Theorem 2.4.3), but also in the more general setting of metric spaces. A proof of this important result is outlined in the Miscellaneous Exercises.

Miscellaneous Exercises

The first two exercises involve the geometric and euclidean metric structure of $\mathbb{R}^n$. For $n \geq 2$, $\mathbb{R}^n = \{(x_1, ..., x_n) : x_i \in \mathbb{R}, i = 1, ..., n\}$. For $\mathbf{p} = (p_1, ..., p_n)$, $\mathbf{q} = (q_1, ..., q_n)$ in $\mathbb{R}^n$ and $c \in \mathbb{R}$, define

$$\mathbf{p} + \mathbf{q} = (p_1 + q_1, ..., p_n + q_n), \qquad and$$
$$c\mathbf{p} = (cp_1, ..., cp_n).$$

*Also, let $\mathbf{0} = (0, ..., 0)$. For $\mathbf{p}, \mathbf{q} \in \mathbb{R}^n$, the **inner product** of $\mathbf{p}$ and $\mathbf{q}$, denoted $\langle \mathbf{p}, \mathbf{q} \rangle$, is defined as*

$$\langle \mathbf{p}, \mathbf{q} \rangle = p_1 q_1 + \cdots p_n q_n.$$

1. Prove each of the following: for $\mathbf{p}, \mathbf{q}, \mathbf{r} \in \mathbb{R}^n$,
   a. $\langle \mathbf{p}, \mathbf{p} \rangle \geq 0$ with equality if and only if $\mathbf{p} = \mathbf{0}$.
   b. $\langle \mathbf{p}, \mathbf{q} \rangle = \langle \mathbf{q}, \mathbf{p} \rangle$.
   c. $\langle a\mathbf{p} + b\mathbf{q}, \mathbf{r} \rangle = a\langle \mathbf{p}, \mathbf{r} \rangle + b\langle \mathbf{q}, \mathbf{r} \rangle$ for all $a, b \in \mathbb{R}$.
   d. $|\langle \mathbf{p}, \mathbf{q} \rangle| \leq \sqrt{\langle \mathbf{p}, \mathbf{p} \rangle} \sqrt{\langle \mathbf{q}, \mathbf{q} \rangle}$.
   *This last inequality is usually called the **Cauchy-Schwarz inequality**. As a hint on how to prove (d), for $\lambda \in \mathbb{R}$, expand $\langle \mathbf{p} - \lambda\mathbf{q}, \mathbf{p} - \lambda\mathbf{q} \rangle$ and then choose λ appropriately. Note that by (a), $\langle \mathbf{p} - \lambda\mathbf{q}, \mathbf{p} - \lambda\mathbf{q} \rangle \geq 0$ for all $\lambda \in \mathbb{R}$.*

2. For $\mathbf{p} = (p_1, ..., p_n) \in \mathbb{R}^n$, set $\|\mathbf{p}\|_2 = \sqrt{\langle \mathbf{p}, \mathbf{p} \rangle} = \sqrt{p_1^2 + \cdots + p_n^2}$. The quantity $\|\mathbf{p}\|_2$ is called the **norm** or the **euclidean length** of the **vector p**.
   a. Use the result of 1(d) to prove that $\|\mathbf{p} + \mathbf{q}\|_2 \leq \|\mathbf{p}\|_2 + \|\mathbf{q}\|_2$ for all $\mathbf{p}, \mathbf{q} \in \mathbb{R}^n$.
   b. Using the result of (a), prove that $d_2(\mathbf{p}, \mathbf{q}) = \|\mathbf{p} - \mathbf{q}\|_2$ is a metric on $\mathbb{R}^n$.

3. If E is an uncountable subset of $\mathbb{R}$, prove that some point of E is a limit point of E.
   The following exercise is designed to prove the converse of Theorem 2.3.7; namely, if K is a subset of a metric space (X, d) having the property that every infinite subset of K has a limit point in K, then K is compact.

4. Let K be a subset of a metric space (X, d) that has the property that every infinite subset of K has a limit point in K.

    a. Prove that there exists a countable subset D of K which is dense in K. (Hint: Fix $n \in \mathbb{N}$. Let $p_1 \in K$ be arbitrary. Choose $p_2 \in K$, if possible, such that $d(p_1, p_2) \geq \frac{1}{n}$. Suppose $p_1, \ldots, p_j$ have been chosen. Choose p_{j+1}, if possible, such that $d(p_1, p_{j+1}) \geq \frac{1}{n}$ for all $i = 1, \ldots, n$. Use the assumption about K to prove that this process must terminate after a finite number of steps. Let $\mathcal{P}_n$ denote this finite collection of points, and let $D = \bigcup_{n \in I} \mathcal{P}_n$. Prove that D is countable and dense in K.)

    b. Let D be as in (a), and let U be an open subset of X such that $U \cap K \neq \emptyset$. Prove that there exists $p \in D$ and $n \in \mathbb{N}$ such that $N_{1/n}(p) \subset U$.

    c. Using the result of (b), prove that for every open cover $\mathcal{U}$ of K, there exists a finite or countable collection $\{U_n\}_n \subset \mathcal{U}$ such that $K \subset \bigcup_n U_n$.

    d. Prove that every countable open cover of K has a finite subcover. (Hint: If $\{U_n\}_{n=1}^{\infty}$ is a countable open cover of K, for each $n \in \mathbb{N}$ let $W_n = \bigcup_{j=1}^{n} U_j$. Prove that $K \subset W_n$ for some $n \in \mathbf{N}$. Assume that the result is false, and obtain an infinite subset of K with no limit point in K which is contradiction.)

Supplemental Reading

Asic, M. D. and Adamovic, D. D., "Limit points of sequences in metric spaces," *Amer. Math. Monthly* **77** (1970) 613–616.

Corazza, P., "Introduction to metric preserving functions," *Amer. Math. Monthly* **106**(1999) 309–323.

Dubeau, F., "Cauchy-Bunyakowski-Schwarz inequality revisited," *Amer. Math. Monthly* **99** (1990) 419–421.

Espelie, M. S. and Joseph, J. E., "Compact subsets of the Sorgenfrey line," *Math. Mag.* **49** (1976) 250–251.

Fleron, Julian F., "A note on the history of the Cantor set and Cantor function," *Math. Mag.* **67** (1994), 136–140.

Geissinger, L., "Pythagoras and the Cauchy Schwarz inequality," *Amer. Math. Monthly* **83** (1976) 40–41.

Kaplansky, I., *Set Theory and Metric Spaces* Chelsea Publ. Co., New York, 1977

Labarre, Jr., A. E., "Structure theorem for open sets of real numbers," *Amer. Math. Monthly* **72** (1965) 1114.

Nathanson, M. B., "Round metric spaces," *Amer. Math. Monthly* **82** (1975) 738–741.

3

Sequences of Real Numbers

Now that we have covered the basic topological concepts required for the study of analysis, we begin with limits of sequences. This topic will be our first serious introduction to the limit process. The notion of convergence of a sequence dates back to the early nineteenth century and the work of Bolzano (1817) and Cauchy (1821). Some of the concepts and results included in this chapter have undoubtedly been encountered previously in the study of calculus. Our presentation however will be considerably more rigorous—emphasizing proofs rather than computations.

Although our primary emphasis will be on sequences of real numbers, these are not the only sequences which are typically encountered. It is not at all unusual to talk about sequences of functions, sequences of vectors, etc. For this reason we will begin our study of sequences in the general setting of metric spaces. Most of the examples however will come from the real numbers. A good understanding of sequences in $\mathbb{R}$ will prove helpful in providing insight into properties of sequences in more general settings.

We begin the chapter by introducing the notion of convergence of a sequence in a metric space, and then by proving the standard limit theorems for sequences of real numbers normally encountered in calculus. In Section 3.3 we will use the least upper bound property of $\mathbb{R}$ to prove that every bounded monotone sequence of real numbers converges in $\mathbb{R}$. The study of subsequences and sub-sequential limits will be the topic of Section 3.4. In this section, we also prove the well known result of Bolzano and Weierstrass that every bounded sequence of real numbers has a convergent subsequence. This result will then be used to provide a short proof of the fact that every Cauchy sequence of real numbers converges. Although the study of series of real numbers is the main topic of Chapter 7, some knowledge of series will be required in the construction of certain examples in Chapters 4 and 6. For this reason we include a brief introduction to series as the last section of this chapter.

3.1 Convergent Sequences

We begin our study of sequences by first considering sequences in arbitrary metric spaces. Throughout this section we let (X, d) be a metric space. When

$X = \mathbb{R}$, unless otherwise specified d will denote the usual euclidean metric on $\mathbb{R}$. Recall that by a **sequence** in X we mean a function $f : \mathbb{N} \to X$. For each $n \in \mathbb{N}$, $p_n = f(n)$ is called the nth **term** of the sequence f, and for convenience, the sequence f is denoted by $\{p_n\}_{n=1}^{\infty}$, or simply $\{p_n\}$.

DEFINITION 3.1.1 *A sequence $\{p_n\}_{n=1}^{\infty}$ in X is said to **converge** if there exists a point $p \in X$ such that for every $\epsilon > 0$, there exists a positive integer $n_o = n_o(\epsilon)$ such that $p_n \in N_\epsilon(p)$ for all $n \geq n_o$. If this is the case, we say that $\{p_n\}$ **converges to** p, or that p is the **limit** of the sequence $\{p_n\}$, and we write*

$$\lim_{n \to \infty} p_n = p \quad or \quad p_n \to p.$$

*If $\{p_n\}$ does not converge, then $\{p_n\}$ is said to **diverge**.*

In the definition, the statement $p_n \in N_\epsilon(p)$ for all $n \geq n_o$ is equivalent to

$$d(p_n, p) < \epsilon \quad \text{for all} \quad n \geq n_o.$$

As a general rule, the integer n_o will depend on the given ϵ. This will be illustrated in the following examples.

EXAMPLES 3.1.2 (a) For our first example we show that the sequence $\{1/n\}_{n=1}^{\infty}$ converges to 0 in $\mathbb{R}$. The proof of this is the remark following Theorem 1.5.1; namely, given $\epsilon > 0$, there exists a positive integer n_o such that $n_o\epsilon > 1$. Thus for all $n \geq n_o$,

$$\left| \tfrac{1}{n} - 0 \right| = \tfrac{1}{n} < \epsilon.$$

Therefore $\lim_{n \to \infty} \tfrac{1}{n} = 0$. In this example, the integer n_o must be chosen so that $n_o > 1/\epsilon$.

(b) If $p \in \mathbb{R}$, the sequence $\{p_n\}$ defined by $p_n = p$ for all $n \in \mathbb{N}$ is called the **constant sequence** p. Since $|p_n - p| = 0$ for all $n \in \mathbb{N}$, we have $\lim_{n \to \infty} p_n = p$.

(c) Consider the sequence $\left\{ \dfrac{2n + 1}{3n + 2} \right\}_{n=1}^{\infty}$. We will show that

$$\lim_{n \to \infty} \frac{2n + 1}{3n + 2} = \frac{2}{3}.$$

Since

$$\left| \frac{2n + 1}{3n + 2} - \frac{2}{3} \right| = \frac{1}{3(3n + 2)} < \frac{1}{9n},$$

given $\epsilon > 0$, choose $n_o \in \mathbb{N}$ such that $n_o > \tfrac{1}{9\epsilon}$. Then for all $n \geq n_o$,

$$\left| \frac{2n + 1}{3n + 2} - \frac{2}{3} \right| < \epsilon.$$

Thus the given sequence converges to 2/3.

(d) The sequence $\{1 - (-1)^n\}_{n=1}^{\infty}$ diverges in $\mathbb{R}$. To prove this, we first note that for this sequence, $|p_n - p_{n+1}| = 2$ for all n. Suppose $p_n \to p$ for some $p \in \mathbb{R}$. Let $0 < \epsilon < 1$. Then by the definition of convergence, there exists an integer n_o such that $|p_n - p| < \epsilon$ for all $n \geq n_o$. But if $n \geq n_o$, then

$$2 = |p_n - p_{n+1}| \leq |p_n - p| + |p - p_{n+1}| < 2\epsilon < 2,$$

which is a contradiction.

(e) Consider the sequence

$$\mathbf{p}_n = \left(1 - \frac{1}{n}, \frac{2n+1}{3n+2}\right)$$

in $\mathbb{R}^2$. In Exercise 10 you will be asked to prove that a sequence $\mathbf{p}_n = (p_n, q_n)$ converges to $\mathbf{p} = (p, q)$ if and only if $p_n \to p$ and $q_n \to q$. Thus the above sequence converges to $(1, \frac{2}{3})$.

(f) Let X denote the set of bounded functions on $[0, 1]$ with metric d given by $d(f, g) = \sup\{|f(x) - g(x)| : x \in [0, 1]\}$. Consider the sequence $\{f_n\}$, where for each $n \in \mathbb{N}$, f_n is the function given by $f_n(x) = x^n/n$. If $f = 0$ denotes the zero function on $[0, 1]$, i.e., $f(x) = 0$ for all $x \in [0, 1]$, then for each $n \in \mathbb{N}$,

$$d(f_n, f) = d(f_n, 0) = \sup\left\{\frac{x^n}{n} : x \in [0, 1]\right\} = \frac{1}{n}.$$

As a consequence the sequence $\{f_n\}$ converges to the zero function in X. $\square$

DEFINITION 3.1.3 *A sequence $\{p_n\}$ in X is said to be* **bounded** *if there exists $p \in X$ and a positive constant M such that $d(p, p_n) \leq M$ for all $n \in \mathbb{N}$.*

Remark. A sequence $\{p_n\}$ in $\mathbb{R}$ is bounded if there exists a positive constant C such that $|p_n| \leq C$ for all $n \in \mathbb{N}$. To see this, suppose $\{p_n\}$ is bounded according to Definition 3.1.3. Then there exists $p \in \mathbb{R}$ such that $|p_n - p| \leq M$ for all n. As a consequence

$$|p_n| \leq |p_n - p| + |p| \leq (M + |p|),$$

which proves the results. Thus with respect to the usual metric a sequence $\{p_n\}$ is bounded if and only if the set $\{p_n : n = 1, 2, \dots\}$ is a bounded subset of $\mathbb{R}$. This however is not always the case for other metrics (see Example 3.1.5(c)).

THEOREM 3.1.4 *Let (X, d) be a metric space.*

(a) *If a sequence $\{p_n\}$ in X converges, then its limit is unique.*

(b) *Every convergent sequence in X is bounded.*

(c) *If $E \subset X$ and p is a limit point of E, then there exists a sequence $\{p_n\}$ in E with $p_n \neq p$ for all n such that*

$$\lim_{n \to \infty} p_n = p.$$

Proof. (a) Suppose the sequence $\{p_n\}$ converges to two distinct points $p, q \in X$. Let $\epsilon = \frac{1}{3}d(p,q)$. Since $p_n \to p$, there exists an integer n_1 such that $d(p_n, p) < \epsilon$ for all $n \geq n_1$. Also, since $p_n \to q$, there exists an integer n_2 such that $d(p_n, q) < \epsilon$ for all $n \geq n_2$. Thus if $n \geq \max\{n_1, n_2\}$, by the triangle inequality

$$d(p,q) \leq d(p_n, p) + d(p_n, q) < 2\epsilon = \tfrac{2}{3}d(p,q)$$

which is a contradiction.

(b) Let $\{p_n\}$ be a convergent sequence in X that converges to $p \in X$. Take $\epsilon = 1$. For this ϵ, there exists an integer n_o such that $d(p_n, p) < 1$ for all $n > n_o$. Let

$$M = \max\{d(p, p_1), \ldots, d(p, p_{n_o}), 1\}.$$

Then $d(p, p_n) \leq M$ for all $n \in \mathbb{N}$. Therefore the sequence is bounded.

(c) We construct the sequence $\{p_n\}$ in E as follows: Since p is a limit point of E, for each positive integer n, by the definition of limit point, there exists $p_n \in E$ with $p_n \neq p$, such that

$$d(p_n, p) < \frac{1}{n}.$$

This sequence clearly satisfies $p_n \to p$. $\square$

EXAMPLES 3.1.5 (a) According to the previous theorem, every convergent sequence is bounded. The converse however is false. The sequence $\{1 - (-1)^n\}_{n=1}^{\infty}$ is bounded, but by Example 3.1.2(d), the sequence does not converge. The sequence is bounded since $|p_n| = |1 - (-1)^n| \leq 2$ for all $n \in \mathbb{N}$.

(b) The sequence $\{n(-1)^n\}$ is not bounded in $\mathbb{R}$, and thus cannot converge.

(c) Let $X = \mathbb{R}$ with the metric

$$\rho(x, y) = \frac{|x - y|}{1 + |x - y|}.$$

With this metric every sequence $\{p_n\}$ in $\mathbb{R}$ satisfies $\rho(p_n, p) < 1$ for any $p \in \mathbb{R}$. Thus every sequence $\{p_n\}$ in $(\mathbb{R}, \rho)$ is bounded according to Definition 3.1.3. Obviously however there exist sequences in $\mathbb{R}$ for which the range is not a bounded subset of $\mathbb{R}$. It is interesting to note that a sequence $\{p_n\}$ converges to p in the usual metric if and only if $p_n \to p$ in the metric ρ (Exercise 11).

(d) In this example, we illustrate part (c) of the previous theorem. Since $\sqrt{2}$ is a limit point of $\mathbb{Q}$, the previous theorem guarantees the existence of a sequence $\{r_n\}$ of rational numbers such that $r_n \to \sqrt{2}$. Note however that this sequence need not be unique. If $r_n \to \sqrt{2}$, then the same is true for the sequence $\{r_n + \frac{1}{n}\}$. $\square$

Exercises 3.1

1. For each of the following sequences, prove, using an ϵ, n_o argument that the sequence converges to the given limit p; that is, given $\epsilon > 0$ determine n_o such that $|p_n - p| < \epsilon$ for all $n \geq n_o$.

   *a. $\left\{ \dfrac{3n+5}{2n+7} \right\}, p = \dfrac{3}{2}$

   b. $\left\{ \dfrac{2n+5}{6n-3} \right\}, p = \dfrac{1}{3}$

   *c. $\left\{ \dfrac{n^2+1}{2n^2} \right\}, p = \dfrac{1}{2}$

   d. $\left\{ 1 - \dfrac{(-1)^n}{n} \right\}, p = 1$

   e. $\left\{ \dfrac{(-1)^n n}{n^2+1} \right\}, p = 0.$

   *f. $\left\{ \sqrt{n+1} - \sqrt{n} \right\}, p = 0$

   g. $\left\{ n\left(\sqrt{1+\frac{1}{n}} - 1 \right) \right\}, p = \dfrac{1}{2}$

2. Show that each of the following sequences diverge in $\mathbb{R}$.

   *a. $\{n (1 + (-1)^n)\}.$

   b. $\left\{ (-1)^n + \dfrac{1}{n} \right\}.$

   *c. $\left\{ \sin \dfrac{n\pi}{2} \right\}.$

   d. $\left\{ n \sin \dfrac{n\pi}{2} \right\}.$

   e. $\left\{ \dfrac{(-1)^n}{n+1} \right\}.$

3. If $b > 0$, prove that $\lim\limits_{n \to \infty} \dfrac{1}{1+nb} = 0.$

4. Prove each of the following.

   *a. If $b > 1$, prove that $\lim\limits_{n \to \infty} \dfrac{1}{b^n} = 0.$

   b. If $0 \leq b < 1$, prove that $\lim\limits_{n \to \infty} b^n = 0.$

5. ***** Let $\{a_n\}$ be a sequence in $\mathbb{R}$ with $\lim\limits_{n \to \infty} a_n = a$. Prove that $\lim\limits_{n \to \infty} a_n^2 = a^2$.

6. ***** If $a_n \geq 0$ for all n and $\lim\limits_{n \to \infty} a_n = a$, prove that $\lim\limits_{n \to \infty} \sqrt{a_n} = \sqrt{a}$.

7. Prove that if $\{a_n\}$ converges to a, then $\{|a_n|\}$ converges to $|a|$. Is the converse true?

8. ***** Let $\{a_n\}$ be a sequence in $\mathbb{R}$ with $\lim\limits_{n \to \infty} a_n = a$. If $a > 0$, prove that there exists $n_o \in \mathbb{N}$ such that $a_n > 0$ for all $n \geq n_o$.

9. Let $\{a_n\}$ be a sequence in $\mathbb{R}$ satisfying $|a_n - a_{n+1}| \geq c$ for some $c > 0$ and all $n \in \mathbb{N}$. Prove that the sequence $\{a_n\}$ diverges.

10. Consider $\mathbb{R}^2$ with the metric d_2 as defined in Example 2.1.7(d). Suppose $\mathbf{p}_n = (a_n, b_n)$ and $\mathbf{p} = (a, b)$. Prove that

    $$\lim\limits_{n \to \infty} \mathbf{p}_n = \mathbf{p} \quad \text{if and only if} \quad \lim\limits_{n \to \infty} a_n = a \text{ and } \lim\limits_{n \to \infty} b_n = b.$$

11. Consider $\mathbb{R}$ with the usual metric, and also with the metric

    $$\rho(x, y) = \dfrac{|x - y|}{1 + |x - y|}.$$

Prove that a sequence $\{p_n\}$ in $\mathbb{R}$ converges to $p \in \mathbb{R}$ in the usual metric if and only if $p_n \to p$ in the metric ρ.

12. Let X be the set of bounded functions on $[0,1]$ with metric d as defined in Example 3.1.2(f). Prove that each of the following sequences of functions converges to the indicated functions f.

    a. $\left\{x + \dfrac{x}{n}\sin nx\right\}_{n=1}^{\infty}$, $f(x) = x$.   **b.** $\left\{\left(\dfrac{x}{b^n}\right)\right\}_{n=1}^{\infty}$, $(b>1)$, $f(x) = 0$.

    c. $\left\{\dfrac{x}{n+x}\right\}_{n=1}^{\infty}$, $f(x) = 0$.

13. Let (X, d) be as in the previous exercise. Does the sequence $\{x^n\}$ converge in (X, d)?

3.2 Sequences of Real Numbers

In this section, we will emphasize some of the important properties of sequences of real numbers, and also investigate the limits of several basic sequences that are frequently encountered in the study of analysis. Our first result involves algebraic operations on convergent sequences.

THEOREM 3.2.1 *If $\{a_n\}$ and $\{b_n\}$ are convergent sequences of real numbers with*

$$\lim_{n\to\infty} a_n = a \qquad and \qquad \lim_{n\to\infty} b_n = b,$$

then

(a) $\displaystyle\lim_{n\to\infty}(a_n + b_n) = a + b,$ *and*

(b) $\displaystyle\lim_{n\to\infty} a_n b_n = ab.$

(c) *Furthermore, if $a \neq 0$, and $a_n \neq 0$ for all n, then* $\displaystyle\lim_{n\to\infty}\frac{b_n}{a_n} = \frac{b}{a}.$

Proof. The proof of (a) is left to the exercises (Exercise 1). To prove (b), we add and subtract the term $a_n b$ to obtain

$$|a_n b_n - ab| = |(a_n b_n - a_n b) + (a_n b - ab)| \leq |a_n||b_n - b| + |b||a_n - a|.$$

Since $\{a_n\}$ converges, by Theorem 3.1.4(b), $\{a_n\}$ is bounded. Thus there exists a constant $M > 0$ such that $|a_n| \leq M$ for all n. Therefore

$$|a_n b_n - ab| \leq M|b_n - b| + |b||a_n - a|.$$

Let $\epsilon > 0$ be given. Since $a_n \to a$, there exists a positive integer n_1 such that

$$|a_n - a| < \frac{\epsilon}{2(|b| + 1)}$$

for all $n \geq n_1$. Also, since $b_n \to b$, there exists a positive integer n_2 such that

$$|b_n - b| < \frac{\epsilon}{2M}$$

for all $n \geq n_2$. Thus if $n \geq \max\{n_1, n_2\}$,

$$|a_n b_n - ab| < M\left(\frac{\epsilon}{2M}\right) + |b|\left(\frac{\epsilon}{2(|b|+1)}\right) < \epsilon.$$

Therefore $\lim\limits_{n \to \infty} a_n b_n = ab$.

To prove (c) it suffices to show that $\lim\limits_{n \to \infty} 1/a_n = 1/a$. The result (c) then follows from (b). Since $a \neq 0$ and $a_n \to a$, there exists a positive integer n_o such that

$$|a_n - a| < \tfrac{1}{2}|a|$$

for all $n \geq n_o$. Also, since

$$|a| \leq |a - a_n| + |a_n| < \tfrac{1}{2}|a| + |a_n|$$

for $n \geq n_o$, we have

$$|a_n| \geq \tfrac{1}{2}|a|$$

for all $n \geq n_o$. Therefore,

$$\left|\frac{1}{a_n} - \frac{1}{a}\right| = \frac{|a - a_n|}{|a_n||a|} < \frac{2}{|a|^2}|a_n - a|.$$

Let $\epsilon > 0$ be given. Since $a_n \to a$, we can choose an integer $n_1 \geq n_o$ so that

$$|a_n - a| < \epsilon\frac{|a|^2}{2}$$

for all $n \geq n_1$. Therefore

$$\left|\frac{1}{a_n} - \frac{1}{a}\right| < \epsilon$$

for all $n \geq n_1$, and as a consequence

$$\lim_{n \to \infty} \frac{1}{a_n} = \frac{1}{a}. \quad \square$$

COROLLARY 3.2.2 *If $\{a_n\}$ is a convergent sequence of real numbers with* $\lim\limits_{n \to \infty} a_n = a$, *then for any $c \in \mathbb{R}$,*

   (a) $\lim\limits_{n \to \infty} (a_n + c) = a + c$, *and*

   (b) $\lim\limits_{n \to \infty} c\, a_n = c\, a$.

Proof. If we define the sequence $\{c_n\}$ by $c_n = c$ for all $n \in \mathbb{N}$, then the conclusions follow by (a) and (b) of the previous theorem. $\quad \square$

THEOREM 3.2.3 *Let* $\{a_n\}$ *and* $\{b_n\}$ *be sequences of real numbers. If* $\{b_n\}$ *is bounded and* $\lim\limits_{n\to\infty} a_n = 0$, *then*

$$\lim_{n\to\infty} a_n b_n = 0.$$

Proof. Exercise 3. □

Remark. Since the sequence $\{b_n\}$ may not converge, Theorem 3.2.1(c) does not apply. The fact that the sequence $\{b_n\}$ is bounded is crucial. For example, consider the sequences $\{\frac{1}{n}\}$ and $\{3n\}$.

THEOREM 3.2.4 *Suppose* $\{a_n\}$, $\{b_n\}$, *and* $\{c_n\}$ *are sequences of real numbers for which there exists* $n_o \in \mathbb{N}$ *such that*

$$a_n \leq b_n \leq c_n \quad \text{for all } n \in \mathbb{N}, n \geq n_o,$$

and that $\lim\limits_{n\to\infty} a_n = \lim\limits_{n\to\infty} c_n = L$. *Then the sequence* $\{b_n\}$ *converges and*

$$\lim_{n\to\infty} b_n = L.$$

Proof. Exercise 4 □

The above result, commonly called the **squeeze theorem**, is very useful in applications. Quite often to show that a given sequence $\{a_n\}$ in $\mathbb{R}$ converges to a, we will first prove that

$$|a_n - a| \leq M b_n$$

for some positive constant M and a nonnegative sequence $\{b_n\}$ with $\lim\limits_{n\to\infty} b_n = 0$. Since the above inequality is equivalent to

$$-M b_n \leq a_n - a \leq M b_n,$$

by Theorem 3.2.4 the sequence $\{a_n - a\}$ converges to 0, or equivalently that $\lim\limits_{n\to\infty} a_n = a$.

Some Special Sequences

We next consider some special sequences of real numbers that occur frequently in the study of analysis. For the proof of Theorem 3.2.6 we require the following result.

THEOREM 3.2.5 (Binomial Theorem) *For* $a \in \mathbb{R}$, $n \in \mathbb{N}$,

$$(1+a)^n = \sum_{k=0}^{n} \binom{n}{k} a^k = \binom{n}{0} + \binom{n}{1} a + \binom{n}{2} a^2 + \cdots + \binom{n}{n} a^n$$

where

$$\binom{n}{k} = \frac{n!}{k!(n-k)!}$$

is the **binomial coefficient**.

In the above, for $n \in \mathbb{N}$, $n!$ (read n **factorial**) is defined by

$$n! = n \cdot (n-1) \cdots 2 \cdot 1,$$

with the usual convention that $0! = 1$. Since the binomial theorem is ancillary to our main topic of discussion, we leave the proof, using mathematical induction, to the exercises (Exercise 13). An alternate proof using Taylor series will be provided in Section 8.7.

THEOREM 3.2.6 .

(a) If $p > 0$, then $\lim\limits_{n \to \infty} \dfrac{1}{n^p} = 0$.

(b) If $p > 0$, then $\lim\limits_{n \to \infty} \sqrt[n]{p} = 1$.

(c) $\lim\limits_{n \to \infty} \sqrt[n]{n} = 1$.

(d) If $p > 1$ and α is real, then $\lim\limits_{n \to \infty} \dfrac{n^{\alpha}}{p^n} = 0$.

(e) If $|p| < 1$, then $\lim\limits_{n \to \infty} p^n = 0$.

(f) For all $p \in \mathbb{R}$, $\lim\limits_{n \to \infty} \dfrac{p^n}{n!} = 0$.

Proof. The proofs of (a) and (b) are left to the exercises (Exercise 5). The proof of (a) is straightforward and the proof of (b) (for $p > 1$) is similar to the proof of (c). For the proof of (c), let $x_n = \sqrt[n]{n} - 1$. Since x_n is positive, by the binomial theorem

$$n = (1 + x_n)^n \geq \binom{n}{2} x_n^2 = \frac{n(n-1)}{2} x_n^2$$

for all $n \geq 2$. Therefore, $x_n^2 \leq 2/(n-1)$ for all $n \geq 2$, and as a consequence

$$0 \leq x_n \leq \sqrt{\frac{2}{n-1}}.$$

Thus by (a) and Theorem 3.2.4, $\lim\limits_{n \to \infty} x_n = 0$, from which the result follows.

(d) Let k be a positive integer so that $k > \alpha$. Since $p > 1$, write $p = (1+q)$, with $q > 0$. By the binomial theorem, for $n > 2k$,

$$p^n = (1+q)^n > \binom{n}{k} q^k = \frac{n(n-1) \cdots (n-k+1)}{k!} q^k.$$

Since $k < \frac{1}{2}n$, $n - k + 1 > \frac{1}{2}n + 1 > \frac{1}{2}n$. Therefore,

$$\frac{n(n-1) \cdots (n-k+1)}{k!} > \frac{n^k}{2^k k!},$$

and as a consequence,

$$0 \le \frac{n^\alpha}{p^n} \le \left(\frac{2^k k!}{q^k}\right) \frac{1}{n^{k-\alpha}}.$$

The result now follows by part (a) and Theorem 3.2.4.

(e) Write p as $p = \frac{\pm 1}{q}$, where $q > 1$. Then

$$|p^n| = |p|^n = \frac{1}{q^n}$$

which by part (d) (with $\alpha = 0$) converges to 0 as $n \to \infty$.

(f) Fix $k \in \mathbb{N}$ such that $k > |p|$. For $n > k$,

$$\left|\frac{p^n}{n!}\right| = \frac{|p|^n}{n!} < \frac{k^{(k-1)}}{(k-1)!}\left(\frac{|p|}{k}\right)^n.$$

Since $|p|/k < 1$, the result follows by (e). $\square$

EXAMPLES 3.2.7 We now provide several examples to illustrate the previous theorems.

(a) As in Example 3.1.2(c), consider the sequence $\left\{\frac{2n+1}{3n+2}\right\}$. We write

$$\frac{2n+1}{3n+2} = \frac{n(2+\frac{1}{n})}{n(3+\frac{2}{n})} = \frac{2+\frac{1}{n}}{3+\frac{2}{n}}.$$

Since $\lim_{n\to\infty} \frac{1}{n} = \lim_{n\to\infty} \frac{2}{n} = 0$, by Corollary 3.2.2(a),

$$\lim_{n\to\infty}\left(2+\frac{1}{n}\right) = 2 \quad\text{and}\quad \lim_{n\to\infty}\left(3+\frac{2}{n}\right) = 3.$$

Therefore by Theorem 3.2.1(b) and (c),

$$\lim_{n\to\infty}\left(\frac{2+\frac{1}{n}}{3+\frac{2}{n}}\right) = \frac{\lim_{n\to\infty}\left(2+\frac{1}{n}\right)}{\lim_{n\to\infty}\left(3+\frac{2}{n}\right)} = 2\cdot\frac{1}{3} = \frac{2}{3}.$$

(b) Consider the sequence $\left\{\frac{(-1)^n}{2\sqrt{n}+7}\right\}$. We first note that

$$0 \le \left|\frac{(-1)^n}{2\sqrt{n}+7}\right| \le \frac{1}{2}\frac{1}{\sqrt{n}}.$$

Thus by Theorems 3.2.4 and 3.2.6(a) with $p = \frac{1}{2}$,

$$\lim_{n\to\infty}\frac{(-1)^n}{2\sqrt{n}+7} = 0.$$

(c) For our next example we consider the sequence $\left\{\dfrac{2^n + n^3}{3^n + n^2}\right\}$. As in (a), we first factor out the dominant power in both the numerator and denominator. By Theorem 3.2.6(d), $\lim\limits_{n\to\infty} n^\alpha/p^n = 0$ for any $\alpha \in \mathbb{R}$ and $p > 1$. This simply states that p^n $(p > 1)$ grows faster than any power of n. Therefore the dominant terms in both the numerator and denominator are 2^n and 3^n, respectively. Thus

$$\frac{2^n + n^3}{3^n + n^2} = \frac{\cdot 2^n(1 + \frac{n^3}{2^n})}{3^n(1 + \frac{n^2}{3^n})} = \left(\frac{2}{3}\right)^n \frac{(1 + \frac{n^3}{2^n})}{(1 + \frac{n^2}{3^n})}.$$

By Theorems 3.2.1 and 3.2.6(d)

$$\lim_{n\to\infty} \frac{(1 + \frac{n^3}{2^n})}{(1 + \frac{n^2}{3^n})} = 1.$$

Finally, since $\lim\limits_{n\to\infty} \left(\frac{2}{3}\right)^n = 0$ (Theorem 3.2.6(e)), we have

$$\lim_{n\to\infty} \frac{2^n + n^3}{3^n + n^2} = 0.$$

(e) As out final example we consider the sequence $\{n((1 + \frac{1}{n})^{-2} - 1)\}$. Before we can evaluate the limit of this sequence we must first simplify the nth term of the sequence. This is accomplished as follows:

$$x_n = n((1 + \tfrac{1}{n})^{-2} - 1) = n\left(\frac{1}{(1 + \frac{1}{n})^2} - 1\right) = n\left(\frac{n^2}{(n+1)^2} - 1\right)$$

$$= n\left(\frac{-2n - 1}{(n+1)^2}\right) = \frac{-2n^2 - n}{(n+1)^2}.$$

Now we can factor out an n^2 from both the numerator and the denominator. This gives

$$x_n = \frac{-2 - \frac{1}{n}}{(1 + \frac{1}{n})^2}.$$

Using the limit theorem we now conclude that $\lim\limits_{n\to\infty} x_n = -2$. $\square$

Exercises 3.2

1. Prove Theorem 3.2.1(a).

2. Let $\{a_n\}$ and $\{b_n\}$ be sequences of real numbers.

   a. If $\{a_n\}$ and $\{a_n + b_n\}$ both converge, prove that the sequence $\{b_n\}$ converges.

   b. Suppose $b_n \neq 0$ for all $n \in \mathbb{N}$. If $\{b_n\}$ and $\{a_n/b_n\}$ both converge, prove that the sequence $\{a_n\}$ also converges.

3. Prove Theorem 3.2.3.

4. Prove Theorem 3.2.4.

5. Prove each of the following.

    a. If $p > 0$, prove that $\lim\limits_{n\to\infty} \dfrac{1}{n^p} = 0$.

    *b. If $p > 0$, prove that $\lim\limits_{n\to\infty} \sqrt[n]{p} = 1$.

6. Find the limit of each of the following sequences.

    *a. $\left\{ \dfrac{3n^2 + 2n + 1}{5n^2 - 2n + 3} \right\}_{n=1}^{\infty}$.

    b. $\left\{ 1 + \dfrac{n^3}{3^n} \right\}_{n=1}^{\infty}$.

    *c. $\left\{ \dfrac{n}{1 + \frac{1}{n}} - n \right\}_{n=1}^{\infty}$.

    d. $\left\{ \dfrac{2\sqrt{n}}{n + 1} \right\}_{n=1}^{\infty}$.

    *e. $\left\{ \dfrac{1}{\sqrt{1 + \frac{1}{n}}} - 1 \right\}_{n=1}^{\infty}$.

    f. $\left\{ \sqrt{n^2 + n} - n \right\}_{n=1}^{\infty}$.

    *g. $\left\{ \sqrt{n} \left(\sqrt{n+a} - \sqrt{n} \right) \right\}_{n=1}^{\infty}$, $a > 0$.

    h. $\left\{ (2^n + 3^n)^{1/n} \right\}_{n=1}^{\infty}$.

7. For each of the following sequences, determine whether the given sequence converges or diverges. If the sequence converges, find its limit; if it diverges, explain why.

    *a. $\left\{ 1 + \dfrac{1 + (-1)^n}{n} \right\}_{n=1}^{\infty}$.

    b. $\left\{ \dfrac{1}{n} \sin \dfrac{n\pi}{2} \right\}_{n=1}^{\infty}$.

    *c. $\left\{ \dfrac{1}{n^2} \left(\dfrac{n^2 + 1}{2n + 3} \right)^2 \right\}_{n=1}^{\infty}$.

    d. $\left\{ \dfrac{3^n}{2^n + n^2} \right\}_{n=1}^{\infty}$.

    *e. $\left\{ \dfrac{\sqrt[3]{8n^3 + 5}}{\sqrt{9n^2 - 4}} \right\}_{n=1}^{\infty}$.

    f. $\left\{ \dfrac{n \cos n\pi}{2n + 3} \right\}_{n=1}^{\infty}$.

8. ***Prove** that $\lim\limits_{n\to\infty} \dfrac{1}{n} \cos n \dfrac{\pi}{2} = 0$.

9. Let $\{x_n\}$ be a sequence in $\mathbb{R}$ with $x_n \to 0$, and $x_n \neq 0$ for all n. Prove that $\lim\limits_{n\to\infty} x_n \sin \dfrac{1}{x_n} = 0$.

10. Let $\{a_n\}$ be a sequence of positive real number such that

$$\lim_{n\to\infty} \frac{a_{n+1}}{a_n} = L.$$

    *a. If $L < 1$, prove that the sequence $\{a_n\}$ converges to zero.

    b. If $L > 1$, prove that the sequence $\{a_n\}$ is unbounded.

    c. Give an example of a convergent sequence $\{a_n\}$ of positive real numbers for which $L = 1$.

    d. Give an example of a divergent sequence $\{a_n\}$ of positive real numbers for which $L = 1$.

11. Use the previous exercise to determine convergence or divergence of each of the following sequences.

    *a. $\{n^2 a^n\}$, $0 < a < 1$.

    b. $\left\{ \dfrac{n^2}{a^n} \right\}$, $0 < a < 1$.

c. $\left\{ \dfrac{a^n}{n!} \right\}$, $0 < a < 1$.   **d.** $\left\{ \dfrac{n!}{n^n} \right\}$.

12. *Suppose $\lim\limits_{n \to \infty} (a_n - 1)/(a_n + 1) = 0$. Prove that $\lim\limits_{n \to \infty} a_n = 1$.

13.   **a.** For $n \in \mathbb{N}$, $1 \le k \le n$, prove that

$$\binom{n}{k} + \binom{n}{k-1} = \binom{n+1}{k}.$$

*b. Use mathematical induction to prove the binomial theorem (Theorem 3.2.5).

14. Let $\{a_k\}_{k=1}^{\infty}$ be a sequence in $\mathbb{R}$. For each $n \in \mathbb{N}$, define
$$s_n = \frac{a_1 + \cdots + a_n}{n}.$$
*a. Prove that if $\lim\limits_{k \to \infty} a_k = a$, then $\lim\limits_{n \to \infty} s_n = a$.

  b. Give an example of a sequence $\{a_k\}$ which diverges, but for which $\{s_n\}$ converges.

3.3   Monotone Sequences

In this section, we will briefly consider monotone sequences of real numbers. As we will see, one of the advantages of such sequences is that they will either converge in $\mathbb{R}$, or diverge to $+\infty$ or $-\infty$.

DEFINITION 3.3.1 *A sequence $\{a_n\}_{n=1}^{\infty}$ of real numbers is said to be*

  (a) **monotone increasing** *(or **nondecreasing**) if $a_n \le a_{n+1}$ for all* $n \in \mathbb{N}$;

  (b) **monotone decreasing** *(or **nonincreasing**) if $a_n \ge a_{n+1}$ for all* $n \in \mathbb{N}$;

  (c) **monotone** *if it is either monotone increasing or monotone decreasing.*

A sequence $\{a_n\}$ is **strictly increasing** if $a_n < a_{n+1}$ for all n. **Strictly decreasing** is defined similarly.

As a general rule, bounded sequences need not converge; e.g. $\{1 - (-1)^n\}$. For monotone sequences however, we have the following convergence result.

THEOREM 3.3.2 *If $\{a_n\}_{n=1}^{\infty}$ is monotone and bounded, then $\{a_n\}_{n=1}^{\infty}$ converges.*

Proof. Suppose $\{a_n\}$ is monotone increasing. Set

$$E = \{\, a_n \,:\, n = 1, 2, \dots \,\}.$$

Then $E \neq \emptyset$ and bounded above. Let $a = \sup E$. We now show that

$$\lim_{n \to \infty} a_n = a.$$

Let $\epsilon > 0$ be given. Since $a - \epsilon$ is not an upper bound of E, there exists a positive integer n_o such that

$$a - \epsilon < a_{n_o} \leq a.$$

Since $\{a_n\}$ is monotone increasing,

$$a - \epsilon < a_n \leq a \qquad \text{for all} \quad n \geq n_o.$$

Thus $a_n \in N_\epsilon(a)$ for all $n \geq n_o$ and therefore $\lim_{n \to \infty} a_n = a$. $\square$

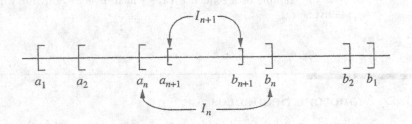

FIGURE 3.1
Nested intervals property

Nested Intervals Property

As an application of the previous theorem we prove the following result usually referred to as the **nested intervals property**. The term *nested* comes from the fact that the sequence $\{I_n\}$ of intervals satisfies $I_n \supset I_{n+1}$ for all $n \in \mathbb{N}$ (see Figure 3.1).

COROLLARY 3.3.3 (Nested Intervals Property) *If $\{I_n\}_{n=1}^{\infty}$ is a sequence of closed and bounded intervals with $I_n \supset I_{n+1}$ for all $n \in \mathbb{N}$, then*

$$\bigcap_{n=1}^{\infty} I_n \neq \emptyset$$

Proof. Suppose $I_n = [a_n, b_n]$, $a_n, b_n \in \mathbb{R}$, $a_n \leq b_n$. Since $I_n \supset I_{n+m}$ for all $m \geq 0$,

$$a_n \leq a_{n+m} \leq b_{n+m} \leq b_m$$

for all $n, m \in \mathbb{N}$. Thus the sequence $\{a_n\}$ is monotone increasing and bounded above by every b_m, $m \in \mathbb{N}$. Thus by the previous theorem, $a = \lim_{n \to \infty} a_n$ exists with $a \leq b_m$ for all $m \in \mathbb{N}$. Therefore $a \in I_m$ for all $m \in \mathbb{N}$ and thus

$$a \in \bigcap_{m=1}^{\infty} I_m,$$

which proves the result. $\square$

Remark. Similarly we can show that if $b = \lim_{n \to \infty} b_n$, then $b \in I_n$ for all n, and thus

$$[a, b] \subset \bigcap_{n=1}^{\infty} I_n.$$

In fact, one can show that equality holds. (See Exercise 3 of Section 2.4.)

EXAMPLES 3.3.4 (a) Our first example shows that the conclusion of Corollary 3.3.3 is false if the intervals I_n are not closed. As in Example 1.7.11(b), for each $n \in \mathbb{N}$ set $I_n = (0, \frac{1}{n})$. Then $I_n \supset I_{n+1}$ for all n, but

$$\bigcap_{n=1}^{\infty} I_n = \emptyset.$$

The conclusion of Corollary 3.3.3 may also be false if the intervals I_n are unbounded (Exercise 1).

(b) Consider the sequence $\{p^n\}$ with $0 < p < 1$. Even though Theorem 3.2.6(e) applies, we use the results of this section to prove that $\lim_{n \to \infty} p^n = 0$. For $n \in \mathbb{N}$ set $s_n = p^n$. Since $p > 0$, $s_n > 0$ for all $n \subset \mathbb{N}$. Thus $\{s_n\}$ is bounded below. Also, since $0 < p < 1$,

$$s_{n+1} = p^{n+1} = ps_n < s_n.$$

Thus the sequence $\{s_n\}$ is monotone decreasing, bounded below, and hence by Theorem 3.3.2, is convergent. Let $s = \lim_{n \to \infty} s_n$. Then

$$s = \lim_{n \to \infty} s_{n+1} = p \lim_{n \to \infty} s_n = ps.$$

Therefore $s = ps$. Since $p \neq 1$, we must have that $s = 0$.

(c) Let $a_1 = 1$ and for $n \geq 1$ set $a_{n+1} = \frac{1}{6}(2a_n + 5)$. The first three terms of the sequence $\{a_n\}$ are $a_1 = 1, a_2 = \frac{7}{6}$, and $a_3 = \frac{11}{9}$. Thus we suspect that the sequence $\{a_n\}$ is monotone increasing and bounded above by 2. Since $a_1 < 2$, if we assume that $a_n \leq 2$, then

$$a_{n+1} = \frac{1}{6}(2a_n + 5) \leq \frac{1}{6}(4 + 9) < 2.$$

Thus by mathematical induction $a_n \leq 2$ for all $n \in \mathbb{N}$. Likewise, since $a_1 < a_2$, if our induction hypothesis is $a_n < a_{n+1}$, then the result is true for $n = 1$, and

$$a_{n+1} = \frac{1}{6}(2a_n + 5) < \frac{1}{6}(2a_{n+1} + 5) = a_{n+2}.$$

Therefore the sequence $\{a_n\}$ is monotone increasing, bounded above, and thus converges. Let $a = \lim_{n \to \infty} a_n$. Then

$$a = \lim_{n \to \infty} a_{n+1} = \lim_{n \to \infty} \frac{1}{6}(2a_n + 5) = \frac{1}{6}(2a + 5).$$

Solving the equation $a = \frac{1}{6}(2a + 5)$ for a gives $a = \frac{5}{4}$.

   (d) Let $a_1 = 1$, and for $n > 1$, set $a_{n+1} = \sqrt{2a_n}$. To investigate the convergence of the sequence $\{a_n\}$, we will establish by induction that

$$1 \leq a_n < a_{n+1} < 2,$$

for all $n \in \mathbb{N}$. When $n = 1$, we have

$$1 = a_1 < \sqrt{2} = a_2 < 2.$$

Thus the statement is true for $n = 1$. Assume that it is true for $n = k$. Then

$$1 \leq a_{k+1} = \sqrt{2a_k} < \sqrt{2a_{k+1}} = a_{k+2},$$

and

$$a_{k+2} = \sqrt{2a_{k+1}} < \sqrt{4} = 2.$$

Thus the sequence $\{a_n\}$ is monotone increasing, bounded above by 2, and hence by Theorem 3.3.2 is convergent. It is possible to prove directly that

$$\sup\{a_n : n = 1, 2, ...\} = 2.$$

However, if we let $a = \lim_{n \to \infty} a_n$, then

$$a = \lim_{n \to \infty} a_{n+1} = \lim_{n \to \infty} \sqrt{2a_n} = \sqrt{2a}.$$

The last equality follows by Exercise 6 of Section 3.1. Therefore a is a solution of $a^2 = 2a$, which since $a \geq 1$, implies $a = 2$. $\square$

Euler's Number e

EXAMPLE 3.3.5 In this example, we consider in detail the very important sequence $\{t_n\}_{n=1}^{\infty}$, where for each $n \in \mathbb{N}$,

$$t_n = \left(1 + \frac{1}{n}\right)^n.$$

We will show that the sequence $\{t_n\}$ is monotone increasing and bounded above, and thus has a limit. The standard notation for this limit is e (in honor of Leonhard Euler); i.e.,

$$e = \lim_{n \to \infty} \left(1 + \frac{1}{n}\right)^n.$$

By the binomial theorem

$$t_n = \left(1 + \frac{1}{n}\right)^n = 1 + n \cdot \frac{1}{n} + \frac{n(n-1)}{1 \cdot 2} \frac{1}{n^2} + \cdots + \frac{n(n-1)\cdots 1}{1 \cdot 2 \cdots n} \cdot \frac{1}{n^n}. \quad (1)$$

For $k = 1, ..., n$, the $(k+1)$-st term on the right side is

$$\frac{n(n-1)\cdots(n-k+1)}{1 \cdot 2 \cdots k} \cdot \frac{1}{n^k}$$

which is equal to

$$\frac{1}{1 \cdot 2 \cdots k} \left(1 - \frac{1}{n}\right)\left(1 - \frac{2}{n}\right) \cdots \left(1 - \frac{k-1}{n}\right). \quad (2)$$

If we expand t_{n+1} in the same way, we obtain $n+2$ terms, and for $k = 1, 2, ..., n$ the $(k+1)$-st term is

$$\frac{1}{1 \cdot 2 \cdots k} \left(1 - \frac{1}{n+1}\right)\left(1 - \frac{2}{n+1}\right) \cdots \left(1 - \frac{k-1}{n+1}\right),$$

which is greater than the corresponding term in (2). Thus $t_n < t_{n+1}$ for all n. From (1) we also obtain

$$t_n \leq 1 + 1 + \frac{1}{1 \cdot 2} + \frac{1}{1 \cdot 2 \cdot 3} + \cdots + \frac{1}{1 \cdot 2 \cdots n}$$

$$\leq 1 + 1 + \frac{1}{2} + \frac{1}{2^2} + \cdots + \frac{1}{2^{n-1}},$$

which by the identity $1 + r + \cdots + r^{n-1} = (r - r^n)/(1 - r)$, $r \neq 1$,

$$= 1 + \frac{1 - (\frac{1}{2})^n}{1 - \frac{1}{2}} < 1 + \frac{1}{1 - \frac{1}{2}} = 3.$$

Thus $\{t_n\}$ is bounded above by 3, and we can apply Theorem 3.3.2. Since $t_n \leq 3$ for all $n \in \mathbb{N}$ we also have that $e \leq 3$. To five decimal places, $e = 2.71828....$ The number e is the base of the **natural logarithm function** which will be defined in Example 6.3.5 as a definite integral. $\square$

Infinite Limits

If a monotone increasing sequence $\{a_n\}$ is bounded above then by Theorem 3.3.2 the sequence converges. If the sequence $\{a_n\}$ is not bounded above, then for each positive real number M there exists $n_o \in \mathbb{N}$ such that $a_n \geq M$ for all $n \geq n_o$. Since the real number M can be taken to be arbitrarily large, this is usually expressed by saying that the sequence $\{a_n\}$ diverges to ∞. We make this concept precise, not only for monotone sequences, but for any sequence of real numbers with the following definition.

DEFINITION 3.3.6 *Let $\{a_n\}$ be a sequence of real numbers. We say that $\{a_n\}$* **approaches infinity**, *or that $\{a_n\}$* **diverges to** ∞, *denoted $a_n \to \infty$, if for every positive real number M, there exists an integer $n_o \in \mathbb{N}$ such that*

$$a_n > M \qquad for\ all\ n \geq n_o.$$

We will also use the notation $\lim\limits_{n\to\infty} a_n = \infty$ to denote that $a_n \to \infty$ as $n \to \infty$. The concept of $a_n \to -\infty$ is defined similarly.

THEOREM 3.3.7 *If $\{a_n\}$ is monotone increasing and not bounded above, then $a_n \to \infty$ as $n \to \infty$.*

As a consequence of Theorems 3.3.2 and 3.3.7, every monotone increasing sequence $\{a_n\}$ either converges to a real number (if the sequence is bounded above) or diverges to ∞. In either case,

$$\lim\limits_{n\to\infty} a_n = \sup\{a_n : n \in \mathbb{N}\}.$$

Remark. Although the definition of diverging to infinity is included in this section on monotone sequences, this should not give the impression that Definition 3.3.6 is applicable only to such sequences. In the following we give an example of a sequence that diverges to infinity but which is not monotone. Also, it is important to remember that when we say that a sequence converges, we mean that it converges to a real number.

EXAMPLE 3.3.8 Consider the sequence $\{n(2+(-1)^n)\}$. If n is even, then $n(2+(-1)^n) = 3n$; if n is odd, then $n(2+(-1)^n) = n$. In either case,

$$n(2+(-1)^n) \geq n,$$

and thus the sequence diverges to ∞. The sequence however is clearly not monotone. □

Exercises 3.3

1. *Show by example that the conclusion of Corollary 3.3.3 is false if the intervals I_n with $I_n \supset I_{n+1}$ are not bounded.

2. Show that each of the following sequences are monotone. find a lower or upper bound if it exists. Find the limit if you can.

    *a. $\left\{ \dfrac{\sqrt{n^2+1}}{n} \right\}$ b. $\left\{ \left(a + \dfrac{1}{n}\right)\left(a - \dfrac{1}{n}\right) \right\}, a > 1.$

    c. $\{a^n\}, a > 1$ d. $\{s_n\}$ where $s_n = \cos^3 \dfrac{\pi}{2} + \cos^2 \dfrac{2\pi}{2} + \cdots + \cos^2 \dfrac{n\pi}{2}$

    e. $\left\{ \dfrac{n^n}{n!} \right\}$

3. Define the sequence $\{a_n\}$ as follows: $a_1 = \sqrt{2}$, and $a_{n+1} = \sqrt{2 + a_n}$.

    a. Show that $a_n \le 2$ for all n.

    b. Show that the sequence $\{a_n\}$ is monotone increasing.

    c. Find $\lim\limits_{n \to \infty} a_n$.

4. *Let $a_1 > 1$, and for $n \in \mathbb{N}, n \ge 1$, define $a_{n+1} = 2 - 1/a_n$. Show that the sequence $\{a_n\}$ is monotone and bounded. Find $\lim\limits_{n \to \infty} a_n$.

5. Let $0 < a < 1$. Set $t_1 = 2$, and for $n \in \mathbb{N}$, set $t_{n+1} = 2 - a/t_n$. Show that the sequence $\{t_n\}$ is monotone and bounded. Find $\lim\limits_{n \to \infty} t_n$.

6. Let $\alpha > 0$. Choose $x_1 > \sqrt{\alpha}$. For $n = 1, 2, 3, ...$, define

    $$x_{n+1} = \frac{1}{2}\left(x_n + \frac{\alpha}{x_n}\right).$$

    *a. Show that the sequence $\{x_n\}$ is monotone and bounded.

    b. Prove that $\lim\limits_{n \to \infty} x_n = \sqrt{\alpha}$

    c. Prove that $0 \le x_n - \sqrt{\alpha} \le (x_n^2 - \alpha)/x_n$.

7. In Exercise 6, let $\alpha = 3$ and $x_1 = 2$. Use part (c) to find x_n such that $|x_n - \sqrt{3}| < 10^{-5}$.

8. For each of the following prove that the sequence $\{a_n\}$ converges and find the limit.

    a. $a_{n+1} = \frac{1}{6}(2a_n + 5), a_1 = 2$ b. $a_{n+1} = \sqrt{2a_n}, a_1 = 3$
    *c. $a_{n+1} = \sqrt{2a_n + 3}, a_1 = 1$ d. $a_{n+1} = \sqrt{2a_n + 3}, a_1 = 4$
    *e. $a_{n+1} = \sqrt{3a_n - 2}, a_1 = 4$ f. $a_{n+1} = \sqrt{3a_n - 2}, a_1 = \frac{3}{2}$

9. Let A be a nonempty subset of $\mathbb{R}$ that is bounded above and let $\alpha = \sup A$. Show that there exists a monotone increasing sequence $\{a_n\}$ in A such that $\alpha = \lim a_n$. Can the sequence $\{a_n\}$ be chosen to be strictly increasing?

10. Use Example 3.3.5 to find the limit of each of the following sequences.

    *a. $\left\{ \left(1 + \dfrac{1}{n}\right)^{2n} \right\}$ b. $\left\{ \left(1 + \dfrac{1}{n}\right)^{n+1} \right\}$

    *c. $\left\{ \left(1 + \dfrac{1}{2n}\right)^{3n} \right\}$ d. $\left\{ \left(1 - \dfrac{1}{n}\right)^{n} \right\}$

11. *For each $n \in \mathbb{N}$, let $s_n = 1 + \dfrac{1}{2} + \cdots + \dfrac{1}{n}$. Show that $\{s_n\}$ is monotone increasing but not bounded above.

12. For each $n \in \mathbb{N}$, let $s_n = 1 + \dfrac{1}{\sqrt{2}} + \cdots + \dfrac{1}{\sqrt{n}}$. Show that the sequence $\{s_n\}$ is monotone increasing but not bounded above.

13. *For each $n \in \mathbb{N}$, let $s_n = \dfrac{1}{1^2} + \dfrac{1}{2^2} + \cdots + \dfrac{1}{n^2}$. Show that the sequence $\{s_n\}$ is monotone increasing and bounded above by 2.

14. Let $0 < b < 1$. For each $n \in \mathbb{N}$, let $s_n = 1 + b + b^2 + \cdots + b^n$. Prove that the sequence $\{s_n\}$ is monotone increasing and bounded above. Find $\lim\limits_{n \to \infty} s_n$.

15. Show that each of the following sequences diverge to $+\infty$

    *a. $\left\{ \dfrac{a^n}{n^a} \right\}$, $a > 1$.   b. $\left\{ \dfrac{n^2 + 1}{n} \right\}$.

    *c. $\left\{ n + \dfrac{(-1)^n}{n} \right\}$ d. $\{ n + (-1)^n \sqrt{n} \}$

16. *Which of the sequences in the previous exercise are not monotone? Explain your answer!

17. If $a_n \to \infty$ and $\{b_n\}$ converges in $\mathbb{R}$, prove that $\{a_n + b_n\}$ diverges to ∞.

18. If $a_n > 0$ for all $n \in \mathbb{N}$ and $\lim\limits_{n \to \infty} a_n = 0$, prove that $\dfrac{1}{a_n} \to \infty$.

19. Suppose $a_1 > a_2 > 0$. For $n \geq 2$ set $a_{n+1} = \frac{1}{2}(a_n + a_{n-1})$. Prove that

    a. $\{a_{2k+1}\}$ is monotone decreasing.   b. $\{a_{2k}\}$ is monotone increasing, and c. $\{a_n\}$ converges.

20. Let $\{s_n\}$ be a bounded sequence of real numbers. For each $n \in \mathbb{N}$ let a_n and b_n be defined as follows; $a_n = \inf\{s_k : k \geq n\}$, $b_n = \sup\{s_k : k \geq n\}$.

    a. Prove that the sequences $\{a_n\}$ and $\{b_n\}$ are monotone and bounded.

    b. Prove that $\lim\limits_{n \to \infty} a_n = \lim\limits_{n \to \infty} b_n$ if and only if the sequence $\{s_n\}$ converges.

21. *In Theorem 3.3.2 we used the supremum property of $\mathbb{R}$ to prove that every bounded monotone sequence converges. Prove that the converse is also true; namely, if every bounded monotone sequence in $\mathbb{R}$ converges, then every nonempty subset of $\mathbb{R}$ that is bounded above has a supremum in $\mathbb{R}$.

22. *Use the nested intervals property to prove that $[0, 1]$ is uncountable.

3.4   Subsequences and the Bolzano-Weierstrass Theorem

In this section, we will consider subsequences and subsequential limits of a given sequence of real numbers. One of the key results of the section is that every bounded sequence of real numbers has a convergent subsequence. This result, also known as the sequential version of the Bolzano-Weierstrass theorem, is one of the fundamental results of real analysis.

DEFINITION 3.4.1 *Let (X, d) be a metric space. Given a sequence $\{p_n\}$ in X, consider a sequence $\{n_k\}_{k=1}^{\infty}$ of positive integers such that $n_1 < n_2 < n_3 < \ldots$. Then the sequence $\{p_{n_k}\}_{k=1}^{\infty}$ is called a* **subsequence** *of the sequence $\{p_n\}$.*

If the sequence $\{p_{n_k}\}$ converges, its limit is called a subsequential limit of the sequence $\{p_n\}$. Specifically, a point $p \subset X$ is a **subsequential limit** of the sequence $\{p_n\}$ if there exists a subsequence $\{p_{n_k}\}$ of $\{p_n\}$ that converges to p. Also, given a sequence $\{p_n\}$ in $\mathbb{R}$, we say that ∞ is a **subsequential limit** of $\{p_n\}$ if there exists a subsequence $\{p_{n_k}\}$ so that $p_{n_k} \to \infty$ as $k \to \infty$. Similarly for $-\infty$.

EXAMPLES 3.4.2 (a) Consider the sequence $\{1 - (-1)^n\}$. If n is even, then $a_n = 0$, and if n is odd, then $a_n = 2$. Thus 0 and 2 are subsequential limits of the given sequence. That these are the only two subsequential limits are left to the exercises (Exercise 1).

(b) As our second example, consider the sequence $\{(-1)^n + \frac{1}{n}\}$. Then both 1 and -1 are subsequential limits. If n is even, i.e., $n = 2k$, then

$$a_n = a_{2k} = 1 + \frac{1}{2k},$$

which converges to 1. On the other hand, if n is odd, i.e., $n = 2k + 1$, then

$$a_n = a_{2k+1} = -1 + \frac{1}{2k+1},$$

which converges to -1. This shows that -1 and 1 are subsequential limits. Suppose $\{a_{n_k}\}$ is any subsequence of $\{a_n\}$. If the sequence $\{n_k\}$ contains an infinite number of both odd and even integers, then the subsequence $\{a_{n_k}\}$ cannot converge. (Why?) On the other hand, if all but a finite number of the n_k are even, then $\{a_{n_k}\}$ converges to 1. Similarly, if all but a finite number of the n_k are odd, then $\{a_{n_k}\}$ converges to -1. Thus -1 and 1 are the only subsequential limits of $\{a_n\}$.

(c) Consider the sequence $\{n(1+(-1)^n)\}$. If n is even, then $n(1+(-1)^n) = 2n$, whereas if n is odd, $n(1+(-1)^n) = 0$. Thus 0 and ∞ are two subsequential limits of the sequence. The same argument as in (b) proves that these are the only two subsequential limits of the sequence. □

Our first result assures us that for convergent sequences, every subsequence also converges to the same limit.

THEOREM 3.4.3 *Let (X, d) be a metric space and let $\{p_n\}$ be a sequence in X. If $\{p_n\}$ converges to p, then every subsequence of $\{p_n\}$ also converges to p.*

Proof. Let $\{p_{n_k}\}$ be a subsequence of $\{p_n\}$, and let $\epsilon > 0$ be given. Since $p_n \to p$, there exists a positive integer n_o such that $d(p_n, p) < \epsilon$ for all $n \geq n_o$. Since $\{n_k\}$ is strictly increasing, $n_k \geq n_o$ for all $k \geq n_o$. Therefore,

$$d(p_{n_k}, p) < \epsilon$$

for all $k \geq n_o$, i.e., $p_{n_k} \to p$. $\square$

EXAMPLES 3.4.4 (a) In this example, we give an application of Theorem 3.4.3. Consider the sequence $\{p^n\}$ where $0 < p < 1$. Since

$$0 < p^{n+1} < p^n < 1$$

for all n, the sequence $\{p^n\}$ is monotone decreasing, bounded below, and hence converges. Let

$$a = \lim_{n \to \infty} p^n.$$

By Theorem 3.4.3 the subsequence $\{p^{2n}\}$ also converges to a. But $p^{2n} = (p^n)^2$, and thus

$$a = \lim_{n \to \infty} p^{2n} = \lim_{n \to \infty} (p^n)^2 = a^2.$$

Thus $a^2 = a$. Since $0 \leq a < 1$, we must have $a = 0$.

(b) In our second example we show how the previous theorem may be used to prove divergence of a sequence. Consider the sequence $\{\sin n\theta\pi\}$, where θ is a rational number with $0 < \theta < 1$. Write $\theta = a/b$, with $a, b \in \mathbb{N}$ and $b \geq 2$. When $a = kb, k \in \mathbb{N}$, then $\sin n\theta\pi = \sin ka\pi = 0$. Therefore 0 is a subsequential limit of the sequence. On the other hand, if $n = 2kb + 1, k \in \mathbb{N}$, then

$$\sin n\theta\pi = \sin (2kb + 1)\frac{a}{b}\pi = \sin \left(2ka\pi + \frac{a}{b}\pi\right)$$
$$= \cos(2ka\pi) \sin \frac{a}{b}\pi = \sin \frac{a}{b}\pi.$$

Since $0 < a/b < 1$, $\sin \frac{a}{b}\pi \neq 0$. Thus $\sin \frac{a}{b}\pi$ is another distinct subsequential limit of $\{\sin n\theta\pi\}$. Hence as a consequence of Theorem 3.4.3 the sequence $\{\sin n\theta\pi\}$ diverges. The result is still true if θ is irrational. The proof however is much more difficult. $\square$

The following result, which is a sequential version of Theorem 2.3.7, will prove useful in subsequent results.

THEOREM 3.4.5 *Let K be a compact subset of a metric space (X, d). Then every sequence in K has a convergent subsequence which converges in K.*

Proof. Let $\{p_n\}$ be a sequence in K, and let $E = \{p_n : n = 1, 2, \ldots\}$. If E is finite, then there exists a point $p \in E$ and a sequence $\{n_k\}$ with $n_1 < n_2 < \cdots$ such that

$$p_{n_1} = p_{n_2} = \cdots = p.$$

The subsequence $\{p_{n_k}\}$ obviously converges to p which is in K.

If E is infinite, then by Theorem 2.3.7, E has a limit point $p \in K$. Choose n_1 such that $d(p, p_{n_1}) < 1$. Having chosen $n_1, \ldots, n_{k-1}$, choose an integer $n_k > n_{k-1}$ so that

$$d(p, p_{n_k}) < \frac{1}{k}.$$

Such an integer n_k exists since every neighborhood of p contains infinitely many points of E. The sequence $\{p_{n_k}\}$ is a subsequence of $\{p_n\}$, which by construction converges to $p \in K$. $\quad\square$

We are now ready to state and prove the sequential version Bolzano-Weierstrass theorem.

COROLLARY 3.4.6 (Bolzano-Weierstrass) *Every bounded sequence in* $\mathbb{R}$ *has a convergent subsequence.*

Proof. Suppose $\{p_n\}$ is a bounded sequence in $\mathbb{R}$. Then there exists a positive integer M such that $\{p_n\}$ is a sequence in the compact set $[-M, M]$. The result now follows by the previous theorem. $\quad\square$

Remark. The converse of Theorem 3.4.5 is also true. If K is a subset of a metric space (X, d) having the property that every sequence in K has a convergent subsequence, then K is compact. For metric spaces, the proof of the converse is similar to Miscellaneous Exercise 4 of the previous chapter.

If K is a subset of $\mathbb{R}$, then the hypothesis can be used to show that K is closed and bounded, and thus by Theorem 2.4.2 is compact.

An argument similar to the one used in the previous Corollary may be used to prove the following.

THEOREM 3.4.7 *Let* $\{p_n\}$ *be a sequence in a metric space* (X, d). *If* p *is a limit point of* $\{p_n : p \in \mathbb{N}\}$, *then there exists a subsequence* $\{p_{n_k}\}$ *of* $\{p_n\}$ *such that* $p_{n_k} \to p$ *as* $k \to \infty$.

Proof. Exercise 8 $\quad\square$

As an application of Theorem 3.4.7 we consider the following example.

EXAMPLE 3.4.8 Let $\{r_n\}_{n=1}^{\infty}$ be an enumeration of the rational numbers in $[0, 1]$. By Example 2.2.13(c), every $p \in [0, 1]$ is a limit point of $\{r_n : n = 1, 2, \ldots\}$. Thus if $p \in [0, 1]$, there exists a subsequence $\{r_{n_k}\}$ of $\{r_n\}$ such that $r_{n_k} \to p$. The sequence $\{r_n\}$ has the property that every $p \in [0, 1]$ is a subsequential limit of the sequence. This sequence also provides an example of a sequence for which the set of subsequential limits of the sequence is uncountable. $\quad\square$

Exercises 3.4

1. **a.** Prove that 0 and 2 are the only subsequential limits of the sequence $\{1 - (-1)^n\}$.

   b. Prove that 0 and ∞ are the only subsequential limits of the sequence $\{n(1 + (-1)^n)\}$.

2. **a.** Construct a sequence $\{s_n\}$ for which the subsequential limits are $\{-\infty, -2, 1\}$.

   b. Construct a sequence $\{s_n\}$ for which the set of subsequential limits is countable.

3. Find all the subsequential limits of the following sequences.

   *a. $\left\{ \sin \dfrac{n\pi}{2} \right\}$.   **b.** $\left\{ n \sin \dfrac{n\pi}{4} \right\}$.

   *c. $\left\{ 1 - \dfrac{(-1)^n}{n} \right\}$ **d.** $\{(1.5 + (-1)^n)^n\}$.

   *e. $\left\{ (-1)^n + 2 \sin \dfrac{n\pi}{2} \right\}$ **f.** $\left\{ n \sin \dfrac{n\pi}{4} \right\}$

4. Use Example 3.3.5 to find the limit of each of the following sequences. Justify your answer.

   *a. $\left\{ \left(1 + \dfrac{1}{3n} \right)^{6n} \right\}$ **b.** $\left\{ \left(1 + \dfrac{1}{2n} \right)^n \right\}$ **c.** $\left\{ \left(1 + \dfrac{2}{n} \right)^n \right\}$

5. Suppose $p > 1$. Use the method of Example 3.4.4 to show that $\lim\limits_{n \to \infty} \sqrt[n]{p} = 1$.

6. For $n \in \mathbb{N}$ set $p_n = n^{1/n}$.

   a. Show that $1 < p_{n+1} < p_n$ for all $n \geq 3$.

   b. Let $p = \lim\limits_{n \to \infty} p_n$. Use the fact that the subsequence $\{p_{2n}\}$ also converges to p to conclude that $p = 1$.

7. Let $\{p_n\}$ be a bounded sequence of real numbers and let $p \in \mathbb{R}$ be such that every convergent subsequence of $\{p_n\}$ converges to p. Prove that the sequence $\{p_n\}$ converges to p.

8. Prove Theorem 3.4.7.

9. Prove that every sequence in $\mathbb{R}$ has a monotone subsequence.

10. *Prove that every bounded sequence in $\mathbb{R}^n$ has a convergent subsequence.

11. Use the Bolzano-Weierstrass theorem to prove the nested intervals property (Corollary 3.3.3).

12. Prove that every uncountable subset of $\mathbb{R}$ has a limit point in $\mathbb{R}$.

3.5   Limit Superior and Inferior of a Sequence

In this section, we define the limit superior and limit inferior of a sequence of real numbers. These two limit operations are important because unlike the limit of a sequence, the limit superior and limit inferior of a sequence always exist. The concept of the limit superior and limit inferior will also be important in our study of both series of real numbers and power series.

Let $\{s_n\}$ be a sequence in $\mathbb{R}$. For each $k \in \mathbb{N}$, we define a_k and b_k as follows:

$$a_k = \inf\{s_n : n \geq k\},$$
$$b_k = \sup\{s_n : n \geq k\}.$$

Recall that for a nonempty subset E of $\mathbb{R}$, $\sup E$ is the least upper bound of E if E is bounded above, and ∞ otherwise.

From the definition, $a_k \leq b_k$ for all k. Furthermore, the sequences $\{a_k\}$ and $\{b_k\}$ satisfy the following:

$$a_k \leq a_{k+1} \qquad \text{and} \qquad b_k \geq b_{k+1} \tag{3}$$

for all k. To prove (3), let $E_k = \{s_n : n \geq k\}$. Then $E_{k+1} \subset E_k$. Therefore, if $b_k = \sup E_k$, $s_n \leq b_k$ for all $n \geq k$. In particular

$$s_n \leq b_k \qquad \text{for all } n \geq k+1.$$

Therefore $b_{k+1} = \sup E_{k+1} \leq b_k$. A similar argument will show that the sequence $\{a_k\}$ is nondecreasing.

As a consequence of (3) the sequence $\{a_k\}$ is monotone increasing and the sequence $\{b_k\}$ is monotone decreasing. Thus by Theorems 3.3.2 and 3.3.7, these two sequences always have limits in $\mathbb{R} \cup \{-\infty, \infty\}$.

DEFINITION 3.5.1 *Let $\{s_n\}$ be a sequence in $\mathbb{R}$. The* **limit superior** *of $\{s_n\}$, denoted $\varlimsup\limits_{n \to \infty} s_n$ or $\varlimsup s_n$, is defined as*

$$\varlimsup_{n \to \infty} s_n = \lim_{k \to \infty} b_k = \inf_{k \in \mathbb{N}} \sup\{s_n : n \geq k\}.$$

The **limit inferior** *of $\{s_n\}$, denoted $\varliminf\limits_{n \to \infty} s_n$ or $\varliminf s_n$, is defined as*

$$\varliminf_{n \to \infty} s_n = \lim_{k \to \infty} a_k = \sup_{k \in \mathbb{N}} \inf\{s_n : n \geq k\}.$$

We now give several examples for which we will compute the limit inferior and limit superior. As will be evident, these computations are very tedious. An easier method will be given in Theorem 3.5.7.

EXAMPLES 3.5.2 (a) $\{1 + (-1)^n\}_{n=1}^{\infty}$. Let $s_n = 1 + (-1)^n$. Then $s_n = 2$ if n is even, 0 otherwise. Thus $a_k = 0$ for all k and $b_k = 2$ for all k. Therefore

$$\varlimsup_{n \to \infty} s_n = 2 \qquad \text{and} \qquad \varliminf_{n \to \infty} s_n = 0.$$

(b) $\{n\,(1+(-1)^n)\}_{n=1}^{\infty}$. In this example,

$$s_n = n\,(1+(-1)^n) = \begin{cases} 0, & \text{if } n \text{ is odd}, \\ 2n, & \text{if } n \text{ is even}. \end{cases}$$

Set $E_k = \{s_n : n \geq k\}$. Then

$$E_k = \{0, 2(k+1), 0, 2(k+3), \cdots\} \qquad \text{if } k \text{ is odd},$$
$$E_k = \{2k, 0, 2(k+2), 0, 2(k+4), \cdots\} \qquad \text{if } k \text{ is even}.$$

Therefore $a_k = \inf E_k = 0$, and $b_k = \sup E_k = \infty$. Thus

$$\lim_{n\to\infty} s_n = 0 \quad \text{and} \quad \overline{\lim_{n\to\infty}} s_n = \infty.$$

(c) $\{(-1)^n + \frac{1}{n}\}_{n=1}^{\infty}$. Set $s_n = (-1)^n + 1/n$. Then

$$s_n = \begin{cases} -1 + \frac{1}{n}, & n \text{ odd}, \\ 1 + \frac{1}{n}, & n \text{ even}. \end{cases}$$

To compute the limit superior and inferior of the sequence $\{s_n\}$, we set $E_k = \{s_n : n \geq k\}$. If k is even, then

$$E_k = \left\{1 + \frac{1}{k}, -1 + \frac{1}{k+1}, 1 + \frac{1}{k+2}, \cdots\right\}.$$

Therefore, for k even,

$$b_k = \sup E_k = 1 + \frac{1}{k} \qquad \text{and} \qquad a_k = \inf E_k = -1.$$

Similarly, for k odd,

$$b_k = \sup E_k = 1 + \frac{1}{k+1} \qquad \text{and} \qquad a_k = \inf E_k = -1.$$

As a consequence,

$$a_k = -1 \text{ for all } k, \qquad b_k = \begin{cases} 1 + \frac{1}{k}, & k \text{ even}, \\ 1 + \frac{1}{k+1}, & k \text{ odd}. \end{cases}$$

Thus

$$\overline{\lim_{n\to\infty}} s_n = 1 \quad \text{and} \quad \lim_{n\to\infty} s_n = -1. \quad \square$$

The following theorem provides an (ϵ, n_o) characterization of the limit superior. An analogous characterization for the limit inferior is given in Theorem 3.5.4.

THEOREM 3.5.3 *Let* $\{s_n\}_{n=1}^{\infty}$ *be a sequence in* $\mathbb{R}$.

(a) *Suppose* $\overline{\lim\limits_{n\to\infty}} \, s_n \in \mathbb{R}$. *Then* $\beta = \overline{\lim\limits_{n\to\infty}} \, s_n$ *if and only if for all* $\epsilon > 0$

   (i) *there exists* $n_o \in \mathbb{N}$ *such that* $s_n < \beta + \epsilon$ *for all* $n \geq n_o$, *and*

   (ii) *given* $n \in \mathbb{N}$, *there exists* $k \in \mathbb{N}$ *with* $k \geq n$ *such that* $s_k > \beta - \epsilon$.

(b) $\varlimsup\limits_{n\to\infty} s_n = \infty$ *if and only if given* M *and* $n \in \mathbb{N}$, *there exists* $k \in \mathbb{N}$ *with* $k \geq n$ *such that* $s_k \geq M$.

(c) $\overline{\lim\limits_{n\to\infty}} \, s_n = -\infty$ *if and only if* $s_n \to -\infty$ *as* $n \to \infty$.

Remark. The statement "$s_n < \beta + \epsilon$ for all $n \geq n_o$" means that $s_n < \beta + \epsilon$ for all but finitely many n. On the other hand, the statement "given n, there exists $k \in \mathbb{N}$ with $k \geq n$ such that $s_k > \beta - \epsilon$" means that $s_n > \beta - \epsilon$ for infinitely many indices n.

THEOREM 3.5.4 *Let* $\{s_n\}$ *be a sequence in* $\mathbb{R}$.

(a') *Suppose* $\varliminf\limits_{n\to\infty} s_n \in \mathbb{R}$. *Then* $\alpha = \varliminf\limits_{n\to\infty} s_n$ *if and only if for all* $\epsilon > 0$

   (i) *there exists* $n_o \in \mathbb{N}$ *such that* $s_n > \alpha - \epsilon$ *for all* $n \geq n_o$, *and*

   (ii) *given* $n \in \mathbb{N}$, *there exists* $k \in \mathbb{N}$ *with* $k \geq n$ *such that* $s_k < \alpha + \epsilon$.

(b') $\varliminf\limits_{n\to\infty} s_n = -\infty$ *if and only if given* M *and* $n \in \mathbb{N}$, *there exists* $k \in \mathbb{N}$ *with* $k \geq n$ *such that* $s_k \leq M$.

(c') $\varliminf\limits_{n\to\infty} s_n = \infty$ *if and only if* $s_n \to \infty$ *as* $n \to \infty$.

Proof of Theorem 3.5.3 We will only proof (a). The proofs of (b) and (c) are left to the exercises (Exercise 5).

(a) Suppose $\beta = \varlimsup s_n = \lim\limits_{k\to\infty} b_k$ where

$$b_k = \sup\{s_n : n \geq k\}.$$

Let $\epsilon > 0$ be given. Since $\lim\limits_{k\to\infty} b_k = \beta$ there exists a positive integer n_o such that $b_k < \beta + \epsilon$ for all $k \geq n_o$. Since $s_n \leq b_k$ for all $n \geq k$,

$$s_n < \beta + \epsilon \qquad \text{for all } n \geq n_o.$$

This proves (i). Suppose $n \in \mathbb{N}$ is given. Since $b_k \to \beta$, and $\{b_k\}$ is monotone decreasing, $b_k \geq \beta$ for all k. In particular, $b_n \geq \beta$. By the definition of b_n however, given $\epsilon > 0$, there exists an integer $k \geq n$ such that

$$s_k > b_n - \epsilon \geq \beta - \epsilon,$$

which proves (ii).

110 — *Introduction to Real Analysis*

Conversely, assume that (i) and (ii) hold. Let $\epsilon > 0$ be given. By (i) there exists $n_o \in \mathbb{N}$ such that $s_n < \beta + \epsilon$ for all $n \geq n_o$. Therefore

$$b_{n_o} = \sup\{s_n : n \geq n_o\} \leq \beta + \epsilon.$$

Since the sequence $\{b_n\}$ is monotone decreasing, $b_n \leq \beta + \epsilon$ for all $n \geq n_o$. Thus

$$\overline{\lim}\, s_n = \lim b_n \leq \beta + \epsilon.$$

Since $\epsilon > 0$ was arbitrary, $\overline{\lim}\, s_n \leq \beta$.

Suppose $\beta' = \overline{\lim}\, s_n < \beta$. Choose $\epsilon > 0$ such that $\beta' < \beta - 2\epsilon$. But then there exists n_o such that

$$s_n < \beta' + \epsilon < \beta - \epsilon \qquad \text{for all } n \geq n_o,$$

which contradicts (ii). Thus $\overline{\lim}\, s_n = \beta$. $\square$

To illustrate the previous two theorems, consider the sequence

$$s_n = (-1)^n + 1/n$$

of Example 3.5.2(c). For this sequence, $\overline{\lim}\, s_n = 1$ and $\underline{\lim}\, s_n = -1$. Given $\epsilon > 0$, then

$$s_n < 1 + \epsilon$$

for all $n \in \mathbb{N}$ with $n \geq 1/\epsilon$. Since the odd terms get close to -1, we can never have the existence of an integer n_o such that $s_n > 1 - \epsilon$ for all $n \geq n_o$. On the other hand, given any $n \in \mathbb{N}$, there exists an even integer $k \geq n$ such that $s_k > 1 - \epsilon$.

An immediate consequence of the previous two theorems is as follows:

COROLLARY 3.5.5 $\overline{\lim}_{n\to\infty} s_n = \underline{\lim}_{n\to\infty} s_n$ *if and only if* $\lim_{n\to\infty} s_n$ *exists in* $\mathbb{R} \cup \{-\infty, \infty\}$.

Proof. Suppose $\overline{\lim}\, s_n = \underline{\lim}\, s_n = \alpha \in \mathbb{R}$. Let $\epsilon > 0$ be given. By (a) and (a') of the previous two theorems, there exist positive integers n_1 and n_2, such that

$$s_n < \alpha + \epsilon \qquad \text{for all} \quad n \geq n_1, \quad \text{and}$$
$$s_n > \alpha - \epsilon \qquad \text{for all} \quad n \geq n_2.$$

Thus if $n_o = \max\{n_1, n_2\}$,

$$\alpha - \epsilon < s_n < \alpha + \epsilon$$

for all $n \geq n_o$; i.e., $\lim_{n\to\infty} s_n = \alpha$. The proofs of the cases $\alpha = \infty$ or $\alpha = -\infty$ are similar.

If $\lim_{n\to\infty} s_n = \alpha$, then it easily follows that both $\overline{\lim}\, s_n = \alpha$ and $\underline{\lim}\, s_n = \alpha$.
$\square$

THEOREM 3.5.6 *Let $\{a_n\}$ and $\{b_n\}$ be bounded sequences in $\mathbb{R}$. Then*

$$\varliminf_{n\to\infty} a_n + \varliminf_{n\to\infty} b_n \le \varliminf_{n\to\infty} (a_n + b_n) \le \varliminf_{n\to\infty} a_n + \varlimsup_{n\to\infty} b_n$$

$$\le \varlimsup_{n\to\infty} (a_n + b_n) \le \varlimsup_{n\to\infty} a_n + \varlimsup_{n\to\infty} b_n.$$

Proof. Exercise 6 $\quad\square$

The following theorem relates the limit superior and inferior of a sequence to the subsequential limits of the sequence, and is in fact very useful for finding $\varlimsup s_n$ and $\varliminf s_n$ of a sequence $\{s_n\}$.

THEOREM 3.5.7 *Let $\{s_n\}_{n=1}^{\infty}$ be a sequence in $\mathbb{R}$ and let*

$$E = \text{ the set of subsequential limits of } \{s_n\} \text{ in } \mathbb{R} \cup \{-\infty, \infty\}.$$

Then $\varlimsup_{n\to\infty} s_n$ and $\varliminf_{n\to\infty} s_n$ are in E and

(a) $\varlimsup_{n\to\infty} s_n = \sup E$, *and*

(b) $\varliminf_{n\to\infty} s_n = \inf E$.

Proof. Let $s = \varlimsup s_n$. Suppose $s \in \mathbb{R}$. To show that $s \in E$, we show the existence of a subsequence $\{s_{n_k}\}$ of $\{s_n\}$ which converges to s. Take $\epsilon = 1$. Let n_1 be the smallest integer such that

$$s - 1 < s_{n_1} < s + 1.$$

Such an integer exists by (i) and (ii) of Theorem 3.5.3(a). Suppose $n_1 < n_2 < \cdots < n_k$ have been chosen. Take $\epsilon = \frac{1}{k+1}$. Let n_{k+1} be the smallest integer greater than n_k such that

$$s - \frac{1}{k+1} < s_{n_{k+1}} < s + \frac{1}{k+1}.$$

Again, such an integer exists by (i) and (ii) of Theorem 3.5.3(a). Then $\{s_{n_k}\}$ is a subsequence of $\{s_n\}$ which clearly converges to s. Therefore $s \in E$. The case $s = \infty$ is treated similarly. If $s = -\infty$, then by (c) of Theorem 3.5.3, $s_n \to -\infty$ as $n \to \infty$.

Since $s \in E$, $s \le \sup E$. It remains to be shown that $\sup E = s$. If $s = \infty$ we are done. Otherwise, suppose $\sup E = \beta > s$. Suppose $\beta \ne \infty$. Then there exists $\alpha \in E$ such that

$$s < \alpha \le \beta.$$

Since $\alpha \in \mathbb{R}$, we can choose $\epsilon > 0$ such that $s + \epsilon < \alpha - \epsilon$. For this ϵ, there exists $n_o \in \mathbb{N}$ such that $s_n < s + \epsilon$ for all $n \ge n_o$. Hence there can exist only finitely many k such that

$$|s_k - \alpha| < \epsilon.$$

Consequently no subsequence of $\{s_n\}$ can converge to α. This contradiction shows that $\sup E = s$. The case $\beta = \infty$ is treated similarly. $\quad\square$

EXAMPLES 3.5.8 In the following examples we use Theorem 3.5.7 to compute $\underline{\lim} \, s_n$ and $\overline{\lim} \, s_n$ for each of the given sequences $\{s_n\}$.

(a) Let $s_n = (-1)^n + 1/n$. By Example 3.4.2(b), the set of subsequential limits of $\{s_n\}$ is $\{-1, 1\}$. Thus by the previous theorem,

$$\underline{\lim} \, s_n = -1 \qquad \text{and} \qquad \overline{\lim} \, s_n = 1.$$

(b) Let $s_n = n(1 + (-1)^n)$. By Example 3.4.2(c) the subsequential limits of $\{s_n\}$ are 0 and ∞. Therefore,

$$\underline{\lim} \, s_n = 0 \qquad \text{and} \qquad \overline{\lim} \, s_n = \infty.$$

(c) Let $s_n = \sin \dfrac{n\pi}{2}$. If n is even, i.e., $n = 2k$, then $s_{2k} = \sin k\pi = 0$. On the other hand, if n is odd, i.e., $n = 2k+1$, then $s_{2k+1} = \sin(2k+1)\frac{\pi}{2} = (-1)^k$. Hence the set of subsequential limits of the sequence $\{s_n\}$ is $\{-1, 0, 1\}$. As a consequence,

$$\underline{\lim} \, s_n = -1 \qquad \text{and} \qquad \overline{\lim} \, s_n = 1. \quad \square$$

Exercises 3.5

1. Find the limit inferior and limit superior of each of the following sequences.

    *a. $\left\{ n \sin \dfrac{n\pi}{4} \right\}$ b. $\left\{ (1 + (-1)^n) \sin \dfrac{n\pi}{4} \right\}$

    *c. $\left\{ \dfrac{n + (-1)^n n^2}{n^2 + 1} \right\}$ d. $\{[1.5 + (-1)^n]^n\}$

    e. $\left\{ \dfrac{1}{n} + n(1 + \cos n\pi) \right\}$ *f. $\left\{ \dfrac{1 - 2(-1)^n n}{3n + 2} \right\}$

2. Let $\{a_n\}$ be a sequence in $\mathbb{R}$. If $\overline{\lim} \, |a_n| = 0$, prove that $\lim\limits_{n \to \infty} a_n = 0$.

3. *Let $\{r_n\}$ be an enumeration of the rationals in $(0, 1)$. Find $\underline{\lim} \, r_n$ and $\overline{\lim} \, r_n$.

4. Let $\{s_n\}$ be a sequence in $\mathbb{R}$. If $s \in \mathbb{R}$ satisfies that for every $\epsilon > 0$, there exists $n_o \in \mathbb{N}$ such $s_n < s + \epsilon$ for all $n \geq n_o$, prove that $\overline{\lim} \, s_n \leq s$.

5. **a.** Prove Theorem 3.5.3(b).

    b. Prove Theorem 3.5.3(c).

6. *a. Let $\{a_n\}$ and $\{b_n\}$ be bounded sequences in $\mathbb{R}$. Prove that

    $$\underline{\lim} \, a_n + \overline{\lim} \, b_n \leq \overline{\lim}(a_n + b_n) \leq \overline{\lim} \, a_n + \overline{\lim} \, b_n.$$

    b. Give an example to show that equality need not hold in (a).

7. **a.** If a_n and b_n are positive for all n, prove that

    $$\overline{\lim}(a_n b_n) \leq (\overline{\lim} \, a_n)(\overline{\lim} \, b_n),$$

    provided the product on the right is not of the form $0 \cdot \infty$.

    b. Need equality hold in (a)?

8. *Let $s_1 = 0$. For $n \in \mathbb{N}$, $n > 1$, let s_n be defined by

$$s_{2m} = \frac{s_{2m-1}}{2}, \qquad s_{2m+1} = \frac{1}{2} + s_{2m}.$$

Find $\overline{\lim}\, s_n$ and $\underline{\lim}\, s_n$.

9. Let $a_n > 0$ for all n. Prove that $\overline{\lim} \sqrt[n]{a_n} \leq \overline{\lim} \dfrac{a_{n+1}}{a_n}$.

10. *Suppose $\{a_n\}$, $\{b_n\}$ are sequences of nonnegative real numbers with $\lim\limits_{n \to \infty} b_n = b \neq 0$, and $\overline{\lim\limits_{n \to \infty}}\, a_n = a$. Prove that $\overline{\lim\limits_{n \to \infty}}\, a_n b_n = a\,b$.

3.6   Cauchy Sequences

In order to apply the definition to prove that a given sequence $\{p_n\}$ converges, it is required that we know the limit of the sequence $\{p_n\}$. For this reason, theorems that provide sufficient conditions for convergence, such as Theorem 3.3.2, are particularly useful. The drawback to Theorem 3.3.2 is that it applies only to monotone sequences of real numbers. In this section, we consider another criterion that for sequences in $\mathbb{R}$ is sufficient to ensure convergence of the sequence.

DEFINITION 3.6.1 *Let (X, d) be a metric space. A sequence $\{p_n\}_{n=1}^{\infty}$ in X is a **Cauchy sequence** if for every $\epsilon > 0$, there exists a positive integer n_o such that*

$$d(p_n, p_m) < \epsilon$$

for all integers $n, m \geq n_o$.

Remark. In the above definition, the criterion $d(p_n, p_m) < \epsilon$ for all integers $n, m \geq n_o$ is equivalent to

$$d(p_{n+k}, p_n) < \epsilon$$

for all $n \geq n_o$ and all $k \in \mathbb{N}$. Thus if $\{p_n\}$ is a Cauchy sequence in X,

$$\lim_{n \to \infty} d(p_{n+k}, p_n) = 0$$

for every $k \in \mathbb{N}$. The converse however is false; namely, if $\{p_n\}$ is a sequence in $\mathbb{R}$ that satisfies $\lim\limits_{n \to \infty} d(p_{n+k}, p_n) = 0$ for every $k \in \mathbb{N}$, this does not imply that the sequence $\{p_n\}$ is a Cauchy sequence (Exercise 4). The hypothesis only implies that for each $k \in \mathbb{N}$, given $\epsilon > 0$, there exists a positive integer n_o such that $d(p_{n+k}, p_n) < \epsilon$ for all $n \geq n_o$.

THEOREM 3.6.2 *Let (X, d) be a metric space*

(a) *Every convergent sequence in X is a Cauchy sequence.*

(b) *Every Cauchy sequence is bounded.*

Proof. (a) Suppose that $\{p_n\}$ converges to $p \in X$. Let $\epsilon > 0$ be given. Then for the given ϵ, there exists a positive integer n_o such that

$$d(p_n, p) < \tfrac{1}{2}\epsilon$$

for all $n \geq n_o$. Thus by the triangle inequality, if $n, m \geq n_o$,

$$d(p_n, p_m) \leq d(p_n, p) + d(p, p_m) < \tfrac{1}{2}\epsilon + \tfrac{1}{2}\epsilon = \epsilon.$$

(b) Take $\epsilon = 1$. By the definition of Cauchy sequence, there exists $n_o \in \mathbb{N}$ such that $d(p_n, p_m) < 1$ for all $n, m \geq n_o$. Let

$$M = \max\{1, d(p_1, p_{n_o}), \ldots, d(p_{n_o-1}, p_{n_o})\}.$$

Then for all n, $d(p_n, p_{n_o}) \leq M$. Thus $\{p_n\}$ is bounded. $\quad\square$

EXAMPLES 3.6.3 (a) Let $X = (0, 1)$ with $d(x, y) = |x - y|$. Consider the sequence $\{1/n\}_{n=1}^{\infty}$. Let $\epsilon > 0$ be given. Choose an integer n_o such that $1/n < \epsilon/2$ for all $n \geq n_o$. Then for all $n, m \geq n_o$,

$$d(\tfrac{1}{n}, \tfrac{1}{m}) = \left| \frac{1}{n} - \frac{1}{m} \right| \leq \frac{1}{n} + \frac{1}{m} < \epsilon.$$

Thus the sequence $\{1/n\}$ is Cauchy but does not converge in X. In $\mathbb{R}$ the sequence converges to 0, but $0 \notin X$. Intuitively, a Cauchy sequence that fails to converge does so because of the absence of a point or element in the space to which it can converge.

(b) Let $X = \mathbb{Q}$ with $d(p, q) = |p - q|$. If $\{p_n\}$ is any sequence of rational numbers that converges to an irrational number, then the sequence $\{p_n\}$ is a Cauchy sequence in $(\mathbb{Q}, d)$, which however does not converge in $\mathbb{Q}$. $\quad\square$

THEOREM 3.6.4 *If $\{p_n\}$ is a Cauchy sequence in a metric space X that has a convergent subsequence, then the sequence $\{p_n\}$ converges.*

Proof. Suppose $\{p_{n_k}\}$ is a convergent subsequence of $\{p_n\}$ with $\lim p_{n_k} = p$. Let $\epsilon > 0$ be given. Since $\{p_n\}$ is Cauchy, there exists an integer N_1 such that

$$d(p_n, p_m) < \tfrac{1}{2}\epsilon \qquad \text{for all } n, m \geq N_1.$$

Since $p_{n_k} \to p$, for the given ϵ, there exists an integer k_1 such that

$$d(p_{n_k}, p) < \tfrac{1}{2}\epsilon \qquad \text{for all } k \geq k_1.$$

Let $n_o = \max\{k_1, N_1\}$, and choose n_k such that $k \geq n_o$. Then $n_k \geq N_1$. Thus if $n \geq n_o$, by the triangle inequality

$$d(p_n, p) \leq d(p_n, p_{n_k}) + d(p_{n_k}, p) < \tfrac{1}{2}\epsilon + \tfrac{1}{2}\epsilon = \epsilon.$$

Therefore $\lim p_n = p$, which proves the result. $\square$

THEOREM 3.6.5 *Every Cauchy sequence of real numbers converges.*

Proof. Let $\{p_n\}$ be a Cauchy sequence in $\mathbb{R}$. By Theorem 3.6.2, the sequence $\{p_n\}$ is bounded. Thus by Corollary 3.4.6, the sequence $\{p_n\}$ has a convergent subsequence. The result now follows by Theorem 3.6.4. $\square$

DEFINITION 3.6.6 *A metric space (X, d) is said to be **complete** if every Cauchy sequence in X converges to a point in X.*

As an example, $\mathbb{R}$ with the usual metric is complete. In the exercises you will be asked to prove that $\mathbb{R}^2$ with the euclidean metric is also complete. Additional examples of complete metric spaces will be encountered in subsequent chapters. Since the proof of Theorem 3.6.5 used the Bolzano-Weierstrass theorem, the *completeness of $\mathbb{R}$ ultimately depends on the least upper bound property of $\mathbb{R}$*. Conversely, if we assume completeness of $\mathbb{R}$, then we can prove that $\mathbb{R}$ satisfies the least upper bound property (Exercise 14). For this reason the least upper bound or supremum property of $\mathbb{R}$ is often called the **completeness property** of $\mathbb{R}$.

EXAMPLES 3.6.7 (a) For our first example we consider the sequence $\{s_n\}$ where for $n \in \mathbb{N}$

$$s_n = 1 + \frac{1}{2^2} + \cdots + \frac{1}{n^2}.$$

For $k \in \mathbb{N}$,

$$|s_{n+k} - s_n| = \frac{1}{(n+1)^2} + \cdots + \frac{1}{(n+k)^2}$$

$$\leq \left[\left(\frac{1}{n} - \frac{1}{n+1}\right) + \cdots + \left(\frac{1}{n+k-1} - \frac{1}{n+k}\right)\right]$$

$$= \frac{1}{n} - \frac{1}{n+k}.$$

In the above we have used the inequality

$$\frac{1}{(n+m)^2} \leq \left(\frac{1}{n+m-1} - \frac{1}{n+m}\right)$$

valid for all $n, m \in \mathbb{N}$. Since the sequence $\{\frac{1}{n}\}$ converges, it is a Cauchy sequence. Thus given $\epsilon > 0$ there exists $n_o \in \mathbb{N}$ such that $|\frac{1}{n} - \frac{1}{n+k}| < \epsilon$ for all $n \geq n_o$ and all $k \in \mathbb{N}$. Therefore the sequence $\{s_n\}$ is a Cauchy sequence and hence converges.

(b) In our second example we give an application to illustrate how the concept of a Cauchy sequence may be used to prove convergence of a given sequence. Additional applications will be given in the exercises. Let a_1, a_2 be arbitrary real numbers with $a_1 \neq a_2$. For $n \geq 3$, define a_n inductively by

$$a_n = \frac{1}{2}(a_{n-1} + a_{n-2}).$$

Our first goal is to show that the sequence $\{a_n\}$ is Cauchy. We first note that

$$a_{n+1} - a_n = -\frac{1}{2}(a_n - a_{n-1}).$$

As a consequence, for $n \geq 2$,

$$a_{n+1} - a_n = (-\tfrac{1}{2})^{n-1}(a_2 - a_1). \tag{4}$$

This last statement is most easily verified by induction (Exercise 5). For $m \geq 1$, consider $|a_{n+m} - a_n|$. By the triangle inequality,

$$|a_{n+m} - a_n| = \left| \sum_{k=0}^{m-1} a_{n+k+1} - a_{n+k} \right| \leq \sum_{k=0}^{m-1} |a_{n+k+1} - a_{n+k}|,$$

which by the above

$$\leq |a_2 - a_1| \sum_{k=0}^{m-1} \frac{1}{2^{n+k-1}} = \frac{1}{2^{n-2}} |a_2 - a_1| \sum_{k=1}^{m} \frac{1}{2^k}.$$

By Example 1.3.2(a)

$$\sum_{k=1}^{m} r^k = \frac{r - r^{m+1}}{1 - r}, \quad r \neq 1. \tag{5}$$

Thus with $r = \frac{1}{2}$,

$$\sum_{k=1}^{m} \frac{1}{2^k} = \frac{\frac{1}{2} - (\frac{1}{2})^{m+1}}{1 - \frac{1}{2}} = 1 - \frac{1}{2^m} < 1.$$

Therefore,

$$|a_{n+m} - a_n| \leq \frac{1}{2^{n-2}} |a_2 - a_1|$$

for all $n \geq 2$ and $m \in \mathbb{N}$. Let $\epsilon > 0$ be given. Choose n_o such that $|a_2 - a_1|/2^{n-2} < \epsilon$ for all $n \geq n_o$. Then by the above,

$$|a_{n+m} - a_n| < \epsilon$$

for all $m \in \mathbb{N}$, $n \geq n_o$. This however is just another way of stating that

$$|a_n - a_m| < \epsilon \qquad \text{for all} \quad m, n \geq n_o.$$

Therefore the sequence $\{a_n\}$ is a Cauchy sequence in $\mathbb{R}$, and thus by Theorem 3.6.5,

$$a = \lim_{n \to \infty} a_n$$

exists in $\mathbb{R}$.

Can we find the limit a here? If we take the same approach as in Example 3.3.4(c), by taking the limit of both sides of equation (5) we only get $a = a$. To find the value of a, let us observe that

$$a_{n+1} - a_1 = (a_{n+1} - a_n) + (a_n - a_{n-1}) + \cdots + (a_2 - a_1)$$

$$= \sum_{k=1}^{n} (a_{k+1} - a_k),$$

then use (4) to get

$$= (a_2 - a_1) \sum_{k=1}^{n} (-\tfrac{1}{2})^{k-1}$$

$$= \tfrac{2}{3}(a_2 - a_1)[1 - (-\tfrac{1}{2})^n].$$

The last equality follows from formula (5). Since $a_{n+1} \to a$ and $(-\tfrac{1}{2})^n \to 0$, upon taking the limit of both sides we obtain

$$a - a_1 = \tfrac{2}{3}(a_2 - a_1) \quad \text{or} \quad a = a_1 + \tfrac{2}{3}(a_2 - a_1). \quad \square$$

Contractive Sequences

One of the key properties of the sequence $\{a_n\}$ of the previous example was that

$$|a_{n+1} - a_n| \leq \tfrac{1}{2}|a_n - a_{n-1}|$$

for all $n \geq 2$. This property was used to show that the sequence $\{a_n\}$ was a Cauchy sequence and thus converged. Sequences that satisfy a criterion such as the above are commonly referred to as contractive sequences. We make this precise in the following definition.

DEFINITION 3.6.8 *A sequence $\{p_n\}$ in a metric space (X, d) is* **contractive** *if there exists a real number b, $0 < b < 1$, such that*

$$d(p_{n+1}, p_n) \leq b \, d(p_n, p_{n-1})$$

for all $n \in \mathbb{N}$, $n \geq 2$.

If $\{p_n\}$ is a contractive sequence, then an argument similar to the one used in the previous example shows that

$$d(p_{n+1}, p_n) \le b^{n-1} d(p_2, p_1)$$

for all $n \ge 1$, and that

$$d(p_{n+m}, p_n) \le b^{n-1} d(p_2, p_1)(1 + b + \cdots + b^{m-1}) < \frac{b^{n-1}}{1-b} d(p_2, p_1)$$

for all $n, m \in \mathbb{N}$. As a consequence, every contractive sequence is a Cauchy sequence. Therefore, if (X, d) is a complete metric space, every contractive sequence in X converges to a point in X. We summarize this in the following theorem.

THEOREM 3.6.9 *Let (X, d) be a complete metric space. Then every contractive sequence in X converges in X. Furthermore, if the sequence $\{p_n\}$ is contractive and $p = \lim p_n$, then*

(a) $d(p, p_n) \le \dfrac{b^{n-1}}{1-b} d(p_2, p_1)$, *and*

(b) $d(p, p_n) \le \dfrac{b}{1-b} d(p_n, p_{n-1})$, *where $0 < b < 1$ is the constant in Definition 3.6.8.*

Proof. We leave the details of the proof to the exercises (Exercise 9). $\quad\square$

Exercises 3.6

1. If $\{a_n\}$ and $\{b_n\}$ are Cauchy sequences in $\mathbb{R}$, prove (without using Theorem 3.6.5) that $\{a_n + b_n\}$ and $\{a_n b_n\}$ are also Cauchy sequences.

2. For each of the following determine whether the given sequence is a Cauchy sequence.

   *a. $\left\{ \dfrac{n+1}{n} \right\}$ b. $\{(-1)^n\}$ c. $\left\{ n + \dfrac{(-1)^n}{n} \right\}$

   *d. $\left\{ \dfrac{1+(-1)^n n}{n^2 + 3} \right\}$ e. $\left\{ \dfrac{1+(-1)^n n^2}{2n^2 + 3} \right\}$ f. $\left\{ \left(1 + \dfrac{1}{\sqrt{n}}\right)^n \right\}$.

3. For $n \in \mathbb{N}$ let $s_n = 1 + \dfrac{1}{2!} + \dfrac{1}{3!} + \cdots + \dfrac{1}{n!}$. Prove that $\{s_n\}$ is a Cauchy sequence.

4. Consider the sequence $\{s_n\}$ defined by $s_n = 1 + \dfrac{1}{2} + \cdots + \dfrac{1}{n}$.

   *a. Show that $\{s_n\}$ is not a Cauchy sequence.

   b. Even though $\{s_n\}$ is not a Cauchy sequence, show that $\lim_{n\to\infty} |s_{n+k} - s_n| = 0$ for all $k \in \mathbb{N}$.

5. Use mathematical induction to prove Identity (4).

6. If K is a compact subset of a metric space (X, d), prove that every Cauchy sequence in K converges to a point in K.

7. Prove that $(\mathbb{R}^2, d_2)$ is complete.

8. Let $\{a_n\}$ be the sequence of Example 3.6.7(b).

   a. Use mathematical induction to prove that
   $$a_{2k+1} = \frac{1}{2^{2k-1}}(a_1 + a_2) + \frac{1}{3}(a_1 + 2a_2)\left(1 - \frac{1}{4^{k-1}}\right).$$

   b. Use the result of (a) to find $\lim a_n$.

9. Prove Theorem 3.6.9.

10. *Let $a_1 > 0$, and for $n \geq 2$, define $a_n = (2 + a_{n-1})^{-1}$. Prove that $\{a_n\}$ is contractive, and find $\lim\limits_{n\to\infty} a_n$.

11. Let $c_1 \in (0, 1)$ be arbitrary, and for $n \in \mathbb{N}$ set $c_{n+1} = \frac{1}{5}(c_n^2 + 2)$.

    a. Show that $\{c_n\}$ is contractive.

    b. Let $c = \lim\limits_{n\to\infty} c_n$ Show that c is a solution of $x^2 - 5x + 2 = 0$.

    c. Let $c_1 = \frac{1}{2}$. Using the result of Theorem 3.6.9, determine the value of n such that $|c_n - c| < 10^{-3}$.

12. Consider the polynomial $p(x) = x^3 + 5x - 1$. It can be shown that $p(x)$ has exactly one root in the open interval $(0, 1)$. Let $a_1 \in (0, 1)$ be arbitrary, and for $n > 1$, set $a_{n+1} = \frac{1}{5}(1 - a_n^3)$.

    a. Prove that the sequence $\{a_n\}$ is contractive.

    b. Show that if $a = \lim\limits_{n\to\infty} a_n$, then $p(a) = 0$.

    c. Let $a_1 = \frac{1}{2}$. Using the result of Theorem 3.6.9(b), determine the value of n such that $|a_n - a| < 10^{-4}$.

13. Let $a_1 \neq a_2$ be real numbers, and let $0 < b < 1$. For $n \geq 3$, set
    $$a_n = b\, a_{n-1} + (1 - b)\, a_{n-2}.$$

    a. Show that the sequence $\{a_n\}$ is contractive.

    *__b.__ Find $\lim\limits_{n\to\infty} a_n$.

14. Prove that if every Cauchy sequence in $\mathbb{R}$ converges, then every nonempty subset of $\mathbb{R}$ that is bounded above has a supremum.

3.7 Series of Real Numbers

In this section, we will give a brief introduction to series of real numbers. Some knowledge of series, especially series with nonnegative terms, will be required in Chapter 4. The topic of series in general, including various convergence tests, alternating series, etc., will be treated in much greater detail in Chapter

7. We begin with some preliminary notation. If $\{a_n\}_{n=1}^{\infty}$ is a sequence in $\mathbb{R}$ and if $p, q \in \mathbb{N}$ with $p \le q$, set

$$\sum_{k=p}^{q} a_k = a_p + a_{p+1} + \cdots + a_q.$$

DEFINITION 3.7.1 *Let $\{a_n\}_{n=1}^{\infty}$ be a sequence of real numbers. Let $\{s_n\}_{n=1}^{\infty}$ be the sequence obtained from $\{a_n\}$, where for each $n \in \mathbb{N}$, $s_n = \sum_{k=1}^{n} a_k$. The sequence $\{s_n\}$ is called an* **infinite series**, *or* **series**, *and is denoted either as*

$$\sum_{k=1}^{\infty} a_k \quad or \ as \quad a_1 + a_2 + \cdots + a_n + \cdots .$$

For each $n \in \mathbb{N}$, s_n is called the nth **partial sum** *of the series and a_n is called the nth* **term** *of the series.*

The series $\sum_{k=1}^{\infty} a_k$ **converges** *if and only if the sequence $\{s_n\}$ of nth partial sums converges in $\mathbb{R}$. If $\lim_{n \to \infty} s_n = s$, then s is called the* **sum of the series**, *and we write*

$$s = \sum_{k=1}^{\infty} a_k.$$

If the sequence $\{s_n\}$ diverges, then the series $\sum_{k=1}^{\infty} a_k$ is said to **diverge**.

EXAMPLES 3.7.2 (a) For $|r| < 1$, consider the **geometric series**

$$\sum_{k=1}^{\infty} r^k.$$

For $n \in \mathbb{N}$,

$$s_n = \sum_{k=1}^{n} r^k = r + r^2 + \cdots + r^n.$$

Thus

$$(1 - r) s_n = s_n - r\, s_n = r - r^{n+1},$$

and as a consequence

$$s_n = \frac{r - r^{n+1}}{1 - r}.$$

Since $|r| < 1$, by Theorem 3.2.6(e), $\lim\limits_{n \to \infty} r^n = 0$. Therefore $\lim\limits_{n \to \infty} s_n = r/(1-r)$, and thus

$$\sum_{k=1}^{\infty} r^k = \frac{r}{1-r} \qquad |r| < 1.$$

For $|r| \geq 1$ the series $\sum\limits_{n=1}^{\infty} r^k$ diverges (Exercise 3).

(b) Consider the series $\sum\limits_{k=1}^{\infty} a_k$, where for each $k \in \mathbb{N}$, $a_k = \left(\dfrac{1}{k} - \dfrac{1}{k+1} \right)$. Then

$$s_n = \sum_{k=1}^{n} a_k$$

$$= \left(1 - \frac{1}{2} \right) + \left(\frac{1}{2} - \frac{1}{3} \right) + \cdots + \left(\frac{1}{n} - \frac{1}{n+1} \right)$$

$$= 1 - \frac{1}{n+1}.$$

Thus $\lim\limits_{n \to \infty} s_n = 1$ and hence $\sum\limits_{k=1}^{\infty} a_k = 1$.

(c) Consider $\sum\limits_{k=1}^{\infty} (-1)^k$. Then

$$s_n = \sum_{k=1}^{n} (-1)^k = \begin{cases} 0, & \text{if } n \text{ is even,} \\ -1, & \text{if } n \text{ is odd.} \end{cases}$$

Thus since $\{s_n\}$ diverges, the series diverges. $\square$

The Cauchy Criterion

The following criterion, which provides necessary and sufficient conditions for the convergence of a series, was formulated by Augustin-Louis Cauchy (1789–1857) in 1821.

THEOREM 3.7.3 (Cauchy Criterion) *The series $\sum\limits_{k=1}^{\infty} a_k$ converges if and only if given $\epsilon > 0$, there exists a positive integer n_o, such that*

$$\left| \sum_{k=n+1}^{m} a_k \right| < \epsilon$$

for all $m > n \geq n_o$.

Proof. Since

$$\left| \sum_{k=n+1}^{m} a_k \right| = |s_m - s_n|,$$

the result is an immediate consequence of Theorems 3.6.2 and 3.6.5. □

Remark. The previous theorem simply states that the series $\sum a_k$ converges if and only if the sequence $\{s_n\}$ of nth partial sums is a Cauchy sequence.

EXAMPLE 3.7.4 In this example, we show that the series $\sum_{k=1}^{\infty} 1/k$ diverges. We accomplish this by showing that the sequence $\{s_n\}$ of partial sums is not a Cauchy sequence. Consider

$$s_{2n} - s_n = \frac{1}{n+1} + \cdots + \frac{1}{2n}, \qquad n \in \mathbb{N}.$$

There are exactly n terms in the sum on the right, and each term is greater than or equal to $1/2n$. Therefore

$$s_{2n} - s_n \geq n\left(\frac{1}{2n}\right) = \frac{1}{2}.$$

The sequence $\{s_n\}$ therefore fails to be a Cauchy sequence and thus the series diverges. The divergence of this series appears to have been first established by Nicole Oresme (1323?–1382) using a method of proof similar to that suggested in the solution of Exercise 11 of Section 3.3 □

COROLLARY 3.7.5 *If* $\sum_{k=1}^{\infty} a_k$ *converges, then* $\lim_{k\to\infty} a_k = 0.$

Proof. Since $a_k = s_k - s_{k-1}$, this is an immediate consequence of the Cauchy criterion. □

Remark. The condition $\lim_{k\to\infty} a_k = 0$ is not sufficient for the convergence of $\sum a_k$. For example, the series $\sum \frac{1}{k}$ diverges, yet $\lim_{k\to\infty} \frac{1}{k} = 0.$

THEOREM 3.7.6 *Suppose* $a_k \geq 0$ *for all* $k \in \mathbb{N}$. *Then* $\sum_{k=1}^{\infty} a_k$ *converges if and only if* $\{s_n\}$ *is bounded above.*

Proof. Since $a_k \geq 0$ for all k, the sequence $\{s_n\}$ is monotone increasing. Thus by Theorem 3.3.2, the sequence $\{s_n\}$ converges if and only if it is bounded above. □

Exercises 3.7

1. *Using the inequality $\dfrac{1}{k^2} \leq \dfrac{1}{k(k-1)} = \dfrac{1}{k-1} - \dfrac{1}{k}$, prove that the series

$$\sum_{k=1}^{\infty} \frac{1}{k^2} \text{ converges.}$$

2. Prove that the series $\displaystyle\sum_{k=1}^{\infty} \frac{1}{k^2 + k}$ converges.

3. If $|r| \geq 1$, show that the series $\displaystyle\sum_{k=1}^{\infty} r^k$ diverges.

4. Prove that the series $\displaystyle\sum_{k=1}^{\infty} \frac{1}{k!}$ converges. (See Exercise 13 of Section 3.6.)

5. *Suppose $a_k \geq 0$ for all k. Prove that if $\sum a_k$ converges, then

$$\sum_{k=1}^{\infty} \frac{\sqrt{a_k}}{k} \text{ converges.}$$

6. If $\displaystyle\sum_{k=1}^{\infty} a_k$ and $\displaystyle\sum_{k=1}^{\infty} b_k$ both converge, prove each of the following:

   a. $\displaystyle\sum_{k=1}^{\infty} c a_k$ converges for all $c \in \mathbb{R}$.

   b. $\displaystyle\sum_{k=1}^{\infty} (a_k + b_k)$ converges.

7. If $\displaystyle\sum_{k=1}^{\infty} (a_k + b_k)$ converges, does this imply that the series

   $$a_1 + b_1 + a_2 + b_2 + \cdots \text{ converges?}$$

8. Suppose $b_k \geq a_k \geq 0$ for all $k \in \mathbb{N}$.

   a. If $\displaystyle\sum_{k=1}^{\infty} b_k$ converges, prove that $\displaystyle\sum_{k=1}^{\infty} a_k$ converges.

   b. If $\displaystyle\sum_{k=1}^{\infty} a_k$ diverges, prove that $\displaystyle\sum_{k=1}^{\infty} b_k$ diverges.

9. Consider the series $\displaystyle\sum_{k=1}^{\infty} \frac{1}{k^p}$, $p \in \mathbb{R}$.

   a. Prove that the series diverges for all $p \leq 1$.

   b. Prove that the series converges for all $p > 1$.

Notes

This chapter provided our first serious introduction to the limit process. In subsequent chapters we will encounter limits of functions, the derivative, and the integral, all of which are further examples of the limit process. Of the many results proved in this chapter, it is difficult to select one or two for special emphasis. They are all important! Many of them will be encountered again—either directly or indirectly—throughout the text.

Some of the concepts and results of this chapter have certainly been encountered previously; others undoubtedly are new. Two concepts which may not have been previously encountered are limit superior (inferior) of a sequence of real numbers and complete metric spaces. The primary importance of the limit superior and inferior of a sequence is that these two limit operations always exist in $\mathbb{R} \cup \{-\infty, \infty\}$. As we will see in Chapter 7, this will allow us to present the correct statements of the root and ratio test for convergence of a series. The limit superior will also be required to define the radius of convergence of a power series. There will be other instances in the text where these two limit operations will be encountered.

In the chapter we have proved several important consequences of the least upper bound property of $\mathbb{R}$. The least upper bound property was used to prove that every bounded monotone sequence converges. This result was subsequently used to prove the nested intervals property, which in turn can be used to provide a proof of the Bolzano-Weierstrass theorem. The nested intervals property can also be used to prove the supremum property of $\mathbb{R}$ (Exercise 21 of Section 3.3). Another property of the real numbers that is equivalent to the least upper bound property is the completeness property of $\mathbb{R}$; namely, every Cauchy sequence of real numbers converges. Other consequences of the least upper bound property will be encountered in subsequent chapters.

Cauchy sequences were originally studied by Cantor in the middle of the nineteenth century. He referred to them as *"fundamental sequences"* and used them in his construction of the real number system $\mathbb{R}$ (See Miscellaneous Exercises 4 – 11). The main reason that these sequences are attributed to Cauchy, rather than Cantor, is due to the fact that his 1821 criterion for convergence of a series (Theorem 3.7.3) is equivalent to the statement that the sequence of partial sums is a Cauchy sequence. The fact that Cauchy was a more prominent mathematician than Cantor may also have been a factor.

Miscellaneous Exercises

The first three exercises involve the concept of an **infinite product**. Let $\{a_k\}$ be a sequence of nonzero real numbers. For each $n = 1, 2, \ldots$, define

$$p_n = \prod_{k=1}^{n} a_k = a_1 \cdot a_2 \cdots a_n.$$

If $p = \lim_{n \to \infty} p_n$ exists, then p is the **infinite product** of the sequence $\{a_k\}_{k=1}^{\infty}$, and we write

$$p = \prod_{k=1}^{\infty} a_k.$$

If the limit does not exist, then the infinite product is said to diverge.

Remark. Some authors require that $p \neq 0$. We will not make this requirement; rather we will specify $p \neq 0$ if this hypothesis is required in a result.

1. Determine whether each of the following infinite products converge. If it converges, find the infinite product.

   a. $\displaystyle\prod_{k=1}^{\infty}(-1)^k.$ **b.** $\displaystyle\prod_{k=2}^{\infty}\left(1-\frac{1}{k}\right).$ **c.** $\displaystyle\prod_{k=2}^{\infty}\left(1-\frac{1}{k^2}\right).$

2. If $\displaystyle\prod_{k=1}^{\infty}a_k = p$ with $p \neq 0$, prove that $\displaystyle\lim_{k\to\infty}a_k = 1$.

3. If $a_n \geq 0$ for all $n \in \mathbb{N}$, prove that

   $\prod_{k=1}^{\infty}(1+a_k)$ converges if and only if $\sum_{k=1}^{\infty}a_k$ converges.

   To prove the result, establish the following inequality:

   $$a_1 + \cdots + a_n \leq (1+a_1)\cdots(1+a_n) \leq e^{a_1+\cdots+a_n}.$$

Construction of the Real Numbers

In the following exercises we outline the construction of the real number system from the rational number system using Cantor's method of Cauchy sequences. Let $\mathbb{Q}$ denote the set of rational numbers. A sequence $\{a_n\}$ in $\mathbb{Q}$ is **Cauchy** if for every $r \in \mathbb{Q}$, $r > 0$, there exists a positive integer n_o such that $|a_n - a_m| < r$ for all $n, m \geq n_o$. A sequence $\{a_n\}$ in $\mathbb{Q}$ is called a **null sequence** if for every $r \in \mathbb{Q}$, $r > 0$, there exists a positive integer n_o such that $|a_n| < r$ for all $n \geq n_o$. Two Cauchy sequences $\{a_n\}$ and $\{b_n\}$ in $\mathbb{Q}$ are said to be **equivalent**, denoted $\{a_n\} \sim \{b_n\}$, provided $\{a_n - b_n\}$ is a null sequence.

4. Let $\{a_n\}$, $\{b_n\}$, $\{c_n\}$, and $\{d_n\}$ be Cauchy sequences in $\mathbb{Q}$. Prove the following:

   a. $\{a_n\} \sim \{a_n\}$.

   b. If $\{a_n\} \sim \{b_n\}$, then $\{b_n\} \sim \{a_n\}$.

   c. If $\{a_n\} \sim \{b_n\}$ and $\{b_n\} \sim \{c_n\}$, then $\{a_n\} \sim \{c_n\}$.

   d. If $\{a_n\} \sim \{b_n\}$, then $\{-a_n\} \sim \{-b_n\}$.

   e. If $\{a_n\} \sim \{c_n\}$ and $\{b_n\} \sim \{d_n\}$, then

   $$\{a_n + b_n\} \sim \{c_n + d_n\} \qquad \text{and} \qquad \{a_n b_n\} \sim \{c_n d_n\}.$$

   Given a Cauchy sequence $\{a_n\}$ in $\mathbb{Q}$, let $[\{a_n\}]$ denote the set of all Cauchy sequences in $\mathbb{Q}$ equivalent to $\{a_n\}$. The set $[\{a_n\}]$ is called the **equivalence class** determined by $\{a_n\}$.

5. Given two Cauchy sequences $\{a_n\}$ and $\{b_n\}$ in $\mathbb{Q}$, prove that $[\{a_n\}] = [\{b_n\}]$ provided $\{a_n\} \sim \{b_n\}$, and $[\{a_n\}] \cap [\{b_n\}] = \emptyset$ otherwise.

   Let $\mathcal{R}$ denote the set of equivalence classes of Cauchy sequences in $\mathbb{Q}$. We denote the elements of $\mathcal{R}$ by lower case Greek letters α, β, γ, ... Thus if $\alpha \in \mathcal{R}$, $\alpha = [\{a_n\}]$ for some Cauchy sequence $\{a_n\}$ in $\mathbb{Q}$. The sequence $\{a_n\}$ is called a representative of the equivalence class α. Suppose $\alpha = [\{a_n\}]$ and $\beta = [\{b_n\}]$. Define $-\alpha$, $\alpha + \beta$, and $\alpha \cdot \beta$ as follows:

   $$-\alpha = [\{-a_n\}],$$
   $$\alpha + \beta = [\{a_n + b_n\}],$$
   $$\alpha \cdot \beta = [\{a_n b_n\}].$$

One needs to show that these operations are well defined; that is, independent of the representative of the equivalence class. For example, to prove that $-\alpha$ is well defined, we suppose that $\{a_n\}$ and $\{b_n\}$ are two representatives of α; i.e., $\{a_n\} \sim \{b_n\}$. But then by 4(d), $\{-a_n\} \sim \{-b_n\}$. Therefore, $[\{-a_n\}] = [\{-b_n\}]$. This shows that $-\alpha$ is well defined.

6. Prove that the operations $+$ and $\cdot$ are well defined on $\mathcal{R}$.

For each $p \in \mathbb{Q}$, let $\{p\}$ denote the sequence all of whose terms are equal to p. If $p \in \mathbb{Q}$, let $\alpha_p = [\{p\}]$. Also, we set

$$\theta = [\{0\}], \qquad \iota = [\{1\}].$$

As we will see, the element θ will be the zero of $\mathcal{R}$ and ι will be the unit of $\mathcal{R}$. A Cauchy sequence $\{b_n\}$ in $\mathbb{Q}$ belongs to θ if and only if $b_n \to 0$. Similarly, $\{a_n\} \in \iota$ if and only if $(a_n - 1) \to 0$. The following problem provides us with the multiplicative inverse of $\alpha \neq \theta$.

7. If $\alpha \neq \theta$, prove that there exists $\{a_n\} \in \alpha$ such that $a_n \neq 0$ for all $n \in \mathbb{N}$, and that $\{\frac{1}{a_n}\}$ is a Cauchy sequence. Define $\alpha^{-1} = [\{\frac{1}{a_n}\}]$.

8. Prove that $\mathcal{R}$ with operations $+$ and $\cdot$ is a field.

We now proceed to define an order relation on $\mathcal{R}$. A Cauchy sequence $\{a_n\}$ in $\mathbb{Q}$ is **positive** if there exists $r \in \mathbb{Q}$, $r > 0$, and $n_o \in \mathbb{N}$ such that $a_n > r$ for all $n \geq n_o$. Let $\mathcal{P}$ be defined by

$$\mathcal{P} = \{[\{a_n\}] : \{a_n\} \text{ is a positive Cauchy sequence }\}.$$

9. Prove that the set $\mathcal{P}$ satisfies the order properties (O1) and (O2) of Section 1.4.

10. Show that the mapping $p \to \alpha_p$ is a one-to-one mapping of $\mathbb{Q}$ into $\mathcal{R}$ which satisfies

$$\alpha_p + \alpha_q = \alpha_{p+q}$$
$$\alpha_p \cdot \alpha_q = \alpha_{pq}$$

for all $p, q \in \mathbb{Q}$. Furthermore, if $p > 0$, then $\alpha_p \in \mathcal{P}$.

11. Prove that every nonempty subset of $\mathcal{R}$ which is bounded above has a least upper bound in $\mathcal{R}$.

The above exercises prove that $\mathcal{R}$ is an ordered field which satisfies the least upper bound property. One can show that any two complete ordered fields are in fact isomorphic, that is, there exists a one-to-one map of one onto the other which preserves the operations of addition, multiplication, and the order properties. Thus $\mathcal{R}$ is isomorphic to the real numbers $\mathbb{R}$.

Supplemental Reading

Aguirre, J. A. F., "A note on Cauchy sequences," *Math. Mag.* **68** (1995), 296–297.

Bell, H. E., "Proof of a fundamental theorem on sequences," *Amer. Math. Monthly* **71** (1964) 665–666.

Goffman, C., "Completeness of the real numbers," *Math. Mag.* **47** (1974) 1–8.

Newman, D. J. and Parsons, T. D., "On monotone subsequences," *Amer. Math. Monthly* **95** (1988) 44–45.

Staib, J. H. and Demos, M. S., "On the limit points of the sequence $\{\sin n\}$," *Math. Mag.* **40** (1967) 210–213.

Wenner, B. R., "The uncountability of the reals," *Amer. Math. Monthly* **76** (1969) 679–680.

4

Limits and Continuity

The concept of limit dates back to the late seventeenth century and the work of Isaac Newton (1642–1727) and Gottfried Leibniz (1646–1716). Both of these mathematicians are given historical credit for inventing the differential and integral calculus. Although the idea of "limit" occurs in Newton's work *Philosophia Naturalis Principia Mathematica* of 1687, he never expressed the concept algebraically; rather he used the phrase "*ultimate ratios of evanescent quantities*" to describe the limit process involved in computing the derivatives of functions.

The subject of limits lacked mathematical rigor until 1821 when Augustin-Louis Cauchy (1789–1857) published his *Cours d'Analyse* in which he offered the following definition of limit: "If the successive values attributed to the same variable approach indefinitely a fixed value, such that finally they differ from it by as little as desired, this latter is called the limit of all the others." Even this statement does not resemble the modern delta-epsilon version of limit given in Section 1. Although Cauchy gave a strictly verbal definition of limit, he did use epsilons, deltas, and inequalities in his proofs. For this reason Cauchy is credited for putting calculus on the rigorous basis with which we are familiar today.

Based on the previous study of calculus, the student should have an intuitive notion of what it means for a function to be continuous. This most likely compares to how mathematicians of the eighteenth century perceived a continuous function; namely one that can be expressed by a single formula or equation involving a variable x. Mathematicians of this period certainly accepted functions that failed to be continuous at a finite number of points. However, even they might have difficulty envisaging a function that is continuous at every irrational number and discontinuous at every rational number in its domain. Such a function is given in Example 4.2.2(g). An example of an increasing function having the same properties will also be given in Section 4 of this chapter.

In Section 1 we define the limit at a point of a real-valued function defined on a subset of a metric space, and provide numerous examples to illustrate this idea. In Sections 2 and 3 we consider the closely related theory of continuity and investigate some of the consequences of this very important concept.

4.1   Limit of a Function

The basic idea underlying the concept of the limit of a function f at a point p is to study the behavior of f at points close to, but not equal to, p. We illustrate this with the following simple examples. Suppose that the velocity v (ft/sec) of a falling object is given as a function $v = v(t)$ of time t. If the object hits the ground in $t = 2$, then $v(2) = 0$. Thus to find the velocity at the time of impact, we investigate the behavior of $v(t)$ as t approaches 2, but is not equal to 2. Neglecting air resistance, the function $v(t)$ is given as follows:

$$v(t) = \begin{cases} -32t, & 0 \le t < 2, \\ 0, & t \ge 2. \end{cases}$$

Our intuition should convince us that $v(t)$ approaches -64 ft/sec as t approaches 2, and that this is the velocity upon impact.

As another example, consider the function $f(x) = x \sin \frac{1}{x}$, $x \ne 0$. Here the function f is not defined at $x = 0$. Thus to investigate the behavior of f at 0 we need to consider the values $f(x)$ for x close to, but not equal to 0. Since

$$|f(x)| = |x \sin \tfrac{1}{x}| \le |x|$$

for all $x \ne 0$, our intuition again should tell us that $f(x)$ approaches 0 as x approaches 0. This indeed is the case as will be shown in Example 4.1.10(c).

We now make this idea of $f(x)$ approaching a value L as x approaches a point p precise. In order that the definition be meaningful, we must require that the point p be a limit point of the domain of the function f.

DEFINITION 4.1.1 *Let (X, d) be a metric space, E be a subset of X and f a real-valued function with domain E. Suppose that p is a limit point of E. The function f has a* **limit** *at p if there exists a number $L \in \mathbb{R}$ such that given any $\epsilon > 0$, there exists a $\delta > 0$ for which*

$$|f(x) - L| < \epsilon$$

for all points $x \in E$ satisfying $0 < d(x, p) < \delta$. If this is the case, we write

$$\lim_{x \to p} f(x) = L \qquad or \qquad f(x) \to L \quad as \quad x \to p.$$

Although we restricted our consideration to the case where f is a real-valued function, we could just as easily have considered the case where f has values in a metric space (Y, ρ). The extension to $f : E \to Y$ is obtained by replacing $|f(x) - L|$ with $\rho(f(x), L)$, where in this case L is an element of Y.

The definition of the limit of a function can also be stated in terms of ϵ

and δ neighborhoods as follows: If $E \subset \mathbb{X}$, $f : E \to \mathbb{R}$, and p is a limit point of E, then

$$\lim_{x \to p} f(x) = L$$

if and only if given $\epsilon > 0$, there exists a $\delta > 0$ such that

$$f(x) \in N_\epsilon(L) \quad \text{for all} \quad x \in E \cap (N_\delta(p) \setminus \{p\}).$$

This is illustrated graphically in Figure 4.1 for the case where E is a subset of $\mathbb{R}$.

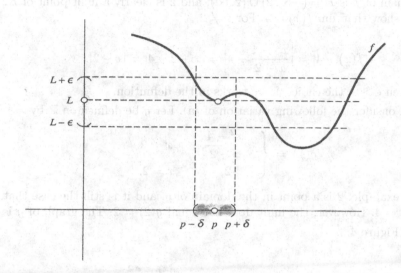

FIGURE 4.1
$$\lim_{x \to p} f(x) = L$$

Remarks. (a) In the definition of limit, the choice of δ for a given ϵ may depend not only on ϵ and the function, but also on the point p. This will be illustrated in Example 4.1.2(f) below.

(b) If p is not a limit point of E, then for δ sufficiently small, there do not exist any $x \in E$ so that $0 < |x - p| < \delta$. Thus if p is an isolated point of E, the concept of the limit of a function at p has no meaning.

(c) In the definition of limit, it is not required that $p \in E$, only that p is a limit point of E. Even if $p \in E$, and f has a limit at p, we may very well have that

$$\lim_{x \to p} f(x) \neq f(p).$$

This will be the case in Example 4.1.2(b) below.

(d) Let $E \subset \mathbb{R}$ and p a limit point of E. To show that a given function f does **not** have a limit at p, we must show that for every $L \in \mathbb{R}$, there exists an $\epsilon > 0$, such that for every $\delta > 0$, there exists an $x \in E$ with $0 < |x - p| < \delta$, for which

$$|f(x) - L| \geq \epsilon.$$

We will illustrate this in Example 4.1.2(d).

EXAMPLES 4.1.2 (a) For $x \neq 2$, let $f(x)$ be defined by

$$f(x) = \frac{x^2 - 4}{x - 2}, \qquad x \neq 2.$$

The domain of f is $E = (-\infty, 2) \cup (2, \infty)$, and 2 is clearly a limit point of E. We now show that $\lim_{x \to 2} f(x) = 4$. For $x \neq 2$,

$$|f(x) - 4| = \left| \frac{x^2 - 4}{x - 2} - 4 \right| = |x + 2 - 4| = |x - 2|.$$

Thus given $\epsilon > 0$, the choice $\delta = \epsilon$ works in the definition.

(b) Consider the following variation of (a). Let g be defined on $\mathbb{R}$ by

$$g(x) = \begin{cases} \dfrac{x^2 - 4}{x - 2}, & x \neq 2, \\ 2, & x = 2. \end{cases}$$

For this example, 2 is a point in the domain of g, and it is still the case that $\lim_{x \to 2} g(x) = 4$. However, the limit does not equal $g(2) = 2$. The graph of g is given in Figure 4.2.

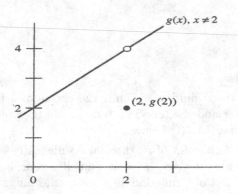

FIGURE 4.2
Graph of g

(c) Let $E = (-1, 0) \cup (1, \infty)$. For $x \in E$, let $h(x)$ be defined by

$$h(x) = \frac{\sqrt{x + 1} - 1}{x}, \qquad x \neq 0.$$

We claim that $\lim_{x \to 0} h(x) = 1/2$. This result is obtained as follows: For $x \neq 0$,

$$\frac{\sqrt{x+1}-1}{x} = \left(\frac{\sqrt{x+1}-1}{x}\right)\left(\frac{\sqrt{x+1}+1}{\sqrt{x+1}+1}\right) = \frac{x}{x(\sqrt{x+1}+1)} = \frac{1}{\sqrt{x+1}+1}.$$

From this last term we now conjecture that $h(x) \to 1/2$ as $x \to 0$. By the above,

$$\left|h(x) - \frac{1}{2}\right| = \left|\frac{1}{\sqrt{x+1}+1} - \frac{1}{2}\right| = \left|\frac{1-\sqrt{x+1}}{2(\sqrt{x+1}+1)}\right|$$

$$= \left|\frac{(1-\sqrt{x+1})(1+\sqrt{x+1})}{2(\sqrt{x+1}+1)^2}\right| = \left|\frac{-x}{2(\sqrt{x+1}+1)^2}\right|$$

$$= \frac{1}{2}\frac{|x|}{(\sqrt{x+1}+1)^2}.$$

For $x \in E$ we have $(\sqrt{x+1}+1)^2 > 1$, and thus

$$\left|h(x) - \frac{1}{2}\right| < \frac{|x|}{2}.$$

Given $\epsilon > 0$, let $\delta = \min\{1, \epsilon\}$. Then for $0 < |x| < \delta$,

$$\left|h(x) - \frac{1}{2}\right| < \frac{|x|}{2} < \frac{\delta}{2} < \epsilon,$$

and thus $\lim_{x \to 0} h(x) = 1/2$.

(d) Let f be defined on $\mathbb{R}$ as follows:

$$f(x) = \begin{cases} 1, & x \in \mathbb{Q}, \\ 0, & x \notin \mathbb{Q}. \end{cases}$$

We will show that for this function, $\lim_{x \to p} f(x)$ fails to exist for every $p \in \mathbb{R}$. Fix $p \in \mathbb{R}$. Let $L \in \mathbb{R}$ and let

$$\epsilon = \max\{|L-1|, |L|\}.$$

Suppose $\epsilon = |L-1|$. By Theorem 1.5.2, for any $\delta > 0$, there exists an $x \in \mathbb{Q}$ such that $0 < |p-x| < \delta$. For such an x,

$$|f(x) - L| = |1 - L| = \epsilon.$$

If $\epsilon = |L|$, then by Exercise 6, Section 1.5, for any $\delta > 0$, there exists an irrational number x with $0 < |x-p| < \delta$. Again, for such an x, $|f(x) - L| = \epsilon$. Thus with ϵ as defined, for any $\delta > 0$, there exists an x with $0 < |x-p| < \delta$ such that $|f(x) - L| \geq \epsilon$. Since this works for every $L \in \mathbb{R}$, $\lim_{x \to p} f(x)$ does not exist.

(e) Let $f : \mathbb{R} \to \mathbb{R}$ be defined by

$$f(x) = \begin{cases} 0, & x \in \mathbb{Q}, \\ x, & x \notin \mathbb{Q}. \end{cases}$$

Then $\lim_{x \to 0} f(x) = 0$. Since $|f(x)| \leq |x|$ for all x, given $\epsilon > 0$, any δ, $0 < \delta \leq \epsilon$, will work in the definition of the limit. A modification of the argument given in (d) shows that for any $p \neq 0$, $\lim_{x \to p} f(x)$ does not exist. An alternate proof will be provided in Example 4.2.2(b)

(f) This example shows dramatically how the choice of δ will in general depend not only on ϵ, but also on the point p. Let $E = (0, 1)$ and let $f : E \to \mathbb{R}$ be defined by

$$f(x) = \frac{1}{x}.$$

We will prove that for $p \in (0, 1]$,

$$\lim_{x \to p} \frac{1}{x} = \frac{1}{p}.$$

If $x > p/2$, then

$$\left| \frac{1}{x} - \frac{1}{p} \right| = \frac{|x - p|}{xp} < \frac{2}{p^2} |x - p|.$$

Therefore, given $\epsilon > 0$, let $\delta = \min\{p/2, p^2\epsilon/2\}$. Then if $0 < |x - p| < \delta$, $x > p/2$, and

$$\left| \frac{1}{x} - \frac{1}{p} \right| < \frac{2}{p^2} \delta < \epsilon.$$

The δ as defined depends both on p and ϵ. This suggests that any δ that works for a given p and ϵ must depend on both p and ϵ. Suppose on the contrary that for a given $\epsilon > 0$ the choice of δ is independent of $p \in (0, 1)$. Then with $\epsilon = 1$, there exists a $\delta > 0$ such that

$$\left| \frac{1}{x} - \frac{1}{p} \right| < 1$$

for all $x, p \in (0, 1)$ with $0 < |x - p| < \delta$. Since any smaller δ will also work, we can assume that $0 < \delta < \frac{1}{2}$. But now if we take $p = \frac{1}{2}\delta$ and $x = \delta$, then $0 < |x - p| < \delta$ and thus

$$|f(x) - f(p)| = \left| \frac{1}{\delta} - \frac{2}{\delta} \right| = \frac{1}{\delta} > 1.$$

This contradiction proves that the choice of δ must depend on both p and ϵ.

(g) For our final example we consider the function f defined on $\mathbb{R}^2 \setminus (0,0)$ given by $f(x,y) = \dfrac{xy}{x^2 + y^2}$. We will prove that the limit $\lim\limits_{(x,y) \to (1,2)} f(x,y) = \frac{2}{5}$. Consider

$$\left| f(x,y) - \frac{2}{5} \right| = \left| \frac{5xy - 2x^2 - 2y^2}{5(x^2 + y^2)} \right|$$

$$= \left| \frac{(x - 2y)(y - 2) + (4y - 2x)(x - 1)}{5(x^2 + y^2)} \right|$$

$$\leq \frac{(|x| + 2|y|)|y - 2| + (4|y| + 2|x|)|x - 1|}{5(x^2 + y^2)}.$$

If $(x,y) \in N_{\frac{1}{2}}(1,2)$, then $\frac{1}{2} < |x| < \frac{3}{2}$ and $\frac{3}{2} < |y| < \frac{5}{2}$. (Verify!) Therefore for $(x,y) \in N_{\frac{1}{2}}(1,2)$ we have $|x| + 2|y| < \frac{13}{2}$, $4|y| + 2|x| < \frac{26}{2}$, and $5(x^2 + y^2) > \frac{25}{2}$. Therefore

$$\left| f(x,y) - \frac{2}{5} \right| < \frac{26}{25}(|y - 2| + |x - 1|).$$

Hence if δ is chosen so that $0 < \delta \leq \frac{1}{2}$, then for $(x,y) \in N_\delta(1,2)$ we have

$$\left| f(x,y) - \frac{2}{5} \right| < \frac{52}{25}\delta.$$

Thus given $\epsilon > 0$, if we choose δ such that $0 < \delta < \min\{\frac{1}{2}, \frac{25}{52}\epsilon\}$, then $(x,y) \in N_\delta(1,2)$ implies that $|f(x,y) - \frac{2}{5}| < \epsilon$. $\quad\square$

Sequential Criterion for Limits

Our first theorem allows us to reduce the question of the existence of the limit of a function to one concerning the existence of limits of sequences. As we will see, this result will be very useful in subsequent proofs, and also in showing that a given function does not have a limit at a point p.

THEOREM 4.1.3 *Let E be a subset of a metric space X, p a limit point of E, and f a real-valued function defined on E. Then*

$$\lim_{x \to p} f(x) = L \qquad \text{if and only if} \qquad \lim_{n \to \infty} f(p_n) = L$$

for every sequence $\{p_n\}$ in E, with $p_n \neq p$ for all n, and $\lim\limits_{n \to \infty} p_n = p$.

Remark. Since p is a limit point of E, Theorem 3.1.4 guarantees the existence of a sequence $\{p_n\}$ in E with $p_n \neq p$ for all $n \in \mathbb{N}$ and $p_n \to p$.

Proof. Suppose $\lim\limits_{x \to p} f(x) = L$. Let $\{p_n\}$ be any sequence in E with $p_n \neq p$ for all n and $p_n \to p$. Let $\epsilon > 0$ be given. Since $\lim\limits_{x \to p} f(x) = L$, there exists a $\delta > 0$ such that

$$|f(x) - L| < \epsilon \qquad \text{for all} \qquad x \in E, \ 0 < |x - p| < \delta. \tag{1}$$

Since $\lim_{n\to\infty} p_n = p$, for the above δ, there exists a positive integer n_o such that

$$0 < |p_n - p| < \delta \quad \text{for all} \quad n \geq n_o.$$

Thus if $n \geq n_o$, by (1), $|f(p_n) - L| < \epsilon$. Therefore $\lim_{n\to\infty} f(p_n) = L$.

Conversely, suppose $f(p_n) \to L$ for every sequence $\{p_n\}$ in E with $p_n \neq p$ for all n and $p_n \to p$. Suppose $\lim_{x\to p} f(x) \neq L$. Then there exists an $\epsilon > 0$ such that for every $\delta > 0$, there exists an $x \in E$ with $0 < |x - p| < \delta$ and $|f(x) - L| \geq \epsilon$. For each $n \in \mathbb{N}$, take $\delta = 1/n$. Then for each n, there exists $p_n \in E$ such that

$$0 < |p_n - p| < \frac{1}{n} \quad \text{and} \quad |f(p_n) - L| \geq \epsilon.$$

Thus $p_n \to p$, but $\{f(p_n)\}$ does not converge to L. This contradiction proves the result. $\square$

An immediate consequence of the previous theorem is the following uniqueness theorem.

COROLLARY 4.1.4 *If f has a limit at p, then it is unique.*

Theorem 4.1.3 is often applied to show that a limit does not exist. If one can find a sequence $\{p_n\}$ with $p_n \to p$, such that $\{f(p_n)\}$ does not converge, then $\lim_{x\to p} f(x)$ does not exist. Alternately, if one can find two sequences $\{p_n\}$ and $\{r_n\}$ both converging to p, but for which

$$\lim_{n\to\infty} f(p_n) \neq \lim_{n\to\infty} f(r_n),$$

then again $\lim_{x\to p} f(x)$ does not exist. We illustrate this with the following two examples.

EXAMPLES 4.1.5 (a) Let $E = (0, \infty)$ and $f(x) = \sin\frac{1}{x}$, $x \in E$. We use the previous theorem to show that

$$\lim_{x\to 0} \sin\frac{1}{x}$$

does not exist. Let $p_n = \dfrac{2}{(2n+1)\pi}$. Then

$$f(p_n) = \sin(2n+1)\frac{\pi}{2} = (-1)^n.$$

Thus $\lim_{n\to\infty} f(p_n)$ does not exist, and consequently by Theorem 4.1.3, $\lim_{x\to 0} f(x)$ also does not exist. The graph of $f(x) = \sin\frac{1}{x}$ is given in Figure 4.3.

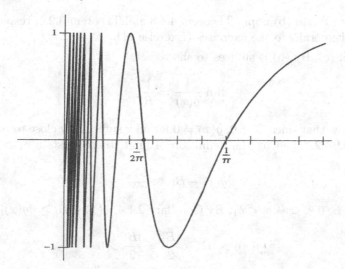

FIGURE 4.3
Graph of $f(x) = \sin(1/x)$, $x > 0$

  (b) As in Example 4.1.2(e) let

$$f(x) = \begin{cases} 0, & x \in \mathbb{Q}, \\ x, & x \notin \mathbb{Q}. \end{cases}$$

Suppose $p \in \mathbb{R}$, $p \neq 0$. Since $\mathbb{Q}$ is dense in $\mathbb{R}$, there exists a sequence $\{p_n\} \subset \mathbb{Q}$ with $p_n \neq p$ for all $n \in \mathbb{N}$ such that $p_n \to p$. Hence $\lim_{n \to \infty} f(p_n) = 0$. On the other hand, since $\mathbb{R} \setminus \mathbb{Q}$ is also dense in $\mathbb{R}$, there exists a sequence $\{q_n\}$ of irrational numbers with $q_n \to p$. But then $\lim_{n \to \infty} f(q_n) = \lim_{n \to \infty} q_n = p$. Thus since $p \neq 0$, by Theorem 4.1.3 $\lim_{x \to p} f(x)$ does not exist. $\square$

Limit Theorems

THEOREM 4.1.6 *Suppose E is a subset of a metric space X, f, $g : E \to \mathbb{R}$, and p is a limit point of E. If*

$$\lim_{x \to p} f(x) = A \qquad and \qquad \lim_{x \to p} g(x) = B,$$

then

  (a) $\lim_{x \to p} [f(x) + g(x)] = A + B,$

  (b) $\lim_{x \to p} f(x)g(x) = AB,$ *and*

  (c) $\lim_{x \to p} \dfrac{f(x)}{g(x)} = \dfrac{A}{B},$ *provided $B \neq 0$.*

Proof. For (a) and (b) apply Theorem 4.1.3 and Theorem 3.2.1, respectively. We leave the details to the exercises (Exercise 11).

Proof of (c). By (b) it suffices to show that

$$\lim_{x \to p} \frac{1}{g(x)} = \frac{1}{B}.$$

We first show that since $B \neq 0$, $g(x) \neq 0$ for all x sufficiently close to p, $x \neq p$. Take $\epsilon = |B|/2$. Then by the definition of limit, there exists a $\delta_1 > 0$ such that

$$|g(x) - B| < \frac{|B|}{2}$$

for all $x \in E$, $0 < |x - p| < \delta_1$. By Corollary 2.1.4 $|g(x) - B| \geq ||g(x)| - |B||$. Thus

$$|g(x)| > |B| - \frac{|B|}{2} = \frac{|B|}{2} > 0$$

for all $x \in E$, $0 < |x - p| < \delta_1$.

We can now apply Theorem 4.1.3 and the corresponding result for sequences of Theorem 3.2.1. Let $\{p_n\}$ be any sequence in E with $p_n \to p$ and $p_n \neq p$ for all n. For the above δ_1, there exists an $n_o \in \mathbb{N}$ such that $0 < |p_n - p| < \delta_1$ for all $n \geq n_o$. Thus $g(p_n) \neq 0$ for all $n \geq n_o$. Therefore by Theorem 3.2.1(c),

$$\lim_{n \to \infty} \frac{1}{g(p_n)} = \frac{1}{B}.$$

Since this holds for every sequence $p_n \to p$, by Theorem 4.1.3,

$$\lim_{x \to p} \frac{1}{g(x)} = \frac{1}{B}. \quad \square$$

The proofs of the following two theorems are easy consequences of Theorem 4.1.3 and the corresponding theorems for sequences (Theorems 3.2.3 and 3.2.4). First however we give the following definition.

DEFINITION 4.1.7 *A real-valued function f defined on a set E is* **bounded** *on E if there exists a constant M such that $|f(x)| \leq M$ for all $x \in E$.*

THEOREM 4.1.8 *Suppose E is a subset of a metric space X, p is a limit point of E, and f, g are real-valued functions on E. If g is bounded on E and $\lim_{x \to p} f(x) = 0$, then*

$$\lim_{x \to p} f(x)g(x) = 0.$$

Proof. Exercise 12. $\square$

THEOREM 4.1.9 *Suppose E is a subset of a metric space X, p is a limit point of E, and f, g, h are functions from E into $\mathbb{R}$ satisfying*

$$g(x) \leq f(x) \leq h(x) \qquad \text{for all} \quad x \in E.$$

If $\lim_{x \to p} g(x) = \lim_{x \to p} h(x) = L$, *then* $\lim_{x \to p} f(x) = L$.

Proof. Exercise 13. □

We now provide examples to illustrate the previous theorems.

EXAMPLES 4.1.10 (a) Using mathematical induction and Theorem 4.1.6(b), $\lim_{x \to c} x^n = c^n$ for all $n \in \mathbb{N}$. If $p(x) = a_n x^n + \cdots + a_1 x + a_o$ is a **polynomial function** of degree n, where n is a nonnegative integer and $a_o, a_1 \ldots, a_n \in \mathbb{R}$ with $a_n \neq 0$, then a repeated application of Theorem 4.1.6(a) gives $\lim_{x \to c} p(x) = p(c)$.

(b) Consider

$$\lim_{x \to -2} \frac{x^3 + 2x^2 - 2x - 4}{x^2 - 4}.$$

By part (a), $\lim_{x \to -2} (x^3 + 2x^2 - 2x - 4) = 0$ and $\lim_{x \to -2} (x^2 - 4) = 0$. Since the denominator has limit zero, Theorem 4.1.6(c) does not apply. In this example however, for $x \neq -2$

$$\frac{x^3 + 2x^2 - 2x - 4}{x^2 - 4} = \frac{(x + 2)(x^2 - 2)}{(x + 2)(x - 2)} = \frac{x^2 - 2}{x - 2}.$$

Since $\lim_{x \to -2} (x - 2) = -4$ which is nonzero, we can now apply Theorem 4.1.6(c) to conclude that

$$\lim_{x \to -2} \frac{x^3 + 2x^2 - 2x - 4}{x^2 - 4} = \lim_{x \to -2} \frac{x^2 - 2}{x - 2} = -\frac{1}{2}.$$

(c) Let $E = \mathbb{R} \setminus \{0\}$, and let $f : E \to \mathbb{R}$ be defined by

$$f(x) = x \sin \frac{1}{x}.$$

Since $|\sin(1/x)| \leq 1$ for all $x \in \mathbb{R}$, $x \neq 0$ and $\lim_{x \to 0} x = 0$, by Theorem 4.1.8

$$\lim_{x \to 0} x \sin \frac{1}{x} = 0.$$

The graph of $f(x) = x \sin \frac{1}{x}$ is given in Figure 4.4.

(d) Let $E = (0, \infty)$ and let f be defined on E by

$$f(t) = \frac{\sin t}{t}.$$

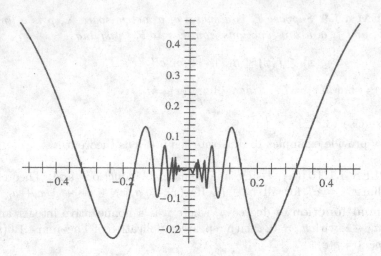

FIGURE 4.4
Graph of $f(x) = x \sin(1/x)$, $x \neq 0$

We now prove that
$$\lim_{t \to 0} \frac{\sin t}{t} = 1.$$

As we will see in the next chapter, this limit will be crucial in computing the derivative of the sine function. From Figure 4.5, we have

$$\text{area } (\triangle OPQ) < \text{ area (sector } OPR) < \text{ area } (\triangle ORS).$$

In terms of t, this gives

$$\frac{1}{2} \sin t \cos t < \frac{1}{2} t < \frac{1}{2} \tan t.$$

Therefore

$$\cos t < \frac{\sin t}{t} < \frac{1}{\cos t}.$$

Using the fact that $\lim_{t \to 0} \cos t = 1$ (Exercise 5), by Theorem 4.1.9 we obtain

$$\lim_{t \to 0} \frac{\sin t}{t} = 1. \quad \square$$

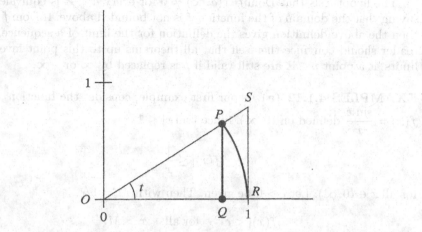

FIGURE 4.5
Triangles and sector of Example 4.1.10(d)

Limits at Infinity

Up to this point we have only considered limits at points $p \in \mathbb{R}$. We now extend the definition to include limits at ∞ or $-\infty$. The definition of the limit at ∞ is very similar to $\lim_{n \to \infty} f(n)$ where $f : \mathbb{N} \to \mathbb{R}$; that is, f is a sequence in $\mathbb{R}$.

DEFINITION 4.1.11 *Let f be a real-valued function such that $\mathrm{Dom}\, f \cap (a, \infty) \neq \emptyset$ for every $a \in \mathbb{R}$. The function f has a **limit at** ∞ if there exists a number $L \in \mathbb{R}$ such that given $\epsilon > 0$, there exists a real number M for which*

$$|f(x) - L| < \epsilon$$

for all $x \in \mathrm{Dom}\, f \cap (M, \infty)$. If this is the case, we write

$$\lim_{x \to \infty} f(x) = L.$$

Similarly, if $\mathrm{Dom}\, f \cap (-\infty, b) \neq \emptyset$ for every $b \in \mathbb{R}$.

$$\lim_{x \to -\infty} f(x) = L$$

if and only if given $\epsilon > 0$, there exists a real number M such that

$$|f(x) - L| < \epsilon$$

for all $x \in \mathrm{Dom}\, f \cap (-\infty, M)$.

The hypothesis that $\text{Dom} f \cap (a, \infty) \neq \emptyset$ for every $a \in \mathbb{R}$ is equivalent to saying that the domain of the function f is not bounded above. If $\text{Dom} f = \mathbb{N}$, then the above definition gives the definition for the limit of a sequence. The reader should convince themself that all theorems up to this point involving limits at a point $p \in \mathbb{R}$ are still valid if p is replaced by ∞ or $-\infty$.

EXAMPLES 4.1.12 (a) As our first example, consider the function $f(x) = \dfrac{\sin x}{x}$ defined on $(0, \infty)$. Since $|\sin x| \leq 1$,

$$|f(x)| \leq \frac{1}{x}$$

for all $x \in (0, \infty)$. Let $\epsilon > 0$ be given. Then with $M = 1/\epsilon$,

$$|f(x)| < \epsilon \qquad \text{for all} \quad x > M.$$

Therefore, $\lim\limits_{x \to \infty} \dfrac{\sin x}{x} = 0$.

(b) For our second example consider $f(x) = x \sin \pi x$. If we set $p_n = (n + \frac{1}{2})$, $n \in \mathbb{N}$, then

$$f(p_n) = (n + \tfrac{1}{2}) \sin(n + \tfrac{1}{2})\pi = (-1)^n (n + \tfrac{1}{2}).$$

Thus the sequence $\{f(p_n)\}_{n=1}^{\infty}$ is unbounded, and as a consequence $\lim\limits_{x \to \infty} x \sin \pi x$ does not exist. $\square$

Exercises 4.1

1. Use the definition to establish each of the following limits.
   *a. $\lim\limits_{x \to 2} (2x - 7) = -3$.
   b. $\lim\limits_{x \to -2} (3x + 5) = -1$
   *c. $\lim\limits_{x \to 1} \dfrac{x}{1 + x} = \dfrac{1}{2}$.
   d. $\lim\limits_{x \to -1} 2x^2 - 3x - 4 = 1$.
   *e. $\lim\limits_{x \to -1} \dfrac{x^3 + 1}{x + 1} = 3$.
   f. $\lim\limits_{x \to 2} \dfrac{x^3 - 2x - 4}{x^2 - 4} = \dfrac{5}{2}$

2. Use the definition to establish each of the following limits.
   a. $\lim\limits_{x \to p} c = c$
   b. $\lim\limits_{x \to p} x = p$.
   *c. $\lim\limits_{x \to p} x^3 = p^3$
   d. $\lim\limits_{x \to p} x^n = p^n$, $n \in \mathbb{N}$
   *e. $\lim\limits_{x \to p} \sqrt{x} = \sqrt{p}$, $p > 0$
   f. $\lim\limits_{x \to p} \dfrac{\sqrt{x + p} - \sqrt{p}}{x} = \dfrac{1}{2\sqrt{p}}$, $p > 0$

3. For each of the following, determine whether the indicated limit exists in $\mathbb{R}$. Justify your answer!
   *a. $\lim\limits_{x \to 0} \dfrac{x}{|x|}$
   b. $\lim\limits_{x \to 1} \dfrac{x^2 - 1}{x + 1}$
   *c. $\lim\limits_{x \to 0} \cos \dfrac{1}{x}$
   d. $\lim\limits_{x \to 0} \sqrt{|x|} \cos \dfrac{1}{x}$.
   *e. $\lim\limits_{x \to 0} \dfrac{(x + 1)^2 - 1}{x}$
   f. $\lim\limits_{x \to 1} \dfrac{x^4 - 2x^2 + 1}{x^3 - x^2 - x + 1}$

4. *Define $f : (-1, 1) \to \mathbb{R}$ by
$$f(x) = \frac{x^2 - x - 2}{x + 1}.$$
Determine the limit L of f at -1 and prove, using ϵ and δ, that f has limit L at -1.

5. *a. Using Figure 4.5, prove that $|\sin h| \leq |h|$ for all $h \in \mathbb{R}$.

   b. Using the trigonometric identity $1 - \cos h = 2\sin^2 \frac{h}{2}$, prove that
   (i) $\lim\limits_{h \to 0} \cos h = 1$.

   (ii) $\lim\limits_{h \to 0} \dfrac{1 - \cos h}{h} = 0$

6. Let E be a subset of a metric space (X, d), p a limit point of E, and $f : E \to \mathbb{R}$. Suppose there exists a constant $M > 0$ and $L \in R$ such that $|f(x) - L| \leq M d(x, p)$ for all $x \in E$. Prove that $\lim\limits_{x \to p} f(x) = L$.

7. Suppose $f : E \to \mathbb{R}$, p is a limit point of E, and $\lim\limits_{x \to p} f(x) = L$.

   *a. Prove that $\lim\limits_{x \to p} |f(x)| = |L|$.

   b. If in addition $f(x) \geq 0$ for all $x \in E$, prove that $\lim\limits_{x \to p} \sqrt{f(x)} = \sqrt{L}$.

   *c. Prove that $\lim\limits_{x \to p} (f(x))^n = L^n$ for each $n \in \mathbb{N}$.

8. Use the limit theorems, examples, and previous exercises to find each of the following limits. State which theorem, examples, or exercises are used in each case.

   *a. $\lim\limits_{x \to -1} \dfrac{5x^2 + 3x - 2}{x - 1}$

   b. $\lim\limits_{x \to -1} \dfrac{x^3 - x^2 + 2}{x + 1}$

   *c. $\lim\limits_{x \to 1} \sqrt{\dfrac{3x + 1}{2x + 5}}$

   d. $\lim\limits_{x \to -2} \dfrac{|x + 2|^{3/2}}{x + 2}$

   *e. $\lim\limits_{x \to 4} \dfrac{\sqrt{x} - 2}{x - 4}$

   f. $\lim\limits_{x \to 0} \dfrac{1}{x} \left[\dfrac{1}{\sqrt{x + p}} - frac1\sqrt{p} \right], p > 0$

   *g. $\lim\limits_{x \to 0} \dfrac{\sin 2x}{x}$

   h. $\lim\limits_{x \to 0} \dfrac{|x - 2| - |x + 2|}{x}$

9. *Suppose $f : (a, b) \to \mathbb{R}$, $p \in [a, b]$, and $\lim\limits_{x \to p} f(x) > 0$. Prove that there exists a $\delta > 0$ such that $f(x) > 0$ for all $x \in (a, b)$ with $0 < |x - p| < \delta$.

10. Suppose E is a subset of a metric space (X, d), p is a limit point of E, and $f : E \to \mathbb{R}$. Prove that if f has a limit at p, then there exists a positive constant M and a $\delta > 0$ such that $|f(x)| \leq M$ for all $x \in E$, $0 < d(x, p) < \delta$.

11. a. Prove Theorem 4.1.6(a).

    b. Prove Theorem 4.1.6(h)

12. *Prove Theorem 4.1.8.

13. Prove Theorem 4.1.9.

14. Let f, g be real-valued functions defined on $E \subset \mathbb{R}$ and let p be a limit point of E.

    *a. If $\lim\limits_{x \to p} f(x)$ and $\lim\limits_{x \to p} (f(x) + g(x))$ exist, prove that $\lim\limits_{x \to p} g(x)$ exists.

b. If $\lim_{x\to p} f(x)$ and $\lim_{x\to p}(f(x)g(x))$ exist, does it follow that $\lim_{x\to p} g(x)$ exists?

15. Let E be a subset of a metric space, p a limit point of E. Suppose f is a bounded real-valued function on E having the property that $\lim_{x\to p} f(x)$ does not exist. Prove that there exist distinct sequences $\{p_n\}$ and $\{q_n\}$ in E with $p_n \to p$ and $q_n \to p$ such that $\lim_{n\to\infty} f(p_n)$ and $\lim_{n\to\infty} f(q_n)$ exist, but are not equal.

16. *Let f be a real-valued function defined on (a,∞) for some $a > 0$. Define g on $(0,\frac{1}{a})$ by $g(t) = f(\frac{1}{t})$. Prove that
$$\lim_{x\to\infty} f(x) = L \quad \text{if and only if} \quad \lim_{t\to 0} g(t) = L.$$

17. Investigate the limits at ∞ of each of the following functions.
    *a. $f(x) = \dfrac{3x^2 + 3x - 1}{2x^2 + 1}$
         b. $f(x) = \dfrac{1}{1 + x^2}$
    *c. $f(x) = \dfrac{\sqrt{4x^2 + 1}}{x}$.
         d. $f(x) = \dfrac{2x + 3}{\sqrt{x+}}$
    *e. $f(x) = \sqrt{x^2 + x} - x$.
         f. $f(x) = \dfrac{\sqrt{x} - 2x}{2\sqrt{x} + 3x}$
    *g. $f(x) = x\cos\frac{1}{x}$
         h. $f(x) = x\sin\frac{1}{x}$.

18. Let $f : (a,\infty) \to \mathbb{R}$ be such that $\lim_{x\to\infty} xf(x) = L$ where $L \in \mathbb{R}$. Prove that $\lim_{x\to\infty} f(x) = 0$.

19. Let $f : \mathbb{R} \to \mathbb{R}$ satisfy $f(x + y) = f(x) + f(y)$ for all $x, y \in \mathbb{R}$. If $\lim_{x\to 0} f(x)$ exists, prove that
    a. $\lim_{x\to 0} f(x) = 0$, and   **b.** $\lim_{x\to p} f(x)$ exists for every $p \in \mathbb{R}$.

4.2 Continuous Functions

The notion of continuity dates back to Leonhard Euler (1707–1783). To Euler, a continuous curve (function) was one that could be expressed by a single formula or equation of the variable x. If the definition of the curve was made up of several parts, it was called discontinuous. This definition was sufficient to convey the concept of continuity if we keep in mind that in Euler's time mathematicians were primarily only concerned with elementary functions; namely functions built up from the trigonometric and exponential functions, and inverses of these functions, using algebraic operations and composition.

The more modern version of continuity is due to Bernhard Bolzano (1817) and Augustin-Louis Cauchy (1821). Both men were motivated to provide a clear and precise definition of continuity in order to be able to prove the intermediate value theorem (Theorem 4.2.11). Cauchy's definition of continuity was as follows: *"The function $f(x)$ will be, between two assigned values* of the variable x, a continuous function of this variable if for each value of x between these limits, the numerical [i.e. absolute value] of the difference $f(x + \alpha) - f(x)$ decreases indefinitely with α"[1]. Even this definition appears strange in comparison with the more modern definition in use today. Both Bolzano and Cauchy were concerned with continuity on an interval, rather than continuity at a point.

DEFINITION 4.2.1 *Let E be a subset of a metric space (X, d) and f a real-valued function with domain E. The function f is* **continuous at a point** *$p \in E$, if for every $\epsilon > 0$, there exists a $\delta > 0$ such that*

$$|f(x) - f(p)| < \epsilon$$

for all $x \in E$ with $d(x, p) < \delta$. The function f is **continuous on E** *if and only if f is continuous at every point $p \in E$.*

The above definition can be rephrased as follows: A function $f : E \to \mathbb{R}$ is continuous at $p \in E$ if and only if given $\epsilon > 0$, there exists a $\delta > 0$ such that

$$f(x) \in N_\epsilon(f(p)) \quad \text{for all} \quad x \in N_\delta(p) \cap E.$$

This is illustrated on the real line in Figure 4.6.

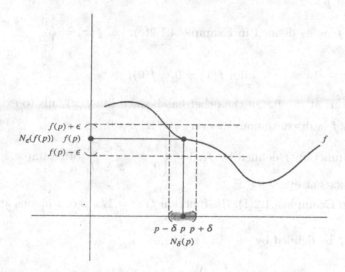

FIGURE 4.6
An illustration of Definition 4.2.1

Remarks. (a) If $p \in E$ is a limit point of E, then f is continuous at p if and only if

$$\lim_{x \to p} f(x) = f(p).$$

---

[1] Cauchy, Cours d'Analyse, p.43

Also, as a consequence of Theorem 4.1.3, f is continuous at p if and only if

$$\lim_{n \to \infty} f(p_n) = f(p)$$

for every sequence $\{p_n\}$ in E with $p_n \to p$.

(b) If $p \in E$ is an isolated point, then **every** function f on E is continuous at p. This follows immediately from the fact that for an isolated point p of E, there exists a $\delta > 0$ such that $N_\delta(p) \cap E = \{p\}$.

We now consider several of the functions given in previous examples, and also some additional new examples.

EXAMPLES 4.2.2 (a) Let g be defined as in Example 4.1.2(b), i.e.,

$$g(x) = \begin{cases} \dfrac{x^2 - 4}{x - 2}, & x \neq 2, \\ 2, & x = 2. \end{cases}$$

At the point $p = 2$, $\lim_{x \to 2} g(x) = 4 \neq g(2)$. Thus g is not continuous at $p = 2$. However, if we redefine g at $p = 2$ so that $g(2) = 4$, then this function is now continuous at $p = 2$.

(b) Let f be as defined in Example 4.1.2(e), i.e. $f(x) = \begin{cases} 0, & x \in \mathbb{Q}, \\ x, & x \notin \mathbb{Q} \end{cases}$.

Since

$$\lim_{x \to 0} f(x) = 0 = f(0),$$

f is continuous at $p = 0$. On the other hand, since $\lim_{x \to p} f(x)$ fails to exists for every $p \neq 0$, f is discontinuous at every $p \in \mathbb{R}$, $p \neq 0$.

(c) The function f defined by $f(x) = \begin{cases} 1, & x \in \mathbb{Q}, \\ 0, & x \notin \mathbb{Q}, \end{cases}$ of Example 4.1.2(d) is discontinuous at every $p \in \mathbb{R}$.

(d) As in Example 4.1.2(f), the function $f(x) = 1/x$ is continuous at every $p \in (0,1)$.

(e) Let f be defined by

$$f(x) = \begin{cases} 0, & x = 0, \\ x \sin \frac{1}{x}, & x \neq 0. \end{cases}$$

By Example 4.1.10(c),

$$\lim_{x \to 0} f(x) = 0 = f(0).$$

Thus f is continuous at $x = 0$.

(f) In this example, we show that $f(x) = \sin x$ is continuous on $\mathbb{R}$. Let $x, y \in \mathbb{R}$. Then

$$|f(y) - f(x)| = |\sin y - \sin x|$$
$$= 2|\cos \tfrac{1}{2}(y+x)\sin \tfrac{1}{2}(y-x)|$$
$$\leq 2|\sin \tfrac{1}{2}(y-x)|.$$

By Exercise 5 of Section 4.1 $|\sin h| \leq |h|$. Therefore

$$|f(y) - f(x)| \leq |y - x|,$$

from which it follows that f is continuous on $\mathbb{R}$.

(g) We now consider a function on $(0,1)$ which is discontinuous at every rational number in $(0,1)$ and continuous at every irrational number in $(0,1)$. For $x \in (0,1)$ define

$$f(x) = \begin{cases} 0, & \text{if } x \text{ is irrational,} \\ \dfrac{1}{n}, & \text{if } x \text{ is rational with } x = \dfrac{m}{n} \text{ in lowest terms.} \end{cases}$$

The graph of f, at least for a few rational numbers, is given in Figure 4.7.

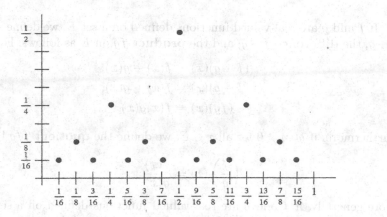

FIGURE 4.7
Graph of the function of Example 4.2.2(g)

To establish our claim we will show that

$$\lim_{x \to p} f(x) = 0$$

for every $p \in (0,1)$. As a consequence, since $f(p) = 0$ for every irrational number $p \in (0,1)$, f is continuous at every irrational number. Also, since

$f(p) \neq 0$ when $p \in \mathbb{Q} \cap (0,1)$, f is discontinuous at every rational number in $(0,1)$.

Fix $p \in (0,1)$ and let $\epsilon > 0$ be given. To prove that $\lim_{x \to p} f(x) = 0$ we need to show that there exists a $\delta > 0$ such that

$$|f(x)| < \epsilon$$

for all $x \in N_\delta(p) \cap (0,1)$, $x \neq p$. This is certainly the case for any irrational number x. On the other hand, if x is rational with $x = \frac{m}{n}$ (in lowest terms), then $f(x) = \frac{1}{n}$. Choose $n_o \in \mathbb{N}$ such that $\frac{1}{n_o} < \epsilon$. There exist only a finite number of rational numbers $\frac{m}{n}$ (in lowest terms) in $(0,1)$ with denominator less than n_o. Denote these by $r_1, ..., r_m$, and let

$$\delta = \min\{|r_i - p| : i = 1, ..., m, \ r_i \neq p\}.$$

Note, since p may be a rational number and thus possibly equal to r_i for some $i = 1, ..., m$, we take the minimum of $\{|r_i - p|\}$ only for those i for which $r_i \neq p$. Thus $\delta > 0$ and if $r \in \mathbb{Q} \cap N_\delta(p) \cap (0,1)$, $r \neq p$, with $r = \frac{m}{n}$ in lowest terms, then $n \geq n_o$. Therefore,

$$|f(r)| = \frac{1}{n} < \epsilon.$$

Thus $|f(x)| < \epsilon$ for all $x \in N_\delta(p) \cap (0,1)$, $x \neq p$. $\quad \square$

If f and g are real-valued functions defined on a set E, we define the **sum** $f + g$, the **difference** $f - g$, and the **product** fg on E as follows: For $x \in E$,

$$(f + g)(x) = f(x) + g(x),$$
$$(f - g)(x) = f(x) - g(x),$$
$$(fg)(x) = f(x)g(x).$$

Furthermore, if $g(x) \neq 0$ for all $x \in E$, we define the **quotient** f/g by

$$\left(\frac{f}{g}\right)(x) = \frac{f(x)}{g(x)}.$$

More generally, if f and g are real-valued functions defined on a set E, the quotient f/g can always be defined on $E_1 = \{x \in E : g(x) \neq 0\}$.

As an application of Theorem 4.1.6 we prove that continuity is preserved under the algebraic operations defined above,

THEOREM 4.2.3 *If E is a subset of a metric space X and $f, g : E \to \mathbb{R}$ are continuous at $p \in E$, then*

   (a) *$f + g$ and $f - g$ are continuous at p, and*

   (b) *fg is continuous at p.*

   (c) *If $g(x) \neq 0$ for all $x \in E$, then $\dfrac{f}{g}$ is continuous at p.*

Proof. If p is an isolated point of E, then the result is true since every function on E is continuous at p. If p is a limit point of E, then the conclusions follow from Theorem 4.1.6. □

Composition of Continuous Functions

In the following theorem we prove that continuity is also preserved under composition of functions.

THEOREM 4.2.4 *Let $A, B \subset \mathbb{R}$ and let $f : A \to \mathbb{R}$ and $g : B \to \mathbb{R}$ be functions such that $\text{Range } f \subset B$. If f is continuous at $p \in A$ and g is continuous at $f(p)$, then $h = g \circ f$ is continuous at p.*

Proof. Let $\epsilon > 0$ be given. Since g is continuous at $f(p)$, there exists a $\delta_1 > 0$ such that

$$|g(y) - g(f(p))| < \epsilon \qquad \text{for all} \quad y \in B \cap N_{\delta_1}(f(p)). \qquad (2)$$

Since f is continuous at p, for this δ_1, there exists a $\delta > 0$ such that

$$|f(x) - f(p)| < \delta_1 \qquad \text{for all} \quad x \in A \cap N_\delta(p).$$

Thus, if $x \in A$ with $|x - p| < \delta$, by (2)

$$|h(x) - h(p)| = |g(f(x)) - g(f(p))| < \epsilon.$$

Therefore h is continuous at p. □

EXAMPLES 4.2.5 (a) If p is a **polynomial function** of degree n, that is

$$p(x) = a_n x^n + a_{n-1} x^{n-1} + \cdots + a_1 x + a_0,$$

where n is a nonnegative integer, $a_0, ..., a_n \in \mathbb{R}$ with $a_n \neq 0$, and $\text{Dom } p = \mathbb{R}$, then by Theorem 4.1.6 (a) and (b), p is continuous on $\mathbb{R}$.

(b) Suppose p and q are polynomials on $\mathbb{R}$ and $E = \{x \in \mathbb{R} : q(x) = 0\}$. Then by Theorem 4.1.6 the **rational function** r defined on $\mathbb{R} \setminus E$ by

$$r(x) = \frac{p(x)}{q(x)}, \qquad x \in \mathbb{R} \setminus E,$$

is continuous on $\mathbb{R} \setminus E$.

(c) By Example 4.2.2(f), $f(x) = \sin x$ is continuous on $\mathbb{R}$. Hence if p is a polynomial function on $\mathbb{R}$, by Theorem 4.2.4 $(f \circ p)(x) = \sin(p(x))$ is also continuous on $\mathbb{R}$. □

Topological Characterization of Continuity

Before considering several consequences of continuity, we provide a strictly topological characterization of a continuous function. In more abstract courses this is often taken as the definition of continuity.

THEOREM 4.2.6 *Let E be a subset of a metric space X and let f be a real-valued function on E. Then f is continuous on E if and only if $f^{-1}(V)$ is open in E for every open subset V of $\mathbb{R}$.*

Proof. Recall (Definition 2.2.21) that a set $U \subset E$ is open in E if for every $p \in U$ there exists a $\delta > 0$ such that $N_\delta(p) \cap E \subset U$.

Suppose f is continuous on E and V is an open subset of $\mathbb{R}$. If $f^{-1}(V) = \emptyset$ we are done. Suppose $p \in f^{-1}(V)$. Then $f(p) \in V$. Since V is open, there exists $\epsilon > 0$ such that $N_\epsilon(f(p)) \subset V$. Since f is continuous at p, there exists a $\delta > 0$ such that $f(x) \in N_\epsilon(f(p))$ for all $x \in N_\delta(p) \cap E$; i.e., $N_\delta(p) \cap E \subset f^{-1}(V)$. Since $p \in f^{-1}(V)$ was arbitrary, $f^{-1}(V)$ is open in E.

Conversely, suppose $f^{-1}(V)$ is open in E for every open subset V of $\mathbb{R}$. Let $p \in E$ and let $\epsilon > 0$ be given. Then by hypothesis $f^{-1}(N_\epsilon(f(p)))$ is open in E. Thus there exists a $\delta > 0$ such that

$$E \cap N_\delta(p) \subset f^{-1}(N_\epsilon(f(p))),$$

that is, $f(x) \in N_\epsilon(f(p))$ for all $x \in N_\delta(p) \cap E$. Therefore f is continuous at p. $\square$

EXAMPLES 4.2.7 (a) We illustrate the previous theorem for the function $f(x) = \sqrt{x}$, $\mathrm{Dom}\, f = [0, \infty)$. Suppose first that V is an open interval (a, b) with $a < b$. Then

$$f^{-1}(V) = \begin{cases} \emptyset, & b \le 0, \\ [0, b^2), & a \le 0 < b, \\ (a^2, b^2), & 0 < a. \end{cases}$$

Clearly $\emptyset$ and (a^2, b^2) are open subsets of $\mathbb{R}$ and hence also of $[0, \infty)$. Although $[0, b^2)$ is not open in $\mathbb{R}$,

$$[0, b^2) = (-b^2, b^2) \cap [0, \infty).$$

Thus by Theorem 2.2.23 $[0, b^2)$ is open in $[0, \infty)$. If V is an arbitrary open subset of $\mathbb{R}$, then by Theorem 2.2.20 $V = \bigcup_n I_n$, where $\{I_n\}$ is a finite or countable collection of open intervals. Since $f^{-1}(V) = \bigcup_n f^{-1}(I_n)$ (Theorem 1.7.14) and each $f^{-1}(I_n)$ is open in $[0, \infty)$, $f^{-1}(V)$ is open in $[0, \infty)$. Therefore $f(x) = \sqrt{x}$ is continuous on $[0, \infty)$.

(b) In this example, we show that if $f : E \to \mathbb{R}$ is continuous on E and

$V \subset E$ is open in E, then $f(V)$ is not necessarily open in Range f. Consider the function $f : \mathbb{R} \to \mathbb{R}$ given by

$$f(x) = \begin{cases} x^2, & x \leq 2, \\ 6 - x, & x > 2. \end{cases}$$

Then f is continuous on $\mathbb{R}$ and Range $f = \mathbb{R}$ (Exercise 10). However, $f((-1, 1)) = [0, 1)$, and this set is not open in $\mathbb{R}$. $\square$

Continuity and Compactness

We now consider several consequences of continuity. In our first result we prove that the continuous image of a compact set is compact. In the proof of the theorem we only use continuity and the definition of a compact set. For subsets of $\mathbb{R}$, an alternate proof using the Heine-Borel-Bolzano-Weierstrass theorem (Theorem 2.4.2) is suggested in the exercises (Exercise 25).

THEOREM 4.2.8 *If K is a compact subset of a metric space X and if $f : K \to \mathbb{R}$ is continuous on K, then $f(K)$ is compact.*

Proof. Let $\{V_\alpha\}_{\alpha \in A}$ be an open cover of $f(K)$. Since f is continuous on K, $f^{-1}(V_\alpha)$ is open in K for every $\alpha \in A$. By Theorem 2.2.23, for each α there exists an open subset U_α of X such that

$$f^{-1}(V_\alpha) = K \cap U_\alpha.$$

We claim that $\{U_\alpha\}_{\alpha \in A}$ is an open cover of K. If $p \in K$, then $f(p) \in f(K)$ and thus $f(p) \in V_\alpha$ for some $\alpha \in A$. But then p is in $f^{-1}(V_\alpha)$ and hence also in U_α. Since each U_α is also open, the collection $\{U_\alpha\}_{\alpha \in A}$ is an open cover of K. Since K is compact, there exists $\alpha_1, ..., \alpha_n \in A$ such that

$$K \subset \bigcup_{j=1}^{n} U_{\alpha_j}.$$

Therefore,

$$K = \bigcup_{j=1}^{n} (U_{\alpha_j} \cap K) = \bigcup_{j=1}^{n} f^{-1}(V_{\alpha_j}),$$

and by Theorem 1.7.14(a),

$$f(K) = \bigcup_{j=1}^{n} f(f^{-1}(V_{\alpha_j})).$$

Since $f(f^{-1}(V_{\alpha_j})) \subset V_{\alpha_j}$, $f(K) \subset \bigcup_{j=1}^{n} V_{\alpha_j}$. Thus $f(K)$ is compact. $\square$

As a corollary of the previous theorem we obtain the following generalization of the usual maximum-minimum theorem normally encountered in calculus.

COROLLARY 4.2.9 *Let K be a compact subset of $\mathbb{R}$ and let $f : K \to \mathbb{R}$ be continuous. Then there exist $p, q \in K$ such that*

$$f(q) \le f(x) \le f(p) \qquad \text{for all} \quad x \in K.$$

Proof. Let $M = \sup\{f(x) : x \in K\}$. By the previous theorem $f(K)$ is compact. Thus, since $f(K)$ is bounded, $M < \infty$. Also, since $f(K)$ is closed, $M \in f(K)$. Thus there exists $p \in K$ such that $f(p) = M$. Similarly for $m = \inf\{f(x) : x \in K\}$. $\square$

We now provide examples to show that the result is false if $K \subset \mathbb{R}$ is not compact; that is, not both closed and bounded.

EXAMPLES 4.2.10 (a) Suppose E is a subset of $\mathbb{R}$. If E is unbounded, consider

$$f(x) = \frac{x^2}{1 + x^2}.$$

Then f is continuous on E, $\sup\{f(x) : x \in E\} = 1$, but $f(x) < 1$ for all $x \in E$. To see that the supremum is 1, we note that since E is unbounded, there exists a sequence $\{x_n\}$ in E such that $x_n^2 \to \infty$ as $n \to \infty$. But then

$$\lim_{n \to \infty} f(x_n) = \lim_{n \to \infty} \frac{1}{1 + \frac{1}{x_n^2}} = 1.$$

Thus if $0 < \beta < 1$, there exists an integer n such that $f(x_n) > \beta$. Hence $\sup\{f(x) : x \in E\} = 1$.

(b) If E is not closed, let x_o be a limit point of E which is not in E. Then

$$g(x) = \frac{1}{1 + (x - x_o)^2}$$

is continuous on E, with $g(x) < 1$ for all $x \in E$. A similar argument as above shows that $\sup\{g(x) : x \in E\} = 1$. $\square$

Intermediate Value Theorem

The following theorem is attributed to both Bolzano and Cauchy. Cauchy however implicitly assumed the completeness of $\mathbb{R}$ in his proof, whereas the proof by Bolzano (given below) uses the least upper bound property. An alternate proof is outlined in the miscellaneous exercises.

THEOREM 4.2.11 (Intermediate Value Theorem) *Let $f : [a,b] \to \mathbb{R}$ be continuous. Suppose $f(a) < f(b)$. If γ is a number satisfying*

$$f(a) < \gamma < f(b),$$

then there exists $c \in (a,b)$ such that $f(c) = \gamma$.

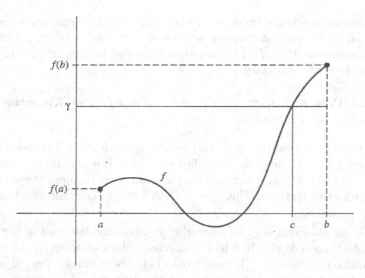

FIGURE 4.8
Intermediate value theorem

The statement and conclusion of the intermediate value theorem is illustrated in Figure 4.8. The theorem simply states that if f is continuous on $[a, b]$ with $f(a) < f(b)$, and $\gamma \in \mathbb{R}$ satisfies that $f(a) < \gamma < f(b)$, then the graph of f crosses the line $y = \gamma$ at least once.

Proof. Let $A = \{x \in [a, b] : f(x) \le \gamma\}$. The set $A \ne \emptyset$ since $a \in A$, and A is bounded above by b. Thus by the least upper bound property A has a supremum in $\mathbb{R}$. Let $c = \sup A$. Since b is an upper bound, $c \le b$.

We now show that $f(c) = \gamma$. Suppose $f(c) < \gamma$. Then $\epsilon = \frac{1}{2}(\gamma - f(c)) > 0$. Since f is continuous at c, for this ϵ there exists a $\delta > 0$ such that

$$f(c) - \epsilon < f(x) < f(c) + \epsilon \qquad \text{for all} \qquad x \in N_\delta(c) \cap [a, b].$$

Since $f(c) < \gamma$, $c \ne b$, and thus $(c, b] \cap N_\delta(c) \ne \emptyset$. But for any $x \in (c, b]$ with $c < x < c + \delta$,

$$f(x) < f(c) + \epsilon = f(c) + \frac{1}{2}\gamma - \frac{1}{2}f(c) = \frac{1}{2}(f(c) + \gamma) < \gamma.$$

But then $x \in A$ and $x > c$ which contradicts $c = \sup A$. Therefore $f(c) \ge \gamma$.

Since $c = \sup A$, either $c \in A$ or c is a limit point of A. If $c \in A$, then $f(c) \le \gamma$. If c is a limit point of A, then by Theorem 3.1.4 there exists a sequence $\{x_n\}$ in A such that $x_n \to c$. Since $x_n \in A$, $f(x_n) \le \gamma$. Since f is continuous,

$$f(c) = \lim_{n \to \infty} f(x_n) \le \gamma.$$

Thus in either case, $f(c) \le \gamma$. Therefore $f(c) = \gamma$. $\square$

The intermediate value theorem is one of the fundamental theorems of calculus. Simply stated, the theorem implies that if I is an interval and $f : I \to \mathbb{R}$ is continuous, then $f(I)$ is an interval. Due to the importance of this result we state it as a corollary.

COROLLARY 4.2.12 *If $I \subset \mathbb{R}$ is an interval and $f : I \to \mathbb{R}$ is continuous on I, then $f(I)$ is an interval.*

Proof. Let $s, t \in f(I)$ with $s < t$, and let $a, b \in I$ with $a \neq b$ be such that $f(a) = s$ and $f(b) = t$. Suppose γ satisfies $s < \gamma < t$. If $a < b$, then since f is continuous on $[a, b]$, by the intermediate value theorem theorem there exists $c \in (a, b)$ such that $f(c) = \gamma$. Thus $\gamma \in f(I)$. A similar argument also holds if $a > b$. $\square$

There is an alternate way to state the previous corollary using the terminology of connected sets. If I is a connected subset of $\mathbb{R}$ and $f : I \to \mathbb{R}$ is continuous on I, then $f(I)$ is connected. This result can be proved using only properties of continuous functions and the definition of a connected set (Exercise 28). The corollary now follows from the fact that a subset of $\mathbb{R}$ is connected if and only if it is an interval (Theorem 2.2.25). The proof of Theorem 2.2.25 however also requires the least upper bound property of $\mathbb{R}$. Consequently, the supremum property of the real numbers cannot be avoided in proving Corollary 4.2.12.

The following two corollaries are additional applications of the intermediate value theorem. Our first result is the proof of Theorem 1.5.3.

COROLLARY 4.2.13 *For every real number $\gamma > 0$ and every positive integer n, there exists a unique positive real number y so that $y^n = \gamma$.*

Proof. That y is unique is clear. Let $f(x) = x^n$, which by Exercise 7 is continuous on $\mathbb{R}$. Let $a = 0$ and $b = \gamma + 1$. Since $(\gamma + 1)^n > \gamma$, f satisfies the hypothesis of Theorem 4.2.11. Thus there exists y, $0 < y < \gamma + 1$, such that

$$f(y) = y^n = \gamma. \quad \square$$

COROLLARY 4.2.14 *If $f : [0, 1] \to [0, 1]$ is continuous, then there exists $y \in [0, 1]$ such that $f(y) = y$.*

Proof. Let $g(x) = f(x) - x$. Then $g(0) = f(0) \geq 0$ and $g(1) = f(1) - 1 \leq 0$. Thus there exists $y \in [0, 1]$ such that $g(y) = 0$; i.e., $f(y) = y$. $\square$

EXAMPLES 4.2.15 (a) In the proof of Theorem 4.2.11 continuity of the function f was required. The following example shows that the converse of Theorem 4.2.11 is false; that is, if a function f satisfies the intermediate value property on an interval $[a, b]$, this does not imply that f is continuous on $[a, b]$. Let f be defined on $[0, \frac{2}{\pi}]$ as follows:

$$f(x) = \begin{cases} \sin \dfrac{1}{x}, & 0 < x \le \frac{2}{\pi}, \\ -1, & x = 0. \end{cases}$$

Then $f(0) = -1$, $f(\frac{2}{\pi}) = 1$, and for every γ, $-1 < \gamma < 1$, there exists an $x \in (0, \frac{2}{\pi})$ such that $f(x) = \gamma$. However, the function f is not continuous at $x = 0$ (see Figure 4.3).

(b) In this example, we show that the conclusion of the intermediate value theorem is false if the interval $[a, b]$ of real numbers is replaced by an interval of rational numbers. Let $E = \{x \in \mathbb{Q} : 0 \le x \le 2\}$, and let $f(x) = x^2$. Then f is continuous on E with $f(0) < 2 < f(2)$. However, there does not exist $r \in E$ such that $f(r) = 2$. □

Exercises 4.2

1. For each of the following, determine whether the given function is continuous at the indicated point x_o.

    a. $f(x) = \begin{cases} \dfrac{2x^2 - 5x - 3}{x - 3}, & x \ne 3, \\ 6, & x = 3, \end{cases}$ at $x_o = 3$

    b. $h(x) = \begin{cases} \dfrac{\sqrt{x} - 2}{x - 4} & x \ne 4 \\ 4 & x = 4 \end{cases}$ at $x_o = 4$

    *c. $g(x) = \begin{cases} \dfrac{1 - \cos x}{x}, & x \ne 0, \\ 0, & x = 0, \end{cases}$ at $x_o = 0$

    *d $k(x) = \begin{cases} x^2, & x \le 2, \\ 4 - x, & x > 2, \end{cases}$ at $x_o = 2$

2. Let $f : \mathbb{R} \to \mathbb{R}$ be defined by

    $$f(x) = \begin{cases} 8x, & \text{when } x \text{ is rational,} \\ 2x^2 + 8, & \text{when } x \text{ is irrational.} \end{cases}$$

    a. Prove, using ϵ and δ, that f is continuous at 2.

    *b. Is f continuous at 1? Justify your answer.

3. Let $f : \mathbb{R} \to \mathbb{R}$ be defined by

    $$f(x) = \begin{cases} x^2, & x \in \mathbb{Q}, \\ x + 2, & x \notin \mathbb{Q}. \end{cases}$$

    Find all points (if any) where f is continuous.

4. *Prove (without using Example 4.2.7) that $f(x) = \sqrt{x}$ is continuous on $[0, \infty)$.

5. Define $f : (0, 1] \to \mathbb{R}$ by $f(x) = \dfrac{1}{\sqrt{x}} - \sqrt{\dfrac{x+1}{x}}$.

    a. Justify that f is continuous on $(0, 1]$.

    ***b.** Can one define $f(0)$ so that f is continuous on $[0, 1]$?

6. Let $E \subset \mathbb{R}$, and suppose $f : E \to \mathbb{R}$ is continuous at $p \in E$.

    ***a.** Prove that $|f|$ is continuous at p. Is the converse true?

    b. Set $g(x) = \sqrt{|fx)|}$. Prove that g is continuous at p.

7. *Let E be a subset of a metric space X, and suppose $f : E \to \mathbb{R}$ is continuous on E. Prove that f^n defined by $f^n(x) = (f(x))^n$ is continuous on E for each $n \in \mathbb{N}$.

8. ***a.** Prove that $f(x) = \cos x$ is continuous on $\mathbb{R}$.

    b. If $E \subset \mathbb{R}$ and $f : E \to \mathbb{R}$ is continuous on E, prove that $g(x) = \cos(f(x))$ is continuous on E.

9. For each of the following equations, determine the largest subset E of $\mathbb{R}$ such that the given equation defines a continuous function on E. In each case state which theorems or examples are used to show that the function is continuous on E.

    ***a.** $f(x) = \dfrac{x^3 + 4x - 5}{x(x^2 - 4)}$.   **b.** $g(x) = \sin \dfrac{1}{x}$.

    ***c.** $h(x) = \dfrac{1}{\sqrt{x^2 + 1}}$.   **d.** $k(x) = \dfrac{\cos x}{\sin x}$.

10. Prove that $f(x) = \begin{cases} x^2, & x \le 2, \\ 6 - x, & x > 2, \end{cases}$ is continuous on $\mathbb{R}$ and that Range $f = \mathbb{R}$.

11. As in Example 4.2.7, use Theorem 4.2.6 to prove that each of the following functions is continuous on the given domain.

    a. $f(x) = \dfrac{1}{x}$,   Dom $f = (0, \infty)$.   **b.** $g(x) = x^2$,   Dom $g = \mathbb{R}$.

12. ***a.** Let $f : \mathbb{R}^2 \to \mathbb{R}^2$ be defined by $f(x, y) = (x + y, x - y)$. Show that f is continuous on $\mathbb{R}^2$.

    b. Let E be a subset of $\mathbb{R}$, and suppose f, g are continuous real-valued functions on E. Prove that $f(x, y) = (f(x), f(y))$ is continuous on $E \times E$.

13. Prove that the function f defined on $D = \mathbb{R}^2 \setminus (0, 0)$ by $f(x, y) = \dfrac{xy}{x^2 + y^2}$ is continuous and bounded on D.

14. Suppose E is a subset of $\mathbb{R}$ and $f, g : E \to \mathbb{R}$ are continuous at $p \in E$. Prove that each of the functions defined below is continuous at p.

    ***a.** $\max\{f, g\}(x) = \max\{f(x), g(x)\}$, $x \in E$.

    b. $\min\{f, g\}(x) = \min\{f(x), g(x)\}$, $x \in E$.

    c. $f^+(x) = \max\{f(x), 0\}$.

15. Prove that there exists $x \in (0, \frac{\pi}{2})$ such that $\cos x = x$.

16. *Use the intermediate value theorem to prove that every polynomial of odd degree has at least one real root.

17. *Suppose $f : [-1, 1] \to \mathbb{R}$ is continuous and satisfies $f(-1) = f(1)$. Prove that there exists $\gamma \in [0, 1]$ such that $f(\gamma) = f(\gamma - 1)$.

18. Suppose $f : [0, 1] \to \mathbb{R}$ is continuous and satisfies $f(0) = f(1)$. Prove that there exists $\gamma \in [0, \frac{1}{2}]$ such that $f(\gamma) = f(\gamma + \frac{1}{2})$.

19. *Let $E \subset \mathbb{R}$ and let $f : E \to \mathbb{R}$ be continuous. Let $F = \{x \in E : f(x) = 0\}$. Prove that F is closed in E. Is F necessarily closed in $\mathbb{R}$?

20. Suppose $f : (0, 1) \to \mathbb{R}$ is continuous and satisfies $f(r) = 0$ for each rational number $r \in (0, 1)$. Prove that $f(x) = 0$ for all $x \in (0, 1)$.

21. Let $E \subset \mathbb{R}$ and let f be a real-valued function on E that is continuous at $p \in E$. If $f(p) > 0$, prove that there exists an $\alpha > 0$ and a $\delta > 0$ such that $f(x) \geq \alpha$ for all $x \in N_\delta(p) \cap E$.

22. *Let $f : E \to \mathbb{R}$ be continuous at $p \in E$. Prove that there exists a positive constant M and $\delta > 0$ such that $|f(x)| \leq M$ for all $x \in E \cap N_\delta(p)$.

23. Let $f : (0, 1) \to \mathbb{R}$ be defined by

$$f(x) = \begin{cases} 0, & \text{if } x \text{ is irrational,} \\ n, & \text{if } x \text{ is rational with } x = \frac{m}{n} \text{ in lowest terms.} \end{cases}$$

a. Prove that f is unbounded on every open interval $I \subset (0, 1)$.

b. Use (a) and the previous exercise to conclude that f is discontinuous at every point of $(0, 1)$.

24. Suppose E is a subset of $\mathbb{R}$ and $f, g : E \to \mathbb{R}$ are continuous on E. Show that $\{x \in E : f(x) > g(x)\}$ is open in E.

25. *Let K be a compact subset of a metric space X and let $f : K \to \mathbb{R}$ be continuous on K. Prove that $f(K)$ is compact by showing that $f(K)$ is closed and bounded.

26. Let $E \subset \mathbb{R}$ and let f be a real-valued function on E. Prove that f is continuous on E if and only if $f^{-1}(F)$ is closed in E for every closed subset F of $\mathbb{R}$.

27. Let $A, B \subset \mathbb{R}$ and let $f : A \to \mathbb{R}$ and $g : B \to \mathbb{R}$ be functions such that Range $f \subset B$.

*a. If $V \subset \mathbb{R}$, prove that $(g \circ f)^{-1}(V) = f^{-1}(g^{-1}(V))$.

b. If f and g are continuous on A and B respectively, use Theorem 4.2.6 to prove that $g \circ f$ is continuous on A.

28. Suppose I is a connected subset of $\mathbb{R}$ and $f : I \to \mathbb{R}$ is continuous on I. Prove, using only the properties of continuity and the definition of connected set, that $f(I)$ is connected.

29. *Let K be a compact subset of a metric space X, and let f be a real-valued function on K. Suppose that for each $x \in K$ there exists $\epsilon_x > 0$ such that f is bounded on $N_{\epsilon_x}(x) \cap K$. Prove that f is bounded on K.

30. Let $A \subset \mathbb{R}$. For $p \in \mathbb{R}$, the **distance from p to the set** A, denoted $d(p, A)$ is defined by $d(p, A) = \inf\{|p - x| : x \in A\}$.

    a. Prove that $d(p, A) = 0$ if and only if $p \in \overline{A}$.

    b. For $x, y \in \mathbb{R}$, prove that $|d(x, A) - d(y, A)| \leq |x - y|$.

    c. Prove that the function $x \to d(x, A)$ is continuous on $\mathbb{R}$.

    d. If A, B are disjoint closed subsets of $\mathbb{R}$, prove that

    $$f(x) = \frac{d(x, A)}{d(x, A) + d(x, B)}$$

    is a continuous function on $\mathbb{R}$ satisfying $0 \leq f(x) \leq 1$ for all $x \in \mathbb{R}$, and

    $$f(x) = \begin{cases} 0, & x \in A, \\ 1, & x \in B. \end{cases}$$

31. Let f be a continuous real-valued function on $\mathbb{R}$ satisfying $f(0) = 1$ and $f(x + y) = f(x)f(y)$ for all $x, y \in \mathbb{R}$. Prove that $f(x) = a^x$ for some $a \in \mathbb{R}$, $a > 0$.

4.3   Uniform Continuity

In the previous section we discussed continuity of a function at a point and on a set. By Definition 4.2.1, a function $f : E \to \mathbb{R}$ is continuous on E if for each $p \in E$, given any $\epsilon > 0$, there exists a $\delta > 0$ such that $|f(x) - f(p)| < \epsilon$ for all $x \in E \cap N_\delta(p)$. In general, for a given $\epsilon > 0$, the choice of δ that works depends not only on ϵ and the function f, but also on the point p. This was illustrated in Example 4.1.2(f) for the function $f(x) = 1/x$, $x \in (0, 1)$. Functions for which a choice of δ independent of p is possible are given a special name.

DEFINITION 4.3.1 *Let E be a subset of a metric space (X, d) and $f : E \to \mathbb{R}$. The function f is* **uniformly continuous** *on E if given $\epsilon > 0$, there exists a $\delta > 0$ such that*

$$|f(x) - f(y)| < \epsilon \qquad for\ all \qquad x, y \in E \quad with \quad d(x, y) < \delta.$$

The key point in the definition of uniform continuity is that the choice of δ must depend only on ϵ, the function f, and the set E; it has to be independent of any $x \in E$. To illustrate this, we consider the following examples.

EXAMPLES 4.3.2 (a) If E is a bounded subset of $\mathbb{R}$, then $f(x) = x^2$ is uniformly continuous on E. Since E is bounded, there exists a positive constant $C > 0$ so that $|x| \leq C$ for all $x \in E$. If $x, y \in E$, then

$$|f(x) - f(y)| = |x^2 - y^2| = |x + y||x - y| \leq (|x| + |y|)|x - y|$$
$$\leq 2C|x - y|.$$

Let $\epsilon > 0$ be given. Take $\delta = \epsilon/2C$. If $x, y \in E$ with $|x - y| < \delta$, then by the above

$$|f(x) - f(y)| \leq 2C\,|x - y| < 2C\delta < \epsilon.$$

Therefore f is uniformly continuous on E. In this example, the choice of δ depends both on ϵ, and the set E. In the exercises you will be asked to show that this result is false if the set E is an unbounded interval.

(b) Let $f(x) = \sin x$. As in Example 4.2.2(f),

$$|f(y) - f(x)| \leq |y - x|$$

for all $x, y \in \mathbb{R}$. Consequently, f is uniformly continuous on $\mathbb{R}$.

(c) In this example, we show that the function $f(x) = 1/x$, $x \in (0, 1)$, is *not* uniformly continuous on $(0, 1)$. Suppose on the contrary that f is uniformly continuous on $(0, 1)$. Then if we take $\epsilon = 1$, there exists a $\delta > 0$ such that

$$|f(x) - f(y)| = \left|\frac{1}{x} - \frac{1}{y}\right| < 1$$

for all $x, y \in (0, 1)$ with $|x - y| < \delta$. Since any smaller δ will also work, we can assume that $\delta < 1$. Then for any $x \in (0, \frac{1}{2})$, $y = x + \frac{1}{2}\delta$ is in $(0, 1)$ and satisfies $|x - y| < \delta$. Thus

$$|f(x) - f(y)| = \frac{\frac{1}{2}\delta}{x(x + \frac{1}{2}\delta)} < 1.$$

Since $x + \frac{1}{2}\delta < 1$ for all $x \in (0, \frac{1}{2})$, we have

$$\tfrac{1}{2}\delta < x(x + \tfrac{1}{2}\delta) < x.$$

Thus $x > \frac{1}{2}\delta$ for all $x \in (0, \frac{1}{2})$, which is a contradiction. The function $f(x) = 1/x$ however is uniformly continuous on $[a, \infty)$ for any fixed $a > 0$ (Exercise 4(a)). $\square$

Lipschitz Functions

Both of the functions in Example 4.3.2 (a) and (b) are examples of an extensive class of functions. If E is a subset of a metric space (X, d), a function $f : E \to \mathbb{R}$ satisfies a **Lipschitz condition** on E if there exists a positive constant M such that

$$|f(x) - f(y)| \leq M\,d(x, y)$$

for all $x, y \in E$. Functions satisfying the above inequality are usually referred to as **Lipschitz functions**. As we will see in the next chapter, functions for which the derivative is bounded are Lipschitz functions. As a consequence of the following theorem, every Lipschitz function is uniformly continuous. However, not every uniformly continuous function is a Lipschitz function. For example, the function $f(x) = \sqrt{x}$ is uniformly continuous on $[0, \infty)$, but f does not satisfy a Lipschitz condition on $[0, \infty)$ (see Exercise 5).

THEOREM 4.3.3 *Suppose E is a subset of a metric space (X, d) and $f :$ $E \to \mathbb{R}$. If there exists a positive constant M such that*

$$|f(x) - f(y)| \leq Md(x, y)$$

for all $x, y \in E$, then f is uniformly continuous on E.

Proof. Exercise 1.   □

Uniform Continuity Theorem

If the function f does not satisfy a Lipschitz condition on E, then to determine whether f is uniformly continuous on E is much more difficult. The following theorem provides a sufficient condition on the set E such that every continuous real-valued function on E is uniformly continuous.

THEOREM 4.3.4 *If K is compact and $f : K \to \mathbb{R}$ is continuous on K, then f is uniformly continuous on K.*

Proof. Let $\epsilon > 0$ be given. Since f is continuous, for each $p \in K$, there exists a $\delta_p > 0$ such that

$$|f(x) - f(p)| < \frac{\epsilon}{2} \tag{3}$$

for all $x \in K \cap N_{\delta_p}(p)$.

The collection $\left\{ N_{\delta_p/2}(p) \right\}_{p \in K}$ is an open cover of K. Since K is compact, a finite number of these will cover K. Thus there exist a finite number of points $p_1, ..., p_n$ in K such that

$$K \subset \bigcup_{i=1}^{n} N_{\delta_{p_i}/2}(p_i).$$

Let

$$\delta = \tfrac{1}{2} \min\{\delta_{p_i} : i = 1, ..., n\}.$$

Then $\delta > 0$. Suppose $x, y \in K$ with $d(x, y) < \delta$. Since $x \in K$, $x \in N_{\delta_{p_i}/2}(p_i)$ for some i. Furthermore, since $d(x, y) < \delta \leq \delta_{p_i}/2$,

$$x, y \in N_{\delta_{p_i}}(p_i).$$

Thus by the triangle inequality and (3),

$$|f(x) - f(y)| \leq |f(x) - f(p_i)| + |f(p_i) - f(y)| < \tfrac{1}{2}\epsilon + \tfrac{1}{2}\epsilon = \epsilon. \quad \square$$

COROLLARY 4.3.5 *A continuous real-valued function on a closed and bounded interval $[a, b]$ is uniformly continuous.*

The definition of uniform continuity as well as the proof of Corollary 4.3.5 appeared in a paper by Eduard Heine in 1872.

EXAMPLE 4.3.6 In this example, we show that both the properties closed and bounded are required. The interval $[0, \infty)$ is closed, but not bounded. The function $f(x) = x^2$ is continuous on $[0, \infty)$, but not uniformly continuous on $[0, \infty)$ (Exercise 2). On the other hand, the interval $(0, 1)$ is bounded, but not closed. The function $f(x) = 1/x$ is continuous on $(0, 1)$, but f is not uniformly continuous on $(0, 1)$. □

Exercises 4.3

1. Prove Theorem 4.3.3.

2. Show that the following functions are not uniformly continuous on the given domain.
    *a. $f(x) = x^2$, Dom $f = [0, \infty)$ b. $g(x) = \dfrac{1}{x^2}$, Dom $g = (0, \infty)$
    *c. $h(x) = \sin \dfrac{1}{x}$, Dom $h = (0, \infty)$ d. $k(x) = \dfrac{1}{\sin x}$, Dom $k = (0, \pi)$

3. Prove that each of the following functions is uniformly continuous on the indicated set.
    *a. $f(x) = \dfrac{x}{1 + x}$, $x \in [0, \infty)$ b. $g(x) = x^2$, $x \in \mathbb{N}$.
    *c. $h(x) = \dfrac{1}{x^2 + 1}$, $x \in \mathbb{R}$.   d. $k(x) = \cos x$, $x \in \mathbb{R}$
    e. $e(x) = \dfrac{x^2}{x + 1}$, $x \in (0, \infty)$ *f. $f(x) = \dfrac{\sin x}{x}$, $x \in (0, 1)$

4. Show that each of the following functions is a Lipschitz function.
    *a. $f(x) = \dfrac{1}{x}$, Dom $f - [u, \infty)$, $a > 0$.
    b. $g(x) = \dfrac{x}{x^2 + 1}$, Dom $g = [0, \infty)$.
    *c. $h(x) = \sin \dfrac{1}{x}$, Dom $h = [a, \infty)$, $a > 0$.
    d. $p(x)$ a polynomial, Dom $p = [-a, a]$, $a > 0$.

5. *a. Show that $f(x) = \sqrt{x}$ satisfies a Lipschitz condition on $[a, \infty)$, $a > 0$.
    b. Prove that $\sqrt{x}$ is uniformly continuous on $[0, \infty)$.
    *c. Show that f does not satisfy a Lipschitz condition on $[0, \infty)$.

6. Suppose $E \subset \mathbb{R}$ and f, g are Lipschitz functions on E.
    a. Prove that $f + g$ is a Lipschitz function on E.
    b. If in addition f and g are bounded on E, prove that fg is a Lipschitz function on E.

7. Suppose $E \subset \mathbb{R}$, and f, g are uniformly continuous real valued functions on E.
    a. Prove that $f + g$ is uniformly continuous on E.

162   *Introduction to Real Analysis*

*b. If in addition f and g are bounded, prove that fg is uniformly continuous on E.

c. Is (b) still true if only one of the two functions is bounded?

8. Suppose E is a subset of a metric space X and $f : E \to \mathbb{R}$ is uniformly continuous. If $\{x_n\}$ is a Cauchy sequence in E, prove that $\{f(x_n)\}$ is a Cauchy sequence.

9. Let $f : (a,b) \to \mathbb{R}$ be uniformly continuous on (a,b). Use the previous exercise to show that f can be defined at a and b such that f is continuous on $[a,b]$

10. ***Suppose** that E is a bounded subset of $\mathbb{R}$ and $f : E \to \mathbb{R}$ is uniformly continuous on E. Prove that f is bounded on E.

11. Suppose $-\infty \le a < c < b \le \infty$, and suppose $f : (a,b) \to \mathbb{R}$ is continuous on (a,b).

a. If f is uniformly continuous on (a,c) and also uniformly continuous on (c,b), prove that f is uniformly continuous on (a,b).

b. Show by example that the conclusion in (a) may be false if f is not continuous on (a,b).

12. Let $a \in \mathbb{R}$. Suppose f is a real-valued function on $[a,\infty)$ satisfying $\lim_{x \to \infty} f(x) = L$, where $L \in \mathbb{R}$. Prove that

*a. f is bounded on $[a,\infty)$, and

b. f is uniformly continuous on $[a,\infty)$

13. Let $E \subset \mathbb{R}$. A function $f : E \to E$ is **contractive** if there exists a constant b, $0 < b < 1$, such that $|f(x) - f(y)| \le b|x - y|$.

*a. If E is closed, and $f : E \to E$ is contractive, prove that there exists a unique point $x_o \in E$ such that $f(x_o) = x_o$. (Such a point x_o is called a **fixed point** of f)

b. Let $E = (0, \frac{1}{3}]$. Show that $f(x) = x^2$ is contractive on E, but that f does not have a fixed point in E.

14. A function $f : \mathbb{R} \to \mathbb{R}$ is **periodic** if there exists $p \in \mathbb{R}$ such that $f(x + p) = f(x)$ for all $x \in \mathbb{R}$. Prove that a continuous periodic function on $\mathbb{R}$ is bounded and uniformly continuous on $\mathbb{R}$.

4.4 Monotone Functions and Discontinuities

In this section, we take a closer look both at limits and continuity for real-valued functions defined on an interval $I \subset \mathbb{R}$. More specifically however, we will be interested in classifying the types of discontinuities which such a function may have. We will also investigate properties of monotone functions defined on an interval I. These functions will play a crucial role in Chapter 6 on Riemann-Stieltjes integration. First however we begin with the right and left limit of a real-valued function defined on a subset E of $\mathbb{R}$.

Right and Left Limits

DEFINITION 4.4.1 *Let $E \subset \mathbb{R}$ and let f be a real-valued function defined on E. Suppose p is a limit point of $E \cap (p, \infty)$. The function f has a* **right limit** *at p if there exists a number $L \in \mathbb{R}$ such that given any $\epsilon > 0$, there exists a $\delta > 0$ for which*

$$|f(x) - L| < \epsilon \text{ for all } x \in E \text{ satisfying } p < x < p + \delta.$$

The right limit of f, if it exists, is denoted by $f(p+)$, and we write

$$f(p+) = \lim_{x \to p^+} f(x) = \lim_{\substack{x \to p \\ x > p}} f(x).$$

Similarly, if p is a limit point of $E \cap (-\infty, p)$, the **left limit** *of f at p, if it exists, is denoted by $f(p-)$, and we write*

$$f(p-) = \lim_{x \to p^-} f(x) = \lim_{\substack{x \to p \\ x < p}} f(x).$$

The hypothesis that p is a limit point of $E \cap (p, \infty)$ guarantees that for every $\delta > 0$, $E \cap (p, p + \delta) \neq \emptyset$. If E is an open interval (a, b), $-\infty < a < b \leq \infty$, then any p satisfying $a \leq p < b$ is a limit point of $E \cap (p, \infty)$. Similarly, if $-\infty \leq a < b < \infty$, then any p satisfying $a < p \leq b$ is a limit point of $(-\infty, p) \cap E$. If I is any interval with $\text{Int}(I) \neq \emptyset$, and $f : I \to \mathbb{R}$, then f has a limit at $p \in \text{Int}(I)$ if and only if

(a) $f(p+)$ and $f(p-)$ both exist, and

(b) $f(p+) = f(p-)$.

The hypothesis that $p \in \text{Int}(I)$ guarantees that p is a limit point of both $(-\infty, p) \cap I$ and also $I \cap (p, \infty)$. If p is a left end point of the interval I, then the right limit of f at p coincides with the limit of f at p. The analogous statement is also true if p is a right endpoint of I.

We also define right and left continuity of a function at a point p as follows:

DEFINITION 4.4.2 *Let $E \subset \mathbb{R}$ and let f be a real-valued function on E. The function f is* **right continuous (left continuous)** *at $p \in E$ if for any $\epsilon > 0$, there exists a $\delta > 0$ such that*

$$|f(x) - f(p)| < \epsilon \text{ for all } x \in E \text{ with } p \leq x < p + \delta \quad (p - \delta < x \leq p).$$

Remarks. If $p \in E$ is an isolated point of E or is not a limit point of $E \cap (p, \infty)$, then there exists a $\delta > 0$ such that $E \cap (p, p + \delta) = \emptyset$. Thus if $f : E \to \mathbb{R}$ and $\epsilon > 0$ is arbitrary, then $|f(x) - f(p)| < \epsilon$ for all $x \in E \cap [p, p + \delta)$. Thus every

$f : E \to \mathbb{R}$ is right continuous at p. In particular, if E is a closed interval $[a, b]$, then every $f : [a, b] \to \mathbb{R}$ is right continuous at b. Also, f is left continuous at b if and only if f is continuous at p.

The following theorem, the proof of which is left to the exercises, is an immediate consequence of the definitions.

THEOREM 4.4.3 *A function $f : (a, b) \to \mathbb{R}$ is right continuous at $p \in (a, b)$ if and only if $f(p+)$ exists and equals $f(p)$. Similarly, f is left continuous at p if and only if $f(p-)$ exists and equals $f(p)$.*

Proof. Exercise 1. □

Types of Discontinuities

By the previous theorem a function f is continuous at $p \in (a, b)$ if and only if

  (a) $f(p+)$ and $f(p-)$ both exist, and

  (b) $f(p+) = f(p-) = f(p)$.

A real-valued function f defined on an interval I can fail to be continuous at a point $p \in \overline{I}$ (the closure of I) for several reasons. One possibility is that $\lim_{x \to p} f(x)$ exists but either does not equal $f(p)$, or f is not defined at p. Such a function can easily be made continuous at p by either defining, or redefining, f at p as follows:

$$f(p) = \lim_{x \to p} f(x).$$

For this reason, such a discontinuity is called a **removable discontinuity**. For example, the function

$$g(x) = \begin{cases} \dfrac{x^2 - 4}{x - 2}, & x \neq 2, \\ 2, & x = 2, \end{cases}$$

of Example 4.2.2(a) is not continuous at 2 since

$$\lim_{x \to 2} g(x) = 4 \neq g(2).$$

By redefining g such that $g(2) = 4$, the resulting function is then continuous at 2. Another example is given by $f(x) = x \sin \frac{1}{x}$, $x \in (0, \infty)$, which is not defined at 0. If we define f on $[0, \infty)$ by

$$f(x) = \begin{cases} 0, & x = 0, \\ x \sin \dfrac{1}{x}, & x > 0, \end{cases}$$

then by Example 4.2.2(e), f is now continuous at 0.

Another possibility is that $f(p+)$ and $f(p-)$ both exist, but are not equal. This type of discontinuity is called a jump discontinuity. (See Figure 4.9)

DEFINITION 4.4.4 *Let f be a real valued function defined on an interval I. The function f has a* **jump discontinuity** *at $p \in \text{Int}(I)$ if $f(p+)$ and $f(p-)$ both exist, but f is not continuous at p. If $p \in I$ is a left (right) endpoint of I, then f has a jump discontinuity at p if $f(p+)$ $(f(p-))$ exists, but f is not continuous at p.*

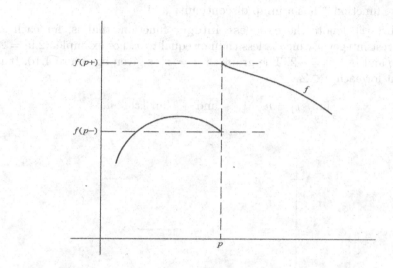

FIGURE 4.9
Jump discontinuity of f at p

Jump discontinuities are also referred to as **simple discontinuities,** or **discontinuities of the first kind.** All other discontinuities are said to be of **second kind**.

If $f(p+)$ and $f(p-)$ both exist, but f is not continuous at p, then either

(a) $f(p+) \neq f(p-)$, or

(b) $f(p+) = f(p-) \neq f(p)$.

In case (a), f has a jump discontinuity at p, whereas in case (b), the discontinuity is removable. All discontinuities for which $f(p+)$ or $f(p-)$ does not exist are discontinuities of the second kind.

EXAMPLES 4.4.5 **(a)** Let f be defined by

$$f(x) = \begin{cases} x, & 0 < x \leq 1, \\ 3 - x^2, & x > 1. \end{cases}$$

If $x < 1$, then $f(x) = x$. Therefore,

$$f(1-) = \lim_{x \to 1^-} f(x) = \lim_{x \to 1} x = 1 = f(1).$$

Likewise, the right limit of f at 1 is

$$f(1+) = \lim_{x \to 1^+} f(x) = \lim_{x \to 1} 3 - x^2 = 2.$$

Therefore $f(1-) = f(1) = 1$, and $f(1+) = 2$. Thus f is left continuous at 1, but not continuous. Since both right and left limits exist at 1, but are not equal, the function f has a jump discontinuity at 1.

(b) Let $[x]$ denote the **greatest integer** function, that is, for each x, $[x] =$ largest integer n which is less than or equal to x. For example, $[2.9] = 2$, $[3.1] = 3$, and $[-1.5] = -2$. The graph of $y = [x]$ is given in Figure 4.10. It is clear that for each $n \in \mathbb{Z}$,

$$\lim_{x \to n^-} [x] = n - 1, \quad \text{and} \quad \lim_{x \to n^+} [x] = n.$$

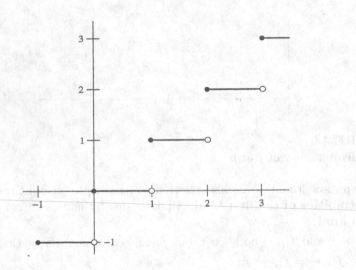

FIGURE 4.10
Graph of $[x]$

Thus f has a jump discontinuity at each $n \in \mathbb{Z}$. Also, since $f(n) = [n] = n$, $f(x) = [x]$ is right continuous at each integer. Finally, since f is constant on each interval $(n - 1, n)$, $n \in \mathbb{Z}$, f is continuous at every $x \in \mathbb{R} \setminus \mathbb{Z}$.

(c) Let f be defined on $\mathbb{R}$ by

$$f(x) = \begin{cases} 0, & \text{if } x \leq 0 \\ \sin \dfrac{1}{x}, & \text{if } x > 0. \end{cases}$$

Then $f(0-) = 0$, but $f(0+)$ does not exist. Thus the discontinuity is of second kind.

(d) Consider the function $g : \mathbb{R} \to \mathbb{R}$ defined by

$$g(x) = \sin(2\pi x[x]).$$

For $x \in (n, n+1), n \in \mathbb{Z}, x[x] = nx$, and thus $g(x)$ is continuous on every interval $(n, n+1), n \in \mathbb{Z}$. On the other hand, for $n \in \mathbb{Z}$,

$$\lim_{x \to n^+} \sin(2\pi x[x]) = \sin(2\pi n^2) = 0, \quad \text{and}$$

$$\lim_{x \to n^-} \sin(2\pi x[x]) = \sin(2\pi n(n-1)) = 0.$$

Since $g(n) = \sin(2\pi n^2) = 0$, g is also continuous at each $n \in \mathbb{Z}$. The function g however, is not uniformly continuous on $\mathbb{R}$ (Exercise 7). □

Monotone Functions

DEFINITION 4.4.6 *Let f be a real-valued function defined on an interval I.*

    (a) *f is **monotone increasing** (increasing, nondecreasing) on I if $f(x) \le f(y)$ for all $x, y \in I$ with $x < y$.*

    (b) *f is **monotone decreasing** (decreasing, nonincreasing) on I if $f(x) \ge f(y)$ for all $x, y \in I$ with $x < y$.*

    (c) *f is **monotone** on I if f is monotone increasing on I or monotone decreasing on I.*

    A function f is **strictly increasing** on I if $f(x) < f(y)$ for all $x, y \in I$ with $x < y$. The concept of **strictly decreasing** is defined similarly. Also, f is **strictly monotone** on I if f is strictly increasing on I or strictly decreasing on I. Our main result for monotone functions is as follows:

THEOREM 4.4.7 *Let $I \subset \mathbb{R}$ be an open interval and let $f : I \to \mathbb{R}$ be monotone increasing on I. Then $f(p+)$ and $f(p-)$ exists for every $p \in I$ and*

$$\sup_{x<p} f(x) = f(p-) \le f(p) \le f(p+) = \inf_{p<x} f(x).$$

Furthermore, if $p < q, p, q \in I$, then $f(p+) \le f(q-)$.

    Although we stated the theorem for monotone increasing functions, a similar statement is also valid for monotone decreasing functions.

Proof. Fix $p \in I$. Since f is increasing on I, $\{f(x) : x < p, x \in I\}$ is bounded above by $f(p)$. Let

$$A = \sup\{f(x) : x < p, x \in I\}.$$

Then $A \le f(p)$. We now show that

$$\lim_{x \to p^-} f(x) = A.$$

The proof of this is similar to the proof of Theorem 3.3.2. Let $\epsilon > 0$ be given. Since A is the least upper bound of $\{f(x) : x < p\}$, there exists $x_o < p$ such that

$$A - \epsilon < f(x_o) \le A.$$

Thus if $x_o < x < p$, $A - \epsilon < f(x_o) \le f(x) \le A$. Therefore,

$$|f(x) - A| < \epsilon \quad \text{for all} \quad x, \; x_o < x < p.$$

Thus by definition, $\lim\limits_{x \to p^-} f(x) = A$. Similarly

$$f(p) \le f(p+) = \inf\{f(x) : p < x, \, x \in I\}.$$

Finally, suppose $p < q$. Then

$$f(p+) = \inf\{f(x) : x > p, \, x \in I\} \le \inf\{f(x) : p < x < q\}$$
$$\le \sup\{f(x) : p < x < q\} \le \sup\{f(x) : x < q, \, x \in I\} = f(q-). \quad \square$$

COROLLARY 4.4.8 *If f is monotone on an open interval I, then the set of discontinuities of f is at most countable.*

Proof. Let $E = \{p \in I : f$ is discontinuous at $p\}$. Suppose f is monotone increasing on I. Then

$$p \in E \qquad \text{if and only if} \qquad f(p-) < f(p+).$$

For each $p \in E$, choose $r_p \in \mathbb{Q}$ such that

$$f(p-) < r_p < f(p+).$$

If $p < q$, then $f(p+) \le f(q-)$. Therefore, if $p, q \in E$, $r_p \ne r_q$, and thus the function $p \to r_p$ is a one-to-one map of E into $\mathbb{Q}$. Therefore E is equivalent to a subset of $\mathbb{Q}$ and thus is at most countable. $\quad \square$

Construction of Monotone Functions with Prescribed Discontinuities

We now proceed to show that given any finite or countable subset A of (a, b), there exists a monotone increasing function f on $[a, b]$ that is discontinuous at each $x \in A$ and continuous on $[a, b] \setminus A$. We first illustrate how this is accomplished for the case where $A = \{a_1, ..., a_n\}$ is a finite subset of (a, b). To facilitate this construction we define the unit jump function I on $\mathbb{R}$ as follows:

DEFINITION 4.4.9 *The **unit jump function** $I : \mathbb{R} \to \mathbb{R}$ is defined by*

$$I(x) = \begin{cases} 0, & \text{when } x < 0, \\ 1, & \text{when } x \ge 0. \end{cases}$$

The function I is right continuous at 0 with $I(0+) = I(0) = 1$ and $I(0-) = 0$. For each $k = 1, ..., n$, let

$$I_k(x) = I(x - a_k) = \begin{cases} 0, & \text{when } x < a_k \\ 1, & \text{when } x \geq a_k. \end{cases}$$

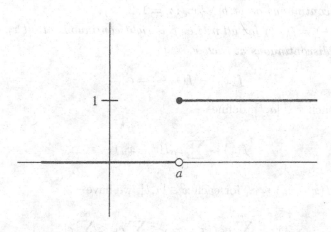

FIGURE 4.11
Graph of $I(x - a)$

Then I_k has a unit jump at each a_k and is right continuous at a_k (see Figure 4.11).

Suppose $\{c_1, ..., c_n\}$ are positive real numbers. Define f on $[a, b]$ by

$$f(x) = \sum_{k=1}^{n} c_k I(x - a_k).$$

The reader should verify that the function f is

(a) monotone increasing on $[a, b]$,

(b) continuous on $[a, b] \setminus \{a_1, a_2, ..., a_n\}$,

(c) right continuous at each a_k, $k = 1, 2, ..., n$, and

(d) discontinuous at each a_k with $f(a_k+) - f(a_k-) = c_k$ for all $k = 1, 2, ..., n$.

That such a function exists for any finite subset $\{a_1, ..., a_n\}$ of (a, b) is not surprising. However, that such a function exists for any countable subset A of (a, b) may take some convincing, especially if one takes A to be dense in $[a, b]$; e.g., the rational numbers in (a, b).

THEOREM 4.4.10 *Let $a, b \in \mathbb{R}$ with $a < b$, and let $\{x_n\}_{n \in \mathbb{N}}$ be a countable subset of (a, b). Let $\{c_n\}_{n=1}^{\infty}$ be any sequence of positive real numbers such that $\sum_{n=1}^{\infty} c_n$ converges. Then there exists a monotone increasing function f on $[a, b]$ such that*

(a) $f(a) = 0$ and $f(b) = \sum_{n=1}^{\infty} c_n$,

(b) f *is continuous on* $[a, b] \setminus \{x_n : n = 1, 2, \ldots\}$,

(c) $f(x_n+) = f(x_n)$ *for all n; i.e. f is right continuous at all x_n, and*

(d) f *is discontinuous at each x_n with*

$$f(x_n) - f(x_n-) = c_n.$$

Proof. For each $x \in [a, b]$, define

$$f(x) = \sum_{n=1}^{\infty} c_n I(x - x_n).$$

Since $0 \le c_n I(x - x_n) \le c_n$ for each $x \in [a, b]$, we have

$$s_n(x) = \sum_{k=1}^{n} c_k I(x - x_k) \le \sum_{k=1}^{n} c_k \le \sum_{k=1}^{\infty} c_k.$$

Thus for each $x \in [a, b]$, the sequence $\{s_n(x)\}$ of partial sums is monotone increasing and bounded above and hence by Theorem 3.7.6 converges. Since

$$I(x - x_n) \le I(y - x_n) \qquad n = 1, 2, \ldots.$$

for all x, y with $x < y$, f is monotone increasing on $[a, b]$. Furthermore, since $x_n > a$ for all n, $I(a - x_n) = 0$ for all n. Therefore $f(a) = 0$. Also, since $I(b - x_n) = 1$ for all n,

$$f(b) = \sum_{k=1}^{\infty} c_k.$$

This proves (a).

We now prove (b). Fix $p \in [a, b]$, $p \ne x_n$ for any n. Let $E = \{x_n : n \in \mathbb{N}\}$. There are two cases to consider.

(i) Suppose p is not a limit point of E. If this is the case, there exists a $\delta > 0$ such that $N_\delta(p) \cap E = \emptyset$. Then

$$I(x - x_k) = I(p - x_k) \qquad \text{for all} \quad x \in (p - \delta, p + \delta)$$

and all $k = 1, 2, \ldots$. Thus f is constant on $(p - \delta, p + \delta)$ and hence continuous.

(ii) Suppose p is a limit point of E. Let $\epsilon > 0$ be given. Since the series

$\sum\limits_{k=1}^{\infty} c_k$ converges, by the Cauchy criterion there exists a positive integer N such that

$$\sum_{k=N+1}^{\infty} c_k < \epsilon.$$

Choose δ such that

$$0 < \delta < \min\{|p - x_n| : n = 1, 2, ..., N\}.$$

With this choice of δ, if $x_k \in N_\delta(p) \cap E$, we have $k > N$. Suppose $p < x < p+\delta$. Then

$$I(p - x_k) = I(x - x_k) \qquad \text{for all} \quad k = 1, 2, ..., N.$$

Furthermore, for any $x > p$, we always have

$$0 \le I(x - x_k) - I(p - x_k) \le 1, \qquad \text{for all} \quad k \in \mathbb{N}.$$

Therefore, for $p < x < p + \delta$,

$$0 \le f(x) - f(p) \le \sum_{k=N+1}^{\infty} c_k(I(x - x_k) - I(p - x_k)) \le \sum_{k=N+1}^{\infty} c_k < \epsilon.$$

Thus f is right continuous at p. Similarly, f is left continuous at p, and therefore f is continuous at p.

For the proof of (c), fix an $x_n \in E$. If x_n is an isolated point of E, then as above, there exists a $\delta > 0$ such that $E \cap (x_n, x_n + \delta) = \emptyset$. Therefore $f(y) = f(x_n)$ for all y, $x_n < y < x_n + \delta$. Thus $f(x_n+) = f(x_n)$. Suppose x_n is a limit point of E. Let $\epsilon > 0$ be given. Again, choose a positive integer N so that

$$\sum_{k=N+1}^{\infty} c_k < \epsilon.$$

As in (b), there exists a $\delta > 0$ such that if $x_k \in (x_n, x_n + \delta) \cap E$, then $k > N$. Thus

$$0 \le f(y) - f(x_n) \le \sum_{k=N+1}^{\infty} c_k < \epsilon \qquad \text{for all} \quad y \in (x_n, x_n + \delta).$$

Therefore $f(x_n+) = f(x_n)$ and f is right continuous at each x_n.

For the proof of (d), suppose $y < x_n$. Again, if x_n is an isolated point of E, there exists a $\delta > 0$ such that $(x_n - \delta, x_n) \cap E = \emptyset$. Therefore, for all $k \ne n$,

$$I(y - x_k) = I(x_n - x_k) \qquad \text{for all} \quad y, \ x_n - \delta < y < x_n,$$

and for all $y < x_n$,

$$0 = I(y - x_n) \le I(x_n - x_n) = I(0) = 1.$$

Therefore,

$$f(x_n) - f(y) = c_n \qquad \text{for all} \quad y,\, x_n - \delta < y < x_n.$$

Suppose x_n is a limit point of E. Given $\epsilon > 0$, choose N such that $\sum\limits_{k=N+1}^{\infty} c_k < \epsilon$. For this N, choose $\delta > 0$ such that if $x_k \in (x_n - \delta, x_n) \cap E$ then $k > N$. Then for all $y \in [a, b]$ with $x_n - \delta < y < x_n$,

$$c_n \le f(x_n) - f(y) \le c_n + \sum_{k=N+1}^{\infty} c_k < c_n + \epsilon.$$

Therefore, $f(x_n) - f(x_n-) = c_n$. $\square$

EXAMPLES 4.4.11 (a) Take $c_n = 2^{-n}$, $x_n = 1 - 1/(n+1)$, $n = 1, 2, ...$, $(a, b) = (0, 1)$. As in Theorem 4.4.10 let

$$f(x) = \sum_{n=1}^{\infty} c_n I(x - x_n).$$

In this example, the sequence $\{x_n\}$ satisfies $0 < x_1 < x_2 < \cdots < 1$. If $0 \le x < x_1 = \frac{1}{2}$, then $I(x - x_n) = 0$ for all n. Thus

$$f(x) = 0, \qquad x \in [0, \tfrac{1}{2}).$$

If $x_1 \le x < x_2 = \frac{2}{3}$, then $I(x - x_1) = 1$ and $I(x - x_k) = 0$ for all $k \ge 2$. Therefore,

$$f(x) = c_1 = \frac{1}{2}, \qquad x \in [\tfrac{1}{2}, \tfrac{1}{3}).$$

If $x_2 \le x < x_3 = \frac{3}{4}$, then $I(x - x_k) = 1$ for $k = 1, 2$ and $I(x - x_k) = 0$ for $k \ge 3$. Therefore,

$$f(x) = c_1 + c_2 = \frac{1}{2} + \frac{1}{2^2} = \frac{3}{4}, \qquad x \in [\tfrac{2}{3}, \tfrac{3}{4}),$$

and so forth. The graph of f is depicted in Figure 4.12.

   (b) Let $c_n = 2^{-n}$ and let $\{x_n\}$ be an enumeration of the rationals in $(0, 1)$. Theorem 4.4.10 guarantees the existence of a nondecreasing function on $[0, 1]$, which is discontinuous at each rational number in $(0, 1)$, and continuous at every irrational number in $(0, 1)$.

   (c) If in the proof of Theorem 4.4.10 we take $\{x_n\}_{n \in \mathbb{N}}$ to be a countable subset of $\mathbb{R}$ and choose $\{c_n\}_{n \in \mathbb{N}}$ ($c_n > 0$) such that $\sum\limits_{n=1}^{\infty} c_n = 1$, then we obtain a nondecreasing real-valued function f on $\mathbb{R}$ satisfying $\lim\limits_{x \to -\infty} f(x) = 0$ and $\lim\limits_{x \to \infty} f(x) = 1$. (See Exercise 21) Such a function is called a **distribution function** on $\mathbb{R}$. Such functions arise naturally in probability theory. $\square$

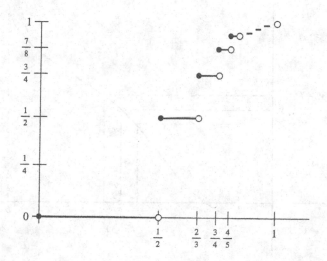

FIGURE 4.12
Graph of $f(x) = \sum c_n I(x - x_n)$

Inverse Functions

Suppose f is a strictly increasing real-valued function on an interval I. Let $x, y \in I$ with $x \neq y$. If $x < y$, then since f is strictly increasing, $f(x) < f(y)$. Similarly, if $x > y$ then $f(x) > f(y)$. Thus $f(x) \neq f(y)$ for any x, $y \in I$ with $x \neq y$. Therefore f is one-to-one and consequently has an inverse function f^{-1} defined on $f(I)$. In the following theorem we prove that if f is continuous on I then f^{-1} is also continuous on $f(I)$.

THEOREM 4.4.12 *Let $I \subset \mathbb{R}$ be an interval and let $f : I \to \mathbb{R}$ be strictly monotone and continuous on I. Then f^{-1} is strictly monotone and continuous on $J = f(I)$.*

Proof. Without loss of generality we consider the case where f is strictly increasing on I. Since f is continuous, by Corollary 4.2.12 $f(I) = J$ is an interval. Furthermore, since f is strictly increasing on I, f is a one-to-one function from I onto J. Hence f^{-1} is a one-to-one function from J onto I.

Suppose $y_1, y_2 \in J$ with $y_1 < y_2$. Then there exist distinct points $x_1, x_2 \in I$ such that $f(x_i) = y_i$, $i = 1, 2$. Since f is strictly increasing, we have $x_1 < x_2$. Thus $f^{-1}(y_1) < f^{-1}(y_2)$, i.e., f^{-1} is strictly increasing.

It remains to be shown that f^{-1} is continuous on J. We first show that f^{-1} is left continuous at each $y_o \in J$ for which $(-\infty, y_o) \cap J \neq \emptyset$. This last assumption only means that y_o is not a left endpoint of J. Let $x_o \in I$ be such that $f(x_o) = y_o$. Then $(-\infty, x_o) \cap I \neq \emptyset$. Let $\epsilon > 0$ be given. Without loss of

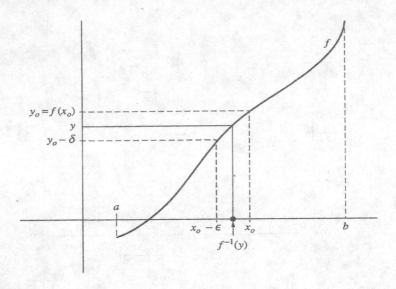

FIGURE 4.13
Continuity of the inverse function

generality, we can assume that ϵ is sufficiently small so that $x_o - \epsilon \in I$. Since f is continuous and strictly increasing,

$$f((x_o - \epsilon, x_o]) = (f(x_o - \epsilon), f(x_o)] = (y_o - \delta, y_o]$$

where $\delta > 0$ is given by $\delta = f(x_o) - f(x_o - \epsilon)$. Thus since f^{-1} is strictly increasing,

$$x_o - \epsilon = f^{-1}(y_o - \delta) < f^{-1}(y) < f^{-1}(y_o) = x_o$$

for all $y \in (y_o - \delta, y_o)$ (see Figure 4.13). Hence,

$$|f^{-1}(y_o) - f^{-1}(y)| < \epsilon \quad \text{for all } y \in (y_o - \delta, y_o].$$

Therefore f^{-1} is left continuous at y_o. A similar argument also proves that f^{-1} is right continuous at each $y_o \in J$ that is not a right endpoint of J. Thus f^{-1} is continuous at each $y_o \in J$. $\square$

For a strictly increasing function f on an open interval I, an alternate proof of the continuity of the inverse function f^{-1} is suggested in the exercises (Exercise 14).

EXAMPLE 4.4.13 The function $f(x) = x^2$ is continuous and strictly increasing on $I = [0, \infty)$ with $J = f(I) = [0, \infty)$. Thus the inverse function $f^{-1}(y) = \sqrt{y}$ is continuous on $[0, \infty)$. As a consequence, the function $g(x) = \sqrt{x}$ is continuous on $[0, \infty)$. Applying the same argument to $f(x) = x^n$ shows that the function $g(x) = \sqrt[n]{x}$ is strictly increasing and continuous on $[0, \infty)$. $\square$

Remark. In the statement of Theorem 4.4.12 we assumed that f was strictly monotone and continuous on the interval I. The fact that f is either strictly increasing or strictly decreasing on I implies that f is one-to-one on the interval I. Conversely, if f is one-to-one and continuous on an interval I, then as a consequence of the intermediate value theorem the function f is strictly monotone on I (Exercise 15). This, however, is false if either f is not continuous on the interval I, or if Dom f is not an interval. (See Exercises 16, 17, and 18).

Exercises 4.4

1. Prove Theorem 4.4.3.

2. For each of the following functions f defined on $\mathbb{R} \setminus \{0\}$, find $\lim\limits_{x \to 0^-} f(x)$ and $\lim\limits_{x \to 0^+} f(x)$, provided the limits exist.

    a. $f(x) = \dfrac{x}{|x|}$.

    *b. $f(x) = \begin{cases} x[x], & x < 0, \\ \dfrac{[x]}{x}, & x > 0. \end{cases}$

    c. $f(x) = [1 - x^2]$

    *d. $f(x) = [x^2 - 1]$

    e. $f(x) = \left[\dfrac{1}{x}\right]$.

    *f. $f(x) = x\left[\dfrac{1}{x}\right]$.

3. For each of the functions f in Exercise 2, determine whether f has a removable discontinuity, a jump discontinuity, or a discontinuity of second kind, at $x = 0$. If f has a removable discontinuity at 0, specify how $f(0)$ should be defined in order that f is continuous at 0.

4. ***** Investigate continuity of $g(x) = (x - 2)[x]$ at $x_o = 2$.

5. Let $f(x) = x - [x]$. Discuss continuity of f. Sketch the graph of f.

6. For each of the following determine the value of b such that f has a removable discontinuity at the indicated point x_o.

    a. $f(x) = \begin{cases} x - 2, & x < 1, \\ bx^3 + 4, & x > 1, \end{cases} \quad x_o = 1$

    *b. $f(x) = \begin{cases} -x^2[x], & x < -2, \\ x + b, & x > -2, \end{cases} \quad x_o = -2$

7. **a.** Sketch the graph of $g(x) = \sin(2\pi x[x])$ for $x \in (-4, 4)$.

    *b. Prove that $g(x)$ is not uniformly continuous on $\mathbb{R}$.

8. Prove that the function f of Example 4.4.11(a) is continuous at $x = 1$.

9. Let $E \subset \mathbb{R}$ and let f be a real-valued function on E. Suppose $p \in \mathbb{R}$ is a limit point of $E \cap (p, \infty)$. Prove that

    $$\lim_{x \to p^+} f(x) = L \quad \text{if and only if} \quad \lim_{n \to \infty} f(p_n) = L$$

    for every sequence $\{p_n\}$ in E with $p_n > p$ for all $n \in \mathbb{N}$ and $p_n \to p$.

10. Let f be a real-valued function $(a, b]$.

*a. If f is continuous on $(a, b]$ and $\lim_{x \to a^+} f(x)$ exists, prove that f is uniformly continuous on $(a, b]$.

b. If f is uniformly continuous on $(a, b]$, prove that $\lim_{x \to a^+} f(x)$ exists.

11. Let $f : [0, 2] \to \mathbb{R}$ be defined by

$$f(x) = \begin{cases} x, & 0 \le x \le 1, \\ 1 + x^2, & 1 < x \le 2. \end{cases}$$

Show that f and f^{-1} are strictly increasing and find $f([0, 2])$. Are f and f^{-1} continuous at every point of their respective domains

12. ***If** $m \in \mathbb{Z}$, $n \in \mathbb{N}$, prove that $f(x) = x^{m/n}$ is continuous on $(0, \infty)$.

13. **a.** If f and g are monotone increasing functions on an interval I, prove that $f + g$ is monotone increasing on I.

b. If in addition f and g are positive, prove that fg is monotone increasing on I.

c. Show by example that the conclusion in part (b) may be false if f and g are not both positive on I.

14. Let $I \subset \mathbb{R}$ be an open interval and let $f : I \to \mathbb{R}$ be strictly increasing and continuous on I.

*a. If $U \subset I$ is open, prove that $f(U)$ is open.

b. Use (a) and Theorem 4.2.6 to prove that f^{-1} is continuous on $f(I)$.

15. Let $I \subset \mathbb{R}$ be an interval and let f be a one-to-one continuous real-valued function on I. Prove that f is strictly monotone on I.

16. Let $f : [0, 1] \to \mathbb{R}$ be defined by

$$f(x) = \begin{cases} 2x, & 0 \le x < \frac{1}{2}, \\ 3 - 2x, & \frac{1}{2} \le x \le 1. \end{cases}$$

a. Sketch the graph of f.

*b. Show that f is one-to-one on $[0, 1]$, but not strictly increasing on $[0, 1]$.

*c. Show that $f([0, 1]) = [0, 2]$.

*d. Find $f^{-1}(y)$. $y \in [0, 2]$, and show that f^{-1} is not continuous at $y_o = 1$.

17. Let $E = [0, 1] \cup [2, 3)$, and for $x \in E$ set

$$f(x) = \begin{cases} x^2, & 0 \le x \le 1, \\ 4 - x, & 2 \le x < 3. \end{cases}$$

a. Sketch the graph of f.

b. Show that f is one-to-one and continuous on E.

c. Show that $f(E) = [0, 2]$.

d. Find $f^{-1}(y)$ for $y \in [0, 2]$, and show that f^{-1} is not continuous at $y_o = 1$.

18. Let $f : [0,1] \to \mathbb{R}$ be defined by

$$f(x) = \begin{cases} x, & x \in \mathbb{Q}, \\ 1 - x, & x \notin \mathbb{Q}. \end{cases}$$

Prove that f is one-to-one, that $f([0,1]) = [0,1]$, but that f is not monotone of any interval $I \subset [0,1]$.

19. Prove that if f is monotone increasing on $[a, \infty)$, $a \in \mathbb{R}$, and bounded above, then $\lim\limits_{x \to \infty} f(x)$ exists.

20. Let $I \subset \mathbb{R}$ be an interval and let $f : I \to \mathbb{R}$ be monotone increasing. For $p \in \text{Int}(I)$, the **jump of f at** p, denoted $J_f(p)$ is defined by

$$J_f(p) = f(p+) - f(p-).$$

If p is a left endpoint, set $J_f(p) = f(p+) - f(p)$, and if p is a right endpoint, set $J_f(p) = f(p) - f(p-)$.

a. Prove that f is continuous at $p \in I$ if and only if $J_f(p) = 0$.

b. If $p \in \text{Int}(I)$, prove that $J_f(p) = \inf\{f(y) - f(x) : x < p < y, \ x, y \in I\}$.

21. Let $\{x_n\}_{n=1}^{\infty}$ be a countable subset of $\mathbb{R}$ and $\{c_n\}_{n=1}^{\infty}$ a sequence of positive real numbers satisfying $\sum c_n = 1$. Let $f : \mathbb{R} \to \mathbb{R}$ be defined by

$$f(x) = \sum_{n=1}^{\infty} c_n I(x - x_n).$$

Prove that $\lim\limits_{x \to -\infty} f(x) = 0$ and $\lim\limits_{x \to \infty} f(x) = 1$.

Notes

The limit of a function at a point is one of the fundamental tools of analysis. Not only is it crucial to continuity, but also to many subsequent topics in the text. The limit process will occur over and over again. We will encounter it in the next chapter in the definition of the derivative. It will occur again in the chapters on integration, series, etc.

Another very important concept that will be encountered on many other occasions in the text is uniform continuity. Uniform continuity is important in that given $\epsilon > 0$, it guarantees the existence of a $\delta > 0$ such that $|f(x) - f(y)| < \epsilon$ for all $x, y \in \text{Dom} f$ with $|x - y| < \delta$. In Chapter 6 we will use this to prove that every continuous real-valued function on $[a, b]$, $a, b \in \mathbb{R}$, is Riemann integrable on $[a, b]$. Other applications of uniform continuity will occur in many other theorems and in the exercises.

One of the most important results of this chapter is the intermediate value theorem (Theorem 4.2.11). The intermediate value theorem has already been used in Corollary 4.2.13 to prove the existence of nth roots; namely, for every positive real number x and $n \in \mathbb{N}$, there exists a unique positive real number y such that $y^n = x$. Even though the existence of nth roots can be proved without the intermediate value theorem, any such proof however is simply the statement that the function

$f(x) = x^n$ satisfies the intermediate value property on $[0, a]$ for every $a > 0$. Other applications of the intermediate value theorem will occur elsewhere in the text.

The proof of the intermediate value theorem depended on the fact that the connected subsets of $\mathbb{R}$ are the intervals (Theorem 2.2.25) and that the continuous image of a connected set is connected (Exercise 28 of Section 4.2). Assuming these two results, the intermediate value theorem is an immediate consequence as follows: Suppose f is continuous on $[a, b]$. Let $I = f([a, b])$. Then I is connected and thus must be an interval. Thus if $f(a) < \gamma < f(b)$, $\gamma \in I$ and hence there exists $c \in [a, b]$ such that $f(c) = \gamma$. That the continuous image of a connected set is connected follows from the definition. However, the proof that the connected subsets of $\mathbb{R}$ are the intervals requires the least upper bound property.

Miscellaneous Exercises

1. Let f be a continuous real-valued function on $[a, b]$ with $f(a) < 0 < f(b)$. Let $c_1 = \frac{1}{2}(a + b)$. If $f(c_1) > 0$, let $c_2 = \frac{1}{2}(a + c_1)$. If $f(c_1) < 0$, let $c_2 = \frac{1}{2}(c_1 + b)$. Continue this process inductively to obtain a sequence $\{c_n\}$ in (a, b) which converges to a point $c \in (a, b)$ for which $f(c) = 0$.

2. Let $E \subset \mathbb{R}$, p a limit point of E, and f a real-valued function defined on E. The **limit superior** of f at p, denoted $\overline{\lim}_{x \to p} f(x)$, is defined by

$$\overline{\lim}_{x \to p} f(x) = \inf_{\delta > 0} \sup\{f(x) : x \in (N_\delta(p) \setminus \{p\}) \cap E\}.$$

Similarly, the **limit inferior** of f at p, denoted $\underline{\lim}_{x \to p} f(x)$, is defined by

$$\underline{\lim}_{x \to p} f(x) = \sup_{\delta > 0} \inf\{f(x) : x \in (N_\delta(p) \setminus \{p\}) \cap E\}.$$

Prove each of the following:

a. $\overline{\lim}_{x \to p} f(x) \leq L$ if and only if given $\epsilon > 0$, there exists a $\delta > 0$ such that $f(x) < L + \epsilon$ for all $x \in E$, $0 < d(x, p) < \delta$.

b. $\overline{\lim}_{x \to p} f(x) \geq L$ if and only if given $\epsilon > 0$ and $\delta > 0$, there exists $x \in E$ with $0 < d(x, p) < \delta$ such that $f(x) > L - \epsilon$.

c. If $\overline{\lim}_{x \to p} f(x) = L$, then for any sequence $\{x_n\}$ in E with $x_n \neq p$ for all $n \in \mathbb{N}$, $\overline{\lim} f(x_n) \leq L$.

d. There exists a sequence $\{x_n\}$ in E with $x_n \neq p$ for all $n \in \mathbb{N}$, such that

$$\overline{\lim}_{x \to \infty} f(x_n) = \overline{\lim}_{x \to p} f(x).$$

3. Let $X \subset \mathbb{R}$ and f a real-valued function on X. For $p \in X$, the **oscillation** of f at p, denoted $\omega(f; p)$, is defined as

$$\omega(f; p) = \inf_{\delta > 0} \sup\{|f(x) - f(y)| : x, y \in N_\delta(p) \cap X\}.$$

Prove each of the following:

a. The function f is continuous at p if and only if $\omega(f; p) = 0$.

b. For every $s \in \mathbb{R}$, the set $\{x \in X : \omega(f; x) < s\}$ is open.

c. The set $\{x \in X : f \text{ is continuous at } x\}$ is the intersection of at most countably many sets that are open in X.

The following set of exercises involve the Cantor ternary function. Let P denote the Cantor ternary set of Section 2.3. For each $x \in (0, 1]$, let $x = .a_1a_2a_3....$ denote the ternary expansion of x. Define N as follows:

$$N = \begin{cases} \infty, & \text{if } a_n \neq 1 \text{ for all } n \in \mathbb{N}, \\ \min\{n : a_n = 1\}, & \text{otherwise.} \end{cases}$$

Define $b_n = \frac{1}{2}a_n$ for $n < N$, and $b_N = 1$, if N is finite. (Note: $b_n \in \{0, 1\}$ for all n.)

4. If $x \in (0, 1]$ has two ternary expansions, show that $\sum_{n=1}^{N} \dfrac{b_n}{2^n}$ is independent of the expansion of x.

The **Cantor ternary function** f on $[0, 1]$ is defined as follows: $f(0) = 0$, and if $x \in (0, 1]$ with ternary expansion $x = .a_1a_2a_3....$, set

$$f(x) = \sum_{n=1}^{N} \frac{b_n}{2^n},$$

where N and b_n are defined as above.

5. Prove each of the following:

   a. f is monotone increasing on $[0, 1]$.

   b. f is constant on each interval in the complement of the Cantor set in $[0, 1]$.

   c. f is continuous on $[0, 1]$.

   d. $f(P) = [0, 1]$.

   e. Sketch the graph of f.

Supplemental Reading

Bryant, J., Kuzmanovich, J. and Pavlichenkov, A., "Functions with compact preimages of compact sets," *Muth. Mag.* **70** (1997), 362–364.

Bumcrot, R. and Sheingorn, M., "Variations on continuity: Sets of infinite limits," *Math. Mag.* **47** (1974), 41–43.

Cauchy, A. L., *Cours d'analyse*, Paris, 1821, in *Oeuvres complétes*

d'Augustin Cauchy, series 2, vol.3, Gauthier-Villars Paris, 1899.

Dupree, E. and Mathes, B., "Functions with dense graphs," *Math. Mag.* **86** (2013), 366–369.

Fleron, Julian F., " A note on the history of the Cantor set and Cantor function," *Math. Mag.* **67** (1994), 136–140.

Grabinger, Judith V., "Who gave you

the epsilon? Cauchy and the origins of rigorous calculus," *Amer. Math. Monthly* **90** (1983), 185–194.

Radcliffe, D. G., "A function that is surjective on every interval," *Amer. Math. Monthly* **123** (2016), 88–89.

Snipes, Ray F., "Is every continuous function uniformly continuous," *Math. Mag.* **57** (1994), 169–173.

Straffin, Jr., Philip, D., "Periodic points of continuous functions," *Math. Mag.* **51** (1978) 99–105.

Velleman, D. J., "Characterizing continuity," *Amer. Math. Monthly* **104** (1997), 318-322.

5

Differentiation

The development of differential and integral calculus by Isaac Newton (1642–1727) and Gottfried Wilhelm Leibniz (1646–1716) in the mid seventeenth century constitutes one of the great advances in mathematics. In the two years following his degree from Cambridge in 1664, Newton invented the method of fluxions (derivatives) and fluents (integrals) to solve problems in physics involving velocity and motion. During the same period he also discovered the laws of universal gravitation and made significant contributions to the study of optics. Leibniz on the other hand, whose contributions came 10 years later, was led to the invention of calculus through the study of tangents to curves and the problem of area. The first published account of Newton's calculus appeared in his 1687 treatise *Philosophia Naturalis Principia Mathematica*. Unfortunately however, much of Newton's work on calculus did not appear until 1737, ten years after his death, in a work entitled *Methodus fluxionum et serierum infinitorum*.

Mathematicians prior to the time of Newton and Leibniz knew how to compute tangents to specific curves and velocities in particular situations. They also knew how to compute areas under elementary curves. What distinguished the work of Newton and Leibniz from that of their predecessors was that they realized that the problems of finding the tangent to a curve and the area under a curve were inversely related. More importantly however, they also developed the notation and a set of techniques (a calculus) to solve these problems for arbitrary functions, whether algebraic or transcendental. In Newton's presentation of his infinitesimal calculus he looked upon y as a flowing quantity, or fluent, of which the quantity $\dot{y}$ was the fluxion or rate of change. The notation of Newton is still in use in physics and differential geometry, whereas every student of calculus is exposed to the d (for difference) and $\int$ (for sum) notation of Leibniz to denote differentiation and integration. Many of the basic rules and formulas of the differential calculus were developed by these two remarkable mathematicians. In the paper *A New Method for Maxima and Minima, and also for Tangents, which is not Obstructed by Irrational Quantities* published in 1684, Leibniz gave correct rules for differentiation of sums, products, quotients, powers, and roots. In addition to his many contributions to the subject, Leibniz also disseminated his results in publications and correspondence with colleagues throughout Europe.

Newton and Leibniz with their invention of the calculus had created a tool of such novel subtlety that its utility was proved for over 150 years before its limitations forced mathematicians to clarify its foundations. The rigorous formulation of the derivative did not occur until 1821 when Cauchy provided a formal definition of limit. This helped to place the theory on a firm mathematical footing. Cauchy's contributions to the rigorous development of calculus will be evident both in this and subsequent chapters.

In this chapter, we develop the theory of differentiation based on the definition of Cauchy, with special emphasis on the mean value theorem and consequences thereof. The first section presents the standard results concerning derivatives of functions obtained by means of algebraic operations and composition. In the examples and exercises we will derive the derivatives of some of the basic algebraic and trigonometric functions. However, throughout the chapter we will assume that the reader is already familiar with standard techniques of differentiation and some of its applications. As a consequence we will concentrate on the mathematical concepts of the derivative, emphasizing many of its more subtle properties.

5.1   The Derivative

In an elementary calculus course the derivative is usually introduced by considering the problem of the tangent line to a curve or as the problem of finding the velocity of an object moving in a straight line. Suppose $y = f(x)$ is a real-valued function defined on an interval $[a, b]$. Fix $p \in [a, b]$. For $x \in [a, b]$, $x \neq p$, the quantity

$$Q(x) = \frac{f(x) - f(p)}{x - p}$$

represents the slope of the straight line (**secant line**) joining the points $(p, f(p))$ and $(x, f(x))$ on the graph of f (see Figure 5.1). The function $Q(x)$ is defined for all values of $x \in [a, b]$, $x \neq p$. The limit of $Q(x)$ as x approaches p, provided this limit exists, is defined as the slope of the **tangent line** to the curve $y = f(x)$ at the point $(p, f(p))$.

A similar type of limit occurs if we consider the problem of defining the velocity of a moving object. Suppose that an object is moving in a straight line and that its distance s from a fixed point P is given as a function of t; namely $s = s(t)$. If t_o is fixed, then the average velocity over the time interval from t to t_o, $t \neq t_o$, is defined as

$$\frac{s(t) - s(t_o)}{t - t_o}.$$

The limit of this quantity as t approaches t_o, again provided that the limit exists, is taken as the definition of the velocity of the object at time t_o.

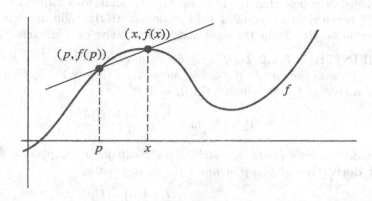

FIGURE 5.1
Secant line between two points on the graph of f

Both of the previous two examples involve identical limits; namely,

$$\lim_{x \to p} \frac{f(x) - f(p)}{x - p} \quad \text{and} \quad \lim_{t \to t_o} \frac{s(t) - s(t_o)}{t - t_o}.$$

These limits, if they exist, are called the derivatives of the functions f and s at p and t_o respectively. The term derivative comes from the French *fonction derivée*.

DEFINITION 5.1.1 *Let $I \subset \mathbb{R}$ be an interval and let f be a real-valued function with domain I. For fixed $p \in I$, the* **derivative** *of f at p, denoted $f'(p)$, is defined to be*

$$f'(p) = \lim_{x \to p} \frac{f(x) - f(p)}{x - p},$$

provided the limit exists. If $f'(p)$ is defined at a point $p \in I$, we say that f is **differentiable** *at p. If the derivative f' is defined at every point of a set $E \subset I$, we say that f is* **differentiable on** E.

If p is an interior point of I, then $p + h \in I$ for all h sufficiently small. If we set $x = p + h$, $h \neq 0$, then the definition of the derivative of f at p can be expressed as

$$f'(p) = \lim_{h \to 0} \frac{f(p + h) - f(p)}{h},$$

provided that the limit exists. This formulation of the derivative is sometimes easier to use.

In the definition of the derivative we do not exclude the possibility that p is an endpoint of I. If $p \in I$ is the left endpoint of I, then

$$f'(p) = \lim_{x \to p^+} \frac{f(x) - f(p)}{x - p} = \lim_{h \to 0^+} \frac{f(p + h) - f(p)}{h},$$

provided of course that the limit exists. The analogous formula also holds if $p \in I$ is the right endpoint of I. In analogy with the right and left limit of a function we also define the right and left derivative of a function.

DEFINITION 5.1.2 *Let $I \subset \mathbb{R}$ be an interval and let f be a real-valued function with domain I. If $p \in I$ is such that $I \cap (p, \infty) \neq \emptyset$, then the* **right derivative** *of f at p, denoted $f'_+(p)$, is defined as*

$$f'_+(p) = \lim_{h \to 0+} \frac{f(p+h) - f(p)}{h},$$

provided the limit exists. Similarly, if $p \in I$ satisfies $(-\infty, p) \cap I \neq \emptyset$, then the **left derivative** *of f at p, denoted $f'_-(p)$, is given by*

$$f'_-(p) = \lim_{h \to 0-} \frac{f(p+h) - f(p)}{h},$$

provided the limit exists.

Remarks. (a) If $p \in \text{Int}(I)$, then $f'(p)$ exists if and only if both $f'_+(p)$ and $f'_-(p)$ exist, and are equal. On the other hand, if $p \in I$ is the left (right) endpoint of I, then $f'(p)$ exists if and only if $f'_+(p)$ ($f'_-(p)$) exists. In this case, $f'(p) = f'_+(p)$ ($f'_-(p)$).

The reader should note the distinction between $f'_+(p)$ and $f'(p+)$. The first denotes the right derivative of f at p, whereas the later is the right limit of the derivative; i.e.,

$$f'(p+) = \lim_{x \to p+} f'(x).$$

Here of course we are assuming that f' is defined for all $x \in (p, p+\delta)$ for some $\delta > 0$.

(b) If f is a differentiable function on an interval I, we will also occasionally use Leibniz's notation

$$\frac{d}{dx} f(x), \quad \frac{df}{dx}, \quad \text{or} \quad \frac{dy}{dx},$$

to denote the derivative of $y = f(x)$.

(c) If f is differentiable on an interval I, then the derivative $f'(x)$ is itself a function on I. Therefore we can consider the existence of the derivative of the function f' at a point $p \in I$. If the function f' has a derivative at a point $p \in I$, we refer to this quantity as the **second derivative** of f at p, that we denote $f''(p)$. Thus

$$f''(p) = \lim_{h \to 0} \frac{f'(p+h) - f'(p)}{h}.$$

In a similar fashion we can define the third derivative of f at p, denoted $f'''(p)$ or $f^{(3)}(p)$. In general, for $n \in \mathbb{N}$, $f^{(n)}(p)$ denotes the nth **derivative** of f at p. In order to discuss the existence of the nth derivative of f at p, we require the existence of the $(n-1)$st derivative of f on an interval containing p.

EXAMPLES 5.1.3 (a) In the exercises (Exercise 2) you will we asked to prove that if $f(x) = x^n$, $n \in \mathbb{Z}$, then $f'(x) = nx^{n-1}$ for all $x \in \mathbb{R}$ ($x \neq 0$ if n is negative). For the function $f(x) = x^2$, the result is obtained as follows:

$$f'(x) = \lim_{h \to 0} \frac{(x+h)^2 - x^2}{h} = \lim_{h \to 0}(2x + h) = 2x.$$

A similar computation shows that $f''(x) = 2$.

(b) Consider $f(x) = \sqrt{x}$, $x > 0$. We first note that for $h \neq 0$.

$$\frac{f(x+h) - f(x)}{h} = \frac{\sqrt{x+h} - \sqrt{x}}{h}$$

$$= \frac{(\sqrt{x+h} - \sqrt{x})}{h} \frac{(\sqrt{x+h} + \sqrt{x})}{(\sqrt{x+h} + \sqrt{x})}$$

$$= \frac{1}{\sqrt{x+h} + \sqrt{x}}.$$

Since $\lim_{h \to 0} \sqrt{x+h} = \sqrt{x}$. we have

$$f'(x) = \lim_{h \to 0} \frac{1}{\sqrt{x+h} + \sqrt{x}} = \frac{1}{2\sqrt{x}}.$$

(c) Consider $f(x) = \sin x$. From the identity

$$\sin(x+h) = \sin x \cos h + \cos x \sin h$$

we obtain

$$\frac{\sin(x+h) - \sin x}{h} = \sin x \left[\frac{\cos h - 1}{h}\right] + \cos x \left[\frac{\sin h}{h}\right].$$

By Example 4.1.10(d) and Exercise 5, Section 4.1,

$$\lim_{h \to 0} \frac{\sin h}{h} = 1 \quad \text{and} \quad \lim_{h \to 0} \frac{\cos h - 1}{h} = 0.$$

Therefore,

$$f'(x) = \lim_{h \to 0} \frac{\sin(x+h) - \sin x}{h}$$

$$= \sin x \lim_{h \to 0}\left[\frac{\cos h - 1}{h}\right] + \cos x \lim_{h \to 0}\left[\frac{\sin h}{h}\right]$$

$$= \cos x.$$

In Exercise 3 you will be asked to prove that $\dfrac{d}{dx}(\cos x) = -\sin x$.

(d) Let f be defined by

$$f(x) = |x| = \begin{cases} x, & x \geq 0, \\ -x, & x < 0. \end{cases}$$

Then

$$f'_+(0) = \lim_{h \to 0^+} \frac{|h|}{h} = \lim_{h \to 0^+} \frac{h}{h} = 1, \quad \text{and}$$

$$f'_-(0) = \lim_{h \to 0^-} \frac{|h|}{h} = \lim_{h \to 0^-} \frac{-h}{h} = -1.$$

Thus $f'_+(0)$ and $f'_-(0)$ both exist, but are unequal. Therefore $f'(0)$ does not exist.

(e) In this example, let $g(x) = x^{3/2}$ with Dom $g = [0, \infty)$. Then for $p = 0$,

$$g'(0) = g'_+(0) = \lim_{x \to 0^+} \frac{x^{3/2}}{x} = \lim_{x \to 0^+} \sqrt{x} = 0.$$

Thus g is differentiable at 0 with $g'(0) = 0$.

(f) Let f be defined by

$$f(x) = \begin{cases} x \sin \dfrac{1}{x}, & x \neq 0, \\ 0, & x = 0. \end{cases}$$

For $x \neq 0$, $f'(x) = -\dfrac{1}{x} \cos \dfrac{1}{x} + \sin \dfrac{1}{x}$ (Exercise 8). When $x = 0$,

$$f'(0) = \lim_{h \to 0} \frac{f(h) - f(0)}{h} = \lim_{h \to 0} \sin \frac{1}{h},$$

which by Example 4.1.5(a) does not exist. Therefore $f'(0)$ does not exist.

(g) Consider the following variation of (f). Let

$$g(x) = \begin{cases} x^2 \sin \dfrac{1}{x}, & x \neq 0, \\ 0, & x = 0. \end{cases}$$

This example is very important! In Exercise 9 you will be asked to show that $g'(x)$ exists for all $x \in \mathbb{R}$ with $g'(0) = 0$, but that the derivative g' is **not** continuous at 0. $\square$

When Cauchy gave the rigorous definition of the derivative, he assumed that the given function was continuous on its domain. As a consequence of the following theorem this requirement is not necessary.

THEOREM 5.1.4 *If $I \subset \mathbb{R}$ is an interval and $f : I \to \mathbb{R}$ is differentiable at $p \in I$, then f is continuous at p.*

Proof. For $t \neq p$,

$$f(t) - f(p) = \left(\frac{f(t) - f(p)}{t - p} \right)(t - p).$$

Since

$$\lim_{t \to p} \frac{f(t) - f(p)}{t - p}$$

exists and equals $f'(p)$, by Theorem 4.1.6(b),

$$\lim_{t \to p}(f(t) - f(p)) = \lim_{t \to p} \left(\frac{f(t) - f(p)}{t - p} \right) \lim_{t \to p}(t - p) = f'(p) \cdot 0 = 0.$$

Therefore, $\lim_{t \to p} f(t) = f(p)$ and thus f is continuous at p. In the above, if p is an endpoint, then the limits are either the right or left limit at p, whichever is appropriate. $\square$

Remark. In both Examples 5.1.3(d) and (f), the given function is continuous at 0, but not differentiable at 0. Given a finite number of points, say $p_1, ..., p_n$, it is easy to construct a function f that is continuous but not differentiable at $p_1, ..., p_n$. For example,

$$f(x) = \sum_{k=1}^{n} |x - p_k|$$

has the desired properties. In 1861, Weierstrass constructed a function f that is continuous at every point of $\mathbb{R}$ but nowhere differentiable. When published in 1874, this example astounded the mathematical community. Prior to this time mathematicians generally believed that continuous functions were differentiable (except perhaps at a finite number of points). In Example 8.5.3 we will consider the function of Weierstrass in detail.

Derivatives of Sums, Products, and Quotients

We now derive the formulas for the derivative of sums, products, and quotients of functions. These rules were discovered by Leibniz in 1675.

THEOREM 5.1.5 *Suppose f, g are real-valued functions defined on an interval I. If f and g are differentiable at $x \in I$, then $f + g$, $f g$ and f/g (if $g(x) \neq 0$) are differentiable at x and*

(a) $(f + g)'(x) = f'(x) + g'(x)$,

(b) $(f g)'(x) = f'(x)g(x) + f(x)g'(x)$,

(c) $\left(\dfrac{f}{g} \right)'(x) = \dfrac{f'(x)g(x) - f(x)g'(x)}{(g(x))^2}$, *provided* $g(x) \neq 0$.

Proof. The proof of (a) is left as an exercise (Exercise 4). For the proof of (b), by adding and subtracting the term $f(x+h)g(x)$, we have for $h \neq 0$,

$$\frac{(f\,g)(x+h) - (f\,g)(x)}{h} = f(x+h)\left(\frac{g(x+h) - g(x)}{h}\right)$$
$$+ \left(\frac{f(x+h) - f(x)}{h}\right)g(x).$$

By Theorem 5.1.4, since f is differentiable at x, $\lim_{h\to 0} f(x+h) = f(x)$. Thus since each of the limits exist, by Theorem 4.1.6

$$(f\,g)'(x) = \lim_{h\to 0} \frac{(f\,g)(x+h) - (f\,g)(x)}{h}$$
$$= \lim_{h\to 0} f(x+h) \lim_{h\to 0}\left(\frac{g(x+h) - g(x)}{h}\right)$$
$$+ g(x) \lim_{h\to 0}\left(\frac{f(x+h) - f(x)}{h}\right)$$
$$= f(x)g'(x) + g(x)f'(x).$$

To prove (c), we first prove that $(1/g)'(x) = -g'(x)/[g(x)]^2$, provided $g(x) \neq 0$. The result then follows by writing f/g as $f \cdot (1/g)$ and applying the product formula (b). If $g(x) \neq 0$, then since g is continuous at x, as in Theorem 4.1.6(c), $g(x+h) \neq 0$ for all h sufficiently small. Thus for h sufficiently small and nonzero,

$$\frac{\dfrac{1}{g(x+h)} - \dfrac{1}{g(x)}}{h} = -\left(\frac{g(x+h) - g(x)}{h}\right)\frac{1}{g(x)g(x+h)}.$$

Again, using the fact that $\lim_{h\to 0} g(x+h) = g(x)$, by Theorem 4.1.6

$$\left(\frac{1}{g}\right)'(x) = \lim_{h\to 0} \frac{\dfrac{1}{g(x+h)} - \dfrac{1}{g(x)}}{h}$$
$$= -\lim_{h\to 0}\left(\frac{g(x+h) - g(x)}{h}\right)\lim_{h\to 0}\frac{1}{g(x)g(x+h)}$$
$$= \frac{-g'(x)}{g^2(x)}. \quad \square$$

The Chain Rule

The previous theorem allows us to compute the derivatives of sums, products, and quotients of differentiable functions. The chain rule on the other hand allows us to compute the derivative of a function obtained from the composition of two or more differentiable functions. Prior to stating and proving the result, we introduce some useful notation.

Suppose f is differentiable at $x \in I$. For $t \in I$, $t \neq x$, set

$$Q(t) = \frac{f(t) - f(x)}{t - x}.$$

Then by Definition 5.1.1, $Q(t) \to f'(x)$ as $t \to x$. If we let $u(t) = Q(t) - f'(x)$, then $u(t) \to 0$ as $t \to x$. Therefore, if f is differentiable at x, for $t \neq x$,

$$f(t) - f(x) = (t - x)\left[f'(x) + u(t)\right], \quad \text{where} \quad u(t) \to 0 \text{ as } t \to x. \quad (1)$$

By setting $u(x) = 0$, the above identity is valid for all $t \in I$.

THEOREM 5.1.6 (Chain Rule) *Suppose f is a real-valued function defined on an interval I and g is a real-valued function defined on some interval J such that* Range $f \subset J$. *If f is differentiable at $x \in I$ and g is differentiable at $f(x)$, then $h = g \circ f$ is differentiable at x and*

$$h'(x) = g'(f(x))\, f'(x).$$

Proof. Let $y = f(x)$. Then by (1)

$$f(t) - f(x) = (t - x)\left[f'(x) + u(t)\right], \qquad (2)$$
$$g(s) - g(y) = (s - y)\left[g'(y) + v(s)\right], \qquad (3)$$

where $t \in I$, $s \in J$ and $u(t) \to 0$ as $t \to x$ and $v(s) \to 0$ as $s \to y$. Let $s = f(t)$. Since f is continuous at x, $s \to y$ as $t \to x$. By identity (3), and then (2),

$$\begin{aligned}
h(t) - h(x) &= g(f(t)) - g(f(x)) \\
&= [f(t) - f(x)]\,[g'(y) + v(s)] \\
&= (t - x)\,[f'(x) + u(t)]\,[g'(y) + v(f(t))].
\end{aligned}$$

Therefore, for $t \neq x$,

$$\frac{h(t) - h(x)}{t - x} = [f'(x) + u(t)]\,[g'(y) + v(f(t))].$$

Since $v(f(t))$ and $u(t)$ both have limit 0 as $t \to x$,

$$\lim_{t \to x} \frac{h(t) - h(x)}{t - x} = f'(x)\, g'(y) = g'(f(x))\, f'(x). \quad \square$$

To illustrate the previous theorem we consider the following examples.

EXAMPLES 5.1.7 (a) By Example 5.1.3(c) the function $f(x) = \sin x$ is differentiable on $\mathbb{R}$. Hence if $g : I \to \mathbb{R}$ is differentiable on the interval I, $h(x) = (f \circ g)(x) = \sin g(x)$ is differentiable on I with

$$h'(x) = f'(g(x))g'(x) = g'(x) \cos g(x).$$

In particular, if $g(x) = 1/x^2$, Dom $g = (0, \infty)$, then by Theorem 5.1.5(c),

$$g'(x) = \frac{-2x}{(x^2)^2} = -\frac{2}{x^3}, \qquad x \in (0, \infty).$$

Therefore,

$$\frac{d}{dx} \sin \frac{1}{x^2} = -\frac{2}{x^2} \cos \frac{1}{x^2}.$$

(b) By Exercise 2

$$\frac{d}{dx} x^n = nx^{n-1} \quad \text{for all } n \in \mathbb{N}.$$

Thus if $f : I \to \mathbb{R}$ is differentiable on the interval I, then by the chain rule $g(x) = [f(x)]^n$, $n \in \mathbb{N}$, is differentiable on I with $g'(x) = n[f(x)]^{n-1} f'(x)$. This formula can also be obtained from Theorem 5.1.5(b) using mathematical induction.

Exercises 5.1

1. Use the definition to find the derivative of each of the following functions.
    *a. $f(x) = x^3, \quad x \in \mathbb{R}$ **b.** $g(x) = \sqrt{x+2}, \quad x > -2$
    *c. $h(x) = \dfrac{1}{x}, \quad x \neq 0$ **d.** $k(x) = \dfrac{1}{\sqrt{x+2}}, \quad x > -2$
    *e. $f(x) = \dfrac{x}{x+1}, \quad x \neq -1$ **f.** $g(x) = \dfrac{x}{x^2+1} \quad x \in \mathbb{R}$

2. ***Prove** that for all integers n, $\dfrac{d}{dx} x^n = n x^{n-1}$ ($x \neq 0$ if n is negative).

3. ***a.** Prove that $\dfrac{d}{dx}(\cos x) = -\sin x$.

    b. Find the derivative of $\tan x = \dfrac{\sin x}{\cos x}$.

4. Prove Theorem 5.1.5(a).

5. For each of the following, determine whether the given function is differentiable at the indicated point x_o. Justify your answer!
    *a. $f(x) = x|x| \quad$ at $x_o = 0$

    b. $f(x) = \begin{cases} x^2, & x \in \mathbb{Q}, \\ 0, & x \notin \mathbb{Q}, \end{cases} \quad$ at $x_o = 0$

    *c. $g(x) = (x-2)[x]$, at $x_o = 2$
    d. $h(x) = \sqrt{x+2}, x \in [-2, \infty)$, at $x_0 = -2$

    *e. $f(x) = \begin{cases} \sin x, & x \in \mathbb{Q}, \\ x, & x \notin \mathbb{Q}, \end{cases} \quad$ at $x_o = 0$

    f. $g(x) = \begin{cases} x^{4/3} \cos x, & x \neq 0, \\ 0, & x = 0, \end{cases} \quad$ at $x_o = 0$

6. Let $f(x) = |x|^3$. Compute $f'(x)$, $f''(x)$, and show that $f'''(0)$ does not exist.

7. Determine where each of the following functions from $\mathbb{R}$ to $\mathbb{R}$ is differentiable and find the derivative.

   *a. $f(x) = x\,[x]$.

   b. $g(x) = |x - 2| + |x + 1|$.

   *c. $h(x) = |\sin x|$.

   d. $k(x) = \begin{cases} x^2 \left[\dfrac{1}{x}\right], & 0 < x \leq 1, \\ 0, & x = 0 \end{cases}$

8. Use the product rule, quotient rule, and chain rule to find the derivative of each of the following.

   a $f(x) = x \sin \frac{1}{x}$, $x \neq 0$

   b. $f(x) = (\cos(\sin x)^n)^m$, $n, m \in \mathbb{N}$

   c. $\sqrt{x + \sqrt{2 + x}}$

   d. $f(x) = x^4(x + \sin \frac{1}{x})$, $x \neq 0$

   e. $f(x) = \dfrac{\sin^2 x}{1 + \sin^2 x}$

   f. $f(x) = \cos^4\left(\frac{x+1}{x-1}\right)$, $x \neq 1$

9. Let g be defined by

   $$g(x) = \begin{cases} x^2 \sin \dfrac{1}{x}, & x \neq 0, \\ 0, & x = 0. \end{cases}$$

   a. Prove that g is differentiable at 0 and that $g'(0) = 0$.

   *b. Show that $g'(x)$ is not continuous at 0.

10. Let f be defined by

    $$f(x) = \begin{cases} x^2 + 2, & x \leq 2, \\ ax + b, & x > 2. \end{cases}$$

    *a. For what values of a and b is f continuous at 2 ?

    *b. For what values of a and b is f differentiable at 2?

11. Let f be defined by

    $$f(x) = \begin{cases} ax + b, & x < -1, \\ x^3 + 1, & -1 \leq x \leq 2, \\ cx + d, & x > 2. \end{cases}$$

    Determine the constants a, b, c, and d such that f is differentiable on $\mathbb{R}$.

12. Assume there exists a function $L : (0, \infty) \to \mathbb{R}$ satisfying $L'(x) = 1/x$ for all $x \in (0, \infty)$. Find the derivative of each of the following.

    *a. $f(x) = L(2x + 1)$, $x > -\frac{1}{2}$

    b. $g(x) = L(x^2)$, $x \neq 0$

    *c. $h(x) = [L(x)]^3$, $x > 0$

    d. $k(x) = L(L(x))$, $x \in \{x > 0 : L(x) > 0\}$

13. Let L be the function of Exercise 12.

    a. Show that L is one-to-one on $(0, \infty)$.

    b. Let $E = L^{-1}$ on $\mathbb{R}$. By considering $L(E(y))$ prove that $E'(y) = E(y)$.

14. For b real, let f be defined by

    $$f(x) = \begin{cases} x^b \sin \dfrac{1}{x}, & x > 0, \\ 0, & x \leq 0. \end{cases}$$

Prove the following:

    a. f is continuous at 0 if and only if $b > 0$.

  *b. f is differentiable at 0 if and only if $b > 1$.

    c. f' is continuous at 0 if and only if $b > 2$.

15.   **a.** If f is differentiable at x_o, prove that

$$\lim_{h \to 0} \frac{f(x_o + h) - f(x_o - h)}{2h} = f'(x_o).$$

  *b. If $\displaystyle\lim_{h \to 0} \frac{f(x_o + h) - f(x_o - h)}{2h}$ exists, is f differentiable at x_o?

16.   If $f : (a, b) \to \mathbb{R}$ is differentiable at $p \in (a, b)$, prove that

$$f'(p) = \lim_{n \to \infty} n[f(p + \tfrac{1}{n}) - f(p)].$$

Show by example that the existence of the limit of the sequence $\{n[f(p + \tfrac{1}{n}) - f(p)]\}$ does not imply the existence of $f'(p)$.

17.   (**Leibniz's Rule**) Suppose f and g have nth order derivatives on (a, b). Prove that

$$(fg)^{(n)}(x) = \sum_{k=0}^{n} \binom{n}{k} f^{(k)}(x) g^{(n-k)}(x).$$

5.2   The Mean Value Theorem

In this section, we will prove the mean value theorem and give several consequences of this important result. Even though the proof itself is elementary, the theorem is one of the most useful results of analysis. Its importance is based on the fact that it allows us to relate the values of a function to values of its derivative. We begin the section with a discussion of local maxima and minima.

Local Maxima and Minima

DEFINITION 5.2.1 *Suppose $E \subset \mathbb{R}$ and f is a real-valued function with domain E. The function f has a **local maximum** at a point $p \in E$ if there exists a $\delta > 0$ such that $f(x) \leq f(p)$ for all $x \in E \cap N_\delta(p)$. The function f has an **absolute maximum** at $p \in E$ if $f(x) \leq f(p)$ for all $x \in E$.*

*Similarly, f has a **local minimum** at a point $q \in E$ if there exists a $\delta > 0$ such that $f(x) \geq f(q)$ for all $x \in E \cap N_\delta(q)$, and f has an **absolute minimum** at $q \in E$ if $f(x) \geq f(q)$ for all $x \in E$.*

Remark. As a consequence of Corollary 4.2.9, every continuous real-valued function defined on a compact subset K of $\mathbb{R}$ has an absolute maximum and minimum on K.

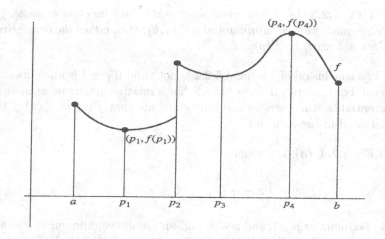

FIGURE 5.2
Absolute maxima and minima on the graph of f

The function f illustrated in Figure 5.2 has a local maximum at a, p_2, and p_4, and a local minimum at p_1, p_3, and b. The points $(p_4, f(p_4))$ and $(p_1, f(p_1))$ are absolute maxima and absolute minima respectively.

The following theorem gives the relationship between local maxima of a function defined on an interval and the values of its derivative.

THEOREM 5.2.2 *Let f be a real-valued function defined on an interval I, and suppose f has either a local minimum or local maximum at $p \in \text{Int}(I)$. If f is differentiable at p, then $f'(p) = 0$.*

Proof. If f is differentiable at $p \in \text{Int}(I)$, then $f'_-(p)$ and $f'_+(p)$ both exist and are equal. Suppose f has a local maximum at p. Then there exists a $\delta > 0$ such that $f(t) \leq f(p)$ for all $t \in I$ with $|t - p| < \delta$. In particular, if $p < t < p + \delta$, $t \in I$, then

$$\frac{f(t) - f(p)}{t - p} \leq 0.$$

Thus $f'_+(p) \leq 0$. Similarly, if $p - \delta < t < p$,

$$\frac{f(t) - f(p)}{t - p} \geq 0,$$

and therefore $f'_-(p) \geq 0$. Finally, since $f'_+(p) = f'_-(p) = f'(p)$, we have $f'(p) = 0$. The proof of the case where f has a local minimum at p is similar. $\square$

As a consequence of the previous theorem we have the following corollary.

COROLLARY 5.2.3 *Let f be a continuous real-valued function on $[a,b]$. If f has a relative maximum or minimum at $p \in (a,b)$, then either the derivative of f at p does not exist, or $f'(p) = 0$.*

Remark. The conclusion of Theorem 5.2.2 is not valid if $p \in I$ is an endpoint of the interval. For example, if $f : [a,b] \to \mathbb{R}$ has a relative maximum at a, and if f is differentiable at a, then we can only conclude that $f'(a) = f'_+(a) \leq 0$. This is illustrated in the following

EXAMPLES 5.2.4 **(a)** The function

$$f(x) = \left(x - \frac{1}{2}\right)^2, \qquad 0 \leq x \leq 2,$$

has a local maximum at $p = 0$ and $p = 2$, and an absolute minimum at $q = \frac{1}{2}$. By computation, we have $f'_+(0) = -1$, $f'_-(2) = 3$, and $f'(\frac{1}{2}) = 0$. In Exercise 1 you will be asked to graph the function f.

(b) The function $f(x) = |x|$, $x \in [-1,1]$, has an absolute minimum at $p = 0$. However, by Example 5.1.3(d) the derivative does not exist at $p = 0$. □

Rolle's theorem

Prior to stating and proving the mean value theorem we first state and prove the following theorem due to Michel Rolle (1652–1719).

THEOREM 5.2.5 (Rolle's Theorem) *Suppose f is a continuous real-valued function on $[a,b]$ with $f(a) = f(b)$, and that f is differentiable on (a,b). Then there exists $c \in (a,b)$ such that $f'(c) = 0$.*

Since the derivative of f at c gives the slope of the tangent line at $(c, f(c))$, a geometric interpretation of Rolle's theorem is that if f satisfies the hypothesis of the theorem, then there exists a least one value of $c \in (a,b)$ where the tangent line to the graph of f is horizontal. For the function f depicted in Figure 5.3, there are exactly two such points.

Proof. If f is constant on $[a,b]$, then $f'(x) = 0$ for all $x \in [a,b]$. Thus, we assume that f is not constant. Since the closed interval $[a,b]$ is compact, by Corollary 4.2.9, f has a maximum and a minimum on $[a,b]$. If $f(t) > f(a)$ for some t, then f has a maximum at some $c \in (a,b)$. Thus by Theorem 5.2.2, $f'(c) = 0$. If $f(t) < f(a)$ for some t, then f has a minimum at some $c \in (a,b)$, and thus again $f'(c) = 0$. □

Remarks. **(a)** Continuity of f on $[a,b]$ is required in the proof of Rolle's theorem. The function

$$f(x) = \begin{cases} x, & 0 \leq x < 1, \\ 0, & x = 1, \end{cases}$$

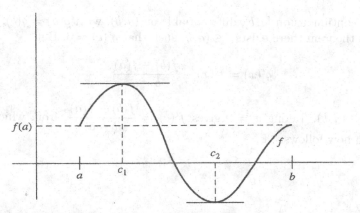

FIGURE 5.3
Rolle's theorem

is differentiable on $(0,1)$ and satisfies $f(0) = f(1) = 0$; yet $f'(x) \neq 0$ for all $x \in (0,1)$. The function f fails to be continuous at 1.

(b) For Rolle's theorem, differentiability of f at a and b is not required. For example, the function $f(x) = \sqrt{4 - x^2}$, $x \in [-2, 2]$ satisfies the hypothesis of Rolle's theorem, yet the derivative does not exist at -2 and 2. For $x \in (-2, 2)$,

$$f'(x) = \frac{-x}{\sqrt{4 - x^2}},$$

and the conclusion of Rolle's theorem is satisfied with $c = 0$.

The Mean Value Theorem

As a consequence of Rolle's theorem we obtain the mean value theorem. This result is usually attributed to Joseph Lagrange (1736–1813).

THEOREM 5.2.6 (Mean Value Theorem) *If $f : [a, b] \to \mathbb{R}$ is continuous on $[a, b]$ and differentiable on (a, b), then there exists $c \in (a, b)$ such that*

$$f(b) - f(a) = f'(c)(b - a).$$

Graphically, the mean value theorem states that there exists at least one point $c \in (a, b)$ such that the slope of the tangent line to the graph of the function f is equal to the slope of the straight line passing through $(a, f(a))$ and $(b, f(b))$. For the function of Figure 5.4, there are two such values of c, namely c_1 and c_2.

Proof. Consider the function g defined on $[a, b]$ by

$$g(x) = f(x) - f(a) - \left[\frac{f(b) - f(a)}{b - a}\right](x - a).$$

Then g is continuous on $[a, b]$, differentiable on (a, b), with $g(a) = g(b)$. Thus by Rolle's theorem there exists $c \in (a, b)$ such that $g'(c) = 0$. But

$$g'(x) = f'(x) - \frac{f(b) - f(a)}{b - a}$$

for all $x \in (a, b)$. Taking $x = c$ gives $f'(c) = \dfrac{f(b) - f(a)}{b - a}$, from which the conclusion now follows. $\square$

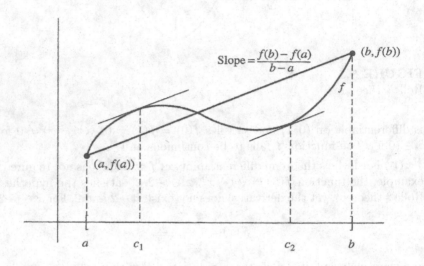

FIGURE 5.4
Mean value theorem

The mean value theorem is one of the fundamental results of differential calculus. Its importance lies in the fact that it enables us to obtain information about a function f from its derivative f'. In Example 5.2.7 we will illustrate how the mean value theorem can be used to derive inequalities. Other applications will be given later in this section and in the exercises. It will also be used in many other instances later in the text.

EXAMPLE 5.2.7 In this example, we illustrate how the mean value theorem may be used in proving elementary inequalities. We will use it to prove that
$$\frac{x}{1 + x} \le \ln(1 + x) \le x \qquad \text{for all} \quad x > -1,$$

where ln denotes the natural logarithm function. This function is considered in greater detail in Example 6.3.5 of the next chapter. Let $f(x) = \ln(1 + x)$, $x \in (-1, \infty)$. Then $f(0) = 0$. If $x > 0$, then by the mean value theorem, there exists $c \in (0, x)$ such that

$$\ln(1 + x) = f(x) - f(0) = f'(c) x.$$

But $f'(c) = (1+c)^{-1}$ and $(1+x)^{-1} < (1+c)^{-1} < 1$ for all $c \in (0, x)$. Therefore

$$\frac{x}{1+x} < f'(c)x < x,$$

and as a consequence

$$\frac{x}{1+x} \le \ln(1 + x) \le x \qquad \text{for all} \quad x \ge 0.$$

Now suppose $-1 < x < 0$. Then again by the mean value theorem there exists $c \in (x, 0)$ such that

$$\ln(1 + x) = f(x) - f(0) = \frac{x}{1+c}.$$

But since $x < c < 0$, we have $1 < (1 + c)^{-1} < (1 + x)^{-1}$, and since x is negative,

$$\frac{x}{1+x} < \ln(1 + x) < x.$$

Hence the desired inequality holds for all $x > -1$, with equality if and only if $x = 0$. $\square$

The following theorem, attributed to Cauchy, is a useful generalization of the mean value theorem.

THEOREM 5.2.8 (Cauchy Mean Value Theorem) *If f, g are continuous real-valued functions on $[a, b]$ that are differentiable on (a, b), then there exists $c \in (a, b)$ such that*

$$[f(b) - f(a)] \, g'(c) = [g(b) - g(a)] \, f'(c).$$

Proof. Let

$$h(x) = [f(b) - f(a)] \, g(x) - [g(b) - g(a)] \, f(x).$$

Then h is continuous on $[a, b]$, differentiable on (a, b) with

$$h(a) = f(b)g(a) - f(a)g(b) = h(b).$$

Thus by Rolle's theorem, there exists $c \in (a, b)$ such that $h'(c) = 0$, which gives the result. $\sqcap$

The geometric interpretation of the Cauchy mean value theorem is very similar to that of the mean value theorem. If $g'(x) \ne 0$ for all $x \in (a, b)$, then $g(a) \ne g(b)$ and the conclusion of Theorem 5.2.8 can be written as

$$\frac{f(b) - f(a)}{g(b) - g(a)} = \frac{f'(c)}{g'(c)}.$$

FIGURE 5.5
Cauchy mean value theorem

Suppose $x = g(t)$, $y = f(t)$, $a \leq t \leq b$, is a parametric representation of a curve C in the plane. As t moves along the interval $[a, b]$ the point (x, y) moves along C from the point $P = (g(a), f(a))$ to $Q = (g(b), f(b))$. The slope of the line joining P to Q is given by $[f(b) - f(a)]/[g(b) - g(a)]$ (see Figure 5.5). On the other hand, the quantity $f'(t)/g'(t)$ is the slope of the curve C at the point $(g(t), f(t))$. Thus one meaning of Theorem 5.2.8 is that there must be a point on the curve C where the slope of the curve is the same as the slope of the line joining P to Q.

Applications of the Mean Value Theorem

We now give several consequences of the mean value theorem. Additional applications are also given in the exercises. In the following, I will denote an arbitrary interval in $\mathbb{R}$.

THEOREM 5.2.9 *Suppose $f : I \to \mathbb{R}$ is differentiable on the interval I.*

  (a) *If $f'(x) \geq 0$ for all $x \in I$, then f is monotone increasing on I.*

  (b) *If $f'(x) > 0$ for all $x \in I$, then f is strictly increasing on I.*

  (c) *If $f'(x) \leq 0$ for all $x \in I$, then f is monotone decreasing on I.*

  (d) *If $f'(x) < 0$ for all $x \in I$, then f is strictly decreasing on I.*

  (e) *If $f'(x) = 0$ for all $x \in I$, then f is constant on I.*

Proof. Suppose $x_1, x_2 \in I$ with $x_1 < x_2$. By the mean value theorem applied to f on $[x_1, x_2]$,
$$f(x_2) - f(x_1) = f'(c)(x_2 - x_1)$$
for some $c \in (x_1, x_2)$. If $f'(c) \geq 0$, then $f(x_2) \geq f(x_1)$. Thus, if $f'(x) \geq 0$ for all $x \in I$, we have $f(x_2) \geq f(x_1)$ for all $x_1, x_2 \in I$ with $x_1 < x_2$. Thus f is monotone increasing on I. The other results follow similarly. $\square$

Remark. It needs to be emphasized that if the derivative of a function f is positive at a point c, then this does **not** imply that f is increasing on an interval containing c. The function f of Exercise 20 satisfies $f'(0) = 1$, but $f'(x)$ assumes both negative and positive values in every neighborhood of 0. Thus f is not monotone on any interval containing 0. If $f'(c) > 0$, the only conclusion that can be reached is that there exists a $\delta > 0$ such that $f(x) < f(c)$ for all $x \in (c-\delta, c)$ and $f(x) > f(c)$ for all $x \in (c, c+\delta)$ (Exercise 17). This however does not mean that f is increasing on $(c-\delta, c+\delta)$. However, if $f'(c) > 0$ **and** f' is continuous at c, then there exists a $\delta > 0$ such that $f'(x) > 0$ for all $x \in (c - \delta, c + \delta)$. Thus f is increasing on $(c - \delta, c + \delta)$.

Theorem 5.2.9 is often used to determine maxima and minima of functions as follows: Suppose f is a real-valued continuous function on (a, b) and $c \in (a, b)$ is such that $f'(c) = 0$ or $f'(c)$ does not exist. Suppose f is differentiable on (a, c) and (c, b). If $f'(x) < 0$ for all $x \in (a, c)$ and $f'(x) > 0$ for all $x \in (c, b)$, then by Theorem 5.2.9, f is decreasing on (a, c) and increasing on (c, b). As a consequence one concludes that f has a relative minimum at c. This method is usually referred to as the **first derivative test** for relative maxima or minima. The natural inclination is to think that the converse is also true; namely, if f has a relative minimum at c, then f is decreasing to the left of c and increasing to the right of c. This however, as the following example shows, is false!

EXAMPLE 5.2.10 Let f be defined by

$$f(x) = \begin{cases} x^4 \left(2 + \sin \dfrac{1}{x} \right), & x \neq 0, \\ 0, & x = 0. \end{cases}$$

The function f has an absolute minimum at $x = 0$; however, $f'(x)$ has both negative and positive values in every neighborhood of 0. The details are left as an exercise (Exercise 21). The graph of $f'(x) = 4x^3(2 + \sin 1/x) - x^2 \cos 1/x)$, $x \neq 0$, is given in Figure 5.6. □

The following theorem, besides being useful in computing right or left derivatives at a point, also states that the derivative (if it exists everywhere on an interval) can only have discontinuities of the second kind.

THEOREM 5.2.11 *Suppose $f : [a, b) \to \mathbb{R}$ is continuous on $[a, b)$ and differentiable on (a, b). If $\lim_{x \to a^+} f'(x)$ exists, then $f'_+(a)$ exists and*

$$f'_+(a) = \lim_{x \to a^+} f'(x).$$

Proof. Let $L = \lim_{x \to a^+} f'(x)$, that is assumed to exist. Given $\epsilon > 0$, there exists a $\delta > 0$ such that

$$|f'(x) - L| < \epsilon \quad \text{for all } x, \, a < x < a + \delta.$$

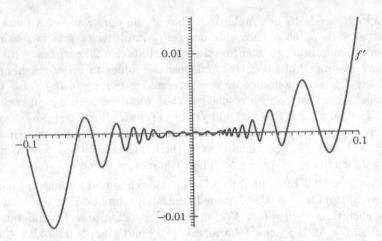

FIGURE 5.6
Graph of $f'(x) = 4x^3(2 + \sin(1/x)) - x^2\cos(1/x)$, $x \neq 0$

Suppose $0 < h < \delta$ is such that $a+h < b$. Since f is continuous on $[a, a+h]$ and differentiable on $(a, a+h)$, by the mean value theorem $f(a+h) - f(a) = f'(\zeta_h)\,h$ for some $\zeta_h \in (a, a+h)$. Therefore

$$\left| \frac{f(a+h) - f(a)}{h} - L \right| = |f'(\zeta_h) - L| < \epsilon$$

for all h, $0 < h < \delta$. Thus $f'_+(a) = L$. $\quad\square$

EXAMPLES 5.2.12 (a) To illustrate the previous theorem, consider the function

$$f(x) = \begin{cases} x^2 + 1, & x < 1, \\ 3 - x^2, & x \geq 1. \end{cases}$$

For $x < 1$, $f'(x) = 2x$ that has a left limit of 2 at $x = 1$. Thus by the theorem,

$$f'_-(1) = \lim_{x \to 1^-} 2x = 2.$$

Similarly,

$$f'_+(1) = \lim_{x \to 1^+} -2x = -2.$$

(b) The converse of Theorem 5.2.11 is false. The function

$$g(x) = \begin{cases} x^2 \sin\dfrac{1}{x}, & x \neq 0, \\ 0, & x = 0, \end{cases}$$

of Example 5.1.3(g) has the property that $g'(0)$ exists but $\lim_{x \to 0} g'(x)$ does not.
$\square$

Intermediate Value Theorem for Derivatives

Our second important result of this section, due to Jean Gaston Darboux (1842–1917), is the intermediate value theorem for derivatives. The remarkable aspect of this theorem is that the hypothesis does not require continuity of the derivative. If the derivative were continuous, then the result would follow from Theorem 4.2.11 applied to f'.

THEOREM 5.2.13 (Intermediate Value Theorem for Derivatives)
Suppose $I \subset \mathbb{R}$ is an interval and $f : I \to \mathbb{R}$ is differentiable on I. Then given a, b in I with $a < b$ and a real number λ between $f'(a)$ and $f'(b)$, there exists $c \in (a, b)$ such that $f'(c) = \lambda$.

Proof. Define g by $g(x) = f(x) - \lambda x$. Then g is differentiable on I with $g'(x) = f'(x) - \lambda$.

Suppose $f'(a) < \lambda < f'(b)$. Then $g'(a) < 0$ and $g'(b) > 0$. As in the remark following Theorem 5.2.9, since $g'(a) < 0$ there exists an $x_1 > a$ such that $g(x_1) < g(a)$. Also, since $g'(b) > 0$, there exists an $x_2 < b$ such that $g(x_2) < g(b)$. As a consequence, g has an absolute minimum at some point $c \in (a, b)$. But then

$$g'(c) = f'(c) - \lambda = 0,$$

i.e., $f'(c) = \lambda$. $\square$

The previous theorem is often used in calculus to determine where a function is increasing or decreasing. Suppose it has been determined that the derivative f' is zero at c_1 and c_2 with $c_1 < c_2$, and that $f'(x) \neq 0$ for all $x \in (c_1, c_2)$. Then by the previous theorem, it suffices to check the sign of the derivative at a single point in the interval (c_1, c_2) to determine whether f' is positive or negative on the whole interval (c_1, c_2). Theorem 5.2.9 then allows us to determine whether f is increasing or decreasing on (c_1, c_2).

Inverse Function Theorem

We conclude this section with the following version of the inverse function theorem.

THEOREM 5.2.14 (Inverse Function Theorem) *Suppose $I \subset \mathbb{R}$ is an interval and $f : I \to \mathbb{R}$ is differentiable on I with $f'(x) \neq 0$ for all $x \in I$. Then f is one-to-one on I, the inverse function f^{-1} is continuous and differentiable on $J = f(I)$ with*

$$\left(f^{-1}\right)'(f(x)) = \frac{1}{f'(x)}$$

for all $x \in I$.

Proof. Since $f'(x) \neq 0$ for all $x \in I$, by Theorem 5.2.13, f' is either positive on I, or negative on I. Assume that $f'(x) > 0$ for all $x \in I$. Then by Theorem

5.2.9, f is strictly increasing on I and by Theorem 4.4.12 f^{-1} is continuous on $J = f(I)$.

It remains to be shown that f^{-1} is differentiable on J. Let $y_o \in J$, and let $\{y_n\}$ be any sequence in J with $y_n \to y_o$, and $y_n \neq y_o$ for all n. For each n, there exists $x_n \in I$ such that $f(x_n) = y_n$. Since f^{-1} is continuous, $x_n \to x_o = f^{-1}(y_o)$. Hence

$$\lim_{n\to\infty} \frac{f^{-1}(y_n) - f^{-1}(y_o)}{y_n - y_o} = \lim_{n\to\infty} \frac{x_n - x_o}{f(x_n) - f(x_o)}$$
$$= \frac{1}{f'(x_o)}.$$

Since this holds for any sequence $\{y_n\}$ with $y_n \to y_o$, $y_n \neq y_o$, by Theorem 4.1.3 and the definition of the derivative

$$\left(f^{-1}\right)'(y_o) = \frac{1}{f'(x_o)}. \quad \square$$

Remark. The hypothesis that $f'(x) \neq 0$ for all $x \in I$ is crucial. For example, the function $f(x) = x^3$ is strictly increasing on $[-1,1]$ with $f'(0) = 0$. The inverse function $f^{-1}(y) = y^{1/3}$ however is not differentiable at $y = 0$.

EXAMPLES 5.2.15 (a) As an application of the previous theorem we show that $f(x) = x^{1/n}$, $x \in (0, \infty)$, $n \in \mathbb{N}$, is differentiable on $(0, \infty)$ with

$$f'(x) = \frac{1}{n} x^{\frac{1}{n}-1}$$

for all $x \in (0, \infty)$. Consider the function $g(x) = x^n$, $n \in \mathbb{N}$, $\text{Dom} \, g = (0, \infty)$. Then $g'(x) = nx^{n-1}$ and $g'(x) > 0$ for all $x \in (0, \infty)$. By the previous theorem g^{-1} is differentiable on $J = g((0, \infty)) = (0, \infty)$ with

$$(g^{-1})'(g(x)) = \frac{1}{g'(x)} = \frac{1}{nx^{n-1}}.$$

If we set $y = g(x) = x^n$, then $x = y^{\frac{1}{n}}$ and

$$(g^{-1})'(y) = \frac{1}{n(y^{\frac{1}{n}})^{n-1}} = \frac{1}{n} y^{\frac{1}{n}-1}.$$

Since $f = g^{-1}$ the desired result follows.

(b) As in Example 5.2.7, let $L(x) = \ln x$ denote the natural logarithm function on $(0, \infty)$. Since $L'(x) = 1/x$ is strictly positive on $(0, \infty)$, the function L is one-to-one, the inverse function L^{-1} is continuous on $\mathbb{R} = \text{Range} \, L$, and by Theorem 5.2.14,

$$(L^{-1})'(L(x)) = \frac{1}{L'(x)} = x.$$

If we set $E = L^{-1}$, then $E'(L(x)) = x$, or $E'(y) = E(y)$ where $y = L(x)$. The function $E(x)$, $x \in \mathbb{R}$ is called the **natural exponential function** on $\mathbb{R}$ and is usually denoted by e^x, where e is Euler's number of Example 3.3.5. The exponential function $E(x)$ is considered in greater detail in Example 8.7.20(d).

(c) In this example, we consider the inverse function of $g(x) = \cos x$, $x \in [0, \pi]$. Since $g'(x) = -\sin x$ is strictly negative for $x \in (0, \pi)$, the function g is strictly decreasing on $[0, \pi]$ with $g([0, \pi]) = [-1, 1]$. Thus for $y \in [-1, 1]$, $x = \arccos y$ if and only if $\cos x = y$. Finally, since $g'(x) \neq 0$ for $x \in (0, \pi)$, by the inverse function theorem

$$\left(g^{-1}\right)'(g(x)) = \frac{1}{g'(x)} = \frac{1}{-\sin x} = \frac{-1}{\sqrt{1 - \cos^2 x}},$$

or since $y = \cos x$,

$$\frac{d}{dy} \arccos y = \frac{-1}{\sqrt{1 - y^2}}. \quad \Box$$

Exercises 5.2

1. Graph the function $f(x) = (x - \frac{1}{2})^2$, $0 \le x \le 2$. Show that $f'_+(0) = -1$ and $f'_-(2) = 3$.

2. Which of the following functions satisfy the hypothesis of the mean value theorem. For those to which the mean value theorem applies, calculate a suitable c.
   a. $f(x) = |x|$, $-2 \le x \le 2$
   b. $f(x) = 2x - x^3$, $0 \le x \le 2$
   c. $f(x) = \dfrac{x}{x+2}$, $-1 \le x \le 2$.
   d. $f(x) = 1 - x^{2/3}$, $-2 \le x \le 1$

3. For each of the following functions determine the interval(s) where the function is increasing, decreasing, and find all local maxima and minima.
   *a. $f(x) = x^3 + 6x - 5$, $x \in \mathbb{R}$.
   b. $g(x) = 4x - x^4$, $x \in \mathbb{R}$.
   *c. $h(x) = \dfrac{x^2}{1 + x^2}$, $x \in \mathbb{R}$.
   d. $k(x) = \sqrt{x} - \frac{1}{2}x$, $x \ge 0$.
   *e. $l(x) = x + \dfrac{4}{x^2}$, $x \neq 0$.
   f. $f(x) = \dfrac{x - a}{x - b}$, $a \neq b$, $x \neq b$

4. Let $f(x) = \sum_{i=1}^{n}(x - a_i)^2$, where $a_1, a_2, ..., a_n$ are constants. Find the value of x where f is a minimum.

5. As in Example 5.2.7 use the mean value theorem to establish each of the following inequalities.
   *a. $\sqrt{1 + x} < 1 + \frac{1}{2}x$, $x > -1$
   b. $e^x \ge 1 + x$, $x \in \mathbb{R}$
   *c. $x^\alpha - a^\alpha < \alpha a^{\alpha-1}(x - a)$, $0 < a < x$, $0 < \alpha < 1$
   d.[1] $(1 + x)^\alpha \ge 1 + \alpha x$, $x > -1$, $\alpha > 1$

---

[1] For $\alpha \in \mathbb{N}$, this inequality was proved by mathematical induction in Example 1.3.2(b). In this exercise and in Exercise 6(b) you may assume that for $\alpha \in \mathbb{R}$, $\dfrac{d}{dx}x^\alpha = \alpha x^{\alpha-1}$.

6. Prove each of the following inequalities.

   *a. $a^{1/n} - b^{1/n} < (a-b)^{1/n}$, $a > b > 0$, $n \in \mathbb{N}$, $n \geq 2$.

   *b. $a^\alpha b^{1-\alpha} \leq \alpha a + (1-\alpha)b$, $a, b > 0$, $0 < \alpha < 1$

7. (**Second Derivative Test**) Let $f : [a, b] \to \mathbb{R}$ be differentiable on (a, b). Suppose $c \in (a, b)$ is such that $f'(c) = 0$, and $f''(c)$ exists.

   *a. If $f''(c) > 0$, prove that f has a local minimum at c.

   b. If $f''(c) < 0$, prove that f has a local maximum at c.

   c. Show by examples that no conclusion can be made if $f''(c) = 0$.

8. *Suppose $f : (a, b) \to \mathbb{R}$ satisfies $|f(x) - f(y)| \leq M |x - y|^\alpha$ for some $\alpha > 1$ and all $x, y \in (a, b)$. Prove that f is a constant function on (a, b).

9. *Find a polynomial $P(x)$ of degree less than or equal to 2 with $P(2) = 0$ such that the function

$$f(x) = \begin{cases} x^2, & x \leq 1, \\ P(x), & x > 1, \end{cases}$$

   is differentiable at $x = 1$.

10. *Let g be defined by

$$g(x) = \begin{cases} 2\sin x + \cos 2x, & x \leq 0, \\ ax^2 + bx + c, & x > 0. \end{cases}$$

   Determine the constants a, b, and c such that $g'(0)$ and $g''(0)$ exist.

11. a. Suppose f is differentiable on an interval I. Prove that f' is bounded on I if and only if there exists a constant M such that $|f(x) - f(y)| \leq M |x - y|$ for all $x, y \in I$.

    b. Prove that $|\sin x - \sin y| \leq |x - y|$ for all $x, y \in \mathbb{R}$.

    c. Prove that $|\sqrt{x} - \sqrt{y}| \leq \frac{1}{2\sqrt{a}}|x - y|$ for all $x, y \in [a, \infty)$, $a > 0$.

12. *a. Show that $\tan x > x$ for $x \in (0, \frac{\pi}{2}]$.

    *b. Set

$$f(x) = \begin{cases} \dfrac{\sin x}{x}, & x \in (0, \frac{\pi}{2}], \\ 1, & x = 0. \end{cases}$$

   Show that f is strictly decreasing on $[0, \frac{\pi}{2}]$.

    c. Using the result of part (b), prove that $\dfrac{2}{\pi} x \leq \sin x \leq x$ for all $x \in [0, \frac{\pi}{2}]$.

13. Give an example of a uniformly continuous function on $[0, 1]$ that is differentiable on $(0, 1)$ but for which f' is not bounded on $(0, 1)$.

14. *a. Suppose $f'(x)$ exists for all $x \in (a, b)$. Let $c \in (a, b)$. Show that there exists a sequence $\{x_n\}$ in (a, b) with $x_n \neq c$ and $x_n \to c$ such that $f'(x_n) \to f'(c)$.

    b. Does $f'(x_n) \to f'(c)$ for every sequence $\{x_n\}$ with $x_n \to c$?

15. Let $f, g : [0, \infty) \to \mathbb{R}$ be continuous on $[0, \infty)$ and differentiable on $(0, \infty)$. If $f(0) = g(0)$ and $f'(x) \geq g'(x)$ for all $x \in (0, \infty)$, prove that $f(x) \geq g(x)$ for all $x \in [0, \infty)$.

16. A differentiable function $f : [a, b] \to \mathbb{R}$ is **uniformly differentiable** on $[a, b]$ if for every $\epsilon > 0$ there exists a $\delta > 0$ such that

$$\left| \frac{f(t) - f(x)}{t - x} - f'(x) \right| < \epsilon$$

for all $t, x \in [a, b]$ with $0 < |t - x| < \delta$. Show that f is uniformly differentiable on $[a, b]$ if and only if f' is continuous on $[a, b]$.

17. *Suppose $f : [a, b] \to \mathbb{R}$ with $f'_+(a) > 0$. Prove that there exists a $\delta > 0$ such that $f(x) > f(a)$ for all x, $a < x < a + \delta$.

18. Prove that the equation $x^3 - 3x + b = 0$ has at most one root in the interval $[-1, 1]$.

19. *Suppose g is differentiable on (a, b) with $|g'(x)| \leq M$ for all $x \in (a, b)$. Prove that there exists an $\epsilon > 0$ such that the function $f(x) = x + \epsilon g(x)$ is one-to-one on (a, b).

20. Let

$$f(x) = \begin{cases} x + 2x^2 \sin \dfrac{1}{x}, & x \neq 0, \\ 0, & x = 0. \end{cases}$$

   a. Show that $f'(0) = 1$.

   *b.** Prove that $f'(x)$ assumes both positive and negative values in every neighborhood of 0.

21. Let f be defined by

$$f(x) = \begin{cases} x^4 \left(2 + \sin \dfrac{1}{x} \right), & x \neq 0, \\ 0, & x = 0. \end{cases}$$

   a. Show that f has an absolute minimum at $x = 0$.

   b. Show that $f'(x)$ assumes both negative and positive values in every neighborhood of 0.

22. Let $f(x) = x^2$, $g(x) = x^3$, $x \in [-1, 1]$.

   a. Find $c \in (-1, 1)$ such that the conclusion of Theorem 5.2.8 holds.

   b. Show that there does not exist any $c \in (-1, 1)$ for which

$$\frac{f(1) - f(-1)}{g(1) - g(-1)} = \frac{f'(c)}{g'(c)}.$$

23. For $r \in \mathbb{Q}$ and $x > 0$, let $f(x) = x^r$. Prove that $f'(x) = r x^{r-1}$.

24. Suppose $L : (0, \infty) \to \mathbb{R}$ is a differentiable function satisfying $L'(x) = 1/x$ with $L(1) = 0$. Prove each of the following:

   *a.** $L(ab) = L(a) + L(b)$ for all $a, b \in (0, \infty)$

   b. $L(1/b) = -L(b), \quad b > 0$

   *c.** $L(b^r) = r L(b), \quad b > 0, r \in \mathbb{R}$

    d. $L(e) = 1$, where e is Euler's number

    e. Range $L = \mathbb{R}$

25. Let $g(x) = \tan x$, $-\frac{\pi}{2} < x < \frac{\pi}{2}$.

    a. Show that g is one-to-one on $(-\frac{\pi}{2}, \frac{\pi}{2})$ with Range $g = \mathbb{R}$.

    *b. Let $\arctan x$, $x \in \mathbb{R}$, denote the inverse function of g. Use Theorem 5.2.14 to prove that

$$\frac{d}{dx} \arctan x = \frac{1}{1 + x^2}.$$

    c. Sketch the graph of $\tan x$ and $\arctan x$.

26. **a.** Show that $f(x) = \sin x$ is one-to-one on $[-\frac{\pi}{2}, \frac{\pi}{2}]$ with $f([-\frac{\pi}{2}, \frac{\pi}{2}]) = [-1, 1]$.

    b. For $x \in [-1, 1]$, let $\arcsin x$ denote the inverse function of f. Show that $\arcsin x$ is differentiable on $(-1, 1)$ and find the derivative of $\arcsin x$.

27. Let $f : (0, \infty) \to \mathbb{R}$ be differentiable on $(0, \infty)$ and suppose that $\lim\limits_{x \to \infty} f'(x) = L$.

    a. Show that for any $h > 0$, $\lim\limits_{x \to \infty} \dfrac{f(x + h) - f(x)}{h} = L$.

    b. Show that $\lim\limits_{x \to \infty} \dfrac{f(x)}{x} = L$.

5.3   L'Hospital's Rule

As another application of the mean value theorem we now prove l'Hospital's rule for evaluating limits. Although the theorem is named after the Marquis de l'Hospital (1661–1704), it should more appropriately be called Bernoulli's rule. The story is that in 1691, l'Hospital asked Johann Bernoulli (1667–1748) to provide, for a fee, lectures on the new subject of calculus. L'Hospital subsequently incorporated these lectures into the first calculus text *L'Analyse des infiniment petis (Analysis of infinitely small quantities)* published in 1696. The initial version (stated without the use of limits) of what is now known as l'Hospital's rule first appeared in this text.

Infinite Limits

Since l'Hospital's rule allows for infinite limits, we provide the following definitions.

DEFINITION 5.3.1 *Let f be a real-valued function defined on a subset E of $\mathbb{R}$ and let p be a limit point of E. We say that f **tends to** ∞ (or **diverges to** ∞) as x approaches p, denoted*

$$\lim_{x \to p} f(x) = \infty,$$

if for every $M \in \mathbb{R}$, there exists a $\delta > 0$ such that

$$f(x) > M \qquad \textit{for all } x \in E \quad \textit{with} \quad 0 < |x - p| < \delta.$$

Similarly,

$$\lim_{x \to p} f(x) = -\infty,$$

if for every $M \in \mathbb{R}$, there exists a $\delta > 0$ such that

$$f(x) < M \qquad \textit{for all } x \in E \quad \textit{with} \quad 0 < |x - p| < \delta.$$

For f defined on an appropriate subset E of $\mathbb{R}$ it is also possible to define each of the following limits:

$$\lim_{x \to p^+} f(x) = \pm\infty, \quad \lim_{x \to p^-} f(x) = \pm\infty, \quad \lim_{x \to \infty} f(x) = \pm\infty, \quad \lim_{x \to -\infty} f(x) = \pm\infty.$$

Since these definitions are similar to Definitions 4.1.11 and 4.4.1 they are left to the exercises (Exercise 1).

Remark. Since we now allow the possibility of a function having infinite limits, it needs to be emphasized that when we say that a function f has a limit at $p \in \mathbb{R}$ (or at $\pm\infty$), we mean a **finite** limit.

L'Hospital's Rule

L'Hospital's rule is useful for evaluating limits of the form

$$\lim_{x \to p} \frac{f(x)}{g(x)}$$

where either (a) $\lim\limits_{x \to p} f(x) = \lim\limits_{x \to p} g(x) = 0$ or (b) f and g tend to $\pm\infty$ as $x \to p$. If (a) holds, then $\lim\limits_{x \to p}(f(x)/g(x))$ is usually referred to as indeterminate of form $0/0$, whereas in (b) the limit is referred to as indeterminate of form ∞/∞. The reason that (a) and (b) are indeterminate are that previous methods may no longer apply.

In (a), if either $\lim\limits_{x \to p} f(x)$ or $\lim\limits_{x \to p} g(x)$ is nonzero, then previous methods discussed in Section 4.1 apply. For example, if both f and g have limits at p and $\lim\limits_{x \to p} g(x) \neq 0$, then by Theorem 4.1.6(c)

$$\lim_{x \to p} \frac{f(x)}{g(x)} = \frac{\lim\limits_{x \to p} f(x)}{\lim\limits_{x \to p} g(x)}.$$

On the other hand, if $\lim\limits_{x \to p} f(x) = A \neq 0$ and $g(x) > 0$ with $\lim\limits_{x \to p} g(x) = 0$, then as $x \to p$, $f(x)/g(x)$ tends to ∞ if $A > 0$, and to $-\infty$ if $A < 0$ (Exercise 5). However, if $\lim\limits_{x \to p} f(x) = \lim\limits_{x \to p} g(x) = 0$, then unless the quotient $f(x)/g(x)$ can somehow be simplified, previous methods may no longer be applicable.

THEOREM 5.3.2 (L'Hospital's Rule) *Suppose f, g are real-valued differentiable functions on (a, b), with $g'(x) \neq 0$ for all $x \in (a, b)$, where $-\infty \leq a < b \leq \infty$. Suppose*

$$\lim_{x \to a^+} \frac{f'(x)}{g'(x)} = L, \qquad where \qquad L \in \mathbb{R} \cup \{-\infty, \infty\}.$$

If

(a) $\lim_{x \to a^+} f(x) = 0$ *and* $\lim_{x \to a^+} g(x) = 0$, *or*

(b) $\lim_{x \to a^+} g(x) = \pm\infty$, *then*

$$\lim_{x \to a^+} \frac{f(x)}{g(x)} = L.$$

Remark. The analogous result where $x \to b^-$ is obviously also true. A more elementary version of l'Hospital's rule that relies only on the definition of the derivative is given in Exercise 2. Also, Exercise 7 provides examples of two functions f and g satisfying (a) for which $\lim_{x \to a}(f(x)/g(x))$ exists but $\lim_{x \to a}(f'(x)/g'(x))$ does not exist.

Proof. Suppose (a) holds. We first prove the case where a is finite. Let $\{x_n\}$ be a sequence in (a, b) with $x_n \to a$ and $x_n \neq a$ for all n. Since we want to apply the generalized mean value theorem to f and g on the interval $[a, x_n]$, we need both f and g continuous at a. This is accomplished by setting

$$f(a) = g(a) = 0.$$

Then by hypothesis (a), f and g are continuous at a. Thus by the generalized mean value theorem, for each $n \in \mathbb{N}$ there exists c_n between a and x_n such that

$$[f(x_n) - f(a)]g'(c_n) = [g(x_n) - g(a)]f'(c_n),$$

or

$$\frac{f(x_n)}{g(x_n)} = \frac{f'(c_n)}{g'(c_n)}.$$

Note, since $g'(x) \neq 0$ for all $x \in (a, b)$, $g(x_n) \neq g(a)$ for all n As $n \to \infty$, $c_n \to a^+$. Thus by Theorem 4.1.3 and the hypothesis,

$$\lim_{n \to \infty} \frac{f(x_n)}{g(x_n)} = \lim_{n \to \infty} \frac{f'(c_n)}{g'(c_n)} = \lim_{x \to a^+} \frac{f'(x)}{g'(x)} = L.$$

Since the above holds for every sequence $\{x_n\}$ with $x_n \to a^+$, the result follows.

Suppose $a = -\infty$. To handle this case, we make the substitution $x = -1/t$. Then as $t \to 0^+$, $x \to -\infty$. Define the functions $\varphi(t)$ and $\psi(t)$ on $(0, c)$ for some $c > 0$ by

$$\varphi(t) = f(-1/t) \qquad \text{and} \qquad \psi(t) = g(-1/t).$$

We leave it as an exercise (Exercise 3) to verify that

$$\lim_{t \to 0^+} \frac{\varphi'(t)}{\psi'(t)} = \lim_{x \to -\infty} \frac{f'(x)}{g'(x)} = L,$$

and that

$$\lim_{t \to 0^+} \varphi(t) = \lim_{t \to 0^+} \psi(t) = 0.$$

Thus by the above,

$$\lim_{x \to -\infty} \frac{f(x)}{g(x)} = \lim_{t \to 0^+} \frac{\varphi(t)}{\psi(t)} = L.$$

Suppose now that (b) holds, i.e., $\lim_{x \to a^+} g(x) = \infty$. The case where $g(x) \to -\infty$ is treated similarly. Rather than treating the finite case and infinite case separately, we provide a proof that works for both.

Suppose first that $-\infty \le L < \infty$, and $\beta \in \mathbb{R}$ satisfies $\beta > L$. Choose r such that $L < r < \beta$. Since

$$\lim_{x \to a^+} \frac{f'(x)}{g'(x)} < r,$$

there exists $c_1 \in (a, b)$ such that

$$\frac{f'(\zeta)}{g'(\zeta)} < r \quad \text{for all} \quad \zeta, \, a < \zeta < c_1.$$

Fix a y, $a < y < c_1$. Since $g(x) \to \infty$ as $x \to a^+$, there exists a c_2, $a < c_2 < y$, such that $g(x) > g(y)$ and $g(x) > 0$ for all x, $a < x < c_2$. Let $x \in (a, c_2)$ be arbitrary. Then by the generalized mean value theorem, there exists $\zeta \in (x, y)$ such that

$$\frac{f(x) - f(y)}{g(x) - g(y)} = \frac{f'(\zeta)}{g'(\zeta)} < r. \tag{4}$$

Multiplying (4) by $(g(x) - g(y))/g(x)$, which is positive, we obtain

$$\frac{f(x) - f(y)}{g(x)} < r \left(1 - \frac{g(y)}{g(x)} \right),$$

or

$$\frac{f(x)}{g(x)} < \frac{f(y)}{g(x)} + r \left(1 - \frac{g(y)}{g(x)} \right) \tag{5}$$

for all x, $a < x < c_2$. Now for fixed y, since $g(x) \to \infty$,

$$\lim_{x \to a^+} \frac{f(y)}{g(x)} = \lim_{x \to a^+} \frac{g(y)}{g(x)} = 0.$$

Therefore

$$\lim_{x \to a^+} \frac{f(y)}{g(x)} + r \left(1 - \frac{g(y)}{g(x)} \right) = r < \beta$$

Thus there exists c_3, $a < c_3 < c_2$, such that

$$\frac{f(y)}{g(x)} + r\left(1 - \frac{g(y)}{g(x)}\right) < \beta$$

for all x, $a < x < c_3$. Thus by (5),

$$\frac{f(x)}{g(x)} < \beta \quad \text{for all} \quad x, a < x < c_3. \tag{6}$$

If $L = -\infty$, then for any $\beta \in \mathbb{R}$, there exists c_3 such that (6) holds for all x, $a < x < c_3$. Thus by definition,

$$\lim_{x \to a^+} \frac{f(x)}{g(x)} = -\infty.$$

If L is finite, then given $\epsilon > 0$, by taking $\beta = L + \epsilon$, there exists c_3 such that

$$\frac{f(x)}{g(x)} < L + \epsilon \quad \text{for all} \quad x, a < x < c_3. \tag{7}$$

Suppose $-\infty < L \le \infty$. Let $\alpha \in \mathbb{R}$, $\alpha < L$ be arbitrary. Then an argument similar to the above gives the existence of $c_3' \in (a, b)$ such that

$$\frac{f(x)}{g(x)} > \alpha \quad \text{for all} \quad x, a < x < c_3'.$$

If $L = \infty$, then this implies that

$$\lim_{x \to a^+} \frac{f(x)}{g(x)} = \infty.$$

On the other hand, if L is finite, taking $\alpha = L - \epsilon$ gives the existence of a c_3' such that

$$\frac{f(x)}{g(x)} > L - \epsilon \quad \text{for all} \quad x, a < x < c_3'.$$

Combining this with (7) proves that

$$\lim_{x \to a^+} \frac{f(x)}{g(x)} = L. \quad \square$$

Remarks. (a) The proof of case (a) could have been done similarly to that of (b), treating the case where a is finite and $-\infty$ simultaneously. I chose not to do so since making the substitution $x = -1/t$ is a useful technique, reducing problems involving limits at $-\infty$ to right limits at 0. Conversely, limits at 0 can be transformed to limits at $\pm\infty$ with the substitution $x = 1/t$. These new limits are in many instances easier to evaluate than the original. This is illustrated in Example 5.3.4(c).

(b) In hypothesis (b) we only required that $\lim\limits_{x \to a^+} g(x) = \pm\infty$. If $\lim\limits_{x \to a^+} f(x)$ is finite, then it immediately follows that

$$\lim_{x \to a^+} \frac{f(x)}{g(x)} = 0,$$

and l'Hospital's rule is not required (Exercise 4). Thus in practice, hypothesis (b) of l'Hospital's rule is used only if both f and g have infinite limits.

For convenience we stated and proved l'Hospital's rule in terms of right limits. Since the analogous results for left limits are also true, combining the two results gives the following corollary.

COROLLARY 5.3.3 (L'Hospital's Rule) *Suppose f, g are real-valued differentiable functions on $(a, p) \cup (p, b)$, with $g'(x) \neq 0$ for all $x \in (a, p) \cup (p, b)$, where $-\infty \leq a < b \leq \infty$. Suppose*

$$\lim_{x \to p} \frac{f'(x)}{g'(x)} = L, \qquad \text{where} \quad L \in \mathbb{R} \cup \{-\infty, \infty\}.$$

If

(a) $\lim\limits_{x \to p} f(x) = \lim\limits_{x \to p} g(x) = 0$, *or*

(b) $\lim\limits_{x \to p} g(x) = \pm\infty$, *then*

$$\lim_{x \to p} \frac{f(x)}{g(x)} = L.$$

EXAMPLES 5.3.4 (a) Consider $\lim\limits_{x \to 0^+} \dfrac{\ln(1 + x)}{x}$, where ln is the natural logarithm function on $(0, \infty)$. This limit is indeterminate of form $0/0$. With $f(x) = \ln(1 + x)$ and $g(x) = x$,

$$\lim_{x \to 0^+} \frac{f'(x)}{g'(x)} = \lim_{x \to 0^+} \frac{1}{1 + x} = 1.$$

Thus by l'Hospital's rule $\lim\limits_{x \to 0^+} \dfrac{\ln(1 + x)}{x} = 1$.

Although l'Hospital's rule provides an easy method for evaluating this limit, the result can also be obtained by using previous techniques. In Example 5.2.7 we proved that

$$\frac{x}{1 + x} \leq \ln(1 + x) \leq x$$

for all $x > -1$. Thus

$$\frac{1}{1 + x} \leq \frac{\ln(1 + x)}{x} \leq 1$$

for all $x > 0$. Thus by Theorem 4.1.9 $\displaystyle\lim_{x \to 0^+} \frac{\ln(1+x)}{x} = 1$.

(b) In this example, we consider $\displaystyle\lim_{x \to 0} \frac{1 - \cos x}{x^2}$. This is indeterminate of form $0/0$. If we apply l'Hospital's rule we obtain

$$\lim_{x \to 0} \frac{\sin x}{2x}$$

which is again indeterminate of form $0/0$. However, applying l'Hospital's rule one more time gives

$$\lim_{x \to 0} \frac{\cos x}{2} = \frac{1}{2}$$

. Therefore $\displaystyle\lim_{x \to 0} \frac{1 - \cos x}{x^2} = \frac{1}{2}$.

(c) Consider

$$\lim_{x \to 0^+} \frac{e^{-\frac{1}{x}}}{x}.$$

Since $\displaystyle\lim_{x \to 0^+} e^{-1/x} = 0$, the above limit is indeterminate of form $0/0$. If we apply l'Hospital's rule we obtain

$$\lim_{x \to 0^+} \frac{e^{-\frac{1}{x}}}{x^2},$$

and this limit is more complicated than the original limit. However, if we let $t = 1/x$, then

$$\lim_{x \to 0^+} \frac{e^{-\frac{1}{x}}}{x} = \lim_{t \to \infty} \frac{t}{e^t}.$$

This limit is indeterminate of form ∞/∞. By l'Hospital's rule

$$\lim_{t \to \infty} \frac{t}{e^t} = \lim_{t \to \infty} \frac{1}{e^t} = 0.$$

Therefore, $\displaystyle\lim_{x \to 0^+} e^{-1/x}/x = 0$. □

Exercises 5.3

1. Provide definitions for each of the following limits:
   a. $\displaystyle\lim_{x \to p^+} f(x) = \infty$.   b. $\displaystyle\lim_{x \to \infty} f(x) = \infty$.

2. *Suppose f, g are differentiable on (a,b), $x_o \in (a,b)$ and $g'(x_o) \neq 0$. If $f(x_o) = g(x_o) = 0$, prove that
$$\lim_{x \to x_o} \frac{f(x)}{g(x)} = \frac{f'(x_o)}{g'(x_o)}.$$
(Hint: apply the definition of the derivative.)

3. Let $h(x)$ be defined on $(-\infty, b)$. Show that there exists a $c > 0$ such that $\varphi(t) = h(-1/t)$ is defined on $(0, c)$, and that $\lim_{x \to -\infty} h(x) = \lim_{t \to 0^+} \varphi(t)$.

4. *Let f, g be real-valued functions defined on (a, b). If $\lim_{x \to a^+} f(x)$ exists in $\mathbb{R}$ and $\lim_{x \to a^+} g(x) = \infty$, prove that

$$\lim_{x \to a^+} \frac{f(x)}{g(x)} = 0.$$

5. Suppose f, g are real-valued functions on (a, b) satisfying $\lim_{x \to a^+} f(x) = A \neq 0$, $\lim_{x \to a^+} g(x) = 0$, and $g(x) > 0$ for all $x \in (a, b)$. If $A > 0$, prove that

$$\lim_{x \to a^+} \frac{f(x)}{g(x)} = \infty.$$

6. Use l'Hospital's rule and any of the differentiation formulas from calculus to find each of the following limits. In the following, $\ln x$, $x > 0$ denotes the natural logarithm function.

   *a. $\lim_{x \to 1} \dfrac{x^5 + 2x - 3}{2x^3 - x^2 - 1}$.
        b. $\lim_{x \to -1} \dfrac{x^5 + 2x - 3}{2x^3 + x^2 + 1}$.

   *c. $\lim_{x \to \infty} \dfrac{\ln x}{x}$.
                  d. $\lim_{x \to 0} \dfrac{1 - \cos 2x}{\sin x}$.

   *e. $\lim_{x \to 0^+} x^a \ln x$ where $a > 0$.
    f. $\lim_{x \to 0} \dfrac{\ln(1 + x)}{\sin x}$.

   *g. $\lim_{x \to \infty} \dfrac{(\ln x)^a}{e^x}$,   where $a > 0$.
    h. $\lim_{x \to \infty} \dfrac{(\ln x)^p}{x^q}$, $p, q \in \mathbb{R}$.

   *i. $\lim_{x \to 0^+} \left(\dfrac{1}{x} - \dfrac{1}{\sin x} \right)$.
    j. $\lim_{x \to \infty} x^{1/x}$

7. Let $f(x) = x^2 \sin(1/x)$ and $g(x) = \sin x$. Show that $\lim_{x \to 0} \dfrac{f(x)}{g(x)}$ exists but that $\lim_{x \to 0} \dfrac{f'(x)}{g'(x)}$ does not exist.

8. Investigate $\lim_{x \to \infty} \dfrac{p(x)}{q(x)}$, where p and q are polynomials of degree n and m, respectively.

9. Let $f(x) = (\sin x)/x$ for $x \neq 0$, and $f(0) = 1$.
   *a. Show that $f'(0)$ exists, and determine its value.
   *b. Show that $f''(0)$ exists, and determine its value.

10. Let f be defined on $\mathbb{R}$ by
$$f(x) = \begin{cases} e^{-\frac{1}{x^2}}, & x \neq 0, \\ 0, & x = 0. \end{cases}$$
Prove that $f^{(n)}(0) = 0$ for all $n = 1, 2, \ldots$.

5.4   Newton's Method[2]

In this section, we consider the iterative method, commonly known as Newton's method, for finding approximations to the solutions of the equation $f(x) = 0$. Although the method is named after Newton, it is actually due to Joseph Raphson (1648–1715) and in many texts the method is referred to as the Newton-Raphson method. Newton did derive an iterative method for finding the roots of a cubic equation; his method however is not the one used in the procedure named after him. That was developed by Raphson.

Suppose f is a continuous function on $[a, b]$ satisfying $f(a)f(b) < 0$. Then f has opposite sign at the endpoints a and b and thus by the intermediate value theorem (Theorem 4.2.11) there exists at least one value $c \in (a, b)$ for which $f(c) = 0$. If in addition f is differentiable on (a, b) with $f'(x) \neq 0$ for all $x \in (a, b)$, then f is either strictly increasing or decreasing on $[a, b]$, and in this case the value c is unique; that is, there is exactly one point where the graph of f crosses the x-axis.

An elementary approach to finding a numerical approximation to the value c is the **method of bisection**. For this method, differentiability of f is not required. To illustrate the method, suppose f satisfies $f(a) < 0 < f(b)$. Let

$$c_1 = \frac{1}{2}(a + b).$$

If $f(c_1) = 0$, we are done. If $f(c_1) \neq 0$, then c belongs to one of the two intervals (a, c_1) or (c_1, b), and thus $|c_1 - c| < \frac{1}{2}(b - a)$. Suppose $f(c_1) > 0$. Then $c \in (a, c_1)$, and in this case we set $c_0 = a$ and

$$c_2 = \frac{1}{2}(c_0 + c_1).$$

If $f(c_2) = 0$, we are done. If not, then suppose $f(c_2) < 0$. Then $c \in (c_2, c_1)$, and as above we set

$$c_3 = \frac{1}{2}(c_1 + c_2).$$

In general, suppose $c_1, c_2, ..., c_n$, $n \geq 2$, have been determined. If by happenstance $f(c_n) = 0$, then we have obtained the exact value. If $f(c_{n-1})f(c_n) < 0$, then c lies between c_{n-1} and c_n, and we define

$$c_{n+1} = \frac{1}{2}(c_n + c_{n-1}).$$

On the other hand, if $f(c_{n-1})f(c_n) > 0$, then c lies between c_n and c_{n-2}, and in this case, we define

$$c_{n+1} = \frac{1}{2}(c_n + c_{n-2}).$$

---

[2]The topics of this section are not required in subsequent chapters.

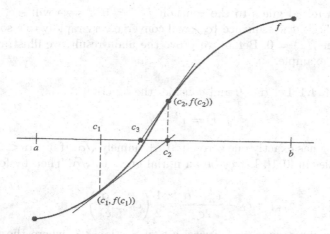

FIGURE 5.7
Newton's method

This gives us a sequence $\{c_n\}$ that satisfies

$$|c_n - c| \le \frac{1}{2^n}(b - a),$$

and thus $\lim_{n \to \infty} c_n = c$.

Although this method provides a sequence of numbers that converges to the zero of f, it has the disadvantage that the convergence is rather slow. An alternate method, due to Raphson, uses tangent lines to the curve to find successive points c_n approximating the zero of f. As we will see, this method will often converge much more rapidly to the solution.

As above, assume that f is differentiable on $[a, b]$ with $f(a)f(b) < 0$ and $f'(x) \ne 0$ for all $x \in [a, b]$. Let c_1 be an initial guess to the value c. The line tangent to the graph of f at $(c_1, f(c_1))$ has equation given by

$$y = f(c_1) + f'(c_1)(x - c_1).$$

Since $f'(c_1) \ne 0$, the line crosses the x-axis at a point that we denote by c_2 (Figure 5.7). Thus

$$0 = f(c_1) + f'(c_1)(c_2 - c_1),$$

that upon solving for c_2 gives

$$c_2 = c_1 - \frac{f(c_1)}{f'(c_1)}.$$

We now replace the point c_1 by the second estimate c_2 to obtain c_3, and so forth. Inductively, we obtain a sequence $\{c_n\}$ given by the formula

$$c_{n+1} = c_n - \frac{f(c_n)}{f'(c_n)}, \qquad n = 1, 2, ..., \tag{8}$$

where c_1 is an initial guess to the solution $f(c) = 0$. As we will see, under suitable hypothesis, the sequence $\{c_n\}$ will converge very rapidly to a solution of the equation $f(x) = 0$. Before we prove the main result, we illustrate the above with an example.

EXAMPLE 5.4.1 Let $\alpha > 0$ and consider the function

$$f(x) = x^2 - \alpha.$$

If $\alpha > 1$, then f has exactly one zero on $[0, \alpha]$, namely $\sqrt{\alpha}$. If $0 < \alpha < 1$, then the zero of f lies in $[0, 1]$. Let c_1 be an initial guess to $\sqrt{\alpha}$. Then by formula (8), for $n \geq 1$,

$$c_{n+1} = c_n - \frac{c_n^2 - \alpha}{2c_n} = \frac{1}{2}\left(c_n + \frac{\alpha}{c_n}\right).$$

This is exactly the sequence of Exercise 6 of Section 3.3, where the reader was asked to prove that the sequence converges to $\sqrt{\alpha}$. With $\alpha = 2$, taking $c_1 = 1.4$ as an initial guess, yields

$$c_2 = 1.4142857,$$
$$c_3 = 1.4142135,$$

which is already correct to at least seven decimal places. $\square$

THEOREM 5.4.2 (Newton's Method) *Let f be a real-valued function on $[a, b]$ that is twice differentiable on $[a, b]$. Suppose that $f(a)f(b) < 0$ and that there exist constants m and M such that $|f'(x)| \geq m > 0$ and $|f''(x)| \leq M$ for all $x \in [a, b]$. Then there exists a subinterval I of $[a, b]$ containing a zero c of f such that for any $c_1 \in I$, the sequence $\{c_n\}$ defined by*

$$c_{n+1} = c_n - \frac{f(c_n)}{f'(c_n)}, \quad n \in \mathbb{N},$$

is in I, and $\lim\limits_{n \to \infty} c_n = c$. Furthermore,

$$|c_{n+1} - c| \leq \frac{M}{2m}|c_n - c|^2. \tag{9}$$

Prior to proving Theorem 5.4.2 we first state and prove the following lemma. The result is in fact a special case of Taylor's theorem (8.7.16) that will be discussed in detail in Chapter 8.

LEMMA 5.4.3 *Suppose $f : [a, b] \to \mathbb{R}$ is such that f and f' are continuous on $[a, b]$ and $f''(x)$ exists for all $x \in (a, b)$. Let $x_0 \in [a, b]$. Then for any $x \in [a, b]$, there exists a real number ζ between x_0 and x such that*

$$f(x) = f(x_0) + f'(x_0)(x - x_0) + \frac{1}{2}f''(\zeta)(x - x_0)^2.$$

Proof. For $x \in [a, b]$, let $\alpha \in \mathbb{R}$ be determined by

$$f(x) = f(x_0) + f'(x_0)(x - x_0) + \alpha(x - x_0)^2.$$

Define g on $[a, b]$ by

$$g(t) = f(t) - f(x_0) - f'(x_0)(t - x_0) - \alpha(t - x_0)^2.$$

If $x = x_0$ then the conclusion is true with $\zeta = x_0$. Assume that $x > x_0$. Then g is continuous and differentiable on $[x_0, x]$ with $g(x_0) = g(x) = 0$. Thus by Rolle's theorem there exists $c_1 \in (x_0, x)$ such that $g'(c) = 0$. But

$$g'(t) = f'(t) - f'(x_0) - 2\alpha(t - x_0).$$

By hypothesis g' is continuous on $[x_0, c]$, differentiable on (x_0, c), and satisfies $g'(x_0) = g'(c) = 0$. Thus by Rolle's theorem again, there exists $\zeta \in (x_0, c)$ such that $g''(\zeta) = 0$. But

$$g''(t) = f''(t) - 2\alpha.$$

Therefore $\alpha = \frac{1}{2} f''(\zeta)$ $\quad\square$

Proof of Theorem 5.4.2 Since $f(a)f(b) < 0$ and $f'(x) \neq 0$ for all $x \in [a, b]$, f has exactly one zero c in the interval (a, b).

Let $x_0 \in [a, b]$ be arbitrary. By Lemma 5.4.3 there exists a point ζ between c and x_0 such that

$$0 = f(c) = f(x_0) + f'(x_0)(c - x_0) + \frac{1}{2} f''(\zeta)(c - x_0)^2,$$

or

$$-f(x_0) = f'(x_0)(c - x_0) + \frac{1}{2} f''(\zeta)(c - x_0)^2. \tag{10}$$

If x_1 is defined by

$$x_1 = x_0 - \frac{f(x_0)}{f'(x_0)},$$

then by equation (10)

$$x_1 = x_0 + (c - x_0) + \frac{1}{2} \frac{f''(\zeta)}{f'(x_0)} (c - x_0)^2.$$

Therefore

$$|x_1 - c| = \frac{1}{2} \frac{|f''(\zeta)|}{|f'(x_0)|} |c - x_0|^2 \leq \frac{M}{2m} |c - x_0|^2. \tag{11}$$

Choose $\delta > 0$ so that $\delta < 2m/M$ and $I = [c - \delta, c + \delta] \subset [a, b]$. If $c_n \in I$, then $|c - c_n| < \delta$. If c_{n+1} is defined by (8), then by (11)

$$|c_{n+1} - c| \leq \frac{M}{2m} \delta^2 < \delta.$$

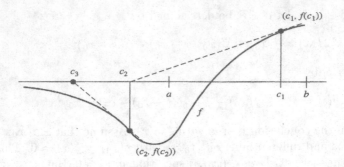

FIGURE 5.8
An illustration of Remark (c)

Therefore $c_{n+1} \in I$. Thus if the initial choice $c_1 \in I$, $c_n \in I$ for all $n = 2, 3, \ldots$.

It remains to be shown that $\lim_{n \to \infty} c_n = c$. If $c_1 \in I$, then by induction

$$|c_{n+1} - c| < \left(\frac{M}{2m} \delta \right)^n |c_1 - c|.$$

But by our choice of δ, $\frac{M}{2m} \delta < 1$, and as a consequence $c_n \to c$. $\square$

Remarks. (a) For a given function f satisfying the hypothesis of the theorem, the constants M and m, and thus δ can be determined. To determine the interval I, one can use the method of bisection to find an approximation x_n to c satisfying $|x_n - x_{n-1}| < \delta$. If c_1 is taken to be x_n, then $|c_1 - c| < \delta$. In practice however, one usually makes a judicious guess for c_1 and proceeds with the computations.

(b) Let $e_n = c - c_n$ be the error in approximating c, and let $K = M/2m$. Then inequality (9) can be expressed as

$$|e_{n+1}| \leq K |e_n|^2.$$

Consequently, if $|e_n| < 10^{-m}$, then $|e_{n+1}| < K 10^{-2m}$. Thus, except for the constant factor K, the accuracy actually doubles at each step. For this reason, Newton's method is usually referred to as a **second order** or **quadratic method**.

(c) Even though Newton's method is very efficient, there are a number of things that can go wrong if c_1 is poorly chosen. For example, in Figure 5.8, the initial choice of c_1 gives a c_2 outside the interval, and the subsequent c_n tend to $-\infty$. Such a function is given by $f(x) = x/(x^2 + 1)$. In Figure 5.9, the initial choice of c_1 causes the subsequent values to oscillate between c_1 and c_2. A function having this property is given by $g(x) = x - \frac{1}{5}x^3$. Taking $c_1 = 1$

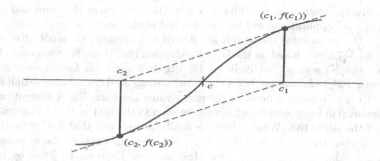

FIGURE 5.9
An example of oscillating c_1 and c_2

gives $c_2 = -1$, $c_3 = 1$, etc. For this reason, the initial choice of c_1 for many functions has to be sufficiently close to c in order to be sure that the method works.

Exercises 5.4

1. *For $\alpha > 0$, apply Newton's method to $f(x) = x^3 - \alpha$ to obtain a sequence $\{c_n\}$ that converges to the cube root of α.

2. Use Newton's method to find approximations to the roots, accurate to six decimal places, of the given functions on the interval $[0, 1]$.
    *a. $f(x) = x^3 - 3x + 1$.   b. $f(x) = 3x^3 - 5x^2 + 1$
    c. $f(x) = 8x^3 - 8x^2 + 1$.

3. Use Newton's method to approximate the real zeros of $f(x) = x^4 - 4x - 3$ accurate to four decimal places

4. Show that $f(x) = \ln x - x + 3$, $x \in (0, \infty)$ has two real zeros. Use Newton's method to approximate them accurate to four decimal places.

5. Let $f : [a, b] \to \mathbb{R}$ be differentiable on $[a, b]$ with $f(a) < 0 < f(b)$. Suppose there exist constants m and M such that $0 < m \le f'(x) \le M$ for all $x \in [a, b]$. Let $c_1 \in [a, b]$ be arbitrary, and define
    $$c_{n+1} = c_n - \frac{f(c_n)}{M}.$$
    Prove that the sequence $\{c_n\}$ converges to the unique zero of f on $[a, b]$. (Hint: Consider the function $g(x) = x - f(x)/M$.)

Notes

Without question the most significant result of this chapter is the mean value theorem. The simplicity of its proof disguises the importance and usefulness of the result. The theorem allows us to obtain information about the function from its derivative.

This has many applications as was illustrated by the subsequent theorems and exercises. Additional applications will be encountered throughout the text.

Although the mean value theorem is attributed to Lagrange, his proof, that appeared around 1772, was based on the false assumption that every function could be expanded in a power series. Cauchy, in his 1823 text *Résumé des Lecons donnees a L'École Royale Polytechnique sur le Calcul Infinitésimal*, used the modern definition of the derivative to provide a proof of the mean value theorem. His statement and proof of the theorem however differs from the version in the text in that he assumed continuity of the derivative. What Cauchy actually proved was that if f' is continuous on $[a, b]$, then the quantity $\{f(b) - f(a)\}/\{b - a\}$ lies between the minimum and maximum values of f' on $[a, b]$. (See Miscellaneous Exercise 4.) Then by the intermediate value theorem (Theorem 4.2.11) applied to the continuous function f', there exists $c \in (a, b)$ such that

$$f(b) - f(a) = f'(c)(b - a).$$

It is worth noting that our proof of the mean value theorem depends ultimately on the completeness property of $\mathbb{R}$ (through Rolle's theorem and Corollary 4.2.9).

The mean value theorem can justifiably be called the fundamental theorem of differential calculus. It allowed the development of rigorous proofs of many results that were previously taken as fact or "proved" from geometric constructions. Although the modern proof of l'Hospital's rule uses the mean value theorem, it should be remembered that the original version for calculating the limit of a quotient where both the numerator and denominator become zero first appeared in 1696, 70 years before Lagrange's proof of the mean value theorem. The original version was stated and "proved" in a purely geometric manner without reference to limits. For further details, including a history of calculus, the reader is referred to the text by Katz listed in the Bibliography.

The Bernoulli brothers, Jakob (1654–1705) and Johann (1667–1748) were among the first mathematicians in Europe to use the new techniques of Newton and Leibniz in the study of curves and related physical problems. Among these were finding the equations of the catenary and isochrone.[3] Both brothers also contributed to the study of differential equations by solving the *Bernoulli equation* $y' + P(x)y = Q(x)y^n$. Through their numerous publications and correspondence with other mathematicians the Bernoulli brothers helped to establish the utility of the new calculus. The first text on differential calculus by l'Hospital also contributed significantly to popularizing the subject.

Leonhard Euler (1707–1783), one of the most prolific mathematicians in history, contributed significantly to establishing calculus as an independent science. Even though the calculus of exponential and logarithmic functions was basically developed by Johann Bernoulli, it was Euler's expositions on these topics in the eighteenth century that brought them into the mainstream of mathematics. Much of what we know today about the exponential, logarithmic and trigonometric functions is due to Euler. He was also among the first mathematicians to define the concept of a function. However to Euler, as for the other mathematicians of that

---

[3]The catenary problem involves finding the equation of a freely hanging cable, whereas the isochrone problem involves finding the equation of a curve along which an object would fall with uniform vertical velocity.

period, a function was one that had a power series expansion. It is important to note that most mathematicians of the eighteenth century, including Euler, were primarily concerned with computations needed for the applications of calculus; proofs did not gain prominence until the nineteenth century. For this reason, numerous results of that era that were assumed to be true were subsequently proved to be true only under more restrictive conditions.

Miscellaneous Exercises

1. Let $f : (a,b) \to \mathbb{R}$ and suppose f'' exists at $x_o \in (a,b)$. Prove that
$$f''(x_o) = \lim_{h \to 0} \frac{f(x_o + h) + f(x_o - h) - 2f(x_o)}{h^2}.$$
Give an example where this limit exists at x_o but f'' does not exist at x_o.

2. Let f be a real-valued differentiable function on $\mathbb{R}$.
   a. If there exists a constant $b < 1$ such that $|f'(x)| < b$ for all $x \in \mathbb{R}$, prove that f has a fixed point in $\mathbb{R}$. (See Exercise 13 of Section 4.3).
   b. Show that the function $f(x) = x + (1 + e^x)^{-1}$ satisfies $|f'(x)| < 1$ for all $x \in \mathbb{R}$ but that f has no fixed point in $\mathbb{R}$.

3. A function f is **convex** (or concave up) on the interval (a,b) if for any $x, y \in (a,b)$, and $0 < t < 1$, $f(tx + (1-t)y) \le t f(x) + (1-t) f(y)$.
   a. If f is convex on (a,b), prove that f is continuous on $(a.b)$.
   b. If f is convex on (a,b), prove that $f'_+(p)$ and $f'_-(p)$ exist for every $p \in (a,b)$. Show by example that a convex function on (a,b) need not be differentiable on (a,b).
   c. Suppose $f''(x)$ exists for all $x \in (a,b)$. Prove that f is convex on (a,b) if and only if $f''(x) \ge 0$ for all $x \in (a,b)$.

4. Suppose f is differentiable on $[a,b]$ and that f' is continuous on $[a,b]$. Without using the mean value theorem, prove that
$$\min\{f'(x) : x \in [a,b]\} \le \frac{f(b) - f(a)}{b - a} \le \max\{f'(x) : x \in [a,b]\}.$$

5. (T.M. Flett) If f is differentiable on $[a,b]$ and $f'(a) = f'(b)$, prove that there exists $\zeta \in (a,b)$ such that
$$\frac{f(\zeta) - f(a)}{\zeta - a} = f'(\zeta).$$

6. Suppose $f : (a,b) \to \mathbb{R}$ is differentiable at $c \in (a,b)$. If $\{s_n\}$ and $\{t_n\}$ are sequences in (a,b) with $s_n < c < t_n$ and $\lim_{n \to \infty} (t_n - s_n) = 0$, prove that
$$\lim_{n \to \infty} \frac{f(t_n) - f(s_n)}{t_n - s_n} = f'(c).$$

Supplemental Reading

Baxley, J. V. and Hayashi, E. K., "Indeterminate forms of exponential type," *Amer. Math. Monthly* **85** (1978), 484–486.

Cajori, F., "Historical note on the Newton-Raphson method of approximation," *Amer. Math. Monthly* **18** (1911), 29–33.

Corless, R. M., "Variations on a theme of Newton," *Math. Mag.* **71** (1998), 34–41.

Flett, T.M., "A mean value theorem," *Math. Gaz.* **42** (1958), 38 – 39.

Hall, W. S. and Newell, M. L., "The mean value theorem for vector valued functions," *Math. Mag.* **52** (1979), 157–158.

Hartig, Donald, "L'Hopitals rule via integration," *Amer. Math. Monthly* **98** (1991), 156–157.

Katznelson, Y. and Stromberg, K., "Everywhere differentiable, nowhere monotone function," *Amer. Math. Monthly* **81** (1974), 349–354.

Langlois, W. E. and Holder, L. I., "The relation of $f'_+(a)$ to "$f'(a+)$," *Math. Mag.* **39** (1966), 112–120.

Lynch, M., "A continuous function that is differentiable only at the rationals," *Math. Mag.* **86** (2013), 132–135.

Miller, A. D. and Vyborny, R., "Some remarks on functions with one-sided derivatives" *Amer. Math. Monthly* **93** (1986), 471–475.

Pan, D., "A maximum principle for high-order derivatives," *Amer. Math. Monthly* **120** (2013), 846–848.

Range, R. M., "Where are limits needed in Calculus," *Amer. Math. Monthly* **118** (2011), 404–417.

Rosenholtz, "There is no differentiable metric for $\mathbb{R}^n$," *Amer. Math. Monthly* **86** (1979), 585–586.

Rotando, L. M. and Korn, H., "The indeterminate form 0^0," *Math. Mag.* **50** (1977), 41–42.

Sahoo, M. R., "Example of a monotonic everywhere differentiable function on $\mathbb{R}$ whose derivative is not continuous," *Amer. Math. Monthly* **120** (2013), 566–568.

Tandra, H., "A yet simpler proof of the chain rule," *Amer. Math. Monthly* **120** (2013), 900.

Thurston, H. A., "On the definition of the tangent line," *Amer. Math. Monthly* **71** (1964), 1099–1103.

Tong, J. and Braza, P. A., "A converse to the mean value theorem," *Amer. Math. Monthly* **104** (1997), 939–942.

6

Integration

When Newton and Leibniz developed the calculus, both considered integration as the inverse operation of differentiation. For example, in the *De analysi*[1], Newton proved that the area under the curve $y = ax^{m/n}$ $(m/n \neq -1)$ is given by

$$\frac{an\, x^{\frac{m}{n}+1}}{m+n}$$

by using his differential calculus to prove that if $A(x)$ represents the area from 0 to x then $A'(x) = ax^{m/n}$. Even though Leibniz arrived at the concept of the integral by using sums to compute the area, integration itself was always the inverse operation of differentiation. Throughout the eighteenth century, the definite integral of a function $f(x)$ on $[a, b]$, denoted $\int_a^b f(x)dx$, was defined as $F(b) - F(a)$ where F was any function whose derivative was $f(x)$. This remained as the definition of the definite integral until the 1820's.

The modern approach to integration is again due to Cauchy, who was the first mathematician to construct a theory of integration based on approximating the area under the curve. Euler had previously used sums of the form $\sum_{k=1}^{n} f(x_{k-1})(x_k - x_{k-1})$ to approximate the integral of a function $f(x)$ in situations where the function $F(x)$ could not be computed. Cauchy however used limits of such sums to develop a theory of integration that was independent of the differential calculus. One of the difficulties with Cauchy's definition of the integral was that it was very restrictive; only functions that were continuous or continuous except at a finite number of points were proved to be integrable. However, one of the key achievements of Cauchy was that using his definition he was able to prove the fundamental theorem of calculus; specifically, if f is continuous on $[a, b]$, then there exists a function F on $[a, b]$ such that $F'(x) = f(x)$ for all $x \in [a, b]$.

The modern definition of integration was developed in 1853 by Georg Bernhard Riemann (1826–1866). Riemann was led to the development of the integral by trying to characterize which functions were integrable according to Cauchy's definition. In the process, he modified Cauchy's definition and developed the theory of integration which now bears his name. One of his achievements was that he was able to provide necessary and sufficient

---

[1] *The Mathematical Works of Isaac Newton*, edited by D. T. Whiteside, Johnson Reprint Corporation, New York, 1964.

conditions for a real-valued bounded function to be integrable. In Section 1, we develop the theory of the Riemann integral using the approach of Jean Gaston Darboux (1842–1917). In this section, we also include the statement of Lebesgue's theorem which provides necessary and sufficient conditions that a bounded real-valued function defined on a closed and bounded interval be Riemann integrable. The equivalence of the Riemann and Darboux approach will be proved in Section 6.2.

In Section 6.5 we will consider the more general Riemann-Stieltjes integral which will give meaning to the following types of integrals:

$$\int_0^1 f(x)\,dx^2, \qquad \int_a^b f(x)\,d[x], \qquad \text{or} \qquad \int_a^b f(x)\,d\alpha(x),$$

where α is a monotone increasing function on $[a, b]$. These types of integrals were developed by Thomas-Jean Stieltjes (1856–1894) and arise in many applications in both mathematics and physics. The theory itself involves only minor modifications in the definition of the Riemann integral; the consequences however are far reaching. The Riemann-Stieltjes integral permits the expression of many seemingly diverse results as a single formula.

6.1   The Riemann Integral

There are traditionally two approaches to the theory of the Riemann integral; namely the original method of Riemann, and the method introduced by Darboux in 1875 using lower and upper sums. I have chosen the latter approach to define the Riemann integral because of its easy adaptability to the Riemann-Stieltjes integral. We will however consider both methods and show that they are in fact equivalent.

Upper and Lower Sums

Let $[a, b]$, $a < b$, be a given closed and bounded interval in $\mathbb{R}$. By a **partition** $\mathcal{P}$ of $[a, b]$ we mean a finite set of points $\mathcal{P} = \{x_0, x_1, ..., x_n\}$ such that

$$a = x_0 < x_1 < \cdots < x_n = b.$$

There is no requirement that the points x_i be equally spaced. For each $i = 1, 2, ..., n$, set

$$\Delta x_i = x_i - x_{i-1},$$

which is equal to the length of the interval $[x_{i-1}, x_i]$.

Suppose f is a bounded real-valued function on $[a, b]$. Given a partition $\mathcal{P} = \{x_0, x_1, ..., x_n\}$ of $[a, b]$, for each $i = 1, 2, ..., n$, let

$$m_i = \inf\{f(t) : x_{i-1} \leq t \leq x_i\},$$
$$M_i = \sup\{f(t) : x_{i-1} \leq t \leq x_i\}.$$

Since f is bounded, by the least upper bound property the quantities m_i and M_i exist in $\mathbb{R}$. If f is a continuous function on $[a, b]$, then by Corollary 4.2.9, for each i there exist points $t_i, s_i \in [x_{i-1}, x_i]$ such that $M_i = f(t_i)$ and $m_i = f(s_i)$.

The **upper sum** $\mathcal{U}(\mathcal{P}, f)$ for the partition $\mathcal{P}$ and function f is defined by

$$\mathcal{U}(\mathcal{P}, f) = \sum_{i=1}^{n} M_i \, \Delta x_i.$$

Similarly, the **lower sum** $\mathcal{L}(\mathcal{P}, f)$ is defined by

$$\mathcal{L}(\mathcal{P}, f) = \sum_{i=1}^{n} m_i \, \Delta x_i.$$

Since $m_i \le M_i$ for all $i = 1, ..., n$, we always have

$$\mathcal{L}(\mathcal{P}, f) \le \mathcal{U}(\mathcal{P}, f)$$

for any partition $\mathcal{P}$ of $[a, b]$. The upper sum for a nonnegative continuous function f is illustrated in Figure 6.1. In this case, $\mathcal{U}(\mathcal{P}, f)$ represents the circumscribed rectangular approximation to the area under the graph of f. Similarly the lower sum represents the inscribed rectangular approximation to the area under the graph of f.

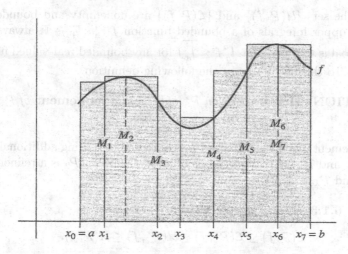

FIGURE 6.1
Upper sum $\mathcal{U}(\mathcal{P}, f)$

Upper and Lower Integrals

If the function f satisfies $m \leq f(t) \leq M$ for all $t \in [a, b]$, then

$$m(b-a) \leq \mathcal{L}(\mathcal{P}, f) \leq \mathcal{U}(\mathcal{P}, f) \leq M(b-a), \tag{1}$$

for any partition $\mathcal{P}$ of $[a, b]$. To see that (1) holds, let $\mathcal{P} = \{x_0, ..., x_n\}$ be any partition of $[a, b]$. Since $M_i \leq M$ for all $i = 1, ..., n$,

$$\mathcal{U}(\mathcal{P}, f) = \sum_{i=1}^{n} M_i \Delta x_i \leq \sum_{i=1}^{n} M(x_i - x_{i-1}) = M(b-a).$$

Similarly $\mathcal{L}(\mathcal{P}, f) \geq m(b-a)$. Thus the set $\{\mathcal{U}(\mathcal{P}, f) : \mathcal{P} \text{ is a partition of } [a, b]\}$ is bounded above and below, as is the set $\{\mathcal{L}(\mathcal{P}, f)\}$.

DEFINITION 6.1.1 *Let f be a bounded real-valued function on the closed and bounded interval $[a, b]$. The* **upper** *and* **lower integrals** *of f, denoted $\overline{\int_a^b} f$ and $\underline{\int_a^b} f$ respectively, are defined by*

$$\overline{\int_a^b} f = \inf\{\mathcal{U}(\mathcal{P}, f) : \mathcal{P} \text{ is a partition of } [a, b]\},$$

$$\underline{\int_a^b} f = \sup\{\mathcal{L}(\mathcal{P}, f) : \mathcal{P} \text{ is a partition of } [a, b]\}.$$

Since the sets $\{\mathcal{U}(\mathcal{P}, f)\}$ and $\{\mathcal{L}(\mathcal{P}, f)\}$ are nonempty and bounded, the lower and upper integrals of a bounded function $f : [a, b] \to \mathbb{R}$ always exist. Our first goal is to prove that $\underline{\int_a^b} f \leq \overline{\int_a^b} f$ for any bounded real-valued function f on $[a, b]$. To this end we make the following definition.

DEFINITION 6.1.2 *A partition $\mathcal{P}^*$ of $[a, b]$ is a* **refinement** *of $\mathcal{P}$ if $\mathcal{P} \subset \mathcal{P}^*$.*

A refinement of a given partition $\mathcal{P}$ is obtained by adding additional points to $\mathcal{P}$. If $\mathcal{P}_1$ and $\mathcal{P}_2$ are two partitions of $[a, b]$, then $\mathcal{P}_1 \cup \mathcal{P}_2$ is a refinement of both $\mathcal{P}_1$ and $\mathcal{P}_2$.

LEMMA 6.1.3 *If $\mathcal{P}^*$ is a refinement of $\mathcal{P}$, then*

$$\mathcal{L}(\mathcal{P}, f) \leq \mathcal{L}(\mathcal{P}^*, f) \leq \mathcal{U}(\mathcal{P}^*, f) \leq \mathcal{U}(\mathcal{P}, f).$$

Proof. Suppose $\mathcal{P} = \{x_0, x_1, ..., x_n\}$ and $\mathcal{P}^* = \mathcal{P} \cup \{x^*\}$, where $x^* \neq x_j$ for any $j = 0, 1, ..., n$. Then there exists an index k such that $x_{k-1} < x^* < x_k$. Let

$$M_k^1 = \sup\{f(t) : t \in [x_{k-1}, x^*]\},$$
$$M_k^2 = \sup\{f(t) : t \in [x^*, x_k]\}.$$

Since $f(t) \le M_k$ for all $t \in [x_{k-1}, x_k]$, we have that $f(t) \le M_k$ for all $t \in [x_{k-1}, x^{\star}]$ and also for all $t \in [x^{\star}, x_k]$. Thus both M_k^1 and M_k^2 are less than or equal to M_k. Now

$$\mathcal{U}(\mathcal{P}^{\star}, f) = \sum_{j=1}^{k-1} M_j \Delta x_j + M_k^1 (x^{\star} - x_{k-1}) + M_k^2 (x_k - x^{\star}) + \sum_{j=k+1}^{n} M_j \Delta x_j.$$

But

$$M_k^1 (x^{\star} - x_{k-1}) + M_k^2 (x_k - x^{\star}) \le M_k \Delta x_k.$$

Therefore

$$\mathcal{U}(\mathcal{P}^{\star}, f) \le \mathcal{U}(\mathcal{P}, f).$$

The proof for the lower sum is similar. If $\mathcal{P}^{\star}$ contains k more points than $\mathcal{P}$, we need only repeat the above argument k times to obtain the result. $\square$

THEOREM 6.1.4 *Let f be a bounded real-valued function on $[a, b]$. Then*

$$\underline{\int_a^b} f \le \overline{\int_a^b} f.$$

Proof. Given any two partitions $\mathcal{P}, \mathcal{Q}$ of $[a, b]$,

$$\mathcal{L}(\mathcal{P}, f) \le \mathcal{L}(\mathcal{P} \cup \mathcal{Q}, f) \le \mathcal{U}(\mathcal{P} \cup \mathcal{Q}, f) \le \mathcal{U}(\mathcal{Q}, f).$$

Thus $\mathcal{L}(\mathcal{P}, f) \le \mathcal{U}(\mathcal{Q}, f)$ for any partitions $\mathcal{P}, \mathcal{Q}$. Hence

$$\underline{\int_a^b} f = \sup_{\mathcal{P}} \mathcal{L}(\mathcal{P}, f) \le \mathcal{U}(\mathcal{Q}, f)$$

for any partition $\mathcal{Q}$. Taking the infimum over $\mathcal{Q}$ gives the result. $\square$

The Riemann Integral

If $f : [a, b] \to \mathbb{R}$ is bounded, then the lower and upper integrals of f on $[a, b]$ always exist and satisfy $\underline{\int_a^b} f \le \overline{\int_a^b} f$. As we will shortly see, there is a large family of functions for which equality holds; such functions are said to be *integrable*.

DEFINITION 6.1.5 *Let f be a bounded real-valued function on the closed and bounded interval $[a, b]$. If*

$$\underline{\int_a^b} f = \overline{\int_a^b} f,$$

*then f is said to be **Riemann integrable** or **integrable** on $[a, b]$. The common value, denoted*

$$\int_a^b f,$$

is called the **Riemann integral** *or* **integral** *of* f *over* $[a, b]$. *We denote by* $\mathcal{R}[a, b]$ *the set of* **Riemann integrable functions** *on* $[a, b]$. *In addition, if* $f \in \mathcal{R}[a, b]$, *we define*

$$\int_b^a f = -\int_a^b f.$$

Alternate notation that is sometimes used to denote the Riemann integral of f is

$$\int_a^b f(x)\,dx.$$

In the above, the variable x could just as easily have been t, or any other convenient letter.

If $f : [a, b] \to \mathbb{R}$ satisfies $m \leq f(t) \leq M$ for all $t \in [a, b]$, then by inequality (1),

$$m\,(b - a) \leq \underline{\int_a^b} f \leq \overline{\int_a^b} f \leq M\,(b - a).$$

If in addition $f \in \mathcal{R}[a, b]$, then

$$m\,(b - a) \leq \int_a^b f \leq M\,(b - a).$$

In particular, if $f(x) \geq 0$ for all $x \in [a, b]$, then $\int_a^b f \geq 0$. If $f \in \mathcal{R}[a, b]$ is non-negative, then the quantity $\int_a^b f$ represents the **area** of the region bounded above by the graph $y = f(x)$, below by the x-axis, and the lines $x = a$ and $x = b$.

EXAMPLES 6.1.6 (a) The following function, attributed to Dirichlet, is the canonical example of a function that is not Riemann integrable on a closed interval. Let f be defined by

$$f(x) = \begin{cases} 1, & x \in \mathbb{Q}, \\ 0, & x \notin \mathbb{Q}. \end{cases}$$

Suppose $a < b$. If $\mathcal{P} = \{x_0, ..., x_n\}$ is any partition of $[a, b]$, then $m_i = 0$ and $M_i = 1$ for all $i = 1, ..., n$. Thus

$$\mathcal{L}(\mathcal{P}, f) = 0 \qquad \text{and} \qquad \mathcal{U}(\mathcal{P}, f) = (b - a)$$

for any partition $\mathcal{P}$ of $[a, b]$. Therefore

$$\underline{\int_a^b} f = 0 \qquad \text{and} \qquad \overline{\int_a^b} f = (b - a).$$

and as a consequence f is *not* Riemann integrable on $[a, b]$.

(b) Let $f : [0, 1] \to \mathbb{R}$ be defined by

$$f(x) = \begin{cases} 0, & 0 \le x < \frac{1}{2}, \\ 1, & \frac{1}{2} \le x \le 1. \end{cases}$$

In this example, we prove that f is integrable on $[0, 1]$ with $\int_0^1 f = \frac{1}{2}$. Let $\mathcal{P} = \{x_0, x_1, ..., x_n\}$ be a partition of $[0, 1]$, and let $k \in \{1, ..., n-1\}$ be such that $x_{k-1} < \frac{1}{2} \le x_k$ (See Figure 6.2). If m_i and M_i denote the infimum and supremum of f on $[x_{i-1}, x_i]$ respectively, then

$$m_i = \begin{cases} 0, & i = 1, ..., k, \\ 1, & i = k+1, ..., n, \end{cases} \quad \text{and} \quad M_i = \begin{cases} 0, & i = 1, ..., k-1, \\ 1, & i = k, ..., n. \end{cases}$$

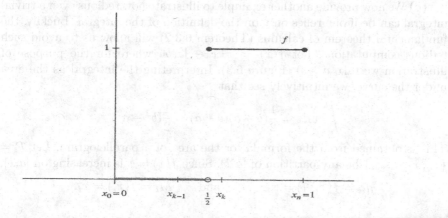

FIGURE 6.2
The function of Example 6.1.6(b)

Thus

$$\mathcal{L}(\mathcal{P}, f) = \sum_{i=k+1}^{n} \Delta x_i = (1 - x_k), \quad \text{and}$$

$$\mathcal{U}(\mathcal{P}, f) = \sum_{i=k}^{n} \Delta x_i = (1 - x_{k-1})$$

Since $1 - x_k \le \frac{1}{2} < 1 - x_{k-1}$, $\mathcal{L}(\mathcal{P}, f) \le \frac{1}{2} \le \mathcal{U}(\mathcal{P}, f)$ for all partitions $\mathcal{P}$ of $[0, 1]$. Thus $\underline{\int_0^1} f \le \frac{1}{2} \le \overline{\int_0^1} f$. Hence if f is integrable on $[0, 1]$, then $\int_0^1 f = 1/2$. We conclude by proving that f is indeed integrable on $[0, 1]$. Since $1 - x_{k-1} = (1 - x_k) + (x_k - x_{k-1})$, we have

$$\mathcal{U}(\mathcal{P}, f) = \mathcal{L}(\mathcal{P}, f) + (x_k - x_{k-1}).$$

Let $\epsilon > 0$ be arbitrary. If $\mathcal{P}$ is any partition of $[0,1]$ with $\Delta x_i < \epsilon$ for all i, then $\mathcal{U}(\mathcal{P},f) < \mathcal{L}(\mathcal{P},f) + \epsilon$. Thus

$$\overline{\int_0^1} f = \inf_{\mathcal{Q}} \mathcal{U}(\mathcal{Q},f) \leq \mathcal{U}(\mathcal{P},f)$$

$$\leq \mathcal{L}(\mathcal{P},f) + \epsilon \leq \sup_{\mathcal{Q}} \mathcal{L}(\mathcal{Q},f) + \epsilon \leq \underline{\int_0^1} f + \epsilon,$$

where the infimum and supremum are taken over all partitions $\mathcal{Q}$ of $[0,1]$. Since $\epsilon > 0$ was arbitrary, we have $\underline{\int_0^1} f = \overline{\int_0^1} f$. Therefore f is integrable on $[0,1]$ with $\int\limits_0^1 f = 1/2$.

(c) We now provide another example to illustrate how tedious even a trivial integral can be if one relies only on the definition of the integral. Luckily, the fundamental theorem of calculus (Theorem 6.3.2) will allow us to avoid such tedious computations. Let $f(x) = x$, $x \in [a,b]$, where for the purpose of illustration we take $a \geq 0$ (Figure 6.3). Interpreting the integral as the area under the curve, we intuitively see that

$$\int_a^b x\, dx = \frac{1}{2}(b-a)(b+a) = \frac{1}{2}(b^2 - a^2).$$

This is obtained from the formula for the area of a parallelogram. Let $\mathcal{P} = \{x_0, x_1, ..., x_n\}$ be any partition of $[a,b]$. Since $f(x) = x$ is increasing on $[a,b]$,

$$m_i = f(x_{i-1}) = x_{i-1}, \qquad \text{and} \qquad M_i = f(x_i) = x_i.$$

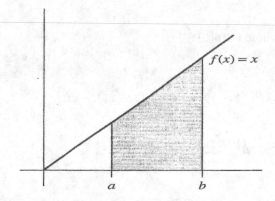

FIGURE 6.3
Example 6.1.6(c)

Therefore

$$\mathcal{L}(\mathcal{P},f) = \sum_{i=1}^n x_{i-1}\Delta x_i \qquad \text{and} \qquad \mathcal{U}(\mathcal{P},f) = \sum_{i=1}^n x_i \Delta x_i.$$

For each index i,

$$x_{i-1} \le \frac{1}{2}(x_{i-1} + x_i) \le x_i.$$

If we multiply this by $\Delta x_i = x_i - x_{i-1}$ and sum from $i = 1$ to n, we obtain

$$\mathcal{L}(\mathcal{P}, f) \le \sum_{i=1}^{n} \frac{1}{2}(x_i^2 - x_{i-1}^2) \le \mathcal{U}(\mathcal{P}, f).$$

But

$$\sum_{i=1}^{n} \frac{1}{2}(x_i^2 - x_{i-1}^2) = \frac{1}{2}(x_n^2 - x_o^2) = \frac{1}{2}(b^2 - a^2).$$

Finally, if we take the supremum and infimum over all partitions $\mathcal{P}$ of $[a, b]$, we have

$$\underline{\int_a^b} x\,dx \le \frac{1}{2}(b^2 - a^2) \le \overline{\int_a^b} x\,dx.$$

This in itself does not prove that

$$\int_a^b x\,dx = \frac{1}{2}(b^2 - a^2).$$

To prove that f is integrable on $[a, b]$, we note that for any partition $\mathcal{P}$ of $[a, b]$,

$$0 \le \overline{\int_a^b} x\,dx - \underline{\int_a^b} x\,dx \le \mathcal{U}(\mathcal{P}, f) - \mathcal{L}(\mathcal{P}, f) = \sum_{i=1}^{n}(x_i - x_{i-1})\Delta x_i$$

$$\le \left(\max_{1 \le i \le n} \Delta x_i\right)(b - a).$$

Let $\epsilon > 0$ be given. If we choose a partition $\mathcal{P}$ such that $\Delta x_i < \epsilon/(b-a+1)$ for all i, then $\overline{\int_a^b} x\,dx - \underline{\int_a^b} x\,dx < \epsilon$. Since this holds for every $\epsilon > 0$, we conclude that $\underline{\int_a^b} x\,dx = \overline{\int_a^b} x\,dx$. Thus $f(x) = x$ is Riemann integrable on $[a, b]$ with $\int\limits_a^b x\,dx = \frac{1}{2}(b^2 - a^2)$.

(d) Consider the function $f(x) = x^2$, $x \in [0, 1]$. For $n \in \mathbb{N}$, let $\mathcal{P}_n$ be the partition $\{0, \frac{1}{n}, \frac{2}{n}, ..., 1\}$. Since f is increasing on $[0, 1]$, its infimum and supremum on each interval $[\frac{i-1}{n}, \frac{i}{n}]$ are attained at the left and right endpoint respectively, with $m_i = (i-1)^2/n^2$ and $M_i = i^2/n^2$. Since $\Delta x_i = \frac{1}{n}$ for all i,

$$\mathcal{L}(\mathcal{P}_n, f) = \frac{1}{n^3}[1^2 + 2^2 + \cdots + (n-1)^2], \qquad \text{and}$$

$$\mathcal{U}(\mathcal{P}_n, f) = \frac{1}{n^3}[1^2 + 2^2 + \cdots + n^2].$$

Using the identity $1^2 + 2^2 + \cdots + m^2 = \frac{1}{6}m(m+1)(2m+1)$ (Exercise 1(c), Section 1.3), we have

$$\mathcal{L}(\mathcal{P}_n, f) = \frac{1}{6}\left(1 - \frac{1}{n}\right)\left(2 - \frac{1}{n}\right) \quad \text{and} \quad \mathcal{U}(\mathcal{P}_n, f) = \frac{1}{6}\left(1 + \frac{1}{n}\right)\left(2 + \frac{1}{n}\right).$$

Thus $\sup_n \mathcal{L}(\mathcal{P}_n, f) = 1/3$ and $\inf_n \mathcal{U}(\mathcal{P}_n, f) = 1/3$. Since the collection $\{\mathcal{P}_n : n \in \mathbb{N}\}$ is a subset of the set of all partitions of $[0,1]$,

$$\frac{1}{3} = \sup_n \mathcal{L}(\mathcal{P}_n, f) \leq \sup_{\mathcal{P}} \mathcal{L}(\mathcal{P}, f) = \underline{\int_0^1} x^2 dx, \qquad \text{and}$$

$$\frac{1}{3} = \inf_n \mathcal{U}(\mathcal{P}_n, f) \geq \inf_{\mathcal{P}} \mathcal{U}(\mathcal{P}, f) = \overline{\int_0^1} x^2 dx.$$

Therefore $f(x) = x^2$ is integrable on $[0,1]$ with $\int_0^1 x^2 \, dx = 1/3$. $\square$

Riemannn's Criterion for Integrability

The following theorem, the original version of which is due to Riemann, provides necessary and sufficient conditions for the existence of the Riemann integral.

THEOREM 6.1.7 *A bounded real-valued function f is Riemann integrable on $[a, b]$ if and only if for every $\epsilon > 0$, there exists a partition $\mathcal{P}$ of $[a, b]$ such that*

$$\mathcal{U}(\mathcal{P}, f) - \mathcal{L}(\mathcal{P}, f) < \epsilon. \tag{2}$$

Furthermore, if $\mathcal{P}$ is a partition of $[a, b]$ for which (2) holds, then the inequality also holds for all refinements of $\mathcal{P}$.

Proof. Suppose (2) holds for a given $\epsilon > 0$. Then

$$0 \leq \overline{\int_a^b} f - \underline{\int_a^b} f \leq \mathcal{U}(\mathcal{P}, f) - \mathcal{L}(\mathcal{P}, f) < \epsilon.$$

Thus f is integrable on $[a, b]$.

Conversely, suppose f is integrable on $[a, b]$. Let $\epsilon > 0$ be given. Then there exist partitions $\mathcal{P}_1$ and $\mathcal{P}_2$ of $[a, b]$ such that

$$\mathcal{U}(\mathcal{P}_2, f) - \int_a^b f < \frac{\epsilon}{2} \quad \text{and} \quad \int_a^b f - \mathcal{L}(\mathcal{P}_1, f) < \frac{\epsilon}{2}.$$

Let $\mathcal{P} = \mathcal{P}_1 \cup \mathcal{P}_2$. Then

$$\mathcal{U}(\mathcal{P}, f) \leq \mathcal{U}(\mathcal{P}_2, f) < \int_a^b f + \frac{\epsilon}{2} < \mathcal{L}(\mathcal{P}_1, f) + \epsilon \leq \mathcal{L}(\mathcal{P}, f) + \epsilon.$$

Therefore, $\mathcal{U}(\mathcal{P}, f) - \mathcal{L}(\mathcal{P}, f) < \epsilon$, which proves (2). If $\mathcal{Q}$ is any refinement of $\mathcal{P}$, then by Lemma 6.1.3

$$0 \leq \mathcal{U}(\mathcal{Q}, f) - \mathcal{L}(\mathcal{Q}, f) \leq \mathcal{U}(\mathcal{P}, f) - \mathcal{L}(\mathcal{P}, f) < \epsilon.$$

Thus (2) is also valid for any refinement of $\mathcal{Q}$ of $\mathcal{P}$. $\square$

Integrability of Continuous and Monotone Functions

As an application of the previous theorem we prove that every continuous real-valued function and every monotone function on $[a, b]$ is Riemann integrable on $[a, b]$. As we will see, both of these results will also follow from Lebesgue's theorem (Theorem 6.1.13).

THEOREM 6.1.8 *Let f be a real-valued function on $[a, b]$.*

(a) *If f is continuous on $[a, b]$, then f is Riemann integrable on $[a, b]$.*

(b) *If f is monotone on $[a, b]$, then f is Riemann integrable on $[a, b]$.*

Proof. (a) Let $\epsilon > 0$ be given. Choose $\eta > 0$ such that $(b - a)\eta < \epsilon$. Since f is continuous on $[a, b]$, by Theorem 4.3.4 f is uniformly continuous on $[a, b]$. Thus there exists a $\delta > 0$ such that

$$|f(x) - f(t)| < \eta \tag{3}$$

for all $x, t \in [a, b]$ with $|x - t| < \delta$. Choose a partition $\mathcal{P}$ of $[a, b]$ such that $\Delta x_i < \delta$ for all $i = 1, 2, ..., n$. Then by (3),

$$M_i - m_i \leq \eta$$

for all $i = 1, 2, ..., n$. Therefore

$$\mathcal{U}(\mathcal{P}, f) - \mathcal{L}(\mathcal{P}, f) = \sum_{i=1}^{n} (M_i - m_i)\Delta x_i \leq \eta \sum_{i=1}^{n} \Delta x_i = \eta(b - a) < \epsilon.$$

Thus by Theorem 6.1.7 f is integrable on $[a, b]$.

(b) Suppose f is monotone increasing on $[a, b]$. For $n \in \mathbb{N}$, set $h = (b-a)/n$. Also for $i = 0, 1, ..., n$, set $x_i = a + i h$. Then $\mathcal{P} = \{x_0, x_1, ..., x_n\}$ is a partition of $[a, b]$ which satisfies $\Delta x_i = h$ for all $i = 1, ..., n$. Since f is monotone increasing on $[a, b]$, $m_i = f(x_{i-1})$ and $M_i = f(x_i)$. Therefore,

$$\mathcal{U}(\mathcal{P}, f) - \mathcal{L}(\mathcal{P}, f) = \sum_{i=1}^{n} [f(x_i) - f(x_{i-1})]\Delta x_i$$

$$= h \sum_{i=1}^{n} [f(x_i) - f(x_{i-1})] = \frac{(b - a)}{n}[f(b) - f(a)].$$

Given $\epsilon > 0$, choose $n \in \mathbb{N}$ such that

$$\frac{(b-a)}{n}[f(b) - f(a)] < \epsilon.$$

For this n and corresponding partition $\mathcal{P}$, $\mathcal{U}(\mathcal{P}, f) - \mathcal{L}(\mathcal{P}, f) < \epsilon$. Thus f is integrable on $[a, b]$. $\square$

The Composition Theorem

We next prove that the composition $\varphi \circ f$ of a continuous function φ with a Riemann integrable function f is again Riemann integrable. As an application of Lebesgue's theorem we will present a much shorter proof of this result later in the section.

THEOREM 6.1.9 *Let f be a bounded Riemann integrable function on $[a, b]$ with $\text{Range} f \subset [c, d]$. If φ is continuous on $[c, d]$, then $\varphi \circ f$ is Riemann integrable on $[a, b]$.*

Proof. Since φ is continuous on the closed and bounded interval $[c, d]$, φ is bounded and uniformly continuous on $[c, d]$. Let $K = \sup\{|\varphi(t)| : t \in [c, d]\}$, and let $\epsilon > 0$ be given. Set $\epsilon' = \epsilon/(b - a + 2K)$.

Since φ is uniformly continuous on $[c, d]$, there exists δ, $0 < \delta < \epsilon'$, such that

$$|\varphi(s) - \varphi(t)| < \epsilon' \tag{4}$$

for all $s, t \in [c, d]$ with $|s - t| < \delta$. Furthermore, since $f \in \mathcal{R}[a, b]$, by Theorem 6.1.7 there exists a partition $\mathcal{P} = \{x_0, x_1, ..., x_n\}$ of $[a, b]$ such that

$$\mathcal{U}(\mathcal{P}, f) - \mathcal{L}(\mathcal{P}, f) < \delta^2.$$

To complete the proof we will show that

$$\mathcal{U}(\mathcal{P}, \varphi \circ f) - \mathcal{L}(\mathcal{P}, \varphi \circ f) \leq \epsilon. \tag{5}$$

By Theorem 6.1.7 it then follows that $\varphi \circ f \in \mathcal{R}[a, b]$.

For each $k = 1, 2, ..., n$, let m_k and M_k denote the infimum and supremum of f on $[x_{k-1}, x_k]$. Also, set

$$m_k^* = \inf\{\varphi(f(t)) : t \in [x_{k-1}, x_k]\} \quad \text{and} \quad M_k^* = \sup\{\varphi(f(t)) : t \in [x_{k-1}, x_k]\}.$$

We partition the set $\{1, 2, ..., n\}$ into disjoint sets A and B as follows:

$$A = \{k : M_k - m_k < \delta\} \quad \text{and} \quad B = \{k : M_k - m_k \geq \delta\}.$$

Since $|f(t) - f(s)| \leq M_k - m_k$ for all $s, t \in [x_{k-1}, x_k]$, if $k \in A$, then by (4)

$$|\varphi(f(t)) - \varphi(f(s))| < \epsilon'$$

for all $s, t \in [x_{k-1}, x_k]$. But

$$M_k^* - m_k^* = \sup\{\varphi(f(t)) - \varphi(f(s)) : s, t \in [x_{k-1}, x_k]\}.$$

Therefore $M_k^* - m_k^* \leq \epsilon'$ for all $k \in A$. On the other hand, if $k \in B$, then $M_k^* - m_k^* \leq 2K$. Thus

$$\mathcal{U}(\mathcal{P}, \varphi \circ f) - \mathcal{L}(\mathcal{P}, \varphi \circ f) = \sum_{k \in A}(M_k^* - m_k^*)\Delta x_k + \sum_{k \in B}(M_k^* - m_k^*)\Delta x_k$$

$$\leq \epsilon' \sum_{k \in A} \Delta x_k + 2K \sum_{k \in B} \Delta x_k$$

$$\leq \epsilon'(b - a) + 2K \sum_{k \in B} \Delta x_k.$$

But for $k \in B$, $\delta \leq M_k - m_k$. Therefore,

$$\sum_{k \in B} \Delta x_k \leq \frac{1}{\delta} \sum_{k \in B}(M_k - m_k)\Delta x_k \leq \frac{1}{\delta}(\mathcal{U}(\mathcal{P}, f) - \mathcal{L}(\mathcal{P}, f)) < \delta < \epsilon',$$

and hence by the above

$$\mathcal{U}(\mathcal{P}, \varphi \circ f) - \mathcal{L}(\mathcal{P}, \varphi \circ f) \leq \epsilon'(b - a) + 2K\epsilon' = \epsilon.$$

This establishes (5), and thus $\varphi \circ f \in \mathcal{R}[a, b]$. $\square$

As a consequence of the previous theorem, if f is Riemann integrable on $[a, b]$, then so are the functions $|f|$ and f^2. For emphasis we state this as a corollary.

COROLLARY 6.1.10 *If $f \in \mathcal{R}[a, b]$, then $|f|$ and f^2 are Riemann integrable on $[a, b]$.*

A natural question to ask is whether the composition of two Riemann integrable functions is Riemann integrable. In Example 6.1.14(b) we will show that the answer is an emphatic *no!*

Lebesgue's Theorem

In Theorem 6.1.8(a) we proved that every continuous function on $[a, b]$ is Riemann integrable on $[a, b]$. By Exercise 16, this is also true for every bounded function on $[a, b]$ that is continuous except at a finite number of points. On the other hand, as a consequence of Theorem 6.1.8(b), every monotone function on $[a, b]$ is Riemann integrable. Hence for example, if $\{r_n\}_{n=1}^{\infty}$ is an enumeration of the rational numbers in $[0, 1]$ and $c_n > 0$ are such that $\sum c_n$ converges, then by Theorem 4.4.10

$$f(x) = \sum_{n=1}^{\infty} c_n I(x - r_n)$$

is monotone increasing on $[0, 1]$, and thus is Riemann integrable on $[0, 1]$. By Theorem 4.4.10, the function f is continuous at every irrational number and discontinuous at every rational number in $[0,1]$.

We now state the beautiful result of Lebesgue which provides necessary and sufficient conditions that a bounded real-valued function on $[a, b]$ be Riemann integrable. To properly state Lebesgue's result we need to introduce the idea of a set of measure zero. The concept of *measure* of a set will be treated in detail in Chapter 10. The basic idea is that the measure of an interval is its length. This is then used to define what we mean by measurable set and the measure of a measurable set. At this point we only need to know what it means for a set to have measure zero.

DEFINITION 6.1.11 *A subset E of $\mathbb{R}$ has* **measure zero** *if given any $\epsilon > 0$, there exists a finite or countable collection $\{I_n\}_{n \in A}$ of open intervals such that*

$$E \subset \bigcup_{n \in A} I_n \qquad \text{and} \qquad \sum_{n \in A} \ell(I_n) < \epsilon,$$

where $\ell(I_n)$ denotes the length of the interval I_n.

EXAMPLES 6.1.12 (a) Every finite set E has measure zero. Suppose $E = \{x_1, ..., x_N\}$ is a finite subset of $\mathbb{R}$. For each $n = 1, 2, ..., N$, as in Figure 6.4 let

$$I_n = \left(x_n - \frac{\epsilon}{2N}, x_n + \frac{\epsilon}{2N} \right).$$

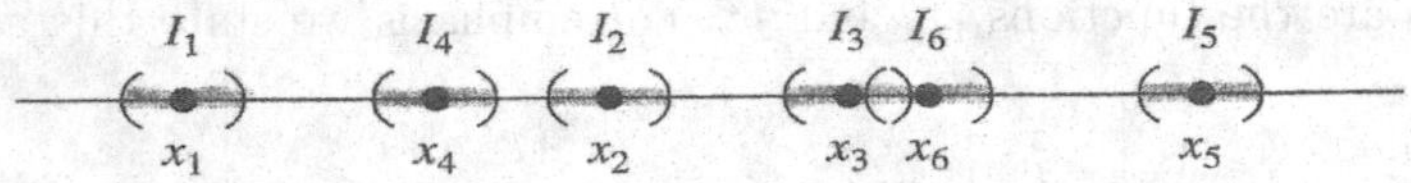

FIGURE 6.4
Example 6.1.12(a)

Then

$$E \subset \bigcup_{n=1}^{N} I_n \qquad \text{and} \qquad \sum_{n=1}^{N} \ell(I_n) = \epsilon.$$

Therefore E has measure zero.

(b) Every countable subset of $\mathbb{R}$ has measure zero. Suppose $E = \{x_n\}_{n=1}^{\infty}$ is a countable subset of $\mathbb{R}$. Let $\epsilon > 0$ be given. For each $n \in \mathbb{N}$, let

$$I_n = \left(x_n - \frac{\epsilon}{2^{n+1}}, x_n + \frac{\epsilon}{2^{n+1}} \right).$$

Since $x_n \in I_n$ for all n, $E \subset \bigcup_{n=1}^{\infty} I_n$. Thus since $\ell(I_n) = \epsilon/2^n$,

$$\sum_{n=1}^{\infty} \ell(I_n) = \epsilon \sum_{n=1}^{\infty} \frac{1}{2^n} = \epsilon.$$

As an example, the set $\mathbb{Q}$ of rational numbers has measure zero.

(c) As we shall see in the exercises of Section 10.2, the Cantor set P in $[0,1]$, which is uncountable, also has measure zero. $\square$

We now state the following theorem of Henri Lebesgue, the proof of which will be given in Section 6.7. This result appeared in 1902 and provides the most succinct form of necessary and sufficient conditions for Riemann integrability.

THEOREM 6.1.13 (Lebesgue) *A bounded real-valued function f on $[a,b]$ is Riemann integrable if and only if the set of discontinuities of f has measure zero.*

Remark. If f is continuous on $[a,b]$, then clearly f satisfies the hypothesis of Theorem 6.1.13 and thus is Riemann integrable. If f is a bounded function which is continuous except at a finite number of points, then by Example 6.1.12(a) the set of discontinuities of f has measure zero. Hence $f \in \mathcal{R}[a,b]$. If f is monotone on $[a,b]$, then by Corollary 4.4.8, the set of discontinuities of f is at most countable, and thus by Example 6.1.12(b), has measure zero. Hence again $f \in \mathcal{R}[a,b]$.

As an application of Lebesgue's theorem we give the following short proof of Theorem 6.1.9.

Proof of Theorem 6.1.9 using Lebesgue's Theorem. As in Theorem 6.1.9, suppose $f \in \mathcal{R}[a,b]$ with Range $f \subset [c,d]$, and suppose $\varphi : [c,d] \to \mathbb{R}$ is continuous. Let

$$E = \{x \in [a,b] : f \text{ is not continuous at } x\} \quad \text{and}$$
$$F = \{x \in [a,b] : \varphi \circ f \text{ is not continuous at } x\}.$$

By Theorem 4.2.4, $F \subset E$. Since f is Riemann integrable on $[a,b]$, the set E has measure zero, and as a consequence so does the set F. Therefore $\varphi \circ f \in \mathcal{R}[a,b]$. $\square$

EXAMPLES 6.1.14 (a) As in Example 4.2.2(g), let f be defined on $[0,1]$ by

$$f(x) = \begin{cases} 1, & x = 0, \\ 0, & \text{if } x \text{ is irrational,} \\ \frac{1}{n}, & \text{if } x = \frac{m}{n} \text{ in lowest terms}, x \neq 0. \end{cases}$$

Since f is continuous except at the rational numbers, which have measure zero, f is Riemann integrable on $[0,1]$. Furthermore, since $\mathcal{L}(\mathcal{P}, f) = 0$ for all partitions $\mathcal{P}$ of $[0,1]$,

$$\int_0^1 f(x) \, dx = 0.$$

(b) Let f be the Riemann integrable function on $[0,1]$ given in (a), and let $g : [0,1] \to \mathbb{R}$ be defined by

$$g(x) = \begin{cases} 0, & x = 0, \\ 1, & x \in (0,1]. \end{cases}$$

Since g is continuous except at 0, $g \in \mathcal{R}[0,1]$. But for $x \in [0,1]$,

$$(g \circ f)(x) = \begin{cases} 1, & \text{if } x \text{ is rational}, \\ 0, & \text{if } x \text{ is irrational}. \end{cases}$$

By Example 6.1.6(a), $g \circ f \notin \mathcal{R}[0,1]$. $\square$

Exercises 6.1

1. Let $f(x) = 1 - x^2$, $x \in [-1,2]$. Find $\mathcal{L}(\mathcal{P}, f)$ and $\mathcal{U}(\mathcal{P}, f)$ for each of the following partitions of $[-1,2]$.
   *a. $\mathcal{P} = \{-1,0,1,2\}$ **b.** $\mathcal{P} = \{-1,-\frac{1}{2},0,\frac{1}{2},1\frac{3}{2},2\}$

2. Show that each of the following functions is Riemann integrable on $[0,2]$ and use the definition to find $\int_0^2 f$.
   *a. $f(x) = \begin{cases} -1, & 0 \le x < 1, \\ 2, & 1 \le x \le 2 \end{cases}$ **b.** $f(x) = \begin{cases} 1 & 0 \le x < \frac{1}{2}, \\ -3, & \frac{1}{2} \le x < \frac{3}{2}, \\ -2, & \frac{3}{2} \le x \le 2 \end{cases}$

3. Show that each of the following functions is Riemann integrable on $[a,b]$, and find $\int_a^b f$.
   *a. $f(x) = c$,   (c a constant)   **b.** $f(x) = \begin{cases} 0, & a \le x < c \\ \frac{1}{2}, & x = c, \\ 1, & c < x \le b \end{cases}$

4. Use one of the methods of Examples 6.1.6 to find $\int_0^1 f$ for each of the following functions f. In the following, $[x]$ denotes the greatest integer function.
   *a. $f(x) = [3x]$ **b.** $f(x) = x[2x]$
   *c. $f(x) = 3x + 1$ **d.** $f(x) = 1 - x^2$

5. Prove that $\int_a^a f = 0$ for any real-valued function f on $[a,a]$.

6. ***** If $f, g \in \mathcal{R}[a,b]$ with $f(x) \le g(x)$ for all $x \in [a,b]$, prove that $\int_a^b f \le \int_a^b g$.

7. **a.** Suppose f is continuous on $[a,b]$ with $f(x) \ge 0$ for all $x \in [a,b]$. If $\int_a^b f = 0$, prove that $f(x) = 0$ for all $x \in [a,b]$.
   b. Show by example that the conclusion may be false if f is not continuous.

8. ***a.** Let $f : [0, 1] \to \mathbb{R}$ be defined by

$$f(x) = \begin{cases} 0, & x \in \mathbb{Q}, \\ x, & x \notin \mathbb{Q} \end{cases}.$$

Compute $\underline{\int_0^1} f$ and $\overline{\int_0^1} f$. Is f integrable on $[0, 1]$.

9. Suppose f is a nonnegative Riemann integrable function on $[a, b]$ satisfying $f(r) = 0$ for all $r \in \mathbb{Q} \cap [a, b]$. Prove that $\int_a^b f = 0$.

10. *****Use the method of Example 6.1.6(c) to show that

$$\int_a^b x^2 \, dx = \frac{1}{3}(b^3 - a^3) \qquad (0 \le a < b).$$

11. Suppose f is monotone increasing on $[a, b]$. For $n \in \mathbb{N}$, set $h = (b - a)/n$. Let $\mathcal{P}_n = \{x_0, x_1, ..., x_n\}$ where for each $k = 0, ..., n$, $x_k = a + kh$.

    a. Prove that

    $$0 \le \mathcal{U}(\mathcal{P}_n, f) - \int_a^b f \le \frac{(b-a)}{n} [f(b) - f(a)].$$

    b. Prove that $\int_a^b f = \lim_{n \to \infty} \mathcal{U}(\mathcal{P}_n, f)$.

12. Use the previous exercise to evaluate the following integrals.

    *a. $\displaystyle\int_0^1 x \, dx$. **b.** $\displaystyle\int_{-2}^1 (3x - 2) \, dx$. ***c.** $\displaystyle\int_0^1 x^3 \, dx$ **d.** $\displaystyle\int_a^b x^3 dx$

13. *****Let f be a bounded function on $[a, b]$. Suppose there exists a sequence $\{\mathcal{P}_n\}$ of partitions of $[a, b]$ such that

    $$\lim_{n \to \infty} \mathcal{L}(\mathcal{P}_n, f) = \lim_{n \to \infty} \mathcal{U}(\mathcal{P}_n, f) = L, \quad L \in \mathbb{R}.$$

    Prove that f is Riemann integrable on $[a, b]$ with $\int_a^b f = L$.

14. ***a.** If $f \in \mathcal{R}[a, b]$, prove directly (without using Theorems 6.1.9 or 6.1.13) that $|f| \in \mathcal{R}[a, b]$.

    b. If $|f| \in \mathcal{R}[a, b]$, is $f \in \mathcal{R}[a, b]$?

    c. If $f \in \mathcal{R}[a, b]$, prove that $\left| \int_a^b f \right| \le \int_a^b |f|$.

15. ***a.** If $f \in \mathcal{R}[a, b]$, prove directly that $f^2 \in \mathcal{R}[a, b]$.

    b. Give an example of a bounded function f on $[a, b]$ for which $f^2 \in \mathcal{R}[a, b]$, but $f \notin \mathcal{R}[a, b]$.

16. *****Suppose f is a bounded real-valued function on $[a, b]$ that has only a finite number of discontinuities. Prove directly that f is Riemann integrable on $[a, b]$.

17. **a.** If E has measure zero, prove that every subset of E has measure zero.

    b. If E_1, E_2 have measure zero, prove that $E_1 \cup E_2$ has measure zero.

c. If each E_n, $n = 1, 2, ...,$ has measure zero, prove that $\bigcup\limits_{n=1}^{\infty} E_n$ has measure zero.

18. Use Theorem 6.1.13 to prove that if $f, g \in \mathcal{R}[a, b]$, then $f + g \in \mathcal{R}[a, b]$.

19. Prove directly that the function f of Example 6.1.14(a) is Riemann integrable on $[a, b]$.

20. Let $f : [a, b] \to \mathbb{R}$ be continuous. Suppose that for every Riemann integrable function $g : [a, b] \to \mathbb{R}$ the product fg is Riemann integrable and $\int_a^b fg = 0$. Prove that $f(x) = 0$ for all $x \in [a, b]$.

21. **a.** Let $\{I_n\}$ be a finite or countable collection of disjoint open intervals in $[a, b]$, and let $U = \bigcup I_n$. Define f on $[a, b]$ by $f(x) = 1$ if $x \in U$, and 0 elsewhere. Prove that f is Riemann integrable on $[a, b]$ and that
$$\int_a^b f = \sum_n \ell(I_n).$$

b. Let P denote the Cantor ternary set in $[0, 1]$ and let f be defined on $[0, 1]$ by $f(x) = 0$ if $x \in P$, and $f(x) = 1$ elsewhere. Prove that f is Riemann integrable on $[0, 1]$ and that $\int_0^1 f = 1$.

6.2 Properties of the Riemann Integral

In this section, we derive some basic properties of the Riemann integral. As in the previous section, $[a, b]$, $a < b$, will be a closed and bounded interval in $\mathbb{R}$, and $\mathcal{R}[a, b]$ denotes the set of Riemann integrable functions on $[a, b]$.

THEOREM 6.2.1 *Let $f, g \in \mathcal{R}[a, b]$. Then*

(a) $f + g \in \mathcal{R}[a, b]$ *with* $\int_a^b (f + g) = \int_a^b f + \int_a^b g$,

(b) $cf \in \mathcal{R}[a, b]$ *for all* $c \in \mathbb{R}$ *with* $\int_a^b cf = c \int_a^b f$, *and*

(c) $fg \in \mathcal{R}[a, b]$.

Proof. (a) The integrability of $f + g$ is actually a consequence of Theorem 6.1.13. However, in establishing the formula for the integral of $(f + g)$, we will also obtain the integrability of $f + g$ as a consequence. Let $\mathcal{P} = \{x_0, ..., x_n\}$ be a partition on $[a, b]$. For each $i = 1, ..., n$ let

$$M_i(f) = \sup\{f(t) : t \in [x_{i-1}, x_i]\},$$
$$M_i(g) = \sup\{g(t) : t \in [x_{i-1}, x_i]\}.$$

Then $f(t) + g(t) \leq M_i(f) + M_i(g)$ for all $t \in [x_{i-1}, x_i]$ and thus

$$\sup\{f(t) + g(t) : t \in [x_{i-1}, x_i]\} \leq M_i(f) + M_i(g).$$

Therefore, for all partitions $\mathcal{P}$ of $[a, b]$,

$$\mathcal{U}(\mathcal{P}, f + g) \leq \mathcal{U}(\mathcal{P}, f) + \mathcal{U}(\mathcal{P}, g).$$

Let $\epsilon > 0$ be given. Since $f, g \in \mathcal{R}[a, b]$, there exist partitions $\mathcal{P}_f$ and $\mathcal{P}_g$ of $[a, b]$ such that

$$\mathcal{U}(\mathcal{P}_f, f) < \int_a^b f + \tfrac{1}{2}\epsilon \quad \text{and} \quad \mathcal{U}(\mathcal{P}_g, g) < \int_a^b g + \tfrac{1}{2}\epsilon.$$

Let $\mathcal{Q} = \mathcal{P}_f \cup \mathcal{P}_g$. Since $\mathcal{Q}$ is a refinement of both $\mathcal{P}_f$ and $\mathcal{P}_g$,

$$\mathcal{U}(\mathcal{Q}, f + g) \leq \mathcal{U}(\mathcal{P}_f, f) + \mathcal{U}(\mathcal{P}_g, g) < \int_a^b f + \int_a^b g + \epsilon.$$

Therefore

$$\overline{\int_a^b} (f + g) < \int_a^b f + \int_a^b g + \epsilon.$$

Since the above holds for all $\epsilon > 0$,

$$\overline{\int_a^b} (f + g) \leq \int_a^b f + \int_a^b g.$$

A similar argument proves that

$$\underline{\int_a^b} (f + g) \geq \int_a^b f + \int_a^b g.$$

Thus the lower integral of $(f + g)$ is equal to the upper integral of $(f + g)$, and as a consequence $(f + g) \in \mathcal{R}[a, b]$ with

$$\int_a^b (f + g) = \int_a^b f + \int_a^b g.$$

(b) The proof of (b) is left as an exercise (Exercise 1).

(c) To prove (c), we first note that

$$fg = \frac{1}{4} \left[(f + g)^2 - (f - g)^2 \right].$$

By (a) the functions $(f + g)$ and $(f - g)$ are integrable on $[a, b]$, and by Corollary 6.1.10, $(f + g)^2$ and $(f - g)^2$ are integrable. Hence by (a) and (b), fg is integrable on $[a, b]$. $\square$

THEOREM 6.2.2 *If $f \in \mathcal{R}[a, b]$, then $|f| \in \mathcal{R}[a, b]$ with*

$$\left| \int_a^b f \right| \leq \int_a^b |f|.$$

Proof. Since $f \in \mathcal{R}[a,b]$, by Corollary 6.1.10 $|f| \in \mathcal{R}[a,b]$. Let $c = \pm 1$ be such that $|\int_a^b f| = c \int_a^b f$. Since $cf(x) \leq |f(x)|$ for all $x \in [a,b]$, $\mathcal{U}(\mathcal{P}, cf) \leq \mathcal{U}(\mathcal{P}, |f|)$ for any partition $\mathcal{P}$ of $[a,b]$. Therefore, since both cf and $|f|$ are integrable, $\int_a^b cf \leq \int_a^b |f|$. Combining the above we have

$$\left| \int_a^b f \right| = c \int_a^b f = \int_a^b cf \leq \int_a^b |f|. \quad \square$$

THEOREM 6.2.3 *Let f be a bounded real-valued function on $[a,b]$, and suppose $a < c < b$. Then $f \in \mathcal{R}[a,b]$ if and only if $f \in \mathcal{R}[a,c]$ and $f \in \mathcal{R}[c,b]$. If this is the case, then*

$$\int_a^b f = \int_a^c f + \int_c^b f. \tag{6}$$

Proof. Let $c \in \mathbb{R}$ satisfy $a < c < b$. We first prove that if f is a bounded real-valued function on $[a,b]$, then

$$\overline{\int_a^b} f = \overline{\int_a^c} f + \overline{\int_c^b} f. \tag{7}$$

Suppose $\mathcal{P}_1$ and $\mathcal{P}_2$ are partitions of $[a,c]$ and $[c,b]$ respectively. Then $\mathcal{P} = \mathcal{P}_1 \cup \mathcal{P}_2$ is a partition of $[a,b]$ with $c \in \mathcal{P}$. Conversely, if $\mathcal{P}$ is any partition of $[a,b]$ with $c \in \mathcal{P}$, then $\mathcal{P} = \mathcal{P}_1 \cup \mathcal{P}_2$ where $\mathcal{P}_1$ and $\mathcal{P}_2$ are partitions of $[a,c]$ and $[c,b]$, respectively. For such a partition $\mathcal{P}$,

$$\mathcal{U}(\mathcal{P}, f) = \mathcal{U}(\mathcal{P}_1, f) + \mathcal{U}(\mathcal{P}_2, f) \geq \overline{\int_a^c} f + \overline{\int_c^b} f.$$

If $\mathcal{Q}$ is any partition of $[a,b]$, then $\mathcal{P} = \mathcal{Q} \cup \{c\}$ is a refinement of $\mathcal{Q}$ containing c. Therefore

$$\mathcal{U}(\mathcal{Q}, f) \geq \mathcal{U}(\mathcal{P}, f) \geq \overline{\int_a^c} f + \overline{\int_c^b} f.$$

Taking the infimum over all partitions $\mathcal{Q}$ of $[a,b]$ gives

$$\overline{\int_a^b} f \geq \overline{\int_a^c} f + \overline{\int_c^b} f.$$

To prove the reverse inequality, let $\epsilon > 0$ be given. Then there exist partitions $\mathcal{P}_1$ and $\mathcal{P}_2$ of $[a,c]$ and $[c,b]$ respectively such that

$$\mathcal{U}(\mathcal{P}_1, f) < \overline{\int_a^c} f + \frac{\epsilon}{2} \quad \text{and} \quad \mathcal{U}(\mathcal{P}_2, f) < \overline{\int_c^b} f + \frac{\epsilon}{2}.$$

Let $\mathcal{P} = \mathcal{P}_1 \cup \mathcal{P}_2$. Then

$$\overline{\int_a^b} f \leq \mathcal{U}(\mathcal{P}, f) = \mathcal{U}(\mathcal{P}_1, f) + \mathcal{U}(\mathcal{P}_2, f) < \overline{\int_a^c} f + \overline{\int_c^b} f + \epsilon.$$

Since $\epsilon > 0$ was arbitrary,

$$\overline{\int_a^b} f \leq \overline{\int_a^c} f + \overline{\int_c^b} f,$$

which when combined with the previous inequality proves (7). A similar argument also proves

$$\underline{\int_a^b} f = \underline{\int_a^c} f + \underline{\int_c^b} f. \tag{8}$$

If f is integrable on $[a, c]$ and $[c, b]$, then by (7) and (8)

$$\int_a^c f + \int_c^b f \leq \underline{\int_a^b} f \leq \overline{\int_a^b} f = \int_a^c f + \int_c^b f.$$

Therefore $f \in \mathcal{R}[a, b]$ and identity (6) holds. Conversely, if $f \in \mathcal{R}[a, b]$, then by (7) and (8),

$$\underline{\int_a^c} f + \underline{\int_c^b} f = \overline{\int_a^c} f + \overline{\int_c^b} f.$$

Since the lower integral of f is always less than or equal to the upper integral of f, the above holds if and only if

$$\underline{\int_a^c} f = \overline{\int_a^c} f \qquad \text{and} \qquad \underline{\int_c^b} f = \overline{\int_c^b} f.$$

Hence $f \in \mathcal{R}[a, c]$ and $f \in \mathcal{R}[c, b]$. $\square$

Riemann's Definition of the Integral

We close this section by comparing the approach of Darboux with the original method of Riemann.

DEFINITION 6.2.4 *Let f be a bounded real-valued function on $[a, b]$ and let $\mathcal{P} = \{x_0, x_1, ..., x_n\}$ be a partition of $[a, b]$. For each $i = 1, 2, ..., n$, choose $t_i \in [x_{i-1}, x_i]$. The sum*

$$\mathcal{S}(\mathcal{P}, f) = \sum_{i=1}^{n} f(t_i)\Delta x_i$$

*is called a **Riemann sum** of f with respect to the partition $\mathcal{P}$ and the points $\{t_i\}$.*

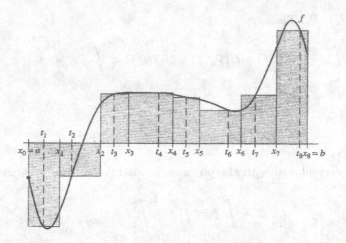

FIGURE 6.5
A Riemann sum $\mathcal{S}(\mathcal{P}, f)$ of f

In the Riemann approach to integration, one defines the integral of a bounded real-valued function f as the *limit* of the Riemann sums of f. Since $\mathcal{S}(\mathcal{P}, f)$ depends not only on the partition $\mathcal{P} = \{x_0, x_1, ..., x_n\}$ but also on the points $t_i \in [x_{i-1}, x_i]$, we first need to clarify what we mean by the limit of the Riemann sums $\mathcal{S}(\mathcal{P}, f)$. For a partition $\mathcal{P} = \{x_0, x_1, ..., x_n\}$ of $[a, b]$, set

$$\|\mathcal{P}\| = \max\{\Delta x_i : i = 1, 2, ..., n\}.$$

The quantity $\|\mathcal{P}\|$ is called the **norm** or the **mesh** of the partition $\mathcal{P}$.

DEFINITION 6.2.5 *Let f be a bounded real-valued function on $[a, b]$. Then*

$$\lim_{\|\mathcal{P}\| \to 0} \mathcal{S}(\mathcal{P}, f) = I$$

if given $\epsilon > 0$, there exists a $\delta > 0$ such that

$$\left| \sum_{i=1}^{n} f(t_i)\Delta x_i - I \right| < \epsilon \qquad (9)$$

for all partitions $\mathcal{P}$ of $[a, b]$ with $\|\mathcal{P}\| < \delta$, and all choices of $t_i \in [x_{i-1}, x_i]$.

THEOREM 6.2.6 *Let f be a bounded real-valued function on $[a, b]$. If*

$$\lim_{\|\mathcal{P}\| \to 0} \mathcal{S}(\mathcal{P}, f) = I,$$

then $f \in \mathcal{R}[a,b]$ and $\int_a^b f = I$. Conversely, if $f \in \mathcal{R}[a,b]$, then $\lim\limits_{\|\mathcal{P}\| \to 0} \mathcal{S}(\mathcal{P},f)$ exists and

$$\lim_{\|\mathcal{P}\| \to 0} \mathcal{S}(\mathcal{P},f) = \int_a^b f.$$

Proof. Suppose $\lim\limits_{\|\mathcal{P}\| \to 0} \mathcal{S}(\mathcal{P},f) = I$. Let $\epsilon > 0$ be given, and let $\delta > 0$ be such that (9) holds for all partitions $\mathcal{P} = \{x_0, x_1, ..., x_n\}$ of $[a,b]$ with $\|\mathcal{P}\| < \delta$, and all $t_i \in [x_{i-1}, x_i]$. By the definition of M_i, for each $i = 1, ..., n$, there exists $\zeta_i \in [x_{i-1}, x_i]$ such that $f(\zeta_i) > M_i - \epsilon$. Thus

$$\mathcal{U}(\mathcal{P},f) = \sum_{i=1}^n M_i \, \Delta x_i$$

$$< \sum_{i=1}^n f(\zeta_i) \, \Delta x_i + \epsilon \sum_{i=1}^n \Delta x_i$$

$$< I + \epsilon + \epsilon \, [b-a] = I + \epsilon \, [1 + b - a].$$

Similarly $\mathcal{L}(\mathcal{P}, f, \alpha) > I - \epsilon \, [1 + b - a]$. Therefore,

$$\mathcal{U}(\mathcal{P},f) - \mathcal{L}(\mathcal{P},f) < 2 \, \epsilon \, [1 + b - a].$$

Thus as a consequence of Theorem 6.1.7, $f \in \mathcal{R}[a,b]$ with $\int_a^b f = I$.

Conversely, suppose $f \in \mathcal{R}[a,b]$. Let $M > 0$ be such that $|f(x)| \le M$ for all $x \in [a,b]$. Let $\epsilon > 0$ be given. Since $f \in \mathcal{R}[a,b]$, by Theorem 6.1.7 there exists a partition $\mathcal{Q}$ of $[a,b]$ such that

$$\int_a^b f - \epsilon < \mathcal{L}(\mathcal{Q},f) \le \mathcal{U}(\mathcal{Q},f) < \int_a^b f + \epsilon.$$

Suppose $\mathcal{Q} = \{x_0, ..., x_N\}$. Let $\delta = \epsilon/NM$, and let $\mathcal{P} = \{y_0, ..., y_n\}$ be any partition of $[a,b]$ with $\|\mathcal{P}\| < \delta$. As in the definition of the integral, let

$$M_i = \sup\{f(x) : x \in [x_{i-1}, x_i]\}, \quad i = 1, ..., N.$$

Consider any interval $[y_{k-1}, y_k]$, $k = 1, ..., n$. This interval may or may not contain points $x_i \in \mathcal{Q}$. Since $\mathcal{Q}$ contains $N+1$ points, there are at most $N-1$ intervals $[y_{k-1}, y_k]$ which contain an $x_i \in \mathcal{Q}$, $i \ne 0, N$. Suppose as in Figure 6.6 $\{x_j, ..., x_{j+m}\} \subset [y_{k-1}, y_k]$. If $x_j \ne y_{k-1}$, set $M_k^1 = \sup\{f(x) : x \in [y_{k-1}, x_j]\}$. Similarly, if $x_{j+m} \ne y_k$, set $M_k^2 = \sup\{f(x) : x \in [x_{j+m}, y_k]\}$.

Let $t_k \in [y_{k-1}, y_k]$ be arbitrary. Since $|f(t) - f(s)| \le 2M$ for all $t, s \in [y_{k-1}, y_k]$,

$$f(t_k) \le 2M + M_{j+s}, \quad s = 1, ..., m, \quad \text{and}$$

$$f(t_k) \le 2M + M_k^i, \quad i = 1, 2.$$

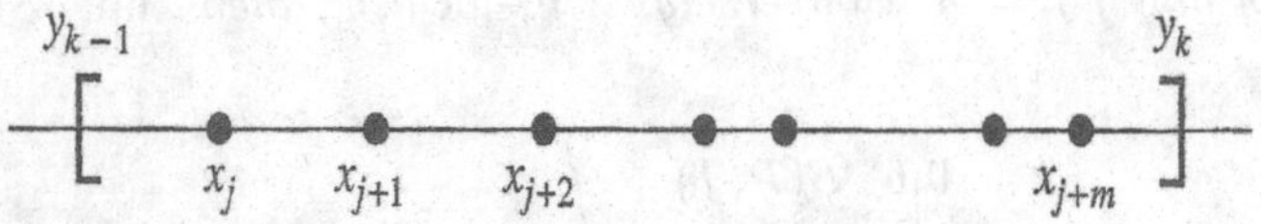

FIGURE 6.6
Partition of the interval $[y_{k-1}, y_k]$

Therefore

$$f(t_k)\Delta y_k = f(t_k)(x_j - y_{k-1}) + \sum_{s=1}^{m} f(t_k)\Delta x_{j+s} + f(t_k)(y_k - x_{j+m})$$

$$\leq 2M\Delta y_k + M_k^1(x_j - y_{k-1}) + \sum_{s=1}^{m} M_{j+s}\Delta x_{j+s} + M_k^2(y_k - x_{j+m})$$

$$< 2M\,\delta + \mathcal{U}(\mathcal{P}_k, f),$$

where $\mathcal{P}_k = \{y_{k-1}, x_j, ..., x_{j+m}, y_k\}$ is a partition of $[y_{k-1}, y_k]$. If the interval $[y_{k-1}, y_k]$ contains no $x_i \in \mathcal{Q}$, $i \neq 0, N$, we simply let $\mathcal{P}_k = \{y_{k-1}, y_k\}$. Let $\mathcal{P}' = \bigcup \mathcal{P}_k$. Then $\mathcal{P}'$ is a partition of $[a, b]$ that is also a refinement of $\mathcal{Q}$. Since at most $N - 1$ intervals $[y_{k-1}, y_k]$ contain a point of $\mathcal{Q}$ other than x_0 and x_N,

$$\mathcal{S}(\mathcal{P}, f) = \sum_{k=1}^{n} f(t_k)\Delta y_k < 2M(N-1)\delta + \sum_{k=1}^{n} \mathcal{U}(\mathcal{P}_k, f)$$

$$< 2\epsilon + \mathcal{U}(\mathcal{P}', f) < 2\epsilon + \mathcal{U}(\mathcal{Q}, f) < 3\epsilon + \int_a^b f.$$

A similar argument proves that

$$\mathcal{S}(\mathcal{P}, f) > \int_a^b f - 3\epsilon.$$

Therefore,

$$\left| \mathcal{S}(\mathcal{P}, f) - \int_a^b f \right| < 3\epsilon.$$

Since this holds for any partition $\mathcal{P} = \{y_0, ..., y_n\}$ of $[a, b]$ with $\|\mathcal{P}\| < \delta$ and any choice of $t_k \in [y_{k-1}, y_k]$, $k = 1, ..., n$,

$$\lim_{\|\mathcal{P}\| \to 0} \mathcal{S}(\mathcal{P}, f) = \int_a^b f. \quad \square$$

EXAMPLE 6.2.7 In this example, we will use the method of Riemann sums to evaluate $\int_a^b x\, dx$. Since $f(x) = x$ is Riemann integrable on $[a, b]$,

$$\int_a^b x\, dx = \lim_{\|\mathcal{P}\| \to 0} S(\mathcal{P}, f).$$

Since the limit exists for any $t_i \in [x_{i-1}, x_i]$, we can take $t_i = (x_{i-1} + x_i)/2$. With this choice of t_i,

$$S(\mathcal{P}, f) = \frac{1}{2} \sum_{i=1}^n (x_{i-1} + x_i)(x_{i-1} - x_i) = \frac{1}{2} \sum_{i=1}^n (x_{i-1}^2 - x_i^2) = \frac{1}{2}(b^2 - a^2),$$

which proves the result. $\square$

Exercises 6.2

1. Prove Theorem 6.2.1(b).

2. ***a.** Use the method of Riemann sums to evaluate $\int_a^b x^2\, dx$.

   b. Use the method of Riemann sums to evaluate $\int_a^b x^n dx$, $n \in \mathbb{N}$, $n \geq 3$.

3. ***a.** Let f be a real-valued function on $[a, b]$ such that $f(x) = 0$ for all $x \neq c_1, ..., c_n$. Prove that $f \in \mathcal{R}[a, b]$ with $\int_a^b f = 0$.

   *b. Let $f, g \in \mathcal{R}[a, b]$ be such that $f(x) = g(x)$ for all but a finite number of points in $[a, b]$. Prove that $\int_a^b f = \int_a^b g$.

   c. Is the result of (a) still true if $f(x) = 0$ for all but countably many points in $[a, b]$?

4. Let $f \in \mathcal{R}[-a, a]$, $a > 0$. Prove each of the following:

   a. If f is **even** (i.e. $f(-x) = f(x)$ for all $x \in [-a, a]$), then $\int_{-a}^a f = 2 \int_0^a f$.

   b. If f is **odd** (i.e. $f(-x) = -f(x)$ for all $x \in [-a, a]$), then $\int_{-a}^a f = 0$.

5. ***Let** f be a bounded real-valued function on $[a, b]$ such that $f \in \mathcal{R}[c, b]$ for every c, $a < c < b$. Prove that $f \in \mathcal{R}[a, b]$ with $\int_a^b f = \lim_{c \to a^+} \int_c^b f$.

6. Let f be continuous on $[0, 1]$. Prove that

$$\lim_{n \to \infty} \frac{1}{n} \sum_{k=1}^n f\left(\frac{k}{n}\right) = \int_0^1 f(x)\, dx.$$

7. Use the previous exercise to evaluate each of the following limits. (You may use any applicable methods from calculus to evaluate the definite integrals.)

    *a. $\displaystyle\lim_{n\to\infty} \frac{1}{n^3} \sum_{k=1}^{n} k^2$ b. $\displaystyle\lim_{n\to\infty} \sum_{k=1}^{n} \frac{k}{n^2 + k^2}$ *c. $\displaystyle\lim_{n\to\infty} \sum_{k=1}^{n} \frac{n}{n^2 + k^2}$

    d. $\displaystyle\lim_{n\to\infty} \frac{1}{n^3} \sum_{k=1}^{m} k\sqrt{k^2 + n^2}$

8. *As in Example 4.4.11(a), let f be a monotone function on $[0,1]$ defined by

    $$f(x) = \sum_{n=1}^{\infty} \frac{1}{2^n} I(x - x_n),$$

    where $x_n = n/(n+1)$, $n \in \mathbb{N}$. Find $\int_0^1 f(x)dx$. (Leave your answer in the form of an infinite series.)

9. Suppose $f \in \mathcal{R}[a,b]$ and $c \in \mathbb{R}$. Define f_c on $[a-c, b-c]$ by $f_c(x) = f(x+c)$. Prove that $f_c \in \mathcal{R}[a-c, b-c]$ with

    $$\int_{a-c}^{b-c} f_c(x)dx = \int_a^b f(x)dx.$$

10. Let f, g, h be bounded real-valued functions on $[a,b]$ satisfying $f(x) \leq g(x) \leq h(x)$ for all $x \in [a,b]$. If $f, h \in \mathcal{R}[a,b]$ with $\int_a^b f = \int_a^b h = I$, prove that $g \in \mathcal{R}[a,b]$ with $\int_a^b g = I$.

6.3   Fundamental Theorem of Calculus

In this section, we will prove two well known and very important results. Collectively they are commonly referred to as the fundamental theorem of calculus. To Newton and Leibniz, integration was the inverse operation of differentiation. In Leibniz's notation this result would simply be expressed as $\int f'(x)\, dx = f(x)$, where here the symbol $\int$ (denoting sum) was Leibniz's notation for the integral. This notation is still used in most calculus texts to denote the antiderivative of a function. Since the modern definition of the integral is based on either Riemann or Darboux sums, we now need to prove that integration is indeed the inverse operation of differentiation. Both versions of the fundamental theorem of calculus presented here are essentially due to Cauchy who proved the results for continuous functions.

As was illustrated in Examples 6.1.6 and 6.2.7, the computation of the integral of a function, using either Darboux's or Riemann's definition, can be extremely tedious. For nontrivial functions these computations are in most instances impossible. The first version of the fundamental theorem of calculus provides a major tool for the evaluation of Riemann integrals. We begin with the following definition.

DEFINITION 6.3.1 *Let f be a real-valued function on an interval I. A function F on I is called an* **antiderivative** *of f on I if $F'(x) = f(x)$ for all $x \in I$.*

Remark. An antiderivative, if it exists is not unique. If F is an antiderivative of f, then so is $F + C$ for any constant C. Conversely, if F and G are antiderivatives of f, then $F'(x) - G'(x) = 0$ for all $x \in [a, b]$. Thus by Theorem 5.2.9, $G(x) = F(x) + C$ for some constant C.

THEOREM 6.3.2 (Fundamental Theorem of Calculus) *If $f \in \mathcal{R}[a, b]$ and if F is an antiderivative of f on $[a, b]$, then*

$$\int_a^b f(x)\, dx = F(b) - F(a).^2$$

Proof. Let $\mathcal{P} = \{x_0, x_1, ..., x_n\}$ be any partition of $[a, b]$. If F is an antiderivative of f, then by the mean value theorem, for each $i = 1, ..., n$, there exists $t_i \in (x_{i-1}, x_i)$ such that

$$F(x_i) - F(x_{i-1}) = f(t_i)\Delta x_i.$$

Therefore,

$$\sum_{i=1}^n f(t_i)\Delta x_i = \sum_{i=1}^n F(x_i) - F(x_{i-1}) = F(b) - F(a).$$

Since

$$\mathcal{L}(\mathcal{P}, f) \leq \sum_{i=1}^n f(t_i)\Delta x_i \leq \mathcal{U}(\mathcal{P}, f),$$

we have

$$\underline{\int_a^b} f(x)\, dx \leq F(b) - F(a) \leq \overline{\int_a^b} f(x)\, dx.$$

Thus if f is Riemann integrable on $[a, b]$,

$$\int_a^b f(x)\, dx = F(b) - F(a). \quad \square$$

Remark. The above version of the fundamental theorem of calculus is considerably stronger than the version given in most elementary calculus texts in that it does not require continuity of the function f. We only need that f is integrable and that it has an antiderivative on $[a, b]$. We will illustrate this in (b) of the following example.

---

[2]The quantity $F(b) - F(a)$ is usually denoted by $F(x)|_a^b$.

EXAMPLES 6.3.3 (a) If $f(x) = x^n$, $n \in \mathbb{N}$, then $F(x) = \frac{1}{n+1}x^{n+1}$ is an antiderivative of f. Thus for any $a, b \in \mathbb{R}$, $a < b$,

$$\int_a^b x^n \, dx = \frac{1}{n+1}(b^{n+1} - a^{n+1}).$$

(b) Consider the function F on $[0, 1]$ defined by

$$F(x) = \begin{cases} x^2 \sin \dfrac{1}{x}, & x \neq 0, \\ 0, & x = 0. \end{cases}$$

A straightforward computation gives

$$f(x) = F'(x) = \begin{cases} -\cos \dfrac{1}{x} + 2x \sin \dfrac{1}{x}, & x \neq 0, \\ 0, & x = 0. \end{cases}$$

Then f is bounded on $[0, 1]$ and continuous everywhere except $x = 0$. Thus $f \in \mathcal{R}[0, 1]$, and by the previous theorem

$$\int_0^1 f = F(1) - F(0) = \sin 1.$$

(c) Let f be defined on $[0, 2]$ by

$$f(x) = \begin{cases} 1, & 0 \leq x < 1, \\ x - 1, & 1 \leq x \leq 2, \end{cases}$$

and for $x \in [0, 2]$ let $F(x)$ be defined by

$$F(x) = \int_0^x f(t) \, dt.$$

If $0 \leq x \leq 1$, then $F(x) = \int\limits_0^x 1 \, dt = x$. On the other hand, if $1 < x \leq 2$, then

$$F(x) = \int_0^1 1 \, dt + \int_1^x (1 - t) \, dt = \frac{1}{2}x^2 - x + \frac{3}{2}.$$

Thus

$$F(x) = \begin{cases} x, & 0 \leq x \leq 1, \\ \frac{1}{2}x^2 - x + \frac{3}{2}, & 1 < x \leq 2. \end{cases}$$

Even though f is not continuous at $x = 1$, the function F is continuous everywhere. This in fact is always the case. (See Exercise 2.) $\quad\square$

The following version of the fundamental theorem of calculus, also due to Cauchy, proves that for continuous functions, integration is the inverse operation of differentiation.

THEOREM 6.3.4 (Fundamental Theorem of Calculus) *Let* $f \in \mathcal{R}[a,b]$. *Define* F *on* $[a,b]$ *by*

$$F(x) = \int_a^x f(t)\, dt.$$

Then F *is continuous on* $[a,b]$. *Furthermore, if* f *is continuous at a point* $c \in [a,b]$, *then* F *is differentiable at* c *and*

$$F'(c) = f(c).$$

Proof. The proof that F is continuous is left as an exercise (Exercise 2). Suppose f is continuous at $c \in [a,b)$. We will show that $F'_+(c) = f(c)$. Let $h > 0$. Then by Theorem 6.2.3,

$$F(c+h) - F(c) = \int_a^{c+h} f(t)\, dt - \int_a^c f(t)\, dt = \int_c^{c+h} f(t)\, dt.$$

Therefore,

$$\frac{F(c+h) - F(c)}{h} - f(c) = \frac{1}{h} \int_c^{c+h} f(t)\, dt - f(c)$$

$$= \frac{1}{h} \int_c^{c+h} [f(t) - f(c)]\, dt.$$

Let $\epsilon > 0$ be given. Since f is continuous at c, there exists a $\delta > 0$ such that

$$|f(t) - f(c)| < \epsilon$$

for all t, $|t - c| < \delta$. Therefore, if $0 < h < \delta$,

$$\left| \frac{F(c+h) - F(c)}{h} - f(c) \right| \leq \frac{1}{h} \int_c^{c+h} |f(t) - f(c)|\, dt$$

$$< \frac{1}{h} \int_c^{c+h} \epsilon\, dt = \epsilon.$$

Thus

$$\lim_{h \to 0^+} \frac{F(c+h) - F(c)}{h} = f(c),$$

i.e., $F'_+(c) = f(c)$. Similarly, if f is continuous at $c \in (a,b]$, $F'_-(c) = f(c)$, which proves the result. $\square$

Remarks. (a) If f is continuous on $[a, b]$, then an antiderivative of f always exists; namely the function F given by

$$F(x) = \int_a^x f(t)\, dt, \qquad a \le x \le b.$$

Since f is continuous, $F'(x) = f(x)$ for all $x \in [a, b]$. As a consequence, we obtain the following more elementary version of Theorem 6.3.2 normally encountered in the study of calculus: If f is continuous on $[a, b]$ and G is any antiderivative of f, then $\int_a^b f = G(b) - G(a)$.

(b) Integrability of a function f on $[a, b]$ does not imply the existence of an antiderivative of f. For example, if f is monotone increasing on $[a, b]$ and $F(x) = \int_a^x f$, then for any $c \in (a, b)$, $F'_+(c) = f(c+)$ and $F'_-(c) = f(c-)$ (Exercise 15). Thus if f is not continuous at c, the derivative of F does not exist at c.

The Natural Logarithm Function

As our first application of the fundamental theorem of calculus, we use the result to define the **natural logarithm function** $\ln x$.

EXAMPLE 6.3.5 For $x > 0$, let $L(x)$ be defined by

$$L(x) = \int_1^x \frac{1}{t}\, dt.$$

Since $f(t) = 1/t$ is continuous on $(0, \infty)$, by Theorem 6.3.4 $L(x)$ satisfies $L'(x) = 1/x$ for all $x > 0$. Furthermore, since $L'(x) > 0$ for all $x \in (0, \infty)$, L is strictly increasing on $(0, \infty)$.

We now prove that the function $L(x)$ satisfies the usual properties of a logarithm function; namely

    (a) $L(ab) = L(a) + L(b)$ for all $a, b > 0$,

    (b) $L(1/b) = -L(b), \quad b > 0$, and

    (c) $L(b^r) = r\, L(b), \quad b > 0, r \in \mathbb{R}$.

To prove (a), consider the function $L(ax)$, $x > 0$. By the chain rule (Theorem 5.1.6).

$$\frac{d}{dx} L(ax) = \frac{1}{ax} a = \frac{1}{x} = L'(x).$$

Thus by Theorem 5.2.9, $L(ax) = L(x) + C$ for some constant C. From the definition of L we have $L(1) = 0$. Therefore,

$$L(a) = L(1) + C = C.$$

Hence $L(ax) = L(a) + L(x)$ for all $x > 0$, which proves (a). The proof of (b) proceeds analogously. It is worth noting that for the proof of (a) and (b) we only used the fact that $L'(x) = 1/x$ and $L(1) = 0$.

To prove (c), if $n \in \mathbb{N}$, then by (a) $L(b^n) = n\,L(b)$. Also by (b),

$$L(b^{-n}) = n\,L\left(\frac{1}{b}\right) = -n\,L(b).$$

Therefore, $L(b^n) = n\,L(b)$ for all $n \in \mathbb{Z}$. Consider $L(\sqrt[n]{b})$ where $n \in \mathbb{N}$. Since $n\,L(\sqrt[n]{b}) = L(b)$, $L(\sqrt[n]{b}) = \frac{1}{n}L(b)$. Therefore

$$L(b^r) = r\,L(b)$$

for all $r \in \mathbb{Q}$. Since L is continuous the above holds for all $r \in \mathbb{R}$.

Our final step will be to prove that $L(e) = 1$, where e is Euler's number of Example 3.3.5. To accomplish this we use the definition to compute the derivative of L at 1. Since $L'(1)$ exists,

$$1 = L'(1) = \lim_{n \to \infty} \frac{L(1 + \frac{1}{n}) - L(1)}{\frac{1}{n}}$$

$$= \lim_{n \to \infty} n\,L(1 + \tfrac{1}{n})$$

$$= \lim_{n \to \infty} L((1 + \tfrac{1}{n})^n) = L(e).$$

The last equality follows by the continuity of L and the definition of e. Therefore $L(e) = 1$ and the function $L(x)$ is the logarithm function to the base e. This function is usually denoted by $\log_e x$ or $\ln x$, and is called the **natural logarithm function**. $\square$

Consequences of the Fundamental Theorem of Calculus

We now prove several other consequences of Theorem 6.3.4. Our first result is the mean value theorem for integrals.

THEOREM 6.3.6 (Mean Value Theorem for Integrals) *Let f be a continuous real-valued function on $[a, b]$. Then there exists $c \in [a, b]$ such that*

$$\int_a^b f = f(c)\,(b - a).$$

Proof. Let $F(x) = \int_a^x f$. Since f is continuous on $[a, b]$, $F'(x) = f(x)$ for all $x \in [a, b]$. Thus by the mean value theorem (Theorem 5.2.6), there exists $c \in [a, b]$ such that

$$\int_a^b f = F(b) - F(a) = F'(c)(b - a) = f(c)(b - a). \quad \square$$

An alternate proof of the above can also be based on the intermediate value theorem using the continuity of f. This alternate method will be used in the proof of the analogous result for the Riemann-Stieltjes integral.

THEOREM 6.3.7 (Integration by Parts Formula) *Let f, g be differentiable functions on $[a, b]$ with f', $g' \in \mathcal{R}[a, b]$. Then*

$$\int_a^b f\, g' = f(b)g(b) - f(a)g(a) - \int_a^b g\, f'.$$

Proof. Since f, g are differentiable on $[a, b]$, they are continuous and thus also integrable on $[a, b]$. Therefore by Theorem 6.2.1(c), fg' and gf' are integrable on $[a, b]$. Since

$$(f\, g)' = g\, f' + f\, g',$$

the function $(f\, g)' \in \mathcal{R}[a, b]$. By the fundamental theorem of calculus (Theorem 6.3.2),

$$f(b)g(b) - f(a)g(a) = \int_a^b (f\, g)' = \int_a^b g\, f' + \int_a^b f\, g',$$

from which the result follows.   □

THEOREM 6.3.8 (Change of Variable Theorem) *Let φ be differentiable on $[a, b]$ with $\varphi' \in \mathcal{R}[a, b]$. If f is continuous on $I = \varphi([a, b])$, then*

$$\int_a^b f(\varphi(t))\varphi'(t)\, dt = \int_{\varphi(a)}^{\varphi(b)} f(x)\, dx.$$

Proof. Since φ is continuous, $I = \varphi([a, b])$ is a closed and bounded interval. Also, since $f \circ \varphi$ is continuous and $\varphi' \in \mathcal{R}[a, b]$, by Theorem 6.2.1(c), $(f \circ \varphi)\varphi' \in \mathcal{R}[a, b]$. If $I = \varphi([a, b])$ is a single point, then φ is constant on $[a, b]$. In this case, $\varphi'(t) = 0$ for all t and both integrals above are zero. Otherwise, for $x \in I$ define

$$F(x) = \int_{\varphi(a)}^x f(s)\, ds.$$

Since f is continuous, $F'(x) = f(x)$ for all $x \in I$. By the chain rule

$$\frac{d}{dt}F(\varphi(t)) = F'(\varphi(t))\varphi'(t) = f(\varphi(t))\varphi'(t)$$

for all $t \in I$. Therefore by Theorem 6.3.2

$$\int_a^b f(\varphi(t))\varphi'(t)\, dt = F(\varphi(b)) - F(\varphi(a)) = \int_{\varphi(a)}^{\varphi(b)} f(s)\, ds.   □$$

Remark. Another version of the change of variable theorem is given in Exercise 11.

EXAMPLES 6.3.9 (a) To illustrate the change of variable theorem, consider $\int_0^2 t/(1+t^2)\, dt$. If we let $\varphi(t) = 1 + t^2$ and $f(x) = 1/x$, then

$$\int_0^2 \frac{t}{1+t^2}\, dt = \frac{1}{2}\int_0^2 f(\varphi(t))\varphi'(t)\, dt,$$

which by Theorem 6.3.8

$$= \frac{1}{2}\int_1^5 \frac{1}{x}\, dx = \frac{1}{2}\ln 5.$$

(b) For integrals involving $\sqrt{a^2 - x^2}$, $\sqrt{x^2 - a^2}$, and $\sqrt{x^2 + a^2}$, an appropriate trigonometric substitution is useful in evaluating the integral. We illustrate this with the following. Find $\int_0^a \sqrt{a^2 - x^2}dx$. We make the substitution

$$x = a\sin t, \quad 0 \le t \le \frac{\pi}{2}.$$

Then $dx = a\cos t\, dt$ and $\sqrt{a^2 - x^2} = a\cos t$. Thus

$$\int_0^a \sqrt{a^2 - x^2}dx = a^2 \int_0^{\frac{\pi}{2}} \cos^2 t\, dt,$$

which by the identity $\cos^2 t = \frac{1}{2}(1 + \cos 2t)$

$$= \frac{a^2}{2}\int_0^{\frac{\pi}{2}}(1 + \cos 2t)dt = \frac{a^2}{2}\left[t + \frac{1}{2}\sin 2t\right]\Big|_0^{\frac{\pi}{2}} = \frac{a^2\pi}{4}. \quad \square$$

Exercises 6.3

1. Sketch both the graph of $f(x)$ and the graph of $F(x)$ of Example 6.3.3 (c).

2. *Let $f \in \mathcal{R}[a, b]$. For $x \in [a, b]$, set $F(x) = \int_a^x f$. Prove that F is continuous on $[a, b]$.

3. For $x \in [0, 1]$, find $F(x) = \int_0^x f(t)dt$ for each of the following functions f defined on $[0, 1]$. In each case verify that F is continuous on $[0, 1]$, and that $F'(x) = f(x)$ at all points where f is continuous.

    a. $f(x) = x^3 - 3x + 5$ ***b.** $f(x) = \begin{cases} 1, & 0 \le x < \frac{1}{2}, \\ -2, & \frac{1}{2} \le x < 1 \end{cases}$

    c. $f(x) = x - [3x]$ ***d** $f(x) = x[3x]$.

4. Let $f(t)$ be defined by $f(t) = \begin{cases} t, & 0 \le t < 1, \\ b - t^2, & 1 \le t \le 2, \end{cases}$ and let $F(x)$ be

   defined by $F(x) = \int\limits_0^x f(t)\,dt, \qquad 0 \le x \le 2.$

   a. Find $F(x)$.

   b. For what value of b in the definition of f is $F(x)$ differentiable for all $x \in [0, 2]$.

5. Let f be a continuous real-valued function on $[a, b]$ and define H on $[a, b]$ by $H(x) = \int\limits_x^b f$. Find $H'(x)$.

6. Find $F'(x)$ where F is defined on $[0, 1]$ as follows:

   a. $F(x) = \int_0^x \dfrac{1}{1 + t^2}\,dt$ ***b.** $F(x) = \int_0^x \cos t^2 \, dt$

   c. $F(x) = \int\limits_x^1 \sqrt{1 + t^3}\,dt$

   *d. $F(x) = \int\limits_0^{x^2} f(t)\,dt$, where f is continuous on $[0, 1]$

7. ***** Let $L(x)$ be defined as in Example 6.3.5. Prove that $L(1/x) = -L(x)$.

8. Let $f : \mathbb{R} \to \mathbb{R}$ be continuous. For $a > 0$ define g on $\mathbb{R}$ by

   $g(x) = \int\limits_{x-a}^{x+a} f(t)dt$

   Show that g is differentiable and find $g'(x)$.

9. Suppose $f : [a, b] \to \mathbb{R}$ is continuous and $g, h : [c, d] \to [a, b]$ are differentiable. For $x \in [c, d]$ define

   $H(x) = \int\limits_{h(x)}^{g(x)} f(t)\,dt.$

   Find $H'(x)$.

10. ***** Let f be a continuous real-valued function on $[a, b]$, $g \in \mathcal{R}[a, b]$ with $g(x) \ge 0$ for all $x \in [a, b]$. Prove that there exists $c \in [a, b]$ such that

    $\int\limits_a^b f(x)g(x)dx = f(c) \int\limits_a^b g(x)dx.$

11. Prove the following change of variables formula: Let φ be a real-valued differentiable function on $[a, b]$ with $\varphi'(x) \ne 0$ for all $x \in [a, b]$. Let ψ be the inverse function of φ on $I = \varphi([a, b])$. If $f : I \to \mathbb{R}$ is continuous on I, then

    $\int\limits_a^b f(\varphi(x))\,dx = \int\limits_{\varphi(a)}^{\varphi(b)} f(t)\psi'(t)\,dt.$

12. Evaluate each of the following integrals. Justify each step.

    *a. $\displaystyle\int_1^2 \frac{\ln x}{x}\,dx.$ ***b.** $\displaystyle\int_1^4 \frac{\sqrt{1 + \sqrt{x}}}{\sqrt{x}}\,dx.$

c. $\displaystyle\int_0^1 x\ln x\, dx.$ **d.** $\displaystyle\int_0^x t\sqrt{at+b}\, dt, \quad a,b>0.$

*e. $\displaystyle\int_0^1 \frac{\sqrt{x}}{1+\sqrt{x}}\, dx.$ **f.** $\displaystyle\int_1^4 \frac{1}{x\sqrt{x+1}}\, dx.$

13. Use the suggested trigonometric substitution to evaluate each of the following integrals.

   a. $\displaystyle\int_{-a}^a \frac{x^2}{\sqrt{a^2-x^2}}\, dx, \quad x=a\sin t, \quad -\tfrac{1}{2}\pi \le t \le \tfrac{1}{2}\pi$

   *b. $\displaystyle\int_0^a \frac{1}{\sqrt{x^2+a^2}}\, dx, \quad x=a\tan t, \quad 0\le t\le \tfrac{\pi}{4}$

   c. $\displaystyle\int_0^{2a} \sqrt{x^2-a^2}\, dx, \quad x=a\sec t, \quad 0\le t\le \tfrac{1}{3}\pi$

14. ***Suppose** $f:[a,b]\to\mathbb{R}$ is continuous. Let $M=\max\{|f(x)| : x\in[a,b]\}$. Show that
$$\lim_{n\to\infty}\left(\int_a^b |f(x)|^n dx\right)^{1/n} = M.$$

15. ***Let** f be a monotone increasing function on $[a,b]$ and let $F(x) = \int_a^b f(t)\, dt$. Prove that $F'_+(c) = f(c+)$ and $F'_-(c) = f(c-)$ for every $c\in(a,b)$.

16. Use Theorem 4.4.10 and the previous exercise to construct a continuous, increasing function F on $[0,1]$ that is differentiable at every irrational number in $[0,1]$, and not differentiable at any rational number in $[0,1]$.

17. (**Cauchy-Schwarz Inequality for Integrals**) Let $f, g \in \mathcal{R}[a,b]$. Prove that
$$\left|\int_a^b f(x)g(x)\, dx\right|^2 \le \left(\int_a^b f^2(x)\, dx\right)\left(\int_a^b g^2(x)\, dx\right).$$

(Hint: For $\alpha, \beta \in \mathbb{R}$ consider $\int_a^b (\alpha f - \beta g)^2$.)

18. (**The Exponential Function**) As in Example 6.3.5, let $L:(0,\infty)\to\mathbb{R}$ be defined by
$$L(x) = \int_1^x \frac{1}{t}\, dt.$$

   a. Show that L is strictly increasing on $(0,\infty)$ with Range $L=\mathbb{R}$.

   b. Let $E:\mathbb{R}\to(0,\infty)$ denote the inverse function of L. Use Theorem 5.2.14 to prove that $E'(x)=E(x)$ for all $x\in\mathbb{R}$. (The function E is called the **natural exponential function**, and is often denoted by $\exp(x)$.)

   c. Prove that $E(x+y)=E(x)E(y)$ for all $x, y\in\mathbb{R}$.

   d. Prove that $E(x)=e^x$ for all $x\in\mathbb{R}$, where e is Euler's number, and e^x is as defined in Miscellaneous Exercise 3 of Chapter 1.

6.4   Improper Riemann Integrals

In the definition of the Riemann integral of a real-valued function f, we required that f be a bounded function defined on a closed and bounded interval $[a, b]$. If these two hypothesis are not satisfied, then the preceding theory does not apply and we have to make some modifications in the definition. In this section, we will briefly consider the changes that are required if the function f is unbounded at some point of its domain, or if the interval on which f is defined is itself unbounded.

Unbounded Functions on Finite Intervals

If $f \in \mathcal{R}[a, b]$ with $|f(x)| \leq M$ for all $x \in [a, b]$, then by Theorem 6.2.3 $f \in \mathcal{R}[c, b]$ for every $c \in (a, b)$, and

$$\left| \int_a^b f - \int_c^b f \right| = \left| \int_a^c f \right| \leq \int_a^c |f| \leq M(c - a).$$

As a consequence, if $f \in \mathcal{R}[a, b]$, then $f \in \mathcal{R}[c, b]$ for every $c \in (a, b)$ and

$$\lim_{c \to a^+} \int_c^b f = \int_a^b f.$$

Suppose now that f is a bounded real-valued function defined only on $(a, b]$ with $f \in \mathcal{R}[c, b]$ for every $c \in (a, b)$. If we define $f(a) = 0$, then f is a bounded function on $[a, b]$ satisfying $f \in \mathcal{R}[c, b]$ for every $c \in (a, b)$. By Exercise 5, Section 6.2, $f \in \mathcal{R}[a, b]$ with

$$\lim_{c \to a^+} \int_c^b f = \int_a^b f.$$

Furthermore, by Exercise 3 of Section 6.2 the answer is independent of how we define $f(a)$.

Using the above as motivation, we extend the definition of the integral to include the case where f becomes unbounded at an endpoint. This extension of the integral is also due to Cauchy.

DEFINITION 6.4.1 *Let f be a real-valued function on $(a, b]$ such that $f \in \mathcal{R}[c, b]$ for every $c \in (a, b)$. The* **improper Riemann integral** *of f on $(a, b]$, denoted $\int_a^b f$, is defined to be*

$$\int_a^b f = \lim_{c \to a^+} \int_c^b f,$$

provided the limit exists. If the limit exists, then the improper integral is said to be **convergent**. *Otherwise, the improper integral is said to be* **divergent**.

A similar definition can also be given if f is defined on $[a, b)$ and becomes unbounded at b. If f becomes unbounded at a point p, $a < p < b$, we consider the improper integrals of f on the intervals $[a, p)$ and $(p, b]$. If each of the improper integrals exist, then we define the improper integral of f on $[a, b]$ to be the sum

$$\int_a^p f + \int_p^b f.$$

EXAMPLES 6.4.2 (a) Consider the function $f(x) = 1/x$ on $(0, 1]$. This function is clearly unbounded at 0. Since f is continuous on $(0, 1]$, $f \in \mathcal{R}[c, 1]$ for every $c \in (0, 1)$. By Example 6.3.5

$$\int_c^1 \frac{1}{x} \, dx = \ln c.$$

To evaluate $\lim_{c \to 0^+} \ln c$, we consider $\ln r^n$, where $0 < r < 1$ and $n \in \mathbb{N}$. Since $\ln r^n = n \ln r$ and $\ln r < 0$,

$$\lim_{n \to \infty} \ln r^n = \lim_{n \to \infty} n \ln r = -\infty.$$

Therefore, since $r^n \to 0$ as $n \to \infty$ and $\ln x$ is monotone increasing $\lim_{c \to 0^+} \ln c = -\infty$. Thus the improper integral of $1/x$ on $(0, 1]$ diverges.

(b) For our second example we consider the improper integral of $L(x) = \ln x$ on $(0, 1]$. Since $L(x)$ is continuous on $(0, 1]$, $L \in \mathcal{R}[c, 1]$ for every $c \in (0, 1)$. Consider $\int_c^1 \ln x \, dx$. If we take $g(x) = x$, then by the integration by parts formula,

$$\int_c^1 \ln x \, dx = \int_c^1 L(x) g'(x) \, dx$$

$$= L(1)g(1) - L(c)g(c) - \int_c^1 L'(x)g(x) \, dx$$

$$= -c \ln c - 1 + c.$$

By the substitution $c = 1/t$, $t > 0$, and l'Hospital's rule,

$$\lim_{c \to 0^+} c \ln c = \lim_{t \to \infty} \frac{-\ln t}{t} = -\lim_{t \to \infty} \frac{1}{t} = 0.$$

Therefore,

$$\lim_{c \to 0^+} \int_c^1 \ln x \, dx = -1.$$

Hence the improper integral of $\ln x$ converges on $(0, 1]$ with

$$\int_0^1 \ln x \, dx = -1.$$

(c) Consider the function f defined on $[-1, 1]$ by

$$f(x) = \begin{cases} 0, & -1 \leq x \leq 0, \\ \dfrac{1}{x}, & 0 < x \leq 1. \end{cases}$$

Here f is unbounded at 0. For this example, the improper integral of f over $[-1, 1]$ fails to exist because the improper integral of f over $(0, 1]$ does not exist.

(d) There are some significant differences between the Riemann integral and the improper Riemann integral. For example, if $f \in \mathcal{R}[a, b]$, then by Theorem 6.2.1 $f^2 \in \mathcal{R}[a, b]$. This however is *false* for the improper Riemann integral! For example, if $f(x) = 1/\sqrt{x}$, $x \in (0, 1]$, then

$$\int_0^1 \frac{1}{\sqrt{x}} \, dx = \lim_{c \to 0^+} \int_c^1 \frac{1}{\sqrt{x}} \, dx = \lim_{c \to 0^+} (2 - 2\sqrt{c}) = 2.$$

Thus the improper integral of f converges on $(0, 1]$. However, $f^2(x) = 1/x$, and the improper integral of $1/x$ on $(0, 1]$ diverges. Also, if $f \in \mathcal{R}[a, b]$, then $|f| \in \mathcal{R}[a, b]$. This again is *false* for improper Riemann integrals. A function f for which the improper integral of f converges, but the improper integral of $|f|$ diverges is given in Exercise 4. □

Infinite Intervals

We now turn our attention to functions defined on infinite intervals.

DEFINITION 6.4.3 *Let f be a real-valued function on $[a, \infty)$ that is Riemann integrable on $[a, c]$ for every $c > a$. The **improper Riemann integral** of f on $[a, \infty)$, denoted $\int_a^\infty f$, is defined to be*

$$\int_a^\infty f = \lim_{c \to \infty} \int_a^c f,$$

*provided the limit exists. If the limit exists, then the improper integral is said to be **convergent**. Otherwise, the improper integral is said to be **divergent**.*

If f is a real-valued function defined on $(-\infty, b]$ satisfying $f \in \mathcal{R}[c, b]$ for every $c < b$, then the improper integral of f on $(-\infty, b]$ is defined as

$$\int_{-\infty}^b f = \lim_{c \to -\infty} \int_c^b f,$$

provided the limit exists. If f is defined on $(-\infty, \infty)$, then the improper integral of f is defined as

$$\int_{-\infty}^p f + \int_p^\infty f,$$

for some fixed $p \in \mathbb{R}$, provided that the improper integrals of f on $(-\infty, p]$ and $[p, \infty)$ are both convergent. For a function f defined on $(-\infty, \infty)$, care must be exercised in computing the improper integral of f. It is <u>incorrect</u> to compute

$$\lim_{c \to \infty} \int_{-c}^{c} f.$$

For example, if $f(x) = x$, then

$$\lim_{c \to \infty} \int_{-c}^{c} x \, dx = \lim_{c \to \infty} \tfrac{1}{2}(c^2 - (-c)^2) = 0.$$

However, the improper integral of f on $(-\infty, \infty)$ is divergent since

$$\int_{0}^{\infty} f = \lim_{c \to \infty} \int_{0}^{c} x \, dx = \lim_{c \to \infty} \tfrac{1}{2}c^2 = \infty.$$

Remark. If f is nonnegative on $[a, \infty)$ with $f \in \mathcal{R}[a, c]$ for every $c > a$, then $\int_a^c f$ is a monotone increasing function of c on $[a, \infty)$. Thus $\lim\limits_{c \to \infty} \int_a^c f$ exists either as a real number or diverges to ∞. For this reason, if $f(x) \geq 0$ for all $x \in [a, \infty)$, we use the notation

$$\int_{a}^{\infty} f(x) \, dx < \infty \quad \text{or} \quad \int_{a}^{b} f(x) \, dx = \infty$$

to denote that the improper integral of f on $[a, \infty)$ converges or diverges respectively.

EXAMPLES 6.4.4 (a) Let $f(x) = 1/x^2$, $x \in [1, \infty)$. Since f is continuous on $[1, \infty)$, $f \in \mathcal{R}[1, c]$ for every $c > 1$. Therefore,

$$\int_{1}^{\infty} f = \lim_{c \to \infty} \int_{1}^{c} \frac{1}{x^2} \, dx = \lim_{c \to \infty} \left(-\frac{1}{c} + 1 \right) = 1.$$

Thus the improper integral converges to the value 1.

(b) In this example, we consider the function $f(x) = (\sin x)/x$, $x \in [\pi, \infty)$. Since f is continuous, $f \in \mathcal{R}[\pi, c]$ for every $c > \pi$. This function has the property that the improper integral of f on $[\pi, \infty)$ converges, but the improper integral of $|f|$ diverges. The proof of the convergence of the improper integral of f is left as an exercise (Exercise 7). Here we will show that

$$\int_{\pi}^{\infty} |f| = \int_{\pi}^{\infty} \frac{|\sin x|}{x} \, dx = \infty.$$

For $n \in \mathbb{N}$, consider

$$\int_{\pi}^{(n+1)\pi} \frac{|\sin x|}{x} \, dx = \sum_{k=1}^{n} \int_{k\pi}^{(k+1)\pi} \frac{|\sin x|}{x} \, dx.$$

Since the integrand is nonnegative,

$$\int_{k\pi}^{(k+1)\pi} \frac{|\sin x|}{x}\, dx \geq \int_{(k+\frac{1}{4})\pi}^{(k+\frac{3}{4})\pi} \frac{|\sin x|}{x}\, dx.$$

On the interval $[(k+\frac{1}{4})\pi, (k+\frac{3}{4})\pi]$, $|\sin x| \geq \sqrt{2}/2$. Also,

$$\frac{1}{x} \geq \frac{1}{(k+1)\pi} \quad \text{for all } x \in [(k+\frac{1}{4})\pi, (k+\frac{3}{4})\pi].$$

Therefore

$$\int_{(k+\frac{1}{4})\pi}^{(k+\frac{3}{4})\pi} \frac{|\sin x|}{x}\, dx \geq \frac{\sqrt{2}}{2} \frac{1}{(k+1)\pi} \left[(k+\tfrac{3}{4})\pi - (k+\tfrac{1}{4})\pi\right] = \frac{\sqrt{2}}{4} \frac{1}{k+1},$$

and as a consequence

$$\int_{\pi}^{(n+1)\pi} \frac{|\sin x|}{x}\, dx \geq \frac{\sqrt{2}}{4} \sum_{k=1}^{n} \frac{1}{k+1}.$$

By Example 3.7.4 the series $\sum_{k=1}^{\infty} 1/k$ diverges. Therefore,

$$\int_{\pi}^{\infty} \frac{|\sin x|}{x}\, dx = \lim_{n \to \infty} \int_{\pi}^{(n+1)\pi} \frac{|\sin x|}{x}\, dx = \infty. \quad \square$$

As the previous example shows, the convergence of the improper integral of f does not imply the convergence of the improper integral of $|f|$. If f is a real-valued function on $[a, \infty)$ such that $f \in \mathcal{R}[a,c]$ for every $c > a$ and the improper integral of $|f|$ exists on $[a, \infty)$, then f is said to be **absolutely integrable** on $[a, \infty)$. An analogous definition can also be given for unbounded functions on a finite interval. We leave it as an exercise to prove that if f is absolutely integrable on $[a, \infty)$, then the improper integral of f also converges on $[a, \infty)$ (Exercise 5).

We conclude this section with the following useful comparison test for improper integrals.

THEOREM 6.4.5 (Comparison Test) *Let $g : [a, \infty) \to \mathbb{R}$ be a nonnegative function satisfying $g \in \mathcal{R}[a,c]$ for every $c > a$ and $\int_a^{\infty} g(x)dx < \infty$. If $f : [a, \infty) \to \mathbb{R}$ satisfies*

(a) $f \in \mathcal{R}[a,c]$ for every $c > a$, and

(b) $|f(x)| \leq g(x)$ for all $x \in [a, \infty)$, then the improper integral of f on $[a, \infty)$ converges, and

$$\left| \int_a^{\infty} f(x)\, dx \right| \leq \int_a^{\infty} g(x)dx.$$

Proof. The proof is left to the exercises (Exercise 6). $\square$

Exercises 6.4

1. For each of the following functions f defined on $(0, 1]$, determine whether the improper integral of f exists. If it exists, find $\int\limits_0^1 f$.

   *a. $f(x) = \dfrac{1}{x^p}$, $0 < p < 1$ **b.** $f(x) = \dfrac{x}{\sqrt{1-x}}$ **c.** $f(x) = \dfrac{\ln x}{x}$

   *d. $f(x) = x \ln x$ **e.** $f(x) = \dfrac{1}{(1+x)\ln(1+x)}$

   *f. $f(x) = \tan\left(\dfrac{\pi}{2}x\right)$

2. For each of the following determine whether the improper integral converges or diverges. If it converges, evaluate the integral.

   *a. $\displaystyle\int_0^\infty e^{-x}\, dx$ **b.** $\displaystyle\int_1^\infty \frac{1}{x}\, dx$ ***c.** $\displaystyle\int_1^\infty x^{-p}\, dx$, $p > 1$

   d. $\displaystyle\int_1^\infty \frac{\ln x}{x}\, dx$ ***e.** $\displaystyle\int_2^\infty \frac{1}{x\ln x}$ **f.** $\displaystyle\int_2^\infty \frac{1}{x(\ln x)^p}$, $p > 1$

   *g. $\displaystyle\int_0^\infty \frac{x}{x^2+1}\, dx$ **h.** $\displaystyle\int_0^\infty \frac{x}{(x^2+1)^p}\, dx$, $p > 1$

3. For each of the following, determine the values of p and q for which the improper integral converges.

   *a. $\displaystyle\int_0^{\frac{1}{2}} x^p |\ln x|^q\, dx$ **b.** $\displaystyle\int_2^\infty x^p (\ln x)^q\, dx$ **c.** $\displaystyle\int_0^\infty x^p [\ln(1+x)]^q\, dx$

4. Let f be defined on $(0, 1]$ by

   $$f(x) = \frac{d}{dx}\left(x^2 \sin\frac{1}{x^2}\right) = 2x \sin\frac{1}{x^2} - \frac{2}{x}\cos\frac{1}{x^2}.$$

   Show that the improper Riemann integral of f converges on $(0, 1]$, but that the improper integral of $|f|$ diverges on $(0, 1]$.

5. ***If** f is absolutely integrable on $[a, \infty)$, and integrable on $[a, c]$ for every $c > a$, prove that the improper Riemann integral of f on $[a, \infty)$ exists.

6. Prove Theorem 6.4.5.

7. Let $f(x) = \dfrac{\cos x}{x^2}$, $x \in [\pi, \infty)$.

   *a. Show that the improper integral of $|f|$ converges on $[\pi, \infty)$.

   *b. Use integration by parts on $[\pi, c]$, $c > \pi$, to show that $\displaystyle\int_\pi^\infty \frac{\sin x}{x}\, dx$ exists.

8. Show that $\displaystyle\int_0^\infty x^{-p}\sin x\, dx$ converges for all p, $0 < p < 2$.

9. For $x > 0$, set

   $$\Gamma(x) = \int_0^\infty e^{-t} t^{x-1}\, dt.$$

   The function Γ is called the **Gamma function**.

   *a. Show that the improper integral converges for all $x > 0$.

b. Use integration by parts to show that $\Gamma(x+1) = x\Gamma(x)$, $x > 0$.

c. Show that $\Gamma(1) = 1$.

d. For $n \in \mathbb{N}$, prove that $\Gamma(n+1) = n!$.

6.5   The Riemann-Stieltjes Integral[3]

In this section, we consider the Riemann-Stieltjes integral, which as we will see is an extension of the Riemann integral. To motivate the Riemann-Stieltjes integral we consider the following example from physics involving the moment of inertia.

EXAMPLE 6.5.1 Consider n-masses, each of mass m_i, $i = 1, ..., n$, located along the x-axis at distances r_i from the origin with $0 < r_1 < \cdots < r_n$ (Figure 6.7). The moment of inertia I about an axis through the origin at right angles to the system of masses is given by

$$I = \sum_{i=1}^{n} r_i^2 \, m_i.$$

On the other hand, if we have a wire of length ℓ along the x-axis with one end at the origin, then the moment of inertia I is given my

$$I = \int_0^\ell x^2 \rho(x) \, dx,$$

where for each $x \in [0, \ell]$, $\rho(x)$ denotes the cross-sectional density at x.   □

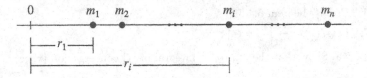

FIGURE 6.7
Example 6.5.1

Although these two problems are totally different, the first being discrete and the second continuous, the Riemann-Stieltjes integral will allow us to express both of these formulas as a single integral. In the definition of the

---

[3]Since the results of this section are not specifically required in subsequent chapters, this topic can be omitted.

Riemann integral we used the length Δx_i of the ith interval to define the upper and lower Riemann sums of a bounded function f. The only difference between the Riemann and Riemann-Stieltjes integral is that we replace Δx_i by

$$\Delta \alpha_i = \alpha(x_i) - \alpha(x_{i-1}),$$

where α is a nondecreasing function on $[a, b]$. Taking $\alpha(x) = x$ will give the usual Riemann integral. Although the modification in the definition is only minor, the consequences however are far reaching. Not only will we obtain a more extensive theory of integration, we also obtain an integral which has broad applications in the mathematical sciences.

Definition of the Riemann-Stieltjes Integral

Let α be a monotone increasing function on $[a, b]$, and let f be a bounded real-valued function on $[a, b]$. For each partition $\mathcal{P} = \{x_0, x_1, ..., x_n\}$ of $[a, b]$, set

$$\Delta \alpha_i = \alpha(x_i) - \alpha(x_{i-1}), \qquad i = 1, ..., n.$$

Since α is monotone increasing, $\Delta \alpha_i \geq 0$ for all i. As in Section 6.1, let

$$m_i = \inf\{f(t) : t \in [x_{i-1}, x_i]\},$$
$$M_i = \sup\{f(t) : t \in [x_{i-1}, x_i]\}.$$

As for the Riemann integral, the **upper Riemann-Stieltjes sum** of f with respect to α and the partition $\mathcal{P}$, denoted $\mathcal{U}(\mathcal{P}, f, \alpha)$, is defined by

$$\mathcal{U}(\mathcal{P}, f, \alpha) = \sum_{i=1}^{n} M_i \Delta \alpha_i.$$

Similarly, the **lower Riemann-Stieltjes sum** of f with respect to α and the partition $\mathcal{P}$, denoted $\mathcal{L}(\mathcal{P}, f, \alpha)$, is defined by

$$\mathcal{L}(\mathcal{P}, f, \alpha) = \sum_{i=1}^{n} m_i \Delta \alpha_i.$$

Since $m_i \leq M_i$ and $\Delta \alpha_i \geq 0$, we always have $\mathcal{L}(\mathcal{P}, f, \alpha) \leq \mathcal{U}(\mathcal{P}, f, \alpha)$. Furthermore, if $m \leq f(x) \leq M$ for all $x \in [a, b]$, then

$$m\,[\alpha(b) - \alpha(a)] \leq \mathcal{L}(\mathcal{P}, f, \alpha) \leq \mathcal{U}(\mathcal{P}, f\alpha) \leq M\,[\alpha(b) - \alpha(a)], \qquad (10)$$

for all partitions $\mathcal{P}$ of $[a, b]$. Let $\mathcal{P}$ be any partition of $[a, b]$. Since $M_i \leq M$ for all i and $\Delta \alpha_i \geq 0$,

$$\sum_{i=1}^{n} M_i \Delta \alpha_i \leq \sum_{i=1}^{n} M \, \Delta \alpha_i = M \sum_{i=1}^{n} \Delta \alpha_i = M[\alpha(b) - \alpha(a)].$$

Thus $\mathcal{U}(\mathcal{P}, f, \alpha) \leq M\left[\alpha(b) - \alpha(a)\right]$. The other inequality follows similarly. In the above we have used the fact that

$$\sum_{i=1}^{n} \Delta\alpha_i = (\alpha(x_1) - \alpha(x_0)) + (\alpha(x_2) - \alpha(x_1)) + \cdots + (\alpha(x_n) - \alpha(x_{n-1}))$$

$$= \alpha(x_n) - \alpha(x_0) = \alpha(b) - \alpha(a).$$

In analogy with the Riemann integral, the **upper** and **lower Riemann-Stieltjes** integrals of f with respect to α over $[a, b]$, denoted $\overline{\int_a^b} f\, d\alpha$ and $\underline{\int_a^b} f\, d\alpha$ respectively, are defined by

$$\overline{\int_a^b} f\, d\alpha = \inf\left\{\mathcal{U}(\mathcal{P}, f, \alpha) : \mathcal{P} \text{ is a partition of } [a, b]\right\},$$

$$\underline{\int_a^b} f\, d\alpha = \sup\left\{\mathcal{L}(\mathcal{P}, f, \alpha) : \mathcal{P} \text{ is a partition of } [a, b]\right\}.$$

By inequality (10) the set $\{\mathcal{U}(\mathcal{P}, f, \alpha) : \mathcal{P} \text{ is a partition of } [a, b]\}$ is bounded below, and thus the upper integral of f with respect to α exists as a real number. Similarly, the lower sums are bounded above and thus the supremum defining the lower integral is also finite. As for the Riemann integral, our first step is to prove that the lower integral is less than or equal to the upper integral.

THEOREM 6.5.2 *Let f be bounded real-valued function on $[a, b]$, and α a monotone increasing function on $[a, b]$. Then*

$$\underline{\int_a^b} f\, d\alpha \leq \overline{\int_a^b} f\, d\alpha.$$

Proof. As in the proof of Lemma 6.1.3, if $\mathcal{P}^*$ is a refinement of the partition $\mathcal{P}$, then

$$\mathcal{L}(\mathcal{P}, f, \alpha) \leq \mathcal{L}(\mathcal{P}^*, f, \alpha) \leq \mathcal{U}(\mathcal{P}^*, f, \alpha) \leq \mathcal{U}(\mathcal{P}, f, \alpha).$$

Thus if $\mathcal{P}, \mathcal{Q}$ are any two partitions of $[a, b]$,

$$\mathcal{L}(\mathcal{P}, f, \alpha) \leq \mathcal{L}(\mathcal{P} \cup \mathcal{Q}, f, \alpha) \leq \mathcal{U}(\mathcal{P} \cup \mathcal{Q}, f, \alpha) \leq \mathcal{U}(\mathcal{Q}, f, \alpha).$$

Therefore $\mathcal{L}(\mathcal{P}, f, \alpha) \leq \mathcal{U}(\mathcal{Q}, f, \alpha)$ for any partitions $\mathcal{P}, \mathcal{Q}$. Hence

$$\underline{\int_a^b} f\, d\alpha = \sup_{\mathcal{P}} \mathcal{L}(\mathcal{P}, f, \alpha) \leq \mathcal{U}(\mathcal{Q}, f, \alpha)$$

for any partition $\mathcal{Q}$. Taking the infimum over $\mathcal{Q}$ gives the result. $\square$

DEFINITION 6.5.3 *Let f be a bounded real-valued function on* $[a, b]$, *and* α *a monotone increasing function on* $[a, b]$. *If*

$$\underline{\int_a^b} f \, d\alpha = \overline{\int_a^b} f \, d\alpha,$$

then f is said to be **Riemann-Stieltjes integrable** *or* **integrable** *with respect to* α *on* $[a, b]$. *The common value is denoted by*

$$\int_a^b f \, d\alpha \quad or \quad \int_a^b f(x) \, d\alpha(x),$$

and is called the **Riemann-Stieltjes integral** *of f with respect to* α.

As was indicated previously, the special case $\alpha(x) = x$ gives the usual Riemann integral on $[a, b]$.

EXAMPLES 6.5.4 (a) Fix $a < c \le b$. As in Definition 4.4.9, let $I_c(x) = I(x - c)$ be the unit jump function at c defined by

$$I_c(x) = \begin{cases} 0, & x < c, \\ 1, & x \ge c. \end{cases}$$

We now prove the following: If f is a bounded real-valued function on $[a, b]$ that is continuous at c, $a < c \le b$, then f is integrable with respect to I_c and

$$\int_a^b f \, dI_c = \int_a^b f(x) \, dI(x - c) = f(c).$$

For convenience, we set $\alpha(x) = I_c(x)$, which is clearly monotone increasing on $[a, b]$. Let $\mathcal{P} = \{x_0, x_1, ..., x_n\}$ be any partition of $[a, b]$. Since $a < c \le b$, there exists an index k, $1 \le k \le n$, such that

$$x_{k-1} < c \le x_k.$$

Then

$$\Delta\alpha_k = \alpha(x_k) - \alpha(x_{k-1}) = 1 - 0 = 1, \quad \text{and}$$
$$\Delta\alpha_i = 0, \quad \text{for all} \quad i \ne k.$$

Therefore

$$\mathcal{U}(\mathcal{P}, f, \alpha) = M_k \Delta\alpha_k = M_k = \sup\{f(t) : x_{k-1} \le t \le x_k\}, \quad \text{and}$$
$$\mathcal{L}(\mathcal{P}, f, \alpha) = m_k \Delta\alpha_k = m_k = \inf\{f(t) : x_{k-1} \le t \le x_k\}.$$

Since f is continuous at c, given $\epsilon > 0$ there exists a $\delta > 0$ such that

$$f(c) - \epsilon < f(t) < f(c) + \epsilon$$

for all $t \in [a, b]$ with $|t - c| < \delta$. If $\mathcal{P}$ is any partition of $[a, b]$ with $x_j - x_{j-1} < \delta$ for all j, then
$$f(c) - \epsilon \le m_k \le M_k \le f(c) + \epsilon.$$
Therefore $f(c) - \epsilon \le \mathcal{L}(\mathcal{P}, f, \alpha) \le \mathcal{U}(\mathcal{P}, f, \alpha) \le f(c) + \epsilon$. As a consequence

$$f(c) - \epsilon \le \underline{\int_a^b} f \, d\alpha \le \overline{\int_a^b} f \, d\alpha \le f(c) + \epsilon.$$

Since $\epsilon > 0$ was arbitrary, the upper and lower integrals of f are equal, and thus f is integrable with respect to α on $[a, b]$ with

$$\int_a^b f \, d\alpha = f(c).$$

(b) The function
$$f(x) = \begin{cases} 1, & x \in \mathbb{Q} \\ 0, & x \notin \mathbb{Q}, \end{cases}$$

is <u>not</u> integrable with respect to any non-constant monotone increasing function α. Suppose α is monotone increasing on $[a, b]$, $a < b$, with $\alpha(a) \ne \alpha(b)$. If $\mathcal{P} = \{x_0, x_1, ..., x_n\}$ is any partition of $[a, b]$, then $m_i = 0$ and $M_i = 1$ for all $i = 1, ..., n$. Therefore $\mathcal{L}(\mathcal{P}, f, \alpha) = 0$ and

$$\mathcal{U}(\mathcal{P}, f, \alpha) = \sum_{i=1}^n \Delta\alpha_i = \alpha(b) - \alpha(a).$$

Thus f is <u>not</u> integrable with respect to α $\square$

Remark: It should be noted that in Example 6.5.4(a) only left continuity of f at c is required (Exercise 2(a)).

The following theorem, which is the analogue of Theorem 6.1.7 for the Riemann integral, provides necessary and sufficient conditions for the existence of the Riemann-Stieltjes integral. The proof of the theorem follows verbatim the proof of Theorem 6.1.7 and thus is omitted.

THEOREM 6.5.5 *Let α be a monotone increasing function on $[a, b]$. A bounded real-valued function f is Riemann-Stieltjes integrable with respect to α on $[a, b]$ if and only if for every $\epsilon > 0$, there exists a partition $\mathcal{P}$ of $[a, b]$ such that*
$$\mathcal{U}(\mathcal{P}, f, \alpha) - \mathcal{L}(\mathcal{P}, f, \alpha) < \epsilon.$$

Furthermore, If $\mathcal{P}$ is a partition of $[a, b]$ for which the above holds, then the inequality also holds for all refinements of $\mathcal{P}$.

We now use the previous theorem to prove the following analogue of Theorem 6.1.8. Except for some minor differences, the two proofs are very similar.

THEOREM 6.5.6 *Let f be a real-valued function on* $[a, b]$ *and* α *a monotone increasing function on* $[a, b]$.

(a) *If f is continuous on* $[a, b]$, *then f is integrable with respect to* α *on* $[a, b]$.

(b) *If f is monotone on* $[a, b]$ *and* α *is continuous on* $[a, b]$, *then f is integrable with respect to* α *on* $[a, b]$.

Proof. (a) The proof of (a) is identical to the proof of Theorem 6.1.8(a) except that given $\epsilon > 0$, we choose $\eta > 0$ such that

$$[\alpha(b) - \alpha(a)]\, \eta < \epsilon.$$

The remainder of the proof now follows verbatim the proof of Theorem 6.1.8(a).

(b) For any positive integer n, choose a partition $\mathcal{P} = \{x_0, x_1, ..., x_n\}$ of $[a, b]$ such that

$$\Delta\alpha_i = \alpha(x_i) - \alpha(x_{i-1}) = \frac{1}{n}[\alpha(b) - \alpha(a)].$$

Since α is continuous, such a choice is possible by the intermediate value theorem. Assume f is monotone increasing on $[a, b]$. Then $M_i = f(x_i)$ and $m_i = f(x_{i-1})$. Therefore,

$$\mathcal{U}(\mathcal{P}, f, \alpha) - \mathcal{L}(\mathcal{P}, f, \alpha) = \sum_{i=1}^{n} [f(x_i) - f(x_{i-1})]\Delta\alpha_i$$

$$= \frac{[\alpha(b) - \alpha(a)]}{n} \sum_{i=1}^{n} [f(x_i) - f(x_{i-1})]$$

$$= \frac{[\alpha(b) - \alpha(a)]}{n} [f(b) - f(a)].$$

Given $\epsilon > 0$, choose $n \in \mathbb{N}$ such that

$$\frac{[\alpha(b) - \alpha(a)]}{n} [f(b) - f(a)] < \epsilon.$$

For this n and corresponding partition $\mathcal{P}$, $\mathcal{U}(\mathcal{P}, f, \alpha) - \mathcal{L}(\mathcal{P}, f, \alpha) < \epsilon$, which proves the result. $\quad\square$

Remark. In part (b) above, the result may be false if α is not continuous. For example, if $a < c \leq b$, then the monotone function I_c is *not* integrable with respect to I_c on $[a, b]$ (Exercise 2(b)).

Properties of the Riemann-Stieltjes Integral

We next consider several basic properties of the Riemann-Stieltjes integral. For notational convenience, we make the following definition.

DEFINITION 6.5.7 *For a given monotone increasing function α on $[a, b]$, $\mathcal{R}(\alpha)$ denotes the set of bounded real-valued functions f on $[a, b]$ which are Riemann-Stieltjes integrable with respect to α.*

THEOREM 6.5.8

(a) *If $f, g \in \mathcal{R}(\alpha)$, then $f + g$ and cf are in $\mathcal{R}(\alpha)$ for every $c \in \mathbb{R}$, and*

$$\int_a^b (f + g) \, d\alpha = \int_a^b f \, d\alpha + \int_a^b g \, d\alpha \quad and \quad \int_a^b cf \, d\alpha = c \int_a^b f \, d\alpha.$$

(b) *If $f \in \mathcal{R}(\alpha_i)$, $i = 1, 2$, then $f \in \mathcal{R}(\alpha_1 + \alpha_2)$ and*

$$\int_a^b f \, d(\alpha_1 + \alpha_2) = \int_a^b f \, d\alpha_1 + \int_a^b f \, d\alpha_2.$$

(c) *If $f \in \mathcal{R}(\alpha)$ and $a < c < b$, then f is integrable with respect to α on $[a, c]$ and $[c, b]$ with*

$$\int_a^b f \, d\alpha = \int_a^c f \, d\alpha + \int_c^b f \, d\alpha.$$

(d) *If $f, g \in \mathcal{R}(\alpha)$ with $f(x) \leq g(x)$ for all $x \in [a, b]$, then*

$$\int_a^b f \, d\alpha \leq \int_a^b g \, d\alpha.$$

(e) *If $|f(x)| \leq M$ on $[a, b]$ and $f \in \mathcal{R}(\alpha)$, then $|f| \in \mathcal{R}(\alpha)$ and*

$$\left| \int_a^b f \, d\alpha \right| \leq \int_a^b |f| \, d\alpha \leq M \, [\alpha(b) - \alpha(a)].$$

Proof. We provide the proofs of (b) and (e). The proofs of (a), (c), and (d), along with other properties of the Riemann-Stieltjes integral are left to the exercises.

(b) Since $f \in \mathcal{R}(\alpha_i)$, given $\epsilon > 0$, there exists a partition $\mathcal{P}_i$, $i = 1, 2$, such that

$$\mathcal{U}(\mathcal{P}_i, f, \alpha_i) - \mathcal{L}(\mathcal{P}_i, f, \alpha_i) < \frac{\epsilon}{2}. \tag{11}$$

Let $\mathcal{P} = \mathcal{P}_1 \cup \mathcal{P}_2$. Since $\mathcal{P}$ is a refinement of both $\mathcal{P}_1$ and $\mathcal{P}_2$, inequality (11) is still valid for the partition $\mathcal{P}$. Thus since

$$\Delta(\alpha_1 + \alpha_2)_i = \Delta(\alpha_1)_i + \Delta(\alpha_2)_i,$$

for all $i = 1, ..., n$,

$$\mathcal{U}(\mathcal{P}, f, \alpha_1 + \alpha_2) - \mathcal{L}(\mathcal{P}, f, \alpha_1 + \alpha_2)$$
$$= \mathcal{U}(\mathcal{P}, f, \alpha_1) - \mathcal{L}(\mathcal{P}, f, \alpha_1) + \mathcal{U}(\mathcal{P}, f, \alpha_2) - \mathcal{L}(\mathcal{P}, f, \alpha_2) < \frac{\epsilon}{2} + \frac{\epsilon}{2} = \epsilon.$$

Therefore by Theorem 6.5.5, $f \in \mathcal{R}(\alpha_1 + \alpha_2)$. Furthermore, for any partition $\mathcal{P}$ of $[a, b]$,

$$\mathcal{L}(\mathcal{P}, f, \alpha_1 + \alpha_2) = \mathcal{L}(\mathcal{P}, f, \alpha_1) + \mathcal{L}(\mathcal{P}, f, \alpha_2)$$
$$\leq \int_a^b f \, d\alpha_1 + \int_a^b f \, d\alpha_2$$
$$\leq \mathcal{U}(\mathcal{P}, f, \alpha_1) + \mathcal{U}(\mathcal{P}, f, \alpha_2) = \mathcal{U}(\mathcal{P}, f, \alpha_1 + \alpha_2).$$

Thus since $f \in \mathcal{R}(\alpha_1 + \alpha_2)$,

$$\int_a^b f \, d(\alpha_1 + \alpha_2) = \int_a^b f \, d\alpha_1 + \int_a^b f \, d\alpha_2.$$

(e) Suppose $f \in \mathcal{R}(\alpha)$ and $\mathcal{P} = \{x_0, x_1, ..., x_n\}$ is a partition of $[a, b]$. For each $i = 1, ..., n$, let

$$M_i = \sup\{f(t) : t \in [x_{i-1}, x_i]\}, \qquad M_i^* = \sup\{|f(t)| : t \subset [x_{i-1}, x_i]\},$$
$$m_i = \inf\{f(t) : t \in [x_{i-1}, x_i]\}, \qquad m_i^* = \inf\{|f(t)| : t \in [x_{i-1}, x_i]\}.$$

If $t, x \in [x_{i-1}, x_i]$, then

$$\big|\,|f(t)| - |f(x)|\,\big| \leq |f(t) - f(x)| \leq M_i - m_i.$$

Thus $M_i^* - m_i^* \leq M_i - m_i$, for all $i = 1, ..., n$, and as a consequence

$$\mathcal{U}(\mathcal{P}, |f|, \alpha) - \mathcal{L}(\mathcal{P}, |f|, \alpha) \leq \mathcal{U}(\mathcal{P}, f, \alpha) - \mathcal{L}(\mathcal{P}, f, \alpha).$$

Therefore, by Theorem 6.5.5, $|f| \in \mathcal{R}(\alpha)$. Choose $c = \pm 1$ such that $c \int f \, d\alpha \geq 0$. Then

$$\left| \int_a^b f \, d\alpha \right| = c \int_a^b f \, d\alpha = \int_a^b c \, f \, d\alpha \leq \int_a^b |f| \, d\alpha$$

$$\leq M \int_a^b d\alpha = M \, [\alpha(b) - \alpha(a)]. \qquad \square$$

As for the Riemann integral, we also have the following mean value theorem and integration by parts formula for the Riemann-Stieltjes integral.

THEOREM 6.5.9 (Mean Value Theorem) *Let f be a continuous real-valued function on $[a, b]$, and α a monotone increasing function on $[a, b]$. Then there exists $c \in [a, b]$ such that*

$$\int_a^b f \, d\alpha = f(c) \left[\alpha(b) - \alpha(a) \right].$$

Proof. Let m and M denote the minimum and maximum of f on $[a, b]$ respectively. Then by Theorem 6.5.8(d),

$$m \left[\alpha(b) - \alpha(a) \right] \le \int_a^b f \, d\alpha \le M \left[\alpha(b) - \alpha(a) \right].$$

If $\alpha(b) - \alpha(a) = 0$, then any $c \in [a, b]$ will work. If $\alpha(b) - \alpha(a) \ne 0$, then by the intermediate value theorem there exists $c \in [a, b]$ such that

$$f(c) = \frac{1}{\alpha(b) - \alpha(a)} \int_a^b f \, d\alpha,$$

which proves the result. $\square$

THEOREM 6.5.10 (Integration by Parts Formula) *Suppose α and β are monotone increasing functions on $[a, b]$. Then $\alpha \in \mathcal{R}(\beta)$ if and only if $\beta \in \mathcal{R}(\alpha)$. If this is the case,*

$$\int_a^b \alpha \, d\beta = \alpha(b)\beta(b) - \alpha(a)\beta(a) - \int_a^b \beta \, d\alpha.$$

Proof. By Exercise 9, for any partition $\mathcal{P}$ of $[a, b]$,

$$\mathcal{U}(\mathcal{P}, \alpha, \beta) = \alpha(b)\beta(b) - \alpha(a)\beta(a) - \mathcal{L}(\mathcal{P}, \beta, \alpha), \quad \text{and}$$
$$\mathcal{L}(\mathcal{P}, \alpha, \beta) = \alpha(b)\beta(b) - \alpha(a)\beta(a) - \mathcal{U}(\mathcal{P}, \beta, \alpha).$$

Therefore,

$$\mathcal{U}(\mathcal{P}, \alpha, \beta) - \mathcal{L}(\mathcal{P}, \alpha, \beta) = \mathcal{U}(\mathcal{P}, \beta, \alpha) - \mathcal{L}(\mathcal{P}, \beta, \alpha).$$

From this identity it immediately follows by Theorem 6.5.5 that $\alpha \in \mathcal{R}(\beta)$ if and only if $\beta \in \mathcal{R}(\alpha)$. Furthermore, if $\beta \in \mathcal{R}(\alpha)$, then given $\epsilon > 0$, there exists a partition $\mathcal{P}$ of $[a, b]$ such that

$$\mathcal{L}(\mathcal{P}, \beta, \alpha) > \int_a^b \beta \, d\alpha - \epsilon.$$

Hence,

$$\int_a^b \alpha \, d\beta \le \mathcal{U}(\mathcal{P}, \alpha, \beta) < \alpha(b)\beta(b) - \alpha(a)\beta(a) - \int_a^b \beta \, d\alpha + \epsilon.$$

Since the above holds for any $\epsilon > 0$,

$$\int_a^b \alpha \, d\beta \leq \alpha(b)\beta(b) - \alpha(a)\beta(a) - \int_a^b \beta \, d\alpha.$$

A similar argument using the lower sum proves the reverse inequality. $\square$

We conclude this section with two results that represent the extremes encountered in Riemann-Stieltjes integration. As in Example 6.5.4(a), let I_c be the unit jump function at $c \in \mathbb{R}$. Suppose $\{s_n\}_{n=1}^N$ is a finite subset of $(a, b]$ and $\{c_n\}_{n=1}^N$ are non-negative real numbers. Define the monotone increasing function α on $[a, b]$ by

$$\alpha(x) = \sum_{n=1}^N c_n I(x - s_n).$$

If f is continuous on $[a, b]$, then by Example 6.5.4(a) and Theorem 6.5.8(b),

$$\int_a^b f \, d\alpha = \sum_{n=1}^N c_n \int_a^b f(x) \, dI(x - s_n) = \sum_{n=1}^N c_n f(s_n). \qquad (12)$$

Suppose $\{s_n\}_{n=1}^\infty$ is a countable subset of $(a, b]$ and $\{c_n\}_{n=1}^\infty$ is a sequence of nonnegative real numbers for which $\sum c_n$ converges. As in Theorem 4.4.10 define α on $[a, b]$ by

$$\alpha(x) = \sum_{n=1}^\infty c_n I(x - s_n). \qquad (13)$$

Since $0 \leq I(x - s_n) \leq 1$ for all n, the series in (13) converges for every $x \in [a, b]$, and α is a monotone increasing function on $[a, b]$. For such a function α we have the following theorem.

THEOREM 6.5.11 *Let f be a continuous real-valued function on $[a, b]$ and let α be the monotone function defined by (13). Then $f \in \mathcal{R}(\alpha)$ and*

$$\int_a^b f \, d\alpha = \sum_{n=1}^\infty c_n f(s_n).$$

Proof. Since f is continuous on $[a, b]$, $f \in \mathcal{R}(\alpha)$ (Theorem 6.5.6(a)). Let $\epsilon > 0$ be given. Choose a positive integer N such that

$$\sum_{n=N+1}^\infty c_n < \epsilon.$$

Define β_1 and β_2 as follows:

$$\beta_1(x) = \sum_{n=1}^N c_n I(x - s_n), \qquad \beta_2(x) = \sum_{n=N+1}^\infty c_n I(x - s_n).$$

Then $\alpha = \beta_1 + \beta_2$, and by identity (12)

$$\int_a^b f \, d\beta_1 = \sum_{n=1}^{N} c_n f(s_n).$$

Let $M = \max\{|f(x)| : x \in [a,b]\}$. Then by Theorem 6.5.8(b) and (e),

$$\left| \int_a^b f \, d\alpha - \sum_{n=1}^{N} c_n f(s_n) \right| = \left| \int_a^b f \, d\beta_2 \right| \leq M \left[\beta_2(b) - \beta_2(a) \right]$$

$$\leq M \sum_{n=N+1}^{\infty} c_n < M \, \epsilon,$$

which proves the result. $\square$

At the other extreme, if the monotone function α is also differentiable, then we have the following result.

THEOREM 6.5.12 *Suppose* $f \in \mathcal{R}[a,b]$ *and* α *is a monotone increasing differentiable function on* $[a,b]$ *with* $\alpha' \in \mathcal{R}[a,b]$. *Then* $f \in \mathcal{R}(\alpha)$ *and*

$$\int_a^b f \, d\alpha = \int_a^b f(x) \, \alpha'(x) \, dx.$$

Proof. Since both f and α' are Riemann integrable on $[a,b]$, by Theorem 6.2.1(c), $f\alpha' \in \mathcal{R}[a,b]$. Let $\epsilon > 0$ be given. Since $\alpha' \in \mathcal{R}[a,b]$, by Theorem 6.1.7 there exists a partition $\mathcal{P}$ of $[a,b]$ such that

$$\mathcal{U}(\mathcal{P}, \alpha') - \mathcal{L}(\mathcal{P}, \alpha') < \epsilon. \tag{14}$$

Let $\mathcal{Q} = \{x_0, ..., x_n\}$ be any refinement of $\mathcal{P}$. As in Theorem 6.2.6, for each $i = 1, ..., n$, we can choose $s_i \in [x_{i-1}, x_i]$ such that

$$\mathcal{U}(\mathcal{Q}, f, \alpha) < \sum_{i=1}^{n} f(s_i) \Delta \alpha_i + \epsilon. \tag{15}$$

By the mean value theorem, for each $i = 1, ..., n$, there exists $t_i \in [x_{i-1}, x_i]$ such that

$$\Delta \alpha_i = \alpha(x_i) - \alpha(x_{i-1}) = \alpha'(t_i) \Delta x_i$$

Therefore,

$$\sum_{i=1}^{n} f(s_i) \Delta \alpha_i = \sum_{i=1}^{n} f(s_i) \alpha'(t_i) \Delta x_i. \tag{16}$$

Let $M = \sup\{|f(x)| : x \in [a,b]\}$, and for $i = 1, ..., n$, let m_i and M_i denote the infimum and supremum respectively of α' over the interval $[x_{i-1}, x_i]$. Then

$$|\alpha'(s_i) - \alpha'(t_i)| \leq M_i - m_i$$

for all $i = 1, ..., n$. Therefore,

$$\left| \sum_{i=1}^{n} f(s_i)\alpha'(t_i)\Delta x_i - \sum_{i=1}^{n} f(s_i)\alpha'(s_i)\Delta x_i \right| \le \sum_{i=1}^{n} |f(s_i)||\alpha'(t_i) - \alpha'(s_i)|$$

$$\le M \sum_{i=1}^{n} (M_i - m_i)\Delta x_i$$

$$= M\left(\mathcal{U}(\mathcal{Q}, \alpha') - \mathcal{L}(\mathcal{Q}, \alpha')\right) < M\epsilon.$$

The last inequality follows by (14) since $\mathcal{Q}$ is a refinement of $\mathcal{P}$. Therefore

$$\sum_{i=1}^{n} f(s_i)\alpha'(t_i)\Delta x_i \le \sum_{i=1}^{n} f(s_i)\alpha'(s_i)\Delta x_i + M\epsilon \le \mathcal{U}(\mathcal{Q}, f\alpha') + M\epsilon.$$

Thus by (15) and (16),

$$\mathcal{U}(\mathcal{Q}, f, \alpha) < \mathcal{U}(\mathcal{Q}, f\alpha') + (M+1)\epsilon.$$

Since $f\alpha' \in \mathcal{R}[a, b]$, there exists a partition $\mathcal{Q}$ of $[a, b]$, which is a refinement of $\mathcal{P}$, such that

$$\mathcal{U}(Q, f\alpha') < \int_a^b f\alpha' + \epsilon.$$

Thus

$$\overline{\int_a^b} f \, d\alpha < \int_a^b f\alpha' + (M+2)\epsilon.$$

Since this holds for any $\epsilon > 0$,

$$\overline{\int_a^b} f \, d\alpha \le \int_a^b f\alpha'.$$

A similar argument using lower sums proves the reverse inequality. Thus $f \in \mathcal{R}(\alpha)$ and

$$\int_a^b f \, d\alpha = \int_a^b f\alpha'. \quad \square$$

We now give several examples to illustrate the previous two theorems.

EXAMPLES 6.5.13 (a) For our first example we illustrate the finite version of Theorem 6.5.11; namely, identity (12). Consider

$$\int_0^2 e^x \, d[x].$$

For $x \in [0, 2]$,

$$[x] = I(x - 1) + I(x - 2).$$

Therefore since e^x is continuous, it is integrable with respect to $[x]$, and by (12)

$$\int_0^2 e^x \, d[x] = e^1 + e^2.$$

(b) To illustrate Theorem 6.5.12, if $\alpha(x) = x^2$ on $[0,1]$, then for any Riemann integrable function f on $[0,1]$,

$$\int_0^1 f(x) \, dx^2 = 2 \int_0^1 f(x)x \, dx.$$

In particular, if $f(x) = \sin \pi x$, then

$$\int_0^1 \sin \pi x \, d \, x^2 = 2 \int_0^1 x \sin \pi x \, dx$$

which by the integration by parts formula

$$= -\frac{1}{\pi} x \cos \pi x \Big|_0^1 + \frac{1}{\pi} \int_0^1 \cos \pi x \, dx$$

$$= -\frac{1}{\pi} \cos \pi + \left[\frac{1}{\pi^2} \sin \pi x \Big|_0^1 \right] = \frac{1}{\pi}.$$

(c) As another illustration of Theorem 6.5.12,

$$\int_0^3 [x] d(e^{2x}) = 2 \int_0^3 [x] e^{2x} dx$$

$$= 2 \int_1^2 e^{2x} dx + 2 \int_2^3 2e^{2x} dx = 2e^6 - e^4 - e^2. \quad \square$$

The Riemann-Stieltjes integral has important applications in a variety of areas, including physics and probability theory. It allows us to express diverse formulas as a single expression. To illustrate this, we again consider the moment of inertia problem of Example 6.5.1.

EXAMPLE 6.5.14 In this example, we will show that both formulas of Example 6.5.1 can be expressed as

$$I = \int_0^\ell x^2 \, dm(x),$$

where $m(x)$ denotes the mass of the wire or system from 0 to x. Clearly the function $m(x)$ is nondecreasing, and thus since x^2 is continuous, the above integral exists.

In the first case, since $0 < r_1 < r_2 < \cdots < r_n$, and the masses m_i are located at r_i (see Figure 6.7), $m(x)$ is given by

$$m(x) = \sum_{i=1}^{n} m_i I(x - r_i).$$

Thus by Theorem 6.5.11,

$$\int_0^\ell x^2 \, dm(x) = \sum_{i=1}^{n} r_i^2 \, m_i.$$

On the other hand, if $m(x)$ is differentiable and $m'(x)$ is Riemann integrable, then by Theorem 6.5.12,

$$\int_0^\ell x^2 \, dm(x) = \int_0^\ell x^2 m'(x) \, dx.$$

It only remains to be shown that $m'(x)$ is the density. The density $\rho(x)$ of a wire (mass per unit length) is defined as the limit of the average density. In the interval $[x, x + \Delta x]$, the average density is

$$\frac{m(x + \Delta x) - m(x)}{\Delta x}.$$

Thus if $m(x)$ is differentiable, $\rho(x) = m'(x)$. $\quad\square$

Riemann-Stieltjes Sums

We conclude this section with a few remarks concerning Riemann-Stieltjes sums. As in Definition 6.2.4, let f be a bounded real-valued function on $[a, b]$, α a monotone increasing function on $[a, b]$, and $\mathcal{P} = \{x_0, ..., x_n\}$ a partition of $[a, b]$. For each $i = 1, 2, ..., n$, choose $t_i \in [x_{i-1}, x_i]$. Then

$$\mathcal{S}(\mathcal{P}, f, \alpha) = \sum_{i=1}^{n} f(t_i) \Delta \alpha_i$$

is called a **Riemann-Stieltjes sum** of f with respect to α and the partition $\mathcal{P}$.

A natural question to ask is whether the analogue of Theorem 6.2.6 is valid for the Riemann-Stieltjes integral. Unfortunately, only part of the Theorem holds. If $\lim_{\|\mathcal{P}\| \to 0} \mathcal{S}(\mathcal{P}, f, \alpha) = I$, then $f \in \mathcal{R}(\alpha)$ on $[a, b]$ and $\int_a^b f \, d\alpha = I$ (Exercise 11). The converse, as the following example will illustrate, is <u>false</u>! However, if f and α satisfy either of the hypothesis of Theorem 6.5.6, then

$$\int_a^b f \, d\alpha = \lim_{\|\mathcal{P}\| \to 0} \mathcal{S}(\mathcal{P}, f, \alpha).$$

The result where f is continuous and α is monotone increasing is given as an exercise (Exercise 10).

EXAMPLE 6.5.15 Let f and α be defined on $[0,2]$ as follows:

$$f(x) = \begin{cases} 0, & 0 \le x \le 1, \\ 1, & 1 < x \le 2, \end{cases}$$

$$\alpha(x) = \begin{cases} 0, & 0 < x < 1, \\ 1, & 1 \le x \le 2. \end{cases}$$

Since f is left continuous at 1, by Example 6.5.4(a) $\int_0^2 f \, d\alpha = f(1) = 0$. Suppose $\mathcal{P} = \{x_0, x_1, ..., x_n\}$ is any partition of $[0,2]$ with $1 \notin \mathcal{P}$. Let k be such that $x_{k-1} < 1 < x_k$. Then $\Delta\alpha_k = 1$ and $\Delta\alpha_i = 0$ for all $i \ne k$. If $t_k \in (1, x_k]$, then $\mathcal{S}(\mathcal{P}, f, \alpha) = 1$. On the other hand, if $t_k = 1$, then $\mathcal{S}(\mathcal{P}, f, \alpha) = 0$. As a consequence,

$$\lim_{\|\mathcal{P}\| \to 0} \mathcal{S}(\mathcal{P}, f, \alpha)$$

does not exist. $\square$

Exercises 6.5

1. *Evaluate $\displaystyle\int_{-1}^{1} f(x) \, d\alpha(x)$ where f is bounded on $[-1, 1]$ and continuous at 0, and α is given by

$$\alpha(x) = \begin{cases} -1, & x < 0, \\ 0, & x = 0, \\ 1, & x > 0. \end{cases}$$

2. **a.** In Example 6.5.4(a), prove that if the function f is left continuous at c, then f is integrable with respect to I_c on $[a, b]$.

   b. Show that I_c is not integrable with respect to I_c

3. Let α be non-decreasing on $[a, b]$. Suppose f is bounded on $[a, b]$ and integrable with respect to α on $[a, b]$. For $x \in [a, b]$ set $F(x) = \int_a^x f \, d\alpha$.

   *a. Prove that $|F(x) - F(y)| \le M |\alpha(x) - \alpha(y)|$ for some positive constant M and all $x, y \in [a, b]$.

   b. Prove that if α is continuous at $x_o \in [a, b]$, then F is also continuous at x_o.

4. **a.** Prove Theorem 6.5.8(a).

   b. Prove Theorem 6.5.8(c).

   c. Prove Theorem 6.5.8(d).

5. Use the theorems from the text to compute each of the following integrals:

   *a. $\displaystyle\int_0^{\pi/2} x \, d(\sin x)$.   **b.** $\displaystyle\int_0^3 [x] \, dx^2$.   *c. $\displaystyle\int_0^3 x^2 \, d[x]$.

   d. $\displaystyle\int_1^3 ([x] + x) \, d(x^2 + e^x)$ **e.** $\displaystyle\int_0^1 \sin \pi x \, d[4x]$.   *f. $\displaystyle\int_1^4 (x - [x]) \, dx^3$

6. Verify the integration by parts formula with (b) and (c) of the previous exercise.

7. Find $\int_0^1 f\, d\alpha$, where f is continuous on $[0,1]$ and

$$\alpha(x) = \sum_{n=1}^{\infty} \frac{1}{2^n} I(x - \tfrac{1}{n}).$$

Leave your answer in the form of an infinite series.

8. Let α be as in Exercise 7. Evaluate each of the following integrals. Leave your answers in the form of an infinite series.

*a. $\int_0^1 x\, d\alpha(x)$ **b.** $\int_0^1 \alpha(x)\, dx$

9. Suppose α and β are monotone increasing on $[a,b]$, and $\mathcal{P}$ is a partition of $[a,b]$. Prove that

$$\mathcal{U}(\mathcal{P},\alpha,\beta) = \alpha(b)\beta(b) - \alpha(a)\beta(a) - \mathcal{L}(\mathcal{P},\beta,\alpha).$$

10. Prove that if f is a continuous real-valued function on $[a,b]$ and α is monotone increasing on $[a,b]$, then $\displaystyle\lim_{\|\mathcal{P}\|\to 0} \mathcal{S}(\mathcal{P},f,\alpha)$ exists and

$$\lim_{\|\mathcal{P}\|\to 0} \mathcal{S}(\mathcal{P},f,\alpha) = \int_a^b f\, d\alpha.$$

11. Let f be a bounded real-valued function on $[a,b]$ and α a monotone increasing function on $[a,b]$. Prove the following: If $\displaystyle\lim_{\|\mathcal{P}\|\to 0} \mathcal{S}(\mathcal{P},f,\alpha) = I$,

then $f \in \mathcal{R}(\alpha)$ and $\int_a^b f\, d\alpha = I$.

12. ***a.** Let α be a monotone increasing function on $[a,b]$. If $f \in \mathcal{R}(\alpha)$, prove that $f^2 \in \mathcal{R}(\alpha)$.

    b. If $f,g \in \mathcal{R}(\alpha)$, prove that $fg \in \mathcal{R}(\alpha)$.

13. If $f \in \mathcal{R}(\alpha)$ on $[a,b]$ with Range $f \subset [c,d]$, and φ is continuous on $[c,d]$, prove that $\varphi \circ f \in \mathcal{R}(\alpha)$ on $[a,b]$.

14. Suppose f is a nonnegative continuous function on $[a,b]$, and α is non-decreasing on $[a,b]$ Define the function β on $[a,b]$ by $\beta(x) = \int_1^x f\, d\alpha$. If g is continuous on $[a,b]$, prove that

$$\int_1^b g\, d\beta = \int_a^b f g\, d\alpha.$$

6.6   Numerical Methods[4]

In this section, we will take a brief look at some elementary numerical methods that are useful in obtaining approximations to Riemann integrals. Even

---

[4]The topics of this section may be omitted on first reading.

though the fundamental theorem of calculus provides an easy method for evaluating definite integrals, it is useful only if we can find an antiderivative of the function being integrated. To illustrate this, in Example 6.3.9 we showed that

$$\int_0^2 \frac{t}{1+t^2} dt = \tfrac{1}{2} \ln 5.$$

This however is not particular useful if we do not know the value of $\ln 5$. By Example 6.3.5,

$$\ln 5 = \int_1^5 \frac{1}{t} dt.$$

To obtain an approximation to $\ln 5$, we can choose from several available methods for obtaining numerical approximations to the definite integral.

Approximations Using Riemann Sums

Since upper and lower sums are in general difficult to evaluate, Darboux's definition of the Riemann integral is not particularly useful in obtaining numerical approximations. The most elementary numerical method is to use Riemann sums. One of the first mathematicians to use numerical methods was Euler, who considered sums of the form

$$\sum_{k=1}^n f(x_{k-1})(x_k - x_{k-1})$$

as an approximation to the integral. This is nothing but the Riemann sum for the partition $\mathcal{P} = \{x_0, x_1, ..., x_n\}$ of $[a, b]$ with $t_k = x_{k-1}$ for all k.

In using Riemann sums to approximate the integral of f it is convenient to take equally spaced partitions. Let $n \in \mathbb{N}$, and set $h = (b-a)/n$. Define

$$x_0 = a, \quad x_1 = a+h, \quad x_2 = a+2h, \cdots, \quad x_n = a+nh = b. \tag{17}$$

Thus if $\mathcal{P}_n = \{x_0, x_1, ..., x_n\}$ with x_i as defined, we always have $\Delta x_i = h$ for all $i = 1, ..., n$, and

$$\mathcal{S}(\mathcal{P}_n, f) = \sum_{i=1}^n f(t_i)\Delta x_i = h \sum_{i=1}^n f(t_i),$$

where for each $i = 1, ..., n$, $t_i \in [x_{i-1}, x_i]$. If we take $t_i = x_{i-1}$ for all i, then we obtain the above formula of Euler. Similarly, we could take t_i to be the right endpoint x_i of the interval $[x_{i-1}, x_i]$. Another choice of t_i would be the midpoint; i.e., $t_i = (x_{i-1} + x_i)/2$. For monotone functions, it is intuitively obvious that the midpoint gives a better approximation than either the right or left endpoint to the integral of f over the interval $[x_{i-1}, x_i]$ (see Figure 6.8). With x_i as defined by (17) and $t_i = (x_{i-1} + x_i)/2$ the above formula becomes

$$M_n(f) = h \sum_{i=1}^n f(a + (i - \tfrac{1}{2})h). \tag{18}$$

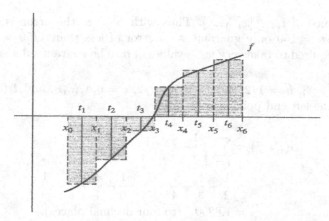

FIGURE 6.8
Midpoint approximation

The quantity $M_n(f)$ is called the *nth* **midpoint approximation** to the integral of f over $[a, b]$.

Regardless of the method used, it is important to be able to estimate the error between the true value and the approximate value. If $f \in \mathcal{R}[a, b]$, then for any partition $\mathcal{P}$ of $[a, b]$, we always have

$$\left| \int_a^b f(x)dx - \mathcal{S}(\mathcal{P}, f) \right| \leq \mathcal{U}(\mathcal{P}, f) - \mathcal{L}(\mathcal{P}, f) \tag{19}$$

for any choice of $t_k \in [x_{k-1}, x_k]$. Inequality (19) follows from the fact that both the Riemann sum of f and the integral of f lie between the lower and upper sum of f. If f is monotone increasing on $[a, b]$, then by the proof of Theorem 6.1.8(b),

$$\mathcal{U}(\mathcal{P}_n, f) - \mathcal{L}(\mathcal{P}_n, f) = h[f(b) - f(a)].$$

Thus by inequality (19),

$$\left| \int_a^b f(x)dx - h \sum_{i=1}^n f(t_i) \right| \leq h|f(b) - f(a)| \tag{20}$$

for any choice of $t_i \in [x_{i-1}, x_i]$. Although we proved (20) for a monotone increasing function, the inequality is also valid for a monotone decreasing function. Thus for monotone functions, inequality (20) provides an estimate on the error between the true value and the approximate value.

EXAMPLE 6.6.1 The function $f(x) = 1/x$ is decreasing on $[1, 5]$. Thus by inequality (20), for $n \in \mathbb{N}$ and $h = (b - a)/n$,

$$\left| \ln 5 - h \sum_{i=1}^n f(t_i) \right| \leq \frac{4}{n} \left| \frac{1}{5} - 1 \right| = \frac{16}{5} \frac{1}{n}$$

for any choice of $t_i \in [x_{i-1}, x_i]$. Thus with $n = 8$, the error is less than $2/5 = 0.4$; $n = 160$ only guarantees an error of less than $1/50 = 0.02$. We would be required to take very large values of n to be guaranteed a sufficiently small error.

With $n = 8$, $h = 1/2$ and $x_i = 1 + i/2$, $i = 0, 1, ..., 8$, and $t_i = x_{i-1} = (1+i)/2$ (the left end point),

$$S(\mathcal{P}_8, f) = \sum_{i=1}^{8} \frac{1}{1+i}$$
$$= \frac{1}{2} + \frac{1}{3} + \frac{1}{4} + \frac{1}{5} + \frac{1}{6} + \frac{1}{7} + \frac{1}{8} + \frac{1}{9}$$
$$= 1.8290 \quad \text{(to four decimal places)}.$$

To four decimal places, $\ln 5 = 1.6094$. Thus the error is 0.2196 which is less than the predicted error of 0.4.

If we use the midpoint approximation (18) (again with $n = 8$), then $t_i = 1 + (i - \frac{1}{2})\frac{1}{2}$ which upon simplification equals $(3 + 2i)/4$. Thus

$$M_8(f) = \frac{1}{2} \sum_{i=1}^{8} \frac{4}{3 + 2i}$$
$$= 2 \left(\frac{1}{5} + \frac{1}{7} + \frac{1}{9} + \frac{1}{11} + \frac{1}{13} + \frac{1}{15} + \frac{1}{17} + \frac{1}{19} \right)$$
$$= 1.5998 \quad \text{(to four decimal places)}.$$

Using the midpoint approximation the error is less than 0.01. This is considerably better than predicted by inequality (20). As we will shortly see, this improved accuracy is not a coincidence. $\square$

Inequality (20) provides an error estimate for monotone functions. For arbitrary real-valued functions, under the additional hypothesis that the derivative is bounded, we have the following theorem.

THEOREM 6.6.2 *Let f be a real-valued differentiable function on $[a, b]$ for which $f'(x)$ is bounded on $[a, b]$. Then for any partition $\mathcal{P}$ of $[a, b]$,*

$$\left| \int_a^b f(x)dx - S(\mathcal{P}, f) \right| \leq \|\mathcal{P}\| M(b - a),$$

where $M = \sup\{|f'(x)| : x \in [a, b]\}$.

Proof. Suppose $\mathcal{P} = \{x_0, x_1, ..., x_n\}$ is a partition of $[a, b]$. Then

$$\mathcal{U}(\mathcal{P}, f) - \mathcal{L}(\mathcal{P}, f) = \sum_{i=1}^{n} (M_i - m_i)\Delta x_i.$$

Since f is continuous on $[a, b]$, for each i there exist $s_i, s_i' \in [x_{i-1}, x_i]$ such that

$$M_i - m_i = f(s_i) - f(s_i').$$

By the mean value theorem, $f(s_i) - f(s_i') = f'(t_i)(s_i - s_i')$ for some t_i between s_i and s_i'. Therefore, since $|f'(t_i)||s_i - s_i'| \leq M \|\mathcal{P}\|$,

$$\mathcal{U}(\mathcal{P}, f) - \mathcal{L}(\mathcal{P}, f) \leq M \|\mathcal{P}\| \sum_{i=1}^{n} \Delta x_i = \|\mathcal{P}\| M(b - a).$$

The result now follows by inequality (19). $\square$

If f satisfies the hypothesis of the theorem, $h = (b - a)/n$, $n \in \mathbb{N}$, and the x_i are defined by (17), then

$$\left| \int_a^b f(x)dx - h \sum_{i=1}^{n} f(t_i) \right| \leq M(b - a)\, h$$

for any choice of $t_i \in [x_{i-1}, x_i]$. A slight improvement of this inequality is given in the exercises (Exercise 6). If $E_n(f)$ denotes the **error** between the true value and the approximate value, that is,

$$E_n(f) = \int_a^b f(x)\, dx - \mathcal{S}(\mathcal{P}_n, f),$$

then the above inequality can simply be written as $|E_n(f)| \leq C\, h$, where C is a fixed constant depending only on f and the interval $[a, b]$. Since the term "h" occurs to the first power, this method is commonly referred to as a **first order method**.

In Example 6.6.1 we saw that the midpoint rule provided much greater accuracy than using the left end point. This is not a coincidence as the following theorem proves.

THEOREM 6.6.3 *If f is twice differentiable on $[a, b]$ and $f''(x)$ is bounded on $[a, b]$, then*

$$\left| \int_a^b f(x) - M_n(f) \right| \leq \frac{M(b - a)}{24}\, h^2,$$

where $M = \sup\{|f''(x)| : x \in [a, b]\}$.

Remark. Since the error between the true value and the approximate value involves the term h^2, the midpoint approximation is a **second order method**.

Proof. To prove the result, we first prove that for each $i = 1, ..., n$,

$$\left| \int_{x_{i-1}}^{x_i} f(x)dx - h\, f(a + (i - \tfrac{1}{2})h) \right| \leq \frac{M}{24} h^3. \tag{21}$$

For $t \in [0, h/2]$, consider the function

$$g_i(t) = \int_{c_i - t}^{c_i + t} f(x)dx - 2t\, f(c_i),$$

where $c_i = (x_{i-1} + x_i)/2$. Then

$$g_i\!\left(\tfrac{h}{2}\right) = \int_{x_{i-1}}^{x_i} f(x)\, dx - h\, f(a + (i - \tfrac{1}{2})h).$$

Since f is continuous, by Theorem 6.3.4,

$$g_i'(t) = f(c_i + t) + f(c_i - t) - 2f'(c_i), \quad \text{and}$$
$$g_i''(t) = f'(c_i + t) - f'(c_i - t).$$

By the mean value theorem

$$f'(c_i + t) - f'(c_i - t) = f''(\zeta)2t$$

for some $\zeta \in (c_k - t, c_k + t)$. Since $|f''(\zeta)| \leq M$, we obtain

$$|g_i''(t)| \leq 2Mt$$

for all $t \in [0, h/2]$. Since $g_i'(0) = 0$, by the fundamental theorem of calculus,

$$|g_i'(t)| = \left| \int_0^t g_i''(x)dx \right| \leq \int_0^t |g_i''(x)|dx \leq 2M \int_0^t x\, dx = Mt^2.$$

Also, since $g_i(0) = 0$, by the fundamental theorem of calculus again,

$$|g_i\!\left(\tfrac{h}{2}\right)| = \left| \int_0^{h/2} g_i'(t)dt \right| \leq \int_0^{h/2} |g_i'(t)|dt \leq \frac{M}{3}\left(\frac{h}{2}\right)^3 = \frac{M}{24}h^3,$$

which proves inequality (21). Therefore,

$$\left| \int_a^b f(x)dx - M_n(t) \right| = \left| \sum_{i=1}^n \int_{x_{i-1}}^{x_i} f(x)dx - hf(a + (i - \tfrac{1}{2})h) \right| = \left| \sum_{i=1}^n g_i\!\left(\tfrac{h}{2}\right) \right|$$

$$\leq \sum_{i=1}^n |g_i\!\left(\tfrac{h}{2}\right)| \leq \frac{M}{24}h^3\, n = \frac{M(b-a)}{24}h^2. \quad \square$$

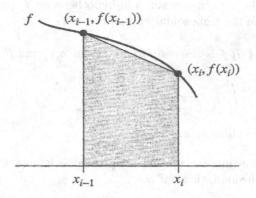

FIGURE 6.9
Trapezoidal approximation

The Trapezoidal Rule

We now consider another common second order approximation method known as the **trapezoidal rule**. In using Riemann sums, regardless of the choice of the points t_i, we used rectangles to approximate the integral of the function f. In our second numerical method we will replace rectangles by trapezoids. Let f be a Riemann integrable function on $[a, b]$ and let $\mathcal{P} = \{x_0, x_1, ..., x_n\}$ be a partition of $[a, b]$. As in Figure 6.9, for each interval $[x_{i-1}, x_i]$, the area of the trapezoid formed by the points $(x_{i-1}, 0)$, $(x_{i-1}, f(x_{i-1}))$, $(x_i, f(x_i))$, $(x_i, 0)$ is given by $\frac{1}{2}[f(x_{i-1}) + f(x_i)]\Delta x_i$. Summing these up gives

$$\frac{1}{2}\sum_{i=1}^{n}[f(x_{i-1}) + f(x_i)]\Delta x_i$$

as an approximation to the integral of f on $[a, b]$. If as previously we set $h = (b - a)/n$ and $x_i = a + ih$, $i = 0, ..., n$, then the above sum becomes

$$\frac{h}{2}\sum_{i=1}^{n}[f(a + (i-1)h) + f(a + ih)] = \frac{h}{2}\left[f(a) + 2\sum_{i=1}^{n-1}f(a + ih) + f(b)\right].$$

If f is Riemann integrable on $[a, b]$ and $n \in \mathbb{N}$, the quantity $T_n(f)$ defined by

$$T_n(f) = \frac{h}{2}\left[f(a) + 2\sum_{i=1}^{n-1}f(a + ih) + f(b)\right] \tag{22}$$

where $h = (b - a)/n$, is called the *n*th **trapezoidal approximation** to the integral of f on $[a, b]$. If we set $y_i = f(a + ih)$, then $T_n(f)$ can be expressed as

$$T_n(f) = \frac{h}{2}\left[y_0 + 2\sum_{i=1}^{n-1}y_i + y_n\right].$$

The following theorem, under suitable hypothesis on f, provides an estimate on the error of the trapezoidal approximation.

THEOREM 6.6.4 *If f is twice differentiable on $[a,b]$ and $f''(x)$ is bounded on $[a,b]$, then*

$$\left| \int_a^b f(x)dx - T_n(f) \right| \le \frac{M(b-a)}{12}h^2,$$

where $M = \sup\{|f''(x)| : x \in [a,b]\}$.

The above error estimate can also be expressed as follows: If f satisfies the hypothesis of the theorem, then for $n \in \mathbb{N}$,

$$\left| \int_a^b f(x)dx - T_n(f) \right| \le \frac{M(b-a)^3}{12}\frac{1}{n^2}. \qquad (23)$$

In this form it is possible to determine the value of n required to guarantee predetermined accuracy.

Proof. The proof of this theorem is very similar to the proof of Theorem 6.6.3. As a consequence we leave most of the details to the exercises (Exercise 5). The first step is to prove that for each $i = 1, ..., n$,

$$\left| \int_{x_{i-1}}^{x_i} f(x)dx - \frac{h}{2}[f(x_{i-1}) + f(x_i)] \right| \le \frac{M}{12}h^3.$$

To accomplish this, consider the function

$$g_i(t) = \int_{x_{i-1}}^{x_{i-1}+t} f(x)dx - \frac{t}{2}[f(x_{i-1}) + f(x_{i-1}+t)]$$

for $t \in [0, h]$. Then $g_i(0) = 0$ and

$$g_i(h) = \int_{x_{i-1}}^{x_i} f(x)dx - \frac{h}{2}[f(x_{i-1}) + f(x_i)].$$

By computing $g_i'(t)$ and $g_i''(t)$ one obtains from the hypothesis on f that $|g_i''(t)| \le \frac{1}{2}tM$. Since $g_i(0) = g_i'(0) = 0$, applying the fundamental theorem of calculus twice we obtain as in Theorem 6.6.3 that $|g_i(h)| \le \frac{1}{12}Mh^3$. The remainder of the proof is identical to Theorem 6.6.3. $\square$

Simpson's Rule

The trapezoidal approximation $T_n(f)$ amounts to approximating the function f with a piecewise linear function g_n that passes through the points $\{(x_i, f(x_i))\}$, $i = 0, \ldots, n$. Our intuition should convince us that one way to obtain a better approximation to the integral of f over $[a, b]$ is to use smoother curves. This is exactly what is done in **Simpson's rule** which uses parabolas to approximate the integral. To use quadratic approximations we will need to use three successive points of the partition of $[a, b]$. This is due to the fact that three points are required to uniquely determine a parabola.

Prior to deriving Simpson's rule, we first establish the following formula: If $p(x) = Ax^2 + Bx + C$ is the quadratic function passing through the points $(0, y_0)$, (h, y_1), $(2h, y_2)$, then

$$\int_0^{2h} p(x)dx = \frac{h}{3}[y_0 + 4y_1 + y_2]. \tag{24}$$

One way to derive this formula would be to first determine the coefficients A, B, C so that $p(0) = y_0$, $p(h) = y_1$, $p(2h) = y_2$, and then integrate $p(x)$. This however is not necessary. By integrating first,

$$\int_0^{2h} p(x)\, dx = \frac{A}{3}(2h)^3 + \frac{B}{2}(2h)^2 + C(2h)$$

$$= \frac{h}{3}[8Ah^2 + 6Bh + 6C]$$

$$= \frac{h}{3}[p(0) + 4p(h) + p(2h)] = \frac{h}{3}[y_0 + 4y_1 + y_2].$$

Let $f \in \mathcal{R}[a, b]$ and let $n \in \mathbb{N}$ be even. Set $h = (b - a)/n$. On each of the intervals $[a + 2(i - 1)h, a + 2ih]$, $i = 1, \ldots, n/2$, we approximate the integral of f by the integral of the quadratic function that agrees with f at the points

$$y_0 = f(a + 2(i - 1)h), \quad y_1 = f(a + (2i - 1)h), \quad y_2 = f(a + 2ih).$$

By identity (24) this gives

$$\frac{h}{3}[f(a + 2(i - 1)h) + 4f(a + (2i - 1)h) + f(a + 2ih)]$$

as an approximation to the integral of f over the interval $[a + 2(i-1)h, a + 2ih]$. Summing these terms from $i = 1$ to $n/2$ gives

$$S_n(f) = \frac{h}{3}[f(a) + 4f(a + h) + 2f(a + 2h) +$$

$$+ 4f(a + 3h) + \cdots + 4f(a + (n - 1)h) + f(b)]$$

as an approximation to the integral of f over $[a, b]$. The quantity $S_n(f)$ is called the *nth* **Simpson approximation** to the integral of f over $[a, b]$. If we set $y_i = f(a + ih)$, then $S_n(f)$ is given by

$$S_n(f) = \frac{h}{3}[y_0 + 4y_1 + 2y_2 + 4y_3 + \cdots + 2y_{n-2} + 4y_{n-1} + y_n]. \qquad (25)$$

The following theorem, again under suitable restrictions on f, provides an error estimate for Simpsons's rule.

THEOREM 6.6.5 *If f is four times differentiable on $[a, b]$ and $f^{(4)}(x)$ is bounded on $[a, b]$, then for $n \in \mathbb{N}$ even,*

$$\left| \int_a^b f(x)dx - S_n(f) \right| \le \frac{M(b-a)}{180} h^4,$$

where $M = \sup\{|f^{(4)}(x)| : x \in [a, b]\}$.

Remarks. Since the error term involves h^4, Simpson's rule is a **fourth order method**. Also, if $f(x)$ is a polynomial of degree less than or equal to 3, then $f^{(4)}(x) = 0$ and thus

$$\int_a^b f(x)\, dx = S_n(f).$$

If f satisfies the hypothesis of the theorem and $n \in \mathbb{N}$ is even, then the above error estimate can be expressed as

$$\left| \int_a^b f(x)dx - S_n(f) \right| \le \frac{M(b-a)^5}{180} \frac{1}{n^4}. \qquad (26)$$

Proof. The proof of the theorem proceeds in an analogous manner as the proofs of the previous two results. We first prove that for $i = 1, ..., n/2$,

$$\left| \int_{c_i - h}^{c_i + h} f(x)dx - \frac{h}{3}[f(c_i - h) + 4f(c_i) + f(c_i + h)] \right| \le \frac{M}{90} h^5, \qquad (27)$$

where $c_i = a + (2i - 1)h$. To accomplish this, define g_i on $[0, h]$ by

$$g_i(t) = \int_{c_i - t}^{c_i + t} f(x)dx - \frac{t}{3}[f(c_i - t) + 4f(c_i) + f(c_i + t)].$$

To prove inequality (27), we are required to show that $|g_i(h)| \le Mh^5/90$. Upon computing the successive derivatives of g_i we obtain

$$g_i'(t) = \frac{t}{3}[f'(c_i - t) + f'(c_i + t)] + \frac{2}{3}[f(c_i - t) + f(c_i + t)] - \frac{4}{3}f(c_i),$$

$$g_i''(t) = \frac{t}{3}[-f''(c_i - t) - f''(c_i + t)] + \frac{1}{3}[-f'(c_i - t) + f'(c_i + t)], \quad \text{and}$$

$$g_i'''(t) = \frac{t}{3}[f'''(c_i - t) - f'''(c_i + t)].$$

By the mean value theorem $f'''(c_i - t) - f'''(c_i + t) = f^{(4)}(\zeta)(2t)$ for some $\zeta \in (c_i - t, c_i + t)$. Therefore

$$|g_i'''(t)| \leq \frac{2}{3} M t^2.$$

As in the previous two theorems, since $g_i(0) = g_i'(0) = g_i''(0) = 0$, upon three integrations we have $|g_i(h)| \leq M h^5 / 90$. This proves inequality (27). Finally,

$$\left| \int_a^b f(x)dx - S_n(f) \right| = \left| \sum_{i=1}^{n/2} g_i(x) \right| \leq \sum_{i=1}^{n/2} |g_i(x)|$$

$$\leq \frac{M h^5}{90} \left(\frac{n}{2} \right) = \frac{M(b-a)}{180} h^4. \quad \square$$

EXAMPLE 6.6.6 In this example, we will use the trapezoidal rule and Simpson's rule with $n = 8$ to obtain approximations to $\ln 5$. For $f(x) = x^{-1}$, $f''(x) = 2x^{-3}$ and $f^{(4)}(x) = 24x^{-5}$. Therefore

$$\sup\{|f''(x)| : x \in [1,5]\} = 2 \quad \text{and} \quad \sup\{|f^{(4)}(x)| : x \in [1,5]\} = 24.$$

By Theorem 6.6.4 with $h = 1/2$, the error E_8 in the trapezoidal approximation satisfies

$$|E_8(f)| \leq \frac{2 \cdot 4}{12} \left(\frac{1}{2} \right)^2 = \frac{1}{6} = 0.16666..$$

On the other hand, the error E_8 in using Simpson's rule is guaranteed to satisfy

$$|E_8(f)| \leq \frac{24 \cdot 4}{180} \left(\frac{1}{2} \right)^4 = \frac{1}{30} = 0.03333...$$

which is considerably better.

With $x_i = 1 + ih$ and $y_i = f(x_i)$, $i = 0, 1, ..., 8$,

$$y_0 = 1, \ y_1 = \frac{2}{3}, \ y_2 = \frac{1}{2}, \ y_3 = \frac{2}{5}, \ y_4 = \frac{1}{3}, \ y_5 = \frac{2}{7}, \ y_6 = \frac{1}{4}, \ y_7 = \frac{2}{9}, \ y_8 = \frac{1}{5}.$$

Therefore,

$$T_8(f) = \frac{1}{4} \left[y_0 + 2 \sum_{i=1}^{7} y_i + y_8 \right]$$

$$= \frac{1}{4} \left[1 + \frac{4}{3} + 1 + \frac{4}{5} + \frac{2}{3} + \frac{4}{7} + \frac{1}{2} + \frac{4}{9} + \frac{1}{5} \right]$$

$$= 1.6290 \quad \text{(to four decimal places)}.$$

Since $\ln 5 = 1.6094$ to four decimal places, the error is less than 0.02, well within the tolerance predicted by the theory. With Simpson's rule,

$$S_8(f) = \frac{1}{6}\left[y_0 + 4y_1 + 2y_2 + 4y_3 + 2y_4 + 4y_5 + 2y_6 + 4y_7 + y_9\right]$$

$$= \frac{1}{6}\left[1 + \frac{8}{3} + 1 + \frac{8}{5} + \frac{2}{3} + \frac{8}{7} + \frac{1}{2} + \frac{8}{9} + \frac{1}{5}\right]$$

$$= 1.6108 \quad \text{(to four decimal places)}.$$

With Simpson's rule the actual error is less than 0.0014.

We can also use the results of the theory to determine how large n must be chosen to guarantee predetermined accuracy. For the trapezoidal rule, since $M = 2$ and $(b - a) = 4$, by inequality (23)

$$|E_n(f)| \leq \frac{2 \cdot 4^3}{12} \frac{1}{n^2}.$$

Thus to obtain accuracy to within 0.001 we need

$$n^2 > \frac{2 \cdot 4^3}{12} 10^3 = 10666.66..$$

which is accomplished with $n \geq 104$. On the other hand, using Simpson's rule,

$$|E_n(f)| \leq \frac{24 \cdot 4^5}{180} \frac{1}{n^4}.$$

Thus to obtain accuracy to within 0.001, we need $n \in \mathbb{N}$ even with

$$n^4 > \frac{24 \cdot 4^5}{180} 10^3.$$

This will be satisfied with $n \geq 20$.   □

Exercises 6.6

1. **a.** Use the midpoint rule, trapezoidal rule, and Simpson's rule to approximate $\ln 2 = \int_1^2 (1/x)\, dx$ with $n = 4$. For each method, determine the estimated error and compare your answer to $\ln 2 = 0.69315$ (to five decimal places).

    b. Repeat (a) with $n = 8$.

    c. For each of the three methods (midpoint rule, trapezoidal rule, Simpson's rule) determine how large n must be chosen to assure accuracy in the approximation of $\ln 2$ to within 0.0001.

2. ***a.** Using the fact that

$$\int_0^1 \frac{dt}{1+t^2} = \frac{\pi}{4},$$

use the midpoint rule, trapezoidal rule, and Simpson's rule to approximate $\pi/4$ with $n = 4$. For each method, determine the estimated error.

    b. Repeat (a) with $n = 8$.

    c. For each of the three methods determine how large n must be chosen to obtain an approximation of π to within 0.0001.

3. Use Simpson's rule to obtain approximations of each of the following integrals accurate to at least four decimal places.

    a. $\displaystyle\int_0^1 \frac{1}{1+x^4}\,dx$ ***b.** $\displaystyle\int_0^2 \sqrt{1+x^2}\,dx$ **c.** $\displaystyle\int_0^1 \sin(x^2)\,dx.$

4. How large must n be chosen so that the trapezoidal approximation T_n approximates $\int_0^2 e^{-x^2}$ with an error less than 10^{-6}.

5. Fill in the details of the proof of Theorem 6.6.4.

6. Let f be a differentiable function on $[a, b]$ with $f'(x)$ bounded on $[a, b]$. Let $n \in \mathbb{N}$. Prove that

$$\left| \int_a^b f(x)dx - h \sum_{i=1}^n f(a + ih) \right| \le \frac{M(b-a)}{2}h,$$

where $h = (b-a)/n$ and $M = \sup\{|f'(x)| : x \in [a, b]\}$.

7. **a.** Show that $T_{2n}(f) = \frac{1}{2}[M_n(f) + T_n(f)]$.

    b. Show that $S_{2n}(f) = \frac{2}{3}M_n(f) + \frac{1}{3}T_n(f)$.

8. Prove the following variation of Theorem 6.6.4: If f is twice differentiable on $[a, b]$ and f'' is continuous on $[a, b]$, then there exists a point $c \in [a, b]$ such that

$$T_n(f) - \int_a^b f(x)\,dx = \frac{(b-a)h^2}{12}f''(c).$$

9. **a.** Use the previous exercise to show that $\dfrac{1}{3072} < T_8 - \ln 2 < \dfrac{1}{384}$.

    b. Using (a) and Exercise 1(b) show that $.6915 < \ln 2 < .6938$.

6.7 Proof of Lebesgue's Theorem

In this final section, we present a self-contained proof of Lebesgue's characterization of the Riemann integrable functions on $[a, b]$. Recall that a subset E of $\mathbb{R}$ has **measure zero** if for any $\epsilon > 0$, there exists a finite or countable collection $\{I_n\}$ of open intervals with $E \subset \bigcup_n I_n$ and $\sum_n \ell(I_n) < \epsilon$, where $\ell(I_n)$ denotes the length of the interval I_n. We begin with several preparatory lemmas.

LEMMA 6.7.1 *A finite or countable union of sets of measure zero has measure zero.*

Proof. We will prove the lemma for the case of a countable sets of measure zero. The result for a finite union is an immediate consequence.

Suppose $\{E_n\}_{n\in\mathbb{N}}$ is a countable collection of sets of measure zero. Set $E = \bigcup_n E_n$ and let $\epsilon > 0$ be given. Since each set E_n is a set of measure zero, for each $n \in \mathbb{N}$ there exists a finite or countable collection $\{I_{n,k}\}_k$ of open intervals such that $E_n \subset \bigcup_k I_{n,k}$ and $\sum_k \ell(I_{n,k}) < \epsilon/2^k$. Since we can always take $I_{n,k}$ to be the empty set, there is no loss of generality in assuming that the collection $\{I_{n,k}\}_k$ is countable. Then $\{I_{n,k}\}_{n,k}$ is again a countable collection of open intervals with $E \subset \bigcup_{n,k} I_{n,k}$.

Since $\mathbb{N} \times \mathbb{N}$ is countable, there exists a one-to-one function f from $\mathbb{N}$ onto $\mathbb{N}\times\mathbb{N}$. For each $m \in \mathbb{N}$, set $J_m = I_{f(m)}$. Then $\{J_m\}_{m\in\mathbb{N}}$ is a countable collection of open intervals with $E \subset \bigcup_m J_m$. Since f is one-to-one, for each $N \in \mathbb{N}$, the set $F_N = f(\{1, \ldots, N\})$ is a finite subset of $\mathbb{N} \times \mathbb{N}$. Hence there exists positive integers N_1 and K_1 such that for all $(n, k) \in F_N$ we have $1 \le n \le N_1$ and $1 \le k \le K_1$. Hence

$$\sum_{m=1}^{N} \ell(J_m) = \sum_{(n,k)\in F_N} \ell(I_{n,k}) \le \sum_{(n,k)\in N_1\times K_1} \ell(I_{n,k}).$$

But

$$\sum_{(n,k)\in N_1\times K_1} \ell(I_{n,k}) = \sum_{n=1}^{N_1}\sum_{k=1}^{K_1} \ell(I_{n,k}) \le \sum_{n=1}^{N_1}\sum_{k=1}^{\infty} \ell(I_{n,k}) < \sum_{n=1}^{N_1} \frac{\epsilon}{2^n} < \epsilon.$$

Thus $\sum_{m=1}^{\infty} \ell(J_m) < \epsilon$. Therefore E has measure zero. $\quad\square$

LEMMA 6.7.2 *Suppose f is a nonnegative Riemann integrable function on $[a,b]$ with $\int_a^b f = 0$. Then $\{x \in [a,b] : f(x) > 0\}$ has measure zero.*

Proof. We first prove that for each $c > 0$, the set $E_c = \{x \in [a,b] : f(x) \ge c\}$ has measure zero. Let $\epsilon > 0$ be given. Since $\int_a^b f = 0$, there exists a partition $\mathcal{P} = \{x_0, x_1, \ldots, x_n\}$ of $[a,b]$ with $\mathcal{U}(\mathcal{P}, f) < c\epsilon$, where

$$\mathcal{U}(\mathcal{P}, f) = \sum_{i=1}^{n} M_i \Delta x_i,$$

and $M_i = \sup\{f(x) : x \in [x_{i-1}, x_i]\}$. Let $I = \{i : E_c \cap [x_{i-1}, x_i] \ne \emptyset\}$. If $i \in I$, then $M_i \ge c$. Hence

$$c\epsilon > \mathcal{U}(\mathcal{P}, f) \ge \sum_{i\in I} M_i \Delta x_i \ge c \sum_{i\in I} \Delta x_i.$$

Thus $\sum_{i\in I}\ell([x_{i-1}, x_1]) = \sum_{i\in I}\Delta x_i < \epsilon$. Finally, since $E \subset \bigcup_{i\in I}[x_{i-1}, x_i]$, it follows that E_c has measure zero.

To conclude the proof we note that $\{x \in [a, b] : f(x) > 0\} = \bigcup_{n\in\mathbb{N}} E_n$, where for each $n \in \mathbb{N}$,

$$E_n = \left\{x \in [a, b] : f(x) \geq \frac{1}{n}\right\}.$$

by the above each E_n has measure zero. Thus by Lemma 6.7.1, the set E has measure zero. $\square$

DEFINITION 6.7.3 *If E is a subset of $\mathbb{R}$, the* **characteristic function** *of E, denoted χ_E, is the function defined by*

$$\chi_E(x) = \begin{cases} 1, & x \in E, \\ 0, & x \notin E. \end{cases}$$

As in Section 6.2, if $\mathcal{P} = \{x_0, x_1, \ldots, x_n\}$ is a partition of $[a, b]$, the **norm** $\|\mathcal{P}\|$ of the partition $\mathcal{P}$ is defined by $\|\mathcal{P}\| = \max_{1\leq i\leq n} \Delta x_i$. If f is a bounded function on $[a, b]$, we denote by m_i and M_i the infimum and supremum respectively of f on $[x_{i-1}, x_i]$. The **lower function** L_f and **upper function** U_f for f and the partition $\mathcal{P}$ are defined respectively by

$$L_f(x) = \sum_{i=1}^{n} m_i\chi_{[x_{i-1}, x_i)}(x), \qquad \text{and}$$

$$U_f(x) = \sum_{i=1}^{n} M_i\chi_{[x_{i-1}, x_i)}(x).$$

The graphs of the lower function L_f and upper function U_f are depicted in Figure 6.10. Since $m_i \leq f(x) \leq M_i$ for all $x \in [x_{i-1}, x_i)$,

$$L_f(x) \leq f(x) \leq U_f(x) \qquad \text{for all } x \in [a, b).$$

Also, since L_f and U_f are continuous except at a finite number of points, they are Riemann integrable on $[a, b]$ (Exercise 1) with

$$\int_a^b L_f = \mathcal{L}(\mathcal{P}, f) \quad \text{and} \quad \int_u^b U_f = \mathcal{U}(\mathcal{P}, f).$$

THEOREM 6.7.4 (Lebesgue) *A bounded real valued function f on $[a, b]$ is Riemann integrable on $[a, b]$ if and only if the set of discontinuities of f has measure zero.*

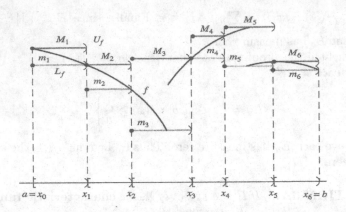

FIGURE 6.10
Graphs of the lower and upper function of f

Proof. Assume first that f is Riemann integrable on $[a, b]$. Then for each $n \in \mathbb{N}$ there exists a partition $\mathcal{P}_n$ of $[a, b]$ with $\|\mathcal{P}_n\| < 1/n$, such that

$$0 \le \int_a^b f - \mathcal{L}(\mathcal{P}_n, f) < \frac{1}{n} \quad \text{and} \quad 0 \le \mathcal{U}(\mathcal{P}_n, f) - \int_a^b f < \frac{1}{n}. \qquad (28)$$

Since $\mathcal{L}(\mathcal{P}_n, f) \le \mathcal{L}(\mathcal{P}, f)$ and $\mathcal{U}(\mathcal{P}, f) \le \mathcal{U}(\mathcal{P}_n, f)$ for any refinement $\mathcal{P}$ of $\mathcal{P}_n$, the partitions $\mathcal{P}_n$, $n = 1, 2, \ldots$ can be chosen so that $\mathcal{P}_{n+1}$ is a refinement of $\mathcal{P}_n$.

For each $n \in \mathbb{N}$, let L_n and U_n respectively denote the lower and upper functions of f for the partition $\mathcal{P}_n$. Then $L_n(x) \le f(x) \le U_n(x)$ for all $x \in [a, b]$ and

$$\int_a^b L_n = \mathcal{L}(\mathcal{P}_n, f) \quad \text{and} \quad \int_a^b U_n = \mathcal{U}(\mathcal{P}_n, f). \qquad (29)$$

Since $\mathcal{P}_{n+1}$ is a refinement of $\mathcal{P}_n$, the functions L_n and U_n satisfy $L_n(x) \le L_{n+1}(x)$ and $U_{n+1}(x) \le U_n(x)$ for all $x \in [a, b]$. Define the function L and U on $[a, b]$ by

$$L(x) = \lim_{n \to \infty} L_n(x) \quad \text{and} \quad U(x) = \lim_{n \to \infty} U_n(x).$$

Then

$$L_n(x) \le L(x) \le f(x) \le U(x) \le U_n(x)$$

for all $x \in [a, b]$. Hence

$$\int_a^b L_n \le \underline{\int_a^b} L \le \overline{\int_a^b} L \le \int_a^b f \le \underline{\int_a^b} U \le \overline{\int_a^b} U \le \int_a^b U_n.$$

But by equations (28) and (29),

$$\lim_{n \to \infty} \int_a^b L_n = \lim_{n \to \infty} \int_a^b U_n = \int_a^b f.$$

Hence L and U are Riemann integrable on $[a, b]$ with

$$\int_a^b L = \int_a^b U = \int_a^b f.$$

Since $U(x) - L(x) \geq 0$ for all $x \in [a, b]$, by Lemma 6.7.2, $\{x \in [a, b] : U(x) \neq L(x)\}$ has measure zero. Furthermore, since each $\mathcal{P}_n$ has measure zero, by Lemma 6.7.1, the set

$$E = \{x \in [a, b] : U(x) \neq L(x)\} \bigcup \left(\bigcup_n \mathcal{P}_n \right)$$

also has measure zero. We conclude by showing that f is continuous on $[a, b] \setminus E$.

Fix $x_o \in [a, b] \setminus E$, and let $\epsilon > 0$ be given. Since $L(x_o) = U(x_o)$, there exists an integer $k \in \mathbb{N}$ such that $U_k(x_o) - L_k(x_o) < \epsilon$. Also, since $x_o \notin \mathcal{P}_k$, the functions U_k and L_k are constant in a neighborhood of x_o. Hence there exists a $\delta > 0$ such that

$$0 \leq U_k(x_o) - L_k(x) = U_k(x) - L_k(x_o) < \epsilon$$

for all $x \in [a, b]$ with $|x - x_o| < \delta$. Finally, since $L_k(x) \leq f(x) \leq U_k(x)$ for all $x \in [a, b)$,

$$-\epsilon < L_k(x) - U_k(x_o) \leq f(x) - f(x_o) < U_k(x) - L_k(x_o) < \epsilon$$

for all x with $|x - x_o| < \delta$. Therefore f is continuous at x_o.

Conversely, suppose f is continuous on $[a, b] \setminus E$, where E is a set of measure zero and $a < b$. Let $M > 0$ be such that $|f(x)| \leq M$ for all $x \in [a, b]$, and let $\epsilon > 0$ be given. Since E has measure zero, there exists a finite or countable collection $\{I_n\}$ of open intervals such that $E \subset \bigcup_n I_n$ and $\sum_n \ell(I_n) < \epsilon/4M$. Also, since f is continuous on $[a, b] \setminus E$, for each $x \in [a, b]$ there exists an open interval J_x such that $|f(z) - f(y)| < \epsilon/2(b - a)$ for all $y, z \in J_x \cap [a, b]$. The collection $\{I_n\} \cup \{J_x : x \in [a, b] \setminus E\}$ is an open cover of $[a, b]$. Thus by compactness, a finite number, say $\{I_k\}_{k=1}^n$ and $\{J_{x_j}\}_{j=1}^m$ also cover $[a, b]$. Let

$$\mathcal{P} = \{a = t_0, t_1, \ldots, t_N = b\}$$

be the partition of $[a, b]$ determined by those endpoints of $I_k, k = 1, \ldots, n$, and $J_{x_j}, j = 1, \ldots, m$, that are in $[a, b]$. For each $j, 1 \leq j \leq N$, the interval (t_{j-1}, t_j) is contained in some I_k or J_{x_j}. Let $J = \{j : (t_{j-1}, t_j) \subset I_k \text{ for some } k\}$. Also, for each $j \in \{1, \ldots, N\}$, let m_j and M_j denote the infimum and supremum of

f on $[t_{j-1}, t_j]$ respectively. Then $M_j - m_j \leq 2M$ for all $j \in J$, and $M_j - m_j < \epsilon/2(b-a)$ for all $j \notin J$. Thus

$$\mathcal{U}(\mathcal{P}, f) - \mathcal{L}(\mathcal{P}, f) = \sum_{j \in J}(M_j - m_j)\Delta t_j + \sum_{j \notin J}(M_j - m_j)\Delta t_j$$

$$\leq 2M \sum_{j \in J} \Delta t_j + \frac{\epsilon}{2(b-a)} \sum_{j \notin J} \Delta t_j$$

$$\leq 2M \sum_k \ell(I_k) + \frac{\epsilon}{2(b-a)} \sum_{j=1}^n \Delta t_j$$

$$\leq 2M \left(\frac{\epsilon}{4M}\right) + \frac{\epsilon}{2(b-a)}(b-a) = \epsilon.$$

Hence by Theorem 6.1.7 the function f is Riemann integrable on $[a, b]$. $\square$

Exercises 6.7

1. Let f be a real-valued function on $[a, b]$ and $\mathcal{P}$ a partition of $[a, b]$. Prove that the lower function L_f and upper function U_f for the partition $\mathcal{P}$ are Riemann integrable on $[a, b]$ with

   $$\int_a^b L_f = \mathcal{L}(\mathcal{P}, f) \quad \text{and} \quad \int_a^b U_f = \mathcal{U}(\mathcal{P}, f).$$

2. *Let P be that Cantor set in $[0, 1]$. Show that $\chi_P \in \mathcal{R}[0, 1]$ and find $\int_0^1 \chi_P(x)dx$.

3. Let $f \in \mathcal{R}[a, b]$, and for $x \in [a, b]$, set $F(x) = \int_a^x f(t)dt$. Prove that there exists a subset E of $[a, b]$ of measure zero such that $F'(x) = f(x)$ for all $x \in [a, b] \setminus E$,

4. *Suppose $f \in \mathcal{R}[a, b]$ and g is a bounded real-valued function on $[a, b]$. If $\{x \in [a, b] : g(x) \neq f(x)\}$ has measure zero, is g Riemann integrable on $[a, b]$?

Notes

The fundamental theorem of calculus is without question the key theorem of calculus; it relates the Cauchy-Riemann theory of integration with differentiation. For Newton, Leibniz, and their successors, integration was the inverse operation of differentiation. When Cauchy however defined integration independent of differentiation, the fundamental theorem of calculus became a necessity. It was crucial in proving that Cauchy's integral was the inverse of differentiation, thereby providing both a convenient tool for the evaluation of definite integrals and proving that every continuous real-valued function defined on an interval I has an antiderivative on I.

Although we stated the fundamental theorem of calculus as two separate theorems, for continuous functions they are the same. Specifically, if f is a continuous real-valued function on $[a, b]$ and F is an antiderivative of f, then for any $x \in [a, b]$,

$$F(x) = F(a) + \int_a^x f(t)\, dt.$$

Conversely, if f is continuous and F is defined as above, then $F'(x) = f(x)$.

Lebesgue's theorem is one of the most beautiful results of analysis. It provides very concise necessary and sufficient conditions for Riemann integrability of a bounded real-valued function f. Although Riemann was the first to provide such conditions, his result lacked the simplicity and elegance of Lebesgue's theorem. Unfortunately for Riemann, the concept of measure had not yet been developed when he stated and proved his result. In Chapter 10 we will develop the theory of measure and Lebesgue's extension of the Riemann integral.

The need for numerical methods for the evaluation of definite integrals was recognized as early as the eighteenth century. Euler and Thomas Simpson (1710–1761), among others, used numerical techniques to approximate the definite integral in problems where an antiderivative could not be found. Even the error estimates developed in Section 6 date back to that era. With the availability of efficient calculators and high speed computers, numerical methods have increased in importance in the past few decades. This has led to the development of very sophisticated numerical algorithms for the evaluation of definite integrals.

Although Newton and Leibniz are credited with inventing the differential and integral calculus, many mathematicians prior to their time knew formulas for computing tangents and areas in particular instances. Archimedes (around 200 B.C.) in his treatise *Quadrature of the Parabola* used the method of exhaustion by inscribed triangles to derive a formula for the area under a parabolic segment. By the mid 1640's, Pierre de Fermat (1601–1665) had determined the formulas for the area under any curve of the form $y = x^k$ ($k \neq -1$), and for finding tangents to such curves. Issac Barrow (1630–1677), who was a professor of geometry at Cambridge, in his 1670 treatise *Lectiones geometriae* developed an algebraic procedure which is virtually identical to the differential calculus for finding tangents to a curve. However, all these early methods were developed using geometric arguments. Newton and Leibniz developed the concepts, the notation, and the algorithms for making these computations for arbitrary functions. Most importantly however, both men realized the inverse nature of the problems of tangents and areas. For these reasons they are credited with the development of the differential and integral calculus. Further information on the historical development of calculus may be found in the text by Katz listed in the Bibliography and the article by Rosenthal listed at the end of the chapter.

Miscellaneous Exercises

A real-valued function f on $[a, b]$ is a **step function** if there exist a finite number of disjoint intervals $\{I_j\}_{j=1}^n$ with $\cup I_j = [a, b]$ such that f is constant on each of the intervals I_j, $j = 1, ..., n$.

1. **a.** If f is a step function on $[a, b]$, prove that $f \in \mathcal{R}[a, b]$ with
   $$\int_a^b f = \sum_{i=1}^n c_i \ell(I_i),$$
   where c_i is the value of f on I_i.

   b. If $f \in \mathcal{R}[a, b]$ and $\epsilon > 0$ is given, prove that there exists a step function h on $[a, b]$ such that
   $$\int_a^b |f - h| < \epsilon.$$

   c. If h is a step function on $[a, b]$ and $\epsilon > 0$ is given, prove that there exists a continuous function g on $[a, b]$ such that
   $$\int_a^b |h - g| < \epsilon.$$

   d. If $f \in \mathcal{R}[a, b]$ and $\epsilon > 0$ is given, prove that there exists a continuous function g on $[a, b]$ such that
   $$\int_a^b |f - g| < \epsilon.$$

   Let f be a real-valued function defined on $[0, \infty)$. The **Laplace transform** of f, denoted $\mathcal{L}(f)$, is the function defined by
   $$\mathcal{L}(f)(s) = \int_0^\infty e^{-st} f(t) \, dt,$$
   whenever the improper Riemann integral exists.

2. Let f be defined on $[0, \infty)$. Prove that there exists $a \in \mathbb{R} \cup \{-\infty, \infty\}$ such that $\mathcal{L}(f)(s)$ is defined for all $s \in (a, \infty)$, and that the integral defining $\mathcal{L}(f)(s)$ diverges for all $s \in (-\infty, a)$.

   (Hint: First show that if $\mathcal{L}(f)(s_o)$ exists for some s_o, then $\mathcal{L}(f)(s)$ exists for all $s > s_o$.)

3. Suppose $f \in \mathcal{R}[0, c]$ for every $c > 0$, and that there exists a positive constant C, and $a \in \mathbb{R}$, such that $|f(t)| \leq C \, e^{at}$ for all $t \geq 0$. Prove that $\mathcal{L}(f)(s)$ exists for all $s > a$.

4. Compute the Laplace transform of each of the following functions. In each case, specify the interval on which $\mathcal{L}(s)$ is defined.

   a. $f(t) = 1$.

   b. $f(t) = e^{at}$, $a \in \mathbb{R}$.

   c. $f(t) = \cos \omega t$.

   d. $f(t) = \sin \omega t$.

   e. $f(t) = t^n$, $n \in \mathbb{N}$. (Use induction.)

   f. $f(t) = I(t - c)$, where $I(t - c)$ is the unit jump function at $t = c$.

   g. $f(t) = t^\alpha$, $\alpha > -1$. (See Exercise 8, Section 6.4.)

5. Suppose f is differentiable on $[0, \infty)$ and $a \in \mathbb{R}$ is such that $\mathcal{L}(f)(s)$ exists for all $s > 1$. If $\lim_{t \to \infty} e^{-st} f(t) = 0$ for all $s > a$, prove that $\mathcal{L}(f')(s) = s \, \mathcal{L}(f)(s) - f(0)$.

Supplemental Reading

Bagby, R. J., "A convergence of limits," *Math. Mag.* **71** (1998), 270–277.

Bao-lin, Z., "A note on the mean-value theorem for integrals," *Amer. Math. Monthly* **104** (1997), 561–562.

Bartle, R. G., "Return to the Riemann integral," *Amer. Math. Monthly* **103** (1996), 625–632.

Botsko, M. W., "An elementary proof that a bounded a.e. continuous function is Riemann integrable," *Amer. Math. Monthly* **95** (1988), 249–252.

Bressoud, D. M., "Historical reflections on teaching the fundamental theorem of calculus," *Amer. Math. Monthly* **118** (2011), 99–115.

Bullock, G. L., "A geometric interpretation of the Riemann-Stieltjes integral," *Amer. Math. Monthly* **95**, (1988), 448-455.

Fazekas, Jr. E. C. and Mercer, P. R., "Elementary proofs of error estimates for the midpoint and Simpson's rules," *Math. Mag.* **82** (2009), 365–370.

Goel, S. K. and Rodriguez, D. M., "A note on evaluating limits using Riemann sums," *Math. Mag.* **60** (1987), 225–228.

Gordon, R. A., "Some integrals involving the Cantor function," *Amer. Math. Monthly* **116** (2009), 218–227.

Gordon, R. A., "A bounded derivative that is not Riemann integrable," *Math. Mag.* **89** (2016), 364–370.

Jacobson, B., "On the mean-value theorem for integrals," *Amer. Math. Montly* **89** (1982) 300–301.

Jones, L. K., "An elementary derivation of the numerical integration bounds in beginning calculus," *Amer. Math. Monthly* **124** (2017), 558–561.

Jones, W. R. and Landau, M. D., "One-sided limits and integrability," *Math. Mag.* **45** (1972), 19–21.

Klippert, J., "On the right-hand derivative of a certain integral function," *Amer. Math. Monthly* **98** (1991), 751–752.

Kristensen, E., Poulsen, E. T., and Reich, E., "A characterization of Riemann integrability," *Amer. Math. Monthly* **69** (1962), 498–505.

Rickey, V. F. and Tuchinsky, P. M., "An application of geography to mathematics: History of the integral of the secant," *Math. Mag.* **53** (1980), 162–166.

Rosenthal, A., "The history of calculus," *Amer. Math. Monthly* **58** (1951), 75–86.

Stein, S. K., "The error of the trapezoidal method for a concave curve," *Amer. Math. Monthly* **83** (1976) 643–645.

Talman, L. A., "Simpson's rule is exact for quintics," *Amer. Math. Monthly* **113** (2006), 141-155.

Tandra, H., "A new proof of the change of variable theorem for the Riemann integral," *Amer. Math. Monthly* **122** (2015), 795–799.

Williams, K. S., "Note on $\int_0^\infty (\sin x/x)dx$," *Math. Mag.* **44** (1971), 9–11.

7

Series of Real Numbers

Although the study of series has a long history in mathematics,[1] the modern definition of convergence dates back only to the beginning of the nineteenth century. In 1821, Cauchy, in his text *Cours d'Analyse*, used his definition of limit to provide the first formal definition of convergence of a series in terms of convergence of the sequence of partial sums. The Cauchy criterion (Theorem 3.7.3) was the first significant result to provide necessary and sufficient conditions for convergence of a series. Cauchy not only stated and proved the result, he also applied his result to prove convergence and divergence of given series. Many of the early convergence tests, such as the root and ratio test, are due to him. Cauchy, with his formal development of series, placed the subject matter on a rigorous mathematical foundation.

In this chapter, we will continue our study of series of real numbers begun earlier in the text. Our primary emphasis in Section 7.1 will be on deriving several tests that are useful in determining the convergence or divergence of a given series. In Section 7.3 we will study the concepts of absolute convergence, conditional convergence, and rearrangements of series. One of the key results of this section is that every rearrangement of an absolutely convergent series not only converges, but converges to the same sum. As we will also see, this fails dramatically if the series converges but fails to converge absolutely.

In Section 7.4 we give a brief introduction to the topic of square summable sequences. These play an important role in the study of Fourier series. One of the main result of this section will be the Cauchy-Schwarz inequality for series. This section also contains a brief introduction to normed linear spaces.

7.1 Convergence Tests

In Section 3.7, we provided a very brief introduction to the subject of infinite series. In the study of infinite series, it is very useful to have tests available by means of which one is able to determine whether a given series converges or diverges. For example, Corollary 3.7.5 is very useful in determining divergence of a series. If the sequence $\{a_k\}$ does not converge to zero, then the series

---

[1]See the notes at the end of the chapter.

$\sum a_k$ diverges. On the other hand however, if $\lim a_k = 0$, then nothing can be ascertained concerning convergence or divergence of the series $\sum a_k$. In this section, we will state and prove several useful results that can be used to establish convergence or divergence of a given series. Additional tests for convergence will also be given in the exercises and subsequent sections. With the exception of Theorem 7.1.1, all of our results in this section will be stated for series of nonnegative terms.

As in Definition 3.7.1, given an infinite series $\sum\limits_{k=1}^{\infty} a_k$ of real numbers, $\{s_n\}_{n=1}^{\infty}$ will denote the associated sequence of **partial sums** defined by

$$s_n = \sum_{k=1}^{n} a_k.$$

The series $\sum a_k$ **converges** if and only if the sequence $\{s_n\}$ of nth partial sums converges. Furthermore, if $\lim\limits_{n\to\infty} s_n = s$, $(s \in \mathbb{R})$, then s is called the **sum of the series**, and we write

$$\sum_{k=1}^{\infty} a_k = s.$$

If the sequence $\{s_n\}$ diverges, then the series $\sum a_k$ is said to **diverge**. Furthermore, if $\lim\limits_{n\to\infty} s_n = \infty$ (or $-\infty$), then we write $\sum a_k = \infty$ $(-\infty)$ to denote that the series **diverges** to ∞ (or $-\infty$).

If $a_k \geq 0$ for all k, then by Theorem 3.7.6 $\sum a_k$ converges if and only if $\lim\limits_{n\to\infty} s_n < \infty$. Thus for series of nonnegative terms we adopt the notation

$$\sum_{k=1}^{\infty} a_k < \infty$$

to denote that the series converges.

Remarks. (a) Although we generally index a series by the positive integers $\mathbb{N}$, it is sometimes more convenient to start with $k = 0$ or with $k = k_o$ for some integer k_o. In this case, the resulting series are denoted as

$$\sum_{k=0}^{\infty} a_k, \qquad \sum_{k=k_o}^{\infty} a_k.$$

Also, from the Cauchy criterion (Theorem 3.7.3) it is clear that $\sum\limits_{k=1}^{\infty} a_k$ converges if and only if $\sum\limits_{k=k_o}^{\infty} a_k$ converges for some, and hence every, $k_o \in \mathbb{N}$.

(b) Given any sequence $\{s_n\}_{n=1}^{\infty}$ of real numbers we can always find a series $\sum a_k$ whose nth partial sum is s_n. If we set $a_1 = s_1$ and $a_k = s_k - s_{k-1}$, $k > 1$, then

$$\sum_{k=1}^{n} a_k = s_1 + (s_2 - s_1) + \cdots + (s_n - s_{n-1}) = s_n.$$

THEOREM 7.1.1 *If* $\sum_{k=1}^{\infty} a_k = \alpha$ *and* $\sum_{k=1}^{\infty} b_k = \beta$, *then*

(a) $\sum_{k=1}^{\infty} c \, a_k = c\alpha,$ *for any* $c \in \mathbb{R}$, *and*

(b) $\sum_{k=1}^{\infty} (a_k + b_k) = \alpha + \beta.$

Proof. The proof of (a) is similar to (b) and thus is omitted. To prove (b), for each $n \in \mathbb{N}$, let

$$s_n = \sum_{k=1}^{n} a_k \quad \text{and} \quad t_n = \sum_{k=1}^{n} b_k.$$

Since the series converge to α and β respectively, $\lim s_n = \alpha$ and $\lim t_n = \beta$. Therefore by Theorem 3.2.1, $\lim(s_n + t_n) = \alpha + \beta$. But

$$s_n + t_n = \sum_{k=1}^{n} a_k + \sum_{k=1}^{n} b_k = \sum_{k=1}^{n} (a_k + b_k).$$

Therefore $s_n + t_n$ is the nth partial sum of the series $\sum (a_k + b_k)$. Since the sequence $\{s_n + t_n\}$ converges to $\alpha + \beta$,

$$\sum_{k=1}^{\infty} (a_k + b_k) = \alpha + \beta. \quad \square$$

Comparison Test

One of the most important convergence tests is the comparison test. Although very elementary, it provides one of the most useful tools in determining convergence or divergence of a series. It is useful both in applications and theory. Several of the proofs of subsequent theorems rely on this test. In applications, by comparing the terms of a given series with the terms of a series for which convergence or divergence is known, we are then able to determine whether the given series converges or diverges.

THEOREM 7.1.2 (Comparison Test) *Suppose* $\sum a_k$ *and* $\sum b_k$ *are two given series of nonnegative real numbers satisfying*

$$0 \leq a_k \leq M \, b_k$$

for some positive constant M and all integers $k \geq k_o$, for some fixed $k_o \in \mathbb{N}$.

(a) *If $\displaystyle\sum_{k=1}^{\infty} b_k < \infty$, then $\displaystyle\sum_{k=1}^{\infty} a_k < \infty$.*

(b) *If $\displaystyle\sum_{k=1}^{\infty} a_k = \infty$, then $\displaystyle\sum_{k=1}^{\infty} b_k = \infty$.*

Proof. Suppose the terms $\{a_k\}$ and $\{b_k\}$ satisfy $a_k \leq M b_k$ for all $k \geq k_o$ and some positive constant M. Then for $n > m \geq k_o$

$$0 \leq \sum_{k=m+1}^{n} a_k \leq M \sum_{k=m+1}^{n} b_k.$$

Suppose $\sum b_k$ converges. Then given $\epsilon > 0$, by the Cauchy criterion (3.7.3) there exists an integer $n_o \geq k_o$ such that

$$\sum_{k=m+1}^{n} b_k < \frac{\epsilon}{M}$$

for all $n > m \geq n_o$. Thus $0 \leq \displaystyle\sum_{k=m+1}^{n} a_k < \epsilon$ for all $n > m \geq n_o$, and hence by the Cauchy criterion $\sum a_k$ converges. On the other hand, if $\sum a_k$ diverges, then $\sum b_k$ must also diverge. $\square$

As a corollary of the previous theorem we also have the following version of the comparison test.

COROLLARY 7.1.3 (Limit Comparison Test) *Suppose $\sum a_k$ and $\sum b_k$ are two given series of positive real numbers.*

(a) *If $\displaystyle\lim_{n \to \infty} \frac{a_n}{b_n} = L$ with $0 < L < \infty$, then $\sum a_k$ converges if and only if $\sum b_k$ converges.*

(b) *If $\displaystyle\lim_{n \to \infty} \frac{a_n}{b_n} = 0$ and $\sum b_k$ converges, then $\sum a_k$ converges.*

Proof. The proof, the details of which are left to the exercises (Exercise 6), follows immediately from the definition of the limit and the comparison test. $\square$

Remark. If $\displaystyle\lim_{n \to \infty} a_n/b_n = 0$ and $\sum a_k$ converges, then nothing can be concluded about the convergence of the series $\sum b_k$. In Example 7.1.4(d), we provide an example of a divergent series $\sum b_k$ and a convergent series $\sum a_k$ for which $\displaystyle\lim_{n \to \infty} a_n/b_n = 0$. On the other hand, in Exercise 23, given a convergent series $\sum a_k$ with $a_k > 0$, you will be asked to construct a convergent series $\sum b_k$ with $b_k > 0$ such that $\displaystyle\lim_{n \to \infty} a_n/b_n = 0$.

EXAMPLES 7.1.4 (a) As an application of the comparison test consider the series

$$\sum_{k=1}^{\infty} \frac{k}{3^k}.$$

We will compare the given series with the convergent geometric series $\sum (1/2)^k$. Thus we wish to show that there exists $k_o \in \mathbb{N}$ such that $k/3^k \leq 1/2^k$ for all $k \geq k_o$. Since $2/3 < 1$, by Theorem 3.2.6(d)

$$\lim_{k \to \infty} k \left(\frac{2}{3} \right)^k = 0.$$

Thus by taking $\epsilon = 1$, there exists an integer k_o such that $k(2/3)^k \leq 1$ for all $k \geq k_o$. As a consequence,

$$\frac{k}{3^k} \leq \frac{1}{2^k} \qquad \text{for all} \quad k \geq k_o.$$

Since $\sum (1/2)^k < \infty$, by the comparison test the given series also converges. A similar argument can be used to prove that $\sum k^n/a^k$ converges for any $n \in \mathbb{Z}$ and $a \in \mathbb{R}$ with $a > 1$ (Exercise 2(l)). This is accomplished by comparing the given series with $\sum b^{-k}$, where $1 < b < a$.

  (b) As our second example consider the series

$$\sum_{k=1}^{\infty} \sqrt{\frac{k+1}{2k^3+1}}.$$

In order to use the comparison test we have to determine what series we want to compare with. Since

$$\sqrt{\frac{k+1}{2k^3+1}} = \frac{1}{k} \sqrt{\frac{1+\frac{1}{k}}{2+\frac{1}{k^3}}} \quad \text{and} \quad \lim_{k \to \infty} \sqrt{\frac{1+\frac{1}{k}}{2+\frac{1}{k^3}}} = \frac{1}{2}\sqrt{2},$$

we will compare the given series with the series $\sum 1/k$. This series is known to diverge (Example 3.7.4). If we take $\epsilon = \frac{1}{4}\sqrt{2}$, then we can conclude that there exists $k_o \in \mathbb{N}$ such that

$$\sqrt{\frac{1+\frac{1}{k}}{2+\frac{1}{k^3}}} \geq \frac{1}{2}\sqrt{2} - \epsilon = \frac{1}{4}\sqrt{2}$$

for all $k \geq k_o$. Thus there exists a positive constant M and $k_o \in \mathbb{N}$ such that

$$\sqrt{\frac{k+1}{2k^3+1}} \geq M\frac{1}{k}$$

for all $k \geq k_o$, and as a consequence, the given series diverges.

(c) The divergence of the series $\sum_{k=1}^{\infty} \sqrt{\dfrac{k+1}{2k^3+1}}$ can also be obtained by the limit comparison test. Comparing the terms of the given series to the terms of the series $\sum 1/k$ we have

$$\lim_{k\to\infty} \frac{\sqrt{\frac{k+1}{2k^3+1}}}{\frac{1}{k}} = \lim_{k\to\infty} \sqrt{\frac{1+\frac{1}{k}}{2+\frac{1}{k^3}}} = \frac{\sqrt{2}}{2}.$$

Thus since the series $\sum 1/k$ diverges, by the limit comparison test the series $\sum_{k=1}^{\infty} \sqrt{\dfrac{k+1}{2k^3+1}}$ also diverges.

(d) Let $a_k = 2^{-k}$ and $b_k = 1/k$. Then $\sum a_k$ converges, $\sum b_k$ diverges, and

$$\lim_{n\to\infty} \frac{a_n}{b_n} = \lim_{n\to\infty} \frac{n}{2^n} = 0.$$

Thus if $\lim_{n\to\infty} \dfrac{a_n}{b_n} = 0$, convergence of the series $\sum a_k$ does not imply convergence of the series $\sum b_k$. $\square$

Integral Test

Our second major convergence test is the integral test. Recall from Section 6.4, if f is a real-valued function defined on $[a,\infty)$ with $f \in \mathcal{R}[a,c]$ for every $c > a$, then the improper Riemann integral of f is defined by

$$\int_a^\infty f(x)\,dx = \lim_{c\to\infty} \int_a^c f(x)\,dx,$$

provided the limit exists. If $f(x) \geq 0$, we use the notation

$$\int_a^\infty f(x)\,dx < \infty \qquad \left(\text{or } \int_a^\infty f(x)\,dx = \infty\right)$$

to denote that the improper integral of f converges (diverges).

THEOREM 7.1.5 (Integral Test) *Let $\{a_k\}_{k=1}^{\infty}$ be a decreasing sequence of nonnegative real numbers, and let f be a nonnegative monotone decreasing function on $[1,\infty)$ satisfying $f(k) = a_k$ for all $k \in \mathbb{N}$. Then*

$$\sum_{k=1}^{\infty} a_k < \infty \qquad \text{if and only if} \qquad \int_1^\infty f(x)\,dx < \infty.$$

Proof. Since f is monotone on $[1,\infty)$, by Theorem 6.1.8 it is Riemann integrable on $[1,c]$ for every $c > 1$. Let $n \in \mathbb{N}$, $n \geq 2$, and consider the partition

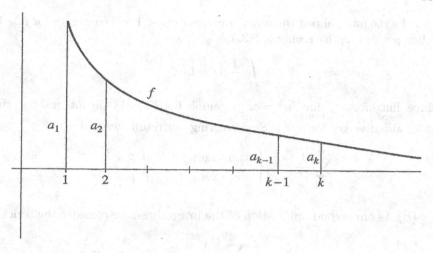

FIGURE 7.1
Integral test

$\mathcal{P} = \{1, 2, ..., n\}$ of $[1, n]$. Since f is decreasing, for each $k = 2, 3, ..., n$, (see Figure 7.1)

$$\sup\{f(t) : t \in [k-1, k]\} = f(k-1) = a_{k-1}, \qquad \text{and}$$
$$\inf\{f(t) : t \in [k-1, k]\} = f(k) = a_k.$$

Therefore

$$\sum_{k=2}^{n} a_k = \mathcal{L}(\mathcal{P}, f) \leq \int_1^n f(x)\, dx \leq \mathcal{U}(\mathcal{P}, f) = \sum_{k=1}^{n-1} a_k,$$

from which the result now follows. $\square$

EXAMPLES 7.1.6 (a) As our first application of the integral test we consider the *p*-series

$$\sum_{k=1}^{\infty} \frac{1}{k^p}.$$

When $p = 1$ this series is referred to as the **harmonic series**. If $p \leq 0$, then $\{k^{-p}\}$ does not converge to zero, and thus by Corollary 3.7.5 the series diverges. Suppose $p > 0$, $p \neq 1$. Let $f(x) = x^{-p}$, which is decreasing on $[1, \infty)$. Then

$$\int_1^c x^{-p}\, dx = \frac{1}{p-1}\left[1 - \frac{1}{c^{p-1}}\right].$$

Therefore,

$$\int_1^{\infty} x^{-p}\, dx = \lim_{c \to \infty} \int_1^c x^{-p}\, dx = \begin{cases} \dfrac{1}{p-1}, & p > 1, \\ \infty, & p < 1. \end{cases}$$

Thus by the integral test the series diverges for $p < 1$ and converges for $p > 1$. When $p = 1$, then by Example 6.3.5,

$$\int_1^c \frac{1}{x}\, dx = \ln c.$$

Since $\lim_{x \to \infty} \ln x = -\lim_{t \to 0+} \ln t = \infty$ (Example 6.4.2(a)), by the integral test the series also diverges for $p = 1$. Summarizing our results we find

$$\sum_{k=1}^{\infty} \frac{1}{k^p} \quad \begin{cases} \text{converges,} & \text{if } p > 1, \\ \text{diverges,} & \text{if } p \le 1. \end{cases}$$

(b) As our second application of the integral test we consider the series

$$\sum_{k=2}^{\infty} \frac{1}{k \ln k}.$$

Let $f(x) = (x \ln x)^{-1}$, $x \in [2, \infty)$. Since

$$f'(x) = -\frac{(1 + \ln x)}{(x \ln x)^2},$$

$f'(x) < 0$ for all $x > 1$. Thus f is monotone decreasing on $[2, \infty)$. But

$$\lim_{c \to \infty} \int_2^c \frac{1}{x \ln x}\, dx = \lim_{c \to \infty} \ln(\ln c) - \ln(\ln 2) = \infty.$$

Thus by the integral test the given series diverges.

(c) As our final example, we consider the series

$$\sum_{k=2}^{\infty} \frac{\ln k}{k^p}, \quad p \in \mathbb{R}.$$

We first consider the case $p = 1$. Since $f(x) = (\ln x)/x$ is decreasing on $[e, \infty)$ and

$$\lim_{c \to \infty} \int_e^c \frac{\ln x}{x}\, dx = \lim_{c \to \infty} (\ln c)^2 - 1 = \infty,$$

by the integral test the series $\sum_{k=2}^{\infty} (\ln k)/k$ diverges.

Suppose now that $p > 1$. Write $p = q + r$, where $q > 1$ and $r > 0$. By l'Hospital's rule, $\lim_{x \to \infty} (\ln x)/x^r = 0$. Thus there exists $k_o \in \mathbb{N}$ such that $(\ln k)/k^r \le 1$ for all $k \ge k_o$. As a consequence

$$\frac{\ln k}{k^p} = \frac{1}{k^q} \frac{\ln k}{k^r} \le \frac{1}{k^q}$$

for all $k \ge k_o$. Since $q > 1$ the series $\sum 1/k^q$ converges. Hence by the comparison test the series $\sum (\ln k)/k^p$ also converges when $p > 1$.

Finally, if $p < 1$, then again the divergence of $\sum 1/k^p$ implies that $\sum(\ln k)/k^p$ also diverges. This follows from the fact that $(\ln k)/k^p \geq (\ln 2)/k^p$ for all $k \in \mathbb{N}$, $k \geq 2$. Summarizing our results we find

$$\sum_{k=2}^{\infty} \frac{\ln k}{k^p} \quad \begin{cases} \text{converges,} & \text{if } p > 1, \\ \text{diverges,} & \text{if } p \leq 1. \end{cases} \quad \square$$

Root and Ratio Test

We now consider the well known root and ratio tests. Although useful in determining the convergence or divergence of certain types of series, both of these tests are also very important in the study of power series. The ratio test, attributed to Jean d'Alembert (1717–1783), is particularly applicable to series involving factorials. The root test, due to Cauchy, will be used in the next chapter to define the interval of convergence of a power series. Because of the close similarity between these two tests we state them together.

Prior to stating the ratio and root tests, we recall the definition of the limit inferior and limit superior of a sequence of real numbers (Definition 3.5.1) If $\{s_n\}$ is a sequence in $\mathbb{R}$, then

$$\varliminf_{n\to\infty} s_n = \sup_{k\in\mathbb{N}} \inf\{s_n : n \geq k\} = \lim_{k\to\infty} \inf\{s_n : n \geq k\}, \quad \text{and}$$

$$\varlimsup_{n\to\infty} s_n = \inf_{k\in\mathbb{N}} \sup\{s_n : n \geq k\} = \lim_{k\to\infty} \sup\{s_n : n \geq k\}.$$

By Theorem 3.5.7, If E denotes the set of subsequential limits of $\{s_n\}$ in $\mathbb{R} \cup \{-\infty, \infty\}$, then

$$\varliminf_{n\to\infty} s_n = \inf E \quad \text{and} \quad \varlimsup_{n\to\infty} s_n = \sup E.$$

In particular, if the sequence $\{s_n\}$ converges (or diverges to either $-\infty$ or ∞), then $\varliminf s_n = \varlimsup s_n = \lim s_n$.

THEOREM 7.1.7 (Ratio Test) *Let $\sum a_k$ be a series of positive terms, and let*

$$R = \varlimsup_{k\to\infty} \frac{a_{k+1}}{a_k}, \quad \text{and} \quad r = \varliminf_{k\to\infty} \frac{a_{k+1}}{a_k}.$$

(a) *If $R < 1$, then $\sum_{k=1}^{\infty} a_k < \infty$.*

(b) *If $r > 1$, then $\sum_{k=1}^{\infty} a_k = \infty$.*

(c) *If $r \leq 1 \leq R$, then the test is inconclusive.*

THEOREM 7.1.8 (Root Test) *Let* $\sum a_k$ *be a series of nonnegative terms, and let*

$$\alpha = \varlimsup_{k \to \infty} \sqrt[k]{a_k}.$$

(a) *If* $\alpha < 1$, *then* $\sum_{k=1}^{\infty} a_k < \infty$.

(b) *If* $\alpha > 1$, *then* $\sum_{k=1}^{\infty} a_k = \infty$.

(c) *If* $\alpha = 1$, *then the test is inconclusive.*

Before proving the theorems we give several examples that illustrate these two convergence tests.

EXAMPLES 7.1.9 (a) Consider the p-series $\sum_{k=1}^{\infty} \dfrac{1}{k^p}$ for $p > 0$. For this series,

$$r = R = \alpha = 1$$

for any $p > 0$; thus both tests are inconclusive. The series however diverges if $0 < p \leq 1$ and converges for $p > 1$.

(b) As our second example we consider

$$\sum_{k=1}^{\infty} \frac{p^k}{k!}, \qquad 0 < p < \infty.$$

Here $a_k = p^k/k!$. Thus

$$\lim_{k \to \infty} \frac{a_{k+1}}{a_k} = \lim_{k \to \infty} \frac{p^{k+1}}{p^k} \frac{k!}{(k+1)!} = p \lim_{k \to \infty} \frac{1}{k+1} = 0.$$

By the ratio test the series converges for all p, $0 < p < \infty$. In this example, the presence of $k!$ makes the root test difficult to use.

(c) Consider $\sum a_n$ where

$$a_n = \begin{cases} \dfrac{1}{2^k}, & \text{if } n = 2k, \\[2mm] \dfrac{1}{3^k}, & \text{if } n = 2k+1. \end{cases}$$

Here

$$\sum_{n=1}^{\infty} a_n = 1 + \frac{1}{2} + \frac{1}{3} + \frac{1}{2^2} + \frac{1}{3^2} + \cdots.$$

By computation,

$$\frac{a_{n+1}}{a_n} = \begin{cases} \left(\dfrac{2}{3}\right)^k, & n = 2k, \\ \dfrac{1}{2}\left(\dfrac{3}{2}\right)^k, & n = 2k+1. \end{cases}$$

For the subsequence of $\{a_{n+1}/a_n\}$ with even n,

$$\lim_{k \to \infty} \frac{a_{2k+1}}{a_{2k}} = \lim_{k \to \infty} \left(\frac{2}{3}\right)^k = 0.$$

For the subsequence with odd n,

$$\lim_{k \to \infty} \frac{a_{2k+2}}{a_{2k+1}} = \frac{1}{2} \lim_{k \to \infty} \left(\frac{3}{2}\right)^k = \infty.$$

Thus $\{0, \infty\}$ is the set of sub-sequential limits of $\{a_{n+1}/a_n\}$. Consequently, $r = 0$ and $R = \infty$. Therefore the ratio test is inconclusive. On the other hand

$$\sqrt[n]{a_n} = \begin{cases} \dfrac{1}{\sqrt{2}}, & n = 2k, \\ \dfrac{1}{\sqrt{3}}\left(\sqrt{3}\right)^{1/(2k+1)}, & n = 2k+1. \end{cases}$$

By Theorem 3.2.6(b), $\lim\limits_{k \to \infty} (\sqrt{3})^{1/(2k+1)} = 1$. Thus the sequence $\{\sqrt[n]{a_n}\}$ has two sub-sequential limits; namely $1/\sqrt{2}$ and $1/\sqrt{3}$. Therefore $\alpha = 1/\sqrt{2}$, and the series converges by the root test. $\square$

Proof of Ratio Test. Suppose

$$\overline{\lim} \frac{a_{n+1}}{a_n} = R < 1.$$

Choose c such that $R < c < 1$. Then by Theorem 3.5.3(a) there exists a positive integer n_o such that

$$\frac{a_{n+1}}{a_n} < c \qquad \text{for all} \quad n \ge n_o.$$

In particular, $a_{n_o+1} < c\,a_{n_o}$. By induction on m, $a_{n_o+m} < c^m a_{n_o}$. Therefore, writing $n = n_o + m$, $m \ge 0$,

$$a_n \le M\,c^n \qquad \text{for all} \quad n \ge n_o,$$

where $M = a_{n_o}/c^{n_o}$. Thus since $0 < c < 1$, by Example 3.7.2(a) the series $\sum c^n$ converges. Therefore $\sum a_n$ also converges by the comparison test.

Suppose

$$\lim_{n \to \infty} \frac{a_{n+1}}{a_n} = r > 1.$$

Again choose c so that $r > c > 1$. As above, there exists a positive integer n_o such that $a_n \geq M\, c^n$ for some constant M and all $n \geq n_o$. But since $c > 1$, $\sum c^n = \infty$, and thus by the comparison test, $\sum a_n = \infty$. $\square$

Proof of Root Test. Let

$$\alpha = \varlimsup_{n \to \infty} \sqrt[n]{a_n}.$$

Suppose $\alpha < 1$. Choose c so that $\alpha < c < 1$. Again by Theorem 3.5.3 there exists a positive integer n_o such that

$$\sqrt[n]{a_n} < c \qquad \text{for all} \quad n \geq n_o.$$

But then $a_n \leq c^n$ for all $n \geq n_o$, and $\sum a_n < \infty$ by the comparison test.

If $\alpha > 1$, then $\sqrt[n]{a_n} > 1$ for infinitely many n. Thus $a_n > 1$ for infinitely many n, and as a consequence $\{a_n\}$ does not converge to zero. Hence by Corollary 3.7.5 the series $\sum a_n$ diverges. $\square$

Example 7.1.9(c) provides an example of a series for which the ratio test is inconclusive, but the root test worked. Thus it appears that the root test is a stronger test, a fact which is confirmed by the following theorem.

THEOREM 7.1.10 *Let* $\{a_n\}$ *be a sequence of positive numbers. Then*

$$\varliminf_{n \to \infty} \frac{a_{n+1}}{a_n} \leq \varliminf_{n \to \infty} \sqrt[n]{a_n} \leq \varlimsup_{n \to \infty} \sqrt[n]{a_n} \leq \varlimsup_{n \to \infty} \frac{a_{n+1}}{a_n}.$$

Remark. As a consequence of the theorem, if a series $\sum a_k$ converges by virtue of the ratio test, i.e., $R < 1$, then we also have $\alpha < 1$. Similarly, if $\sum a_k$ diverges by virtue of the ratio test, i.e., $r > 1$, then we also have $\alpha > 1$. Thus if the ratio test proves convergence or divergence of the series, so does the root test. The converse however, as indicated above, is false.

Proof. Let
$$R = \varlimsup_{n \to \infty} \frac{a_{n+1}}{a_n}.$$

If $R = \infty$, then there is nothing to prove. Thus assume that $R < \infty$, and let $\beta > R$ be arbitrary. Then there exists a positive integer n_o such that

$$\frac{a_{n+1}}{a_n} \leq \beta \qquad \text{for all} \quad n \geq n_o.$$

As in the proof of the ratio test, this gives $a_n \leq M\,\beta^n$ for all $n \geq n_o$, where $M = a_{n_o}/\beta^{n_o}$. Hence

$$\sqrt[n]{a_n} \leq \beta \sqrt[n]{M} \qquad \text{for all} \quad n \geq n_o.$$

Since $\lim_{n \to \infty} \sqrt[n]{M} = 1$, we have

$$\varlimsup_{n \to \infty} \sqrt[n]{a_n} \leq \beta.$$

Since $\beta > R$ was arbitrary, $\varlimsup_{n \to \infty} \sqrt[n]{a_n} \leq R$, which proves the result. The inequality for the limit inferior is proved similarly. $\square$

EXAMPLE 7.1.11 If $a_k = k!$, then $\lim(a_{k+1}/a_k) = \infty$. Thus as a consequence of the previous theorem,

$$\lim_{k \to \infty} \sqrt[k]{k!} = \infty. \quad \square$$

Exercises 7.1

1. If a and b are positive real numbers, prove that

$$\sum_{k=1}^{\infty} \frac{1}{(ak+b)^p}$$

    converges if $p > 1$ and diverges if $p \leq 1$.

2. Test each of the following series for convergence:

    *a. $\displaystyle\sum_{k=1}^{\infty} \frac{k}{k^2 + 1}$ b. $\displaystyle\sum_{k=1}^{\infty} \frac{\sqrt{k}}{k^2 + 2k - 1}$ *c. $\displaystyle\sum_{k=1}^{\infty} \frac{k^2}{2^k}$

    d. $\displaystyle\sum_{k=1}^{\infty} k^3 e^{-k}$ *e. $\displaystyle\sum_{k=1}^{\infty} \frac{3^k}{k^3}$ f. $\displaystyle\sum_{k=1}^{\infty} \frac{3^k}{k!}$

    *g. $\displaystyle\sum_{k=1}^{\infty} \frac{(k!)^2}{(2k)!}$ h. $\displaystyle\sum_{k=1}^{\infty} \frac{k!}{k^k}$ *i. $\displaystyle\sum_{k=1}^{\infty} \frac{\sqrt{k+1} - \sqrt{k}}{k}$

    j. $\displaystyle\sum_{k=1}^{\infty} (\sqrt[k]{k} - 1)^k$ *k. $\displaystyle\sum_{k=2}^{\infty} \frac{1}{(\ln k)^2}$

    l. $\displaystyle\sum_{k=1}^{\infty} \frac{k^n}{a^k}$, $a > 1$, $n \in \mathbb{Z}$ *m. $\displaystyle\sum_{k=1}^{\infty} \sin(1/k^p)$, $p > 0$

    n. $\displaystyle\sum_{k=1}^{\infty} \cos(1/k^p)$, $p > 0$ *o. $\displaystyle\sum_{k=1}^{\infty} \frac{1}{k} \ln\left(1 + \frac{1}{k}\right)$

3. For each of the following, determine all values of $p \in \mathbb{R}$ for which the given geometric series converges, and find the sum of the series.

    *a. $\displaystyle\sum_{k=0}^{\infty} (\sin p)^k$ b. $\displaystyle\sum_{k=1}^{\infty} \left(\frac{p}{3}\right)^{2k}$ c $\displaystyle\sum_{k=0}^{\infty} \left(\frac{1+p}{1-p}\right)^k$, $p \neq 1$

4. Suppose $a_k \geq 0$ for all $k \in \mathbb{N}$ and $\sum a_k < \infty$. For each of the following, either prove that the given series converges or provide an example for which the series diverges.

    a. $\displaystyle\sum_{k=1}^{\infty} \frac{a_k}{1 + a_k}$ *b. $\displaystyle\sum_{k=1}^{\infty} a_k^2$ c. $\displaystyle\sum_{k=1}^{\infty} \sqrt{a_k}$

    *d. $\displaystyle\sum_{k=1}^{\infty} k a_k$ e. $\displaystyle\sum_{k=1}^{\infty} \sqrt[k]{k}\, a_k$ *f. $\displaystyle\sum_{k=1}^{\infty} \sqrt{\frac{a_k}{k}}$

    g. $\sum a_{n_k}$ where $\{a_{n_k}\}$ is a subsequence of $\{a_n\}$

5. *Determine all values of p and q for which the following series converges:

$$\sum_{k=2}^{\infty} \frac{1}{k^q (\ln k)^p}.$$

    (Hint: Consider the three cases $q > 1$, $q = 1$, $q < 1$)

6. *Prove Corollary 7.1.3.

7. If $\sum a_k$ converges and $\sum b_k = \infty$, prove that $\sum(a_k + b_k) = \infty$.

8. *Suppose $\{a_n\}$ is a sequence in $\mathbb{R}$ with $a_n > 0$ for all $n \in \mathbb{N}$. For each $k \in \mathbb{N}$ set

$$b_k = \frac{1}{k} \sum_{n=1}^{k} a_n.$$

Prove that $\sum_{k=1}^{\infty} b_k$ diverges.

9. Determine all values of p for which the following series converges:

$$\sum_{k=1}^{\infty} \frac{1}{k^p} \left(\sum_{n=1}^{k} \frac{1}{n^p} \right)$$

10. Suppose that the series $\sum a_k$ converges and $\{n_j\}$ is a strictly increasing sequence of positive integers. Define the sequence $\{b_k\}$ as follows.

$$b_1 = a_1 + \cdots + a_{n_1}$$
$$b_2 = a_{n_1+1} + \cdots + a_{n_2}$$
$$\vdots$$
$$b_k = a_{n_{k-1}+1} + \cdots + a_{n_k}.$$

Prove that $\sum b_k$ converges and that $\sum_{k=1}^{\infty} b_k = \sum_{k=1}^{\infty} a_k$. (*This exercise proves that if the series $\sum a_k$ converges, then any series obtained from $\sum a_k$ by inserting parentheses also converges to the same sum. The following exercise shows that removing parentheses may lead to difficulties.*)

11. *Give an example of a series $\sum a_k$ such that $\sum_{k=1}^{\infty} (a_{2k-1} + a_{2k})$ converges, but $\sum a_k$ diverges.

12. *Suppose that the series $\sum a_k$ of positive real numbers converges by virtue of the root or ratio test. Show that the series $\sum_{k=1}^{\infty} k^n a_k$ converges for all $n \in \mathbb{N}$.

13. *Show that the series

$$\frac{1}{1^2} + \frac{1}{2^3} + \frac{1}{3^2} + \frac{1}{4^3} + \cdots$$

converges, but that both the ratio and root tests are inconclusive.

14. Apply the root and ratio tests to the series $\sum a_k$ where

$$a_k = \begin{cases} \dfrac{1}{2^k}, & \text{when } k \text{ is even,} \\ \dfrac{1}{2^{k+2}}, & \text{when } k \text{ is odd.} \end{cases}$$

15. Suppose $a_k \geq 0$ for all $k \in \mathbb{N}$. Prove that the series $\sum_{k=1}^{\infty} a_k$ converges if and only if some subsequence $\{s_{n_k}\}$ of the sequence $\{s_n\}$ of partial sums converges.

16. *(**Cauchy Condensation Test**) Suppose that $a_1 \geq a_2 \geq a_3 \geq \cdots \geq 0$. Use the previous exercise to prove that $\sum_{k=1}^{\infty} a_k$ converges if and only if

$$\sum_{k=0}^{\infty} 2^k a_{2^k} \text{ converges.}$$

17. Use the Cauchy condensation test to show that $\sum_{n=1}^{\infty} \frac{1}{n^p}$ converges for all $p > 1$, and diverges for all p, $0 < p \leq 1$.

18. Use the Cauchy condensation test to determine the convergence or divergence of each of the following series.

    *a. $\sum_{k=2}^{\infty} \frac{1}{k \ln k}$ **b.** $\sum_{k=2}^{\infty} \frac{1}{k(\ln k)^p}$ $(p > 1)$.   **c.** $\sum_{k=3}^{\infty} \frac{1}{k(\ln k)(\ln(\ln k))}$

19. ***For** $k \in \mathbb{N}$ let c_k be defined by
    $$c_k = 1 + \frac{1}{2} + \cdots + \frac{1}{k} - \ln k.$$
    Prove that $\{c_k\}$ is a monotone decreasing sequence of positive numbers that is bounded below. The limit c of the sequence is called **Euler's constant**. Show that c is approximately 0.577.

20. **(Raabe's Test)** Let $a_k > 0$ for all $k \in \mathbb{N}$. Prove the following:

    a. If $a_{k+1}/a_k \leq 1 - r/k$ for some $r > 1$ and all $k \geq k_o$, $k_o \in \mathbb{N}$, then $\sum a_k < \infty$. (Hint: Show that $(k-1)a_k - ka_{k+1} \geq (r-1)a_k$ for all $k \geq k_o$.)

    b. If $a_{k+1}/a_k \geq 1 - 1/k$ for all $k \geq k_o$, $k_o \in \mathbb{N}$, then $\sum a_k = \infty$. (Hint: Show that $\{ka_{k+1}\}$ is monotone increasing for $k \geq k_o$.)

21. ***If** $p, q > 0$, show that the series
    $$\sum_{k=1}^{\infty} \frac{(p+1)(p+2)\cdots(p+k)}{(q+1)(q+2)\cdots(q+k)}$$
    converges for $q > p+1$ and diverges for $q \leq p+1$.

22. For $p > 0$ consider the series
    $$\left(\frac{1}{2}\right)^p + \left(\frac{1\cdot3}{2\cdot4}\right)^p + \cdots + \left(\frac{1\cdot3\cdots(2k-1)}{2\cdot4\cdots(2k)}\right)^p + \cdots$$

    a. Show that the ratio test fails for this series.

    b. Use Raabe's test to show that the series converges for $p > 2$, diverges for $p < 2$, and that the test is inconclusive when $p = 2$.

    *c. Prove that the series diverges for $p = 2$.

23. Let $\{a_n\}$ be a sequence of positive real numbers.

    *a. Suppose $\sum a_n$ converges. Construct a convergent series $\sum b_n$ with $b_n > 0$ such that $\lim_{n \to \infty} a_n/b_n = 0$.

    b. Suppose $\sum a_n$ diverges. Construct a divergent series $\sum b_n$ with $b_n > 0$ such that $\lim_{n \to \infty} b_n/a_n = 0$.

7.2 The Dirichlet Test

In this section, we prove the Dirichlet convergence test, named after Peter Lejeune Dirichlet (1805–1859), and then apply it to both alternating series and trigonometric series. The key to the Dirichlet test is the following summation by parts formula of Neils Abel (1802–1829). This formula is the analogue for series of the integration by parts formula.

THEOREM 7.2.1 (Abel Partial Summation Formula) *Let $\{a_k\}$ and $\{b_k\}$ be sequences of real numbers. Set*

$$A_0 = 0 \quad and \quad A_n = \sum_{k=1}^{n} a_k, \quad if\ n \geq 1.$$

Then if $1 \leq p \leq q$,

$$\sum_{k=p}^{q} a_k\, b_k = \sum_{k=p}^{q-1} A_k(b_k - b_{k+1}) + A_q\, b_q - A_{p-1}\, b_p.$$

Proof. Since $a_k = A_k - A_{k-1}$,

$$\sum_{k=p}^{q} a_k\, b_k = \sum_{k=p}^{q}(A_k - A_{k-1})b_k$$

$$= \sum_{k=p}^{q} A_k\, b_k - \sum_{k=p-1}^{q-1} A_k\, b_{k+1}$$

$$= \sum_{k=p}^{q-1} A_k(b_k - b_{k+1}) + A_q\, b_q - A_{p-1}b_p. \quad \square$$

As an application of the partial summation formula we prove the following theorem of Dirichlet. Another application is given in Exercise 1.

THEOREM 7.2.2 (Dirichlet Test) *Suppose $\{a_k\}$ and $\{b_k\}$ are sequences of real numbers satisfying*

(a) *the partial sums $A_n = \sum_{k=1}^{n} a_k$ form a bounded sequence,*

(b) *$b_1 \geq b_2 \geq b_3 \geq \cdots \geq 0$, and*

(c) *$\lim_{k \to \infty} b_k = 0$.*

Then $\sum_{k=1}^{\infty} a_k\, b_k$ converges.

Proof. Since $\{A_n\}$ is a bounded sequence, we can choose $M > 0$ such that $|A_n| \leq M$ for all n. Also, since $b_n \to 0$, given $\epsilon > 0$, there exists a positive integer n_o such that $b_n < \epsilon/2M$ for all $n \geq n_o$. Thus, if $n_o \leq p \leq q$, by the partial summation formula

$$\left| \sum_{k=p}^{q} a_k b_k \right| = \left| \sum_{k=p}^{q-1} A_k(b_k - b_{k+1}) + A_q b_q - A_{p-1} b_p \right|$$

$$\leq \sum_{k=p}^{q-1} |A_k|(b_k - b_{k+1}) + |A_q| b_q + |A_{p-1}| b_p$$

$$\leq M \left(\sum_{k=p}^{q-1} (b_k - b_{k+1}) + b_q + b_p \right)$$

$$\leq 2M b_p < \epsilon.$$

Hence by the Cauchy criterion (Theorem 3.7.3), $\sum_{k=1}^{\infty} a_k b_k$ converges. $\square$

Alternating Series

Our first application of the Dirichlet test is to alternating series. An **alternating series** is a series of the form $\sum (-1)^k b_k$ or $\sum (-1)^{k+1} b_k$, with $b_k \geq 0$ for all k.

THEOREM 7.2.3 (Alternating Series Test) *If $\{b_k\}$ is a sequence of real numbers satisfying*

(a) $b_1 \geq b_2 \geq \cdots \geq 0$, *and*

(b) $\lim_{k \to \infty} b_k = 0$,

then $\sum_{k=1}^{\infty} (-1)^{k+1} b_k$ *converges.*

Proof. Let $a_k = (-1)^{k+1}$. Then $|A_n| \leq 1$ for all n, and the Dirichlet test applies. $\square$

Remark. The hypothesis (a) that $\{b_k\}$ is decreasing is required. If we only assume that $b_k \geq 0$ and $\lim_{k \to \infty} b_k = 0$, then the conclusion is false (Exercise 2).

For an alternating series satisfying the hypothesis of Theorem 7.2.3, we can actually do better than just prove convergence. The following theorem provides an estimate on the rate of convergence of the partial sums $\{s_n\}$ to the sum of the series.

THEOREM 7.2.4 *Consider the series* $\sum_{k=1}^{\infty} (-1)^{k+1} b_k$, *where the sequence $\{b_k\}$ satisfies the hypothesis of Theorem 7.2.3. Let*

$$s_n = \sum_{k=1}^{n} (-1)^{k+1} b_k \quad \text{and} \quad s = \sum_{k=1}^{\infty} (-1)^{k+1} b_k.$$

Then $|s - s_n| \le b_{n+1}$ *for all* $n \in \mathbb{N}$.

Proof. Consider the sequence $\{s_{2n}\}$. Since

$$s_{2n} = \sum_{k=1}^{2n} (-1)^{k+1} b_k = (b_1 - b_2) + \cdots + (b_{2n-1} - b_{2n}),$$

and $(b_{k-1} - b_k) \ge 0$ for all k, the sequence $\{s_{2n}\}$ is monotone increasing. Similarly $\{s_{2n+1}\}$ is monotone decreasing. Since $\{s_n\}$ converges to s, so do the subsequences $\{s_{2n}\}$ and $\{s_{2n+1}\}$. Therefore $s_{2n} \le s \le s_{2n+1}$ for all $n \in \mathbb{N}$. As a consequence $|s - s_k| \le |s_{k+1} - s_k| = b_{k+1}$ for all $k \in \mathbb{N}$.   □

EXAMPLE 7.2.5 Since the sequence $\{1/(2k-1)\}$ decreases to zero, by Theorem 7.2.3 the series

$$\sum_{k=1}^{\infty} \frac{(-1)^{k+1}}{2k-1}$$

converges. As we will see in Chapter 9 (Example 9.5.7(a)), this series converges to $\pi/4$. Thus by Theorem 7.2.4, if s_n is the nth partial sum of the series,

$$\left| \frac{\pi}{4} - s_n \right| \le \frac{1}{2n+1}$$

for all $n \in \mathbb{N}$. Although this can be used to obtain an approximation to π, the convergence is very slow. We would have to take $n = 50$ to be guaranteed accuracy to two decimal places.   □

Trigonometric Series

Our next application of the Dirichlet test is to the convergence of trigonometric series. These types of series will be studied in much greater detail in Chapter 9.

THEOREM 7.2.6 (Trigonometric Series) *Suppose* $\{b_k\}$ *is a sequence of real numbers satisfying* $b_1 \ge b_2 \ge \cdots \ge 0$, *and* $\lim_{k \to \infty} b_k = 0$. *Then*

(a) $\sum_{k=1}^{\infty} b_k \sin kt$ *converges for all* $t \in \mathbb{R}$, *and*

(b) $\sum_{k=1}^{\infty} b_k \cos kt$ *converges for all* $t \in \mathbb{R}$, *except perhaps* $t = 2p\pi$, $p \in \mathbb{Z}$.

Proof. To prove the result, we require the following two identities: For $t \ne 2p\pi$, $p \in \mathbb{Z}$,

$$\sum_{k=1}^{n} \sin kt = \frac{\cos \frac{1}{2}t - \cos(n + \frac{1}{2})t}{2 \sin \frac{1}{2}t}, \tag{1}$$

$$\sum_{k=1}^{n} \cos kt = \frac{\sin(n + \frac{1}{2})t - \sin \frac{1}{2}t}{2 \sin \frac{1}{2}t}. \tag{2}$$

We will prove (1), leaving (2) for the exercises (Exercise 4). Set $A_n = \sum\limits_{k=1}^{n} \sin kt$.
Using the trigonometric identity $\sin A \sin B = \frac{1}{2}[\cos(A - B) - \cos(A + B)]$ we obtain

$$(\sin \tfrac{1}{2}t)\, A_n = \sum_{k=1}^{n} \sin \tfrac{1}{2}t \, \sin kt$$

$$= \frac{1}{2} \sum_{k=1}^{n} [\cos(k - \tfrac{1}{2})t - \cos(k + \tfrac{1}{2})t]$$

$$= \frac{1}{2} \left[\cos \tfrac{1}{2}t - \cos(n + \tfrac{1}{2})t\right].$$

Thus for $t \neq 2p\pi$, $p \in \mathbb{Z}$,

$$A_n = \frac{\cos \tfrac{1}{2}t - \cos(n + \tfrac{1}{2})t}{2 \sin \tfrac{1}{2}t},$$

and therefore

$$|A_n| \leq \frac{|\cos \tfrac{1}{2}t| + |\cos(n + \tfrac{1}{2})t|}{2|\sin \tfrac{1}{2}t|} \leq \frac{1}{|\sin \tfrac{1}{2}t|},$$

which is finite provided $t \neq 2p\pi$, $p \in \mathbb{Z}$. Consequently, by the Dirichlet test the series in (a) converges for all $t \neq 2p\pi$, $p \in \mathbb{Z}$. If $t = 2p\pi$, $p \in \mathbb{Z}$, then $\sin kt = 0$ for all k. Thus the series in (a) converges for all $t \in \mathbb{R}$.

The proof of the convergence of the series in (b) is similar. However in this case, when $t = 2p\pi$, $\cos kt = 1$ for all $k \in \mathbb{N}$ and the given series may or may not converge. $\square$

EXAMPLE 7.2.7 By Theorem 7.2.6, the series

$$\sum_{k=1}^{\infty} \frac{1}{k} \cos kt$$

converges for all $t \neq 2p\pi$, $p \in \mathbb{Z}$. When $t = 2p\pi$, $p \in \mathbb{Z}$, then $\cos kt = 1$ for all k and the series $\sum 1/k$ diverges. On the other hand, the series

$$\sum_{k-1}^{\infty} \frac{1}{k^2} \cos kt$$

converges for all $t \in \mathbb{R}$. $\square$

Exercises 7.2

1. *(**Abel's Test**) Prove that if $\sum a_k$ converges, and $\{b_k\}$ is monotone and bounded, then $\sum a_k b_k$ converges.

2. *Show by example that the hypothesis of Theorem 7.2.3 cannot be replaced by $b_k \geq 0$ and $\lim_{k \to \infty} b_k = 0$.

3. If $\sum a_k$ converges, does $\sum a_k^2$ always converge?

4. *Prove that
$$\sum_{k=1}^{n} \cos kt = \frac{\sin(n + \frac{1}{2})t - \sin \frac{1}{2}t}{2 \sin \frac{1}{2}t}, \quad t \neq 2p\pi, p \in \mathbb{Z}.$$

5. Test each of the following series for convergence:

   *a. $\sum_{k=1}^{\infty} \frac{(-1)^{k+1}}{k^p}, p > 0$ b. $\sum_{k=2}^{\infty} \frac{(-1)^k \ln k}{k}$ *c. $\sum_{k=2}^{\infty} \frac{(-1)^k}{k \ln k}$

   *d. $\sum_{k=1}^{\infty} (-1)^{k+1} \frac{k^k}{(k+1)^k}$ e. $\sum_{k=1}^{\infty} (-1)^{k+1} \frac{k^k}{(k+1)^{k+1}}$ *f. $\sum_{k=2}^{\infty} \frac{\sin k}{\ln k}$

   g. $\sum_{k=1}^{\infty} \frac{\sin kt}{k^p}, t \in \mathbb{R}, p > 0$ *h. $\sum_{k=1}^{\infty} \frac{\cos kt}{k^p}, t \in \mathbb{R}, p > 0$

   i. $\sum_{k=1}^{\infty} (-1)^{k+1} \sin(\pi/k)$

6. Given that
$$\sum_{k=1}^{\infty} \frac{(-1)^{k+1}}{k^2} = \pi^2/12,$$
determine how large $n \in \mathbb{N}$ must be chosen so that $|\pi^2/12 - s_n| < 10^{-4}$, where s_n is the nth partial sum of the series.

7. If p and q are strictly positive real numbers, show that
$$\sum_{k=2}^{\infty} (-1)^k \frac{(\ln k)^p}{k^q}$$
converges.

8. *Suppose that $\sum a_k$ converges. Prove that
$$\lim_{n \to \infty} \frac{1}{n} \sum_{k=1}^{n} k a_k = 0.$$

9. As in Exercise 19 of Section 7.1, let $c_k = 1 + \frac{1}{2} + \cdots + \frac{1}{k} - \ln k$. Set
$$b_n = 1 - \frac{1}{2} + \frac{1}{3} - \cdots - \frac{1}{2n} = \sum_{k=1}^{2n} \frac{(-1)^{k+1}}{k}.$$
Show that $\lim_{n \to \infty} b_n = \ln 2$. (Hint: $b_n = c_{2n} - c_n + \ln 2$.)

7.3   Absolute and Conditional Convergence

In this section, we introduce the concept of absolute convergence of a series. As we will see in this and subsequent sections of the text, the notion of absolute convergence is very important in the study of series. We begin with the definition of absolute and conditional convergence.

DEFINITION 7.3.1 *A series* $\sum a_k$ *of real numbers is said to be* **absolutely convergent** *(or converges absolutely) if* $\sum |a_k|$ *converges. The series is said to be* **conditionally convergent** *if it is convergent but not absolutely convergent.*

We illustrate these two definitions with the following examples.

EXAMPLES 7.3.2 (a) Since the sequence $\{1/k\}$ decreases to zero, by Theorem 7.2.3 the series $\sum (-1)^{k+1}/k$ converges. However,

$$\sum_{k=1}^{\infty} \left| \frac{(-1)^{k+1}}{k} \right| = \sum_{k=1}^{\infty} \frac{1}{k} = \infty.$$

Thus the series $\sum (-1)^{k+1}/k$ is conditionally convergent.

(b) Consider the series $\sum (-1)^{k+1}/k^2$. By Theorem 7.2.3 the alternating series converges. Furthermore, since $\sum 1/k^2 < \infty$, the series is absolutely convergent. $\square$

Our first result for absolutely convergent series is as follows:

THEOREM 7.3.3 *Every absolutely convergent series of real numbers converges.*

Proof. Suppose $\sum a_k$ converges absolutely; i.e. $\sum |a_k| < \infty$. By the triangle inequality, for $1 \leq p \leq q$,

$$\left| \sum_{k=p}^{q} a_k \right| \leq \sum_{k=p}^{q} |a_k|,$$

and the result now follows by the Cauchy criterion (Theorem 3.7.3). $\square$

Remark. To test a series $\sum a_k$ for absolute convergence we can apply any of the appropriate convergence tests of Section 7.1 to the series $\sum |a_k|$. There is however one important fact which needs to be emphasized. If the series $\sum |a_k|$ diverges by virtue of the ratio or root test, i.e.,

$$r = \lim_{n \to \infty} \frac{|a_{n+1}|}{|a_n|} > 1 \qquad \text{or} \qquad \alpha = \overline{\lim_{n \to \infty}} \sqrt[n]{|a_n|} > 1,$$

then not only does $\sum |a_k|$ diverge, but $\sum a_k$ also diverges.

To see this, suppose $\alpha > 1$. Then as in the proof of the root test, $|a_k| > 1$ for infinitely many k. Hence the sequence $\{a_k\}$ does not converge to zero, and thus by Corollary 3.7.5, $\sum a_k$ diverges. Similarly, if $r > 1$ and if $1 < c < r$, then as in the proof of Theorem 7.1.7(b), there exists a positive integer n_o and constant M such that

$$|a_n| \geq M c^n$$

for all $n \geq n_o$. Thus again, since $c > 1$, $\{a_n\}$ does not converge to zero, and the series $\sum a_k$ diverges. We summarize this as follows:

THEOREM 7.3.4 (Root and Ratio Test) *Let $\sum a_k$ be a series of real numbers, and let*

$$\alpha = \varlimsup_{k \to \infty} \sqrt[k]{|a_k|}.$$

Also, if $a_k \neq 0$ for all $k \in \mathbb{N}$, let

$$R = \varlimsup_{k \to \infty} \left| \frac{a_{k+1}}{a_k} \right| \qquad and \qquad r = \varliminf_{k \to \infty} \left| \frac{a_{k+1}}{a_k} \right|.$$

(a) *If $\alpha < 1$ or $R < 1$, then the series $\sum a_k$ is absolutely convergent.*

(b) *If $\alpha > 1$ or $r > 1$, then the series $\sum a_k$ is divergent.*

(c) *If $\alpha = 1$ or $r \leq 1 \leq R$, then the test is inconclusive.*

EXAMPLE 7.3.5 To illustrate the previous theorem we consider the series $\sum\limits_{k=1}^{\infty} p(k) b^k$, where p is a polynomial and $b \in \mathbb{R}$ with $|b| < 1$. In this example, $a_k = p(k) b^k$ and thus

$$\left| \frac{a_{k+1}}{a_k} \right| = |b| \left| \frac{p(k+1)}{p(k)} \right|.$$

Since $\lim\limits_{k \to \infty} p(k+1)/p(k) = 1$ (Exercise 8), $\lim\limits_{k \to \infty} |a_{k+1}/a_k| = |b| < 1$. Thus by the ratio test the series is absolutely convergent. □

Rearrangements of Series

We next take up the topic of rearrangements of series. Loosely speaking, a series $\sum a_k'$ is a *rearrangement* of the series $\sum a_k$ if all the terms in the original series $\sum a_k$ appear exactly once in the series $\sum a_k'$, but not necessarily in the same order. For example, the series

$$\frac{1}{1^2} + \frac{1}{3^2} + \frac{1}{2^2} + \frac{1}{4^2} + \frac{1}{5^2} + \frac{1}{7^2} + \frac{1}{6^2} + \cdots$$

is a rearrangement of the series

$$\sum_{k=1}^{\infty} \frac{1}{k^2}.$$

The following provides a formal definition of this concept.

DEFINITION 7.3.6 *A series* $\sum a'_k$ *is a* **rearrangement** *of the series* $\sum a_k$ *if there exists a one-to-one function j from $\mathbb{N}$ onto $\mathbb{N}$ such that $a'_k = a_{j(k)}$ for all $k \in \mathbb{N}$.*

A natural question to ask is the following: If the series $\sum a_k$ converges and $\sum a'_k$ is a rearrangement of $\sum a_k$, does the series $\sum a'_k$ converge? If it converges, does it converge to the same sum. As the following example illustrates, the answer to both of these questions is no!

EXAMPLE 7.3.7 Consider the series

$$\sum_{k=1}^{\infty} \frac{(-1)^{k+1}}{k} = 1 - \frac{1}{2} + \frac{1}{3} - \frac{1}{4} + \frac{1}{5} \cdots$$

which converges, but not absolutely. Consider also the following series which is a rearrangement of the above:

$$1 + \frac{1}{3} - \frac{1}{2} + \frac{1}{5} + \frac{1}{7} - \frac{1}{4} + \frac{1}{9} + \frac{1}{11} - \frac{1}{6} + \cdots . \tag{3}$$

Let

$$s = \sum_{k=1}^{\infty} (-1)^{k+1} \frac{1}{k}.$$

As in the proof of Theorem 7.2.4, $s < s_{2n+1}$ for all $n \in \mathbb{N}$. In particular,

$$s < s_3 = 1 - \frac{1}{2} + \frac{1}{3} = \frac{5}{6}.$$

Let s'_n denote the nth partial sum of the series (3). Then

$$s'_{3n} = \sum_{k=1}^{n} \left(\frac{1}{4k-3} + \frac{1}{4k-1} - \frac{1}{2k} \right).$$

Since

$$\frac{1}{4k-3} + \frac{1}{4k-1} - \frac{1}{2k} = \frac{8k-3}{2k(4k-1)(4k-3)}$$

we have

$$0 < \frac{1}{4k-3} + \frac{1}{4k-1} - \frac{1}{2k} \le M \frac{1}{k^2}$$

for some constant M. Thus the sequence $\{s'_{3n}\}$ is strictly increasing, and by the comparison test converges. Let $s' = \lim_{n \to \infty} s'_{3n}$. Since

$$s'_{3n+1} = s'_{3n} + \frac{1}{4n+1}, \quad \text{and}$$
$$s'_{3n+2} = s'_{3n} + \frac{1}{4n+1} + \frac{1}{4n+3},$$

the sequences $\{s'_{3n+1}\}$ and $\{s'_{3n+2}\}$ also converge to s'. Therefore $\lim_{n\to\infty} s'_n = s'$. Thus the series (3) also converges. However, since

$$\frac{5}{6} = s'_3 < s'_6 < s'_9 < \cdots,$$

$s' = \lim_{n\to\infty} s'_n > \frac{5}{6}$. Thus the series (3) does not converge to the same sum as the original series. This, as we will see, is due to the fact that the given series does not converge absolutely.   □

THEOREM 7.3.8 *If the series $\sum a_k$ converges absolutely, then every rearrangement of $\sum a_k$ converges to the same sum.*

Proof. Let $\sum a'_k$ be a rearrangement of $\sum a_k$. Since $\sum |a_k| < \infty$, given $\epsilon > 0$, there exists a positive integer N such that

$$\sum_{k=n}^{m} |a_k| < \epsilon \tag{4}$$

for all $m \geq n \geq N$. Suppose $a'_k = a_{j(k)}$, where j is a one-to-one function of $\mathbb{N}$ onto $\mathbb{N}$. Choose an integer $p\,(p \geq N)$ such that

$$\{1, 2, ..., N\} \subset \{j(1), j(2), ..., j(p)\}.$$

Such a p exists since the function j is one-to-one and onto. Let

$$s_n = \sum_{k=1}^{n} a_k \qquad \text{and} \qquad s'_n = \sum_{k=1}^{n} a'_k.$$

If $n \geq p$,

$$s_n - s'_n = \sum_{k=1}^{n} a_k - \sum_{k=1}^{n} a_{j(k)}.$$

By the choice of p, the numbers $a_1, ..., a_N$ appear in both sums and consequently will cancel. Thus the only terms remaining will have index k or $j(k)$ greater than or equal to N. Therefore by (4),

$$|s_n - s'_n| \leq 2\epsilon$$

for all $n \geq p$. Therefore $\lim_{n\to\infty} s'_n = \lim_{n\to\infty} s_n$, and thus the rearrangement converges to the same sum as the original series.   □

For conditionally convergent series we have the following very interesting result of Riemann.

THEOREM 7.3.9 *Let $\sum a_k$ be a conditionally convergent series of real numbers. Suppose $\alpha \in \mathbb{R}$. Then there exists a rearrangement $\sum a'_k$ of $\sum a_k$ which converges to α.*

Before proving the theorem, we illustrate the method of proof with the alternating series

$$\sum_{k=1}^{\infty} \frac{(-1)^{k+1}}{k} = 1 - \frac{1}{2} + \frac{1}{3} - \frac{1}{4} + \cdots.$$

This series converges, but fails to converge absolutely. The positive terms of this series are $P_k = 1/(2k-1)$, $k \in \mathbb{N}$, and the absolute value of the negative terms are $Q_k = 1/2k$, $k \in \mathbb{N}$. Since $\sum \frac{1}{k} = \infty$, we also have $\sum P_k = \sum Q_k = \infty$. Suppose for purposes of illustration $\alpha = 1.5$. Our first step is to add "just enough" positive terms to exceed α. More precisely, we let m_1 be the smallest integer such that $P_1 + \cdots + P_{m_1} > \alpha$. For $\alpha = 1.5$, $m_1 = 3$; i.e., $1 + \frac{1}{3} + \frac{1}{5} > 1.5$ whereas $1 + \frac{1}{3} < 1.5$. Our next step is to go in the other direction; namely, we let n_1 be the smallest integer such that

$$P_1 + \cdots + P_{m_1} - Q_1 - \cdots - Q_{n_1} < \alpha.$$

Again, for $\alpha = 1.5$, $n_1 = 1$, and $1 + \frac{1}{3} + \frac{1}{5} - \frac{1}{2} < 1.5$. We are able to do this since the series $\sum P_k$ and $\sum Q_k$ both diverge to ∞. We continue this process inductively. Choosing the smallest integers m_k and n_k at each stage of the construction will be the key in proving that the resulting series converges to α.

Proof. Without loss of generality we assume that $a_k \neq 0$ for all k. Let p_k and q_k be defined by

$$p_k = \frac{1}{2}(|a_k| + a_k) \qquad \text{and} \qquad q_k = \frac{1}{2}(|a_k| - a_k).$$

Then $p_k - q_k = a_k$ and $p_k + q_k = |a_k|$. Furthermore, if $a_k > 0$, then $q_k = 0$ and $p_k = a_k$; if $a_k < 0$, then $p_k = 0$ and $q_k = |a_k|$.

We first prove that the series $\sum p_k$ and $\sum q_k$ both diverge. Since

$$\sum_{k=1}^{\infty}(p_k + q_k) = \sum_{k=1}^{\infty} |a_k|,$$

they cannot both converge. Also, since

$$\sum_{k=1}^{n} a_k = \sum_{k=1}^{n} p_k - \sum_{k=1}^{n} q_k,$$

the convergence of one implies the convergence of the other. Thus they both must diverge.

Let P_1, P_2, P_3, ... denote the positive terms of $\sum a_k$ in the original order, and let Q_1, Q_2, Q_3, ... be the absolute values of the negative terms, also in the original order. The series $\sum P_k$ and $\sum Q_k$ differ from $\sum p_k$ and $\sum q_k$ only by zero terms and thus are also divergent.

We will inductively construct sequences $\{m_k\}$ and $\{n_k\}$ such that the series

$$P_1 + \cdots + P_{m_1} - Q_1 - \cdots - Q_{n_1} + P_{m_1+1} + \cdots P_{m_2} - Q_{n_1+1} - \cdots - Q_{n_2} + \cdots \quad (5)$$

has the desired property. Clearly (5) is a rearrangement of the original series.

Let m_1 be the smallest integer such that

$$X_1 = P_1 + \cdots + P_{m_1} > \alpha.$$

Such an m_1 exists since $\sum P_k = \infty$. Similarly, let n_1 be the smallest integer such that

$$Y_1 = X_1 - Q_1 - \cdots - Q_{n_1} < \alpha.$$

Suppose $\{m_1, ..., m_k\}$ and $\{n_1, ..., n_k\}$ have been chosen. Let m_{k+1} and n_{k+1} be the smallest integers greater than m_k and n_k respectively such that

$$X_{k+1} = Y_k + P_{m_k+1} + \cdots + P_{m_{k+1}} > \alpha, \qquad \text{and}$$

$$Y_{k+1} = X_{k+1} - Q_{n_k+1} - \cdots - Q_{n_{k+1}} < \alpha.$$

Such integers always exist due to the divergence of the series $\sum P_k$ and $\sum Q_k$. Since m_{k+1} was chosen to be the smallest integer such that the above holds,

$$X_{k+1} - P_{m_{k+1}} \leq \alpha.$$

Therefore

$$0 < X_{k+1} - \alpha \leq P_{m_{k+1}}.$$

Similarly,

$$0 < \alpha - Y_{k+1} \leq Q_{n_{k+1}}.$$

Since $\sum a_n$ converges, we have $\lim_{k \to \infty} P_k = \lim_{k \to \infty} Q_k = 0$. Therefore,

$$\lim_{k \to \infty} X_k = \lim_{k \to \infty} Y_k = \alpha.$$

Let S_n be the nth partial sum of the series (5). If the last term of S_n is a P_n, then there exists a k such that

$$Y_k < S_n \leq X_{k+1}.$$

If the last term of S_n is $-Q_n$, then there exists a k such that

$$Y_{k+1} \leq S_n < X_{k+1}.$$

In either case we obtain $\lim_{n \to \infty} S_n = \alpha$, which proves the result. $\quad \square$

Remark. By a variation of the above proof one can show that if the series $\sum a_k$ is conditionally convergent, then given α, β with

$$-\infty \leq \alpha \leq \beta \leq \infty,$$

there exists a rearrangement $\sum a_k'$ of $\sum a_k$ such that

$$\underline{\lim}_{n \to \infty} S_n = \alpha, \qquad \text{and} \qquad \overline{\lim}_{n \to \infty} S_n = \beta,$$

where S_n is the nth partial sum of $\sum a_k'$ (Exercise 13).

Exercises 7.3

1. Prove the following:

    a. If $\lim\limits_{k \to \infty} k^p a_k = A$ for some $p > 1$, then $\sum a_k$ converges absolutely.

    b. If $\lim\limits_{k \to \infty} k\, a_k = A \neq 0$, then $\sum a_k$ diverges.

    c. If $\lim\limits_{k \to \infty} k\, a_k = 0$, then the test is inconclusive.

2. *Suppose $\sum a_k^2 < \infty$ and $\sum b_k^2 < \infty$. Prove that the series $\sum a_k b_k$ converges absolutely.

3. **a.** Prove that if $\sum a_k$ converges and $\sum b_k$ converges absolutely, then $\sum a_k b_k$ converges.

    b. Show by example that the conclusion may be false if one of the two series does not converge absolutely.

4. *Suppose that the sequence $\{b_n\}$ is monotone decreasing with $\lim b_n = 0$. If $\{a_n\}$ is a sequence in $\mathbb{R}$ satisfying $|a_n| \leq b_n - b_{n+1}$ for all $n \in \mathbb{N}$, prove that $\sum a_k$ converges absolutely.

5. **a.** If $\sum a_k$ converges absolutely, prove that $\sum \epsilon_k a_k$ converges for every choice of $\epsilon_k \in \{-1, 1\}$.

    b. If $\sum \epsilon_k a_k$ converges for every choice of $\epsilon_k \in \{-1, 1\}$, prove that $\sum a_k$ converges absolutely.

6. Test each of the following series for absolute and conditional convergence.

    *a. $\displaystyle\sum_{k=1}^{\infty} \frac{(-1)^{k+1}}{\sqrt{k}}$
    b. $\displaystyle\sum_{k=3}^{\infty} \frac{(-1)^k}{\sqrt{k}\ln(\ln k)}$
    *c. $\displaystyle\sum_{k=1}^{\infty} \frac{(-1)^{k+1}}{k^p},\ p > 0$

    d. $\displaystyle\sum_{k=3}^{\infty} \frac{(-1)^k \ln(\ln k)}{\sqrt{\ln k}}$
    *e. $\displaystyle\sum_{k=2}^{\infty} \frac{(-1)^k}{k(\ln k)^p},\ p > 0$
    f. $\displaystyle\sum_{k=1}^{\infty} \frac{(-1)^{k+1}k^k}{(k+1)^{k+1}}$

    *g. $\displaystyle\sum_{k=1}^{\infty} \frac{(-1)^{k+1}k^k}{(k+1)^k}$
    h. $\displaystyle\sum_{k=1}^{\infty} (-1)^{k+1} \sin\left(\frac{1}{k}\right)$
    *i. $\displaystyle\sum_{k=2}^{\infty} \frac{(-1)^k}{k^2 + (-1)^k}$

7. Test the series $\displaystyle\sum_{k=1}^{\infty} \frac{p^k}{k^p}$, $p \in \mathbb{R}$, for absolute and conditional convergence.

8. Prove that $\lim\limits_{k \to \infty} \dfrac{|p(k+1)|}{|p(k)|} = 1$ for any polynomial p.

9. *Show that the series $1 + \dfrac{1}{2} - \dfrac{1}{3} + \dfrac{1}{4} + \dfrac{1}{5} - \dfrac{1}{6} + \cdots$ diverges.

10. Determine whether the series $1 - \dfrac{1}{2} - \dfrac{1}{3} + \dfrac{1}{4} + \dfrac{1}{5} - \dfrac{1}{6} - \dfrac{1}{7} + \cdots$ converges or diverges.

11. *Prove that the series $\displaystyle\sum_{k=1}^{\infty} \frac{\sin k}{k}$ is conditionally convergent.

12. If $a_k \geq 0$ for all $k \in \mathbb{N}$, and $\sum a_k = \infty$, prove that $\sum a_k' = \infty$ for any rearrangement $\sum a_k'$ of $\sum a_k$.

13. Suppose that the series $\sum a_k$ is conditionally convergent. Given α, β with $-\infty \leq \alpha \leq \beta \leq \infty$, show that there exists a rearrangement $\sum a'_k$ of $\sum a_k$ such that $\varliminf_{n\to\infty} S_n = \alpha$ and $\varlimsup_{n\to\infty} S_n = \beta$, where S_n is the nth partial sum of $\sum a'_k$.

14. Suppose every rearrangement of the series $\sum a_k$ converges, prove that $\sum a_k$ converges absolutely.

7.4   Square Summable Sequences[2]

In this section, we introduce the set ℓ^2 of square summable sequences of real numbers and derive several useful inequalities. This set occurs naturally in the study of Fourier series.

DEFINITION 7.4.1 *A sequence* $\{a_k\}_{k=1}^{\infty}$ *of real numbers is said to be in* ℓ^2, *or to be* **square summable**, *if*

$$\sum_{k=1}^{\infty} a_k^2 < \infty.$$

For $\{a_k\} \in \ell^2$ *set*

$$\|\{a_k\}\|_2 = \sqrt{\sum_{k=1}^{\infty} a_k^2}.$$

The set ℓ^2 *is called the space of* **square summable sequences**, *and the quantity* $\|\{a_k\}\|_2$ *is called the* **norm** *of the sequence* $\{a_k\}$.

Remark. Since a sequence $\{a_k\}$ in $\mathbb{R}$ is by definition a function $\mathbf{a}$ from $\mathbb{N}$ into $\mathbb{R}$ with $a_k = \mathbf{a}(k)$, it is sometimes convenient to think of ℓ^2 as the set of all functions $\mathbf{a} : \mathbb{N} \to \mathbb{R}$ for which

$$\|\mathbf{a}\|_2 = \sqrt{\sum_{k=1}^{\infty} |\mathbf{a}(k)|^2} < \infty.$$

---

[2]The topic of square summable sequences, although important in the study of Fourier series, is not specifically required in Chapter 9 and thus can be omitted on first reading. The concept of a normed linear space occurs on several occassions in the discussion of subsequent topics in the text. At that point the reader can study this topic more carefully.

EXAMPLES 7.4.2 (a) For the sequence $\{1/k\}_{k=1}^{\infty}$,

$$\|\{1/k\}\|_2^2 = \sum_{k=1}^{\infty} \frac{1}{k^2}.$$

Since this is a p-series with $p = 2$, the series $\sum 1/k^2$ converges and thus $\{1/k\} \in \ell^2$. On the other hand, since

$$\|\{1/\sqrt{k}\}\|_2^2 = \sum_{k=1}^{\infty} \frac{1}{k} = \infty,$$

we have $\{1/\sqrt{k}\} \notin \ell^2$.

(b) For fixed q, $0 < q < \infty$, consider the sequence $\{1/k^q\}$. Then

$$\|\{1/k^q\}\|_2^2 = \sum_{k=1}^{\infty} \frac{1}{k^{2q}}.$$

By Example 7.1.9(a) this series converges for all $q > 1/2$ and diverges for all $q \le 1/2$. Thus $\{1/k^q\} \in \ell^2$ if and only if $q > 1/2$. $\quad\square$

Cauchy-Schwarz Inequality

Our main goal in this section is to prove the Cauchy-Schwarz inequality for sequences in ℓ^2. First however we prove the finite version of this inequality.

THEOREM 7.4.3 (Cauchy-Schwarz Inequality) *If $n \in \mathbb{N}$ and $a_1, ..., a_n$ and $b_1, ..., b_n$ are real numbers, then*

$$\sum_{k=1}^{n} |a_k b_k| \le \sqrt{\sum_{k=1}^{n} a_k^2} \sqrt{\sum_{k=1}^{n} b_k^2}.$$

Proof. Let $\lambda \in \mathbb{R}$ and consider

$$0 \le \sum_{k=1}^{n}(|a_k| - \lambda|b_k|)^2 = \sum_{k=1}^{n} a_k^2 - 2\lambda \sum_{k=1}^{n} |a_k b_k| + \lambda^2 \sum_{k=1}^{n} b_k^2.$$

The above can be written as

$$0 \le A - 2\lambda C + \lambda^2 B,$$

where $A = \sum_{k=1}^{n} a_k^2$, $C = \sum_{k=1}^{n} |a_k b_k|$, and $B = \sum_{k=1}^{n} b_k^2$. If $B = 0$, then $b_k = 0$ for all $k = 1, ..., n$ and the Cauchy-Schwarz inequality certainly holds. If $B \ne 0$, we take $\lambda = C/B$ which gives

$$0 \le A - \frac{C^2}{B}.$$

or $C^2 \leq AB$; that is,

$$\left(\sum_{k=1}^{n} |a_k b_k|\right)^2 \leq \left(\sum_{k=1}^{n} a_k^2\right)\left(\sum_{k=1}^{n} b_k^2\right).$$

Taking the square root of both sides gives the desired result. $\square$

As a consequence of the previous result we have the following corollary.

COROLLARY 7.4.4 (Cauchy-Schwarz Inequality) *If $\{a_k\}$, $\{b_k\} \in \ell^2$, then $\sum_{k=1}^{\infty} a_k b_k$ is absolutely convergent and*

$$\sum_{k=1}^{\infty} |a_k b_k| \leq \|\{a_k\}\|_2 \, \|\{b_k\}\|_2.$$

Proof. For each $n \in \mathbb{N}$, by the previous theorem

$$\sum_{k=1}^{n} |a_k b_k| \leq \sqrt{\sum_{k=1}^{n} a_k^2} \sqrt{\sum_{k=1}^{n} b_k^2} \leq \|\{a_k\}\|_2 \|\{b_k\}\|_2.$$

Letting $n \to \infty$ gives the desired result. $\square$

For $\mathbf{a}, \mathbf{b} \in \ell^2$ the **inner product** of $\mathbf{a}$ and $\mathbf{b}$, denoted $\langle \mathbf{a}, \mathbf{b} \rangle$, is defined as

$$\langle \mathbf{a}, \mathbf{b} \rangle = \sum_{k=1}^{\infty} a_k b_k.$$

As a consequence of the Cauchy-Schwarz inequality we have

$$|\langle \mathbf{a}, \mathbf{b}, \rangle| \leq \|\mathbf{a}\|_2 \|\mathbf{b}\|_2.$$

Minkowski's Inequality

Our next result shows that the norm $\| \; \|_2$ satisfies the "triangle inequality" on ℓ^2.

THEOREM 7.4.5 (Minkowski's Inequality) *If $\{a_k\}$ and $\{b_k\}$ are in ℓ^2, then $\{a_k + b_k\} \in \ell^2$ and*

$$\|\{a_k + b_k\}\|_2 \leq \|\{a_k\}\|_2 + \|\{b_k\}\|_2.$$

Proof. By hypothesis, each of the series $\sum a_k^2$ and $\sum b_k^2$ converge. Also, by Corollary 7.4.4 the series $\sum a_k b_k$ converges absolutely. Since

$$(a_k + b_k)^2 = a_k^2 + 2a_k b_k + b_k^2 \leq a_k^2 + 2|a_k b_k| + b_k^2,$$

we have

$$\|\{a_k + b_k\}\|_2^2 = \sum_{k=1}^{\infty}(a_k + b_k)^2 \le \|\{a_k\}\|_2^2 + 2\sum_{k=1}^{\infty}|a_k b_k| + \|\{b_k\}\|_2^2,$$

which by the Cauchy Schwarz inequality

$$\le \|\{a_k\}\|_2^2 + 2\|\{a_k\}\|_2\|\{b_k\}\|_2 + \|\{b_k\}\|_2^2$$
$$= (\|\{a_k\}\|_2 + \|\{b_k\}\|_2)^2.$$

Taking the square root of both sides gives the desired result. □

Although not specifically stated, Minkowski's inequality is also true for finite sums. In particular if $n \in \mathbb{N}$ and $a_1, ..., a_n, b_1, ..., b_n$, are real numbers, then

$$\sqrt{\sum_{k=1}^{n}(a_k + b_k)^2} \le \sqrt{\sum_{k=1}^{n}a_k^2} + \sqrt{\sum_{k=1}^{n}b_k^2}. \tag{6}$$

In the following theorem we summarize some of the properties of the norm $\| \ \|_2$ on ℓ^2. These are very similar to the properties of the absolute value function on $\mathbb{R}$. As for the absolute value, the inequality

$$\|\{a_k + b_k\}\|_2 \le \|\{a_k\}\|_2 + \|\{b_k\}\|_2$$

is also referred to as the **triangle inequality** for ℓ^2.

THEOREM 7.4.6 *The norm* $\| \ \|_2$ *on* ℓ^2 *satisfies the following properties:*

(a) $\|\{a_k\}\|_2 \ge 0$ *for all* $\{a_k\} \in \ell^2$.

(b) $\|\{a_k\}\|_2 = 0$ *if and only if* $a_k = 0$ *for all* $k \in \mathbb{N}$.

(c) *If* $\{a_k\} \in l^2$ *and* $c \in \mathbb{R}$, *then* $\{ca_k\} \in \ell^2$ *and*

$$\|\{ca_k\}\|_2 = |c| \ \|\{a_k\}\|_2.$$

(d) *If* $\{a_k\}, \{b_k\} \in \ell^2$, *then* $\{a_k + b_k\} \in \ell^2$ *and*

$$\|\{a_k + b_k\}\|_2 \le \|\{a_k\}\|_2 + \|\{b_k\}\|_2.$$

Proof. The results (a) and (b) are obvious from the definition, and (d) is just a restatement of Minkowski's inequality. The verification of (c) is left as an exercise (Exercise 5) □

Normed Linear Spaces

The space ℓ^2 as well as $\mathbb{R}^n$ are both examples of vector spaces over $\mathbb{R}$. For completeness we include the definition of a vector space.

DEFINITION 7.4.7 *A set X with two operations "$+$", vector addition, and "$\cdot$", scalar multiplication, satisfying*

$$\mathbf{x} + \mathbf{y} \in X \qquad \text{for all} \quad \mathbf{x}, \mathbf{y} \in X, \quad \text{and}$$
$$c \cdot \mathbf{x} \in X \qquad \text{for all} \quad \mathbf{x} \in X, c \in \mathbb{R}$$

is a **vector space** *over $\mathbb{R}$ if the following are satisfied:*

(a) $\mathbf{x} + \mathbf{y} = \mathbf{y} + \mathbf{x}$. *(commutative law)*

(b) $\mathbf{x} + (\mathbf{y} + \mathbf{z}) = (\mathbf{x} + \mathbf{y}) + \mathbf{z}$. *(associative law)*

(c) *There is a unique element in X called the* **zero element**, *denoted* $\mathbf{0}$, *such that*

$$\mathbf{x} + \mathbf{0} = \mathbf{x} \qquad \text{for all} \quad \mathbf{x} \in X.$$

(d) *For each $\mathbf{x} \in X$, there exists a unique element $-\mathbf{x} \in X$ such that* $\mathbf{x} + (-\mathbf{x}) = \mathbf{0}$.

(e) $(ab) \cdot \mathbf{x} = a \cdot (b \cdot \mathbf{x})$ *for all $a, b \in \mathbb{R}$, $\mathbf{x} \in X$.*

(f) $a \cdot (\mathbf{x} + \mathbf{y}) = a \cdot \mathbf{x} + b \cdot \mathbf{y}$.

(g) $(a + b) \cdot \mathbf{x} = a \cdot \mathbf{x} + b \cdot \mathbf{x}$.

(h) $1 \cdot \mathbf{x} = \mathbf{x}$.

It is clear that $\mathbb{R}^n$ with $+$ and $\cdot$ defined by

$$\mathbf{a} + \mathbf{b} = (a_1 + b_1, ..., a_n + b_n),$$
$$c \cdot \mathbf{a} = (ca_1, ..., ca_n)$$

is a vector space over $\mathbb{R}$. Similarly if addition and scalar multiplication of sequences $\{a_k\}$, $\{b_k\}$ in ℓ^2 are defined by

$$\{a_k\} + \{b_k\} = \{a_k + b_k\},$$
$$c \cdot \{a_k\} = \{ca_k\}, \quad c \in \mathbb{R},$$

then it is easily shown that ℓ^2 is a vector space over $\mathbb{R}$. The fact that $\{a_k + b_k\}$ and $\{ca_k\}$ are in ℓ^2 is part of the statement of Theorems 7.4.5 and 7.4.6. The zero element $\mathbf{0}$ in ℓ^2 is the sequence $\{a_k\}$ where $a_k = 0$ for all $k \in \mathbb{N}$.

In addition to being vector spaces, the spaces $\mathbb{R}^n$ and ℓ^2 are also examples of normed linear spaces. The concepts of a "normed linear space" as well as that of a "norm" are defined as follows:

DEFINITION 7.4.8 *Let X be a vector space of $\mathbb{R}$. A function $\| \ \| : X \to \mathbb{R}$ satisfying*

(a) $\|\mathbf{x}\| \geq 0$ *for all $\mathbf{x} \in X$,*

(b) $\|\mathbf{x}\| = 0$ *if and only if $\mathbf{x} = \mathbf{0}$,*

(c) $\|c\mathbf{x}\| = |c| \, \|\mathbf{x}\|$ *for all $c \in \mathbb{R}$, $\mathbf{x} \in X$, and*

(d) $\|\mathbf{x} + \mathbf{y}\| \leq \|\mathbf{x}\| + \|\mathbf{y}\|$ *for all $\mathbf{x}, \mathbf{y} \in X$,*

is called a **norm** *on X. The pair $(X, \| \ \|)$ is called a* **normed linear space**.

Inequality (d) is called the **triangle inequality** for the norm $\| \ \|$. By Theorem 7.4.6, $(\ell^2, \| \ \|_2)$ is a normed linear space. Additional examples of normed linear spaces will occur elsewhere in the book, both in the text and the exercises.

If $(X, \| \ \|)$ is a normed linear space, the **distance** $d(\mathbf{x}, \mathbf{y})$ between two points $\mathbf{x}, \mathbf{y} \in X$ is defined as $d(\mathbf{x}, \mathbf{y}) = \|\mathbf{x} - \mathbf{y}\|$. From the definition of the norm $\| \ \|$, it immediately follows that

(a) $d(\mathbf{x}, \mathbf{y}) \geq 0$ for all $\mathbf{x}, \mathbf{y} \in X$,

(b) $d(\mathbf{x}, \mathbf{y}) = 0$ if and only if $\mathbf{x} = \mathbf{y}$,

(c) $d(\mathbf{x}, \mathbf{y}) = d(\mathbf{y}, \mathbf{x})$, and

(d) $d(\mathbf{x}, \mathbf{y}) \leq d(\mathbf{x}, \mathbf{z}) + d(\mathbf{z}, \mathbf{y})$ for all $\mathbf{x}, \mathbf{y}, \mathbf{z} \in X$.

Thus d is a **metric** on the vector space X. Since the notions of "limit point of a set" (Definition 2.2.12) and "convergence of a sequence" (Definition 3.1.1) are defined in terms of ϵ-neighborhoods, both of these concepts have analogous definitions in a normed linear space $(X, \| \ \|)$. Thus for example, a sequence $\{\mathbf{x}_n\}$ in X converges to $\mathbf{x} \in X$, if given $\epsilon > 0$, there exists $n_o \in \mathbb{N}$ such that $\|\mathbf{x}_n - \mathbf{x}\| < \epsilon$ for all $n \in \mathbb{N}$, $n \geq n_o$. This type of convergence, referred to as **norm convergence**, will be encountered again later in the text.

As a general rule, many of the results involving sequences of real numbers are still valid in the setting of normed linear space. This is especially true for those theorems whose proofs relied only on properties of the absolute value. Great care however must be exercised with theorems that rely on the supremum property of $\mathbb{R}$. For example, the Bolzano-Weierstrass theorem fails for $(\ell^2, \| \ \|_2)$ (Exercise 6). On the other hand however, the Bolzano-Weierstrass theorem still holds in $(\mathbb{R}^n, \| \ \|_2)$ (Miscellaneous Exercise 5).

Exercises 7.4

1. Determine which of the following sequences are in ℓ^2.

   *a. $\left\{ \dfrac{1}{\ln k} \right\}_{k=2}^{\infty}$ b. $\left\{ \dfrac{1}{\sqrt{k}\ln k} \right\}_{k=2}^{\infty}$ *c. $\left\{ \dfrac{\ln k}{\sqrt{k}} \right\}_{k=2}^{\infty}$ a. $\left\{ \dfrac{\sin k}{k} \right\}_{k=1}^{\infty}$.

2. Determine all values of $p \in \mathbb{R}$ such that the given sequence is in ℓ^2.

   *a. $\{p^k\}_{k=1}^{\infty}$ b. $\left\{ \dfrac{k^p}{p^k} \right\}_{k=1}^{\infty}$ *c. $\left\{ \dfrac{1}{k^p \ln k} \right\}_{k=2}^{\infty}$ d. $\left\{ \dfrac{1}{\sqrt{k}(\ln k)^p} \right\}_{k=2}^{\infty}$.

3. *If $\{a_k\} \in \ell^2$, prove that $\sum\limits_{k=1}^{\infty} a_k/k$ converges absolutely.

4. Give an example of a sequence $\{a_k\} \in \ell^2$ for which $\sum\limits_{k=1}^{\infty} |a_k| = \infty$.

5. If $\{a_k\} \in \ell^2$ and $c \in \mathbb{R}$, prove that $\{ca_k\} \in \ell^2$ and $\|\{ca_k\}\|_2 = |c|\|\{a_k\}\|_2$.

6. For each $n \in \mathbb{N}$, let $\mathbf{e}_n$ be the sequence in ℓ^2 defined by

$$e_n(k) = \begin{cases} 0, & k \neq n, \\ 1, & k = n. \end{cases}$$

Show that $\|\mathbf{e}_n - \mathbf{e}_m\|_2 = \sqrt{2}$ if $n \neq m$. (**Remark.** The sequence $\{\mathbf{e}_n\}$ is a bounded sequence in ℓ^2 with no convergent subsequence. Thus the Bolzano-Weierstrass theorem (3.4.6) fails in ℓ^2.)

7. Show that equality holds in the Cauchy-Schwarz inequality if and only if for all $k \in \mathbb{N}$, $b_k = ca_k$ for some $c \in \mathbb{R}$.

8. For $\mathbf{a}, \mathbf{b} \in \mathbb{R}^n$, let $\langle \mathbf{a}, \mathbf{b} \rangle$ denote the inner product of $\mathbf{a}$ and $\mathbf{b}$. Prove each of the following:

   a. $\langle \mathbf{a}, \mathbf{a} \rangle \geq 0$.

   b. $\langle \mathbf{a}, \mathbf{a} \rangle = 0$ if and only if $\mathbf{a} = (0, ..., 0)$.

   c. $\langle \mathbf{a}, \mathbf{b} + \mathbf{c} \rangle = \langle \mathbf{a}, \mathbf{b} \rangle + \langle \mathbf{a}, \mathbf{c} \rangle$.

   d. $\langle \mathbf{a}, \mathbf{b} \rangle = \langle \mathbf{b}, \mathbf{a} \rangle$.

   e. $\langle \mathbf{a}, \mathbf{b} \rangle = \frac{1}{2} \left[\|\mathbf{a}\|_2^2 + \|\mathbf{b}\|_2^2 - \|\mathbf{a} - \mathbf{b}\|_2^2 \right]$.

9. *For $\mathbf{a}, \mathbf{b} \in \mathbb{R}^n$, use the law of cosines to prove that $\langle \mathbf{a}, \mathbf{b} \rangle = \|\mathbf{a}\|_2 \|\mathbf{b}\|_2 \cos\theta$, where θ is the angle between the vectors $\mathbf{a}$ and $\mathbf{b}$.

10. Let $(X, \| \ \|)$ be a normed linear space. Prove that $|\, \|\mathbf{x}\| - \|\mathbf{y}\| \,| \leq \|\mathbf{x} - \mathbf{y}\|$ for all $\mathbf{x}, \mathbf{y} \in X$.

11. Let ℓ^1 denote the set of all sequences $\{a_k\}$ satisfying $\|\{a_k\}\|_1 = \sum_{k=1}^{\infty} |a_k| < \infty$.

    a. Prove that $(\ell^1, \| \ \|_1)$ is a normed linear space.

    b. Prove that $\ell^1 \subsetneq \ell^2$.

12. Determine all values of $p \in \mathbb{R}$ for which each of the sequences in Exercise 2 is in ℓ^1

13. Let X be a non-empty set, and let $\mathcal{B}(X)$ denote the set of all bounded real-valued functions on X. For $f \in \mathcal{B}(x)$, set

    $$\|f\|_\infty = \sup\{|f(x)| : x \in X\}.$$

    Prove that $(\mathcal{B}(X), \| \ \|_\infty)$ is a normed linear space.

Notes

The geometric series is perhaps one of the most important series in analysis. In the seventeenth and eighteenth century, the convergence of many series was established by comparison with the geometric series. It forms the basis for the proof of the root and ratio test, and thus also for the study of convergence of power series. The geometric series dates back to Euclid in the 3rd century B.C. The formula for a finite sum of a geometric progression appeared in Euclid's *Elements*, and Archimedes, in his treatise *Quadrature of the Parabola*, indirectly used the geometric series to find the area under a parabolic arc.[3] Even though Greek mathematicians knew how to

---

[3] See p. 105 of the text by Katz.

sum a finite geometric progression, they used *reducto ad absurdum* arguments to avoid dealing with infinite quantities.

Infinite series first appeared in the middle ages. In the fourteenth century, Nicole Oresme (1323?–1382) of Italy provided a geometric proof to the effect that the infinite series

$$\frac{1}{2} + \frac{2}{2^2} + \cdots + \frac{n}{2^n} + \cdots$$

had sum equal to 2, a result now quite easy to prove using power series. He was also the first to prove that the harmonic series diverged, whereas the Italian mathematician Pietro Mengoli (1624–1686) was the first to show that the sum of the alternating harmonic series $1 - \frac{1}{2} + \frac{1}{3} - \frac{1}{4} + \cdots$ is $\ln 2$. The results of Oresme and Mengoli, as well as others of this era, were stated verbally; the infinite sum notation did not appear until the seventeenth century.

With the development of the calculus in the mid seventeenth century, the emphasis shifted to the study of power series expansions of functions, and computations involving power series. The expansion of functions in power series was for Newton, and his successors, an indispensable tool. In his treatise *Of the Method of Fluxions and Infinite series*[4], Newton includes a discussion of infinite series techniques for the solution of both algebraic and differential equations. Both the geometric series and the binomial series were employed in many of his computations.

During the seventeenth and eighteenth century, mathematicians operated with power series in the same way as with polynomials; often ignoring questions of convergence. An excellent illustration of these eighteenth century techniques is Euler's derivation of the sum of the series

$$\frac{1}{1^2} + \frac{1}{2^2} + \frac{1}{3^2} + \cdots.$$

The convergence of this series had already been established earlier by Johann Bernoulli. Using results from the theory of equations and applying them to the power series expansion of $\sin x$, Euler was able to show that the sum of the given series was $\pi^2/6$ [5].

As power series were used more frequently to approximate mathematical quantities, the emphasis turned to deriving precise error estimates in approximating a function by a finite sum of the terms of its power series. In 1768, d'Alembert made a careful examination of the binomial series $(1 + r)^p$, $p \in \mathbb{Q}$, discovered earlier by Newton[6]. D'Alembert computed the value of n for which the absolute value of the ratio of successive terms is less than 1, thereby ensuring that the successive terms decrease. More importantly, d'Alembert computed the bounds on the error made in approximating $(1 + r)^p$ by a finite number of terms of this series. His argument relied on a term by term comparison with a geometric series, similar to the proof used in establishing the ratio test. D'Alembert's result can easily be converted into a proof of the convergence of the series.

Computations analogous to those of d'Alembert were also undertaken by Lagrange in this book *Théorie des Fonctions Analytiques* published in 1797. The remainder term for Taylors formula (to be discussed in the next chapter) first appeared

---

[4] *The Mathematical Works of Issac Newton*, edited by D. T. Whiteside, Johnson Reprint Corporation, New York, 1964.

[5] For details the reader is referred to the article by J. Grabiner.

[6] *Réflexions sur les suites ét sur les racines imaginaires*, Opuscules Mathématiques, vol. 5 (1768), pp. 171–215.

in this text. Cauchy was undoubtedly influenced by the works of d'Alembert and Lagrange in his definition of convergence of a series. However, unlike the results of d'Alembert, which applied only to the binomial series, and those of Lagrange for Taylor series, Cauchy's definition of convergence was entirely general; applicable to any series of real numbers.

Using his definition of convergence, Cauchy proved that the statement now known as the **Cauchy criterion** was necessary for the convergence of a series of real numbers. This was was also known earlier to Bolzano. However, without the completeness property of the real number system, neither Bolzano nor Cauchy were able to prove that the Cauchy criterion was also sufficient for convergence of the series.

Miscellaneous Exercises

1. Given two series $\sum\limits_{k=0}^{\infty} a_k$ and $\sum\limits_{k=0}^{\infty} b_k$, set $c_n = \sum\limits_{k=0}^{n} a_k b_{n-k}$, $\quad n = 0, 1, 2, \dots$.

   The series $\sum\limits_{n=0}^{\infty} c_n$ is called the **Cauchy product** of $\sum a_k$ and $\sum b_k$.

   a. If $\sum a_k$ and $\sum b_k$ converge absolutely, prove that $\sum c_n$ converges absolutely and that

   $$\sum_{n=0}^{\infty} c_n = \left(\sum_{k=0}^{\infty} a_k \right) \left(\sum_{k=0}^{\infty} b_k \right).$$

   b. Let $a_k = b_k = (-1)^k / \sqrt{k+1}$, $k = 0, 1, 2\dots$ Prove that the Cauchy product of $\sum a_k$ and $\sum b_k$ diverges.

   c. Prove that the result of (a) is still true if only one of the two series converge absolutely; the other series must still converge.

2. Let X be a non-empty set. If f is a real-valued function on X, define

   $$\|f\|_1 = \sup\{ \sum_{x \in F} |f(x)| : F \text{ is a finite subset of } X\}.$$

   In the above, the supremum is taken over all finite subsets F of X. Denote by $\ell^1(X)$ the set of all real-valued functions f on X for which $\|f\|_1 < \infty$.

   a. Suppose X is infinite. If $f \in \ell^1(X)$, prove that $\{x \in X : f(x) \neq 0\}$ is at most countable.

   b. If $\{x \in X : f(x) \neq 0\} = \{x_n : n \in A\}$, where $A \subset \mathbb{N}$, prove that $\|f\|_1 = \sum\limits_{n \in A} |f(x_n)|$.

   c. Prove that $(\ell^1(X), \| \ \|_1)$ is a normed linear space.

3. For $f \in \mathcal{R}[a,b]$, set

   $$\|f\|_2 = \left[\int\limits_a^b (f(x))^2 \, dx \right]^{1/2}.$$

   a. Prove that for $f, g \in \mathcal{R}[a,b]$, $\int\limits_a^b |f(x)g(x)| \, dx \leq \|f\|_2 \|g\|_2$.

b. For $f, g \in \mathcal{R}[a,b]$, prove that $\|f+g\|_2 \leq \|f\|_2 + \|g\|_2$.

c. Prove that $\| \ \|_2$ defines a norm on $\mathcal{C}[a,b]$, the space of continuous real-valued functions on $[a,b]$.

d. Is $\| \ \|_2$ a norm on $\mathcal{R}[a,b]$?

4. As in the previous exercise, let $\mathcal{C}[a,b]$ denote the vector space of continuous real-valued functions on $[a,b]$. For $f \in \mathcal{C}[a,b]$, set $\|f\|_u = \max\{|f(x)| : x \in [a,b]\}$.

   a. Prove that $\| \ \|_u$ is a norm on $\mathcal{C}[a,b]$.

   b. Prove that a sequence $\{f_n\}$ in $\mathcal{C}[a,b]$ converges to $f \in \mathcal{C}[a,b]$ if and only if given $\epsilon > 0$, there exists $n_o \in \mathbb{N}$ such that $|f(x) - f_n(x)| < \epsilon$ for all $x \in [a,b]$ and all $n \in \mathbb{N}$, $n \geq n_o$.

Definition (a) Let $(X, \| \ \|)$ be a normed linear space, and let $E \subset X$. A function $f : E \to \mathbb{R}$ is **continuous** at $\mathbf{p} \in E$ if given $\epsilon > 0$, there exists a $\delta > 0$ such that

$$|f(\mathbf{x}) - f(\mathbf{p})| < \epsilon \quad \text{for all } \mathbf{x} \in E \text{ with } \|\mathbf{x} - \mathbf{p}\| < \delta.$$

(b) Let X be a vector space over $\mathbb{R}$. A function $f : X \to \mathbb{R}$ is **linear** if

$$f(a\mathbf{x} + b\mathbf{y}) = af(\mathbf{x}) + bf(\mathbf{y})$$

for all $\mathbf{x}, \mathbf{y} \in X$ and $a, b \in \mathbb{R}$.

5. Let $(X, \| \ \|)$ be a normed linear space and let $f : X \to \mathbb{R}$ be a linear function. Prove that the following are equivalent.

   a. f is continuous at some $\mathbf{p} \in X$.

   b. f is continuous at $\mathbf{0}$.

   c. There exists a positive constant M such that $|f(\mathbf{x})| \leq M\|\mathbf{x}\|$ for all $\mathbf{x} \in X$.

   d. There exists a positive constant M such that $|f(\mathbf{x}) - f(\mathbf{y})| \leq M\|\mathbf{x} - \mathbf{y}\|$ for all $\mathbf{x}, \mathbf{y} \in X$.

   e. f is uniformly continuous on X.

6. For fixed $\mathbf{b} \in \ell^2$, define $\Gamma : \ell^2 \to \mathbb{R}$ by $\Gamma(\mathbf{a}) = \langle \mathbf{a}, \mathbf{b} \rangle = \sum\limits_{k=1}^{\infty} \mathbf{a}(k)\mathbf{b}(k)$. Prove that Γ is a continuous linear function on ℓ^2.

7. As in Exercise 3, let $\mathcal{C}[a,b]$ denote the vector space of continuous real valued functions on $[a,b]$ with norm $\| \ \|_2$. For fixed $g \in \mathcal{R}[a,b]$, prove that Γ defined by

$$\Gamma(f) = \int_a^b f(x)g(x)\,dx$$

is a continuous linear function on $\mathcal{C}[a,b]$.

Supplemental Reading

Ali, S. A., "The mth ratio test: new convergence tests for series," *Amer. Math. Monthly* **115** (2008), 514–524.

Behforooz, G. H., "Thinning out the harmonic series," *Math. Mag.* **68** (1995), 289–293.

Boas, R. P., "Estimating remainders," *Math. Mag.* **51** (1978), 83–89.

Boas, R. P., "Partial sums of infinite series and how they grow," *Amer. Math. Monthly* **84** (1977), 237–258.

Bradley, D. M., "An infinite series that displays the concavity of the natural logarithm," *Math. Mag.* **90** (2017), 353–354.

Cowen, C. C., Davidson, K. R., and Kaufman, R.P., "Rearranging the alternating harmonic series," *Amer. Math. Monthly* **87** (1980), 817–819.

Creswell, S. H.. "A continuous bijection of ℓ^2 onto a subset of ℓ^2 whose inverse is discontinuous everywhere," *Amer. Math. Monthly* **117** (2010), 823–828.

Daners, D., "A short elementary proof of $\sum 1/k^2 = \pi^2/6$," *Math. Mag.* **85** (2012), 361–364.

Dence, T. P. and Dence, J. P., "A survey of Euler's constant," *Math. Mag.* **82** (2009), 255–265.

Efthimiou, C. J., "Finding exact values for infinite series," *Math. Mag.* **72** (1999), 45–51.

Goar, M., "Olivier and Abel on series convergence: An episode from early 19th century analysis," *Math. Mag.* **72** (1999), 347–355.

Grabinger, Judith V., *The Origins of Cauchy's Rigorous Calculus,* The MIT Press, Cambridge, Massachusetts, 1981.

Hoang, N. S., "A limit comparison test for general series," *Amer. Math. Monthly* **122** (2015), 893–896.

Jungck, G. "An alternative to the integral test," *Math. Mag.* **56** (1983), 232–235.

Katz, V. J., "Ideas of Calculus in Islam and India," *Math. Mag.* **68** (1995), 163–174.

Kortram, R. A., "Simple proofs for $\sum\limits_{k=1}^{\infty} 1/k^2 = \frac{1}{6}\pi^2$ and $\sin x = x \prod\limits_{k=1}^{\infty}\left(1 - \dfrac{x^2}{k^2\pi^2}\right)$," *Math. Mag.* **69** (1996), 122-125.

Krantz, S. G. and McNeal, J. D., "Creating more convergent series," *Amer. Math. Monthly* **111** (2004), 32–38.

Maher, P., "Jensens inequality by differentiation," *Math. Gaz.* **73** (1989), 139–140.

Passare, M., "How to compute $\sum 1/n^2$ by solving triangles," *Amer. Math. Monthly* **115** (2008), 745–752.

Tolsted, E., "An elementary derivation of the Cauchy, Hölder, and Minkowski inequalities from Young's inequality," *Math. Mag.* **37** (1964), 2–12.

Young, R. M., "Euler's constant," *Math. Gazette* **75** (1991), 187–190.

8

Sequences and Series of Functions

In this chapter, we will study convergence properties of sequences and series of real-valued functions defined on a set E. In most instances E will be a subset of $\mathbb{R}$. Since we are dealing with sequences and series of functions, there naturally arise questions involving preservation of continuity, differentiability, and integrability. Specifically, is the limit function of a convergent sequence of continuous, differentiable, or integrable functions again continuous, differentiable, or integrable? We will discuss these questions in detail in Section 1 and show by examples that the answer to all of these questions is in general no! Convergence by itself is not sufficient for preservation of either continuity, differentiability, or integrability. Additional hypotheses will be required.

In the 1850's Weierstrass made a careful distinction between convergence of a sequence or series of numbers and that of a sequence or series of functions. It is to him that we are indebted for the concept of uniform convergence which is the additional hypothesis required for the preservation of continuity and integrability. It was also Weierstrass who constructed a continuous but nowhere differentiable function, and who proved that every continuous real-valued function on a closed and bounded interval can be uniformly approximated by a polynomial. As prominently as Cauchy is associated with the study of sequences and series of numbers, Weierstrass is likewise associated with the study of sequences and series of functions. For his many contributions to the subject area, Weierstrass is often referred to as the father of modern analysis.

The study of sequences and series of functions has its origins in the study of power series representation of functions. The power series of $\ln(1 + x)$ was known to Nicolaus Mercator (1620–1687) by 1668, and the power series for many of the transcendental functions such as $\arctan x$, $\arcsin x$, among others, were discovered around 1670 by James Gregory (1625–1683). All of these series were obtained without any reference to calculus. Newton's first discoveries, dating back to the early months of 1665, resulted from his ability to express functions in terms of power series. His treatise on calculus, published posthumously in 1737, was appropriately entitled *A treatise of the method of fluxions and infinite series*. Among his many accomplishments, Newton derived the power series expansion of $(1 + x)^{m/n}$ using algebraic techniques. This series and the geometric series were crucial in many of his computations. Newton also displayed the power of his calculus by deriving the power series expansion of $\ln(1 + x)$ using term-by-term integration of the expansion of $1/(1 + x)$. The mathematicians Colin Maclaurin (1698–1746) and Brooks Taylor (1685–1731) are prominent for being the first mathematicians to use the methods of the new calculus in determining the coefficients in the power series expansion of a function.

8.1   Pointwise Convergence and Interchange of Limits

In this section, we consider a number of questions involving sequences and series of functions and interchange of limits. Some of these questions were actually believed to be true by many mathematicians prior to the nineteenth century. Even Cauchy in his text *Cours d'Analyse* "proved" a theorem to the effect that the limit of a convergent sequence of continuous functions was again continuous. As we will shortly see, this result is false!

To begin our study of sequences and series of functions we first define what we mean by pointwise convergence of a sequence of functions. If E is a nonempty set and if for each $n \in \mathbb{N}$, f_n is a real-valued function on E, then we say that $\{f_n\}$ is a **sequence of real-valued functions** on E. For each $x \in E$, such a sequence gives rise to the sequence $\{f_n(x)\}$ of real numbers, which may or may not converge. If the sequence $\{f_n(x)\}$ converges for all $x \in E$, then the sequence $\{f_n\}$ is said to converge pointwise on E, and by the uniqueness of the limit

$$f(x) = \lim_{n \to \infty} f_n(x)$$

defines a function f from E into $\mathbb{R}$. We summarize this in the following definition.

DEFINITION 8.1.1 *Let (X, d) be a metric space and let $E \subset X$. Let $\{f_n\}_{n=1}^{\infty}$ be a sequence of real-valued functions defined on E. The sequence $\{f_n\}$ **converges pointwise** on E if $\{f_n(x)\}_{n=1}^{\infty}$ converges for every $x \in E$. If this is the case, then f defined by*

$$f(x) = \lim_{n \to \infty} f_n(x), \qquad x \in E,$$

*defines a function on E. The function f is called the **limit of the sequence** $\{f_n\}$.*

In terms of ϵ and n_o, the sequence $\{f_n\}$ converges pointwise to f if for each $x \in E$, given $\epsilon > 0$, there exists a positive integer $n_o = n_o(x, \epsilon)$ such that

$$|f_n(x) - f(x)| < \epsilon$$

for all $n \geq n_o$. The expression $n_o = n_o(x, \epsilon)$ indicates that the positive integer n_o may depend both on ϵ and $x \in E$.

If as above, $\{f_n\}_{n=1}^{\infty}$ is a sequence of real-valued functions on a nonempty set E, then with the sequence $\{f_n\}$ we can associate the sequence $\{S_n\}$ of nth **partial sums**, where for each $n \in \mathbb{N}$, S_n is the real-valued function on E defined by

$$S_n(x) = f_1(x) + \cdots + f_n(x) = \sum_{k=1}^{n} f_k(x).$$

The sequence $\{S_n\}$ is called a **series of functions** on E denoted by $\sum_{k=1}^{\infty} f_k$ or simply $\sum f_k$. The series $\sum f_k$ **converges pointwise** on E if for each $x \in E$ the sequence $\{S_n(x)\}$ of partial sums converges; that is, the series $\sum_{k=1}^{\infty} f_k(x)$ converges for each $x \in E$. If the sequence $\{S_n\}$ converges pointwise to the function S on E, then S is called the **sum of the series** $\sum f_k$ and we write $S = \sum_{k=1}^{\infty} f_k$, or if we wish to emphasize the variable x,

$$S(x) = \sum_{k=1}^{\infty} f_k(x), \qquad x \in E.$$

Suppose $f_n : [a, b] \to \mathbb{R}$ for all $n \in \mathbb{N}$, and $f_n(x) \to f(x)$ for each $x \in [a, b]$. Among the questions we want to consider are the following:

(a) If each f_n is continuous at $p \in [a, b]$, is the function f continuous at p? Recall that the function f is continuous at p if and only if

$$\lim_{t \to p} f(t) = f(p).$$

Since $f(x) = \lim_{n \to \infty} f_n(x)$, what we are really asking is whether

$$\lim_{t \to p} \left(\lim_{n \to \infty} f_n(t) \right) = \lim_{n \to \infty} \left(\lim_{t \to p} f_n(t) \right) ?$$

(b) If for each $n \in \mathbb{N}$ the function f_n is differentiable at $p \in [a, b]$, is f differentiable at p? If so, does

$$f'(p) = \lim_{n \to \infty} f_n'(p) ?$$

(c) If for each $n \in \mathbb{N}$ the function f_n is Riemann integrable on $[a, b]$, is f Riemann integrable? If so, does

$$\int_a^b f = \lim_{n \to \infty} \int_a^b f_n ?$$

We now provide a number of examples to show that the answer to all of the above questions is in general **no**.

EXAMPLES 8.1.2 (a) Let $E = [0, 1]$, and for each $x \in E$, $n \in \mathbb{N}$, let $f_n(x) = x^n$. Clearly each f_n is continuous on E. Since $f_n(1) = 1$ for all n, $\lim_{n \to \infty} f_n(1) = 1$. If $0 \le x < 1$, then by Theorem 3.2.6(e), $\lim_{n \to \infty} f_n(x) = 0$. Therefore

$$\lim_{n \to \infty} f_n(x) = f(x) = \begin{cases} 0, & 0 \le x < 1, \\ 1, & x = 1. \end{cases}$$

The function f however is not continuous on $[0, 1]$. (In Exercise 1 you will be asked to sketch the graphs of f_1, f_2, f_4.)

(b) Consider the sequence $\{f_k\}_{k=0}^{\infty}$ defined by

$$f_k(x) = \frac{x^2}{(1 + x^2)^k}, \qquad x \in \mathbb{R}.$$

For each $k = 0, 1, 2, ..$, f_k is continuous on $\mathbb{R}$. Consider the series $\sum_{k=0}^{\infty} f_k$ which for each $x \in \mathbb{R}$ is given by

$$\sum_{k=0}^{\infty} f_k(x) = \sum_{k=0}^{\infty} x^2 \left(\frac{1}{1 + x^2} \right)^k.$$

We now show that this series converges for all $x \in \mathbb{R}$ and also find its sum f. If $x = 0$, then $f_k(0) = 0$ for all k, and thus $f(0) = 0$. If $x \ne 0$, then $1/(1 + x^2) < 1$ and hence by Example 3.7.2(a)

$$x^2 \sum_{k=0}^{\infty} \left(\frac{1}{1 + x^2} \right)^k = x^2 \left[\frac{1}{1 - \frac{1}{1+x^2}} \right] = 1 + x^2 = f(x).$$

Therefore

$$f(x) = \begin{cases} 0, & x = 0, \\ 1 + x^2, & x \ne 0, \end{cases}$$

which again is not continuous on $\mathbb{R}$.

(c) Let $\{x_k\}$ be an enumeration of the rational numbers in $[0, 1]$. For each $n \in \mathbb{N}$, define f_n as follows:

$$f_n(x) = \begin{cases} 0, & \text{if } x = x_k, \ 1 \le k \le n, \\ 1, & \text{otherwise.} \end{cases}$$

Since each f_n is continuous except at $x_1, ..., x_n$, f_n is Riemann integrable on $[0, 1]$ with $\int_0^1 f_n(x)dx = 1$. On the other hand,

$$\lim_{n \to \infty} f_n(x) = f(x) = \begin{cases} 0, & \text{if } x \text{ is rational,} \\ 1, & \text{if } x \text{ is irrational,} \end{cases}$$

which by Example 6.1.6(a) is not Riemann integrable on $[0, 1]$.

(d) For $x \in [0, 1]$, $n \in \mathbb{N}$, let $f_n(x) = nx(1 - x^2)^n$. Since each f_n is continuous, f_n is Riemann integrable on $[0, 1]$. If $0 < x < 1$, then $0 < 1 - x^2 < 1$, and thus by Theorem 3.2.6

$$\lim_{n \to \infty} nx(1 - x^2)^n = 0, \qquad \text{if} \quad 0 < x < 1.$$

Finally, since $f_n(0) = f_n(1) = 0$, we have

$$f(x) = \lim_{n \to \infty} f_n(x) = 0 \qquad \text{for all } x \in [0, 1].$$

Thus f is also Riemann integrable on $[0, 1]$. On the other hand,

$$\int_0^1 f_n(x)\, dx = n \int_0^1 x(1 - x^2)^n\, dx = \frac{1}{2}\frac{n}{n+1}.$$

Therefore,

$$\lim_{n \to \infty} \int_0^1 f_n(x)\, dx = \frac{1}{2} \neq 0 = \int_0^1 f(x)\, dx.$$

(e) As our final example, consider $f_n(x) = \dfrac{\sin nx}{n}$, $x \in \mathbb{R}$. Since $|\sin nx| \leq 1$ for all $x \in \mathbb{R}$ and $n \in \mathbb{N}$,

$$f(x) = \lim_{n \to \infty} f_n(x) = 0 \quad \text{for all } x \in \mathbb{R}.$$

Therefore $f'(x) = 0$ for all x. On the other hand,

$$f_n'(x) = \cos nx.$$

In particular $f_n'(0) = 1$ so that $\lim\limits_{n \to \infty} f_n'(0) = 1 \neq f'(0)$. This example shows that in general

$$\frac{d}{dx}\left(\lim_{n \to \infty} f_n(x)\right) \neq \lim_{n \to \infty} f_n'(x).$$

Additional examples are also given in the exercises. $\quad\square$

Exercises 8.1

1. Let f_n be as in Example 8.1.2. Sketch the graphs of $f_1, f_2,$ and f_4.

2. Find the pointwise limits of each of the following sequences of functions on the given set.

   *a. $\left\{\dfrac{nx}{1 + nx}\right\}$, $\quad x \in [0, \infty)$
   b. $\left\{\dfrac{\sin nx}{1 + nx}\right\}$, $\quad x \in [0, \infty)$

   *c. $\{(\cos x)^{2n}\}$, $\quad x \in \mathbb{R}$.
   d. $\left\{nxe^{-nx^2}\right\}$, $\quad x \in \mathbb{R}$

3. Determine the values of x for which each of the following series converge.

   *a. $\displaystyle\sum_{n=1}^{\infty} \frac{nx^n}{2^n}$
   b. $\displaystyle\sum_{n=1}^{\infty} \frac{x^n}{(1 - x)^n}$, $\quad x \neq 1$

   *c. $\displaystyle\sum_{n=1}^{\infty} \frac{1}{3^{nx}}$
   d. $\displaystyle\sum_{n=1}^{\infty} \frac{2^n(\sin x)^n}{n}$

344   *Introduction to Real Analysis*

4. For $n \in \mathbb{N}$, define $f_n : \mathbb{N} \to \mathbb{R}$ by $f_n(m) = n/(m+n)$. Prove that

$$\lim_{m \to \infty} \left(\lim_{n \to \infty} f_n(m) \right) \neq \lim_{n \to \infty} \left(\lim_{m \to \infty} f_n(m) \right).$$

5. Consider the sequence $\{f_n\}$ with $n \geq 2$, defined on $[0,1]$ by

$$f_n(x) = \begin{cases} n^2 x, & 0 \leq x \leq 1/n, \\ 2n - n^2 x, & 1/n < x \leq 2/n, \\ 0, & 2/n < x \leq 1. \end{cases}$$

    a. Sketch the graph of f_n for $n = 2$, 3, and 4.

    b. Prove that $\lim_{n \to \infty} f_n(x) = 0$ for each $x \in [0,1]$.

    *c. Show that $\int_0^1 f_n(x)\,dx = 1$ for all $n = 2, 3, \dots$.

6. Let $g_n(x) = e^{-nx}/n$, $x \in [0, \infty)$, $n \in \mathbb{N}$. Find $\lim_{n \to \infty} g_n(x)$ and $\lim_{n \to \infty} g_n'(x)$.

7. Let $f_n(x) = (x/n)e^{-x/n}$, $x \in [0, \infty)$.

    *a. Show that $\lim_{n \to \infty} f_n(x) = 0$ for all $x \in [0, \infty)$.

    *b. Given $\epsilon > 0$, does there exist an integer $n_o \in \mathbb{N}$ such that $|f_n(x)| < \epsilon$ for all $x \in [0, \infty)$ and all $n \geq n_o$. (Hint: determine the maximum of f_n on $[0, \infty)$)

    c. Answer the same question as in (b) for $x \in [0, a]$, $a > 0$.

8. ***If** $a_{n,m} \geq 0$, $n, m \in \mathbb{N}$, prove that

$$\sum_{n=1}^{\infty} \sum_{m=1}^{\infty} a_{n,m} = \sum_{m=1}^{\infty} \sum_{n=1}^{\infty} a_{n,m},$$

with the convention that if one of the sums is finite, the other is also and equality holds, and if one is infinite, so is the other.

8.2 Uniform Convergence

All of the examples of the previous section show that pointwise convergence by itself is not sufficient to allow the interchange of limit operations; additional hypotheses are required. It was Weierstrass who realized in the 1850's what additional assumptions were needed to insure that the limit function of a convergent sequence of continuous functions was again continuous.

Recall from Definition 8.1.1, a sequence $\{f_n\}$ of real-valued functions defined on a set E converges pointwise to a function f on E if for each $x \in E$, given $\epsilon > 0$, there exists a positive integer $n_o = n_o(x, \epsilon)$ such that

$$|f(x) - f_n(x)| < \epsilon$$

for all $n \geq n_o$. The key here is that the choice of the integer n_o may depend not only on ϵ, but also on $x \in E$. If this dependence on x can be removed, then we have the following:

DEFINITION 8.2.1 *A sequence of real-valued functions $\{f_n\}$ defined on a set E **converges uniformly** to f on E, if for every $\epsilon > 0$, there exists a positive integer n_o such that*

$$|f_n(x) - f(x)| < \epsilon$$

for all $x \in E$ and all $n \geq n_o$. Similarly, a series $\sum_{k=1}^{\infty} f_k$ of real-valued functions **converges uniformly** *on a set E if and only if the sequence $\{S_n\}$ of partial sums converges uniformly on E.*

The inequality in the definition can also be expressed as

$$f(x) - \epsilon < f_n(x) < f(x) + \epsilon$$

for all $x \in E$ and $n \geq n_o$. If E is a subset of $\mathbb{R}$, then the geometric interpretation of the above inequality is that for $n \geq n_o$ the graph of $y = f_n(x)$ lies between the graphs of $y = f(x) - \epsilon$ and $y = f(x) + \epsilon$.

EXAMPLES 8.2.2 (a) For $x \in [0,1]$, $n \in \mathbb{N}$, let $f_n(x) = x^n$. By Example 8.1.2(a), the sequence $\{f_n\}$ converges pointwise to the function

$$f(x) = \begin{cases} 0, & 0 \leq x < 1, \\ 1, & x = 1. \end{cases}$$

We now show that the convergence is not uniform. If the convergence were uniform, then given $\epsilon > 0$, there would exist a positive integer n_o such that $|f_n(x) - f(x)| < \epsilon$ for all $n \geq n_o$. In particular,

$$x^{n_o} < \epsilon \qquad \text{for all} \quad x \in [0,1).$$

This however is a contradiction if $\epsilon < 1$. Even though the convergence is not uniform on $[0,1]$, the sequence does converge uniformly to 0 on $[0,a]$ for every a, $0 < a < 1$. This follows immediately from the fact that for $x \in [0,a]$, $|f_n(x)| = |x^n| \leq a^n$.

(b) Consider the series

$$\sum_{k=1}^{\infty} \left[kxe^{-kx^2} - (k-1)xe^{-(k-1)x^2} \right], \qquad 0 \leq x \leq 1.$$

Since the series is a telescoping series, the nth partial sum $S_n(x)$ is given by

$$S_n(x) = nxe^{-nx^2}.$$

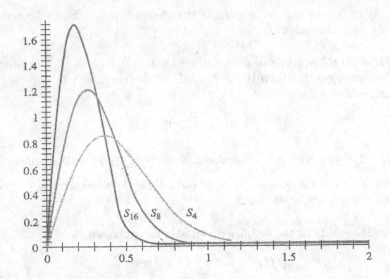

FIGURE 8.1
Graphs of S_4, S_8, S_{16}

It is easily shown (Exercise 2(d), Section 8.1) that

$$S(x) = \lim_{n\to\infty} S_n(x) = 0 \qquad \text{for all} \quad x \in [0,1].$$

The graphs of S_4, S_8, S_{16} are given in Figure 8.1.

We now show that the convergence is not uniform. Suppose that the sequence $\{S_n\}$ converges uniformly to 0 on $[0,1]$. Then if we take $\epsilon = 1$, there exists a positive integer n_o such that

$$|S_n(x) - S(x)| = S_n(x) = nxe^{-nx^2} < 1$$

for all $n \geq n_o$ and $x \in [0,1]$. However, for each $n \in \mathbb{N}$, by the first derivative test S_n has a maximum at $x_n = \sqrt{1/2n}$ with

$$M_n = \max_{0 \leq x \leq 1} S_n(x) = \sqrt{\frac{n}{2e}}.$$

This term however is greater than 1 for $n \geq 6$. Thus the convergence is not uniform. Since the maximum of each S_n moves along the x-axis as $n \to \infty$, such functions are often referred to as "sliding-hump" functions. For this example $M_n \to \infty$ as $n \to \infty$. $\square$

The Cauchy Criterion

Our first criterion for uniform convergence is the Cauchy criterion. The statement of this result is very similar to the definition of Cauchy sequence.

THEOREM 8.2.3 (Cauchy Criterion) *A sequence $\{f_n\}$ of real-valued functions defined on a set E converges uniformly on E if and only if for every $\epsilon > 0$, there exists an integer $n_o \in \mathbb{N}$ such that*

$$|f_n(x) - f_m(x)| < \epsilon \tag{1}$$

for all $x \in E$ and all $n, m \geq n_o$.

Proof. If $\{f_n\}$ converges uniformly to f on E, then the proof that (1) holds is similar to the proof that every convergent sequence is Cauchy. Conversely, suppose that the sequence $\{f_n\}$ satisfies (1). Then for each $x \in E$, the sequence $\{f_n(x)\}$ is a Cauchy sequence in $\mathbb{R}$, and hence converges (Theorem 3.6.5). Therefore,

$$f(x) = \lim_{n \to \infty} f_n(x)$$

exists for every $x \in E$.

We now show that the sequence $\{f_n\}$ converges uniformly to f on E. Let $\epsilon > 0$ be given. By hypothesis, there exists $n_o \in \mathbb{N}$ such that (1) holds for all $x \in E$ and all $n, m \geq n_o$. Fix an $m \geq n_o$. Then

$$|f(x) - f_m(x)| = \lim_{n \to \infty} |f_n(x) - f_m(x)| \leq \epsilon$$

for all $x \in E$. Since the above holds for all $m \geq n_o$, the sequence $\{f_n\}$ converges uniformly to f on E. $\square$

The analogous result for series is as follows:

COROLLARY 8.2.4 *The series $\sum\limits_{k=1}^{\infty} f_k$ of real-valued functions on E converges uniformly on E if and only if given $\epsilon > 0$, there exists a positive integer n_o, such that*

$$\left| \sum_{k=n+1}^{m} f_k(x) \right| < \epsilon$$

for all $x \in E$ and all integers $m > n \geq n_o$.

Proof. The proof of the Corollary follows by applying the previous theorem to the partial sums $S_n(x)$ of the series $\sum f_k(x)$. $\square$

THEOREM 8.2.5 *Suppose the sequence $\{f_n\}$ of real-valued functions on the set E converges pointwise to f on E. For each $n \in \mathbb{N}$, set*

$$M_n = \sup_{x \in E} |f_n(x) - f(x)|.$$

Then $\{f_n\}$ converges uniformly to f on E if and only if $\lim\limits_{n \to \infty} M_n = 0$.

Proof. Exercise 1. $\square$

EXAMPLES 8.2.6 (a) To illustrate the previous theorem we consider the sequence

$$S_n(x) = n x e^{-nx^2}, \qquad n = 1, 2, \dots$$

of Example 8.2.2(b). For this sequence, $\lim_{n\to\infty} S_n(x) = 0$ for all x, $0 \le x < \infty$. However,

$$M_n = \sup_{x\in[0,\infty)} S_n(x) = \sqrt{\frac{n}{2e}},$$

which diverges to ∞. Thus the convergence is not uniform on $[0,\infty)$. However, the sequence $\{S_n\}$ does converge uniformly to the zero function on $[a,\infty)$ for every fixed $a > 0$. (Exercise 6)

(b) Consider the sequence $\{f_n\}$ of Example 8.2.2(a) given by $f_n(x) = x^n$, $x \in [0,1]$. This sequence converges pointwise to the function $f(x) = 0$, $0 \le x < 1$, and $f(1) = 1$. Since

$$|f_n(x) - f(x)| = \begin{cases} x^n, & 0 \le x < 1, \\ 0, & x = 1, \end{cases}$$

we have

$$M_n = \sup_{x\in[0,1]} |f_n(x) - f(x)| = 1.$$

Thus since $\{M_n\}$ does not converge to zero, the sequence $\{f_n\}$ does not converge uniformly to f on $[0,1]$. On the other hand, if $0 < a < 1$ is fixed, then

$$M_n = \sup_{x\in[0,a]} |f_n(x)| = a^n.$$

Since $\lim_{n\to\infty} a^n = 0$, by Theorem 8.2.5 the sequence $\{f_n\}$ converges uniformly to the zero function on $[0,a]$ for every fixed a, $0 < a < 1$, $\square$

The Weierstrass M-Test

The following theorem of Weierstrass provides a very useful test for uniform convergence of a series of functions.

THEOREM 8.2.7 (Weierstrass M-Test) *Suppose $\{f_k\}$ is a sequence of real-valued functions defined on a set E, and $\{M_k\}$ is a sequence of real numbers satisfying*

$$|f_k(x)| \le M_k, \qquad \text{for all } x \in E \text{ and } k \in \mathbb{N}.$$

If $\sum_{k=1}^{\infty} M_k < \infty$, then $\sum_{k=1}^{\infty} f_k(x)$ converges uniformly and absolutely on E.

Proof. Let $S_n(x) = \sum\limits_{k=1}^{n} f_k(x)$. Then for $n > m$,

$$|S_n(x) - S_m(x)| = \left| \sum_{k=m+1}^{n} f_k(x) \right| \leq \sum_{k=m+1}^{n} |f_k(x)| \leq \sum_{k=m+1}^{n} M_k.$$

Uniform convergence now follows by the Cauchy Criterion. That $\sum |f_k(x)|$ also converges is clear. $\square$

EXAMPLES 8.2.8 (a) If $\sum a_k$ converges absolutely, then since $|a_k \cos kx| \leq |a_k|$ for all $x \in \mathbb{R}$, by the Weierstrass M-test the series $\sum a_k \cos kx$ converges uniformly on $\mathbb{R}$. Similarly for the series $\sum a_k \sin kx$. In particular, the series

$$\sum_{k=1}^{\infty} \frac{\cos kx}{k^p}, \quad \sum_{k=1}^{\infty} \frac{\sin kx}{k^p}, \quad p > 1,$$

converge uniformly on $\mathbb{R}$.

(b) Consider the series $\sum\limits_{k=1}^{\infty} (x/2)^k$. This is a geometric series that converges for all $x \in \mathbb{R}$ satisfying $|x| < 2$. If $0 < a < 2$ and $|x| \leq a$, then

$$\left| \left(\frac{x}{2} \right)^k \right| \leq \left(\frac{a}{2} \right)^k.$$

Since $a/2 < 1$ the series $\sum (a/2)^k$ converges. Thus by the Weierstrass M-test, the series $\sum\limits_{k=1}^{\infty} (x/2)^k$ converges uniformly on $[-a, a]$ for any a, $0 < a < 2$. The series however does not converge uniformly on $(-2, 2)$ (Exercise 11). $\square$

Although the Weierstrass M-test automatically implies absolute convergence, the following example shows that uniform convergence as a general rule does not imply absolute convergence.

EXAMPLES 8.2.9 (a) Consider the series

$$\sum_{k=1}^{\infty} (-1)^{k+1} \frac{x^k}{k}, \qquad 0 \leq x \leq 1.$$

For each $k \in \mathbb{N}$, set $a_k(x) = x^k/k$. For $x \in [0, 1]$, we have

$$a_1(x) \geq a_2(x) \geq \cdots \geq 0, \qquad \text{and} \qquad \lim_{k \to \infty} a_k(x) = 0.$$

Thus by Theorem 7.2.3, the series $\sum (-1)^{k+1} a_k(x)$ converges for all $x \in [0, 1]$. Let

$$S(x) = \sum_{k+1}^{\infty} (-1)^{k+1} a_k(x).$$

If $S_n(x)$ is the nth partial sum of the series, then by Theorem 7.2.4

$$|S(x) - S_n(x)| \le a_{n+1}(x) \le \frac{1}{n+1}, \qquad \text{for all } x \in [0,1].$$

Thus $\{S_n\}$ converges uniformly to S on $[0,1]$. However, the given series does not converge absolutely when $x = 1$. The series $\sum(-1)^{k+1}x^k/k$, $x \in [0,1]$, also provides an example of a series that converges uniformly on $[0,1]$ but for which the Weierstrass M-test fails.

(b) The converse is also false; absolute convergence need not imply uniform convergence! As an example, consider the series $\sum\limits_{k=1}^{\infty} x^2(1+x^2)^{-k}$ of Example 8.1.2(b). Since all the terms are nonnegative, the series converges absolutely to

$$f(x) = \begin{cases} 0, & x = 0, \\ 1 + x^2, & x \ne 0, \end{cases}$$

on $\mathbb{R}$. However, as a consequence of Corollary 8.3.2 of the next section, since f is not continuous at 0, the convergence cannot be uniform on any interval containing 0. $\square$

Exercises 8.2

1. Prove Theorem 8.2.5.

2. **a.** If $\{f_n\}$ and $\{g_n\}$ converge uniformly on a set E, prove that $\{f_n + g_n\}$ converges uniformly on E.

   *b. If $\{f_n\}$ and $\{g_n\}$ converge uniformly on a set E, and if in addition there exist constants M and N such that $|f_n(x)| \le M$ and $|g_n(x)| \le N$ for all $n \in \mathbb{N}$ and all $x \in E$, prove that $\{f_n g_n\}$ converges uniformly on E.

   c. Find examples of sequences $\{f_n\}$ and $\{g_n\}$ that converge uniformly on a set E, but for which $\{f_n g_n\}$ does not converge uniformly on E.

3. Show that if $\{f_n\}$ converges uniformly on (a, b) and $\{f_n(a)\}$ and $\{f_n(b)\}$ converge, then $\{f_n\}$ converges uniformly on $[a, b]$.

4. ***Let** $f_n(x) = n\,x(1 - x^2)^n$, $0 \le x \le 1$. Show that $\{f_n\}$ does not converge uniformly to 0 on $[0, 1]$.

5. Let $f_n(x) = \dfrac{x^n}{1 + x^n}$, $0 \le x \le 1$.

   *a. Show that $\{f_n\}$ converges uniformly to 0 on $[0, a]$ for any a, $0 < a < 1$.

   *b. Does $\{f_n\}$ converge uniformly on $[0, 1]$?

6. Show that the sequence $\{nxe^{-nx^2}\}$ converges uniformly to 0 on $[a, \infty)$ for every $a > 0$.

7. For each $n \in \mathbb{N}$, set $f_n(x) = x + \dfrac{x}{n}\sin nx$, $x \in \mathbb{R}$. Show that the sequence $\{f_n\}$ converges uniformly to $f(x) = x$ for all $x \in [-a, a]$, $a > 0$. Does $\{f_n\}$ converge uniformly to f on $\mathbb{R}$?

8. Show that each of the following series converge uniformly on the indicated interval.

   *a. $\displaystyle\sum_{k=1}^{\infty} \frac{1}{k^2 + x^2}$, $0 \le x < \infty$.

   b. $\displaystyle\sum_{k=1}^{\infty} e^{-kx} x^k$, $0 \le x < \infty$.

   *c. $\displaystyle\sum_{k=1}^{\infty} k^2 e^{-kx}$, $1 \le x < \infty$.

   d. $\displaystyle\sum_{k=1}^{\infty} \frac{(-1)^{k+1}}{k+x}$, $0 \le x < \infty$.

9. Test each of the following series for uniform convergence on the indicated interval.

   *a. $\displaystyle\sum_{k=1}^{\infty} \frac{\sin 2kx}{(2k+1)^{3/2}}$, $x \in \mathbb{R}$.

   b. $\displaystyle\sum_{k=2}^{\infty} \frac{x^k}{k(\ln k)^2}$, $|x| \le 1$.

   c. $\displaystyle\sum_{k=0}^{\infty} \frac{(-1)^{k+1} x^{2k+1}}{2k+1}$, $|x| \le 1$.

   *d. $\displaystyle\sum_{k=1}^{\infty} \sin(x/k^p)$, $p > 1$, $|x| \le 2$.

   *e. $\displaystyle\sum_{k=0}^{\infty} \left(\frac{1}{kx+2} - \frac{1}{(k+1)x+2} \right)$, $0 \le x \le 1$.

10. Show that each of the following series converge uniformly on $[a, \infty)$ for any $a > 0$, but do not converge uniformly on $(0, \infty)$.

    *a. $\displaystyle\sum_{k=0}^{\infty} \frac{1}{1 + k^2 x}$.

    b. $\displaystyle\sum_{k=1}^{\infty} \frac{1}{k^{1+x}}$.

11. Show that the series $\displaystyle\sum_{k=1}^{\infty} (x/2)^k$ does not converges uniformly on $(-2, 2)$.

12. *If $\sum_{k=0}^{\infty} a_k$ converges absolutely, prove that $\sum_{k=0}^{\infty} a_k x^k$ converges uniformly on $[-1, 1]$.

13. Let $\{f_n\}$ be a sequence of functions that converges uniformly to a continuous function f on $(-\infty, \infty)$. Prove that

    $$\lim_{n \to \infty} f_n \left(x + \tfrac{1}{n} \right) = f(x) \quad \text{for all } x \in (-\infty, \infty).$$

14. Let $\{c_k\}$ be a sequence of real numbers satisfying $\sum |c_k| < \infty$, and let $\{x_k\}$ be a countable subset of $[a, b]$. Prove that the series

    $$\sum_{k=1}^{\infty} c_k I(x - x_k)$$

    converges uniformly on $[a, b]$. Here I is the unit jump function defined in Definition 4.4.9.

15. (**Dirichlet Test for Uniform Convergence**) Suppose $\{f_k\}$ and $\{g_k\}$ are sequences of functions on a set E satisfying

    (a) the partial sums $S_n(x) = \displaystyle\sum_{k=1}^{n} g_k(x)$ are **uniformly bounded** on E, i.e., there exists $M > 0$ such that $|S_n(x)| \le M$ for all $n \in \mathbb{N}$ and $x \in E$.

    (b) $f_k(x) \ge f_{k+1}(x) \ge 0$ for all $k \in \mathbb{N}$ and $x \in E$, and

    (c) $\displaystyle\lim_{k \to \infty} f_k(x) = 0$ uniformly on E.

    Prove that $\sum f_k(x) g_k(x)$ converges uniformly on E.

16. *Prove that $\displaystyle\sum_{k=1}^{\infty} \frac{\sin kx}{k^p}$, $\displaystyle\sum_{k=1}^{\infty} \frac{\cos kx}{k^p}$ $(p > 0)$ converge uniformly on any closed interval which does not contain an integer multiple of 2π.

17. Define a sequence of functions $\{f_n\}$ on $[0, 1]$ by

$$f_n(x) = \begin{cases} \dfrac{1}{n}, & \text{if } \dfrac{1}{2^{n+1}} < x \le \dfrac{1}{2^n}, \\ 0, & \text{elsewhere.} \end{cases}$$

Prove that $\sum\limits_{n=1}^{\infty} f_n(x)$ converges uniformly on $[0, 1]$, but that the Weierstrass M-test fails.

18. *Let F_0 be a bounded Riemann integrable function on $[0, 1]$. For $n \in \mathbb{N}$, define $F_n(x)$ on $[0, 1]$ by

$$F_n(x) = \int_0^x F_{n-1}(t)\,dt.$$

Prove that $\sum\limits_{k=0}^{\infty} F_k(x)$ converges uniformly on $[0, 1]$.

8.3   Uniform Convergence and Continuity

In this section, we will prove that the limit of a uniformly convergent sequence of continuous functions is again continuous. Prior to proving this result, we first prove a stronger result that will have additional applications later.

THEOREM 8.3.1 *Suppose $\{f_n\}$ is a sequence of real-valued functions that converges uniformly to a function f on a subset E of a metric space (X, d). Let p be a limit point of E, and suppose that for each $n \in \mathbb{N}$,*

$$\lim_{x \to p} f_n(x) = A_n.$$

Then the sequence $\{A_n\}$ converges and

$$\lim_{x \to p} f(x) = \lim_{n \to \infty} A_n.$$

Remark. The last statement can be rewritten as

$$\lim_{x \to p} \left(\lim_{n \to \infty} f_n(x) \right) = \lim_{n \to \infty} \left(\lim_{x \to p} f_n(x) \right).$$

It should be noted that p is not required to be a point of E; only a limit point of E.

Proof. Let $\epsilon > 0$ be given. Since the sequence $\{f_n\}$ converges uniformly to f on E, there exists a positive integer n_o such that

$$|f_n(x) - f_m(x)| < \epsilon \tag{2}$$

for all $n, m \geq n_o$ and all $x \in E$. Since (2) holds for all $x \in E$, letting $x \to p$ gives

$$|A_n - A_m| \leq \epsilon, \qquad \text{for all} \quad n, m \geq n_o.$$

Thus $\{A_n\}$ is a Cauchy sequence in $\mathbb{R}$, which as a consequence of Theorem 3.6.5 converges. Let $A = \lim_{n \to \infty} A_n$.

It remains to be shown that $\lim_{x \to p} f(x) = A$. Again, let $\epsilon > 0$ be given. First, by the uniform convergence of the sequence $\{f_n(x)\}$ and the convergence of the sequence $\{A_n\}$, there exists a positive integer m such that

$$|f(x) - f_m(x)| < \frac{\epsilon}{3}$$

for all $x \in E$, and also that

$$|A - A_m| < \frac{\epsilon}{3}.$$

Since $\lim_{x \to p} f_m(x) = A_m$, there exists a $\delta > 0$ such that

$$|f_m(x) - A_m| < \frac{\epsilon}{3} \qquad \text{for all} \quad x \in E, 0 < d(x, p) < \delta.$$

By the triangle inequality,

$$|f(x) - A| \leq |f(x) - f_m(x)| + |f_m(x) - A_m| + |A_m - A|$$
$$< \frac{2\epsilon}{3} + |f_m(x) - A_m| < \epsilon.$$

Thus if $x \in E$ with $0 < d(x, p) < \delta$,

$$|f(x) - A| < \epsilon;$$

i.e., $\lim_{x \to p} f(x) = A$. $\square$

COROLLARY 8.3.2 *Let E be a subset of a metric space X.*

(a) If $\{f_n\}$ is a sequence of continuous real-valued functions on E, and if $\{f_n\}$ converges uniformly to f on E, then f is continuous on E.

(b) If $\{f_n\}$ is a sequence of continuous real-valued functions on E, and if $\sum_{n=1}^{\infty} f_n$ converges uniformly on E, then

$$S(x) = \sum_{n=1}^{\infty} f_n(x)$$

is continuous on E.

Proof. (a) If $p \in E$ is an isolated point, then f is automatically continuous at p. If $p \in E$ is a limit point of E, then since f_n is continuous for each $n \in \mathbb{N}$,

$$\lim_{x \to p} f_n(x) = f_n(p).$$

Thus by the previous theorem,

$$\lim_{x \to p} f(x) = \lim_{n \to \infty} f_n(p) = f(p).$$

Therefore f is continuous at p.
(b) For the proof of (b), let

$$S_n(x) = \sum_{k=1}^{n} f_k(x).$$

Then for each $n \in \mathbb{N}$, S_n is continuous on E. Since $\{S_n\}$ converges uniformly to S on E, by part (a) S is also continuous on E. $\square$

EXAMPLE 8.3.3 The sequence $\{x^n\}_{n=1}^{\infty}$, $x \in [0,1]$, of Example 8.1.2(a) does not converge uniformly on $[0,1]$ since the limit function

$$f(x) = \begin{cases} 0 & 0 \le x < 1, \\ 1, & x = 1, \end{cases}$$

is not continuous on $[0,1]$. Likewise, the series

$$\sum_{k=0}^{\infty} x^2 \left(\frac{1}{1+x^2} \right)^k = \begin{cases} 0, & x = 0, \\ 1+x^2, & x \ne 0, \end{cases}$$

of Example 8.1.2(b) cannot converge uniformly on any interval containing 0, since the sum of the series is not continuous at 0. $\square$

Dini's Theorem[1]

A natural question to ask is whether the converse of Corollary 8.3.2 is true. Namely, if f and f_n are continuous for all n and $f_n \to f$ pointwise, is the convergence necessarily uniform? The following example shows that this need not be the case. However, in Theorem 8.3.5 we will prove that with the additional assumption that the sequence $\{f_n(x)\}$ is monotone for all x, then the convergence is indeed uniform.

[1]This topic is not required in subsequent sections and thus can be omitted on first reading.

EXAMPLE 8.3.4 As in Example 8.2.2(b), for each $n \in \mathbb{N}$, let

$$S_n(x) = nxe^{-nx^2}, \qquad x \in [0, 1].$$

Then S_n is continuous on $[0, 1]$ for each n, and $\lim_{n \to \infty} S_n(x) = S(x) = 0$, which is also continuous. However, since

$$\max_{0 \le x \le 1} S_n(x) = \sqrt{\frac{n}{2e}},$$

by Theorem 8.2.5 the convergence cannot be uniform. $\square$

THEOREM 8.3.5 (Dini's Theorem) *Suppose K is a compact subset of a metric space X and $\{f_n\}$ is a sequence of continuous real-valued functions on K satisfying,*

(a) *$\{f_n\}$ converges pointwise on K to a continuous function f, and*

(b) *$f_n(x) \ge f_{n+1}(x)$ for all $x \in K$ and $n \in \mathbb{N}$.*

Then $\{f_n\}$ converges uniformly to f on K.

Proof. For each $n \in \mathbb{N}$ let $g_n = f_n - f$. Then g_n is continuous on K, $g_n(x) \ge g_{n+1}(x) \ge 0$ for all $x \in K$ and all $n \in \mathbb{N}$, and

$$\lim_{n \to \infty} g_n(x) = 0 \qquad \text{for each} \quad x \in K.$$

Let $\epsilon > 0$ be given. For each $n \in \mathbb{N}$, let

$$K_n = \{x \in K : g_n(x) \ge \epsilon\}.$$

We first prove that K_n is closed. Let p be a limit point of K_n. By Theorem 3.1.4 there exists a sequence $\{x_k\}$ in K_n which converges to p. Since g_n is continuous and $g_n(x_k) \ge \epsilon$ for all $k \in \mathbb{N}$,

$$g_n(p) = \lim_{k \to \infty} g_n(x_k) \ge \epsilon.$$

Therefore $p \in K_n$. Thus K_n is closed and as a consequence of Theorem 2.3.5 also compact. Furthermore, since $g_n(x) \ge g_{n+1}(x)$ for all $x \in K$,

$$K_{n+1} \subset K_n \qquad \text{for all } n.$$

Finally, since $g_n(x) \to 0$ for each $x \in K$,

$$\bigcap_{n=1}^{\infty} K_n = \emptyset.$$

However, by Theorem 2.3.8 this can only be the case if $K_{n_o} = \emptyset$ for some $n_o \in \mathbb{N}$. Thus for all $n \ge n_o$,

$$0 \le g_n(x) < \epsilon \qquad \text{for all } x \in K.$$

Therefore the sequence $\{g_n\}$ converges uniformly to 0 on K. $\square$

EXAMPLE 8.3.6 We now provide an example to show that compactness is required. For each $n \in \mathbb{N}$, set

$$f_n(x) = \frac{1}{nx + 1}, \qquad 0 < x < 1.$$

Then $\{f_n(x)\}$ monotonically decreases to $f(x) = 0$ for each $x \in (0,1)$. However, since $\lim_{x \to 0} f_n(x) = 1$, by Theorem 8.3.1 the convergence cannot be uniform. $\square$

The Space $\mathcal{C}(K)^2$

We conclude this section with a brief discussion of the space $\mathcal{C}(K)$ of all continuous real-valued functions on a compact set K. If $f, g \in \mathcal{C}(K)$ and $c \in \mathbb{R}$, then by Theorem 4.2.3 the functions $f + g$ and cf are also continuous on K. Thus $\mathcal{C}(K)$ is a vector space over $\mathbb{R}$ where for the zero element we take the constant function 0; that is, the function given by $f(x) = 0$ for all $x \in K$. We define a norm on $\mathcal{C}(K)$ as follows.

DEFINITION 8.3.7 *For each $f \in \mathcal{C}(K)$, set*

$$\|f\|_u = \max\{|f(x)| : x \in K\}.$$

The quantity $\|f\|_u$ is called the **uniform norm** *of f on K.*

That $\| \ \|_u$ is indeed a norm on $\mathcal{C}(K)$ is left to the exercises (Exercise 12). We now take a closer look at uniform convergence of a sequence of continuous real-valued functions and also introduce the concept of convergence in norm. We first note that a sequence $\{f_n\}$ of continuous real-valued functions on K is nothing else but a sequence in the set $\mathcal{C}(K)$. Suppose that the sequence $\{f_n\}$ in $\mathcal{C}(K)$ converges uniformly to f on K. Then by Corollary 8.3.2 the function $f \in \mathcal{C}(K)$. By the definition of uniform convergence, given $\epsilon > 0$ there exists a positive integer n_o such that

$$|f_n(x) - f(x)| < \epsilon$$

for all $x \in K$ and all $n \geq n_o$. Since $f_n - f$ is continuous, if $n \geq n_o$, by Corollary 4.2.9

$$\|f_n - f\|_u = \max\{|f_n(x) - f(x)| : x \in [a,b]\} < \epsilon.$$

Therefore $\|f_n - f\|_u < \epsilon$ for all $n \geq n_o$

Conversely, suppose $f, f_n \in \mathcal{C}(K)$ satisfy the following: For each $\epsilon > 0$ there exists a positive integer n_o such that $\|f - f_n\|_u < \epsilon$ for all $n \geq n_o$. But then $|f(x) - f_n(x)| < \epsilon$ for all $x \in K$ and all $n \geq n_o$; i.e., $\{f_n\}$ converges uniformly to f on K. This proves the following theorem.

---

[2]This topic can also be omitted on first reading. The concept of a complete normed linear space (Definition 8.3.10) is only required in Section 10.9.

THEOREM 8.3.8 *A sequence $\{f_n\}$ in $\mathcal{C}(K)$ converges uniformly to $f \in \mathcal{C}(K)$ if and only if given $\epsilon > 0$, there exists $n_o \in \mathbb{N}$ such that $\|f - f_n\|_u < \epsilon$ for all $n \geq n_o$.*

Using the above as motivation, we define convergence in a normed linear space as follows:

DEFINITION 8.3.9 *Let $(X, \| \ \|)$ be a normed linear space. A sequence $\{x_n\}$ in X **converges in norm** if there exists $x \in X$ such that for every $\epsilon > 0$, there exists a positive integer n_o such that $\|x - x_n\| < \epsilon$ for all $n \geq n_o$. If this is the case, we say that $\{x_n\}$ converges in norm to x and denote this by $x_n \overset{\| \ \|}{\to} x$ as $n \to \infty$.*

From the definition it is clear that a sequence $\{x_n\}$ in X converges in norm to $x \in X$ if and only if $\lim_{n \to \infty} \|x_n - x\| = 0$. Also, as in the proof of Theorem 3.1.4 if $\{x_n\}$ converges in norm then its limit is unique. Using the norm it is also possible to define what we mean by a Cauchy sequence in a normed linear space.

DEFINITION 8.3.10 **(a)** *A sequence $\{x_n\}$ in a normed linear space $(X, \| \ \|)$ is a **Cauchy sequence** if for every $\epsilon > 0$ there exists a positive integer n_o such that*

$$\|x_n - x_m\| < \epsilon$$

for all integers $n, m \geq n_o$.

(b) *A normed linear space $(X, \| \ \|)$ is **complete** if every Cauchy sequence in X converges in norm to an element of X.*

As for sequences of real numbers, every sequence $\{x_n\}$ in X that converges in norm to $x \in X$ is a Cauchy sequence. In Theorem 3.6.5 we proved that the normed linear space $(\mathbb{R}, | \ |)$ is complete. The following theorem proves that $(\mathcal{C}(K), \| \ \|_u)$ is also complete.

THEOREM 8.3.11 *If K is a compact subset of a metric space X, then the normed linear space $(\mathcal{C}(K), \| \ \|_u)$ is complete.*

Proof. Let $\{f_n\}$ be a Cauchy sequence in $\mathcal{C}(K)$; i.e., given $\epsilon > 0$, there exists a positive integer n_o such that $\|f_n - f_m\|_u < \epsilon$ for all $n, m \geq n_o$. But then

$$|f_n(x) - f_m(x)| \leq \|f_n - f_m\|_u < \epsilon$$

for all $x \in [a, b]$ and all $n, m \geq n_o$. Thus by Theorem 8.2.3 and Corollary 8.3.2 the sequence $\{f_n\}$ converges uniformly to a continuous function f on K. Finally, since the convergence is uniform, given $\epsilon > 0$, there exists an integer n_o such that

$$|f_n(x) - f(x)| < \epsilon \qquad \text{for all } x \in K \quad \text{and } n \geq n_o.$$

As a consequence we have $\|f_n - f\|_u < \epsilon$ for all $n \geq n_o$. Therefore the sequence $\{f_n\}$ converges to f in the norm $\| \ \|_u$. $\quad \square$

Contraction Mappings

In Exercise 13 of Section 4.3 we defined the notion of a contractive function on a subset E of $\mathbb{R}$. We now extend this to normed linear spaces.

DEFINITION 8.3.12 *Let* $(X, \| \ \|)$ *be a normed linear space. A mapping (function)* $T : X \to X$ *is called a* **contraction mapping (function)** *if there exists a constant* c, $0 < c < 1$, *such that*

$$\|T(x) - T(y)\| \leq c\|x - y\|$$

for all $x, y \in X$.

Clearly every contraction mapping on X is continuous, in fact uniformly continuous on X. As in Exercise 13 of Section 4.3, we now prove that if T is a contraction mapping on a complete normed linear space $(X, \| \ \|)$, then T has a unique fixed point in X.

THEOREM 8.3.13 *Let* $(X, \| \ \|)$ *be a complete normed linear space and let* $T : X \to X$ *be a contraction mapping. Then there exists a unique point* $x \in X$ *such that* $T(x) = x$.

Proof. Suppose $T : X \to X$ satisfies $\|T(x) - T(y)\| \leq c\|x - y\|$ for fixed c, $0 < c < 1$, and all $x, y \in X$. We now define a sequence $\{x_n\}$ in X as follows: Let $x_o \in X$ be arbitrary. For $n \in \mathbb{N}$ set $x_n = T(x_{n-1})$. That is, $x_1 = T(x_o)$, $x_2 = T(x_1)$, etc. Since

$$\|x_{n+1} - x_n\| = \|T(x_n) - T(x_{n-1})\| \leq c\|x_n - x_{n-1}\|,$$

the sequence $\{x_n\}$ is a contractive sequence in X (see Definition 3.6.8). An argument similar to the one used for a contractive sequence in Section 3.6 shows that

$$\|x_{n+m} - x_n\| \leq \frac{c^n}{1-c}\|x_1 - x_o\|$$

for all $n, m \in \mathbb{N}$. The details are left as an exercise (Exercise 10). Since $c^n \to 0$, the sequence $\{x_n\}$ is a Cauchy sequence in X. By completeness, the sequence $\{x_n\}$ converges (in the norm) to some $x \in X$. But by continuity of the mapping T,

$$x = \lim_{n \to \infty} x_{n+1} = \lim_{n \to \infty} T(x_n) = T(x),$$

i.e., $T(x) = x$. Suppose $y \in X$ also satisfies $T(y) = y$. But then

$$\|y - x\| = \|T(y) - T(x)\| \leq c\|y - x\|.$$

Since $0 < c < 1$, the above can be true if and only if $\|y - x\| = 0$, that is $y = x$. Thus x is unique. $\quad \square$

Exercises 8.3

1. *Show that the series $\sum_{k=0}^{\infty} x(1-x)^k$ cannot converge uniformly for $0 \le x \le 1$.

2. For $n \in \mathbb{N}$, let $f_n(x) = x^n/(1+x^n)$, $x \in [0,1]$. Prove that the sequence $\{f_n\}$ does not converge uniformly on $[0,1]$.

3. Give an example of a sequence of functions that are not continuous at any point but which converges uniformly to a continuous function.

4. *Suppose that f is uniformly continuous on $\mathbb{R}$. For each $n \in \mathbb{N}$, set $f_n(x) = f(x + \frac{1}{n})$. Prove that the sequence $\{f_n\}$ converges uniformly to f on $\mathbb{R}$.

5. Let $\{f_n\}$ be a sequence of continuous real-valued functions that converges uniformly to a function f on a set $E \subset \mathbb{R}$. Prove that $\lim_{n \to \infty} f_n(x_n) = f(x)$ for every sequence $\{x_n\} \subset E$ such that $x_n \to x \in E$.

6. * Let $E \subset \mathbb{R}$ and let D be a dense subset of E. If $\{f_n\}$ is a sequence of continuous real-valued functions on E, and if $\{f_n\}$ converges uniformly on D, prove that $\{f_n\}$ converges uniformly on E. (Recall that D is dense in E if every point of E is either a point of D or a limit point of D)

7. Find a sequence $\{f_n\}$ in $\mathcal{C}[0,1]$ with $\|f_n\|_u = 1$ such that no subsequence of $\{f_n\}$ converges (in norm) in $\mathcal{C}[0,1]$.

8. Suppose $\{f_n\}$ is a sequence of continuous functions on $[a,b]$ that converges uniformly on $[a.b]$. For each $x \in [a,b]$, set $g(x) = \sup_{n}\{f_n(x)\}$.

    a. Prove that g is continuous on $[a,b]$.

    b. Show by example that the conclusion may be false if the sequence $\{f_n\}$ converges only pointwise on $[a,b]$.

9. *For each $n \in \mathbb{N}$ and $x \in \mathbb{R}$, set $f_n(x) = (1 + \frac{x}{n})^n$. Use Dini's theorem to prove that the sequence $\{f_n\}$ converges uniformly to e^x on $[a,b]$ for any fixed $a, b \in \mathbb{R}$.

10. Let $(X, \|\ \|)$ be a normed linear space and let $T : X \to X$ be a contraction mapping with constant c, $0 < c < 1$. If $\{x_n\}$ is the sequence in X as defined in the proof of Theorem 8.3.13, prove that

    $$\|x_{n+m} - x_n\| \le (c^n/(1-c))\|x_1 - x_0\| \text{ for all } n, m \in \mathbb{N}.$$

11. Define $T : \mathcal{C}[0,1] \to \mathcal{C}[0,1]$ by $(T\varphi)(x) = \int_0^x \varphi(t)\, dt$, $0 \le x \le 1$, $\varphi \in \mathcal{C}[0,1]$, and set $T^2 = T \circ T$.

    a. Prove that $|(T^2\varphi)(x)| \le \frac{1}{2}x^2\|\varphi\|_u$.

    b. Show that T^2 is a contraction mapping on $\mathcal{C}[0,1]$ and thus has a fixed point in $\mathcal{C}[0,1]$.

    c. Prove that T has a fixed point in $\mathcal{C}[0,1]$.

12. Prove that $(\mathcal{C}(K), \|\ \|_u)$ is a normed linear space.

13. Prove that $(\ell^2, \|\ \|_2)$ is a complete normed linear space.

8.4   Uniform Convergence and Integration

In Example 8.1.2(c) we provided an example of a sequence of Riemann integrable functions that converges pointwise, but for which the limit function is not Riemann integrable. Furthermore, in Example 8.1.2(d) we provided an example of a seuence of continuous function on $[0,1]$ for which $\lim\limits_{n\to\infty} f_n(x) = 0$ for all $x \in [0,1]$ but for which

$$\int_0^1 f_n(x)dx = \frac{1}{2}\frac{n}{n+1}.$$

Thus $\lim\limits_{n\to\infty} \int\limits_0^1 f_n(x) \neq \int\limits_0^1 \lim\limits_{n\to\infty} f_n(x)dx$. Hence, pointwise convergence, even if the limit function is Riemann integrable, is also not sufficient for the interchange of limits.

In this section, we will prove that uniform convergence of a sequence $\{f_n\}$ of Riemann integrable functions is again sufficient for the limit function f to be Riemann integrable, and for convergence of the definite integrals of f_n to the definite integral of f. The analogous result for the Riemann-Stieltjes integral is left to the exercises (Exercise 2).

THEOREM 8.4.1 *Suppose $f_n \in \mathcal{R}[a,b]$ for all $n \in \mathbb{N}$, and suppose that the sequence $\{f_n\}$ converges uniformly to f on $[a,b]$. Then $f \in \mathcal{R}[a,b]$ and*

$$\int_a^b f(x)\,dx = \lim_{n\to\infty} \int_a^b f_n(x)\,dx.$$

Proof. For each $n \in \mathbb{N}$, set

$$\epsilon_n = \max_{x\in[a,b]} |f_n(x) - f(x)|.$$

Since $f_n \to f$ uniformly on $[a,b]$, by Theorem 8.2.5, $\lim\limits_{n\to\infty} \epsilon_n = 0$. Also, for all $x \in [a,b]$,

$$f_n(x) - \epsilon_n \leq f(x) \leq f_n(x) + \epsilon_n.$$

Hence

$$\int_a^b (f_n - \epsilon_n) \leq \underline{\int_a^b} f \leq \overline{\int_a^b} f \leq \int_a^b (f_n + \epsilon_n). \tag{3}$$

Therefore

$$0 \leq \overline{\int_a^b} f - \underline{\int_a^b} f \leq 2\epsilon_n[b-a].$$

Since $\epsilon_n \to 0$, $f \in \mathcal{R}[a, b]$. Also by (3),

$$\left| \int_a^b f(x)\, dx - \int_a^b f_n(x)\, dx \right| < \epsilon_n\, [b - a],$$

and thus

$$\lim_{n \to \infty} \int_a^b f_n(x)\, dx = \int_a^b f(x)\, dx. \quad \square$$

COROLLARY 8.4.2 *If $f_k \in \mathcal{R}[a, b]$ for all $k \in \mathbb{N}$, and if*

$$f(x) = \sum_{k=1}^{\infty} f_k(x), \qquad x \in [a, b],$$

where the series converges uniformly on $[a, b]$, then $f \in \mathcal{R}[a, b]$ and

$$\int_a^b f(x)\, dx = \sum_{k=1}^{\infty} \int_a^b f_k(x)\, dx.$$

Proof. Apply the previous theorem to $S_n(x) = \sum_{k=1}^{n} f_k(x)$, which by Theorem 6.2.1 is integrable for each $n \in \mathbb{N}$. $\quad \square$

Although uniform convergence is sufficient for the conclusion of Theorem 8.4.1; it is not necessary. For example, if $f_n(x) = x^n$, $x \in [0, 1]$, then $\{f_n\}$ converges pointwise, but not uniformly, to the function

$$f(x) = \begin{cases} 0, & 0 \le x < 1, \\ 1, & x = 1. \end{cases}$$

The function $f \in \mathcal{R}[0, 1]$ and

$$\lim_{n \to \infty} \int_0^1 x^n\, dx = \lim_{n \to \infty} \frac{1}{n + 1} = 0 = \int_0^1 f(x) dx.$$

In Section 10.6, using results from the Lebesgue theory of integration, we will be able to prove a stronger convergence result that does not require uniform convergence of the sequence $\{f_n\}$. However, it does require that the limit function f is Riemann integrable. For completeness we include a statement of that result at this point.

THEOREM 8.4.3 (Bounded Convergence Theorem) *Suppose f and f_n, $n \in \mathbb{N}$, are Riemann integrable functions on $[a, b]$ with $\lim_{n \to \infty} f_n(x) = f(x)$ for all $x \in [a, b]$. Suppose also that there exists a positive constant M such that $|f_n(x)| \le M$ for all $x \in [a, b]$ and all $n \in \mathbb{N}$. Then*

$$\lim_{n \to \infty} \int_a^b f_n(x)\, dx = \int_a^b f(x)\, dx.$$

It is easily checked that the sequence $\{f_n\}$ on $[0, 1]$, where for each $n \in \mathbb{N}$ $f_n(x) = x^n$, satisfies the hypothesis of the previous theorem. Also, since the limit function f is continuous except at $x = 1$, $f \in \mathcal{R}[0, 1]$. On the other hand, the sequence $\{f_n\}$ of Example 8.1.2(d) does not satisfy the hypothesis of the theorem.

Exercises 8.4

1. *If $\sum |a_k| < \infty$, prove that

$$\int_0^1 \left(\sum_{k=0}^\infty a_k x^k \right) dx = \sum_{k=0}^\infty \frac{a_k}{k+1}.$$

2. Let α be a monotone increasing function on $[a, b]$. Suppose $f_n \in \mathcal{R}(\alpha)$ on $[a, b]$ for all $n \in \mathbb{N}$, and suppose that the sequence $\{f_n\}$ converges uniformly to f on $[a, b]$. Then $f \in \mathcal{R}(\alpha)$ and

$$\int_a^b f \, d\alpha = \lim_{n \to \infty} \int_a^b f_n \, d\alpha.$$

3. For each $n \in \mathbb{N}$, let $f_n(x) = nx/(1 + nx)$, $x \in [0, 1]$. Show that the sequence $\{f_n\}$ converges pointwise, but not uniformly, to an integrable function f on $[0, 1]$, and that

$$\lim_{n \to \infty} \int_0^1 f_n(x) \, dx = \int_0^1 f(x) \, dx.$$

4. *If f is Riemann integrable on $[0, 1]$, use the bounded convergence theorem to prove that

$$\lim_{n \to \infty} \int_0^1 x^n f(x) \, dx = 0.$$

5. Let $\{f_n\}$ be a sequence in $\mathcal{R}[a, b]$ that converges uniformly to $f \in \mathcal{R}[a, b]$. For $n \in \mathbb{N}$ set $F_n(x) = \int_0^x f_n$, and let $F(x) = \int_0^x f$, $x \in [a, b]$. Prove that $\{F_n\}$ converges uniformly to F on $[a, b]$.

6. *Suppose $f : [0, 1] \to \mathbb{R}$ is continuous. Prove that $\lim_{n \to \infty} \int_0^1 f(x^n) \, dx = f(0)$.

7. *Let $\{r_n\}$ be an enumeration of the rational numbers in $[0, 1]$, and let $f : [0, 1] \to \mathbb{R}$ be defined by

$$f(x) = \sum_{k=1}^\infty \frac{1}{2^k} I(x - x_k),$$

where I is the unit jump function of Definition 4.4.9. Prove that $f \in \mathcal{R}[0, 1]$.

8. Define g on $\mathbb{R}$ by $g(x) = x - [x]$, where $[x]$ denotes the greatest integer function. Prove that the function

$$f(x) = \sum_{n=1}^\infty \frac{g(nx)}{n^2}$$

is Riemann integrable on $[0, 1]$. (This function was given by Riemann as an example of a function that is not integrable according to Cauchy's definition.)

9. *Let $g \in \mathcal{R}[a, b]$ and let $\{f_n\}$ be a sequence of Riemann integrable functions on $[a, b]$ that converges uniformly to f on $[a, b]$. Prove that

$$\lim_{n \to \infty} \int_a^b f_n g = \int_a^b f g.$$

10. *Let g be a nonnegative real-valued function on $[0, \infty)$ for which $\int_0^\infty g(x)dx$ is finite. Suppose $\{f_n\}$ is a sequence of real-valued functions on $[0, \infty)$ satisfying $f_n \in \mathcal{R}[0, c]$ for every $c > 0$ and $|f_n(x)| \le g(x)$ for all $x \in [0, \infty)$ and $n \in \mathbb{N}$. If the sequence $\{f_n\}$ converges uniformly to f on $[0, c]$ for every $c > 0$, prove that

$$\lim_{n \to \infty} \int_0^\infty f_n(x)\,dx = \int_0^\infty f(x)\,dx.$$

11. For $f \in \mathcal{C}[a, b]$, define $\|f\|_1 = \int_a^b |f(x)|\,dx$.

    a. Prove that $(\mathcal{C}[a, b], \| \ \|_1)$ is a normed linear space.

    *b. Show by example that the normed linear space $(\mathcal{C}[a, b], \| \ \|_1)$ is not complete.

8.5 Uniform Convergence and Differentiation

In this section, we consider the question of interchange of limits and differentiation. Example 8.1.2(e) shows that even if the sequence $\{f_n\}$ converges uniformly to f, this is not sufficient for convergence of the sequence $\{f_n'\}$ of derivatives. Example 8.5.3 will further demonstrate very dramatically the failure of the interchange of limits and differentiation. There we will give an example of a series, each of whose terms has derivatives of all orders, that converges uniformly to a continuous function f, but for which f' fails to exist at every point of $\mathbb{R}$. Clearly, uniform convergence of the sequence $\{f_n\}$ is not sufficient. What is required is uniform convergence of the sequence $\{f_n'\}$.

THEOREM 8.5.1 *Suppose $\{f_n\}$ is a sequence of differentiable functions on* $[a, b]$. *If*

   (a) $\{f_n'\}$ *converges uniformly on* $[a, b]$, *and*

   (b) $\{f_n(x_o)\}$ *converges for some* $x_o \in [a, b]$,

then $\{f_n\}$ converges uniformly to a function f on $[a, b]$, with

$$f'(x) = \lim_{n \to \infty} f_n'(x).$$

Remarks. (a) Convergence of $\{f_n(x_o)\}$ at some $x_o \in [a, b]$ is required. For example, if we let $g_n(x) = f_n(x) + n$, then $g_n'(x) = f_n'(x)$, but $\{g_n(x)\}$ need not converge for any $x \in [a, b]$. In Exercise 1 you will be asked to show that uniform convergence of $\{f_n'\}$ is also required; pointwise convergence is not sufficient.

(b) If in addition to the hypotheses we assume that f_n' is continuous on $[a, b]$, then a much shorter and easier proof can be provided using the fundamental theorem of calculus. Since f_n' is continuous, by Theorem 6.3.2

$$f_n(x) = f_n(x_o) + \int_{x_o}^{x} f_n'(t)dt$$

for all $x \in [a, b]$. The result can now be proved using Corollary 8.3.2 and Theorem 8.4.1. The details are left to the exercises (Exercise 2).

Proof. Let $\epsilon > 0$ be given. Since $\{f_n(x_o)\}$ converges and $\{f_n'\}$ converges uniformly, there exists $n_o \in \mathbb{N}$ such that

$$|f_n(x_o) - f_m(x_o)| < \frac{\epsilon}{2}, \qquad \text{for all} \quad n, m \geq n_o, \tag{4}$$

and

$$|f_n'(t) - f_m'(t)| < \frac{\epsilon}{2(b-a)}, \qquad \text{for all } t \in [a, b], \text{ and all} \quad n, m \geq n_o. \tag{5}$$

Apply the mean value theorem to the functions $f_n - f_m$ with $n, m \geq n_o$ fixed. Then for $x, y \in [a, b]$, there exists t between x and y such that

$$|(f_n(x) - f_m(x)) - (f_n(y) - f_m(y))| = |[f_n'(t) - f_m'(t)](x - y)|.$$

Thus by (5),

$$|(f_n(x) - f_m(x)) - (f_n(y) - f_m(y))| \leq \frac{\epsilon}{2(b-a)}|x - y| < \frac{\epsilon}{2}. \tag{6}$$

Take $y = x_o$ in (6). Then by (4) and (6), for all $x \in [a, b]$ and $n, m \geq n_o$,

$$|f_n(x) - f_m(x)| \leq |(f_n(x) - f_m(x)) - (f_n(x_o) - f_m(x_o))| + |f_n(x_o) - f_m(x_o)|$$
$$< \frac{\epsilon}{2} + \frac{\epsilon}{2} = \epsilon.$$

Hence by Theorem 8.2.3 the sequence $\{f_n\}$ converges uniformly on $[a, b]$. Let

$$f(x) = \lim_{n \to \infty} f_n(x).$$

It remains to be shown that f is differentiable and that

$$f'(x) = \lim_{n \to \infty} f_n'(x)$$

for all $x \in [a, b]$. Fix $p \in [a, b]$, and for $t \neq p$, $t \in [a, b]$, define

$$g_n(t) = \frac{f_n(t) - f_n(p)}{t - p}, \qquad g(t) = \frac{f(t) - f(p)}{t - p}.$$

Then $g_n(t) \to g(t)$ for each $t \in [a, b]$, $t \neq p$, and for each n

$$\lim_{t \to p} g_n(t) = f_n'(p).$$

Let $E = [a, b] \setminus \{p\}$. Take $y = p$ in inequality (6). Then for all $t \in E$,

$$|g_n(t) - g_m(t)| \leq \frac{\epsilon}{2(b-a)}, \qquad \text{for all } n, m \geq n_o.$$

Therefore $\{g_n\}$ converges uniformly to g on E. Hence by Theorem 8.3.1,

$$f'(p) = \lim_{t \to p} g(t) = \lim_{n \to \infty} f_n'(p). \quad \Box$$

EXAMPLE 8.5.2 To illustrate the previous theorem consider the series

$$\sum_{k=1}^{\infty} \frac{\sin kx}{2^k}.$$

Since $\left| 2^{-k} \sin kx \right| \leq 2^{-k}$ for all $x \in \mathbb{R}$, by the Weierstrass M-test this series converges uniformly to a function S on $\mathbb{R}$. For $n \in \mathbb{N}$, let

$$S_n(x) = \sum_{k=1}^{n} \frac{\sin kx}{2^k}.$$

Then

$$S_n'(x) = \sum_{k=1}^{n} \frac{k \cos kx}{2^k}.$$

Since $\sum k 2^{-k}$ converges, by the Weierstrass M-test the sequence $\{S_n'\}$ converges uniformly on $\mathbb{R}$. Thus by Theorem 8.5.1,

$$S'(x) = \lim_{n \to \infty} S_n'(x) = \sum_{k=1}^{\infty} \frac{k \cos kx}{2^k}. \quad \Box$$

A Continuous Nowhere Differentiable Function

We conclude this section with the following example of Weierstrass of a continuous function which is nowhere differentiable. When this example first appeared in 1874, it astounded the mathematical community.

EXAMPLE 8.5.3 Consider the function f on $\mathbb{R}$ defined by

$$f(x) = \sum_{k=0}^{\infty} \frac{\cos a^k \pi x}{2^k}, \qquad (7)$$

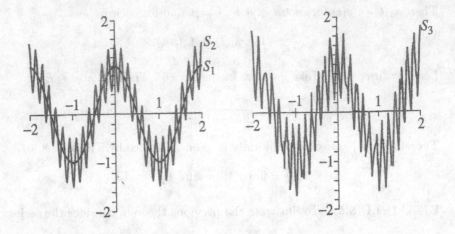

FIGURE 8.2
Graphs of S_1, S_2, S_3

where a is an odd positive integer satisfying $a > 3\pi + 2$. Since

$$\left| \frac{\cos a^k \pi x}{2^k} \right| \le \frac{1}{2^k},$$

the series (7) converges uniformly on $\mathbb{R}$, and hence f is continuous. The graphs of the partial sums (with $a = 13$) $S_1(x) = \cos(\pi x)$, $S_2(x) = S_1(x) + \frac{1}{2}\cos(13\pi x)$, and $S_3(x) = S_2(x) + \frac{1}{2^2}\cos(13^2\pi x)$ are illustrated in Figure 8.2.

We prove that f is nowhere differentiable by showing that for each $x \in \mathbb{R}$, there exists a sequence $h_n \to 0$ such that

$$\lim_{n \to \infty} \left| \frac{f(x + h_n) - f(x)}{h_n} \right| = \infty.$$

For $n \in \mathbb{N}$, set

$$S_n(x) = \sum_{k=0}^{n-1} \frac{\cos a^k \pi x}{2^k},$$

$$R_n(x) = \sum_{k=n}^{\infty} \frac{\cos a^k \pi x}{2^k}.$$

Then for $h > 0$,

$$\frac{f(x+h) - f(x)}{h} = \frac{S_n(x+h) - S_n(x)}{h} + \frac{R_n(x+h) - R_n(x)}{h}.$$

Our first step will be to estimate

$$\left| \frac{S_n(x+h) - S_n(x)}{h} \right|$$

from above. By the mean value theorem,

$$\frac{\cos a^k \pi(x+k) - \cos a^k \pi x}{h} = -a^k \pi \sin[a^k \pi(x+\zeta)],$$

for some ζ, $0 < \zeta < h$. Since

$$|-a^k \pi \sin[a^k \pi(x+\zeta)]| \le a^k \pi,$$

we obtain

$$\left| \frac{S_n(x+h) - S_n(x)}{h} \right| \le \sum_{k=0}^{n-1} \frac{1}{2^k} \left| \frac{\cos a^k \pi(x+h) - \cos a^k \pi x}{h} \right| \qquad (8)$$

$$\le \pi \sum_{k=0}^{n-1} \left(\frac{a}{2}\right)^k = \frac{\pi \left[1 - \left(\frac{a}{2}\right)^n\right]}{1 - \frac{a}{2}}$$

$$< \frac{2\pi}{a-2} \left(\frac{a}{2}\right)^n.$$

We now proceed to obtain a lower estimate on the term involving R_n. To do so we write

$$a^n x = k_n + \delta_n,$$

where k_n is an integer, and $-\frac{1}{2} \le \delta_n < \frac{1}{2}$. Set

$$h_n = \frac{1 - \delta_n}{a^n}.$$

Since $-\frac{1}{2} \le \delta_n < \frac{1}{2}$, we have $\frac{3}{2} \ge 1 - \delta_n > \frac{1}{2}$. Therefore,

$$\frac{2}{3} a^n \le \frac{1}{h_n} < 2 a^n. \qquad (9)$$

For $k \ge n$,

$$a^k \pi(x + h_n) = a^{k-n} a^n \pi(x + h_n)$$

$$= a^{k-n} \pi(a^n x + (1 - \delta_n)) = a^{k-n} \pi(k_n + 1).$$

Since a^{k-n} is odd and k_n is an integer,

$$\cos[a^k \pi(x + h_n)] = \cos[a^{k-n} \pi(k_n + 1)] = (-1)^{k_n+1}.$$

Also, $a^k \pi x = a^{k-n} a^n \pi x = a^{k-n} \pi(k_n + \delta_n)$. Using the trigonometric identity

$$\cos(A + B) = \cos A \cos B - \sin A \sin B$$

and the fact that $\sin(k_n a^{k-n} \pi) = 0$, we have

$$\cos a^k \pi x = \cos(a^{k-n} k_n \pi) \cos(a^{k-n} \delta_n \pi)$$

$$= (-1)^{k_n} \cos(a^{k-n} \delta_n \pi).$$

Therefore,

$$\cos a^k \pi(x + h_n) - \cos a^k \pi x = (-1)^{k_n+1} - (-1)^{k_n} \cos(a^{k-n}\delta_n\pi)$$
$$= (-1)^{k_n+1}[1 + \cos(a^{k-n}\delta_n\pi)].$$

As a consequence,

$$\left| \frac{R_n(x + h_n) - R_n(x)}{h_n} \right| = \left| \sum_{k=n}^{\infty} \frac{1}{2^k} \frac{\cos a^k \pi(x + h_n) - \cos a^k \pi x}{h_n} \right|$$

$$= \left| \sum_{k=n}^{\infty} \frac{(-1)^{k_n+1}}{h_n} \frac{[1 + \cos a^{k-n}\delta_n\pi]}{2^k} \right|$$

$$= \frac{1}{h_n} \sum_{k=n}^{\infty} \frac{1 + \cos a^{k-n}\delta_n\pi}{2^k} \geq \frac{1}{h_n} \frac{1 + \cos \delta_n\pi}{2^n}.$$

Since $-\frac{1}{2} \leq \delta_n < \frac{1}{2}$, $\cos \delta_n\pi \geq 0$. Therefore by (9) and the above,

$$\left| \frac{R_n(x + h_n) - R_n(x)}{h_n} \right| \geq \frac{1}{h_n} \frac{1}{2^n} \geq \frac{2}{3} \left(\frac{a}{2} \right)^n. \tag{10}$$

Using the reverse triangle inequality, $|a + b| \geq |a| - |b|$, we have

$$\left| \frac{f(x + h_n) - f(x)}{h_n} \right| \geq \left| \frac{R_n(x + h_n) - R_n(x)}{h_n} \right| - \left| \frac{S_n(x + h_n) - S_n(x)}{h_n} \right|,$$

which by (8) and (10),

$$\geq \frac{2}{3} \left(\frac{a}{2} \right)^n - \frac{2\pi}{a - 2} \left(\frac{a}{2} \right)^n$$

$$= \left(\frac{a}{2} \right)^n \left[\frac{2}{3} - \frac{2\pi}{a - 2} \right].$$

Since $a/2 > 1$, we obtain

$$\left| \frac{f(x + h_n) - f(x)}{h_n} \right| \to \infty \qquad \text{as } n \to \infty,$$

provided a is an odd positive integer satisfying

$$\frac{2}{3} - \frac{2\pi}{a - 2} > 0;$$

i.e., $a > 3\pi + 2$. Since $\pi < 3.15$, we need $a \geq 13$.   □

Remark. The above proof is based on the proof of a more general result given in the text by E. Hewitt and K. Stromberg. There it is proved (Theorem 17.7) that

$$f(x) = \sum_{k=0}^{\infty} \frac{\cos a^k \pi x}{b^k}$$

has the desired property if a is an odd positive integer, and b is any real number with $b > 1$ satisfying

$$\frac{a}{b} > 1 + \tfrac{3}{2}\pi.$$

The above function was carefully examined by G.H. Hardy [Trans. Amer. Math. Soc., **17** (1916), 301–325] who proved that the above f has the stated properties provided $1 < b \leq a$.

These are by no means the only examples of such functions. A slightly easier construction of a continuous function which is nowhere differentiable is given in Exercise 7.

Exercises 8.5

1. For $n \in \mathbb{N}$, set $f_n(x) = x^n/n$, $x \in [0, 1]$. Prove that the sequence $\{f_n\}$ converges uniformly to $f(x) = 0$ on $[0, 1]$, that the sequence $\{f'_n(x)\}$ converges pointwise on $[0, 1]$, but that $\{f'_n(1)\}$ does not converge to $f'(1)$.

2. *Let $\{f_n\}$ be a sequence of differentiable functions on $[a, b]$ for which f'_n is continuous on $[a, b]$ for all $n \in \mathbb{N}$. If $\{f'_n\}$ converges uniformly on $[a, b]$, and $\{f_n(x_o)\}$ converges for some $x_o \in [a, b]$, use the fundamental theorem of calculus to prove that $\{f_n\}$ converges uniformly to a function f on $[a, b]$ and that $f'(x) = \lim_{n \to \infty} f'_n(x)$ for all $x \in [a, b]$.

3. Let $\{a_k\}_{k=0}^{\infty}$ be a sequence of real numbers satisfying $\sum k|a_k| < \infty$. Show that the series $\sum_{k=0}^{\infty} a_k x^k$ converges uniformly to a function f on $|x| \leq 1$ and that

$$f'(x) = \sum_{k=1}^{\infty} k a_k x^{k-1}$$

for all x, $|x| \leq 1$.

4. *Let $\{f_n\}$ be a sequence of differentiable real-valued functions on (a, b) that converges pointwise to a function f on (a, b). Suppose the sequence $\{f'_n\}$ converges uniformly on every compact subset of (a, b). Prove that f is differentiable on (a, b) and that $f'(x) = \lim_{n \to \infty} f'_n(x)$ for all $x \in (a, b)$.

5. State and prove an analogue of Theorem 8.5.1 for a series of functions $\sum f_k(x)$.

6. Show that each of the following series converge on the indicated interval and that the derivative of the sum can be obtained by term by term differentiation of the series:

   *a. $\sum_{k=1}^{\infty} \dfrac{1}{(1 + kx)^2}$, $x \in (0, \infty)$ b. $\sum_{k=1}^{\infty} e^{-kx}$, $x \in (0, \infty)$

   *c. $\sum_{k=0}^{\infty} x^k$, $|x| < 1$ d. $\sum_{k=0}^{\infty} \dfrac{x^k}{k!}$, $e \in (-\infty, \infty)$

7. This exercise provides another construction of a continuous function f on $\mathbb{R}$ which is nowhere differentiable. Set $g(x) = |x|$, $-1 \leq x \leq 1$, and extend g to $\mathbb{R}$ to be periodic of period 2 by setting $g(x+2) = g(x)$. Define f on $\mathbb{R}$ by

$$f(x) = \sum_{k=0}^{\infty} \left(\frac{3}{4}\right)^k g(4^k x).$$

a. Prove that f is continuous on $\mathbb{R}$.

b. Fix $x_o \in \mathbb{R}$ and $m \in \mathbb{N}$. Set $\delta_m = \pm \frac{1}{2} 4^{-m}$, where the sign is chosen so that no integer lies between $4^m x_o$ and $4^m (x_o + \delta_m)$. Show that

$$g(4^n(x_o + \delta_m)) - g(4^n x_o) = 0 \text{ for all } n > m.$$

c. Show that

$$\left| \frac{f(x_o + \delta_m) - f(x_o)}{\delta_m} \right| \geq \frac{1}{2}(3^m + 1).$$

Since $\delta_m \to 0$ as $m \to \infty$, it now follows that $f'(x_o)$ does not exist.

8.6   The Weierstrass Approximation Theorem

In this section, we will prove the following well known theorem of Weierstrass.

THEOREM 8.6.1 (Weierstrass) *If f is a continuous real-valued function on $[a, b]$, then given $\epsilon > 0$, there exists a polynomial P such that*

$$|f(x) - P(x)| < \epsilon$$

for all $x \in [a, b]$.

An equivalent version, and what we will actually prove, is the following:

If f is a continuous real-valued function on $[a, b]$, then there exists a sequence $\{P_n\}$ of polynomials such that

$$f(x) = \lim_{n \to \infty} P_n(x) \qquad \text{uniformly on } [a, b].$$

Before we prove Theorem 8.6.1, we state and prove a more fundamental result that will also have applications later. Prior to doing so, we need the following definitions.

DEFINITION 8.6.2 *A real-valued function f on $\mathbb{R}$ is **periodic** with period p if*

$$f(x + p) = f(x) \qquad \text{for all } x \in \mathbb{R}.$$

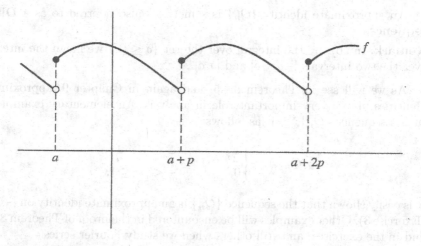

FIGURE 8.3
Graph of a periodic function

The canonical examples of periodic functions are the functions $\sin x$ and $\cos x$, both of which are periodic of period 2π. The graph of a periodic function of period p is illustrated in Figure 8.3. The graphs of a periodic function of period p on any two successive intervals of length p are identical. It is clear that if f is periodic of period p, then

$$f(x + kp) = f(x) \qquad \text{for all } k \in \mathbb{Z}.$$

Another useful property of periodic functions is as follows:

THEOREM 8.6.3 *If f is periodic of period p and Riemann integrable on $[0, p]$, then f is Riemann integrable on $[a, a + p]$ for every $a \in \mathbb{R}$, and*

$$\int_{a}^{a+p} f(x)\, dx = \int_{0}^{p} f(x)\, dx.$$

Proof. Exercise 2. □

Approximate Identities

DEFINITION 8.6.4 *A sequence $\{Q_n\}$ of nonnegative Riemann integrable functions on $[-a, a]$ satisfying*

(a) $\displaystyle \int_{-a}^{a} Q_n(t)\, dt = 1,$ *and*

(b) $\displaystyle \lim_{n \to \infty} \int_{\{\delta \le |t|\}} Q_n(t)\, dt = 0$ *for every $\delta > 0$, is called an **approximate***

identity *on $[-a, a]$.*

An approximate identity $\{Q_n\}$ is sometimes also referred to as a **Dirac sequence**.

Remark. In (b), by the integral over the set $\{\delta \le |t|\}$ we mean the integral over the two intervals $[-a, -\delta]$ and $[\delta, a]$.

As we will see in Theorem 8.6.5, and again in Chapter 9, approximate identities play a very important role in analysis. An elementary example of such a sequence $\{Q_n\}_{n=1}^{\infty}$ is as follows:

$$Q_n(t) = \begin{cases} \frac{n}{2}, & -\frac{1}{n} \le x \le \frac{1}{n}, \\ 0, & \frac{1}{n} < |x| \le 1. \end{cases}$$

It is easily shown that the sequence $\{Q_n\}$ is an approximate identity on $[-1, 1]$ (Exercise 3). Other examples will be encountered in the proof of Theorem 8.6.1 and in the exercises, and still others when we study Fourier series.

As a general rule, the Q_n are usually taken to be even functions; i.e. $Q_n(-x) = Q_n(x)$. The fact that the integrals over the set $\{t : |t| \ge \delta\}$ become small as $n \to \infty$ seems to suggest that in some sense the functions themselves become small as n becomes large. On the other hand, since the integrals over $[-a, a]$ are always 1, by property (b) of Definition 8.6.4

$$\lim_{n \to \infty} \int_{-\delta}^{\delta} Q_n(t) \, dt = 1$$

for every $\delta > 0$. This seems to indicate that the functions are concentrated near 0 and must become very large near 0 (see Exercise 6). The graphs of the first few functions $Q_1, Q_2,$ and Q_3 of a typical approximate identity $\{Q_n\}$ are given in Figure 8.4.

THEOREM 8.6.5 *Let $\{Q_n\}$ be an approximate identity on $[-1, 1]$, and let f be a bounded real-valued periodic function on $\mathbb{R}$ of period 2 with $f \in \mathcal{R}[-1, 1]$. For $n \in \mathbb{N}$, $x \in \mathbb{R}$, define*

$$S_n(x) = \int_{-1}^{1} f(x+t) Q_n(t) \, dt. \tag{11}$$

If f is continuous at $x \in \mathbb{R}$, then

$$\lim_{n \to \infty} S_n(x) = f(x).$$

Furthermore, if f is continuous on $[-1, 1]$, then

$$\lim_{n \to \infty} S_n(x) = f(x) \qquad uniformly \ on \ \mathbb{R}.$$

Proof. We first note that since f is periodic and integrable on $[-1, 1]$, f is integrable on every finite subinterval of $\mathbb{R}$. Thus the integral in (11) is defined

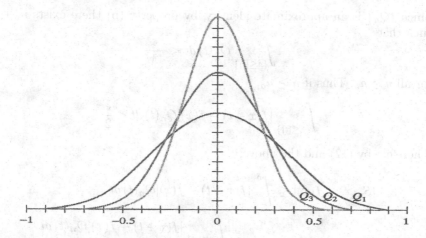

FIGURE 8.4
Graphs of Q_1, Q_2, Q_3

for all $x \in \mathbb{R}$. Also, since f is bounded, there exists a constant $M > 0$ such that $|f(x)| \leq M$ for all $x \in \mathbb{R}$.

Suppose first that f is continuous at $x \in \mathbb{R}$. By (a) of Definition 8.6.4

$$f(x) = \int_{-1}^{1} f(x) Q_n(t)\, dt.$$

Therefore,

$$|S_n(x) - f(x)| = \left| \int_{-1}^{1} [f(x+t) - f(x)] Q_n(t)\, dt \right|$$

$$\leq \int_{-1}^{1} |f(x+t) - f(x)| Q_n(t)\, dt. \tag{12}$$

Let $\epsilon > 0$ be given. Since f is continuous at x, there exists a $\delta > 0$ such that

$$|f(x+t) - f(x)| < \frac{\epsilon}{2}$$

for all t, $|t| < \delta$. Therefore

$$\int_{-\delta}^{\delta} |f(x+t) - f(x)| Q_n(t)\, dt < \frac{\epsilon}{2} \int_{-\delta}^{\delta} Q_n(t)\, dt \leq \frac{\epsilon}{2} \int_{-1}^{1} Q_n(t)\, dt = \frac{\epsilon}{2} \tag{13}$$

On the other hand,

$$\int_{\{\delta \leq |t|\}} |f(x+t) - f(x)| Q_n(t)\, dt \leq 2M \int_{\{\delta \leq |t|\}} Q_n(t)\, dt.$$

Since $\{Q_n\}$ is an approximate identity, by property (b) there exists $n_o \in \mathbb{N}$ such that

$$\int_{\{\delta \leq |t|\}} Q_n(t)\, dt < \frac{\epsilon}{4M}$$

for all $n \geq n_o$. Thus if $n \geq n_o$,

$$\int_{\{\delta \leq |t|\}} |f(x+t) - f(x)| Q_n(t)\, dt < \frac{\epsilon}{2}.$$

Therefore by (12) and the above,

$$|S_n(x) - f(x)| \leq \int_{-\delta}^{\delta} |f(x+t) - f(x)| Q_n(t)\, dt$$

$$+ \int_{\{\delta \leq |t|\}} |f(x+t) - f(x)| Q_n(t)\, dt$$

$$< \frac{\epsilon}{2} + \frac{\epsilon}{2} = \epsilon$$

for all $n \geq n_o$. Thus $\lim_{n\to\infty} S_n(x) = f(x)$.

Suppose f is continuous on $[-1, 1]$. Since f is periodic, this implies that $f(-1) = f(1)$. By Theorem 4.3.4, f is uniformly continuous on $[-1, 1]$, and hence by periodicity, also on $\mathbb{R}$ (Exercise 1). Thus given $\epsilon > 0$, there exists a $\delta > 0$ such that

$$|f(x+t) - f(x)| < \frac{\epsilon}{2}$$

for all $x \in \mathbb{R}$ and all t, $|t| < \delta$. As a consequence, inequality (13) holds for all $x \in \mathbb{R}$. Therefore, as above, there exists $n_o \in \mathbb{N}$ such that

$$|S_n(x) - f(x)| < \epsilon$$

for all $x \in \mathbb{R}$ and all $n \geq n_o$. This proves that the sequence $\{S_n\}$ converges uniformly to f on $\mathbb{R}$. $\square$

Proof of the Weierstrass Approximation Theorem.

We now use Theorem 8.6.5 to prove the Weierstrass approximation theorem (Theorem 8.6.1). Suppose f is a continuous real-valued function on $[a, b]$. By making a change of variable, i.e.,

$$g(x) = f((b - a)x + a), \qquad x \in [0, 1],$$

we can assume that f is continuous on $[0, 1]$. Also, if we let

$$g(x) = f(x) - f(0) - x[f(1) - f(0)], \qquad x \in [0, 1],$$

then $g(0) = g(1) = 0$ and $g(x) - f(x)$ is a polynomial. If we can approximate g by a polynomial Q and set

$$P(x) = Q(x) + [f(x) - g(x)],$$

then P is also a polynomial with $|f(x) - P(x)| = |g(x) - Q(x)|$. Therefore, without loss of generality, we can assume that f is defined on $[0,1]$ satisfying

$$f(0) = f(1) = 0.$$

Extend f to $[-1,1]$ by defining $f(x) = 0$ for all $x \in [-1,0)$. Then f is continuous on $[-1,1]$. Finally, we extend f to all of $\mathbb{R}$ by defining

$$f(x) = f(x - 2k), \qquad k \in \mathbb{Z},$$

where $k \in \mathbb{Z}$ is chosen so that $x - 2k \in (-1,1]$

Our next step is to find an approximate identity $\{Q_n\}$ on $[-1,1]$ such that the corresponding functions S_n of Theorem 8.6.5 defined by equation (11) are polynomials. To accomplish this we let

$$Q_n(t) = c_n(1 - t^2)^n,$$

where $c_n > 0$ is chosen such that

$$\int_{-1}^{1} Q_n(t)\, dt = 1.$$

Thus the sequence $\{Q_n\}$ satisfies hypothesis (a) of Definition 8.6.4. To show that it also satisfies (b) we need an estimate on the magnitude of c_n. Since

$$1 = c_n \int_{-1}^{1} (1 - t^2)^n\, dt = 2c_n \int_{0}^{1} (1 - t^2)^n\, dt$$

$$\geq 2c_n \int_{0}^{1/\sqrt{n}} (1 - t^2)^n\, dt$$

$$\geq 2c_n \int_{0}^{1/\sqrt{n}} (1 - nt^2)\, dt = 2c_n \left(\frac{1}{\sqrt{n}} - \frac{1}{3\sqrt{n}}\right)$$

$$= \frac{4c_n}{3\sqrt{n}},$$

we obtain

$$c_n \leq \tfrac{3}{4}\sqrt{n} < \sqrt{n}.$$

In the above we have used the inequality $(1 - t^2)^n \geq 1 - nt^2$ valid for all $t \in [0,1]$ (Example 1.3.2(b)). Finally, for any δ, $0 < \delta < 1$,

$$Q_n(t) = c_n(1 - t^2)^n \leq \sqrt{n}(1 - \delta^2)^n, \qquad \text{for all } t, \delta \leq |t| \leq 1.$$

Thus since $0 < (1 - \delta^2) < 1$, by Theorem 3.2.6(d), $\lim_{n\to\infty} Q_n(t) = 0$ uniformly in $\delta \leq |t| \leq 1$. Therefore,

$$\lim_{n\to\infty} \int_{\{\delta \leq |t|\}} Q_n(t)\, dt = 0.$$

For $x \in [0,1]$, set

$$P_n(x) = \int_{-1}^{1} f(x+t)Q_n(t)\,dt.$$

This is the function $S_n(x)$ of Theorem 8.6.5 except restricted to $x \in [0,1]$. Let $x \in [0,1]$. Since $f(t) = 0$ for $t \in [-1,0] \cup [1,2]$,

$$\int_{-1}^{1} f(x+t)Q_n(t)\,dt = \int_{-x}^{1-x} f(x+t)Q_n(t)\,dt,$$

which by the change of variables $s = t + x$ gives

$$P_n(x) = \int_{0}^{1} f(s)Q_n(s-x)\,ds.$$

Therefore $P_n(x)$, for $x \in [0,1]$, is a polynomial of degree less than or equal to $2n$. As a consequence of Theorem 8.6.5,

$$\lim_{n \to \infty} P_n(x) = f(x) \qquad \text{uniformly on} \qquad [0,1],$$

thereby proving the result. $\quad \square$

Remarks. (a) The above proof of the Weierstrass approximation theorem is a variation of a proof found in the text by Walter Rudin listed in the Bibliography.

  (b) The Weierstrass approximation theorem proves that the set of polynomials is dense in $\mathcal{C}([0,1])$. A natural question is the following. Do we need all polynomials? Let $N = \{n_j\}_{j=1}^{\infty}$ be a stricly increasing sequence of positive integers, and let $\mathcal{P}_N$ be the set of all polynomials of the form

$$P(x) = a_0 + a_1 x^{n_1} + \cdots + a_k x^{n_k}.$$

A very interesting result, whose proof is beyond the scope of the text, is the **Müntz-Szasz Theorem**[3] as follows: The set $\mathcal{P}_n$ is dense in $\mathcal{C}([0,1])$ if and only if

$$\sum_{k=1}^{\infty} \frac{1}{n_k} = \infty.$$

Hence, the set of all polynomials with even exponents is dense, whereas the set of all polynomials of the form

$$P(x) = a_0 + a_1 x^2 + a_2 x^{2^2} + a_3 x^{3^2} + \cdots + a_n x^{n^2}$$

is not.

---

[3] A proof of the Müntz-Szasz theorem may by found in the following text by Walter Rudin: *Real and Complex Analysis*, McGraw–Hill, New York, 1966.

Exercises 8.6

1. If $f : \mathbb{R} \to \mathbb{R}$ is periodic of period 2 and continuous on $[-1, 1]$, prove that f is uniformly continuous on $\mathbb{R}$.

2. *Prove Theorem 8.6.3.

3. For $n \in \mathbb{N}$, define Q_n on $[-1, 1]$ as follows:

$$Q_n(x) = \begin{cases} \frac{n}{2}, & -\frac{1}{n} \le x \le \frac{1}{n}, \\ 0, & \frac{1}{n} < |x| \le 1. \end{cases}$$

   Show that $\{Q_n\}$ is an approximate identity on $[-1, 1]$.

4. For $n \in \mathbb{N}$, set $Q_n(x) = c_n(1 - |x|)^n$, $x \in [-1, 1]$.

   *a. Determine $c_n > 0$ so that $\int_{-1}^{1} Q_n(t) dt = 1$.

   b. Prove that with the above choice of c_n the sequence $\{Q_n\}$ is an approximate identity on $[-1, 1]$.

   c. Sketch the graph of $Q_n(x)$ for $n = 2, 4, 8$.

5. For $n \in \mathbb{N}$, set $Q_n(x) = c_n |x| e^{-nx^2}$, $x \in [-1, 1]$.

   a. Determine $c_n > 0$ so that $\int_{-1}^{1} Q_n(t) dt = 1$.

   b. Prove that with the above choice of c_n the sequence $\{Q_n\}$ is an approximate identity on $[-1, 1]$.

6. *If $\{Q_n\}$ is an approximate identity on $[-1, 1]$, prove that

$$\overline{\lim_{n \to \infty}} \sup\{Q_n(x) : x \in [-\delta, \delta]\} = \infty$$

   for every $\delta > 0$

7. Let f be a continuous real-valued function on $[0, 1]$. Prove that given $\epsilon > 0$, there exists a polynomial P with <u>rational coefficients</u> such that $|f(x) - P(x)| < \epsilon$ for all $x \in [0, 1]$.

8. Suppose f is a continuous real-valued function on $[0, 1]$ satisfying

$$\int_0^1 f(x)x^n \, dx = 0 \qquad \text{for all } n = 0, 1, 2, ...$$

   Prove that $f(x) = 0$ for all $x \in [0, 1]$. (Hint: First show that $\int_0^1 f(x)P(x)dx = 0$ for every polynomial P, then use the Weierstrass theorem to show that $\int_0^1 f^2(x)dx = 0$.)

8.7 Power Series Expansions

In this section, we turn our attention to the study of power series and the representation of functions by means of power series. Because of their special nature, power series possess certain properties which are not valid for series of functions. We begin with the following definition.

DEFINITION 8.7.1 *Let* $\{a_k\}_{k=0}^{\infty}$ *be a sequence of real numbers, and let* $c \in \mathbb{R}$. *A series of the form*

$$\sum_{k=0}^{\infty} a_k(x-c)^k = a_0 + a_1(x-c) + a_2(x-c)^2 + a_3(x-c)^3 + \cdots$$

is called a **power series** *in* $(x-c)$. *When* $c = 0$, *the series is called a power series in* x. *The numbers* a_k *are called the* **coefficients** *of the power series.*

Even though the study of representation of functions by means of power series dates back to the mid seventeenth century, the rigorous study of convergence is much more recent. Certainly Newton and his successors were concerned with questions involving the convergence of a power series to its defining functions. It was Cauchy however who with his formal development of series brought mathematical rigor to the subject. As an application of his root and ratio test, Cauchy was among the first to use these tests to determine the interval of convergence of a power series. This is accomplished as follows: Consider a power series $\sum_{k=0}^{\infty} a_k(x-c)^k$. Applying the root test to this series gives

$$\varlimsup_{k \to \infty} \sqrt[k]{|a_k||x-c|^k} = |x-c| \varlimsup_{k \to \infty} \sqrt[k]{|a_k|} = |x-c|\,\alpha,$$

where $\alpha = \varlimsup_{k \to \infty} \sqrt[k]{|a_k|}$. Thus by Theorem 7.3.4 the series converges absolutely if $\alpha|x-c| < 1$, and diverges if $\alpha|x-c| > 1$. If $\alpha = 0$, then $\alpha|x-c| < 1$ for all $x \in \mathbb{R}$. If $0 < \alpha < \infty$, then

$$\alpha|x-c| < 1 \quad \text{if and only if} \quad |x-c| < \frac{1}{\alpha}.$$

DEFINITION 8.7.2 *Given a power series* $\sum a_k(x-c)^k$, *the* **radius of convergence** R *is defined by*

$$\frac{1}{R} = \varlimsup_{n \to \infty} \sqrt[n]{|a_n|}.$$

If $\varlimsup \sqrt[n]{|a_n|} = \infty$ *we take* $R = 0$, *and if* $\lim \sqrt[n]{|a_n|} = 0$ *we set* $R = \infty$.

When $R = 0$, the power series $\sum a_k(x-c)^k$ converges only for $x = c$. On the other hand, if $R = \infty$, then the power series converges for all $x \in \mathbb{R}$.

Remark. If $a_k \neq 0$ for all k and $\lim\limits_{k\to\infty} |a_{k+1}|/|a_k|$ exists, then by Theorem 7.1.10 the radius of convergence of $\sum a_k x^k$ is also given by

$$\frac{1}{R} = \lim_{k\to\infty} \frac{|a_{k+1}|}{|a_k|}.$$

This formulation is particularly useful if the coefficients involve factorials.

THEOREM 8.7.3 *Given a power series $\sum\limits_{k=0}^{\infty} a_k(x-c)^k$ with radius of convergence R, $0 < R \leq \infty$, then the series*

(a) *converges absolutely for all x with $|x - c| < R$, and*

(b) *diverges for all x with $|x - c| > R$.*

(c) *Furthermore, if $0 < \rho < R$, then the series converges uniformly for all x with $|x - c| \leq \rho$.*

Proof. Statements (a) and (b) were proved in the discussion preceding the statement of the theorem. Suppose $0 < \rho < R$. Choose β such that $\rho < \beta < R$. Since

$$\varlimsup_{k\to\infty} \sqrt[k]{|a_k|} = \frac{1}{R} < \frac{1}{\beta},$$

there exists $n_o \in \mathbb{N}$ such that

$$\sqrt[k]{|a_k|} < \frac{1}{\beta} \qquad \text{for all } k \geq n_o.$$

Hence for $k \geq n_o$ and $|x - c| \leq \rho$,

$$|a_k(x-c)^k| \leq |a_k|\rho^k < \left(\frac{\rho}{\beta}\right)^k.$$

But $(\rho/\beta) < 1$ and thus $\sum (\rho/\beta)^k < \infty$. Therefore by the Weierstrass M-test, the series converges uniformly on $|x - c| \leq \rho$. $\square$

The previous theorem provides no suggestion as to what happens when $|x - c| = R$. As the following examples (with $c = 0$) illustrate, the series may either converge or diverge when $|x| = R$.

EXAMPLES 8.7.4 (a) The series $\sum\limits_{k=0}^{\infty} x^k$ has radius of convergence $R = 1$. This series diverges at both $x = 1$ and -1.

(b) The series $\sum\limits_{k=1}^{\infty} \dfrac{x^k}{k}$ also has radius of convergence $R = 1$. In this case, when $x = 1$ the series diverges; whereas when $x = -1$, the series is an alternating series that converges by Theorem 7.2.3.

(c) Consider the series $\sum\limits_{k=1}^{\infty} \dfrac{x^k}{k^2}$. Again the radius of convergence is $R = 1$. In this example, the series converges at both $x = 1$ and -1.

(d) Consider the series

$$1 + 2x + 3^2 x^2 + 2^3 x^3 + 3^4 x^4 + \cdots = \sum_{k=0}^{\infty} a_k x^k$$

where

$$a_k = \begin{cases} 3^k, & \text{if } k \text{ is even,} \\ 2^k, & \text{if } k \text{ is odd.} \end{cases}$$

Hence $\overline{\lim}\, \sqrt[k]{|a_k|} = 3$, and therefore $R = 1/3$. The series diverges at both $x = 1/3$ and $x = -1/3$.

(e) Finally, consider the series $\sum k! x^k$. Here $a_k = k!$, and

$$\lim_{k \to \infty} \frac{a_{k+1}}{a_k} = \lim_{k \to \infty} (k+1) = \infty.$$

Thus by Theorem 7.1.10 $\sqrt[k]{|a_k|} \to \infty$, and $R = 0$. Therefore the power series converges only for $x = 0$. $\square$

Abel's Theorem

Suppose we are given a power series $\sum a_k (x - c)^k$ with radius of convergence $R > 0$. By setting

$$f(x) = \sum_{k=0}^{\infty} a_k (x - c)^k, \tag{14}$$

we obtain a function that is defined for all x, $|x - c| < R$. Functions that are defined in terms of a power series (as in (14)) are usually referred to as **real analytic** functions. Since the series converges uniformly for all x with $|x - c| \leq \rho$, for any ρ, $0 < \rho < R$, the function f is continuous on $|x - c| \leq \rho$. Since this holds for all $\rho < R$, the function f is continuous on $|x - c| < R$. If the series (14) also converges at an endpoint, say at $x = c + R$, then f is continuous not only in $(c - R, c + R)$ but also at $x = c + R$. This follows from the following theorem of Abel. For convenience, we take $c = 0$ and $R = 1$.

THEOREM 8.7.5 (Abel's Theorem) *Suppose* $f(x) = \sum\limits_{k=0}^{\infty} a_k x^k$ *has radius of convergence* $R = 1$, *and that* $\sum\limits_{k=0}^{\infty} a_k$ *converges. Then*

$$\lim_{x \to 1^-} f(x) = \sum_{k=0}^{\infty} a_k.$$

Proof. Set $s_{-1} = 0$, and for $n = 0, 1, 2, \ldots$ let $s_n = \sum\limits_{k=0}^{n} a_k$. Then by the partial summation formula (7.2.1)

$$\sum_{k=0}^{n} a_k x^k = \sum_{k=0}^{n-1} s_k (x^k - x^{k+1}) + s_n x^n$$

$$= (1 - x) \sum_{k=0}^{n-1} s_k x^k + s_n x^n.$$

Since the sequence $\{s_n\}$ converges, if we let $n \to \infty$, then for all x, $|x| < 1$,

$$f(x) = (1 - x) \sum_{k=0}^{\infty} s_k x^k.$$

Let $s = \lim\limits_{n \to \infty} s_n$, and let $\epsilon > 0$ be given. Choose $n_o \in \mathbb{N}$ such that $|s - s_n| < \epsilon/2$ for all $n \geq n_o$. Since

$$(1 - x) \sum_{k=0}^{\infty} x^k = 1, \qquad |x| < 1,$$

we have for all x, $0 < x < 1$,

$$|f(x) - s| = \left| (1 - x) \sum_{k=0}^{\infty} (s_k - s) x^k \right| \leq (1 - x) \sum_{k=0}^{\infty} |s_k - s| x^k$$

$$\leq (1 - x) \sum_{k=0}^{n_o} |s_k - s| + \frac{\epsilon}{2}(1 - x) \sum_{k=n_o+1}^{\infty} x^k$$

$$\leq (1 - x)M + \frac{\epsilon}{2},$$

where $M = \sum\limits_{k=0}^{n_o} |s_k - s|$. If we now choose $\delta > 0$ such that $1 - \delta < x < 1$ implies that $(1 - x)M < \epsilon/2$, then $|f(x) - s| < \epsilon$ for all x, $1 - \delta < x < 1$. Thus $\lim\limits_{x \to 1^-} f(x) = s$. $\square$

EXAMPLE 8.7.6 To illustrate Abel's theorem, consider the series $\sum\limits_{k=0}^{\infty} (-1)^k t^k$. This series has radius of convergence $R = 1$. Furthermore, the series converges to $f(t) = 1/(1 + t)$ for all t, $|t| < 1$. Since the convergence is uniform on $|t| \leq |x|$, where $|x| < 1$, by Corollary 8.4.2

$$\ln(1 + x) = \int_0^x \frac{dt}{1 + t} = \sum_{k=0}^{\infty} (-1)^k \int_0^x t^k \, dt$$

$$= \sum_{k=0}^{\infty} \frac{(-1)^k}{k + 1} x^{k+1} = \sum_{k=1}^{\infty} \frac{(-1)^{k+1}}{k} x^k$$

for all x, $|x| < 1$. The series $\sum\limits_{k=1}^{\infty} \dfrac{(-1)^{k+1}}{k} x^k$ has radius of convergence $R = 1$, and also converges when $x = 1$. Thus by Abel's theorem,

$$\ln 2 = \sum_{k=1}^{\infty} \frac{(-1)^{k+1}}{k} = 1 - \tfrac{1}{2} + \tfrac{1}{3} - \tfrac{1}{4} + \cdots . \quad \square$$

Differentiation of Power Series

Suppose the power series $\sum\limits_{k=0}^{\infty} a_k(x-c)^k$ has radius of convergence $R > 0$. If we differentiate the series term-by-term we obtain the new power series

$$\sum_{k=1}^{\infty} k\, a_k(x-c)^{k-1} = \sum_{k=0}^{\infty} (k+1)a_{k+1}(x-c)^k. \tag{15}$$

The obvious question to ask is, what is the radius of convergence of the differentiated series (15)? Furthermore, if f is defined by $f(x) = \sum\limits_{k=0}^{\infty} a_k(x-c)^k$, $|x-c| < R$, does the series (15) converge to $f'(x)$? The answers to both of these questions are provided by the following theorem.

THEOREM 8.7.7 *Suppose* $\sum\limits_{k=0}^{\infty} a_k(x-c)^k$ *has radius of convergence* $R > 0$, *and*

$$f(x) = \sum_{k=0}^{\infty} a_k(x-c)^k, \qquad |x-c| < R.$$

Then

(a) $\sum\limits_{k=1}^{\infty} k\, a_k(x-c)^{k-1}$ *has radius of convergence* R, *and*

(b) $f'(x) = \sum\limits_{k=1}^{\infty} k\, a_k(x-c)^{k-1}$, *for all* x, $|x-c| < R$.

Proof. For convenience we take $c = 0$. Consider the differentiated series $\sum k\, a_k x^{k-1}$. By Theorem 3.2.6, $\lim\limits_{k\to\infty} \sqrt[k]{k} = 1$, and for $x \neq 0$,

$$\lim_{k\to\infty} \sqrt[k]{|x|^{k-1}} = \lim_{k\to\infty} \frac{|x|}{\sqrt[k]{|x|}} = |x|.$$

Therefore $\lim\limits_{k\to\infty} \sqrt[k]{k|x|^{k-1}} = |x|$. By Exercise 10 of Section 3.5, for $x \neq 0$,

$$\varlimsup_{k\to\infty} \sqrt[k]{|a_k|\, k\, |x|^{k-1}} = \varlimsup_{k\to\infty} \sqrt[k]{|a_k|} \lim_{k\to\infty} \sqrt[k]{k\, |x|^{k-1}}$$

$$= |x| \varlimsup_{k\to\infty} \sqrt[k]{|a_k|}.$$

Therefore, if $R = \infty$, the differentiated series (15) converges for all x, and if $0 < R < \infty$, the differentiated series converges for all x, $|x| < R$, and diverges for all x, $|x| > R$. Thus the radius of convergence of $\sum\limits_{k=1}^{\infty} k\, a_k x^{k-1}$ is also R.

Furthermore, for any ρ, $0 < \rho < R$, by Theorem 8.7.3 the series $\sum k\, a_k x^{k-1}$ converges uniformly for all x, $|x| \le \rho$. Thus by Theorem 8.5.1, the series (15) obtained by term-by-term differentiation converges to $f'(x)$, i.e.,

$$f'(x) = \sum_{k=1}^{\infty} k\, a_k x^{k-1}, \qquad \text{for all} \quad x, \ |x| < R. \quad \square$$

COROLLARY 8.7.8 *Suppose* $\sum\limits_{k=0}^{\infty} a_k(x-c)^k$ *has radius of convergence* $R > 0$, *and*

$$f(x) = \sum_{k=0}^{\infty} a_k(x - c)^k, \qquad |x - c| < R.$$

Then f *has derivatives of all orders in* $|x - c| < R$, *and for each* $n \in \mathbb{N}$,

$$f^{(n)}(x) = \sum_{k=n}^{\infty} k(k-1)\cdots(k-n+1)\, a_k(x-c)^{k-n}. \tag{16}$$

In particular,

$$f^{(n)}(c) = n!\, a_n. \tag{17}$$

Proof. The result is obtained by successively applying the previous theorem to f, f', f'', etc. Equation (17) follows by setting $x = c$ in (16). $\quad \square$

DEFINITION 8.7.9 *A real-valued function* f *defined on an open interval* I *is said to be* **infinitely differentiable** *on* I *if* $f^{(n)}(x)$ *exists on* I *for all* $n \in \mathbb{N}$. *The set of infinitely differentiable functions on an open interval* I *is denoted by* $C^{\infty}(I)$.

As a consequence of Corollary 8.7.8, if $\sum a_k(x-c)^k$ has radius of convergence $R > 0$ and if f is defined by $f(x) = \sum\limits_{k=0}^{\infty} a_k(x-c)^k$ for $|x - c| < R$, then the function f is infinitely differentiable on $(c - R, c + R)$ and its nth derivative is given by (16). We illustrate this with the following example.

EXAMPLE 8.7.10 For $|x| < 1$,

$$\frac{1}{1-x} = \sum_{k=0}^{\infty} x^k.$$

Thus by the previous corollary,

$$\frac{1}{(1-x)^2} = \sum_{k=1}^{\infty} k\, x^{k-1} = \sum_{k=0}^{\infty} (k+1)x^k,$$

$$\frac{2}{(1-x)^3} = \sum_{k=2}^{\infty} k(k-1)\, x^{k-2} = \sum_{k=0}^{\infty} (k+2)(k+1)x^k,$$

and for arbitrary $n \in \mathbb{N}$,

$$\frac{(n-1)!}{(1-x)^n} = \sum_{k=0}^{\infty} (k+n-1)\cdots(k+1)\, x^k. \quad \square$$

Uniqueness Theorem for Power Series

The following uniqueness result for power series is another consequence of Corollary 8.7.8 .

COROLLARY 8.7.11 *Suppose $\sum a_k(x-c)^k$ and $\sum b_k(x-c)^k$ are two power series which converge for all x, $|x - c| < R$, for some $R > 0$. Then*

$$\sum_{k=0}^{\infty} a_k(x - c)^k = \sum_{k=0}^{\infty} b_k(x - c)^k, \qquad |x - c| < R,$$

if and only if $a_k = b_k$ for all $k = 0, 1, 2, \ldots$.

Proof. Clearly, if $a_k = b_k$ for all k, then the two power series are equal and converge to the same function. Conversely, set

$$f(x) = \sum_{k=0}^{\infty} a_k(x - c)^k \qquad \text{and} \qquad g(x) = \sum_{k=0}^{\infty} b_k(x - c)^k.$$

If $f(x) = g(x)$ for all x, $|x-c| < R$, then $f^{(n)}(x) = g^{(n)}(x)$ for all $n = 0, 1, 2, \ldots$, and all x, $|x - c| < R$. In particular, $f^{(n)}(c) = g^{(n)}(c)$ for all $n = 0, 1, 2, \ldots$. Thus by (17), $a_n = b_n$ for all n. $\quad \square$

Representation of a Function by a Power Series

Up to this point we have shown that if a function f is defined by a power series, that is

$$f(x) = \sum_{k=0}^{\infty} a_k(x - c)^k, \qquad |x - c| < R,$$

with radius of convergence $R > 0$, then by Corollary 8.7.8, f is infinitely differentiable on $(c - R, c + R)$ and the coefficients a_k are given by $a_k = f^{(k)}(c)/k!$.

We now consider the converse question. Given an infinitely differentiable function on an open interval I and $c \in I$, can f be expressed as a power series in a neighborhood of the point c. Specifically, does there exist an $\epsilon > 0$ such that

$$f(x) = \sum_{k=0}^{\infty} a_k(x - c)^k$$

for all x, $|x - c| < \epsilon$, with $a_k = f^{(k)}(c)/k!$ for all $k = 0, 1, 2, ...$ The following example of Cauchy shows that this is not always possible.

EXAMPLE 8.7.12 Let f be defined on $\mathbb{R}$ by

$$f(x) = \begin{cases} e^{-1/x^2}, & x \neq 0, \\ 0, & x = 0. \end{cases}$$

Since $\lim_{x \to 0} e^{-1/x^2} = \lim_{t \to \infty} e^{-t^2} = 0$, f is continuous at 0. For $x \neq 0$,

$$f'(x) = \frac{2\,e^{-1/x^2}}{x^3}.$$

When $x = 0$, we have

$$f'(0) = \lim_{h \to 0} \frac{f(h) - f(0)}{h} = \lim_{h \to 0} \frac{e^{-1/h^2}}{h} = \lim_{t \to \infty} \frac{t}{e^{t^2}} = 0.$$

The last step follows from l'Hospital's rule. Thus,

$$f'(x) = \begin{cases} \dfrac{2}{x^3} e^{-1/x^2}, & x \neq 0, \\ 0, & x = 0. \end{cases}$$

By induction, it follows as above, that for each $n \in \mathbb{N}$,

$$f^{(n)}(x) = \begin{cases} P(1/x)e^{-1/x^2}, & x \neq 0, \\ 0, & x = 0, \end{cases}$$

where P is a polynomial of degree $3n$. The details are left to the exercises (Exercise 16). Thus the function f is infinitely differentiable on $\mathbb{R}$. If there exists $R > 0$ such that $f(x) = \sum_{k=0}^{\infty} a_k x^k$ for all x, $|x| < R$, then $a_k = 0$ for all k. As a consequence, f cannot be presented by a power series which converges to f in a neighborhood of 0.

Taylor Polynomials and Taylor Series

We now consider the problem of representing a function f in terms of a power series in greater detail. Newton derived the power series expansion of many of the elementary functions by algebraic techniques or term-by-term integration. For example, the series expansion of $1/(1+x)$ can easily be obtained by long division, which upon term-by-term integration gives the power series expansion of $\ln(1+x)$. Maclaurin and Taylor were among the first mathematicians to use Newton's calculus in determining the coefficients in the power series expansion of a function. Both realized that if a function $f(x)$ had a power series expansion $\sum a_k(x-c)^k$, then the coefficients a_k had to be given by $f^{(k)}(c)/k!$.

DEFINITION 8.7.13 *Let f be a real-valued function defined on an open interval I, and let $c \in I$ and $n \in \mathbb{N}$. Suppose $f^{(n)}(x)$ exists for all $x \in I$. The polynomial*

$$T_n(f,c)(x) = \sum_{k=0}^{n} \frac{f^{(k)}(c)}{k!}(x-c)^k$$

*is called the **Taylor polynomial** of order n of f at the point c. If f is infinitely differentiable on I, the series*

$$\sum_{k=0}^{\infty} \frac{f^{(k)}(c)}{k!}(x-c)^k$$

*is called the **Taylor series** of f at c.*

For the special case $c = 0$, the Taylor series of a function f is often referred to as the **Maclaurin series**. The first three Taylor polynomials T_0, T_1, T_2, are given specifically by

$$T_0(f,c)(x) = f(c),$$
$$T_1(f,c)(x) = f(c) + f'(c)(x-c),$$
$$T_2(f,c)(x) = f(c) + f'(c)(x-c) + \frac{f''(c)}{2!}(x-c)^2,$$

The Taylor polynomial $T_1(f,c)$ is the **linear approximation** to f at c; that is, the equation of the straight line passing through $(c, f(c))$ with slope $f'(c)$.

In general, the Taylor polynomial T_n of f is a polynomial of degree less than or equal to n, that satisfies

$$T_n^{(k)}(f,c)(c) = f^{(k)}(c),$$

for all $k = 0, 1, ..., n$. Since $f^{(n)}(c)$ might possibly be zero, T_n (as the next example shows) could very well be a polynomial of degree strictly less than n.

EXAMPLES 8.7.14 In the following examples we compute the Taylor series of several functions. At this stage nothing is implied about the convergence of the series to the function.

(a) Let $f(x) = \sin x$ and take $c = \frac{\pi}{2}$. Then

$$f(\tfrac{\pi}{2}) = \sin\frac{\pi}{2} = 1,$$
$$f'(\tfrac{\pi}{2}) = \cos\frac{\pi}{2} = 0,$$
$$f''(\tfrac{\pi}{2}) = -\sin\frac{\pi}{2} = -1,$$
$$f^{(3)}(\tfrac{\pi}{2}) = -\cos\frac{\pi}{2} = 0.$$

Thus
$$T_3(f, \tfrac{\pi}{2})(x) = 1 - \frac{1}{2!}(x - \tfrac{\pi}{2})^2,$$

which is a polynomial of degree 2. In general, if n is odd, $f^{(n)}(\frac{\pi}{2}) = 0$, and if $n = 2k$ is even, $f^{(2k)}(\frac{\pi}{2}) = (-1)^k$. Therefore, if n is even,

$$T_n(f, \tfrac{\pi}{2})(x) = T_{n+1}(f, \tfrac{\pi}{2}) = \sum_{k=0}^{n/2} \frac{(-1)^k}{(2k)!}(x - \tfrac{\pi}{2})^{2k}.$$

The Taylor expansion of $f(x) = \sin x$ about $c = \frac{\pi}{2}$ is given by

$$\sum_{k=0}^{\infty} \frac{(-1)^k}{(2k)!}(x - \tfrac{\pi}{2})^{2k}.$$

(b) For the function $f(x) = e^{-1/x^2}$, by Example 8.7.12

$$T_n(f, 0)(x) = 0 \qquad \text{for all } n \in \mathbb{N}.$$

Thus the Taylor series of f at $c = 0$ converges for all $x \in \mathbb{R}$; namely to the zero function. It however does not converge to f.

(c) In many instances, the Taylor expansion of a given function can be computed from a know series. As an example, we find the Taylor series expansion of $f(x) = 1/x$ about $c = 2$. This could be done by computing the derivatives of f and evaluating them at $c = 2$. However, it would still remain to be shown that the given series converges to $f(x)$. An easier method is as follows: We first write

$$\frac{1}{x} = \frac{1}{2 - (2 - x)} = \frac{1}{2}\frac{1}{1 - (\frac{2-x}{2})}.$$

For $|w| < 1$,

$$\frac{1}{1-w} = \sum_{k=0}^{\infty} w^k.$$

Setting $w = (2-x)/2$, we have

$$\frac{1}{x} = \frac{1}{2} \sum_{k=0}^{\infty} \frac{(2-x)^k}{2^k} = \sum_{k=0}^{\infty} \frac{(-1)^k}{2^{k+1}} (x-2)^k,$$

for all x, $|x-2| < 2$. By uniqueness, the given series must be the Taylor series of $f(x) = 1/x$. In this instance, the power series also converges to the function $f(x)$ for all x satisfying $|x-2| < 2$. $\quad\square$

Remainder Estimates

To investigate when the Taylor series of a function f converges to $f(x)$, we consider

$$R_n(x) = R_n(f,c)(x) = f(x) - T_n(f,c)(x). \tag{18}$$

The function R_n is called the **remainder** or **error function** between f and $T_n(f,c)$. Clearly,

$$f(x) = \lim_{n \to \infty} T_n(f,c)(x) \qquad \text{if and only if} \qquad \lim_{n \to \infty} R_n(f,c)(x) = 0.$$

Since the Taylor polynomial T_n is the nth partial sum of the Taylor series of f, the Taylor series converges to f at a point x if and only if $\lim_{n \to \infty} R_n(f,c)(x) = 0$. To emphasize this fact, we state it as a theorem.

THEOREM 8.7.15 *Suppose f is an infinitely differentiable real-valued function on the open interval I and $c \in I$. Then for $x \in I$,*

$$f(x) = \sum_{k=0}^{\infty} \frac{f^{(k)}(c)}{k!} (x-c)^k,$$

if and only if $\lim_{n \to \infty} R_n(f,c)(x) = 0$.

The formula

$$f(x) = f(c) + f'(c)(x-c) + \frac{f''(c)}{2!}(x-c)^2 + \cdots$$
$$+ \frac{f^{(n)}(c)}{n!}(x-c)^n + R_n(f,x)(x)$$

is known as Taylor's formula with remainder. We now proceed to derive several formulas for the remainder term R_n. These can be used to show convergence of T_n to f.

Lagrange Form of the Remainder

Our first result, due to Joseph Lagrange (1736–1813), is called the **Lagrange form of the remainder**. This result, sometimes also referred to as Taylor's theorem, was previously proved for the special case $n = 2$ in Lemma 5.4.3.

THEOREM 8.7.16 *Suppose f is a real-valued function on an open interval I, $c \in I$ and $n \in \mathbb{N}$. If $f^{(n+1)}(t)$ exists for every $t \in I$, then for any $x \in I$, there exists a ζ between x and c such that*

$$R_n(x) = R_n(f, c)(x) = \frac{f^{(n+1)}(\zeta)}{(n+1)!}(x - c)^{n+1}. \tag{19}$$

Remark. Continuity of $f^{(n+1)}$ is not required.

Proof. Fix $x \in I$, and let M be defined by

$$f(x) = T_n(f, c)(x) + M (x - c)^{n+1}.$$

To prove the result we need to show that $(n+1)! \, M = f^{(n+1)}(\zeta)$ for some ζ between x and c. To accomplish this, set

$$
\begin{aligned}
g(t) &= f(t) - T_n(f, c)(t) - M (t - c)^{n+1} \\
&= R_n(t) - M (t - c)^{n+1}.
\end{aligned}
$$

First, since T_n is a polynomial of degree less than or equal to n,

$$g^{(n+1)}(t) = f^{(n+1)}(t) - (n+1)! \, M.$$

Also, since $T_n^{(k)}(f, c)(c) = f^{(k)}(c)$, $k = 0, 1, ..., n$,

$$g(c) = g'(c) = \cdots = g^{(n)}(c) = 0.$$

For convenience, let's assume $x > c$. By the choice of M, $g(x) = 0$. By the mean value theorem applied to g on the interval $[c, x]$, there exits x_1, $c < x_1 < x$, such that

$$0 = g(x) - g(c) = g'(x_1)(x - c).$$

Thus $g'(x_1) = 0$ for some x_1, $c < x_1 < x$. Since $g'(c) = 0$, by the mean value theorem applied to g' on the interval $[c, x_1]$, $g''(x_2) = 0$ for some x_2, $c < x_2 < x_1$. Continuing in this manner we obtain a point x_n satisfying $c < x_n < x$, such that $g^{(n)}(x_n) = 0$. Applying the mean value theorem one more time to the function $g^{(n)}$ on the interval $[c, x_n]$, we obtain the existence of a $\zeta \in (c, x_n)$ such that

$$0 = g^{(n)}(x_n) - g^{(n)}(c) = g^{(n+1)}(\zeta)(x_n - c).$$

Thus $g^{(n+1)}(\zeta) = 0$; i.e., $f^{(n+1)}(\zeta) - (n+1)! \, M = 0$, for some ζ between x and c. $\square$

In 8.7.20 we will give several examples to show how the remainder estimates may be used to prove convergence of the Taylor series to its defining function. In the following example we show how the previous theorem may be used to derive simple estimates and inequalities.

EXAMPLES 8.7.17 (a) In this example, we use Theorem 8.7.16 with $n = 2$ to approximate $f(x) = \sqrt{1+x}$, $x > -1$. With $c = 0$ we find that

$$f(0) = 1, \quad f'(0) = \frac{1}{2}, \quad f''(0) = -\frac{1}{4}.$$

Therefore $T_2(f, 0)(x) = 1 + \frac{1}{2}x - \frac{1}{8}x^2$, and thus

$$\sqrt{1+x} = 1 + \frac{1}{2}x - \frac{1}{8}x^2 + R_2(f, x).$$

By formula (19),

$$R_2(f, 0)(x) = \frac{f^{(3)}(\zeta)}{3!}x^3 = \frac{1}{16}(1+\zeta)^{-5/2}x^3$$

for some ζ between 0 and x. If $x > 0$, then $\zeta > 0$ and in this case $(1+\zeta)^{-5/2} < 1$. Therefore we have

$$|\sqrt{1+x} - T_2(f, 0)(x)| < \frac{1}{16}x^3$$

for any $x > 0$. If we let $x = 0.4$, then $T_2(f, 0)(.4) = 1.18$, and by the above $|\sqrt{1.4} - 1.18| < 0.004$, so that two decimal place accuracy is assured. In fact, to five decimal places $\sqrt{1.4} = 1.18322$.

(b) The error estimates can also be used to derive inequalities. As in the previous example,

$$\sqrt{1+x} = 1 + \frac{1}{2}x - \frac{1}{8}x^2 + R_2(f, x).$$

For $x > 0$ we have $0 < R_2(f, 0)(x) < \frac{1}{16}x^3$. Thus

$$1 + \frac{1}{2}x - \frac{1}{8}x^2 < \sqrt{1+x} < 1 + \frac{1}{2}x - \frac{1}{8}x^2 + \frac{1}{16}x^3$$

for all $x > 0$. $\square$

Integral Form of the Remainder

Another formula for $R_n(f, c)$ is given by the following **integral form of the remainder**. This however does require the additional hypothesis that the $(n+1)st$ derivative is Riemann integrable.

THEOREM 8.7.18 *Suppose f is a real-valued function on an open interval I, $c \in I$ and $n \in \mathbb{N}$. If $f^{(n+1)}(t)$ exists for every $t \in I$ and is Riemann integrable on every closed and bounded subinterval of I, then*

$$R_n(x) = R_n(f,c)(x) = \frac{1}{n!} \int_c^x f^{(n+1)}(t)(x-t)^n \, dt, \qquad x \in I. \qquad (20)$$

Proof. The result is proved by induction on n. Suppose $n = 1$. Then

$$R_1(x) = f(x) - f(c) - f'(c)(x-c),$$

which by the fundamental theorem of calculus

$$= \int_c^x f'(t) \, dt - f'(c) \int_c^x dt = \int_c^x [f'(t) - f'(c)] \, dt.$$

From the integration by parts formula (Theorem 6.3.7) with

$$u(t) = f'(t) - f'(c), \qquad v'(t) = 1,$$
$$u'(t) = f''(t), \qquad v(t) = (t - x),$$

we obtain

$$\int_c^x [f'(t) - f'(c)] \, dt = [f'(t) - f'(c)](t - x) \Big|_c^x - \int_c^x f''(t)(t-x) dt$$

$$= \int_c^x (x - t) f''(t) \, dt.$$

To complete the proof, we assume that the result holds for $n = k$, and prove that this implies the result for $n = k + 1$. Thus assume $R_k(x)$ is given by (20). Then

$$R_{k+1}(x) = f(x) - T_{k+1}(f,c)(x)$$

$$= f(x) - T_k(f,c)(x) - \frac{f^{(k+1)}(c)}{(k+1)!}(x-c)^{k+1}$$

$$= R_k(x) - \frac{f^{(k+1)}(c)}{(k+1)!}(x-c)^{k+1}$$

$$= \frac{1}{k!} \int_c^x (x-t)^k f^{(k+1)}(t) \, dt - \frac{1}{k!} f^{(k+1)}(c) \int_c^x (x-t)^k \, dt$$

$$= \frac{1}{k!} \int_c^x (x-t)^k [f^{(k+1)}(t) - f^{(k+1)}(c)] \, dt.$$

As for the case $n = 1$, we again use integration by parts with

$$u(t) = f^{(k+1)}(t) - f^{(k+1)}(c) \quad \text{and} \quad v(t) = -\frac{1}{k+1}(x - t)^{k+1},$$

which upon simplification gives,

$$R_{k+1}(x) = \frac{1}{(k+1)!} \int_c^x (x-t)^{k+1} f^{(k+2)}(t) \, dt. \qquad \square$$

Cauchy's Form for the Remainder

Under the additional assumption of continuity of $f^{(n+1)}$ we obtain **Cauchy's form for the remainder** as follows:

COROLLARY 8.7.19 *Let f be a real-valued function on an open interval I, $c \in I$ and $n \in \mathbb{N}$. If $f^{(n+1)}$ is continuous on I, then for each $x \in I$, there exists a ζ between c and x such that*

$$R_n(x) = R_n(f,c)(x) = \frac{f^{(n+1)}(\zeta)}{n!}(x-c)(x-\zeta)^n. \tag{21}$$

Proof. Since $f^{(n+1)}(t)(x-t)^n$ is continuous on the interval from c to x, by the mean value theorem for integrals (Theorem 6.3.6), there exists a ζ between c and x such that

$$\int_c^x f^{(n+1)}(t)(x-t)^n dt = (x-c)f^{(n+1)}(\zeta)(x-\zeta)^n.$$

The result now follows by (20). $\square$

We now compute the Taylor series for several elementary functions and use the previous formulas for the remainder to show that the series converges to the function.

EXAMPLES 8.7.20 (a) As our first example we prove the **binomial theorem** (Theorem 3.2.5). For $n \in \mathbb{N}$ let $f(x) = (1+x)^n$, $x \in \mathbb{R}$. Since f is a polynomial of degree n, if $k > n$ then $f^{(k)}(x) = 0$ for all $x \in \mathbb{R}$. Therefore by Theorem 8.7.16,

$$f(x) = \sum_{k=0}^n \frac{f^{(k)}(0)}{k!}x^k.$$

By computation, $f^{(k)}(0) = n!/(n-k)!$ for $k = 0, 1, ..., n$. Therefore

$$(1+x)^n = \sum_{k=0}^n \frac{n!}{k!(n-k)!}x^k.$$

The series expansion of $(1+x)^\alpha$ for $\alpha \in \mathbb{R}$ with $\alpha < 0$ is given in Theorem 8.8.4, whereas the expansion for $\alpha > 0$ is given in Exercise 7 of Section 8.8. For rational numbers α, the expansion of $(1+x)^\alpha$ was known to Newton as early as 1664.

(b) Let $f(x) = \sin x$ with $c = 0$. Then

$$f^{(n)}(x) = \begin{cases} (-1)^k \cos x, & n = 2k+1, \\ (-1)^k \sin x, & n = 2k. \end{cases}$$

Thus $f^{(n)}(0) = 0$ for all even $n \in \mathbb{N}$, and $f^{(n)}(0) = (-1)^k$, whenever $n = 2k + 1$, $k = 0, 1, 2, \dots$. Therefore the Taylor series of f at $c = 0$ is given by

$$\sum_{k=0}^{\infty} \frac{(-1)^k}{(2k+1)!} x^{2k+1}.$$

To show convergence of the series to $\sin x$ we consider the remainder term $R_n(x)$. By Theorem 8.7.16, for each $x \in \mathbb{R}$ there exists a ζ such that

$$R_n(x) = \frac{f^{(n+1)}(\zeta)}{(n+1)!} x^{n+1}.$$

Since $|f^{(n+1)}(x)| \leq 1$ for all x, we have

$$|R_n(x)| \leq \frac{|x|^{n+1}}{(n+1)!}.$$

By Theorem 3.2.6(f), $\lim_{n \to \infty} |x|^{n+1}/(n+1)! = 0$ for any $x \in \mathbb{R}$. As a consequence, $\lim_{n \to \infty} R_n(x) = 0$ for all $x \in \mathbb{R}$, and thus

$$\sin x = \sum_{k=0}^{\infty} \frac{(-1)^k}{(2k+1)!} x^{2k+1}, \qquad x \in \mathbb{R}.$$

The sine function, as well as the cosine function, can be defined strictly in terms of power series. For further details, the reader is encouraged to look at Miscellaneous Exercise 3.

(c) As our third example we derive the Taylor series for $f(x) = \ln(1+x)$, where as in Example 6.3.5

$$\ln x = \int_1^x \frac{1}{t} dt, \qquad x > 0,$$

denotes the natural logarithm function on $(0, \infty)$. Then $f(0) = \ln(1) = 0$, and by the fundamental theorem of calculus $f'(x) = 1/(1+x)$. Thus for $n = 1, 2, \dots$,

$$f^{(n)}(x) = (-1)^{n+1} \frac{(n-1)!}{(1+x)^n}.$$

In particular, $f^{(n)}(0) = (-1)^{n+1}(n-1)!$, and the Taylor series of f at 0 becomes

$$\sum_{n=1}^{\infty} \frac{(-1)^{n+1}}{n} x^n.$$

Although we have already proved that this series converges to $\ln(1+x)$ for all x, $-1 < x \leq 1$ (Example 8.7.6), we will prove this again to illustrate the use of the remainder formulas.

Suppose first that $0 < x \leq 1$. By Theorem 8.7.16,

$$R_n(x) = R_n(f,0)(x) = \frac{(-1)^{n+2}x^{n+1}}{(n+1)(1+\zeta)^{n+1}}$$

for some ζ, $0 < \zeta < x$. In this case, $(1+\zeta) > 1$, and thus

$$|R_n(x)| \leq \frac{1}{n+1}x^{n+1} \leq \frac{1}{n+1},$$

for all x, $0 \leq x \leq 1$. Therefore,

$$\lim_{n\to\infty} R_n(x) = 0, \qquad \text{for all } x \in [0,1].$$

We next consider the more difficult case $-1 < x < 0$. By the Cauchy form of the remainder, if $-1 < x < 0$, there exists ζ, $x \leq \zeta \leq 0$, such that

$$R_n(x) = \frac{1}{n!}f^{(n+1)}(\zeta)(x-\zeta)^n\,x = (-1)^{n+2}\frac{(x-\zeta)^n\,x}{(1+\zeta)^{n+1}}.$$

Therefore,

$$|R_n(x)| \leq \frac{|x|}{(1+\zeta)}\left[\frac{(\zeta-x)}{1+\zeta}\right]^n.$$

Consider the function $\varphi(t)$ defined on $[x,0]$ by

$$\varphi(t) = \frac{t-x}{1+t}.$$

Then $\varphi'(t) = (1+x)/(1+t)^2$ which is positive on $[x,0]$ provided $x > -1$. Therefore

$$\varphi(t) \leq \varphi(0) = -x = |x|$$

for all t, $x \leq t \leq 0$. Thus

$$|R_n(x)| \leq \frac{|x|}{(1+\zeta)}|x|^n.$$

Since $|x| < 1$, $\lim_{n\to\infty} R_n(x) = 0$. Therefore the Taylor series converges to $\ln(1+x)$ for all x, $-1 < x \leq 1$; i.e.,

$$\ln(1+x) = \sum_{n=1}^{\infty}\frac{(-1)^{n+1}}{n}x^n, \qquad -1 < x \leq 1.$$

The Taylor expansion of $\ln(1+x)$ was first obtained in 1668 by Nicolaus Mercator. Newton shortly afterward also obtained the same expansion by term by term integration of the series expansion of $1/(1+x)$ (see Example 8.7.6).

(d) As our final example, we consider the **natural exponential function** $E(x) = \exp x$ which is defined as the inverse function of the natural logarithm function $L(x) = \ln x$. The domain of L is $(0, \infty)$ with range $(-\infty, \infty)$. Since $L'(x) = 1/x$ is strictly positive on $(0, \infty)$, L is a strictly increasing function on $(0, \infty)$. The inverse function $E(x)$ is defined by

$$y = E(x) \quad \text{if and only if} \quad x = \ln y.$$

By the inverse function theorem (Theorem 5.2.14), E is differentiable on $\mathbb{R}$ with

$$E'(x) = E'(L(y)) = \frac{1}{L'(y)} = y = E(x).$$

Thus $E'(x) = E(x)$ for all $x \in \mathbb{R}$. Since $\ln ab = \ln a + \ln b$, it immediately follows that

$$E(x + y) = E(x)\, E(y) \qquad \text{and} \qquad E(-x) = \frac{1}{E(x)}.$$

Also, $E(0) = 1$, and by Example 6.3.5 $E(1) = e$, where e is Euler's number defined in Example 3.3.5.. Therefore $E(n) = e^n$ for every integer n, and if $r = m/n$, m, $n \in \mathbb{Z}$ with $n \neq 0$, then $E(nr) = E(m) = e^m$. But $E(nr) = (E(r))^n$. Therefore, if $r \in \mathbb{Q}$,

$$E(r) = e^r.$$

For arbitrary $x \in \mathbb{R}$ we define e^x by $e^x = E(x)$. This definition of e^x is consistent with the definition given in Miscellaneous Exercise 3 of Chapter 1 (See Exercise 17).

Since $E^{(n)}(x) = E(x)$ for all $x \in \mathbb{R}$, $E^{(n)}(0) = E(0) = 1$. Thus the Taylor series expansion of $E(x)$ about $c = 0$ is given by

$$\sum_{k=0}^{\infty} \frac{1}{k!} x^k.$$

It is left as an exercise (Exercise 11) to show that this series converges to e^x for all $x \in \mathbb{R}$. $\square$

There is a more subtle question involving power series representation of functions which we have not touched upon. The question concerns the following: How is the radius of convergence R of the Taylor series of f related to the function f? The full answer to this question requires a knowledge of complex analysis and thus is beyond the scope of this text. However, we will illustrate the question and provide a hint of the answer with the following examples. If $f(x) = 1/(1 + x)$, then the Taylor expansion of f about $x = 0$ is given by

$$\sum_{k=0}^{\infty} (-1)^k x^k,$$

which has radius of convergence $R = 1$. This is expected since the given function f is not defined at $x = -1$, and thus the series could not have radius of convergence $R > 1$. If it did, this would imply that f would then have a finite limit at $x = -1$.

On the other hand, the function $g(x) = 1/(1+x^2)$ is infinitely differentiable on all of $\mathbb{R}$. However, the Taylor series expansion of g about $c = 0$ is given by

$$\sum_{k=0}^{\infty} (-1)^k x^{2k},$$

which again has only radius of convergence $R = 1$. The reason for this is that even though g is well behaved on $\mathbb{R}$, if we extend g to the complex plane $\mathbb{C}$ by

$$g(z) = \frac{1}{1 + z^2},$$

then g is not defined when $z^2 = -1$; i.e., $z = \pm i$, where i is the complex number that satisfies $i^2 = -1$.

Exercises 8.7

1. Find the radius of convergence of each of the following power series:

   a. $\sum_{k=1}^{\infty} \frac{3^k}{k^3} x^k$ ***b.** $\sum_{k=0}^{\infty} \frac{1}{4^k} (x+1)^{2k}$

   c. $\sum_{k=1}^{\infty} \left(1 - \frac{1}{k}\right)^k x^k$ ***d.** $\sum_{k=1}^{\infty} \left(1 - \frac{1}{k}\right)^{k^2} x^k$

   e. $\sum_{k=1}^{\infty} k \left(\frac{x}{2}\right)^k$ ***f.** $\sum_{k=0}^{\infty} a_k x^k$ where $a_k = \begin{cases} \dfrac{1}{2^k}, & \text{when } k \text{ is even,} \\ \dfrac{1}{2^{k+2}}, & \text{when } k \text{ is odd.} \end{cases}$

2. For each of the following, determine all values of x for which the given series converges:

   a. $\sum_{k=0}^{\infty} \frac{k}{x^k}, (x \neq 0)$ ***b.** $\sum_{k=1}^{\infty} \frac{3^k x^k}{2^k (1-x)^k}, (x \neq 1)$ **c.** $\sum_{k=0}^{\infty} \left(\frac{1+x}{1-x}\right)^k, (x \neq 1)$

3. Using the power series expansion of $1/(1-x)$ and its derivatives, find

   *a. $\sum_{k=1}^{\infty} k x^k, |x| < 1$ **b.** $\sum_{k=1}^{\infty} k^2 x^k, |x| < 1.$ **c.** $\sum_{k=1}^{\infty} \frac{k}{2^k}.$

4. **a.** Use Theorem 8.7.16 to show that

   $$\left| \sqrt[3]{1+x} - \left(1 + \frac{1}{3}x - \frac{1}{9}x^2\right) \right| < \frac{5}{81} x^3$$

   for all $x > 0$.

   b. Use the above inequality to approximate $\sqrt[3]{1.2}$ and $\sqrt[3]{2}$, and provide an estimate of the error.

5. Determine how large n must be chosen so that $|\sin x - T_n(\sin, 0)(x)| < .001$ for all x, $|x| \leq 1$.

6. Use the Taylor series and remainder estimate of Example 8.7.20(c) to compute $\ln 1.2$ accurate to four decimal places.

7. Suppose $f(x) = \sum\limits_{k=0}^{\infty} a_k(x-c)^k$ has radius of convergence $R > 0$. For $|x - c| < R$, set $F(x) = \int\limits_{c}^{x} f(t)dt$.

   Prove that
   $$F(x) = \sum_{k=0}^{\infty} \frac{a_k}{k+1}(x-c)^{k+1}, \qquad |x-c| < R.$$

8. *a. Use the previous exercise and the fact that $\dfrac{d}{dx}\arctan x = \dfrac{1}{1+x^2}$ to obtain the Taylor series expansion of $\arctan x$ about $c = 0$.

   b. Use part (a) to obtain a series expansion for π.

   *c. How large must n be chosen so that the nth partial sum of the series in (b) provides an approximation of π correct to four decimal places?

9. Use Exercise 8 and Abel's theorem to prove that $\sum\limits_{k=0}^{\infty} \dfrac{(-1)^k}{2k+1} = \dfrac{\pi}{4}$.

10. *Find constants a_0, a_1, a_2, a_3, and a_4 such that
    $$x^4 + 3x^2 - 2x + 5 = a_4(x-1)^4 + a_3(x-1)^3 + a_2(x-1)^2 + a_1(x-1) + a_0.$$

11. Prove that the Taylor series of e^x (with $c = 0$) converges to e^x for all $x \in \mathbb{R}$.

12. Using any applicable method, find the Taylor series of each of the following functions at the indicated point, and specify the interval on which the series converges to the function
    a. $f(x) = \cos x$, $\quad c = 0$.   *b. $f(x) = \ln x$, $\quad c = 1$
    c. $f(x) = \ln\left(\dfrac{1+x}{1-x}\right)$, $\quad c = 0$ *d. $f(x) = (1-x)^{-1/2}$, $\quad c = 0$
    e. $f(x) = \dfrac{x^2}{1-x^2}$, $\quad c = 0$ *f. $f(x) = \arcsin x$, $\quad c = 0$
    g. $f(x) = \sqrt{x}$, $\quad c = 1$ *h. $f(x) = (1+x)^p$, $\quad c = 0$, p real

13. Suppose $f(x) = \sum\limits_{k=0}^{\infty} a_k x^k$, $|x| < R$, where $R > 0$. Prove the following:

    a. $f(x)$ is even if and only if $a_k = 0$ for all odd k.

    b. $f(x)$ is odd if and only if $a_k = 0$ for all even k.

14. If $\sum a_k$ converges, prove that $\sum\limits_{k=0}^{\infty} a_k x^k$ converges uniformly on $[0, 1]$.

15. Suppose $f(x) = \sum\limits_{k=0}^{\infty} a_k x^k$, $|x| < R_1$, and $g(x) = \sum\limits_{k=0}^{\infty} b_k x^k$, $|x| < R_2$. Prove that
    $$f(x)g(x) = \sum_{k=0}^{\infty} c_k x^k, \qquad |x| < \min\{R_1, R_2\}, \text{ where } c_k - \sum_{j=0}^{k} a_j b_{k-j}.$$

16. Let $f : \mathbb{R} \to \mathbb{R}$ be defined by $f(x) = e^{-1/x^2}$ for $x \neq 0$, and $f(0) = 0$. Prove that for each $n \in \mathbb{N}$,
    $$f^{(n)}(x) = \begin{cases} P(1/x)e^{-1/x^2}, & x \neq 0, \\ 0, & x = 0, \end{cases}$$

where P is a polynomial of degree $3n$.

17. Suppose $b > 1$. For $x \in \mathbb{R}$ define $b(x) = E(x \ln b)$, where E is the natural exponential function.

    a. Prove that $b(r) = b^r$ for all $r \in \mathbb{Q}$.

    b. For $x \in \mathbb{R}$, prove that $b(x) = \sup\{b^r : r \in \mathbb{Q}, r < x\}$.

8.8   The Gamma Function

We close this chapter with a brief discussion of the Beta and Gamma functions, both of which are due to Euler. The Gamma function is closely related to factorials and arises in many areas of mathematics. The origin, history, and the development of the Gamma function are described very nicely in the article by Philip Davis listed in the supplemental reading. Our primary application of the Gamma function will be in the Taylor expansion of $(1-x)^{-\alpha}$, where $\alpha > 0$ is arbitrary.

DEFINITION 8.8.1 *For $0 < x < \infty$, the* **Gamma function** $\Gamma(x)$ *is defined by*

$$\Gamma(x) = \int_0^\infty t^{x-1} e^{-t} dt. \tag{22}$$

When $0 < x < 1$, the integral in (22) is an improper integral not only at ∞, but also at 0. The convergence of the improper integral defining $\Gamma(x)$, $x > 0$, was given as Exercise 9 in Section 6.4. The Graph of $\Gamma(x)$ for $0 < x < 5$ is given in Figure 8.5. The following properties of the Gamma function show that it is closely related to factorials.

THEOREM 8.8.2 **(a)** *For each x, $0 < x < \infty$, $\Gamma(x+1) = x\,\Gamma(x)$.*

(b) *For $n \in \mathbb{N}$, $\Gamma(n+1) = n!$.*

Proof. Let $0 < c < R < \infty$. We apply integration by parts to

$$\int_c^R t^x e^{-t}\, dt.$$

With $u = t^x$ and $v' = e^{-t}$,

$$\int_c^R t^x e^{-t}\, dt = -t^x e^{-t}\Big|_c^R + x \int_c^R t^{x-1} e^{-t}\, dt$$

$$= -\frac{R^x}{e^R} + c^x e^{-c} + x \int_c^R t^{x-1} e^{-t}\, dt.$$

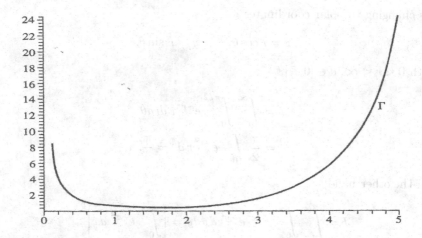

FIGURE 8.5
Graph of $\Gamma(x)$, $0 < x < 5$

Since $\lim\limits_{c \to 0^+} c^x e^{-c} = 0$ and $\lim\limits_{R \to \infty} R^x e^{-R} = 0$, taking the appropriate limits in the above yields

$$\Gamma(x+1) = \int_0^\infty t^x e^{-t}\, dt = x \int_0^\infty t^{x-1} e^{-t}\, dt = x\,\Gamma(x).$$

This proves (a). For the proof of (b) we first note that

$$\Gamma(1) = \int_0^\infty e^{-t}\, dt = 1.$$

Thus by induction, $\Gamma(n+1) = n!$. $\square$

EXAMPLE 8.8.3 Since the value of $\Gamma(\frac{1}{2})$ occurs frequently, we now show that $\Gamma(\frac{1}{2}) = \sqrt{\pi}$. By definition,

$$\Gamma(\tfrac{1}{2}) = \int_0^\infty t^{-\frac{1}{2}} e^{-t}\, dt.$$

With the substitution $t = s^2$,

$$\int_0^\infty t^{-\frac{1}{2}} e^{-t}\, dt = 2 \int_0^\infty e^{-s^2}\, ds.$$

To complete the result, we need to evaluate the so-called probability integral $\int_0^\infty e^{-s^2}\, ds$. This can be accomplished by the following trick using the change of variables theorem from multivariable calculus. Consider the double integral

$$J = \int_0^\infty \int_0^\infty e^{-x^2 - y^2}\, dx\, dy.$$

By changing to polar coordinates

$$x = r\cos\theta, \qquad y = r\sin\theta,$$

with $0 < r < \infty$, $\theta \in [0, \frac{\pi}{2}]$,

$$J = \int_0^\infty \int_0^{\pi/2} e^{-r^2} r \, dr \, d\theta$$
$$= \frac{\pi}{2} \int_0^\infty e^{-r^2} r \, dr = \frac{\pi}{4}.$$

On the other hand,

$$J = \int_0^\infty \int_0^\infty e^{-x^2} e^{-y^2} \, dx \, dy = \left[\int_0^\infty e^{-x^2} \, dx \right]^2.$$

Therefore,

$$\int_0^\infty e^{-x^2} \, dx = \frac{\sqrt{\pi}}{2},$$

from which the result follows. □

The Binomial Series

As an application of the Gamma function we will derive the power series expansion of $f(x) = (1-x)^{-\alpha}$, where $\alpha > 0$ is real. The coefficients of this expansion are expressed very nicely in terms of the Gamma function. By Example 8.7.10, for $n \in \mathbb{N}$,

$$(1-x)^{-n} = \frac{1}{(n-1)!} \sum_{k=0}^\infty (k+n-1)\cdots(k+1)\, x^k$$
$$= \frac{1}{(n-1)!} \sum_{k=0}^\infty \frac{(k+n-1)!}{k!}\, x^k,$$

which in terms of the Gamma function gives

$$\frac{1}{(1-x)^n} = \frac{1}{\Gamma(n)} \sum_{k=0}^\infty \frac{\Gamma(k+n)}{k!}\, x^k.$$

We will now prove that this formula is still valid for all $\alpha \in \mathbb{R}$ with $\alpha > 0$.

THEOREM 8.8.4 (Binomial Series) *For $\alpha > 0$,*

$$\frac{1}{(1-x)^\alpha} = \frac{1}{\Gamma(\alpha)} \sum_{n=0}^\infty \frac{\Gamma(n+\alpha)}{n!}\, x^n, \qquad |x| < 1. \tag{23}$$

Proof. We first show that the radius of convergence of the series (23) is $R = 1$. Set $a_n = \Gamma(n + \alpha)/n!$. Then

$$\frac{a_{n+1}}{a_n} = \frac{\Gamma(n + 1 + \alpha)}{(n + 1)!} \frac{n!}{\Gamma(n + \alpha)}.$$

But by Theorem 8.8.2, $\Gamma(n + 1 + \alpha) = (n + \alpha)\Gamma(n + \alpha)$. Therefore,

$$\lim_{n \to \infty} \frac{a_{n+1}}{a_n} = \lim_{n \to \infty} \frac{n + \alpha}{n + 1} = 1,$$

and as a consequence of Theorem 7.1.10 we have $R = 1$.

To show that the series actually converges to $(1 - x)^{-\alpha}$, we set

$$f_\alpha(x) = \frac{1}{\Gamma(\alpha)} \sum_{n=0}^{\infty} \frac{\Gamma(n + \alpha)}{n!} x^n, \qquad |x| < 1.$$

Since a power series can be differentiated term-by-term,

$$f_\alpha'(x) = \frac{1}{\Gamma(\alpha)} \sum_{n=1}^{\infty} \frac{n\,\Gamma(n + \alpha)}{n!} x^{n-1}.$$

Multiplying by $(1 - x)$ gives

$$(1 - x)\,f_\alpha(x) = \frac{1}{\Gamma(\alpha)} \sum_{n=1}^{\infty} \frac{n\,\Gamma(n + \alpha)}{n!} (1 - x)\, x^{n-1}$$

$$= \frac{1}{\Gamma(\alpha)} \left[\sum_{n=1}^{\infty} \frac{\Gamma(n + \alpha)}{(n - 1)!} x^{n-1} - \sum_{n=1}^{\infty} \frac{n\,\Gamma(n + \alpha)}{n!} x^n \right]$$

$$= \frac{1}{\Gamma(\alpha)} \sum_{n=0}^{\infty} \left[\frac{\Gamma(n + 1 + \alpha)}{n!} - \frac{n\,\Gamma(n + \alpha)}{n!} \right] x^n.$$

But $\Gamma(n + 1 + \alpha) - n\,\Gamma(n + \alpha) = \alpha\,\Gamma(n + \alpha)$. Therefore,

$$(1 - x)f_\alpha'(x) = \alpha f_\alpha(x).$$

As a consequence,

$$\frac{d}{dx}\left[(1 - x)^\alpha f_\alpha(x) \right] = -\alpha\,(1 - x)^{\alpha-1} f_\alpha(x) + (1 - x)^\alpha f_\alpha'(x)$$

$$= -\alpha\,(1 - x)^{\alpha-1} f_\alpha(x) + \alpha\,(1 - x)^{\alpha-1} f_\alpha(x) = 0.$$

Therefore $(1-x)^\alpha f_\alpha(x)$ is equal to a constant for all x, $|x| < 1$. But $f_\alpha(0) = 1$. Thus $(1 - x)^\alpha f_\alpha(x) = 1$; that is

$$f_\alpha(x) = (1 - x)^{-\alpha},$$

which proves the result. $\square$

The Beta Function

There are a number of important integrals that can be expressed in terms of the Gamma function. Some of these, which can be obtained by a change of variables, are given in the exercises. There is one integral however which is very important and thus we state it as a theorem. Since the proof is nontrivial and would take us too far astray, we state the result without proof. For a proof of the theorem the reader is referred to Theorem 8.20 in the text by Rudin.

THEOREM 8.8.5 *For $x > 0$, $y > 0$,*

$$\int_0^1 t^{x-1}(1-t)^{y-1}\,dt = \frac{\Gamma(x)\Gamma(y)}{\Gamma(x+y)}. \tag{24}$$

The function

$$B(x,y) = \frac{\Gamma(x)\Gamma(y)}{\Gamma(x+y)}, \qquad x, y > 0,$$

is called the **Beta function**.

Exercises 8.8

1. ***a.** Compute $\Gamma(\frac{3}{2})$, $\quad \Gamma(\frac{7}{2})$.

   b. Prove that for $n \in \mathbb{N}$, $\Gamma(n + \frac{1}{2}) = \dfrac{(2n)!\sqrt{\pi}}{4^n n!}$.

2. ***** By making a change of variable, prove that

$$\Gamma(x) = \int_0^1 \left(\ln\frac{1}{t}\right)^{x-1}\,dt, \qquad 0 < x < \infty.$$

3. Evaluate each of the following definite integrals:

   *a. $\displaystyle\int_0^\infty e^{-t}t^{\frac{3}{2}}\,dt$ \qquad **b.** $\displaystyle\int_0^1 \left(\ln\frac{1}{t}\right)^n\,dt, \quad n \in \mathbb{N}.$

4. By making the change of variable $t = \sin^2 u$ in Theorem 8.8.5, prove that

$$\int_0^{\frac{\pi}{2}} (\sin u)^{2n-1}(\cos u)^{2m-1}\,du = \frac{1}{2}\frac{\Gamma(n)\Gamma(m)}{\Gamma(n+m)}, \qquad n, m > 0.$$

5. Evaluate each of the following integrals:

   *a. $\displaystyle\int_0^{\frac{\pi}{2}} (\sin x)^{2n}\,dx, \quad n \in \mathbb{N}.$ \qquad **b.** $\displaystyle\int_0^{\frac{\pi}{2}} (\sin x)^{2n+1}\,dx, \quad n \in \mathbb{N}.$

6. Use the binomial series and term-by-term integration to find the power series expansion of

$$\arcsin x = \int_0^x (1-t^2)^{-1/2}\,dt.$$

7. Let $\alpha > 0$. Set $\dbinom{\alpha}{0} = 1$ and for $k \in \mathbb{N}$ set

$$\binom{\alpha}{k} = \frac{\alpha(\alpha-1)(\alpha-2)\cdots(\alpha-k+1)}{k!}.$$

Note, if $m \in \mathbb{N}$,

$$\binom{m}{k} = \frac{m!}{k!(m-k)!}, \ k \le m, \text{ and } \binom{m}{k} = 0 \text{ for } k > m.$$

a. Prove that the series $\sum_{k=0}^{\infty} \binom{\alpha}{k} x^k$ converges uniformly and absolutely for $x \in [-1, 1]$.

b. Prove that $\sum_{k=0}^{\infty} \binom{\alpha}{k} x^k = (1+x)^\alpha$, $x \in [-1, 1]$.

Notes

Without question the most important concept of this chapter is that of uniform convergence of a sequence or series of functions. It is the additional hypothesis required in proving that the limit function of a sequence of continuous or integrable functions is again continuous or integrable. As was shown by numerous examples, pointwise convergence is not sufficient. For differentiation, uniform convergence of $\{f_n\}$ is not sufficient; what is also required is uniform convergence of the sequence of derivatives $\{f'_n\}$.

The example of Weierstrass (Example 8.5.3) is interesting for several reasons. First, it provides an example of a continuous function which is nowhere differentiable on $\mathbb{R}$. Furthermore, it also provides an example of a sequence of infinitely differentiable functions which converges uniformly on $\mathbb{R}$, but for which the limit function is nowhere differentiable. Exercise 7 of Section 8.5 provides another construction of a continuous function f that is nowhere differentiable. Although this construction is much easier, the partial sums of the series defining the function f are themselves not differentiable everywhere. Thus it is not so surprising that f itself is not differentiable anywhere on $\mathbb{R}$.

The proof of the Weierstrass approximation theorem presented in the text is only one of the many poofs available. A constructive proof by S. N. Bernstein using the so-called Bernstein polynomials can be found on p. 107 of the text by Natanson listed in the Bilbiography. The proof in the text using approximate identities was chosen because the technique involved is very important in analysis and will be encountered later in the text. In Theorem 9.4.5 we will prove a variation of the Weierstrass approximation theorem. At that point we will show that every continuous real-valued function on $[-\pi, \pi]$ with $f(-\pi) = f(\pi)$ can be uniformly approximated to within a given ϵ by a finite sum of a trigonometric series.

Miscellaneous Exercises

1. Using Miscellaneous Exercise 1 of Chapter 6 and the Weierstrass approximation theorem, prove the following: If $f \in \mathcal{R}[a,b]$ and $\epsilon > 0$ is given, then there exists a polynomial P such that

$$\int_a^b |f - P| < \epsilon.$$

2. Define f on $\mathbb{R}$ by

$$f(x) = \begin{cases} c \exp\left(\dfrac{-1}{1 - x^2}\right), & |x| < 1, \\ 0, & |x| \geq 1, \end{cases}$$

where $\exp(x) = e^x$, and $c > 0$ is chosen so that $\int_{-\infty}^{\infty} f(x)\, dx = 1$. For $\lambda > 0$, set $f_\lambda(x) = \dfrac{1}{\lambda} f(\tfrac{1}{\lambda} x)$.

   a. Prove that $f_\lambda \in C^\infty(\mathbb{R})$ for all $\lambda > 0$.

   b. Prove that $f_\lambda(x) = 0$ for all $x \in \mathbb{R}$, $|x| \geq \lambda$, and that $\int_{-\infty}^{\infty} f_\lambda(x)dx = 1$.

   c. Prove that for every $\delta > 0$, $\displaystyle\lim_{\lambda \to 0^+} \int_{\{\delta \leq |t|\}} f_\lambda(t)\, dt = 0$.

3. In this exercise, we show how the trigonometric functions may be defined by means of power series. Define the functions S and C on $\mathbb{R}$ by

$$S(x) = \sum_{k=0}^{\infty} \frac{(-1)^k}{(2k+1)!} x^{2k+1}, \quad C(x) = \sum_{k=0}^{\infty} \frac{(-1)^k}{(2k)!} x^{2k}.$$

   a. Show that the power series defining S and C converge for all $x \in \mathbb{R}$.

   b. Show that $S'(x) = C(x)$ and $C'(x) = -S(x)$, $x \in \mathbb{R}$.

   c. Show that $S''(x) = -S(x)$ and $C''(x) = -C(x)$.

   d. Show that if $f : \mathbb{R} \to \mathbb{R}$ satisfies $f''(x) = -f(x)$ with $f(0) = 0$, $f'(0) = 1$, then $f(x) = S(x)$ for all $x \in \mathbb{R}$.

   e. If $f : \mathbb{R} \to \mathbb{R}$ satisfies $f''(x) = -f(x)$, prove that there exist constants c_1, c_2 such that $f(x) = c_1 S(x) + c_2 C(x)$.

   f. Show that $(S(x))^2 + (C(x))^2 = 1$. (Hint: Consider the function $f(x) = (S(x))^2 + (C(x))^2$.)

   g. Show that $C(x+y) = C(x)C(y) - S(x)S(y)$ and $S(x+y) = S(x)C(y) + C(x)S(y)$ for all $x, y \in \mathbb{R}$.

Supplemental Reading

Andrushkiw, J. W. , "A note on multiple series of positive terms," *Amer. Math. Monthly* **68** (1961), 253–258.

Billingsly, P., "Van der Waerden's continuous nowhere differentiable function," *Amer. Math. Monthly* **89** (1982), 691.

Blank, A. A., "A simple example of a Weierstrass function," *Amer. Math. Monthly* **73** (1966), 515–519.

Boas, Jr., R. P., "Partial sums of infinite series and how they grow," *Amer. Math. Monthly* **84** (1977), 237–258.

Boas, Jr., R. P. and Pollard, H., "Continuous analogues of series," *Amer. Math. Monthly* **80** (1973), 18–25.

Cunningham, Jr., F., "Taking limits under the integral sign," *Math. Mag.* **40** (1967), 179–186.

Davis, P. J., "Leonhard Euler's integral; a historical profile of the Gamma function," *Amer. Math. Monthly* **66** (1959), 849–869.

de Silva, N., "A concise elementary proof of Arzelá's bounded convergence theorem," *Amer. Math. Monthly* **117** (2010), 918–920.

Ferguson, Le Baron O., "What can be approximated by polynomials with integer coefficients," *Amer. Math. Monthly* **113** (2006), 403–414.

French, A. P., "The integral definition of the logarithm and the logarithmic series," *Amer. Math. Monthly* **85** (1978), 580–582.

Garcia-Caballero, E. M. and Moreno, S. G., "Yet another generalization of a celebrated inequality of the Gamma function," *Amer. Math. Monthly* **120** (2013), 821.

Kestleman, H., "Riemann integration of limit functions," *Amer. Math. Monthly* **77** (1970), 182–187.

Lewin, J. W., "Some applications of the bounded convergence theorem for an introductory course in analysis," *Amer. Math. Monthly* **94** (1987), 988–993.

Mathé, P., "Approximation of Hölder continuous functions by Bernstein polynomials," *Amer. Math. Monthly* **106** (1999), 568–725.

McLoughlin. P. F., "A simple proof of Taylor's Theorem," *Amer. Math. Monthly* **120** (2013), 767–768.

Miller, K. S., "Derivatives of noninteger order," *Math. Mag.* **68** (1995), 183–192.

Minassian, D. P. and Gaisser, J. W., "A simple Weierstrass function," *Amer. Math. Monthly* **91** (1984), 254–256.

Neuschel, T., "A new proof of Stirling's formula," *Amer. Math. Monthly* **121** (2014), 350–352.

Patin, J. M., "A very short proof of Stirling's formula," *Amer. Math. Monthly* **96** (1989), 41–42.

Roy, R., "The discovery of the series formula for π by Leibniz, Gregory and Nilakantha," *Math. Mag.* **63** (1990), 291–306.

Sagan, H., "An elementary proof that Schoenberg's space filling curve is nowhere differentiable," *Math. Mag.* **65** (1992), 125–128.

Schenkman, E., "The Weierstrass approximation theorem," *Amer. Math. Monthly* **79** (1972), 65–66.

Weinstock, R., "Elementary evaluations of $\int_0^\infty e^{-x^2}\, dx$, $\int_0^\infty \cos x^2\, dx$, and $\int_0^\infty \sin x^2\, dx$," *Amer. Math. Monthly* **97** (1990), 30–42.

Wen, L., "A nowhere differentiable continuous function constructed by infinite products," *Amer. Math. Monthly* **109** (2002), 378–380.

9

Fourier Series

In this chapter, we consider the problem of expressing a real-valued periodic function of period 2π in terms of a trigonometric series

$$\tfrac{1}{2}a_0 + \sum_{n=1}^{\infty}(a_n \cos nx + b_n \sin nx),$$

where the a_n and b_n are real numbers. As we will see, such series afford much greater generality in the type of functions that can be represented as opposed to Taylor series. The study of trigonometric series has its origins in the monumental work of Joseph Fourier (1768–1830) on heat conduction in solids. His 1807 presentation to the French Academy introduced a whole new subject area in mathematics while at the same time providing very useful techniques for solving physical problems.

Fourier's work is the source of all modern methods in mathematical physics involving boundary value problems and has been a source of new ideas in mathematical analysis for the past two centuries. To see how greatly mathematics has been influenced by the studies of Fourier one only needs to look at the two volume work *Trigonometric Series* by A. Zygmund (Cambridge University Press, 1968). In addition to trigonometric series, Fourier's original method of separation of variables leads very naturally to the study of orthogonal functions and the representation of functions in terms of a series of orthogonal functions. All of these have many applications in mathematical physics and engineering.

Fourier initially claimed and tried to show, with no success, that the Fourier series expansion of a function actually represented the function. Although his claim is false, in view of the eighteenth century concept of a function this was not an unrealistic expectation. Fourier's claim had an immediate impact on nineteenth century mathematics. It caused mathematicians to reconsider the definition of "function." The question of what type of function has a Fourier series expansion also led Riemann to the development of the theory of the integral and the notion of an integrable function. The first substantial progress on the convergence of a Fourier series to its defining function is due to Dirichlet in 1829. Instead of trying to prove like Fourier that the Fourier series always converged to its defining function, Dirichlet considered the more restrictive problem of trying to find sufficient conditions on the function f for which the Fourier series converges pointwise to the function.

In the first section we provide a brief introduction to the theory of orthogonal functions and to the concept of approximation in the mean. In Section 9.2 we also introduce the notion of a complete sequence of orthogonal functions and show that this is equivalent to the convergence in the mean of the sequence of partial sums of the Fourier series to its defining function. The proof of the completeness of the trigonometric system $\{1, \sin nx, \cos nx\}_{n=1}^{\infty}$ will be presented in Section 9.4. In this section, we also prove Fejér's theorem on the uniform approximation of a continuous function by the nth partial sum of a trigonometric series. In the final section, we present Dirichlet's contributions to the pointwise convergence problem.

9.1   Orthogonal Functions

In this section, we provide a brief introduction to the topic of orthogonal functions and the question of representing a function by means of a series of orthogonal functions. Although these topics have their origins in the study of partial differential equations and boundary value problems[1], they are closely related to concepts normally encountered in the study of vector spaces.

If X is a vector space over $\mathbb{R}$, a function $\langle \, , \, \rangle : X \times X \to \mathbb{R}$ is an **inner product** on X if

(a) $\langle \mathbf{x}, \mathbf{x} \rangle \geq 0$ for all $\mathbf{x} \in X$,
(b) $\langle \mathbf{x}, \mathbf{x} \rangle = 0$ if and only if $\mathbf{x} = \mathbf{0}$,
(c) $\langle \mathbf{x}, \mathbf{y} \rangle = \langle \mathbf{y}, \mathbf{x} \rangle$ for all $\mathbf{x}, \mathbf{y} \in X$, and
(d) $\langle a\mathbf{x} + b\mathbf{y}, \mathbf{z} \rangle = a\langle \mathbf{x}, \mathbf{z} \rangle + b\langle \mathbf{y}, \mathbf{z} \rangle$ for all $\mathbf{x}, \mathbf{y}, \mathbf{z} \in X$ and $a, b \in \mathbb{R}$.

In $\mathbb{R}^n$, the usual inner product is given by

$$\langle \mathbf{a}, \mathbf{b} \rangle = \sum_{j=1}^{n} a_j b_j$$

for $\mathbf{a} = (a_1, ..., a_n)$ and $\mathbf{b} = (b_1, ..., b_n)$ in $\mathbb{R}^n$. If $\langle \, , \, \rangle$ is an inner product on X, then two non-zero vectors $\mathbf{x}, \mathbf{y} \in X$ are **orthogonal** if $\langle \mathbf{x}, \mathbf{y} \rangle = 0$. The term "orthogonal" is synonymous with "perpendicular" and comes from geometric considerations in $\mathbb{R}^n$. Two non-zero vectors $\mathbf{a}$ and $\mathbf{b}$ in $\mathbb{R}^n$ are orthogonal if and only if they are mutually perpendicular; that is, the angle θ between the two vectors $\mathbf{a}$ and $\mathbf{b}$ is $\pi/2$ or $90°$ (see Exercise 9, Section 7.4).

In the study of analysis we typically encounter vector spaces whose elements are functions. For example, in previous sections we have shown that the space ℓ^2 of square summable sequences as well as the space $\mathcal{C}[a, b]$ of continuous real-valued functions on $[a, b]$ are vector spaces over $\mathbb{R}$. With the usual

---

[1] For a detailed treatment of this subject the reader is referred to the texts by Berg and McGregor or by Weinberger listed in the Bibliography.

rules of addition and scalar multiplication, $\mathcal{R}[a,b]$, the set of Riemann integrable functions on $[a,b]$, is also a vector space over $\mathbb{R}$. If for $f,\ g \in \mathcal{R}[a,b]$ we define

$$\langle f, g \rangle = \int_a^b f(x)g(x)\,dx,$$

then it is easily shown that $\langle\ ,\ \rangle$ satisfies (a), (c), and (d) of the definition of an inner product. It however does not satisfy (b). If $a < b$ and $c_1, ..., c_n$ are a finite number of points in $[a,b]$, then the function

$$f(x) = \begin{cases} 0, & x \neq c_i, \\ 1, & x = c_i, \end{cases}$$

is in $\mathcal{R}[a,b]$ satisfying $\langle f, f \rangle = 0$, but f is not the zero function. Thus technically $\langle\ ,\ \rangle$ is not an inner product on $\mathcal{R}[a,b]$, a minor difficulty which can easily be overcome by defining two Riemann integrable functions f and g to be equal if $f(x) = g(x)$ for all $x \in [a,b]$ except on a set of measure zero. This will be explored in greater detail in Chapter 10. Alternately, if we restrict ourself to the subset $\mathcal{C}[a,b]$ of $\mathcal{R}[a,b]$, then $\langle f, g \rangle$ as defined above is an inner product on $\mathcal{C}[a,b]$ (Exercise 11).

Orthogonal Functions

We now define orthogonality with respect to the above *inner product* on $\mathcal{R}[a,b]$.

DEFINITION 9.1.1 *A finite or countable collection of Riemann integrable functions $\{\phi_n\}$ on $[a,b]$ satisfying $\int_a^b \phi_n^2 \neq 0$ is* **orthogonal** *on $[a,b]$ if*

$$\langle \phi_n, \phi_m \rangle = \int_a^b \phi_n(x)\phi_m(x)\,dx = 0, \qquad \text{for all} \quad n \neq m.$$

EXAMPLES 9.1.2 (a) For our first example we consider the two functions $\phi(x) = 1$ and $\psi(x) = x$, $x \in [-1,1]$. Since

$$\int_{-1}^1 \phi(x)\psi(x)\,dx = \int_{-1}^1 x\,dx = 0,$$

the functions ϕ and ψ are orthogonal on the interval $[-1,1]$.

(b) In this example, we show that the sequence of functions $\{\sin nx\}_{n=1}^{\infty}$ is orthogonal on $[-\pi, \pi]$. By the trigonometric identity

$$\sin A \sin B = \tfrac{1}{2}[\cos(A - B) - \cos(A + B)],$$

for $n \neq m$,

$$\int_{-\pi}^{\pi} \sin nx \sin mx \, dx = \frac{1}{2} \int_{-\pi}^{\pi} [\cos(n-m)x - \cos(n+m)x] \, dx$$

$$= \frac{1}{2} \left[\frac{\sin(n-m)x}{(n-m)} - \frac{\sin(n+m)x}{(n+m)} \right] \bigg|_{-\pi}^{\pi} = 0.$$

For future reference, when $n = m$,

$$\int_{-\pi}^{\pi} \sin^2 nx \, dx = \frac{1}{2} \int_{-\pi}^{\pi} (1 - \cos 2nx) \, dx$$

$$= \frac{1}{2} \left(x - \frac{\sin 2nx}{2n} \right) \bigg|_{-\pi}^{\pi} = \pi.$$

(c) As our final example we consider the collection

$$\{1, \sin \tfrac{n\pi x}{L}, \cos \tfrac{n\pi x}{L} \}_{n=1}^{\infty}$$

on the interval $[-L, L]$ where $L > 0$. As in (b), if $n \neq m$, then

$$\int_{-L}^{L} \sin \tfrac{n\pi x}{L} \sin \tfrac{m\pi x}{L} \, dx = 0.$$

Thus the collection $\{\sin \tfrac{n\pi x}{L}\}$ is orthogonal on $[-L, L]$. Also, by the trigonometric identities

$$\cos A \cos B = \tfrac{1}{2}[\cos(A - B) + \cos(A + B)]$$
$$\sin A \cos B = \tfrac{1}{2}[\sin(A - B) + \sin(A + B)],$$

we have for $n \neq m$

$$\int_{-L}^{L} \cos \tfrac{n\pi x}{L} \cos \tfrac{m\pi x}{L} \, dx = \int_{-L}^{L} \sin \tfrac{n\pi x}{L} \cos \tfrac{m\pi x}{L} \, dx = 0.$$

Thus the functions in the collection $\{\cos \tfrac{n\pi x}{L}\}$ are all orthogonal on $[-L, L]$ as are the function $\sin \tfrac{n\pi x}{L}$ and $\cos \tfrac{m\pi x}{L}$ for all $n, m \in \mathbb{N}$ with $n \neq m$. For $m = n$

$$\int_{-L}^{L} \sin \tfrac{n\pi x}{L} \cos \tfrac{n\pi x}{L} \, dx = \tfrac{L}{2n\pi} \sin^2 \tfrac{n\pi x}{L} \bigg|_{-L}^{L} = 0.$$

This last identity shows that the functions $\sin \tfrac{n\pi x}{L}$ and $\cos \tfrac{n\pi x}{L}$ are also orthogonal on $[-L, L]$ for all $n \in \mathbb{N}$. Finally, since

$$\int_{-L}^{L} \sin \tfrac{n\pi x}{L} \, dx = \int_{-L}^{L} \cos \tfrac{n\pi x}{L} \, dx = 0$$

for all $n \in \mathbb{N}$, the constant function 1 is orthogonal to $\sin \tfrac{n\pi x}{L}$ and $\cos \tfrac{n\pi x}{L}$ for all $n \in \mathbb{N}$. In this example, we also have

$$\int_{-L}^{L} \sin^2 \tfrac{n\pi x}{L} \, dx = \int_{-L}^{L} \cos^2 \tfrac{n\pi x}{L} \, dx = L. \quad \square$$

If in Example 9.1.2(b) we define $\phi_n(x) = \frac{1}{\sqrt{\pi}} \sin nx$, then the sequence $\{\phi_n(x)\}_{n=1}^{\infty}$ satisfies

$$\int_{-\pi}^{\pi} \phi_n(x)\phi_m(x)\, dx = \begin{cases} 0, & \text{when } n \neq m, \\ 1, & \text{when } n = m. \end{cases}$$

Such a sequence of orthogonal functions is given a special name.

DEFINITION 9.1.3 *A finite or countable collection of Riemann integrable functions $\{\phi_n\}$ is* **orthonormal** *on $[a,b]$ if*

$$\int_{a}^{b} \phi_n(x)\,\phi_m(x)\, dx = \begin{cases} 0, & \text{when } n \neq m, \\ 1, & \text{when } n = m. \end{cases}$$

Given a collection $\{\phi_n\}$ of orthogonal functions on $[a,b]$, we can always construct a family $\{\psi_n\}$ of orthonormal functions on $[a,b]$ by setting

$$\psi_n(x) = \frac{1}{c_n}\,\phi_n(x),$$

where c_n is defined by

$$c_n^2 = \int_{a}^{b} \phi_n^2(x)\, dx.$$

Approximation in the Mean

Let $\{\phi_n\}$ be a finite or countable family of orthogonal functions defined on an interval $[a,b]$. For each $N \in \mathbb{N}$ and $c_1, ..., c_N \subset \mathbb{R}$, consider the Nth partial sum

$$S_N(x) = \sum_{n=1}^{N} c_n\phi_n(x). \tag{1}$$

A natural question is, given a real-valued function f on $[a,b]$, how must the coefficients c_n be chosen so that S_N gives the *best approximation* to f on $[a,b]$? In the Weierstrass approximation theorem we have already encountered one form of approximation; namely uniform approximation or approximation in the uniform norm. However, for the study of orthogonal functions there is another type of norm approximation that turns out to be more useful.

If X is a vector space over $\mathbb{R}$ with inner product $\langle\,,\,\rangle$, then there is a natural **norm** on X associated with this inner product. If for $\mathbf{x} \in X$ we define

$$\|\mathbf{x}\| = \sqrt{\langle \mathbf{x}, \mathbf{x}\rangle},$$

then $\|\ \|$ is a norm on X as defined in Definition 7.4.8. The details that $\|\ \|$ is a norm is left to the exercises (Exercise 12). The crucial step in proving the triangle inequality for $\| \ \|$ is the following version of the **Cauchy-Schwarz inequality**: For all $\mathbf{x}, \mathbf{y} \in X$,

$$|\langle \mathbf{x}, \mathbf{y} \rangle| \leq \|\mathbf{x}\| \, \|\mathbf{y}\|.$$

The proof of this inequality follows verbatim the proof of Theorem 7.4.3. For the vector space $\mathcal{R}[a,b]$ with inner product $\langle f, g \rangle = \int_a^b f(x)g(x)\, dx$, the **norm** of a function f, denoted $\|f\|_2$, is given by

$$\|f\|_2 = \left[\int_a^b (f(x))^2 dx \right]^{1/2}.$$

Thus for $f \in \mathcal{R}[a,b]$, the natural problem to consider is how must the constants c_n be chosen in order to minimize the quantity

$$\|f - S_N\|_2^2 = \int_a^b [f(x) - S_N(x)]^2 \, dx\,?$$

This type of norm approximation is referred to as **approximation in the mean** or **least squares approximation**. The following theorem specifies the choice of $\{c_n\}$ so that S_N provides the best approximation to f in the mean.

THEOREM 9.1.4 *Let $f \in \mathcal{R}[a,b]$ and let $\{\phi_n\}$ be a finite or countable collection of orthogonal functions on $[a,b]$. For $N \in \mathbb{N}$, let S_N be defined by (1). Then the quantity*

$$\int_a^b [f(x) - S_N(x)]^2 \, dx,$$

is minimal if and only if

$$c_n = \frac{\int_a^b f(x)\phi_n(x)dx}{\int_a^b \phi_n^2(x)dx}, \quad n = 1, 2, \ldots N. \tag{2}$$

Furthermore, for this choice of c_n,

$$\int_a^b [f(x) - S_n(x)]^2 dx = \int_a^b f^2(x)\, dx - \sum_{n=1}^N c_n^2 \int_a^b \phi_n^2(x)dx. \tag{3}$$

Prior to proving the result, we give the following alternate statement of the previous theorem.

COROLLARY 9.1.5 *Let $f \in \mathcal{R}[a,b]$ and let $S_N(x) = \sum_{n=1}^N c_n\phi_n(x)$ where the c_n are defined by (2). If $T_N(x) = \sum_{n=1}^N a_n\phi_n(x)$, $a_n \in \mathbb{R}$, then*

$$\int_a^b [f(x) - S_N(x)]^2 dx \leq \int_a^b [f(x) - T_N(x)]^2 dx,$$

for any choice of a_n, $n = 1, 2, \ldots, N$.

Proof of Theorem 9.1.4 For fixed $N \in \mathbb{N}$,

$$0 \le \int_a^b [f(x) - S_N(x)]^2 dx$$

$$= \int_a^b f^2(x)\, dx - 2 \int_a^b f(x)S_N(x)\, dx + \int_a^b S_N^2(x)\, dx. \qquad (4)$$

By linearity of the integral,

$$\int_a^b f(x)S_N(x)\, dx = \sum_{n=1}^N c_n \int_a^b f(x)\phi_n(x)\, dx.$$

Also,

$$\int_a^b S_N^2(x)\, dx = \int_a^b S_N(x) \left(\sum_{n=1}^N c_n\phi_n(x) \right)\, dx = \sum_{n=1}^N c_n \int_a^b S_N(x)\phi_n(x)\, dx.$$

But

$$\int_a^b S_N(x)\phi_n(x) = \sum_{k=1}^N c_k \int_a^b \phi_k(x)\phi_n(x)\, dx,$$

which by orthogonality,

$$= c_n \int_a^b \phi_n^2(x)\, dx.$$

Therefore,

$$\int_a^b S_N^2(x)\, dx = \sum_{n=1}^N c_n^2 \int_a^b \phi_n^2(x)\, dx.$$

Upon substituting into (4) we obtain

$$0 \le \int_a^b [f(x) - S_N(x)]^2 dx$$

$$= \int_a^b f^2(x)\, dx - 2 \sum_{n=1}^N c_n \int_a^b f(x)\phi_n(x)\, dx + \sum_{n=1}^N c_n^2 \int_a^b \phi_n^2(x)\, dx,$$

which upon completing the square

$$= \int_a^b f^2(x)dx + \sum_{n=1}^N \int_a^b \phi_n^2(x)dx \left[c_n - \frac{\int_a^b f\phi_n}{\int_a^b \phi_n^2} \right]^2 - \sum_{n=1}^N \frac{\left[\int_a^b f\,\phi_n \right]^2}{\int_a^b \phi_n^2}.$$

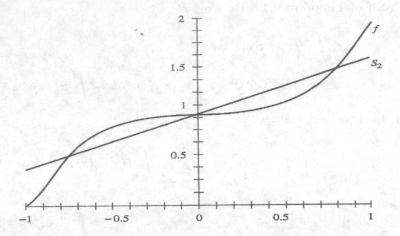

FIGURE 9.1
Graphs of f and S_2

The coefficients c_n occur only in the middle term. Since this term is nonnegative, the right side is a minimum if and only if

$$c_n = \frac{\int_a^b f \phi_n}{\int_a^b \phi_n^2}, \qquad n = 1, 2, ..., N.$$

With this choice of c_n, we also obtain formula (3) upon substitution.   □

EXAMPLE 9.1.6 As was previously shown, the functions $\phi_1(x) = 1$ and $\phi_2(x) = x$ are orthogonal on $[-1, 1]$. Let $f(x) = x^3 + 1$. Then

$$c_1 = \frac{\int_{-1}^1 f(x)\phi_1(x)dx}{\int_{-1}^1 \phi_1^2(x)dx} = \frac{1}{2}\int_{-1}^1 (x^3 + 1)\, dx = 1,$$

and

$$c_2 = \frac{\int_{-1}^1 f(x)\phi_2(x)dx}{\int_{-1}^1 \phi_2^2(x)dx} = \frac{3}{2}\int_{-1}^1 (x^4 + x)\, dx = \frac{3}{5}.$$

Therefore, $S_2(x) = 1 + \frac{3}{5}x$ is the best approximation in the mean to $f(x) = 1 + x^3$ on $[-1, 1]$. The graphs of f and S_2 are given in Figure 9.1.   □

DEFINITION 9.1.7 *Let* $\{\phi_n\}_{n=1}^{\infty}$ *be a sequence of orthogonal functions on* $[a, b]$ *and let* $f \in \mathcal{R}[a, b]$. *For each* $n \in \mathbb{N}$, *the number*

$$c_n = \frac{\int_a^b f(x)\phi_n(x)\, dx}{\int_a^b \phi_n^2(x)\, dx} \qquad (5)$$

is called the **Fourier coefficient** of f with respect to $\{\phi_n\}$. The series $\sum_{n=1}^{\infty} c_n \phi_n(x)$ is called the **Fourier series** of f. This is denoted by

$$f(x) \sim \sum_{n=1}^{\infty} c_n \phi_n(x). \tag{6}$$

Remark. The notation "$\sim$" in (6) only means that the coefficients $\{c_n\}$ in the series are given by formula (5). Nothing is implied about convergence of the series!

EXAMPLE 9.1.8 In Example 9.1.2(b) it was shown that the sequence of functions $\{\sin nx\}_{n=1}^{\infty}$ is orthogonal on $[-\pi, \pi]$. Since

$$\int_{-\pi}^{\pi} \sin^2 nx \, dx = \pi, \qquad n = 1, 2, \ldots,$$

if $f \in \mathcal{R}[-\pi, \pi]$, the Fourier coefficients c_n, $n = 1, 2, \ldots$, of f with respect to the orthogonal system $\{\sin nx\}$ are given by

$$c_n = \frac{1}{\pi} \int_{-\pi}^{\pi} f(x) \sin nx \, dx,$$

and the Fourier series of f becomes

$$f(x) \sim \sum_{n=1}^{\infty} c_n \sin nx.$$

As indicated above, nothing is implied about convergence. Even if the series should converge, it need not converge to the function f. Since the terms of the series are odd functions of x, the series, if it converges, defines an odd function on $[-\pi, \pi]$. Thus unless f itself is odd, the series could not converge to f. For example, if $f(x) = 1$, then

$$c_n = \frac{1}{\pi} \int_{-\pi}^{\pi} \sin nx \, dx = \frac{-1}{n\pi} \cos nx \Big|_{-\pi}^{\pi} = 0.$$

In this case, the series converges for all x, but clearly not to $f(x) = 1$. $\square$

Bessel's Inequality

For each $N \in \mathbb{N}$, let $S_N(x)$ denote the Nth partial sum of the Fourier series of f; i.e.,

$$S_N(x) = \sum_{n=1}^{N} c_n \phi_n(x),$$

where the c_n are the Fourier coefficients of f with respect to the sequence $\{\phi_n\}$ of orthogonal functions on $[a, b]$. Then by identity (3) of Theorem 9.1.4,

$$0 \le \int_a^b f^2(x)\, dx - \sum_{n=1}^N c_n^2 \int_a^b \phi_n^2(x)\, dx.$$

Therefore

$$\sum_{n=1}^N c_n^2 \int_a^b \phi_n^2(x)\, dx \le \int_a^b f^2(x)\, dx.$$

Since this holds for every $N \in \mathbb{N}$, by letting $N \to \infty$ we obtain the following inequality.

THEOREM 9.1.9 (Bessel's Inequality) *If $f \in \mathcal{R}[a, b]$ and $\{c_n\}_{n=1}^\infty$ are the Fourier coefficients of f with respect to the sequence of orthogonal functions $\{\phi_n\}_{n=1}^\infty$, then*

$$\sum_{n=1}^\infty c_n^2 \int_a^b \phi_n^2(x)\, dx \le \int_a^b f^2(x)\, dx.$$

In Example 9.1.8 with $f(x) = 1$, $\int_{-\pi}^{\pi} f^2(x)dx = 2\pi$, and $c_n = 0$ for all $n = 1, 2, \dots..$ Thus it is clear that equality need not hold in Bessel's inequality. However, there is one consequence of Theorem 9.1.9 which will prove useful later.

COROLLARY 9.1.10 *Suppose $\{\phi_n\}_{n=1}^\infty$ is a sequence of orthogonal functions on $[a, b]$. If $f \in \mathcal{R}[a, b]$, then*

$$\lim_{n \to \infty} \frac{\int_a^b f(x)\phi_n(x)\, dx}{\sqrt{\int_a^b \phi_n^2(x)\, dx}} = 0.$$

Proof. Since $f \in \mathcal{R}[a, b]$, $\int_a^b f^2(x)\, dx$ is finite. Thus by Bessel's inequality, the series $\sum c_n^2 \int_a^b \phi_n^2$ converges. As a consequence,

$$\lim_{n \to \infty} c_n^2 \int_a^b \phi_n^2(x)\, dx = 0.$$

and thus

$$\lim_{n \to \infty} c_n \sqrt{\int_a^b \phi_n^2(x)\, dx} = \lim_{n \to \infty} \frac{\int_a^b f(x)\phi_n(x)\, dx}{\sqrt{\int_a^b \phi_n^2(x)\, dx}} = 0. \quad \square$$

Exercises 9.1

1. *Let $f(x) = \sin \pi x$, $\phi_1(x) = 1$, and $\phi_2(x) = x$. Find c_1 and c_2 so that $S_2(x) = c_1\phi_1(x) + c_2\phi_2(x)$ gives the best approximation in the mean to f on $[-1, 1]$.

2. **a.** Show that the polynomials $P_0(x) = 1$, $P_1(x) = x$, and $P_2(x) = \frac{3}{2}x^2 - \frac{1}{2}$ are orthogonal on $[-1, 1]$.

   b. Let
   $$f(x) = \begin{cases} 0, & -1 \le x < 0, \\ 1, & 0 \le x \le 1 \end{cases}$$

   Find the constants c_0, c_1, and c_2, such that $S_2(x) = c_0 P_0(x) + c_1 P_1(x) + c_2 P_2(x)$ gives the best approximation in the mean to f on $[-1, 1]$.

3. *a. Let $\phi_0(x) = 1$, $\phi_1(x) = x - a_1$, $\phi_2(x) = x^2 - a_2 x - a_3$. Determine the constants a_1, a_2, and a_3, so that $\{\phi_0, \phi_1, \phi_2\}$ are orthogonal on $[0, 1]$.

   *b. Find the polynomial of degree less than or equal to 2 that best approximates $f(x) = \sin \pi x$ in the mean on $[0, 1]$.

4. Let $\{\phi_n\}_{n=1}^{\infty}$ be a sequence of orthogonal functions on $[a, b]$. For $f, g \in \mathcal{R}[a, b]$ with $f \sim \sum a_n \phi_n$ and $g \sim \sum b_n \phi_n$, show that for $\alpha, \beta \in \mathbb{R}$,
   $$(\alpha f + \beta g) \sim \sum_{n=1}^{\infty} (\alpha a_n + \beta b_n)\phi_n.$$

5. **a.** Show that the sequence $\{\sin nx\}_{n=1}^{\infty}$ is orthogonal on $[0, \pi]$.

   b. For $f \in \mathcal{R}[0, \pi]$, show that the Fourier series of f with respect to the sequence $\{\sin nx\}$ is given by
   $$\sum_{n=1}^{\infty} b_n \sin nx \quad \text{where} \quad b_n = \frac{2}{\pi} \int_0^{\pi} f(x) \sin nx \, dx.$$

   *c. Find the Fourier series of $f(x) = x$ on $[0, \pi]$ in terms of the orthogonal sequence $\{\sin nx\}$.

6. **a.** Show that the sequence $\{1, \cos nx\}_{n=1}^{\infty}$ is orthogonal on $[0, \pi]$.

   b. For $f \in \mathcal{R}[0, \pi]$, show that the Fourier series of f with respect to the above orthogonal sequence is given by
   $$\frac{1}{2}a_0 + \sum_{n=1}^{\infty} a_n \cos nx$$
   where $a_0 = \frac{2}{\pi} \int_0^{\pi} f(x) \, dx$ and $a_n = \frac{2}{\pi} \int_0^{\pi} f(x) \cos nx \, dx$, $n = 1, 2, \dots$.

   *c. Find the Fourier series of $f(x) = x$ on $[0, \pi]$ in terms of the orthogonal sequence $\{1, \cos nx\}$.

7. If $f \in \mathcal{R}[0, \pi]$, prove that
   $$\lim_{n \to \infty} \int_0^{\pi} f(x) \sin nx \, dx = \lim_{n \to \infty} \int_0^{\pi} f(x) \cos nx \, dx = 0.$$

8. Let $\{\phi_n\}$ be a sequence of orthogonal functions on $[a, b]$. If the series $\sum_{n=1}^{\infty} a_n \phi_n(x)$ converges uniformly to a function $f(x)$ on $[a, b]$, prove that for each $n \in \mathbb{N}$, a_n is the Fourier coefficient of f.

9. Let $\{a_n\}$ be a sequence in $(0,1)$ satisfying $0 < a_{n+1} < a_n < 1$ for all $n \in \mathbb{N}$. Define ϕ_n on $[0,1]$ by

$$\phi_n(x) = \begin{cases} 0, & 0 \le x < a_{n+1}, \\ \dfrac{2(x - a_{n+1})}{a_n - a_{n+1}}, & a_{n+1} \le x < \tfrac{1}{2}(a_{n+1} + a_n) \\ \dfrac{-2(x - a_n)}{a_n - a_{n+1}}, & \tfrac{1}{2}(a_{n+1} + a_n) \le x \le a_n \\ 0, & a_n < x \le 1. \end{cases}$$

Show that $\{\phi_n\}$ is orthogonal in $\mathcal{R}[0,1]$ and compute $\|\phi_n\|_2$ for each $n \in \mathbb{N}$.

10. Let $P_0(x) = 1$ and for $n \in \mathbb{N}$ let

$$P_n(x) = \frac{1}{2^n n!} \frac{d^n}{dx^n} (1 - x^2)^n, \ x \in [-1,1].$$

The polynomials P_n are called the **Legendre polynomials** on $[-1,1]$.

    a. Find P_1, P_2, and P_3.

    b. Show that the sequence $\{P_n\}_{n=0}^{\infty}$ is orthogonal on $[-1,1]$. (Hint: Use repeated integration by parts.)

11. For $f, g \in \mathcal{C}[a,b]$ define $\langle f, g \rangle = \int_a^b f(x)g(x)dx$. Prove that $\langle \ , \ \rangle$ is an inner product on $\mathcal{C}[a,b]$.

12. Let X be a vector space over $\mathbb{R}$ with inner product $\langle \ , \ \rangle$. Prove each of the following:

    *a. $|\langle \mathbf{x}, \mathbf{y} \rangle| \le \|\mathbf{x}\| \, \|\mathbf{y}\|$ for all $\mathbf{x}, \mathbf{y} \in X$.

    b. $\| \ \|$ is a norm on X.

13. **(Cauchy-Schwarz Inequality)** Use the previous exercise to prove that if $f, g \in \mathcal{R}[a,b]$, then

$$\left| \int_a^b f(x)g(x)dx \right|^2 \le \left(\int_a^b f^2(x)\,dx \right) \left(\int_a^b g^2(x)\,dx \right).$$

14. Let X be a vector space over $\mathbb{R}$ with inner product $\langle \ , \ \rangle$. If $\{\mathbf{y}_1, ..., \mathbf{y}_n\}$ are non-zero orthogonal vectors in X and $\mathbf{x} \in X$, prove that the quantity $\|\mathbf{x} - (c_1\mathbf{y}_1 + \cdots c_n\mathbf{y}_n)\|$ is a minimum if and only if

$$c_i = \frac{\langle \mathbf{x}, \mathbf{y}_i \rangle}{\|\mathbf{y}_i\|^2}$$

for all $i = 1, ..., n$. (Imitate the proof of Theorem 9.1.4)

9.2   Completeness and Parseval's Equality

In this section, we look for necessary and sufficient conditions on the sequence $\{\phi_n\}$ of orthogonal functions on $[a,b]$ for which equality holds in Bessel's inequality. To accomplish this it will be useful to introduce the notion of convergence in the mean.

DEFINITION 9.2.1 *A sequence* $\{f_n\}$ *of Riemann integrable function on* $[a, b]$ **converges in the mean** *to* $f \in \mathcal{R}[a, b]$ *if*

$$\lim_{n\to\infty} \int_a^b [f(x) - f_n(x)]^2 dx = 0.$$

If we consider $\mathcal{R}[a, b]$ as a normed linear space with norm

$$\|f\|_2 = \left[\int_a^b (f(x))^2 dx \right]^{1/2},$$

then convergence in the mean is nothing else but **convergence in norm** as defined in Definition 8.3.9. Thus a sequence $\{f_n\}$ in $\mathcal{R}[a, b]$ converges to $f \in \mathcal{R}[a, b]$ in the mean if and only if $\lim_{n\to\infty} \|f - f_n\|_2 = 0$. Convergence in the mean is sometimes also referred to as **mean-square convergence**.

It is natural to ask how convergence is the mean is related to pointwise or uniform convergence. Our first theorem proves that uniform convergence implies convergence in the mean. As should be expected, pointwise convergence is not sufficient (Exercise 2). In the other direction, we will show in Example 9.2.3 that convergence in the mean does not imply pointwise convergence, and thus certainly not uniform convergence. There we construct a sequence $\{f_n\}$ of Riemann integrable functions on $[0, 1]$ such that $\|f_n\|_2 \to 0$, but for which $\{f_n(x)\}$ fails to converge for any $x \in [0, 1]$.

THEOREM 9.2.2 *If* f, f_n, $n = 1, 2, ...$, *are Riemann integrable on* $[a, b]$, *and* $\{f_n\}$ *converges uniformly to* f *on* $[a, b]$, *then* $\{f_n\}$ *converges in the mean to* f *on* $[a, b]$.

Proof. Since the proof of this result is similar to the proof of Theorem 8.4.1, we leave it as an exercise (Exercise 1). $\square$

EXAMPLE 9.2.3 In this example, we construct a sequence $\{f_n\}$ on $[0, 1]$ that converges to zero in the mean, but for which $\{f_n(x)\}$ does not converge for any $x \in [0, 1]$. This sequence is constructed as follows: For each $n \in \mathbb{N}$, write $n = 2^k + j$ where $k = 0, 1, 2..$ and $0 \le j < 2^k$. For example, $1 = 2^0 + 0$, $2 = 2^1 + 0$, $3 = 2^1 + 1$, etc. Define f_n on $[0, 1]$ by

$$f_n(x) = \begin{cases} 1, & \dfrac{j}{2^k} \le x \le \dfrac{j+1}{2^k}, \\ 0, & \text{otherwise.} \end{cases}$$

The first four functions f_1, f_2, f_3, and f_4 are given as follows: $f_1(x) = 1$, and

$$f_2(x) = \begin{cases} 1, & 0 \le x \le \frac{1}{2}, \\ 0, & \frac{1}{2} < x \le 1, \end{cases}$$

$$f_3(x) = \begin{cases} 0, & 0 \le x < \frac{1}{2} \\ 1 & \frac{1}{2} \le x \le 1, \end{cases}$$

$$f_4(x) = \begin{cases} 1, & 0 \le x \le \frac{1}{4}, \\ 0, & \frac{1}{4} < x \le 1. \end{cases}$$

For each $n \in \mathbb{N}$, $f_n \in \mathcal{R}[0,1]$ with

$$\int_0^1 f_n^2(x)\, dx = \int_{j/2^k}^{(j+1)/2^k} 1\, dx = \frac{1}{2^k}.$$

Thus $\lim\limits_{n \to \infty} \int_0^1 f_n^2(x)\, dx = 0$. On the other hand, if $x \in [0,1]$, then the sequence $\{f_n(x)\}$ contains an infinite number of 0's and 1's, and thus does not converge. $\square$

   In the following theorem we prove that convergence in the mean of the partial sums of the Fourier series is equivalent to equality in Bessel's inequality.

THEOREM 9.2.4 *Let $\{\phi_n\}_{n=1}^{\infty}$ be a sequence of orthogonal functions on $[a,b]$. Then the following are equivalent:*

   (a) *For every $f \in \mathcal{R}[a,b]$,*

$$\lim_{N \to \infty} \int_a^b [f(x) - S_N(x)]^2\, dx = 0,$$

where S_N is the Nth partial sum of the Fourier series of f.

   (b) *For every $f \in \mathcal{R}[a,b]$,*

$$\sum_{n=1}^{\infty} c_n^2 \int_a^b \phi_n^2(x)\, dx = \int_a^b f^2(x)\, dx, \qquad \textbf{(Parseval's equality)}$$

where the c_n are the Fourier coefficients of f.

Proof. Suppose $S_N(x) = \sum\limits_{n=1}^{N} c_n \phi_n(x)$ is the Nth partial sum of the Fourier series of f. Then by Theorem 9.1.4

$$\int_a^b [f(x) - S_N(x)]^2 dx = \int_a^b f^2(x)\, dx - \sum_{n=1}^{N} c_n^2 \int_a^b \phi_n^2(x)\, dx.$$

From this it follows immediately that $\{S_N\}$ converges in the mean to f if and only if Parseval's equality holds. $\square$

DEFINITION 9.2.5 *A sequence* $\{\phi_n\}_{n=1}^{\infty}$ *of orthogonal functions on* $[a, b]$ *is said to be* **complete** *if for every* $f \in \mathcal{R}[a, b]$,

$$\sum_{n=1}^{\infty} c_n^2 \int_a^b \phi_n^2(x) \, dx = \int_a^b f^2(x) \, dx.$$

As a consequence of the previous theorem, the orthogonal sequence $\{\phi_n\}$ is complete on $[a, b]$ if and only if for every $f \in \mathcal{R}[a, b]$, the sequence $\{S_N\}$ of the partial sums of the Fourier series of f converges in the mean to f. We now prove some additional consequences of completeness of an orthogonal sequence.

THEOREM 9.2.6 *If the sequence* $\{\phi_n\}_{n=1}^{\infty}$ *of orthogonal functions on* $[a, b]$ *is complete, and if* f *is a continuous real-valued function on* $[a, b]$ *satisfying*

$$\int_a^b f(x)\phi_n(x) \, dx = 0, \qquad \text{for all} \quad n = 1, 2, ...,$$

then $f(x) = 0$ *for all* $x \in [a, b]$.

Proof. The hypothesis implies that the Fourier coefficients c_n of f are zero for all $n \in \mathbb{N}$. Thus by Parseval's equality,

$$\int_a^b f^2(x) \, dx = 0.$$

Since f^2 is continuous and nonnegative, by Exercise 7, Section 6.1, this holds if and only if $f^2(x) = 0$ for all $x \in [a, b]$. Thus $f(x) = 0$ for all $x \in [a, b]$. $\square$

There is a converse to Theorem 9.2.6. Since the proof of the converse requires a knowledge of the Lebesgue integral, we only state the result. A sketch of the proof is provided in the miscellaneous exercises (Exercise 3) of Chapter 10.

THEOREM 9.2.7 *If* $\{\phi_n\}_{n=1}^{\infty}$ *is a sequence of orthogonal functions on* $[a, b]$ *having the property that the only real-valued continuous function* f *on* $[a, b]$ *satisfying*

$$\int_a^b f(x)\phi_n(x) \, dx = 0 \qquad \text{for all} \quad n = 1, 2, ...$$

is the zero function, then the system $\{\phi_n\}_{n=1}^{\infty}$ *is complete.*

For the orthogonal system $\{\sin nx\}_{n=1}^{\infty}$ on $[-\pi, \pi]$ and $f(x) = 1$, we have $c_n = 0$ for all $n = 1, 2, \dots$. Thus as a consequence of Theorem 9.2.6 the orthogonal system $\{\sin nx\}$ is not complete on $[-\pi, \pi]$. However, as we will see in Exercise 3 of Section 9.4, this system will be complete on $[0, \pi]$.

Another consequence of completeness is the following: Suppose $\{\phi_n\}$ is complete on $[a, b]$ and f, g are continuous real-valued functions on $[a, b]$ satisfying

$$\int_a^b f(x)\phi_n(x)\,dx = \int_a^b g(x)\phi_n(x)\,dx$$

for all $n = 1, 2, ...$, then $f(x) = g(x)$ for all $x \in [a, b]$. The above assumption simply means that f and g have the same Fourier coefficients. To prove the result, apply Theorem 9.2.6 to $h(x) = f(x) - g(x)$.

THEOREM 9.2.8 *Suppose* $\{\phi_n\}_{n=1}^\infty$ *is complete on* $[a, b]$, f, $g \in \mathcal{R}[a, b]$ *with*

$$f(x) \sim \sum_{n=1}^\infty c_n\phi_n(x) \qquad and \qquad g(x) \sim \sum_{n=1}^\infty b_n\phi_n(x).$$

Then,

$$\int_a^b f(x)g(x)\,dx = \sum_{n=1}^\infty c_n b_n \int_a^b \phi_n^2(x)\,dx.$$

Proof. Exercise 5.   □

Exercises 9.2

1. Prove Theorem 9.2.2.

2. For $n \in \mathbb{N}$, define the function f_n on $[0, 1]$ by

$$f_n(x) = \begin{cases} \sqrt{n}, & 0 < x < \dfrac{1}{n}, \\ 0, & \text{elsewhere.} \end{cases}$$

Show that $\{f_n\}$ converges to 0 pointwise, but not in the mean.

3. Consider the orthogonal system $\{1, \cos\frac{n\pi x}{L}, \sin\frac{n\pi x}{L}\}_{n=1}^\infty$ on $[-L, L]$.

    a. Show that if $f \in \mathcal{R}[-L, L]$, then the Fourier series of f with respect to the above orthogonal system is given by

$$f(x) \sim \tfrac{1}{2}a_o + \sum_{n=1}^\infty (a_n \cos\tfrac{n\pi x}{L} + b_n \sin\tfrac{n\pi x}{L}),$$

where

$$a_n = \frac{1}{L} \int_{-L}^L f(x) \cos\frac{n\pi x}{L}\,dx, \ n = 0, 1, 2, ..., \text{ and}$$

$$b_n = \frac{1}{L} \int_{-L}^L f(x) \sin\frac{n\pi x}{L}\,dx, \ n = 1, 2, ...$$

    b. Show that Bessel's inequality becomes

$$\tfrac{1}{2}a_o^2 + \sum_{n=1}^\infty (a_n^2 + b_n^2) \le \frac{1}{L} \int_{-L}^L f^2(x)\,dx.$$

4. ***a.** Assuming that the orthogonal system $\{\sin nx\}$ is complete on $[0, \pi]$, show that Parseval's equality becomes

$$\sum_{n=1}^{\infty} b_n^2 = \frac{2}{\pi} \int_0^{\pi} [f(x)]^2 dx \quad \text{where} \quad b_n = \frac{2}{\pi} \int_0^{\pi} f(x) \sin nx\, dx.$$

*b. Use Parseval's equality and the indicated function to find the sum of the given series.

(i) $\displaystyle\sum_{k=1}^{\infty} \frac{1}{(2k-1)^2}$, $f(x) = 1$.   (ii) $\displaystyle\sum_{k=1}^{\infty} \frac{1}{k^2}$, $f(x) = x$.

5. ***Prove Theorem 9.2.8.**

6. ***Show by example that continuity of f is required in Theorem 9.2.6.**

9.3   Trigonometric and Fourier Series

In Section 9.1, we introduced Fourier series with respect to any system $\{\phi_n\}_{n=1}^{\infty}$ of orthogonal functions on $[a, b]$. In this section, we will emphasize the trigonometric system

$$\left\{1, \cos \frac{n\pi x}{L}, \sin \frac{n\pi x}{L}\right\}_{n=1}^{\infty},$$

which by Example 9.1.2(c) is orthogonal on $[-L, L]$. For convenience we will take $L = \pi$.

Any series of the form

$$\frac{1}{2}A_0 + \sum_{n=1}^{\infty}(A_n \cos nx + B_n \sin nx),$$

where the A_n and B_n are real numbers, is called a **trigonometric series**. For example, the series

$$\sum_{n=2}^{\infty} \frac{\sin nx}{\ln n} \quad \text{and} \quad \sum_{n=1}^{\infty} \frac{\cos nx}{n}$$

are both examples of trigonometric series. Since the coefficients

$$\left\{\frac{1}{\ln n}\right\}_{n=2}^{\infty} \quad \text{and} \quad \left\{\frac{1}{n}\right\}_{n=1}^{\infty}$$

are nonnegative and decrease to zero, by Theorem 7.2.6 the first series converges for all $x \in \mathbb{R}$, whereas the second converges for all $x \in \mathbb{R}$, except $x = 2p\pi$, $p \in \mathbb{Z}$.

Fourier Series

For the orthogonal system $\{1, \cos nx, \sin nx\}_{n=1}^{\infty}$ on $[-\pi, \pi]$ we have

$$\int_{-\pi}^{\pi} \sin^2 nx\, dx = \int_{-\pi}^{\pi} \cos^2 nx\, dx = \pi, \quad \text{and} \quad \int_{-\pi}^{\pi} 1^2\, dx = 2\pi,$$

Thus by Definition 9.1.7, the Fourier coefficients of a function $f \in \mathcal{R}[-\pi, \pi]$ with respect to the orthogonal system are defined as follows.

DEFINITION 9.3.1 *Let $f \in \mathcal{R}[-\pi, \pi]$. The* **Fourier coefficients** *of f with respect to the orthogonal system $\{1, \cos nx, \sin nx\}$ are defined by*

$$a_0 = \frac{1}{\pi} \int_{-\pi}^{\pi} f(x)\, dx,$$

$$a_n = \frac{1}{\pi} \int_{-\pi}^{\pi} f(x) \cos nx\, dx, \qquad n = 1, 2, \ldots$$

$$b_n = \frac{1}{\pi} \int_{-\pi}^{\pi} f(x) \sin nx\, dx, \qquad n = 1, 2, \ldots$$

Also, the **Fourier series** *of f is given by*

$$f(x) \sim \frac{1}{2} a_0 + \sum_{n=1}^{\infty} (a_n \cos nx + b_n \sin nx).$$

Remark. For the constant function $\phi_0 = 1$, since $\int_{-\pi}^{\pi} (\phi_0)^2 = 2\pi$, the term a_0 according to Definition 9.1.7 should be defined as $\frac{1}{2\pi} \int_{-\pi}^{\pi} f(x) dx$. However, for notational convenience it is easier to define a_0 as $\frac{1}{\pi} \int_{-\pi}^{\pi} f(x) dx$ and to include the constant $\frac{1}{2}$ in the definition of the Fourier series.

For the orthogonal system $\{1, \cos nx, \sin nx\}$, by Exercise 3(b) of the previous section, Bessel's inequality of Theorem 9.1.9 becomes

$$\tfrac{1}{2} a_0^2 + \sum_{n=1}^{\infty} (a_n^2 + b_n^2) \leq \frac{1}{\pi} \int_{-\pi}^{\pi} [f(x)]^2 dx. \qquad \textbf{(Bessel's Inequality)}$$

Thus for $f \in \mathcal{R}[a, b]$ the sequences $\{a_n\}$ and $\{b_n\}$ of Fourier coefficients of f are square summable sequences. In Theorem 10.8.7 we will prove that if $\{a_n\}$ and $\{b_n\}$ are square summable sequences, then there exists a Lebesgue integrable function f such that $\{a_n\}$ and $\{b_n\}$ are the Fourier coefficients of f.

Remark on Notation. In subsequent sections we will primarily be interested in real-valued functions defined on all of $\mathbb{R}$ that are periodic of period 2π. As a consequence, in the examples and exercises, rather than defining our functions on $[-\pi, \pi]$, we only define the function f on $[-\pi, \pi)$ with the convention that $f(\pi) = f(-\pi)$. This then allows us to extend f to all of $\mathbb{R}$ as a 2π periodic function according to the following definition.

DEFINITION 9.3.2 *For a real-valued function f defined on $[-\pi, \pi)$, the* **periodic extension** *(of period 2π) of f to $\mathbb{R}$ is obtained by defining $f(x) = f(x - 2k\pi)$, where $k \in \mathbb{Z}$ is such that $x - 2k\pi \in [-\pi, \pi)$.*

EXAMPLES 9.3.3 (a) Let $f(x) = \begin{cases} 0, & -\pi \le x < 0, \\ 1, & 0 \le x < \pi \end{cases}$. Then

$$a_0 = \frac{1}{\pi} \int_{-\pi}^{\pi} f(x)\,dx = \frac{1}{\pi} \int_{0}^{\pi} 1\,dx = 1,$$

and for n=1,2,...,

$$a_n = \frac{1}{\pi} \int_{0}^{\pi} \cos nx\,dx = \frac{1}{n\pi} \sin nx \Big|_{0}^{\pi} = 0,$$

$$b_n = \frac{1}{\pi} \int_{0}^{\pi} \sin nx\,dx = \frac{-1}{n\pi} \cos nx \Big|_{0}^{\pi} = \frac{1}{n\pi}[1 - \cos n\pi] = \frac{1}{n\pi}[1 - (-1)^n].$$

In the above we have used the fact that $\cos n\pi = (-1)^n$. Thus the Fourier series of f is given by

$$f(x) \sim \frac{1}{2} + \sum_{n=1}^{\infty} \frac{1}{n\pi}[1 - (-1)^n] \sin nx = \frac{1}{2} + \frac{2}{\pi} \sum_{k=0}^{\infty} \frac{1}{2k+1} \sin(2k+1)x.$$

If $S_N(x)$ denotes the Nth partial sum of the Fourier series, then S_1 and S_3 are given by

$$S_1(x) = \frac{1}{2} + \frac{2}{\pi} \sin x \quad \text{and} \quad S_3(x) = \frac{1}{2} + \frac{2}{\pi}\left[\sin x + \frac{1}{3}\sin 3x\right].$$

The graph of f, S_1, S_3, S_5, and S_{15} are given in Figure 9.2.

(b) Let $f(x) = x$. Before we compute the Fourier coefficients, we will make several observations which simplify this task. Recall that a function $g(x)$ is **even** on $[-a, a]$ if $g(-x) = g(x)$ for all x, and $g(x)$ is **odd** if $g(-x) = -g(x)$ for all x. By Exercise 4 of Section 6.2, if $g(x)$ is even on $[-a, a]$, then

$$\int_{-a}^{a} g(x)\,dx = 2\int_{0}^{a} g(x)\,dx,$$

whereas if $g(x)$ is odd,

$$\int_{-a}^{a} g(x)\,dx = 0.$$

The functions $\sin nx$ are all odd, whereas $\cos nx$ are even for all n. Therefore since $f(x) = x$ is odd, $x\cos nx$ is odd and $x\sin nx$ is even. Thus, $a_n = 0$ for all $n = 0, 1, 2, ...,$ and

$$b_n = \frac{2}{\pi} \int_{0}^{\pi} x \sin nx\,dx,$$

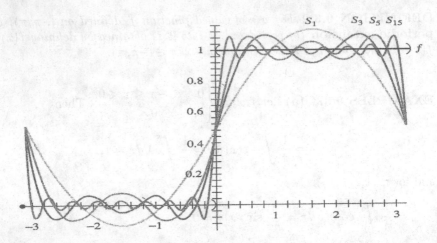

FIGURE 9.2
Graphs of f, S_1, S_3, S_5, and S_{15}

which by an integration by parts

$$= \frac{2}{\pi} \left[\frac{-x}{n} \cos nx \Big|_0^\pi + \frac{1}{n} \int_0^\pi \cos nx \, dx \right] = -\frac{2}{n} \cos n\pi = \frac{2}{n}(-1)^{n+1}.$$

Therefore

$$x \sim 2 \sum_{n=1}^\infty \frac{(-1)^{n+1}}{n} \sin nx. \quad \square$$

Riemann-Lebesgue Lemma

There is one additional result from the general theory that will be needed later.
For the orthogonal system $\{1, \cos nx, \sin nx\}$, Corollary 9.1.10 is as follows:

THEOREM 9.3.4 (Riemann-Lebesgue Lemma) *If $f \in \mathcal{R}[-\pi, \pi]$, then*

$$\lim_{n \to \infty} \int_{-\pi}^\pi f(x) \cos nx \, dx = \lim_{n \to \infty} \int_{-\pi}^\pi f(x) \sin nx \, dx = 0.$$

The following example shows that integrability of the function f is required.

EXAMPLE 9.3.5 Let f be defined on $[-\pi, \pi)$ as follows:

$$f(x) = \begin{cases} 0, & -\pi \le x \le 0, \\ \dfrac{1}{x}, & 0 < x < \pi. \end{cases}$$

Then

$$\int_{-\pi}^{\pi} f(x) \sin nx \, dx = \int_{0}^{\pi} \frac{\sin nx}{x} \, dx = \int_{0}^{n\pi} \frac{\sin x}{x} \, dx.$$

Hence,

$$\lim_{n \to \infty} \int_{-\pi}^{\pi} f(x) \sin nx \, dx = \lim_{n \to \infty} \int_{0}^{n\pi} \frac{\sin x}{x} \, dx = \int_{0}^{\infty} \frac{\sin x}{x} \, dx = \frac{\pi}{2}.^2 \quad \square$$

Is a Trigonometric Series a Fourier Series?

Since every Fourier series is a trigonometric series, an obvious question to ask is whether every trigonometric series is a Fourier series? More specifically, given a trigonometric series

$$\tfrac{1}{2}A_0 + \sum_{n=1}^{\infty} (A_n \cos nx + B_n \sin nx),$$

with $\{A_n\}$ and $\{B_n\}$ converging to zero, does there exist a function f on $[-\pi, \pi]$ such that the coefficients A_n and B_n are given by Definition 9.3.1? As we will see, the answer is **no!** First however, in the positive direction, we prove the following.

THEOREM 9.3.6 *If the trigonometric series*

$$\frac{1}{2}A_0 + \sum (A_n \cos nx + B_n \sin nx)$$

converges uniformly on $[-\pi, \pi]$, *then it is the Fourier series of a continuous real-valued function on* $[-\pi, \pi]$.

Proof. For $n \in \mathbb{N}$, let

$$S_n(x) = \frac{1}{2}A_0 + \sum_{k=1}^{n} (A_k \cos kx + B_k \sin kx).$$

Since the series converges uniformly on $[-\pi, \pi]$, and S_n is continuous for each n,

$$f(x) = \lim_{n \to \infty} S_n(x)$$

---

[2]The value of $\pi/2$ for the improper integral is most easily obtained by contour integration and the theory of residues of complex analysis. A real variables approach that computes the value of this integral is given in the article by K. S. Williams listed in the supplemental readings.

is a continuous function on $[-\pi, \pi]$. For $m \in \mathbb{N}$, consider

$$\int_{-\pi}^{\pi} f(x) \cos mx\, dx = \int_{-\pi}^{\pi} \left(\lim_{n \to \infty} S_n(x) \right) \cos mx\, dx$$

$$= \lim_{n \to \infty} \int_{-\pi}^{\pi} S_n(x) \cos mx\, dx.$$

Since for each m, the sequence $\{S_n(x) \cos mx\}$ converges uniformly to $f(x) \cos mx$ on $[-\pi, \pi]$, the above interchange of limits and integration is valid by Theorem 8.4.1. If $n > m$, then

$$\int_{-\pi}^{\pi} S_n(x) \cos mx = \tfrac{1}{2} A_0 \int_{-\pi}^{\pi} \cos mx\, dx + \sum_{k=1}^{n} A_k \int_{-\pi}^{\pi} \cos kx \cos mx\, dx$$

$$+ \sum_{k=1}^{n} B_k \int_{-\pi}^{\pi} \sin kx \cos mx\, dx.$$

Thus by orthogonality,

$$\int_{-\pi}^{\pi} S_n(x) \cos mx\, dx = A_m \pi.$$

Letting $n \to \infty$ gives

$$A_m = \frac{1}{\pi} \int_{-\pi}^{\pi} f(x) \cos mx\, dx.$$

The analogous formula also holds for B_m, and thus the given series is the Fourier series of f. $\square$

Remarks. (a) If $\sum |A_k|$ and $\sum |B_k|$ both converge, then by the Weierstrass M-test the series

$$\frac{1}{2} A_0 + \sum_{k=1}^{\infty} (A_k \cos kx + B_k \sin kx)$$

converges uniformly on $\mathbb{R}$, and thus is the Fourier series of a continuous function on $[-\pi, \pi]$. Convergence of the series $\sum |A_k|$ and $\sum |B_k|$ is however not necessary for uniform convergence of the trigonometric series. For example, the trigonometric series

$$\sum_{n=2}^{\infty} \frac{\sin nx}{n \ln n}$$

converges uniformly on $\mathbb{R}$ (Exercise 12), yet $\displaystyle\sum_{n=2}^{\infty} \frac{1}{n \ln n} = \infty$.

(b) In 1903 Lebesgue proved the following stronger version of Theorem 9.3.6: If $f(x) = \tfrac{1}{2} A_0 + \sum_{k=1}^{\infty} (A_k \cos kx + B_k \sin kx)$ for all $x \in (-\pi, \pi)$, and if f

is continuous (in fact, measurable) then A_k and B_k are the Fourier coefficients of the function f.[3] (See also Miscellaneous Exercise 1.)

We now turn to the negative results. Consider the series

$$\sum_{n=2}^{\infty} \frac{\sin nx}{\ln n}, \tag{7}$$

which by Theorem 7.2.6 converges for all $x \in \mathbb{R}$. However, there does not exist a <u>Riemann</u> integrable function f on $[-\pi, \pi]$ such that

$$\frac{1}{\ln n} = b_n = \frac{1}{\pi} \int_{-\pi}^{\pi} f(x) \sin nx \, dx.$$

If such a function f exists, then by Bessel's inequality we would have

$$\sum_{n=2}^{\infty} b_n^2 = \sum_{n=2}^{\infty} \frac{1}{(\ln n)^2} \leq \frac{1}{\pi} \int_{-\pi}^{\pi} f^2(x) \, dx.$$

But, since f is Riemann integrable on $[-\pi, \pi]$, so is f^2, and thus the integral is finite. On the other hand however,

$$\sum_{n=2}^{\infty} \frac{1}{(\ln n)^2} = \infty,$$

which gives a contradiction.

The above argument only shows that the series (7) is not obtained by means of Definition 9.3.1 from a Riemann integrable function. There remains however the question whether this is still the case if we extend our definition to allow the class of Lebesgue integrable functions to be introduced in Chapter 10? As we will see in Section 10.8, the answer to this is still **no!**

Fourier Sine and Cosine Series

We close this section with a brief discussion of Fourier sine and cosine series. As we have seen in Exercise 5 of Section 9.1, the sequence $\{\sin nx\}_{n=1}^{\infty}$ is orthogonal on $[0, \pi]$. Also, by Exercise 6 of Section 9.1 the same is true of the sequence $\{1, \cos nx\}_{n=1}^{\infty}$. For $f \in \mathcal{R}[0, \pi]$, the Fourier series with respect to each of these two orthogonal systems are called the Fourier sine and cosine series of f respectively. Since

$$\int_0^{\pi} \sin^2 nx \, dx = \int_0^{\pi} \cos^2 nx \, dx = \tfrac{1}{2}\pi,$$

the formulas of Definition 9.1.7 give the following.

---

[3] *Sur les series trigonometric*, Annales Scientifiques de l'École Normale Supérieure, (3) 20 (1903), 453–485.

DEFINITION 9.3.7 *For $f \in \mathcal{R}[0,\pi]$, the* **Fourier sine series** *of f is given by*

$$f(x) \sim \sum_{n=1}^{\infty} b_n \sin nx,$$

where

$$b_n = \frac{2}{\pi} \int_0^{\pi} f(x) \sin nx \, dx, \qquad n = 1, 2, \ldots$$

are the **Fourier sine coefficients** *of f. Similarly, the* **Fourier cosine series** *of $f \in \mathcal{R}[0,\pi]$ is given by*

$$f(x) \sim \tfrac{1}{2}a_0 + \sum_{n=1}^{\infty} a_n \cos nx,$$

where

$$a_0 = \frac{2}{\pi} \int_0^{\pi} f(x) \, dx \qquad and \qquad a_n = \frac{2}{\pi} \int_0^{\pi} f(x) \cos nx \, dx, \qquad n = 1, 2, \ldots$$

are the **Fourier cosine coefficients** *of f.*

There is a simple connection between Fourier series and the Fourier sine and cosine series. As in Example 9.3.3(b), we first note that if f is an even function on $[-\pi, \pi]$, then by Exercise 1,

$$f(x) \sim \tfrac{1}{2}a_0 + \sum_{n=1}^{\infty} a_n \cos nx, \quad \text{where } a_n = \frac{2}{\pi} \int_0^{\pi} f(x) \cos nx \, dx$$

Similarly, if f is an odd function on $[-\pi, \pi]$, then

$$f(x) \sim \sum_{n=1}^{\infty} b_n \sin nx, \quad \text{where } b_n = \frac{2}{\pi} \int_0^{\pi} f(x) \sin nx \, dx.$$

Thus the coefficients $\{a_n\}$ and $\{b_n\}$ depend only on the values of the function f on $[0, \pi]$. Conversely, given a function f on $[0, \pi)$, the following definition extends f both as an even and an odd function to the interval $(-\pi, \pi)$:

DEFINITION 9.3.8 *Let f be a real-valued function defined on $[0, \pi)$. The* **even extension** *of f to $(-\pi, \pi)$, denoted f_e, is the function defined by*

$$f_e(x) = \begin{cases} f(-x), & -\pi < x < 0, \\ f(x), & 0 \le x < \pi. \end{cases}$$

Similarly, the **odd extension** *of f to $(-\pi, \pi)$, denoted f_o, is the function defined by*

$$f_o(x) = \begin{cases} -f(-x), & -\pi < x < 0, \\ 0, & x = 0, \\ f(x), & 0 < x < \pi. \end{cases}$$

It is easily seen that the functions $f_e(x)$ and $f_o(x)$ are even and odd respectively on $(-\pi, \pi)$, and agree with the given function f on $(0, \pi)$. From the above discussion it is easily seen that if $f \in \mathcal{R}[0, \pi]$, then the Fourier sine series of f is equal to the Fourier series of f_o, and the Fourier cosine series of f is equal to the Fourier series of f_e (Exercise 2).

Remark. In the definition of the even and odd extension we only assumed that f was defined on $[0, \pi)$, and then defined f_e and f_o on $(-\pi, \pi)$. If

$$f(\pi-) = \lim_{x \to \pi^-} f(x)$$

exists, then for the even extension we define $f_e(-\pi) = f(\pi-)$. Thus f_e is now defined on $[-\pi, \pi)$ and hence can be extended to all of $\mathbb{R}$ as a 2π-periodic function. For the odd extension f_o, we set $f_o(-\pi) = -f(\pi-)$, thereby defining f_o on $[-\pi, \pi)$.

Exercises 9.3

1. Let $f \in \mathcal{R}[-\pi, \pi]$. Prove the following:

   *a. If f is even on $[-\pi, \pi]$, then the Fourier series of f is given by

   $$f(x) \sim \tfrac{1}{2}a_0 + \sum_{n=1}^{\infty} a_n \cos nx, \quad \text{where } a_n = \frac{2}{\pi} \int_0^\pi f(x) \cos nx \, dx,$$

   $n = 0, 1, 2, \ldots$

   b. If f is odd on $[-\pi, \pi]$, then the Fourier series of f is given by

   $$f(x) \sim \sum_{n=1}^{\infty} b_n \sin nx \quad \text{where } b_n = \frac{2}{\pi} \int_0^\pi f(x) \sin nx \, dx, \quad n = 1, 2, \ldots$$

2. Let $f \in \mathcal{R}[0, \pi]$. Prove the following:

   a. The Fourier sine series of f is equal to the Fourier series of the odd extension f_o of f.

   b. The Fourier cosine series of f is equal to the Fourier series of the even extension f_e of f.

3. Find the Fourier series of each of the following functions f on $[-\pi, \pi)$.

   *a. $f(x) = \begin{cases} -1, & -\pi \le x < 0, \\ 1, & 0 \le x < \pi. \end{cases}$ b. $f(x) = \begin{cases} 0, & -\pi \le x \le 0, \\ x, & 0 < x < \pi. \end{cases}$

   *c. $f(x) = |x|$.   d. $f(x) = x^2$.

   *e. $f(x) = 1 + x$.

   f. $f(x) = \begin{cases} 0, & -\pi \le x < -\frac{\pi}{2}, \\ -1, & -\frac{\pi}{2} \le x < 0, \\ 1, & 0 \le x < \frac{\pi}{2}, \\ 0, & \frac{\pi}{2} \le x < \pi. \end{cases}$

4. On the interval $[-2\pi, 2\pi]$ sketch the graph of the 2π-periodic extension of each of the functions in Exercise 3.

5. Find the Fourier sine and cosine series on $[0, \pi)$ of each of the following functions.

a. $f(x) = 1$ **b.** $f(x) = x$

c. $f(x) = x^2$ **d.** $f(x) = e^{-x}$

e. $f(x) = \begin{cases} 1, & 0 \le x < \frac{\pi}{2}, \\ -1, & \frac{\pi}{2} \le x < \pi \end{cases}$ ***f.** $f(x) = \begin{cases} 0, & 0 \le x < \frac{\pi}{2}, \\ 1, & \frac{\pi}{2} \le x < \pi \end{cases}$

6. Let h be defined on $[0, \pi)$ as follows:

$$h(x) = \begin{cases} cx, & 0 \le x \le \frac{\pi}{2}, \\ c(\pi - x), & \frac{\pi}{2} < x < \pi, \end{cases}$$

where $c > 0$ is a constant.

   a. Sketch the graph of h on $[0, \pi)$.

   b. Sketch the graph of the even extension h_e on $(-\pi, \pi)$.

   *c. Find the Fourier series of h_e.

   d. Sketch the graph of the odd extension h_o on $(-\pi, \pi)$.

   e. Find the Fourier series of h_o.

7.   **a.** Find the Fourier series of $f(x) = \sin x$ on $[-\pi, \pi]$.

   *b. Find the Fourier cosine series of $f(x) = \sin x$ on $[0, \pi]$.

8. ***Suppose** f, f' are continuous on $[-\pi, \pi]$, and $f'' \in \mathcal{R}[-\pi, \pi]$. Also, suppose that

$$f(-\pi) = f(\pi) \quad \text{and} \quad f'(-\pi) = f'(\pi).$$

   Prove that the Fourier series of f converges uniformly to f on $[-\pi, \pi]$. (Hint: Apply integration by parts to show that $|a_k|$ and $|b_k|$ are less than $C k^{-2}$ for some positive constant C.)

9. Show that the trigonometric series

$$\sum_{n=1}^{\infty} \frac{1}{\sqrt{n}} \sin nx$$

   converges for all $x \in \mathbb{R}$ but is not the Fourier series of a Riemann integrable function on $[-\pi, \pi]$.

10. If f is absolutely integrable on $[-\pi, \pi]$ (see Section 6.4), prove that

$$\lim_{n \to \infty} \int_{-\pi}^{\pi} f(x) \cos nx \, dx = \lim_{n \to \infty} \int_{-\pi}^{\pi} f(x) \sin nx \, dx = 0.$$

11. Using the previous exercise, show that each of the following hold.

   a. $\displaystyle \lim_{n \to \infty} \int_{0}^{\pi} \ln x \sin nx \, dx = \lim_{n \to \infty} \int_{0}^{\pi} \ln x \cos nx \, dx = 0.$

   b. $\displaystyle \lim_{n \to \infty} \int_{0}^{\pi} x^{\alpha} \sin nx \, dx = \lim_{n \to \infty} \int_{0}^{\pi} x^{\alpha} \cos nx \, dx = 0$ for all $\alpha > -1$.

12. Suppose $a_n \ge 0$ for all $n \in \mathbb{N}$, and $\{na_n\}$ is monotone decreasing with $\lim_{n \to \infty} na_n = 0$. Prove that $\sum_{n=1}^{\infty} a_n \sin nx$ converges uniformly. (Hint: Write $\sum a_n \sin nx$ as $\sum (na_n)(\frac{\sin nx}{n})$ and use the fact that the partial sums of $\sum \frac{\sin nx}{n}$ are uniformly bounded. See also Exercise 15 of Section 8.2.)

9.4   Convergence in the Mean of Fourier Series

Let f be a real-valued function defined on $[-\pi, \pi)$, and extend f to all of $\mathbb{R}$ to be periodic of period 2π. Throughout this section we will assume that $f \in \mathcal{R}[-\pi, \pi]$. Let

$$f(x) \sim \frac{1}{2}a_0 + \sum_{n=1}^{\infty}(a_n \cos nx + b_n \sin nx)$$

be the Fourier Series of f, and for each $n \in \mathbb{N}$, let

$$S_n(x) = \frac{1}{2}a_0 + \sum_{k=1}^{n}(a_k \cos kx + b_k \sin kx)$$

be the nth partial sum of the Fourier series.

Our goal in this section will be to prove that if $f \in \mathcal{R}[-\pi, \pi]$, then the sequence $\{S_n\}$ converges in the mean to the function f on $[-\pi, \pi]$. By Theorem 9.2.4 this is equivalent to completeness of the trigonometric system $\{1, \cos nx, \sin nx\}$. To investigate mean-square convergence, and also pointwise convergence in the next section, it will be useful to obtain an integral expression for S_n.

The Dirichlet Kernel

THEOREM 9.4.1 *Let* $f \in \mathcal{R}[-\pi, \pi]$. *Then for each* $n \in \mathbb{N}$ *and* $x \in \mathbb{R}$,

$$S_n(x) = \frac{1}{\pi}\int_{-\pi}^{\pi} f(t) D_n(x - t)\, dt,$$

where D_n *is the* **Dirichlet kernel,** *given by*

$$D_n(t) = \frac{1}{2} + \sum_{k=1}^{n}\cos kt = \begin{cases} \dfrac{\sin(n + \frac{1}{2})t}{2\sin\frac{t}{2}}, & t \neq 2p\pi,\ p \in \mathbb{Z} \\[2mm] n + \frac{1}{2}, & t = 2p\pi. \end{cases}$$

Proof. By the definition of the Fourier coefficients a_k and b_k (Definition 9.3.1),

$$S_n(x) = \frac{1}{2}a_0 + \sum_{k=1}^{n}(a_k \cos kx + b_k \sin kx)$$

$$= \frac{1}{\pi}\int_{-\pi}^{\pi} \frac{1}{2}f(t)\,dt + \sum_{k=1}^{n}\left(\frac{1}{\pi}\int_{-\pi}^{\pi} f(t)\cos kt\,dt\right)\cos kx$$

$$+ \sum_{k=1}^{n}\left(\frac{1}{\pi}\int_{-\pi}^{\pi} f(t)\sin kt\,dt\right)\sin kx$$

$$= \frac{1}{\pi}\int_{-\pi}^{\pi} f(t)\left[\frac{1}{2} + \sum_{k=1}^{n}(\cos kt \cos kx + \sin kt \sin kx)\right]dt$$

$$= \frac{1}{\pi}\int_{-\pi}^{\pi} f(t)\left[\frac{1}{2} + \sum_{k=1}^{n}\cos k(x-t)\right]dt.$$

In the last step we have used the trigonometric identity

$$\cos k(x-t) = \cos kx \cos kt + \sin kx \sin kt.$$

Set $D_n(s) = \frac{1}{2} + \sum_{k=1}^{n}\cos ks$. Then by the above,

$$S_n(x) = \frac{1}{\pi}\int_{-\pi}^{\pi} f(t)D_n(x-t)\,dt.$$

To conclude the proof, it remains to derive the formula for $D_n(s)$. If $s = 2p\pi$, $p \in \mathbb{Z}$, then

$$D_n(2p\pi) = \frac{1}{2} + \sum_{k=1}^{n}1 = n + \frac{1}{2}.$$

By identity (2) of Theorem 7.2.6, for $s \neq 2p\pi$,

$$D_n(s) = \frac{1}{2} + \frac{\sin(n+\frac{1}{2})s - \sin\frac{s}{2}}{2\sin\frac{s}{2}} = \frac{\sin(n+\frac{1}{2})s}{2\sin\frac{s}{2}},$$

which establishes the result. $\square$

If the sequence $\{D_n\}_{n=1}^{\infty}$ were an approximate identity, convergence results would follow easily from Theorem 8.6.5. Unfortunately however, the functions D_n are neither nonnegative nor do they satisfy (b) of Definition 8.6.4. They however still satisfy

$$\frac{1}{\pi}\int_{-\pi}^{\pi} D_n(t)\,dt = 1, \tag{8}$$

a fact which will prove useful later. To see that (8) holds, it suffices to integrate

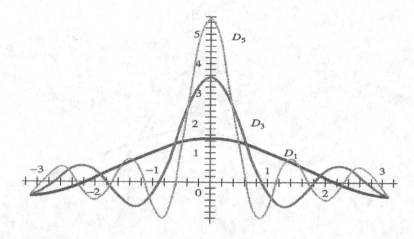

FIGURE 9.3
Graphs of D_1, D_3, and D_5

the sum defining $D_n(s)$ term by term. The only nonzero term will be the integral involving $1/2$, and thus

$$\frac{1}{\pi} \int_{-\pi}^{\pi} D_n(t)\, dt = \frac{1}{\pi} \int_{-\pi}^{\pi} \frac{1}{2}\, dt = 1.$$

The graphs of D_1, D_3, and D_5 are illustrated in Figure 9.3.

The Fejér Kernel

To prove mean-square convergence, it turns out to be more useful to first consider the arithmetic means of the partial sums S_n. For each $n \in \mathbb{N}$, set

$$\sigma_n(x) = \frac{S_0(x) + \cdots + S_n(x)}{n+1}.$$

By Exercise 14 of Section 3.2, if $\lim_{n\to\infty} S_n(x) = S(x)$, then we also have

$$\lim_{n\to\infty} \sigma_n(x) = S(x).$$

However, it is possible that the sequence $\{S_n(x)\}$ diverges for a particular x, whereas the sequence $\{\sigma_n(x)\}$ may converge.

LEMMA 9.4.2 *For $n \in \mathbb{N}$, let $S_n(x) = \frac{1}{2}a_0 + \sum\limits_{k=1}^{n} a_k \cos ks + b_k \sin kx$. Then*

(a) $\sigma_n(x) = \dfrac{1}{2}a_0 + \sum\limits_{j=1}^{n} \left(1 - \dfrac{j}{n+1}\right)(a_j \cos jx + b_j \sin jx)$, *and*

(b) $\displaystyle\int_{-\pi}^{\pi} [f(x) - S_n(x)]^2 dx \le \int_{-\pi}^{\pi} [f(x) - \sigma_n(x)]^2 dx.$

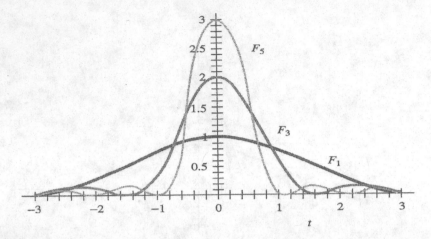

FIGURE 9.4
Graphs of F_1, F_3, and F_5

Proof. The proof of (a) is left as an exercise (Exercise 1). Since σ_n is itself the partial sum of a trigonometric series, the result (b) follows by Corollary 9.1.5. $\square$

Our next step is to express $\sigma_n(x)$ as an integral analogous to that of $S_n(x)$ in the previous theorem.

THEOREM 9.4.3 *Let $f \in \mathcal{R}[-\pi, \pi]$. Then for each $n \in \mathbb{N}$ and $x \in \mathbb{R}$,*

$$\sigma_n(x) = \frac{1}{\pi} \int_{-\pi}^{\pi} f(t) F_n(x - t)\, dt,$$

where F_n is the **Fejér kernel,** *given by*

$$F_n(t) = \frac{1}{n+1} \sum_{k=0}^{n} D_k(t) = \begin{cases} \dfrac{1}{2(n+1)} \left[\dfrac{\sin(n+1)\frac{t}{2}}{\sin \frac{t}{2}} \right]^2, & t \neq 2p\pi, \\[2ex] \dfrac{n+1}{2}, & t = 2p\pi. \end{cases}$$

The graphs of F_1, F_3, and F_5 are illustrated in Figure 9.4.

Proof. By Theorem 9.4.1,

$$S_n(x) = \frac{1}{\pi} \int_{-\pi}^{\pi} f(s) D_n(x - s)\, ds,$$

where D_n is the Dirichlet kernel. Therefore,

$$\sigma_n(x) = \frac{1}{\pi} \int_{-\pi}^{\pi} f(s) F_n(x - s)\, ds,$$

where

$$F_n(t) = \frac{1}{n+1} \sum_{k=0}^{n} D_k(t).$$

If $t = 2p\pi$, $p \in \mathbb{Z}$, then $D_k(2p\pi) = k + \frac{1}{2}$, and thus

$$F_n(2p\pi) = \frac{1}{n+1} \sum_{k=0}^{n} (k + \tfrac{1}{2}) = \frac{(n+1)}{2}.$$

If $t \neq 2p\pi$, $p \in \mathbb{Z}$, then

$$F_n(s) = \frac{1}{2(n+1)\sin\frac{t}{2}} \sum_{k=0}^{n} \sin(k + \tfrac{1}{2})t$$

$$= \frac{1}{2(n+1)\sin^2\frac{t}{2}} \sum_{k=0}^{n} \sin\tfrac{t}{2}\sin(k + \tfrac{1}{2})t.$$

By the identity $\sin A \sin B = \frac{1}{2}[\cos(B-A) - \cos(B+A)]$,

$$\sum_{k=0}^{n} \sin\tfrac{t}{2}\sin(k + \tfrac{1}{2})t = \frac{1}{2} \sum_{k=0}^{n} (\cos kt - \cos(k+1)t)$$

$$= \frac{1}{2}(1 - \cos(n+1)t) = \sin^2(n+1)\tfrac{t}{2}.$$

Therefore, for $t \neq 2p\pi$,

$$F_n(t) = \frac{1}{2(n+1)} \left[\frac{\sin(n+1)\frac{t}{2}}{\sin\frac{t}{2}} \right]^2. \quad \square$$

We now prove the following properties of the Fejér kernel.

THEOREM 9.4.4

(a) F_n *is periodic of period 2π with $F_n(-t) = F_n(t)$.*

(b) $F_n(t) \geq 0$ *for all t.*

(c) $\dfrac{1}{\pi} \displaystyle\int_{-\pi}^{\pi} F_n(t)dt = 1.$

(d) *For $0 < \delta < \pi$, $\displaystyle\lim_{n\to\infty} F_n(t) = 0$ uniformly for all t, $\delta \leq |t| \leq \pi$.*

Proof. (a) Clearly $F_n(-t) = F_n(t)$. Also, since

$$\sin\left(\frac{(n+1)}{2}(t+2\pi) \right) = \sin\left((n+1)\frac{t}{2} \right)\cos(n+1)\pi$$

$$= (-1)^{n+1}\sin\left((n+1)\frac{t}{2} \right),$$

Introduction to Real Analysis

and $\sin \frac{1}{2}(t + 2\pi) = -\sin \frac{t}{2}$, substituting into the formula for F_n gives $F_n(t + 2\pi) = F_n(t)$. Therefore F_n is periodic of period 2π. The proof of (b) is obvious.

(c) Since $\frac{1}{\pi} \int_{-\pi}^{\pi} D_k(t)dt = 1$, we have

$$\frac{1}{\pi} \int_{-\pi}^{\pi} F_n(t)dt = \frac{1}{n+1} \sum_{k=0}^{n} \frac{1}{\pi} \int_{-\pi}^{\pi} D_k(t)\, dt = 1.$$

To prove (d), we first note that $\sin^2 \frac{t}{2} \geq \sin^2 \frac{\delta}{2}$ for all t, $0 < \delta \leq |t| \leq \pi$. Also, since $|\sin(n+1)\frac{t}{2}| \leq 1$ for all t,

$$F_n(t) \leq \frac{1}{2(n+1)\sin^2 \frac{\delta}{2}} \qquad \text{for all } t,\, \delta \leq |t| \leq \pi.$$

Therefore, for any δ, $0 < \delta < \pi$, $\lim_{n \to \infty} F_n(t) = 0$ uniformly on $\delta \leq |t| \leq \pi$. $\quad\square$

As a consequence of (b), (c), and (d) above, the sequence $\{\frac{1}{\pi}F_n(t)\}_{n=1}^{\infty}$ is an approximate identity on $[-\pi, \pi]$. If f is a real-valued function on $\mathbb{R}$ that is periodic of period 2π, then

$$\sigma_n(x) = \frac{1}{\pi} \int_{-\pi}^{\pi} f(t) F_n(x - t)dt,$$

which by the change of variable $s = t - x$,

$$= \frac{1}{\pi} \int_{-\pi-x}^{\pi-x} f(s + x) F_n(s)\, ds.$$

Since the function $s \to f(s + x)F_n(s)$ is periodic of period 2π, by Theorem 8.6.3

$$\sigma_n(x) = \frac{1}{\pi} \int_{-\pi}^{\pi} f(s + x) F_n(s)\, ds.$$

Thus if f is continuous on $[-\pi, \pi]$ with $f(-\pi) = f(\pi)$, by Theorem 8.6.5,

$$\lim_{n \to \infty} \sigma_n(x) = f(x)$$

uniformly on $[-\pi, \pi]$. We summarize this in the following theorem of L. Fejér (1880–1959).

THEOREM 9.4.5 (Fejér) *If f is a continuous real-valued function on $[-\pi, \pi]$ with $f(-\pi) = f(\pi)$, then*

$$\lim_{n \to \infty} \sigma_n(x) = f(x)$$

uniformly on $[-\pi, \pi]$.

COROLLARY 9.4.6 *If f is a continuous real-valued function on $[-\pi, \pi]$ with $f(-\pi) = f(\pi)$, then*

$$\lim_{n \to \infty} \int_{-\pi}^{\pi} [f(x) - S_n(x)]^2 \, dx = 0.$$

Proof. The proof of the Corollary is an immediate consequence of Lemma 9.4.2(b). $\square$

Remark. There is a similarity between Theorem 9.4.5 and the Weierstrass approximation theorem. An alternate way of expressing Theorem 9.4.5 is as follows: Let f be continuous on $[-\pi, \pi]$ with $f(-\pi) = f(\pi)$. Given $\epsilon > 0$, there exists a **trigonometric polynomial**

$$T_n(x) = \frac{1}{2}A_0 + \sum_{k=1}^{n} (A_k \cos kx + B_k \sin kx)$$

such that

$$|f(x) - T_n(x)| < \epsilon$$

for all $x \in [-\pi, \pi]$. The function $T_n(x)$ is called a trigonometric polynomial since in complex form it can be rewritten as

$$T_n(x) = \sum_{k=-n}^{n} c_k e^{ikx},$$

where by De Moivre's formula $e^{ikx} = \cos kx + i \sin kx$.

Convergence in the Mean

Corollary 9.4.6 only proves convergence in the mean for the case where f is a continuous real-valued function on $[-\pi, \pi]$. We are now ready to prove the main result of this section.

THEOREM 9.4.7 *If $f \in \mathcal{R}[-\pi, \pi]$, then*

$$\lim_{n \to \infty} \int_{-\pi}^{\pi} [f(x) - S_n(x)]^2 dx = 0,$$

where S_n is the nth partial sum of the Fourier series of f.

For the proof of the theorem we require the following lemma.

LEMMA 9.4.8 *Let $f \in \mathcal{R}[a, b]$ with $|f(x)| \leq M$ for all $x \in [a, b]$. Then given $\epsilon > 0$, there exists a continuous function g on $[a, b]$ with $g(a) = g(b)$ and $|g(x)| \leq M$ for all $x \in [a, b]$, such that*

$$\int_{a}^{b} |f(x) - g(x)| \, dx < \epsilon.$$

Before proving the lemma, we will first use the result to prove Theorem 9.4.7, then consider some consequences of this theorem.

Proof of Theorem 9.4.7 Suppose $|f(x)| \leq M$ with $M > 0$, and let $\epsilon > 0$ be given. By the lemma, there exists a continuous function g on $[-\pi, \pi]$ with $g(-\pi) = g(\pi)$ and $|g(x)| \leq M$ for all $x \in [-\pi, \pi]$ such that

$$\int_{-\pi}^{\pi} |f(x) - g(x)| \, dx < \frac{\epsilon}{8M}. \tag{9}$$

Let $S_n(g)(x)$ be the function defined by

$$S_n(g)(x) = \frac{1}{\pi} \int_{-\pi}^{\pi} g(t) D_n(x - t) \, dt,$$

where D_n is the Dirichlet kernel. Then since g is continuous, by Corollary 9.4.6 the sequence $\{S_n(g)\}$ converges in the mean to g on $[-\pi, \pi]$. Thus there exists $n_o \in \mathbb{N}$ such that

$$\int_{-\pi}^{\pi} [g(x) - S_n(g)(x)]^2 dx < \frac{\epsilon}{4}$$

for all $n \geq n_o$. Consider $\int_{-\pi}^{\pi} [f(x) - S_n(g)(x)]^2 dx$. Since

$$|f(x) - S_n(g)(x)| \leq |f(x) - g(x)| + |g(x) - S_n(g)(x)|,$$

we have

$$|f(x) - S_n(g)(x)|^2 \leq 2 \left[|f(x) - g(x)|^2 + |g(x) - S_n(g)(x)|^2 \right].$$

But by inequality (9),

$$\int_{-\pi}^{\pi} |f(x) - g(x)|^2 \, dx \leq 2M \int_{-\pi}^{\pi} |f(x) - g(x)| \, dx < \frac{\epsilon}{4}$$

Therefore, for all $n \geq n_o$,

$$\int_{-\pi}^{\pi} [f(x) - S_n(g)(x)]^2 dx < \frac{\epsilon}{2} + 2 \int_{-\pi}^{\pi} [g(x) - S_n(g)(x)]^2 \, dx < \epsilon.$$

However, for each $n \in \mathbb{N}$, $S_n(g)$ is the nth partial sum of a trigonometric series. Thus if S_n is the nth partial sum of the Fourier series of f, by Corollary 9.1.5

$$\int_{-\pi}^{\pi} [f(x) - S_n(x)]^2 \, dx \leq \int_{-\pi}^{\pi} [f(x) - S_n(g)(x)]^2 \, dx.$$

Thus, given $\epsilon > 0$, there exists an integer n_o, such that

$$\int_{-\pi}^{\pi} [f(x) - S_n(x)]^2 \, dx < \epsilon$$

for all $n \geq n_o$; i.e., $\{S_n\}$ converges in the mean to f on $[-\pi, \pi]$. $\square$

As a consequence of Theorems 9.2.4 and 9.4.7, we have the following corollary:

COROLLARY 9.4.9 (Parseval's equality) *If $f \in \mathcal{R}[-\pi, \pi]$, then*

$$\frac{1}{2}a_0^2 + \sum_{n=1}^{\infty}(a_n^2 + b_n^2) = \frac{1}{\pi}\int_{-\pi}^{\pi}f^2(x)\,dx.$$

EXAMPLE 9.4.10 By Example 9.3.3(b), the Fourier series of $f(x) = x$ is given by

$$x \sim \sum_{n=1}^{\infty}\frac{2(-1)^{n+1}}{n}\sin nx.$$

Here $a_n = 0$ for all n, and $b_n = 2(-1)^{n+1}/n$. Thus by Parseval's equality,

$$\sum_{n=1}^{\infty}\frac{4}{n^2} = \frac{1}{\pi}\int_{-\pi}^{\pi}x^2\,dx = \frac{2}{3}\pi^2,$$

which gives

$$\sum_{n=1}^{\infty}\frac{1}{n^2} = \frac{\pi^2}{6}. \quad \square$$

Proof of Lemma 9.4.8. Let $f \in \mathcal{R}[a, b]$ with $|f(x)| \leq M$ for all $x \in [a, b]$. Since $f_1(x) = f(x) + M$ is nonnegative, if we can prove the result for the function f_1, the result also follows for the function f. Thus we can assume that f satisfies $0 \leq f(x) \leq M$ for all $x \in [a, b]$.

Let $\epsilon > 0$ be given. Since f is Riemann integrable on $[a, b]$, there exists a partition $\mathcal{P} = \{x_0, x_1, ..., x_n\}$ of $[a, b]$ such that

$$0 \leq \int_a^b f(x)\,dx - \mathcal{L}(\mathcal{P}, f) < \frac{\epsilon}{2},$$

where

$$\mathcal{L}(\mathcal{P}, f) = \sum_{i=1}^{n}m_i\Delta x_i,$$

and $m_i = \inf\{f(t) : t \in [x_{i-1}, x_i]\}$, $i = 1, 2, ..., n$. For each $i = 1, 2, ..., n$, define the functions h_i on $[a, b]$ by

$$h_i(x) = \begin{cases} m_i, & x_{i-1} \leq x < x_i, \\ 0, & \text{elsewhere,} \end{cases}$$

and let $h(x) = \sum_{i=1}^{n}h_i(x)$ (see Figure 9.5). The function h is called a **step function** on $[a, b]$. Since h is continuous except at a finite number of points, h is Riemann integrable and

$$\int_a^b h(x)\,dx = \sum_{i=1}^{n}m_i\Delta x_i = \mathcal{L}(\mathcal{P}, f).$$

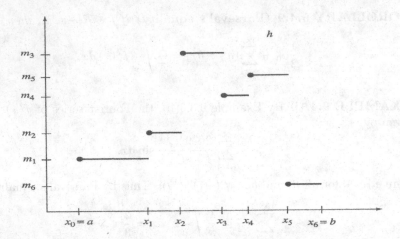

FIGURE 9.5
Graph of the Step Function h

Therefore,

$$0 \le \int_a^b [f(x) - h(x)]\, dx < \frac{\epsilon}{2}.$$

Also, since $m_i = \inf\{f(t) : t \in [x_{i-1}, x_i]\}$, $0 \le h(x) \le f(x)$ for all $x \in [a, b]$. By taking slightly shorter intervals, and connecting the endpoints with straight line segments, we leave it as an exercise (Exercise 9) to show that there exists a continuous function g on $[a, b]$ with $g(a) = g(b) = 0$, $0 \le g(x) \le h(x)$, such that

$$\int_a^b [h(x) - g(x)]\, dx < \frac{\epsilon}{2}.$$

Then $0 \le g(x) \le f(x)$, and

$$0 \le \int_a^b [f(x) - g(x)]\, dx = \int_a^b [f(x) - h(x)]\, dx + \int_a^b [h(x) - g(x)]\, dx$$

$$< \frac{\epsilon}{2} + \frac{\epsilon}{2} = \epsilon. \quad \square$$

Exercises 9.4

1. Prove Lemma 9.4.2(a).

2. *Using the Fourier series of $f(x) = x^2$ and Parseval's equality, find
$$\sum_{n=1}^{\infty} \frac{1}{n^4}.$$

3. *Prove that the orthogonal systems $\{\sin nx\}_{n=1}^{\infty}$ and $\{1, \cos nx\}_{n=1}^{\infty}$ are both complete on $[0, \pi]$.

4. Show that the set $\{\sin \frac{1}{2}x, \sin \frac{3}{2}x, \sin \frac{5}{2}x, \dots\}$ is complete on $[0, \pi]$

5. Let $f \in \mathcal{R}[-\pi, \pi]$, and let S_n denote the nth partial sum of the Fourier series of f.

   a. Using the Cauchy-Schwarz inequality for integrals (Exercise 13, Section 9.1), prove that

$$\int_{-\pi}^{\pi} |S_n(x) - f(x)| \, dx \le \sqrt{2\pi} \left(\int_{-\pi}^{\pi} |S_n(x) - f(x)|^2 \, dx \right)^{1/2}.$$

   b. Prove that $\displaystyle\lim_{n \to \infty} \int_{-\pi}^{\pi} S_n(x) \, dx = \int_{-\pi}^{\pi} f(x) dx$.

6. * Use the previous exercise to prove the following: Suppose $f \in \mathcal{R}[-\pi, \pi]$ with

$$f(x) \sim \tfrac{1}{2}a_o + \sum_{n=1}^{\infty} (a_n \cos nx + b_n \sin nx).$$

   For $x \in [-\pi, \pi]$, set $F(x) = \int_{-\pi}^{x} f(t)dt - \frac{1}{2}a_o x$.

   Then F is continuous on $[-\pi, \pi]$ with $F(-\pi) = F(\pi)$, and for all $x \in [-\pi, \pi]$,

$$F(x) = \frac{1}{2}a_o\pi + \sum_{n=1}^{\infty} \left(\frac{a_n}{n} \sin nx - \frac{b_n}{n}(\cos nx - \cos n\pi) \right),$$

   where the convergence is uniform on $[-\pi, \pi]$.

7. By integrating the Fourier series of $f(x) = x$ (Example 9.3.3(b)), find the Fourier series of $g(x) = x^2$.

8. Suppose f is a real-valued periodic function of period 2π with $f \in \mathcal{R}[-\pi, \pi]$, and $x_o \in \mathbb{R}$ is such that $f(x_o-)$ and $f(x_o+)$ both exist. Prove that

$$\lim_{n \to \infty} \sigma_n(x_o) = \tfrac{1}{2}[f(x_o-) + f(x_o+)].$$

9. Complete the proof of Lemma 9.4.8; namely, given the step function h on $[a, b]$ and $\epsilon > 0$, there exists a continuous function g on $[a, b]$ with

$$0 \le g(x) \le h(x), \; g(a) = g(b) = 0, \text{ and } \int_a^b [h(x) - g(x)] \, dx < \epsilon.$$

10. * Let $f \in \mathcal{R}[a, b]$. Given $\epsilon > 0$, prove that there exists a polynomial $p(x)$ such that $\int_a^b |f(x) - p(x)|dx < \epsilon$.

9.5 Pointwise Convergence of Fourier Series

We now turn our attention to the question of pointwise convergence of a Fourier series to its defining function. It is known that even if the function is continuous, this is not always possible; additional hypothesis on the function

f will be required. As was indicated in the introduction, Dirichlet was the first to find sufficient conditions on the function f so that the Fourier series of f converges to f. For the proof of his result we need several preliminary lemmas.

LEMMA 9.5.1 *Let f be a periodic real-valued function on $\mathbb{R}$ of period 2π with $f \in \mathcal{R}[-\pi, \pi]$. Fix $x \in \mathbb{R}$. Then for any real number A,*

$$S_n(x) - A = \frac{2}{\pi} \int_0^\pi \left[\frac{f(x+t) + f(x-t)}{2} - A \right] D_n(t)\, dt,$$

where D_n is the Dirichlet kernel and S_n is the nth partial sum of the Fourier series of f.

Proof. By Theorem 9.4.1,

$$S_n(x) = \frac{1}{\pi} \int_{-\pi}^\pi f(t) D_n(x-t)\, dt,$$

which by the change of variable $s = x - t$,

$$= -\frac{1}{\pi} \int_{x+\pi}^{x-\pi} f(x-s) D_n(s)\, ds = \frac{1}{\pi} \int_{x-\pi}^{x+\pi} f(x-s) D_n(s)\, ds.$$

Since both f and D_n are periodic of period 2π, by Theorem 8.6.3

$$\frac{1}{\pi} \int_{x-\pi}^{x+\pi} f(x-s) D_n(s)\, ds = \frac{1}{\pi} \int_{-\pi}^\pi f(x-s) D_n(s)\, ds.$$

Therefore,

$$S_n(x) = \frac{1}{\pi} \int_{-\pi}^0 f(x-s) D_n(s)\, ds + \frac{1}{\pi} \int_0^\pi f(x-s) D_n(s)\, ds.$$

In the first integral set $u = -s$. Then since $D_n(-u) = D_n(u)$,

$$\frac{1}{\pi} \int_{-\pi}^0 f(x-s) D_n(s)\, ds = -\frac{1}{\pi} \int_\pi^0 f(x+u) D_n(u)\, du$$

$$= \frac{1}{\pi} \int_0^\pi f(x+s) D_n(s)\, ds.$$

Therefore,

$$S_n(x) = \frac{1}{\pi} \int_0^\pi [f(x+s) + f(x-s)] D_n(s)\, ds.$$

Finally, since $\frac{2}{\pi} \int_0^\pi D_n(s)\, ds = 1$,

$$S_n(x) - A = \frac{2}{\pi} \int_0^\pi \left[\frac{f(x+s) - f(x-s)}{2} - A \right] D_n(s)\, ds,$$

which is the desired identity. $\square$

LEMMA 9.5.2 *If $g \in \mathcal{R}[0, \pi]$, then*

$$\lim_{n \to \infty} \int_0^\pi g(t) \sin \left(n + \frac{1}{2} \right) t \, dt = 0.$$

Proof. Extend g to $[-\pi, \pi]$ by defining $g(x) = 0$ for all x, $-\pi \le x < 0$. Since $\sin(n + \frac{1}{2})t = \cos \frac{t}{2} \sin nt + \sin \frac{t}{2} \cos nt$, we have

$$\int_0^\pi g(t) \sin \left(n + \tfrac{1}{2} \right) t \, dt = \int_{-\pi}^\pi g(t) \sin \left(n + \tfrac{1}{2} \right) t \, dt$$

$$= \int_{-\pi}^\pi g_1(t) \sin nt \, dt + \int_{-\pi}^\pi g_2(t) \cos nt \, dt,$$

where $g_1(t) = g(t) \cos \frac{t}{2}$ and $g_2(t) = g(t) \sin \frac{t}{2}$. Since g is Riemann integrable on $[-\pi, \pi]$, so are both g_1 and g_2. Thus by Theorem 9.3.4,

$$\lim_{n \to \infty} \int_{-\pi}^\pi g_1(t) \sin nt \, dt = \lim_{n \to \infty} \int_{-\pi}^\pi g_2(t) \cos nt \, dt = 0,$$

from which the result follows. $\square$

Dirichlet's Theorem

Before we state and prove Dirichlet's theorem, we briefly review some notation introduced in Chapter 4. For a real-valued function f defined in a neighborhood of a given point p, the **right** and **left limits** of f at p, denoted $f(p+)$ and $f(p-)$ respectively, are defined by

$$f(p+) = \lim_{x \to p^+} f(x) \quad \text{and} \quad f(p-) = \lim_{x \to p^-} f(x),$$

provided of course that the limits exist.

THEOREM 9.5.3 (Dirichlet) *Let f be a real-valued periodic function on $\mathbb{R}$ of period 2π with $f \in \mathcal{R}[-\pi, \pi]$. Suppose $x_o \in \mathbb{R}$ is such that*

(a) *$f(x_o+)$ and $f(x_o-)$ both exist, and*

(b) *there exists a constant M and a $\delta > 0$ such that*

$$|f(x_o + t) - f(x_o+)| \le M t \quad \text{and} \quad |f(x_o - t) - f(x_o-)| \le M t \tag{10}$$

for all t, $0 < t \le \delta$, then

$$\lim_{n \to \infty} S_n(x_o) = \frac{1}{2}[f(x_o+) + f(x_o-)].$$

Remark. If f is continuous at x_o, then $\frac{1}{2}[f(x_o+) + f(x_o-)] = f(x_o)$, and thus

$$\lim_{n \to \infty} S_n(x_o) = f(x_o),$$

provided of course that the inequalities (10) hold.

Proof. Set $A = \frac{1}{2}[f(x_o+) + f(x_o-)]$. By Lemma 9.5.1

$$S_n(x_o) - A = \frac{2}{\pi} \int_0^\pi \left[\frac{f(x_o+t) + f(x_o-t)}{2} - A \right] D_n(t)\, dt$$

$$= \frac{1}{\pi} \int_0^\pi [(f(x_o+t) - f(x_o+)) + (f(x_o-t) - f(x_o-))] \frac{\sin(n+\frac{1}{2})t}{2\sin\frac{t}{2}}\, dt$$

$$= \frac{1}{\pi} \int_0^\pi g_1(t) \sin(n+\tfrac{1}{2})t\, dt + \frac{1}{\pi} \int_0^\pi g_2(t) \sin(n+\tfrac{1}{2})t\, dt,$$

where

$$g_1(t) = \frac{f(x_o+t) - f(x_o+)}{2\sin\frac{t}{2}} \qquad \text{and} \qquad g_2(t) = \frac{f(x_o-t) - f(x_o-)}{2\sin\frac{t}{2}}.$$

To prove the result, it suffices to show that the functions g_1 and g_2 are Riemann integrable on $[0, \pi]$. If this is the case, then by Lemma 9.5.2,

$$\lim_{n\to\infty} \int_0^\pi g_i(t) \sin(n+\tfrac{1}{2})t\, dt = 0, \qquad i = 1, 2.$$

Therefore,

$$\lim_{n\to\infty} S_n(x_o) = A = \frac{1}{2}[f(x_o+) + f(x_o-)].$$

To finish the proof we still have to show that $g_i \in \mathcal{R}[0, \pi]$, $i = 1, 2$. We will prove the result for the function g_1, the proof for g_2 being similar. Set $h(t) = f(x_o+t) - f(x_o+)$. Since $f \in \mathcal{R}[-\pi, \pi]$ and is periodic of period 2π, h is Riemann integrable on $[0, \pi]$. Also, $(\sin\frac{t}{2})^{-1}$ is continuous on $(0, \pi]$ and thus Riemann integrable on $[c, \pi]$ for any c, $0 < c < \pi$. Therefore,

$$g_1(t) = \frac{h(t)}{2\sin\frac{t}{2}}$$

is Riemann integrable on $[c, \pi]$ for any c, $0 < c < \pi$. Let $\delta > 0$ be as in hypothesis (b). Then for $0 < t < \delta$, using inequality (10),

$$|g_1(t)| = \frac{|f(x_o+t) - f(x_o+)|}{|2\sin\frac{t}{2}|}$$

$$= \frac{|f(x_o+t) - f(x_o+)|}{|t|} \left| \frac{\frac{1}{2}t}{\sin\frac{1}{2}t} \right| \le M \left| \frac{\frac{1}{2}t}{\sin\frac{1}{2}t} \right|.$$

Since $\lim_{x\to 0+} \dfrac{x}{\sin x} = 1$, there exists a constant C such that

$$\left| \frac{\frac{1}{2}t}{\sin\frac{1}{2}t} \right| \le C$$

for all t, $0 < t \le \pi$. Therefore, g_1 is bounded on $(0, \delta)$, and hence also on $(0, \pi]$. Thus since g_1 is Riemann integrable on $[c, \pi]$ for any c, $0 < c < \pi$, by Exercise 5 of Section 6.2, g_1 is Riemann integrable on $(0, \pi]$. $\square$

Piecewise Continuous Functions

Dirichlet's theorem required that $f(x_o-)$ and $f(x_o+)$ both exist at x_o, and that f satisfy the inequalities (10) at the point x_o. The existence of both $f(x_o-)$ and $f(x_o+)$ means that either f is continuous at x_o, or in the terminology of Section 4.4, has a removable or jump discontinuity at x_o. Functions that have only simple discontinuities, and at most a finite number on an interval $[a, b]$, are said to be piecewise continuous on $[a, b]$. We make this precise with the following definition.

DEFINITION 9.5.4 *A real-valued function f is **piecewise continuous** on $[a, b]$ if there exist finitely many points $a = x_0 < x_1 < \cdots < x_n = b$ such that*

(a) *f is continuous on (x_{i-1}, x_i) for all $i = 1, 2, ..., n$, and*

(b) *for each $i = 0, 1, 2, ..., n$, $f(x_i+)$ and $f(x_i-)$ exist as finite limits.*

For $i = 0$ and n, we of course only require the existence of the right and left limit respectively. Also, we **do not** require that f be defined at $x_0, x_1, ..., x_n$.

In addition to piecewise continuous functions, it will also be useful to consider functions for which the derivative is piecewise continuous. If f is piecewise continuous on $[a, b]$, then the derivative f' is **piecewise continuous** on $[a, b]$ if there exist a finite number of points $a = x_0 < x_1 < \cdots < x_n = b$ such that

(a) $f'(x)$ exists and is continuous on each interval (x_{i-1}, x_i), $i = 1, 2, ..., n$, and

(b) for each $i = 0, 1, 2, ..., n$, the quantities $f'(x_i+)$ and $f'(x_i-)$ both exist as finite limits. Again, for the endpoints a and b we obviously only require the existence of $f'(a+)$ and $f'(b-)$.

EXAMPLES 9.5.5 (a) Consider the function $f(x) = \begin{cases} x, & 0 < x < 1, \\ x^2 + 2, & 1 < x < 2. \end{cases}$
Since f is continuous on $(0, 1)$ and $(1, 2)$ and $f(0+)$, $f(1-)$, $f(1+)$, and $f(2-)$ all exist, f is piecewise continuous on $[0, 2]$. Also, since

$$f'(x) = \begin{cases} 1, & 0 < x < 1, \\ 2x, & 1 < x < 2, \end{cases}$$

and $f'(0+)$, $f'(1-)$, $f'(1+)$, and $f'(2-)$ all exist, f' is also piecewise continuous on $[0, 2]$.

(b) As our second example, consider the function

$$f(x) = \begin{cases} 0, & x = 0, \\ x^2 \sin \frac{1}{x}, & 0 < x \leq 1. \end{cases}$$

The function f is both continuous and differentiable on $[0, 1]$ with

$$f'(x) = \begin{cases} 0, & x = 0, \\ 2x \sin \frac{1}{x} - \cos \frac{1}{x}, & 0 < x < 1. \end{cases}$$

However, since $f'(0+)$ does not exist, f' is not piecewise continuous on $[0, 1]$. □

Returning to Dirichlet's theorem, suppose f is such that both $f'(x_o-)$ and $f'(x_o+)$ exist at the point x_o. If f' is piecewise continuous on $[-\pi, \pi]$, then there exists $x_1 > x_o$ such that f and f' are continuous on (x_o, x_1). If f is not continuous at x_o, redefine f at x_o as $f(x_o+)$. Then f (redefined if necessary) is continuous on $[x_o, x_1)$, and thus by Theorem 5.2.11

$$\lim_{t \to 0^+} \frac{f(x_o + t) - f(x_o+)}{t} = f'(x_o+).$$

As a consequence, there exists a constant M and a $\delta > 0$ such that

$$|f(x_o + t) - f(x_o+)| \leq M t$$

for all t, $0 < t < \delta$. Similarly, the existence of $f'(x_o-)$ implies the existence of a constant M and a $\delta > 0$ such that

$$|f(x_o - t) - f(x_o-)| \leq M t$$

for all t, $0 < t < \delta$. Thus f satisfies the hypothesis of Dirichlet's theorem. On combining the above discussion with Theorem 9.5.3, we obtain the following corollary.

COROLLARY 9.5.6 *Suppose f is a real-valued periodic function on $\mathbb{R}$ of period 2π. If f and f' are piecewise continuous on $[-\pi, \pi)$, then*

$$\lim_{n \to \infty} S_n(x) = \tfrac{1}{2}[f(x+) + f(x-)]$$

for all $x \in \mathbb{R}$.

EXAMPLES 9.5.7 (a) Consider the function f defined on $[-\pi, \pi)$ by

$$f(x) = \begin{cases} 0, & -\pi \leq x < -\frac{\pi}{2}, \\ 3, & -\frac{\pi}{2} \leq x \leq \frac{\pi}{2}, \\ 0, & \frac{\pi}{2} < x < \pi. \end{cases}$$

Extend f to all of $\mathbb{R}$ as a periodic function of period 2π. The function f is then continuous except at $x = \frac{1}{2}\pi + k\pi$, $k \in \mathbb{Z}$. At each discontinuity x_o,

$$\tfrac{1}{2}[f(x_o+) + f(x_o-)] = \tfrac{3}{2}.$$

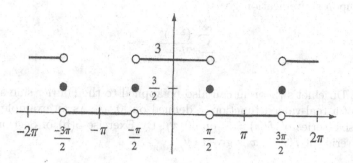

FIGURE 9.6
Graph of $\frac{1}{2}[f(x+) + f(x-)]$

The graph of $\frac{1}{2}[f(x+) + f(x-)]$ on the interval $[-2\pi, 2\pi]$ is given in Figure 9.6.

To discuss convergence of the Fourier series of f, we first note that f is piecewise continuous on $[-\pi, \pi]$ and differentiable on $(-\pi, -\frac{\pi}{2})$, $(-\frac{\pi}{2}, \frac{\pi}{2})$, and $(\frac{\pi}{2}, \pi)$ with $f'(x) = 0$ in the respective intervals. Thus f' is also piecewise continuous on $[-\pi, \pi]$. As a consequence of Corollary 9.5.6 the Fourier series of f converges to $\frac{1}{2}[f(x+) + f(x-)]$ for all $x \in \mathbb{R}$.

The Fourier series of f is obtained as follows:

$$a_0 = \frac{1}{\pi} \int_{-\frac{\pi}{2}}^{\frac{\pi}{2}} 3 \, dx = 3,$$

$$a_n = \frac{1}{\pi} \int_{-\frac{\pi}{2}}^{\frac{\pi}{2}} 3 \cos nx \, dx = \frac{6}{n\pi} \sin \frac{n\pi}{2}, \quad n = 1, 2, \ldots$$

and since f is an even function, $b_n = 0$ for all $n \in \mathbb{N}$. Therefore,

$$f(x) \sim \frac{3}{2} + \sum_{n=1}^{\infty} \frac{6}{n\pi} \sin \frac{n\pi}{2} \cos nx.$$

If n is even, $\sin \frac{n\pi}{2} = 0$. If n is odd, i.e., $n = 2k + 1$,

$$\sin(2k + 1)\frac{\pi}{2} = \sin\left(k\pi + \frac{\pi}{2}\right) = \cos k\pi \sin \frac{\pi}{2} = (-1)^k.$$

Thus

$$f(x) \sim \frac{3}{2} + \frac{6}{\pi} \sum_{k=0}^{\infty} \frac{(-1)^k}{2k + 1} \cos(2k + 1)x.$$

When $x = 0$, the series converges to $f(0) = 3$. As a consequence

$$3 = \frac{3}{2} + \frac{6}{\pi} \sum_{k=0}^{\infty} \frac{(-1)^k}{2k + 1},$$

which upon simplification gives

$$\sum_{k=0}^{\infty} \frac{(-1)^k}{2k+1} = \frac{\pi}{4}.$$

(b) Dirichlet's theorem can also be applied to the Fourier sine and cosine series of a real-valued function f defined on $[0, \pi)$. As an example, consider the cosine series of $f(x) = x$ on $(0, \pi)$. By Exercise 5(b) of Section 9.3, the cosine series of $f(x) = x$ is given by

$$x \sim \frac{1}{2}\pi - \frac{4}{\pi}\sum_{k=1}^{\infty} \frac{1}{(2k-1)^2}\cos(2k-1)x.$$

Since the even extension f_e of f to $[-\pi, \pi]$ is given by $f_e(x) = |x|$, the cosine series of x is the Fourier series of $|x|$ on $[-\pi, \pi]$. Since both f_e and f_e' are piecewise continuous on $[-\pi, \pi]$, the Fourier series converges to the 2π-periodic extension of $|x|$ for all $x \in \mathbb{R}$.

By Dirichlet's theorem, the Fourier series converges to $|x|$ for all $x \in [-\pi, \pi]$. Taking $x = 0$ gives

$$0 = \frac{\pi}{2} - \frac{4}{\pi}\sum_{k=1}^{\infty} \frac{1}{(2k-1)^2},$$

or $\displaystyle\sum_{k=1}^{\infty} \frac{1}{(2k-1)^2} = \frac{\pi^2}{8}$. □

Remarks. Although the inequalities (10) of Theorem 9.5.3 are sufficient, they are by no means necessary. There are variations of Dirichlet's theorem which provide sufficient conditions on f to guarantee convergence of the series to $\frac{1}{2}[f(x_0+) + f(x_0-)]$. For example, the inequalities of Theorem 9.5.3 can be replaced by the following:

$$|f(x_0 + t) - f(x_0+)| \le M t^\alpha \quad \text{and} \quad |f(x_0 + t) - f(x_0-)| \le M t^\alpha \qquad (11)$$

for all t, $0 < t < \delta$, and some α, $0 < \alpha \le 1$. If f satisfies the above at x_o, then the conclusion of Dirichlet's theorem is still valid. An even more general condition is due to Ulisse Dini (1845–1918) who proved that if $f \in \mathcal{R}[-\pi, \pi]$ satisfies

$$\int_0^\delta \frac{|f(x_o + t) - f(x_o+)|}{t}\, dt < \infty \quad \text{and} \quad \int_0^\delta \frac{|f(x_o - t) - f(x_o-)|}{t}\, dt < \infty$$

for some $\delta > 0$, then the Fourier series of f converges to $\frac{1}{2}[f(x_o+) + f(x_o-)]$ at x_o. Both of the above hold if f satisfies (11) at x_o.

Differentiation of Fourier Series

Our final result of this section involves the derivative of a Fourier series. Consider the function $f(x) = x$. Since both f and f' are continuous on $[-\pi, \pi]$, the Fourier series of f converges to f for all $x \in (-\pi, \pi)$. Therefore, by Example 9.3.3(b),

$$x = \sum_{n=1}^{\infty} \frac{2(-1)^{n+1}}{n} \sin nx, \qquad x \in (-\pi, \pi).$$

The differentiated series

$$\sum_{n=1}^{\infty} 2(-1)^{n+1} \cos nx$$

does not converge since its nth term fails to approach zero as $n \to \infty$. The 2π-periodic extension of f has discontinuities at $x = \pm(2k-1)\pi$, $k \in \mathbb{N}$. It turns out that continuity of the periodic function is important for differentiability of the Fourier series. Sufficient conditions are given by the following theorem.

THEOREM 9.5.8 *Let f be a continuous function on $[-\pi, \pi]$ with $f(-\pi) = f(\pi)$, and let f' be piecewise continuous on $[-\pi, \pi]$. If*

$$f(x) = \frac{1}{2}a_0 + \sum_{n=1}^{\infty}(a_n \cos nx + b_n \sin nx), \qquad x \in [-\pi, \pi],$$

is the Fourier series of f, then at each $x \in (-\pi, \pi)$ where $f''(x)$ exists,

$$f'(x) = \sum_{n=1}^{\infty}(-na_n \sin nx + nb_n \cos nx). \qquad (12)$$

Remark. At a point x where $f''(x)$ does not exist but $f''(x-)$ and $f''(x+)$ both exist, the series (12) converges to $\frac{1}{2}[f'(x-) + f'(x+)]$.

Proof. Suppose $x \in (-\pi, \pi)$ is such that $f''(x)$ exists. Since f' is continuous at x, by Dirichlet's theorem

$$f'(x) = \frac{1}{2}\alpha_0 + \sum_{n=1}^{\infty}(\alpha_n \cos nx + \beta_n \sin nx).$$

Since f is continuous with $f(-\pi) = f(\pi)$, and $f' \in \mathcal{R}[-\pi, \pi]$,

$$\alpha_0 = \frac{1}{2\pi}\int_{-\pi}^{\pi} f'(t)\,dt = \frac{1}{2\pi}[f(\pi) - f(-\pi)] = 0.$$

Also, by integration by parts,

$$\alpha_n = \frac{1}{2\pi}\int_{-\pi}^{\pi} f'(t) \cos nt$$

$$= \frac{1}{2\pi}[f(\pi)\cos n\pi - f(-\pi)\cos(-n\pi)] + \frac{n}{2\pi}\int_{-\pi}^{\pi} f(t)\sin nt\,dt$$

$$= nb_n.$$

Similarly, $\beta_n = -na_n$, which proves the result. □

EXAMPLE 9.5.9 Consider the function $f(x) = x^2$. Since f is even on $(-\pi, \pi)$,

$$f(x) = \tfrac{1}{2}a_0 + \sum_{n=1}^{\infty} a_n \cos nx.$$

Also, since f satisfies the hypothesis of Theorem 9.5.8,

$$2x = \sum_{n=1}^{\infty} (-na_n \sin nx).$$

On the other hand, by Example 9.3.3(b),

$$2x = \sum_{n=1}^{\infty} \frac{4(-1)^{n+1}}{n} \sin nx.$$

Therefore $a_n = 4(-1)^n/n^2$ for $n = 1, 2, 3, \dots$. To find a_0 we use the definition. This gives $a_0 = 2\pi^2/3$, and thus

$$x^2 = \frac{\pi^2}{3} + 4\sum_{n=1}^{\infty} \frac{(-1)^n}{n^2} \cos nx, \quad x \in [-\pi, \pi]. \quad □$$

Remarks. In closing this section, it should be mentioned that there exist continuous functions f for which the Fourier series of f fails to converge at a given point. This was first shown by P. du Bois Reymond. Other examples were subsequently constructed by L. Fejér and Lebesgue. The example of Fejér can be found on p. 416 of the text by E.C. Titchmarsh. Given any countable set E in $(-\pi, \pi)$, it is possible to construct a continuous function f on $[-\pi, \pi]$ such that the Fourier series of f diverges on E and converges on $(-\pi, \pi) \setminus E$.[4] In fact, it is possible to construct a continuous function on $[-\pi, \pi]$ whose Fourier series diverges on an uncountable subset of $(-\pi, \pi)$.[5] The existence of functions having such pathological behavior is due to the fact that for the Dirichlet kernel D_n,

$$\lim_{n \to \infty} \int_0^{\pi} |D_n(t)| \, dt = \infty.$$

The details are left to the miscellaneous exercises.

For all of the above examples it is still the case that the Fourier series of f converges to f except on a set of measure zero. This in fact is the case for the Fourier series of every Riemann integrable function f on $[-\pi, \pi]$. In 1926, Kolmogorov showed that there exist Lebesgue integrable functions whose Fourier

---

[4] See Chapter VIII of the text by A. Zygmund.
[5] Ibid.

series diverge everywhere.[6] The biggest question about the convergence of Fourier series was asked by Lusin: If a function f satisfies the hypothesis that f^2 is Lebesgue integrable on $[-\pi, \pi]$, does the Fourier series of f converge to f, except perhaps on a set of measure zero? This problem remained unanswered for 50 years. The first proof that this was indeed the case was provided by L. Carleson in 1966.[7]

Exercises 9.5

1. For each of the functions of Exercise 3, Section 9.3, sketch the graph of the function on the interval $[-2\pi, 2\pi]$ to which the Fourier series converges.

2. *Use the Fourier series of $f(x) = x^2$ to find each of the following sums:

   a. $\displaystyle\sum_{n=1}^{\infty} \frac{(-1)^{n+1}}{n^2}$,   b. $\displaystyle\sum_{n=1}^{\infty} \frac{1}{n^2}$.

3. Using the Fourier cosine series of $f(x) = \sin x$ on $[0, \pi]$ (Exercise 7, Section 9.3), find each of the following sums:

   *a. $\displaystyle\sum_{n=1}^{\infty} \frac{1}{4n^2 - 1}$,   *b. $\displaystyle\sum_{n=1}^{\infty} \frac{(-1)^n}{4n^2 - 1}$.

4. a. Find the Fourier series of $f(x) = (\pi - |x|)^2$ on $[-\pi, \pi]$.

   b. On the interval $[-2\pi, 2\pi]$ sketch the graph of the function to which the series in (a) converges.

   c. Use the results of (a) to find the sums of the following series:

   i. $\displaystyle\sum_{n=1}^{\infty} \frac{1}{n^2}$,   ii. $\displaystyle\sum_{n=1}^{\infty} \frac{1}{n^4}$.

5. a. Show that for $-\pi < x < \pi$,

   $$e^x = \frac{e^{\pi} - e^{-\pi}}{2\pi} \left[1 - 2\sum_{n=1}^{\infty} \frac{(-1)^{n+1}}{n^2 + 1}(\cos nx - n\sin nx) \right].$$

   b. Use the results of (a) to find the sums of the following series:

   i. $\displaystyle\sum_{n=1}^{\infty} \frac{(-1)^{n+1}}{n^2 + 1}$,   ii. $\displaystyle\sum_{n=1}^{\infty} \frac{1}{n^2 + 1}$.

6. *a. Find the Fourier cosine series of e^{ax} on $[0, \pi)$.

   *b. On the interval $[-2\pi, 2\pi]$, sketch the graph of the function to which the series in (a) converges.

7. Let f be a continuous function on $[-\pi, \pi]$ with $f(-\pi) = f(\pi)$. If in addition f' is piecewise continuous on $[-\pi, \pi]$, prove that the Fourier series of f converges uniformly to f on $[-\pi, \pi]$.

---

[6] "Une série de Fourier–Lebesgue divergente partout," *Compte Rendus*, 183 (1926), 1327–1328.

[7] "On convergence and growth of partial sums of Fourier Series," *Acta Math.*, 116 (1966), 135–157.

8. Suppose $f \in \mathcal{R}[-\pi, \pi]$, and $x_o \in (-\pi, \pi)$ is such that $f(x_o-)$ and $f(x_o+)$ exist, and that

$$\lim_{t \to 0^+} \frac{f(x_o + t) - f(x_o+)}{t}, \quad \lim_{t \to 0^+} \frac{f(x_o - t) - f(x_o-)}{t}$$

both exist as finite limits. Prove that f satisfies the hypothesis of Dirichlet's theorem (9.5.3).

9. Suppose $f \in \mathcal{R}[-\pi, \pi]$ and $x_o \in (-\pi, \pi)$ is such that $f(x_o-)$ and $f(x_o+)$ both exist. If f satisfies the inequalities (11) at x_o for some α, $0 < \alpha \leq 1$, prove that the Fourier series of f at x_o converges to $\frac{1}{2}[f(x_o+) + f(x_o-)]$.

Notes

There is no doubt about the significance of Fourier's contributions to the areas of mathematical physics and applied mathematics; one only needs to consult a text on partial differential equations. The methods which he developed in connection with the theory of heat conduction are applicable to a large class of physical phenomena, including problems in acoustics, elasticity, optics, and the theory of electrical networks, among others. Fourier's work however is even more significant in that it inaugurated a new area of mathematics.

The study of Fourier series led to the development of the fundamental concepts and methods of what is now called real analysis. The study of the concept of a function by Dirichlet and others was directly linked to their interest in Fourier series. The study of Fourier series by Riemann led to his development of the Riemann integral. He was concerned with the question of finding sufficient conditions for the existence of the integrals which gave the Fourier coefficients of a function f, that is

$$a_n = \frac{1}{\pi} \int_{-\pi}^{\pi} f(x) \cos nx \, dx \quad \text{and} \quad b_n = \frac{1}{\pi} \int_{-\pi}^{\pi} f(x) \sin nx \, dx.$$

The quest for an understanding of what types of functions possessed Fourier series is partly responsible for the development of Lebesgue's theory of integration. The need for a more extensive theory of integration was illustrated by Lebesgue in 1903 in the paper "Sur les series trigonometric."[8] In this paper, he constructed an example of a function that is not Riemann integrable but that is representable everywhere by its Fourier series. Such a function is $f(x) = -\ln|2\sin(x/2)|$ whose Fourier series is

$$\sum_{k=1}^{\infty} \frac{\cos kx}{k}.$$

This series converges everywhere to the function f on $[-\pi, \pi]$, but since f is unbounded, it is not Riemann integrable on $[-\pi, \pi]$. It is however integrable in the sense of Lebesgue. The article by Alan Gluchoff provides an excellent exposition on the influence of trigonometric series to the theories of integration of Cauchy, Riemann, and Lebesgue.

---

[8] *Annales Scientifiques de l'École Normale Supérieure*, (3) 20 (1903), 453–485.

There are several important topics that we have not touched upon in the course of this chapter. One of these is the **Gibbs phenomenon**, named after Josiah Gibbs (1839–1903). To explain this phenomenon, we consider the Fourier series of $f(x) = 0$, $x \in [-\pi, 0)$, and $f(x) = 1$, $x \in [0, \pi)$ of Example 9.3.3(a). By Dirichlet's theorem we have

$$\frac{1}{2} + \frac{2}{\pi} \sum_{k=1}^{\infty} \frac{1}{(2k-1)} \sin(2k-1)x = \begin{cases} 0, & -\pi < x < 0, \\ \frac{1}{2}, & x = 0, \\ 1, & 0 < x < \pi. \end{cases}$$

A careful examination of the graphs (see Figure 9.2) of the partial sums S_3, S_5, and S_{15}, shows that near 0, each of the functions S_i, $i = 3, 5, 15$, has an absolute maximum at a point t_i, where the t_i get closer to 0, but $S_i(t_i)$ is bounded away from 1.

We now consider the behavior of the partial sums S_n more closely. For $n = (2k-1)$, $k \in \mathbb{N}$,

$$S_{2k-1}(x) = \frac{1}{2} + \frac{2}{\pi}\left[\sin x + \frac{1}{3}\sin 3x + \cdots + \frac{1}{(2k-1)}\sin(2k-1)x\right],$$

and

$$S'_{2k-1}(x) = \frac{2}{\pi}\left[\cos x + \cos 3x + \cdots + \cos(2k-1)x\right].$$

If we multiply $S'_{2k-1}(x)$ by $\sin x$ and use the identity

$$\sin x \cos jx = \tfrac{1}{2}[\sin(j+1)x - \sin(j-1)x],$$

we obtain

$$\sin x\, S'_{2k-1}(x) = \frac{1}{\pi}\sin kx.$$

From this it now follows that $S_{2k-1}(x)$ has relative maxima and minima at the points

$$2kx = \pm\pi, \pm 2\pi, ..., \pm 2(k-1)\pi.$$

These points are equally spaced in $(-\pi, \pi)$. Consider the points $x_k = \pi/(2k)$, at which each S_{2k-1} has an (absolute) maximum with

$$S_{2k-1}(x_k) = \frac{1}{2} + \frac{2}{\pi}\left[\sin\frac{\pi}{2k} + \frac{1}{3}\sin\frac{3\pi}{2k} + \cdots + \frac{1}{2k-1}\sin\frac{(2k-1)\pi}{2k}\right].$$

To find $\lim\limits_{k\to\infty} S_{2k-1}(x_k)$, we write the above sum as a Riemann sum of the function $g(x) = (\sin x)/x$. This function is Riemann integrable on $[0, \pi]$. With the partition $\mathcal{P}$ of $[0, \pi]$ given by $y_j = j\frac{\pi}{k}$, $j = 0, 1, 2, ..., k$, and $t_j = \frac{1}{2}(y_{j-1} + y_j)$,

$$\frac{2}{\pi}\left[\sin\frac{\pi}{2k} + \frac{1}{3}\sin\frac{3\pi}{2k} + \cdots + \frac{1}{2k-1}\sin\frac{(2k-1)\pi}{2k}\right] = \frac{1}{\pi}\sum_{j=1}^{k} g(t_j)\Delta y_j.$$

Therefore,

$$\lim_{k\to\infty} S_{2k-1}(x_k) = \frac{1}{2} + \frac{1}{\pi}\int_0^{\pi}\frac{\sin x}{x}dx.$$

To approximate the integral we use the Taylor series expansion of $\sin x$. This gives

$$\frac{\sin x}{x} = 1 - \frac{x^2}{3!} + \frac{x^4}{5!} - \frac{x^6}{7!} + \frac{x^8}{9!} - \cdots, \qquad x \in \mathbb{R}.$$

Therefore,

$$\frac{1}{\pi}\int_0^\pi \frac{\sin x}{x}\,dx = \frac{1}{\pi}\left(\pi - \frac{\pi^3}{3\cdot 3!} + \frac{\pi^5}{5\cdot 5!} - \frac{\pi^7}{7\cdot 7!} + \frac{\pi^9}{9\cdot 9!} - \cdots\right)$$

$$= 1 - \frac{\pi^2}{3\cdot 3!} + \frac{\pi^4}{5\cdot 5!} - \frac{\pi^6}{7\cdot 7!} + \frac{\pi^8}{9\cdot 9!} - \cdots$$

$$\approx 1 - .54831 + .16235 - .02725 + .00291$$

$$\approx .59 \quad \text{(to two decimal places)}.$$

The notation "$\approx$" denotes approximately equal to. Therefore $\lim_{k\to\infty} S_{2k-1}(x_k) \approx 1.09$. Even though $\lim_{n\to\infty} S_n(x) = 1$ for all $x \in (0,\pi)$, if we approach 0 along the points x_k, then $S_{2k-1}(x_k)$ *overshoots* the value 1 by approximately .09, i.e.,

$$\lim_{k\to\infty} |S_{2k-1}(x_k) - f(x_k)| \approx 0.09.$$

If the Fourier series converges uniformly, this cannot happen.

The above behavior, known as the Gibbs phenomenon, is due to the fact that f has a jump discontinuity at $x_0 = 0$, and is typical of the behavior of the Fourier series of a piecewise continuous function at a jump discontinuity. Furthermore, if f and f' are piecewise continuous on $[-\pi,\pi]$, then the amount of overshoot at a discontinuity x_0, due to the Gibbs phenomenon, is approximately equal to $0.09[f(x_0+) - f(x_0-)]$. The article by Shelupsky in the supplemental readings provides a very nice discussion of this phenomenon.

Another important question involves the uniqueness of the representation of a function by a trigonometric series. Specifically, suppose

$$f(x) = \sum_{k=0}^\infty (A_k \cos kx + B_k \sin kx) = \sum_{k=0}^\infty (C_k \cos kx + D_k \sin kx)$$

for all $x \in [-\pi,\pi]$, must $A_k = C_k$ and $B_k = D_k$ for all $k = 0,1,2,...$? Alternately, if

$$\sum_{k=0}^\infty (A_k \cos kx + B_k \sin kx) = 0 \tag{13}$$

for all $x \in [-\pi,\pi]$, must $A_k = B_k = 0$ for all $k = 0,1,2,...$? By (13) we mean that $\lim_{n\to\infty} S_n(x) = 0$ for all $x \in [-\pi,\pi]$, where S_n is the nth partial sum of the series.

In 1870 Eduard Heine proved that if the sequence $\{S_n\}$ converges uniformly to 0 on $[-\pi,\pi]$ then $A_k = B_k = 0$ for all k. This is Theorem 9.3.6. The general uniqueness problem was solved by Cantor in the early 1870's. He was also able to prove uniqueness if (13) holds for all but a finite number of x in $[-\pi,\pi]$. This then led Cantor to consider the uniqueness problem for infinite subsets of $[-\pi,\pi]$; specifically, if $E \subset [-\pi,\pi]$ is infinite and (13) holds for all $x \in [-\pi,\pi] \setminus E$, must $A_k = B_k = 0$ for all k? Since point set theory was undeveloped at this time, this question also led Cantor to devote much of his time and effort to studying point subsets of $\mathbb{R}$. For a thorough discussion of the uniqueness problem the reader is referred to the article by Marshall Ash listed in the supplemental readings or to the text by A. Zygmund listed in the Bibliography.

Micellaneous Exercises

1. Suppose f is continuous on $[-\pi, \pi)$ and $f(x) = \frac{1}{2}A_0 + \sum_{k=1}^{\infty} (A_k \cos kx + B_k \sin kx)$. Let S_n be the nth partial sum of the series. If there exists a positive constant M such that $|S_n(x)| \le M$ for all $x \in [-\pi, \pi]$ and $n \in \mathbb{N}$, prove that A_k and B_k are the Fourier coefficients of f.

2. Prove that $\lim_{n \to \infty} \int_0^{\pi} |D_n(t)| \, dt = \infty$, where D_n is the Dirichlet kernel. (See Example 6.4.4(b).)

3. Let $f(x) = -\ln|2\sin(x/2)|$. Show that the Fourier series of f is given by $\sum_{k=1}^{\infty} (\cos kx)/k$. (Note: Since f is unbounded at $x = 0$, the integrals defining a_k and b_k have to be interpreted as improper integrals.)

Supplemental Reading

Ash, M. J., "Uniqueness of representation by trigonometric series," *Amer. Math. Monthly* **96** (1989), 873–885.

Askey, R. and Haimo, D. T., "Similarities between Fourier series and power series," *Amer. Math. Manthly* **103** (1996), 297–304.

Gluchoff, A. D., "Trigonometric series and theories of integration," *Math. Mag.* **67** (1994), 3–20.

Gonzalez-Velasco, E. A., "Connections in mathematical analysis: The case of Fourier series," *Amer. Math. Monthly* **99** (1992), 427–441.

Halmos, P., "Fourier series," *Amer. Math. Monthly* **85** (1978), 33–34.

Lanczos, C., *Discourse on Fourier Series,* Hafner Publ. Co., New York, NY, 1966.

Shelupsky, D., "Derivation of the Gibbs phenomenen," *Amer. Math. Monthly* **87** (1980), 210–212.

Shepp, L. A. and Kruskal, J. B., "Computerized tomography: The new medical X-ray technology," *Amer. Math. Monthly* **85** (1978), 420–439.

Simon, B., "Uniform convergence of Fourier series," *Amer. Math. Monthly* **67** (1969), 55–56.

Williams, K. S., "Note on $\int_0^{\infty} (\sin x/x) \, dx$," *Math. Mag.* **44** (1971), 9–11.

10

Lebesgue Measure and Integration

The concept of measure plays a very important role in the theory of real analysis. On the real line the idea of measure generalizes the length of an interval, in the plane, the area of a rectangle, and so forth. It allows us to talk about the measure of a set in the same way that we talk about the length of an interval. The development of the Riemann integral of a bounded function on a closed and bounded interval $[a, b]$ depended very much on the fact that we partitioned $[a, b]$ into intervals. The notion of measure and measurable set will play a prominent role in the development of the Lebesgue integral in that we will partition $[a, b]$ not into intervals, but instead into pairwise disjoint measurable sets.

The theory of measure is due to Henri Lebesgue (1875–1941) who in his famous 1902 thesis defined measure of subsets of the line and the plane, and also the Lebesgue integral of a nonnegative function. Like Riemann, Lebesgue was also led to the development of his theory of integration by the problem of finding sufficient conditions on a function f for which the integrals defining the Fourier coefficients of f exist. In this chapter, we will develop the theory of Lebesgue measure of subsets of $\mathbb{R}$ following the original approach of Lebesgue using inner and outer measure. Although this approach is somewhat more tedious than the modern approach due to Constantin Carathéodory (1873–1950), it has the advantage of being more intuitive and conceptually easier to visualize.

In the first section we will illustrate the need for the concept of *measure* of a set and *measurable* function by considering an alternate approach to integration developed by Lebesgue in 1928. Although this ultimately will not be how we define the Lebesgue integral, the approach is instructive in emphasizing the concepts required for the development of the Lebesgue theory of integration. In Section 10.2, we use the fact that every open subset of $\mathbb{R}$ can be expressed as a finite or countable union of disjoint open intervals to define the measure of open sets, and then of compact sets. These are then used to define the inner and outer measure of subsets of $\mathbb{R}$. A bounded subset of $\mathbb{R}$ is then said to be measurable if these two quantities are the same.

In Section 10.6, we develop the theory of the Lebesgue integral of a bounded real-valued function using upper and lower sums as in the development of the Riemann integral. However, rather than using point partitions of the interval, we will use measurable partitions. They key result of the section is that a bounded real-valued function on $[a, b]$ is Lebesgue integrable if and only if it is measurable. As we will see, the class of Lebesgue integrable functions contains the class of Riemann integrable functions, and for Riemann integrable functions, the two integrals coincide. One of the advantages of the Lebesgue theory of integration involves the interchange of limits of sequences of functions and integration. We will prove several very important and useful convergence theorems, including the well known bounded convergence theorem and Lebesgue's dominated convergence theorem.

10.1   Introduction to Measure

In Definition 6.1.11, we defined what it means for a subset of $\mathbb{R}$ to have measure zero. In this chapter, we will consider the concept of *measure* of a set in greater detail. When introducing a new concept it is of course very natural to ask both "why," and "does it lead to something useful?" Both of these questions were answered by Lebesgue in 1903 when he exhibited a trigonometric series that converged everywhere to a nonnegative function f that was not Riemann integrable.[1] The function f however is integrable according to Lebesgue's definition and the trigonometric series is the Fourier series of f.

In this section, we will illustrate why it is necessary to consider the concept of measure of a set by considering an alternate approach to integration. As we will see later in the chapter, this approach leads to greater generality in the types of functions that can be integrated. In addition, Lebesgue's theory of integration also allows us to prove interchange of limit and integration theorems without requiring uniform convergence of the sequence of functions.

Let f be a bounded function on $[a, b]$. In developing the theory of the Riemann integral we partitioned the interval $[a, b]$ and defined the upper and lower sums of f corresponding to the partition. An alternate approach to integration, due to Lebesgue, is to partition the range of the function, rather than the domain. For purposes of illustration suppose f is nonnegative with Range $f \subset [0, \beta)$. Let $n \in \mathbb{N}$, and partition $[0, \beta)$ into n disjoint subintervals

$$\left[(j-1)\frac{\beta}{n}, j\frac{\beta}{n}\right), \qquad j = 1, 2, ..., n.$$

See Figure 10.1 with $n = 8$.

For each $j = 1, 2, ..., n$, we let

$$E_j = \left\{x \in [a, b] : (j-1)\frac{\beta}{n} \leq f(x) < j\frac{\beta}{n}\right\}.$$

---

[1] "Sur les series trigonometric," *Annales Scientifiques de l'École Normale Supérieure,* (3) 20 (1903), 453–485.

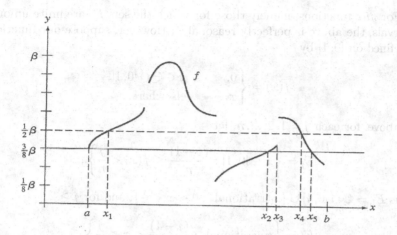

FIGURE 10.1
Partition of the range of f

In Figure 10.1,

$$E_4 = \left\{ x \in [a,b] : 3\frac{\beta}{8} \le f(x) < 4\frac{\beta}{8} \right\} = [a, x_1) \cup [x_2, x_3) \cup [x_4, x_5).$$

For a set such as E_4, which is a finite union of disjoint intervals, it is reasonable to define the measure of E_4, denoted $m(E_4)$, as the sum of the length of the intervals, i.e.,

$$m(E_4) = (x_1 - a) + (x_3 - x_2) + (x_5 - x_4).$$

Assuming that we can do this for each of the sets E_j, a lower approximation to the area under the graph of f is given by

$$\sum_{j=1}^{n} (j-1)\frac{\beta}{n} m(E_j),$$

and an upper approximation would be given by

$$\sum_{j=1}^{n} j\frac{\beta}{n} m(E_j).$$

Taking limits as $n \to \infty$, assuming that they exist and are equal, would in fact provide another approach to integration. That this indeed is the case for a large class of functions, including the Riemann integrable functions, will be proved in Section 10.6.

For *nice* functions, namely those for which the sets E_j are finite unions of intervals, the above is perfectly reasonable. However, suppose our function f is defined on $[0,1]$ by

$$f(x) = \begin{cases} 0, & x \in \mathbb{Q} \cap [0,1], \\ x, & \text{elsewhere.} \end{cases}$$

As above, for each $j = 1, 2, ..., n$, let

$$E_j = \left\{ x \in [0,1] : \frac{(j-1)}{n} \leq f(x) < \frac{j}{n} \right\}.$$

Thus $E_1 = \mathbb{Q} \cap [0,1] \cup \{x \text{ irrational} : 0 < x < \frac{1}{n}\}$, and for $j \geq 2$,

$$E_j = \left\{ x \text{ irrational} : \frac{(j-1)}{n} < x < \frac{j}{n} \right\}.$$

Here the sets E_j are no longer unions of intervals, and so what is meant by the measure of the set is by no means obvious.

Our goal in the next two sections is to define a function λ defined on a family $\mathcal{M}$ of subsets of $\mathbb{R}$, called the *measurable sets*, which contains all the intervals, and has the property that

(a) for an interval J, $\lambda(J) = $ length of J,

(b) $\lambda(E + x) = \lambda(E)$, for all $E \in \mathcal{M}$ and $x \in \mathbb{R}$, where

$$E + x = \{y + x : y \in E\}, \quad \text{and}$$

(c) $\lambda(\bigcup E_k) = \sum \lambda(E_k)$ for any finite or countable family $\{E_k\}$ of pairwise disjoint sets in $\mathcal{M}$.

10.2   Measure of Open Sets: Compact Sets

We begin our discussion of measure theory by first defining the measure of open and compact subsets of $\mathbb{R}$.

DEFINITION 10.2.1 *If J is an interval, we define the **measure** of J, denoted $m(J)$, to be the length of J.*

Thus if J is (a,b), $(a,b]$, $[a,b)$, or $[a,b]$, $a,b \in \mathbb{R}$, then

$$m(J) = b - a.$$

If J is $\mathbb{R}$, (a,∞), $[a,\infty)$, $(-\infty,b)$ or $(-\infty,b]$, we set $m(J) = \infty$. In dealing with the symbols ∞ and $-\infty$, it is customary to adopt the following conventions:

(a) If x is real, then $x + \infty = \infty, \quad x - \infty = -\infty$.

(b) If $x > 0$ then $x \cdot \infty = \infty, \; x \cdot (-\infty) = -\infty$.

(c) If $x < 0$ then $x \cdot \infty = -\infty, \; x \cdot (-\infty) = \infty$.

(d) Also, $\infty + \infty = \infty, \; -\infty - \infty = -\infty, \; \infty \cdot (\pm\infty) = \pm\infty, \; -\infty \cdot (\pm\infty) = \mp\infty$.

The symbols $\infty - \infty$ and $-\infty + \infty$ are undefined, but we shall adopt the arbitrary convention that $0 \cdot \infty = 0$.

Measure of Open Sets

Our first step will be to extend the set function m to the open subsets of $\mathbb{R}$. Since this extension relies on Theorem 2.2.20, we restate that result at this point.

THEOREM 2.2.20 If U is an open subset of $\mathbb{R}$, then there exists a finite or countable collection $\{I_n\}$ of pairwise disjoint open intervals such that

$$U = \bigcup_n I_n.$$

Recall, the family $\{I_n\}$ is **pairwise disjoint** if and only if $I_n \cap I_m = \emptyset$ whenever $n \neq m$.

DEFINITION 10.2.2 *If U is an open subset of $\mathbb{R}$ with $U = \bigcup_n I_n$, where $\{I_n\}$ is a finite or countable collection of pairwise disjoint open intervals, we define the* **measure** *of U, denoted $m(U)$, by*

$$m(U) = \sum_n m(I_n).$$

Remarks. (a) For the empty set $\emptyset$, we set $m(\emptyset) = 0$.

(b) The sum defining $m(U)$ may be either finite or infinite. If any of the intervals are of infinite length, then $m(U) = \infty$. On the other hand, if

$$U = \bigcup_{n=1}^{\infty} I_n,$$

where the I_n are pairwise disjoint bounded open intervals, we may still have

$$m(U) = \sum_{n=1}^{\infty} m(I_n) = \infty,$$

due to the divergence of the series to ∞. Since $m(I_n) \geq 0$ for all n, the sequence of partial sums is monotone increasing and thus will either converge to a real number or diverge to ∞.

EXAMPLES 10.2.3 (a) For each $n = 1, 2, ...,$ set

$$I_n = \left(n - \frac{1}{2^n}, n + \frac{1}{2^n}\right).$$

Then $I_1 = (1-\frac{1}{2}, 1+\frac{1}{2})$, $I_2 = (2-\frac{1}{4}, 2+\frac{1}{4})$, etc. Since $n+2^{-n} < (n+1)-2^{-(n+1)}$ for all $n \in \mathbb{N}$, the collection $\{I_n\}_{n=1}^{\infty}$ is pairwise disjoint. Let $U = \bigcup_{n=1}^{\infty} I_n$. Then

$$m(U) = \sum_{n=1}^{\infty} m(I_n) = \sum_{n=1}^{\infty} 2\frac{1}{2^n} = \sum_{n=0}^{\infty} \frac{1}{2^n} = \frac{1}{1-\frac{1}{2}} = 2.$$

The set U is an example of an unbounded set with finite measure.

(b) Let $J_n = (n, n+1/n)$, $n = 1, 2, ...,$ and let $V = \bigcup_{n=1}^{\infty} J_n$. Then

$$m(V) = \sum_{n=1}^{\infty} m(J_n) = \sum_{n=1}^{\infty} \frac{1}{n} = \infty.$$

(c) As in Section 2.5, let U denote the complement of the Cantor set in $[0, 1]$. Since U is the union of the open intervals that have been removed, by Property 4 of the Cantor set,

$$m(U) = 1. \quad \square$$

We now state and prove several results concerning the measure of open sets.

THEOREM 10.2.4 *If U and V are open subsets of $\mathbb{R}$ with $U \subset V$, then*

$$m(U) \leq m(V).$$

Proof. The statement of the theorem appears to be so obvious that no proof seems to be required. However, it is important to keep in mind how the measure of an open set is defined. Suppose

$$U = \bigcup_n I_n \quad \text{and} \quad V = \bigcup_m J_m,$$

where $\{I_n\}_n$ and $\{J_m\}_m$ are finite or countable collections of pairwise disjoint open intervals. Since $U \subset V$, each interval $I_n \subset J_m$ for some m. For each m, let

$$N_m = \{n : I_n \subset J_m\}.$$

Since the collection $\{J_m\}_m$ is pairwise disjoint, so is the collection $\{N_m\}_m$, and

$$U = \bigcup_n I_n = \bigcup_m \bigcup_{n \in N_m} I_n.$$

Therefore

$$m(U) = \sum_m \sum_{n \in N_m} m(I_n).$$

But by Exercise 1,

$$\sum_{n \in N_m} m(I_n) \le m(J_m),$$

from which the result follows. $\square$

Remark. As a consequence of the previous theorem, if U is an open subset of (a, b), $a, b \in \mathbb{R}$, then

$$m(U) \le b - a.$$

Thus every bounded open set has finite measure.

THEOREM 10.2.5 *If U is an open subset of $\mathbb{R}$, then*

$$m(U) = \lim_{k \to \infty} m(U_k),$$

where for each $k \in \mathbb{N}$, $U_k = U \cap (-k, k)$.

Proof. For each k, U_k is open, with

$$U_k \subset U_{k+1} \subset U$$

for all $k \in \mathbb{N}$. By Theorem 10.2.4, the sequence $\{m(U_k)\}$ is monotone increasing with $m(U_k) \le m(U)$ for all k. Therefore

$$\lim_{k \to \infty} m(U_k) \le m(U). \tag{1}$$

If U is bounded, then there exists $k_o \in \mathbb{N}$ such that

$$U \cap (-k, k) = U$$

for all $k \ge k_o$. Hence $m(U_k) = m(U)$ for all $k \ge k_o$, and thus equality holds in (1).

Suppose that U is an unbounded open subset of $\mathbb{R}$ with

$$U = \bigcup_n I_n,$$

where $\{I_n\}$ is a finite or countable collection of pairwise disjoint open intervals. If $m(I_n) = \infty$ for some n, then $m(U) = \infty$, and for that n, either $I_n = \mathbb{R}$ or I_n is an interval of the form

$$(-\infty, a_n) \quad \text{or} \quad (a_n, \infty)$$

for some $a_n \in \mathbb{R}$. Suppose $I_n = (a_n, \infty)$. Choose $k_o \in \mathbb{N}$ such that $k_o \ge |a_n|$. Then for all $k \ge k_o$,

$$I_n \cap (-k, k) = (a_n, k),$$

and thus

$$\infty = \lim_{k\to\infty} m(I_n \cap (-k,k)) \leq \lim_{k\to\infty} m(U_k) \leq m(U).$$

Therefore equality holds in (1). The other two cases follow similarly.

Suppose $m(I_n) < \infty$ for all n. Since U is unbounded, the collection $\{I_n\}$ must be infinite. If the collection were finite, then since each interval has finite length, each interval is bounded, and as a consequence U must also be bounded. Let $\alpha \in \mathbb{R}$ with $\alpha < m(U)$. Since

$$\sum_{n=1}^{\infty} m(I_n) = m(U) > \alpha,$$

there exists a positive integer N such that

$$\sum_{n=1}^{N} m(I_n) > \alpha.$$

Let $V = \bigcup_{n=1}^{N} I_n$. Then V is a bounded open set, and thus by the above,

$$m(V) = \lim_{k\to\infty} m(V \cap (-k,k)).$$

Since $m(V) > \alpha$, there exists $k_o \in \mathbb{N}$ such that

$$m(V \cap (-k,k)) > \alpha \qquad \text{for all } k \geq k_o.$$

But $V \cap (-k,k) \subset U_k$ for all $k \in \mathbb{N}$. Hence by Theorem 10.2.4, $m(V \cap (-k,k)) \leq m(U_k)$ and as a consequence

$$m(U_k) > \alpha \qquad \text{for all } k \geq k_o.$$

If $m(U) = \infty$, then since $\alpha < m(U)$ was arbitrary, we have $m(U_k) \to \infty$ as $k \to \infty$. If $m(U) < \infty$, then given $\epsilon > 0$, take $\alpha = m(U) - \epsilon$. By the above, there exists $k_o \in \mathbb{N}$ such that

$$m(U) - \epsilon < m(U_k) \leq m(U) \qquad \text{for all } k \geq k_o.$$

Therefore, $\lim_{k\to\infty} m(U_k) = m(U)$. $\square$

Remark. In proving results about open sets, the previous theorem allows us to first prove the result for the case where U is a bounded open set, and then to use the limit process to extend the result to the unbounded case.

Our next goal is to prove the following:

THEOREM 10.2.6 *If $\{U_n\}_n$ is a finite or countable collection of open subsets of $\mathbb{R}$, then*

$$m\left(\bigcup_n U_n\right) \leq \sum_n m(U_n).$$

For the proof of the theorem we require the following lemma.

LEMMA 10.2.7 *If* $\{I_n\}_{n=1}^N$ *is a finite collection of bounded open intervals, then*

$$m\left(\bigcup_{n=1}^N I_n\right) \leq \sum_{n=1}^N m(I_n).$$

It should be noted that the collection $\{I_n\}$ is not assumed to be pairwise disjoint. The lemma is most easily proved by resorting to the theory of the Riemann integral.

DEFINITION 10.2.8 *If* E *is a subset of* $\mathbb{R}$, *the* **characteristic function** *of* E, *denoted* χ_E, *is the function defined by*

$$\chi_E(x) = \begin{cases} 1, & x \in E, \\ 0, & x \notin E. \end{cases}$$

Suppose I is a bounded open interval. Choose $a, b \in \mathbb{R}$ such that $I \subset [a, b]$. Since χ_I is continuous on $[a, b]$ except at the two endpoints of I, $\chi_I \in \mathcal{R}[a, b]$ with

$$\int_a^b \chi_I(x)\, dx = m(I).$$

It should be clear that this is independent of the interval $[a, b]$ containing I. If U is an open subset of $[a, b]$ with

$$U = \bigcup_{k=1}^m J_k, \qquad m \leq N,$$

where the $\{J_k\}$ are pairwise disjoint open intervals, then

$$\chi_U(x) = \sum_{k=1}^m \chi_{J_k}(x),$$

and thus

$$m(U) = \sum_{k=1}^m m(J_k) = \sum_{k=1}^m \int_a^b \chi_{J_k}(x)\, dx = \int_a^b \chi_U(x)\, dx. \qquad (2)$$

Proof of Lemma 10.2.7 Let $\{I_n\}_{n=1}^N$ be a finite collection of bounded open intervals and let $U = \bigcup_{n=1}^N I_n$. Choose $a, b \in \mathbb{R}$ such that $U \subset [a, b]$. Then χ_U is continuous except at a finite number of points, and

$$\chi_U(x) \leq \sum_{n=1}^N \chi_{I_n}(x).$$

Therefore by (2) and the above,

$$m(U) = \int_a^b \chi_U(x)\,dx \le \int_a^b \sum_{n=1}^N \chi_{I_n}(x)\,dx = \sum_{n=1}^N \int_a^b \chi_{I_n}(x)\,dx = \sum_{n=1}^N m(I_n). \quad \square$$

Proof of Theorem 10.2.6 Since the result for a finite collection follows obviously from that of a countable collection, we suppose $\{U_n\}_{n=1}^\infty$ is a countable collection of open sets and

$$U = \bigcup_{n=1}^\infty U_n.$$

Since U is open, $U = \bigcup_m J_m$ where $\{J_m\}_m$ is a finite or countable collection of pairwise disjoint open intervals. Also, for each n,

$$U_n = \bigcup_k I_{n,k},$$

where for each n, $\{I_{n,k}\}_k$ is a finite or countable collection of pairwise disjoint open intervals.

We assume first that U is bounded with $m(U) < \infty$. Let $\epsilon > 0$ be given. If the collection $\{J_m\}_m$ is infinite, then since $\sum_m m(J_m) < \infty$, there exists a positive integer N such that

$$\sum_{m=N+1}^\infty m(J_m) < \epsilon.$$

Thus

$$m(U) < \sum_{m=1}^N m(J_m) + \epsilon.$$

If the collection $\{J_m\}_m$ is finite, the previous step is not necessary.

Consider the collection $\{J_m\}_{m=1}^N$. For each $m = 1, 2, ..., N$, let K_m be an open interval such that

$$\overline{K}_m \subset J_m \quad \text{and} \quad m(J_m) < m(K_m) + \frac{\epsilon}{N}.$$

Then

$$m(U) < \sum_{m=1}^N m(J_m) + \epsilon < \sum_{m=1}^N m(K_m) + 2\epsilon. \tag{3}$$

Let $A = \bigcup_{m=1}^N \overline{K}_m$. Since each $\overline{K}_m$ is closed and bounded, so is the set A. Thus A is compact, and since

$$A \subset U = \bigcup_{n=1}^\infty \left(\bigcup_k I_{n,k} \right),$$

the collection $\{I_{n,k}\}$ is an open cover of A. Hence by compactness there exists a finite number $I_{n_i,k_{i_j}}$, $i = 1, ..., J$, $j = 1, ..., m_i$, such that

$$A \subset \bigcup_{i,j} I_{n_i,k_{i_j}}.$$

Since the intervals $\{K_m\}$ are pairwise disjoint

$$\sum_{m=1}^{N} m(K_m) = m\left(\bigcup_{m=1}^{N} K_m\right) \leq m\left(\bigcup_{i,j} I_{n_i,k_{i_j}}\right),$$

which by Lemma 10.4

$$\leq \sum_{i,j} m(I_{n_i,k_{i_j}}) = \sum_{i=1}^{J} \sum_{j=1}^{m_i} m(I_{n_i,k_{i_j}})$$

$$\leq \sum_{i=1}^{J} m(U_{n_i}) \leq \sum_{n=1}^{\infty} m(U_n).$$

Combining this with inequality (3) gives

$$m(U) < \sum_{n=1}^{\infty} m(U_n) + 2\epsilon.$$

Since $\epsilon > 0$ was arbitrary, the result follows for the case where U is bounded.

If U is unbounded, then for each $k \in \mathbb{N}$,

$$m(U \cap (-k, k)) \leq \sum_{n=1}^{\infty} m(U_n \cap (-k, k)) \leq \sum_{n=1}^{\infty} m(U_n).$$

The result now follows by Theorem 10.2.5. $\square$

THEOREM 10.2.9 *If U and V are open subsets of $\mathbb{R}$, then*

$$m(U) + m(V) = m(U \cup V) + m(U \cap V).$$

Proof. (a) If both U and V are unions of a finite number of bounded open intervals, then so are $U \cap V$ and $U \cup V$. Thus the functions

$$\chi_U, \quad \chi_V, \quad \chi_{U \cup V}, \quad \chi_{U \cap V}$$

are all Riemann integrable on some interval $[a, b]$ with

$$\chi_U(x) + \chi_V(x) = \chi_{U \cup V}(x) + \chi_{U \cap V}(x)$$

for all $x \in [a, b]$. Therefore by identity (2),

$$m(U) + m(V) = m(U \cup V) + m(U \cap V).$$

(b) For the general case, suppose

$$U = \bigcup_{n=1}^{\infty} I_n \quad \text{and} \quad V = \bigcup_{n=1}^{\infty} J_n,$$

where the collections $\{I_n\}$ and $\{J_n\}$ consist of pairwise disjoint open intervals respectively. If one of $m(U)$ or $m(V)$ is ∞, then by Theorem 10.2.4, $m(U \cup V) = \infty$ and the conclusion holds. Thus we can assume that both $m(U)$ and $m(V)$ are finite.

Let $\epsilon > 0$ be given. Choose $N \in \mathbb{N}$ such that

$$\sum_{n=N+1}^{\infty} m(I_n) < \epsilon \quad \text{and} \quad \sum_{n=N+1}^{\infty} m(J_n) < \epsilon.$$

Let U^* and U^{**} be defined by

$$U^* = \bigcup_{n=1}^{N} I_n, \qquad U^{**} = \bigcup_{n=N+1}^{\infty} I_n.$$

Also let V^* and V^{**} be defined analogously. Then

$$m(U) = m(U^*) + m(U^{**}) \quad \text{and} \quad m(V) = m(V^*) + m(V^{**}).$$

Since $m(U^{**}) < \epsilon$ and $m(V^{**}) < \epsilon$,

$$m(U) + m(V) < m(U^*) + m(V^*) + 2\epsilon.$$

Since the sets U^* and V^* are finite unions of open intervals, by part (a)

$$m(U^*) + m(V^*) = m(U^* \cup V^*) + m(U^* \cap V^*),$$

which by Theorem 10.2.4

$$\leq m(U \cup V) + m(U \cap V).$$

The last inequality follows since both $U^* \cup V^*$ and $U^* \cap V^*$ are subsets of $U \cup V$ and $U \cap V$ respectively. Hence

$$m(U) + m(V) < m(U \cup V) + m(U \cap V) + 2\epsilon.$$

Since $\epsilon > 0$ was arbitrary, we have

$$m(U) + m(V) \leq m(U \cup V) + m(U \cap V).$$

We now proceed to prove the reverse inequality. We first note that

$$U \cup V = (U^* \cup V^*) \cup (U^{**} \cup V^{**}),$$

and as a consequence

$$m(U \cup V) \le m(U^* \cup V^*) + 2\epsilon.$$

Also by the distributive law,

$$\begin{aligned}
U \cap V &= (U^* \cup U^{**}) \cap (V^* \cup V^{**}) \\
&= (U^* \cap V^*) \cup (U^{**} \cap V^*) \cup (U^* \cap V^{**}) \cup (U^{**} \cap V^{**}) \\
&\subset (U^* \cap V^*) \cup U^{**} \cup V^{**}.
\end{aligned}$$

Therefore,

$$m(U \cap V) < m(U^* \cap V^*) + 2\epsilon.$$

Combining the above gives

$$m(U \cup V) + m(U \cap V) < m(U^* \cup V^*) + m(U^* \cap V^*) + 4\epsilon,$$

which since U^* and V^* are finite unions of intervals

$$\begin{aligned}
&= m(U^*) + m(V^*) + 4\epsilon \\
&\le m(U) + m(V) + 4\epsilon.
\end{aligned}$$

Again since $\epsilon > 0$ was arbitrary, this proves the reverse inequality. $\square$

Measure of Compact Sets

We now define the measure of a compact subset of $\mathbb{R}$. If K is a compact subset of $\mathbb{R}$ and U is any bounded open set containing K, then

$$U = K \cup (U \setminus K).$$

Using the fact that $U \setminus K$ is also open and bounded, and thus has finite measure, we define the measure of K as follows:

DEFINITION 10.2.10 *Let K be a compact subset of $\mathbb{R}$. The* **measure** *of K, denoted $m(K)$, is defined to be*

$$m(K) = m(U) - m(U \setminus K),$$

where U is any bounded open subset of $\mathbb{R}$ containing K.

We first show that the definition of $m(K)$ is independent of the choice of U.

THEOREM 10.2.11 *If K is compact, then $m(K)$ is well defined.*

Proof. Suppose U and V are any two bounded open sets containing K. Then by Theorem 10.2.9

$$m(U) + m(V \setminus K) = m(U \cup (V \setminus K)) + m(U \cap (V \setminus K))$$
$$= m(U \cup V) + m((U \cap V) \setminus K).$$

In the above we have used the fact that

$$U \cup (V \setminus K) = U \cup V \quad \text{and} \quad U \cap (V \setminus K) = (U \cap V) \setminus K.$$

Similarly

$$m(U \setminus K) + m(V) = m(U \cup V) + m((U \cap V) \setminus K).$$

Therefore

$$m(U) + m(V \setminus K) = m(V) + m(U \setminus K).$$

Since all the terms are finite,

$$m(U) - m(U \setminus K) = m(V) - m(V \setminus K).$$

Thus the definition of $m(K)$ is independent of the choice of U; i.e., $m(K)$ is well defined. $\square$

EXAMPLES 10.2.12 (a) In our first example we show that for a closed and bounded interval $[a, b]$, $a, b \in \mathbb{R}$, Definition 10.2.10 is consistent with Definition 10.2.1; namely

$$m([a, b]) = b - a.$$

Let $U = (a - \epsilon, b + \epsilon)$, $\epsilon > 0$. Then

$$U \setminus [a, b] = (a - \epsilon, a) \cup (b, b + \epsilon).$$

Therefore,

$$m([a, b]) = m(U) - m(U \setminus [a, b])$$
$$= (b - a) + 2\epsilon - 2\epsilon = b - a.$$

(b) If $K = \{x_1, ..., x_n\}$ with $x_i \in \mathbb{R}$, then $m(K) = 0$. Choose $\delta > 0$ such that the intervals

$$I_j = (x_j - \delta, x_j + \delta), \qquad j = 1, 2, ..., n,$$

are pairwise disjoint, and let $U = \bigcup_{j=1}^n I_j$. Then

$$U \setminus K = \bigcup_{j=1}^n [(x_j - \delta, x_j) \cup (x_j, x_j + \delta)].$$

Thus $m(U) = m(U \setminus K)$; i.e., $m(K) = 0$. $\square$

THEOREM 10.2.13

(a) *If K is compact and U is open with $K \subset U$, then $m(K) \leq m(U)$.*

(b) *If K_1 and K_2 are compact with $K_1 \subset K_2$, then $m(K_1) \leq m(K_2)$.*

Proof. The proof of (a) is an immediate consequence of the definition. For the proof of (b), if U is a bounded open set containing K_2, then since $U \setminus K_2 \subset U \setminus K_1$,

$$m(K_1) = m(U) - m(U \setminus K_1) \leq m(U) - m(U \setminus K_2) = m(K_2) \quad \square$$

If I is an open interval and $a, b \in \mathbb{R}$, then $I \cap [a, b]$ is an interval and thus $m(I \cap [a, b])$ is defined with

$$m(I \cap [a, b]) = m(I \cap (a, b)).$$

We extend this to open subsets of $\mathbb{R}$ as follows:

DEFINITION 10.2.14 *If U is an open subset of $\mathbb{R}$ with $U = \bigcup_n I_n$ where $\{I_n\}_n$ is a finite or countable collection of pairwise disjoint open intervals, and $a, b \in \mathbb{R}$, we define*

$$m(U \cap [a, b]) = \sum_n m(I_n \cap [a, b]).$$

Since $m(I_n \cap [a, b]) = m(I_n \cap (a, b))$ for all n, we have

$$m(U \cap [a, b]) = m(U \cap (a, b)).$$

Remark. What the above definition really defines is the measure of relatively open subsets of $[a, b]$ (see Definition 2.2.21). By Theorem 2.2.23, a subset G of $[a, b]$ is open in $[a, b]$ if and only if there exists an open subset U of $\mathbb{R}$ such that $G = U \cap [a, b]$. Since the set U may not be unique, we leave it as an exercise (Exercise 4) to show that if U, V are open subsets of $\mathbb{R}$ with

$$U \cap [a, b] = V \cap [a, b],$$

then $m(U \cap [a, b]) = m(V \cap [a, b])$.

THEOREM 10.2.15 *If U is an open subset of $\mathbb{R}$ and $a, b \in \mathbb{R}$, then*

$$m(U \cap [a, b]) + m(U^c \cap [a, b]) = b - a.$$

Recall from Chapter 1 that $U^c = \mathbb{R} \setminus U = \{x \in \mathbb{R} : x \notin U\}$. If U is open, then U^c is closed and thus $U^c \cap [a, b]$ is a compact subset of $[a, b]$.

Proof. Suppose $V \supset [a, b]$ is open. Let $K = U^c \cap [a, b]$. Then since $V \supset K$,

$$m(K) = m(V) - m(V \setminus K).$$

But
$$V \setminus K = V \cap (U^c \cap [a,b])^c = (V \cap U) \cup (V \cap [a,b]^c) \supset V \cap U.$$
Therefore, $m(V \cap U) \le m(V \setminus K)$. Since $U \cap [a,b] \subset U \cap V$,
$$m(U \cap [a,b]) + m(K) \le m(U \cap V) + m(V) - m(V \setminus K) \le m(V).$$
Given $\epsilon > 0$, take $V = (a - \epsilon, b + \epsilon)$. Then
$$m(U \cap [a,b]) + m(U^c \cap [a,b]) \le b - a + 2\epsilon.$$
Since $\epsilon > 0$ is arbitrary, this proves
$$m(U \cap [a,b]) + m(U^c \cap [a,b]) \le b - a.$$

To prove the reverse inequality, let $I_\epsilon = [a+\epsilon, b-\epsilon]$, where $0 < \epsilon < \frac{1}{2}(b-a)$. Then
$$m(U \cap [a,b]) + m(U^c \cap [a,b]) \ge m(U \cap (a,b)) + m(U^c \cap I_\epsilon).$$
Since (a,b) is an open set containing $U^c \cap I_\epsilon$,
$$m(U^c \cap I_\epsilon) = b - a - m((a,b) \setminus (U^c \cap I_\epsilon)).$$
But
$$m((a,b) \setminus (U^c \cap I_\epsilon)) = m(((a,b) \cap U) \cup ((a,b) \cap I_\epsilon^c))$$
$$= m(((a,b) \cap U) \cup (a, a - \epsilon) + (b - \epsilon, b)),$$
which by Theorem 10.2.6
$$\le m(U \cap (a,b)) + 2\epsilon.$$
Therefore
$$m(U \cap [a,b]) + m(U^c \cap [a,b]) \ge b - a - 2\epsilon.$$
Since $\epsilon > 0$ was arbitrary, the reverse inequality follows. $\square$

Exercises 10.2

1. If $\{I_n\}_n$ is a finite or countable collection of disjoint open intervals with $\bigcup_n I_n \subset (a,b)$, prove that $\sum_n m(I_n) \le m((a,b))$.

2. *If $U \ne \emptyset$ is an open subset of $\mathbb{R}$, prove that $m(U) > 0$.

3. *Let P denote the Cantor set in $[0,1]$. Prove that $m(P) = 0$.

4. Suppose U, V are open subsets of $\mathbb{R}$, $a,b \in \mathbb{R}$ with $U \cap [a,b] = V \cap [a,b]$. Prove that $m(U \cap [a,b]) = m(V \cap [a,b])$.

5. If A, B are subsets of $\mathbb{R}$, prove that
   $\chi_{A \cap B}(x) = \chi_A(x)\chi_B(x),$
   $\chi_{A \cup B}(x) = \chi_A(x) + \chi_B(x) - \chi_{A \cap B}(x),$
   $\chi_{A^c}(x) = 1 - \chi_A(x).$

6. *If K_1 and K_2 are disjoint compact subsets of $\mathbb{R}$, prove that
   $$m(K_1 \cup K_2) = m(K_1) + m(K_2).$$

10.3 Inner and Outer Measure: Measurable Sets

Our goal in this section is to define a function λ on a large family $\mathcal{M}$ of subsets of $\mathbb{R}$, called the measurable sets, which agrees with the function m on the open and compact subsets of $\mathbb{R}$. We begin with the definition of inner and outer measure of a set.

DEFINITION 10.3.1 *Let E be a subset of $\mathbb{R}$. The* **outer measure** *of E, denoted $\lambda^*(E)$, is defined by*

$$\lambda^*(E) = \inf\{m(U) : U \text{ is open with } E \subset U\}.$$

The **inner measure** *of E, denoted $\lambda_*(E)$, is defined by*

$$\lambda_*(E) = \sup\{m(K) : K \text{ is compact with } K \subset E\}.$$

THEOREM 10.3.2

(a) *For any subset E of $\mathbb{R}$,*

$$0 \leq \lambda_*(E) \leq \lambda^*(E).$$

(b) *If E_1 and E_2 are subsets of $\mathbb{R}$ with $E_1 \subset E_2$, then*

$$\lambda_*(E_1) \leq \lambda_*(E_2) \qquad and \qquad \lambda^*(E_1) \leq \lambda^*(E_2).$$

Proof. (a) If K is compact and U is open with $K \subset E \subset U$, then

$$0 \leq m(K) \leq m(U).$$

If we fix K, then $m(K) \leq m(U)$ for all open sets U containing E. Taking the infimum over all such U gives

$$0 \leq m(K) \leq \lambda^*(E).$$

Taking the supremum over all compact subsets K of E proves (a). The proof of (b) is similar and is left as an exercise (Exercise 7). $\square$

EXAMPLES 10.3.3 (a) If E is any countable subset of $\mathbb{R}$, then

$$\lambda_*(E) = \lambda^*(E) = 0.$$

Suppose $E = \{x_n\}_{n=1}^{\infty}$. Let $\epsilon > 0$ be arbitrary. For each n, let

$$I_n = \left(x_n - \frac{\epsilon}{2^n}, x_n + \frac{\epsilon}{2^n}\right),$$

and set $U = \bigcup_{n=1}^{\infty} I_n$. Then U is open with $E \subset U$. By Theorem 10.2.6,

$$m(U) \leq \sum_{n=1}^{\infty} m(I_n) = \sum_{n=1}^{\infty} \frac{\epsilon}{2^{n-1}} = 2\epsilon.$$

Therefore, $\lambda^*(E) < 2\epsilon$. Since $\epsilon > 0$ was arbitrary, $\lambda^*(E) = 0$. As a consequence, we also have $\lambda_*(E) = 0$.

(b) If I is any bounded interval, then

$$\lambda_*(I) = \lambda^*(I) = m(I).$$

Suppose $I = (a, b)$ with $a, b \in \mathbb{R}$. Since I itself is open,

$$\lambda^*(I) \leq m(I) = b - a.$$

On the other hand, if $0 < \epsilon < (b - a)$, then $[a + \epsilon/2, b - \epsilon/2]$ is a compact subset of I, and as a consequence

$$b - a - \epsilon = m\left(\left[a + \frac{\epsilon}{2}, b - \frac{\epsilon}{2}\right]\right) \leq \lambda_*(I).$$

Therefore,

$$b - a - \epsilon \leq \lambda_*(I) \leq \lambda^*(I) \leq b - a.$$

Since $\epsilon > 0$ was arbitrary, equality holds. A similar argument proves that if I is any closed and bounded interval, then

$$\lambda_*(I) = \lambda^*(I) = m(I).$$

As a consequence of Theorem 10.3.2(b), the result holds for any bounded interval I.

(c) For any open set U,

$$\lambda_*(U) = \lambda^*(U) = m(U).$$

By definition, $\lambda^*(U) = m(U)$. But $m(U) = \sum_N m(I_n)$, where $\{I_n\}$ is a pairwise disjoint collection of open intervals with $U = \bigcup_n I_n$. Suppose $\alpha \in \mathbb{R}$ satisfies $\alpha < \lambda^*(U)$. Since $m(U) > \alpha$, there exists a finite number of intervals $I_1, \ldots, I_N$ such that $\sum_{j=1}^{N} m(I_j) > \alpha$. For each j, choose a closed and bounded interval $J_j \subset I_j$ such that $\sum_{j=1}^{N} m(J_j) > \alpha$. Let $K = \bigcup_{j=1}^{N} J_j$. Then K is a compact subset of U and thus $\lambda_*(U) < m(K)$ Finally, since the intervals $\{J_j\}_{j=1}^{N}$ are pairwise disjoint, by Exercise 6 of Section 10.2,

$$m(K) = \sum_{j=1}^{N} m(J_j) > \alpha.$$

Therefore $\lambda_*(U) > \alpha$. If $\lambda^*(U) = \infty$, then by the above $\lambda_*(U) > \alpha$ for every $\alpha \in \mathbb{R}$; that is, $\lambda_*(U) = \infty$. On the other hand, if $\lambda^*(U)$ is finite, take $\alpha = \lambda^*(U) - \epsilon$, where $\epsilon > 0$ is arbitrary. But then $\lambda^*(U) \geq \lambda_*(U) > \lambda^*(U) - \epsilon$ for every $\epsilon > 0$. From this it now follows that $\lambda_*(U) = \lambda^*(U) = m(U)$. $\square$

Measurable Sets

In both of the previous examples, the inner and outer measure of the sets are equal. As we shall see, all subsets of $\mathbb{R}$ build out of open sets or closed sets by countable unions, intersections, and complementation will have this property, and this includes most sets encountered in practice. In fact, the explicit construction of a set whose inner and outer measure are different requires use of an axiom from set theory, the *Axiom of Choice*, which we have not discussed. The construction of such a set is outlined in the miscellaneous exercises.

DEFINITION 10.3.4

(a) *A bounded subset E of $\mathbb{R}$ is said to be* **Lebesgue measurable** *or* **measurable** *if*

$$\lambda_*(E) = \lambda^*(E).$$

If this is the case, then the **measure** *of E, denoted $\lambda(E)$, is defined to be*

$$\lambda(E) = \lambda_*(E) = \lambda^*(E).$$

(b) *An unbounded set E is* **measurable** *if $E \cap [a,b]$ is measurable for every closed and bounded interval $[a, b]$. If this is the case, we define*

$$\lambda(E) = \lim_{k \to \infty} \lambda(E \cap [-k, k]).$$

Remarks. (a) As a consequence of Example 10.3.3(c) every open set U is measurable with

$$\lambda(U) = m(U)$$

(b) If E is unbounded and $E \cap I$ is measurable for every closed and bounded interval I, then by Theorem 10.3.2 the sequence $\{\lambda(E \cap [-k, k])\}_{k=1}^{\infty}$ is non-decreasing, and as a consequence

$$\lambda(E) = \lim_{k \to \infty} \lambda(E \cap [-k, k])$$

exists.

(c) There is no discrepancy between the two parts of the definition. We will shortly prove in Theorem 10.4.1 that if E is a bounded measurable set, then $E \cap I$ is measurable for every interval I. Conversely, if E is a bounded set for which

$$\lambda_*(E \cap [a, b]) = \lambda^*(E \cap [a, b])$$

for all $a, b \in \mathbb{R}$, then by choosing a and b sufficiently large such that $E \subset [a, b]$, we have $\lambda_*(E) = \lambda^*(E)$. The two separate definitions are required due to the existence of unbounded nonmeasurable sets E for which

$$\lambda_*(E) = \lambda^*(E) = \infty.$$

An example of such a set will be given in Exercise 5 of Section 10.4.

THEOREM 10.3.5 *Every set E of outer measure zero is measurable with* $\lambda(E) = 0$.

Proof. Suppose $E \subset \mathbb{R}$ with $\lambda^*(E) = 0$. Then for any closed and bounded interval I,
$$\lambda^*(E \cap I) \leq \lambda^*(E) = 0$$
Thus $\lambda_*(E \cap I) = \lambda^*(E \cap I) = 0$, and hence $E \cap I$ is measurable for every closed and bounded interval I. Since $\lambda(E \cap [-k, k]) = 0$ for every $k \in \mathbb{N}$, $\lambda(E) = 0$. $\square$

As a consequence of the previous theorem and Example 10.3.3(a), every countable set E is measurable with $\lambda(E) = 0$. In particular, $\mathbb{Q}$ is measurable with $\lambda(\mathbb{Q}) = 0$. Another consequence of Theorem 10.3.5 is that every subset of a set of measure zero is measurable.

THEOREM 10.3.6 *Every interval I is measurable with* $\lambda(I) = m(I)$.

Proof. By Example 10.3.3(b), if I is a bounded interval, $\lambda_*(I) = \lambda^*(I) = m(I)$. Thus I is measurable with $\lambda(I) = m(I)$, On the other hand, if I is unbounded, then $I \cap [a, b]$ is a bounded interval for every $a, b \in \mathbb{R}$, and thus measurable. In this case,

$$\lambda(I) = \lim_{k \to \infty} \lambda(I \cap [-k, k]) = \lim_{k \to \infty} m(I \cap [-k, k]) = \infty. \quad \square$$

THEOREM 10.3.7 *For any $a, b \in \mathbb{R}$ and $E \subset \mathbb{R}$,*

$$\lambda^*(E \cap [a, b]) + \lambda_*(E^c \cap [a, b]) = b - a.$$

Proof. Let U be any open subset of $\mathbb{R}$ with $E \cap [a, b] \subset U$. Then $U^c \cap [a, b]$ is compact with $U^c \cap [a, b] \subset E^c \cap [a, b]$. Therefore,

$$m(U) + \lambda_*(E^c \cap [a, b]) \geq m(U \cap [a, b]) + m(U^c \cap [a, b]) = b - a.$$

The last equality follows by Theorem 10.2.15. Taking the infimum over all open sets U containing $E \cap [a, b]$ gives

$$\lambda^*(E \cap [a, b]) + \lambda_*(E^c \cap [a, b]) \geq b - a.$$

To prove the reverse inequality, let K be a compact subset of $E^c \cap [a, b]$. Then K^c is open with $K^c \cap [a, b] \supset E \cap [a, b]$. Therefore

$$\lambda^*(E \cap [a, b]) + m(K \cap [a, b]) \leq m(K^c \cap [a, b]) + m(K \cap [a, b]) = b - a.$$

The last equality again follows by Theorem 10.2.15 with $U = K^c$. Thus taking the supremum over all compact subsets K of $E \cap [a, b]$ gives

$$\lambda^*(E \cap [a, b]) + \lambda_*(E^c \cap [a, b]) \leq b - a,$$

which combined with the above, proves the result. $\square$

LEMMA 10.3.8 *For any subset E of $\mathbb{R}$,*

$$\lambda_*(E) = \lim_{k \to \infty} \lambda_*(E \cap [-k, k]).$$

Proof. Since the proof is similar to that of Theorem 10.2.5, we leave it as an exercise (Exercise 8). $\square$

THEOREM 10.3.9 *Suppose E_1, E_2 are subsets of $\mathbb{R}$. Then*
   (a) $\lambda^*(E_1 \cup E_2) + \lambda^*(E_1 \cap E_2) \leq \lambda^*(E_1) + \lambda^*(E_2)$, *and*
   (b) $\lambda_*(E_1 \cup E_2) + \lambda_*(E_1 \cap E_2) \geq \lambda_*(E_1) + \lambda_*(E_2)$.

Proof. (a) If $\lambda^*(E_i) = \infty$ for some i, $i = 1, 2$, then inequality (a) certainly holds. Thus suppose $\lambda^*(E_i) < \infty$ for $i = 1, 2$. Let $\epsilon > 0$ be given. By the definition of outer measure, for each i, we can choose an open set U_i containing E_i such that

$$m(U_i) < \lambda^*(E_i) + \frac{\epsilon}{2}.$$

Therefore

$$\epsilon + \lambda^*(E_1) + \lambda^*(E_2) > m(U_1) + m(U_2),$$

which by Theorem 10.2.9

$$= m(U_1 \cup U_2) + m(U_1 \cap U_2)$$
$$\geq \lambda^*(E_1 \cup E_2) + \lambda^*(E_1 \cap E_2).$$

The last inequality follows from the definition of outer measure. Since $\epsilon > 0$ was arbitrary, inequality (a) follows.

   (b) Let $a, b \in \mathbb{R}$ be arbitrary. By (a) applied to $[a, b] \cap E_i^c$, we have

$$\lambda^*([a, b] \cap E_1^c) + \lambda^*([a, b] \cap E_2^c)$$
$$\geq \lambda^*([a, b] \cap (E_1^c \cup E_2^c)) + \lambda^*([a, b] \cap (E_1^c \cap E_2^c))$$
$$= \lambda^*([a, b] \cap (E_1 \cap E_2)^c) + \lambda^*([a, b] \cap (E_1 \cup E_2)^c).$$

But by Theorem 10.3.7, for any $E \subset \mathbb{R}$,

$$\lambda^*([a,b] \cap E^c) = (b-a) - \lambda_*(E \cap [a,b]).$$

Therefore,

$$\lambda_*(E_1 \cap [a,b]) + \lambda_*(E_2 \cap [a,b])$$
$$\leq \lambda_*([a,b] \cap (E_1 \cap E_2)) + \lambda_*([a,b] \cap (E_1 \cup E_2)).$$

For each $k \in \mathbb{N}$, set $I_k = [-k,k]$. By the above,

$$\lambda_*(E_1 \cap I_k) + \lambda_*(E_2 \cap I_k) \leq \lambda_*((E_1 \cap E_2) \cap I_k) + \lambda_*((E_1 \cup E_2) \cap I_k)$$
$$\leq \lambda_*(E_1 \cap E_2) + \lambda_*(E_1 \cup E_2).$$

The result now follows by Lemma 10.3.8. $\square$

Exercises 10.3

1. **a.** If $E \subset \mathbb{R}$, $a, b \in \mathbb{R}$, prove that $\lambda^*(E \cap (a,b)) = \lambda^*(E \cap [a,b])$, and $\lambda_*(E \cap (a,b)) = \lambda_*(E \cap [a,b])$.

   *b. If $E \subset \mathbb{R}$, prove that $\lambda^*(E+x) = \lambda^*(E)$ and $\lambda_*(E+x) = \lambda_*(E)$ for every $x \in \mathbb{R}$, where $E + x = \{a + x : a \in E\}$.

2. Prove that every subset of a set of measure zero is measurable.

3. ***Let** $E_1 \subset \mathbb{R}$ with $\lambda^*(E_1) = 0$. If E_2 is a measurable subset of $\mathbb{R}$, prove that $E_1 \cap E_2$ and $E_1 \cup E_2$ are measurable.

4. Let $E = [0,1] \setminus \mathbb{Q}$. Prove that E is measurable and $\lambda(E) = 1$.

5. Let P denote the Cantor set in $[0,1]$.

   a. Prove that $\lambda_*(P^c \cap [0,1]) = 1$.

   b. Prove that $\lambda^*(P) = 0$.

6. ***If** $E \subset \mathbb{R}$, prove that there exists a sequence $\{U_n\}$ of open sets with $E \subset U_n$ for all $n \in \mathbb{N}$ such that $\lambda^*(E) = \lambda^*(\bigcap_n U_n)$.

7. Prove Theorem 10.3.2(b).

8. ***Prove Lemma 10.3.8.**

9. **a.** Prove that every compact set K is measurable with $\lambda(K) = m(K)$.

   b. Prove that every closed set is measurable.

10.4 Properties of Measurable Sets

In this section, we will study some of the basic properties of measurable sets. Our first result proves that the union and intersection of two measurable sets are again measurable.

THEOREM 10.4.1 *If E_1 and E_2 are measurable subsets of $\mathbb{R}$, then*

$$E_1 \cap E_2 \quad and \quad E_1 \cup E_2$$

are measurable with

$$\lambda(E_1) + \lambda(E_2) = \lambda(E_1 \cup E_2) + \lambda(E_1 \cap E_2).$$

Proof. (a) We consider first the case where both E_1 and E_2 are bounded measurable sets, in which case, $E_1 \cap E_2$ and $E_1 \cup E_2$ are also bounded. Since

$$\lambda(E_i) = \lambda_*(E_i) = \lambda^*(E_i), \qquad i = 1, 2,$$

by Theorem 10.3.9,

$$\lambda(E_1) + \lambda(E_2) \leq \lambda_*(E_1 \cup E_2) + \lambda_*(E_1 \cap E_2)$$
$$\leq \lambda^*(E_1 \cup E_2) + \lambda^*(E_1 \cap E_2) \leq \lambda(E_1) + \lambda(E_2).$$

Therefore,

$$\lambda_*(E_1 \cup E_2) + \lambda_*(E_1 \cap E_2) = \lambda^*(E_1 \cup E_2) + \lambda^*(E_1 \cap E_2),$$

and as a consequence,

$$\lambda_*(E_1 \cup E_2) - \lambda^*(E_1 \cup E_2) = \lambda^*(E_1 \cap E_2) - \lambda_*(E_1 \cap E_2).$$

But for any bounded set E, $\lambda^*(E) - \lambda_*(E) \geq 0$. Thus equality can hold in the above if and only if both sides are zero; namely,

$$\lambda_*(E_1 \cup E_2) = \lambda^*(E_1 \cup E_2) \quad and \quad \lambda_*(E_1 \cap E_2) = \lambda^*(E_1 \cap E_2).$$

Therefore $E_1 \cap E_2$ and $E_1 \cup E_2$ are measurable, with

$$\lambda(E_1) + \lambda(E_2) = \lambda(E_1 \cup E_2) + \lambda(E_1 \cap E_2).$$

(b) Suppose one or both of the measurable sets E_1 and E_2 are unbounded. Let $I = [a, b]$ with $a, b \in \mathbb{R}$. If both E_1 and E_2 are unbounded, then $E_1 \cap I$ and $E_2 \cap I$ are measurable by definition. If one of the two sets, say E_1, is bounded, then by part (a) $E_1 \cap I$ is measurable. Thus in both cases, $E_1 \cap I$ and $E_2 \cap I$ are bounded measurable sets. But then

$$(E_1 \cap E_2) \cap I \quad and \quad (E_1 \cup E_2) \cap I$$

are measurable for every closed and bounded interval I with

$$\lambda(E_1 \cap I) + \lambda(E_2 \cap I) = \lambda((E_1 \cap E_2) \cap I) + \lambda((E_1 \cup E_2) \cap I). \qquad (4)$$

Since $E_1 \cup E_2$ is unbounded, $E_1 \cup E_2$ is measurable by definition. Also, if

$E_1 \cap E_2$ is unbounded, then it is measurable by definition. On the other hand, if $E_1 \cap E_2$ is bounded, choose I such that $E_1 \cap E_2 \subset I$. In this case,

$$(E_1 \cap E_2) \cap I = E_1 \cap E_2.$$

Therefore, $E_1 \cap E_2$ is measurable. Finally, to prove that

$$\lambda(E_1) + \lambda(E_2) = \lambda(E_1 \cup E_2) + \lambda(E_1 \cap E_2),$$

take $I_k = [-k, k]$ in (4) and let $k \to \infty$. $\square$

Remark. As a consequence of part (a) of the previous theorem, if E is a bounded measurable subset of $\mathbb{R}$, then $E \cap I$ is measurable for every bounded interval I. Also, if $E_1, ..., E_n$ are measurable sets, then by induction

$$\bigcap_{k=1}^{n} E_k \quad \text{and} \quad \bigcup_{k=1}^{n} E_k$$

are measurable.

THEOREM 10.4.2 *A subset E of $\mathbb{R}$ is measurable if and only if*

$$\lambda^*(E \cap [a, b]) + \lambda^*(E^c \cap [a, b]) \le b - a$$

for every $a, b \in \mathbb{R}$.

Proof. Let $E \subset \mathbb{R}$. Interchanging the roles of E and E^c in Theorem 10.3.7 gives

$$\lambda_*(E \cap [a, b]) + \lambda^*(E^c \cap [a, b]) = b - a$$

for any $a, b \in \mathbb{R}$. If E is measurable, then $E \cap [a, b]$ is measurable for every $a, b \in \mathbb{R}$, and thus

$$\lambda_*(E \cap [a, b]) = \lambda^*(E \cap [a, b]).$$

Thus if E is measurable, $\lambda^*(E \cap [a, b]) + \lambda^*(E^c \cap [a, b]) = b - a$.

Conversely, suppose E satisfies $\lambda^*(E \cap [a, b]) + \lambda^*(E^c \cap [a, b]) \le b - a$ for every $a, b \in \mathbb{R}$. Since we always have

$$\lambda_*(E \cap [a, b]) + \lambda^*(E^c \cap [a, b]) = b - a,$$

we obtain

$$\lambda^*(E \cap [a, b]) - \lambda_*(E \cap [a, b]) \le 0.$$

Since $\lambda_*(E \cap [a, b]) \le \lambda^*(E \cap [a, b])$, the above can hold if and only if

$$\lambda_*(E \cap [a, b]) = \lambda^*(E \cap [a, b])$$

Thus $E \cap [a, b]$ is measurable for every $a, b \in \mathbb{R}$. Hence E is measurable. $\square$

COROLLARY 10.4.3 *A set E is measurable if and only if E^c is measurable.*

Proof. This is an immediate consequence of the fact that E satisfies Theorem 10.4.2 if and only if E^c satisfies Theorem 10.4.2. $\square$

Our next goal is to show that the union and intersection of a countable collection of measurable sets is again measurable. For the proof of this result we require to following theorem.

THEOREM 10.4.4

(a) *If $\{E_n\}_{n=1}^{\infty}$ is a sequence of subsets of $\mathbb{R}$, then*

$$\lambda^* \left(\bigcup_{n=1}^{\infty} E_n \right) \leq \sum_{n=1}^{\infty} \lambda^*(E_n).$$

(b) *If $\{E_n\}_{n=1}^{\infty}$ is a sequence of pairwise disjoint subsets of $\mathbb{R}$, then*

$$\lambda_* \left(\bigcup_{n=1}^{\infty} E_n \right) \geq \sum_{n=1}^{\infty} \lambda_*(E_n).$$

Proof. (a) If $\sum_{n=1}^{\infty} \lambda^*(E_n) = \infty$, then the conclusion in (a) is certainly true. Thus we assume that $\sum_{n=1}^{\infty} \lambda^*(E_n) < \infty$. Let $\epsilon > 0$ be given. For each $n \in \mathbb{N}$, there exists an open set U_n with $E_n \subset U_n$ such that

$$m(U_n) < \lambda^*(E_n) + \frac{\epsilon}{2^n}.$$

Let $U = \bigcup_{n=1}^{\infty} U_n$. Then U is an open subset of $\mathbb{R}$ with $E = \bigcup_{n=1}^{\infty} E_n \subset U$. Thus

$$\lambda^*(E) \leq m \left(\bigcup_{n=1}^{\infty} U_n \right),$$

which by Theorem 10.2.6

$$\leq \sum_{n=1}^{\infty} m(U_n) < \sum_{n=1}^{\infty} \left(\lambda^*(E_n) + \frac{\epsilon}{2^n} \right)$$

$$= \sum_{n=1}^{\infty} \lambda^*(E_n) + \epsilon.$$

Since this holds for all $\epsilon > 0$, the result follows.

(b) Suppose the sets E_n, $n = 1, 2, \ldots$ are pairwise disjoint. Since $E_1 \cap E_2 = \emptyset$, by Theorem 10.3.9(b)

$$\lambda_*(E_1) + \lambda_*(E_2) \leq \lambda_*(E_1 \cup E_2).$$

By induction,

$$\sum_{k=1}^{n} \lambda_*(E_k) \leq \lambda_* \left(\bigcup_{k=1}^{n} E_k \right) \leq \lambda_* \left(\bigcup_{k=1}^{\infty} E_k \right).$$

Since the above holds for all $n \in \mathbb{N}$, letting $n \to \infty$ gives the desired result. $\square$

THEOREM 10.4.5 *Let $\{E_n\}_{n=1}^{\infty}$ be a sequence of measurable sets. Then $\bigcup_{n=1}^{\infty} E_n$ and $\bigcap_{n=1}^{\infty} E_n$ are measurable with*

(a) $\quad \lambda \left(\bigcup_{n=1}^{\infty} E_n \right) \leq \sum_{n=1}^{\infty} \lambda(E_n).$

(b) *If in addition the sets E_n, $n = 1, 2, \ldots$, are pairwise disjoint, then*

$$\lambda \left(\bigcup_{n=1}^{\infty} E_n \right) = \sum_{n=1}^{\infty} \lambda(E_n).$$

Proof. Let $E = \bigcup_{n=1}^{\infty} E_n$. Without loss of generality we can assume that E (and hence all the sets E_n) is bounded. If not, we consider

$$E \cap [a, b] = \bigcup_{n=1}^{\infty} (E_n \cap [a, b])$$

for $a, b \in \mathbb{R}$.

Set $A_1 = E_1$, and for each $n \in \mathbb{N}$, $n \geq 2$, set

$$A_n = E_n \setminus \left(\bigcup_{k=1}^{n-1} E_k \right) = E_n \cap \left(\bigcup_{k=1}^{n-1} E_k \right)^c.$$

Since finite unions, intersections, and complements of measurable sets are again measurable, A_n is measurable for each $n \in \mathbb{N}$. Furthermore, the sets A_n, $n \in \mathbb{N}$, are pairwise disjoint with

$$\bigcup_{n=1}^{\infty} A_n = E.$$

Thus by Theorem 10.4.4(a) and (b),

$$\sum_{n=1}^{\infty} \lambda(A_n) \leq \lambda_*(E) \leq \lambda^*(E) \leq \sum_{n=1}^{\infty} \lambda(A_n).$$

Therefore $\lambda_*(E) = \lambda^*(E)$, and thus since E is bounded, E is measurable. Furthermore, by Theorem 10.4.4

$$\lambda(E) \leq \sum_{n=1}^{\infty} \lambda(E_n),$$

with equality if the sets E_n, $n = 1, 2 \ldots$, are pairwise disjoint.

By Corollary 10.4.3, E_n^c is measurable for each n. Thus by the above, $\bigcup_{n=1}^{\infty} E_n^c$ is measurable. Since

$$\bigcap_{n=1}^{\infty} E_n = \left(\bigcup_{n=1}^{\infty} E_n^c \right)^c,$$

the intersection is also measurable. □

THEOREM 10.4.6

   (a) *If* $\{E_n\}_{n=1}^{\infty}$ *is a sequence of measurable sets with* $E_1 \subset E_2 \subset \cdots$, *then*

$$\lambda \left(\bigcup_{n=1}^{\infty} E_n \right) = \lim_{n \to \infty} \lambda(E_n).$$

   (b) *If* $\{E_n\}_{n=1}^{\infty}$ *is a sequence of measurable sets with* $E_1 \supset E_2 \supset \cdots$ *and* $\lambda(E_1) < \infty$, *then*

$$\lambda \left(\bigcap_{n=1}^{\infty} E_n \right) = \lim_{n \to \infty} \lambda(E_n).$$

Proof. (a) Let $E = \bigcup_{n=1}^{\infty} E_n$. By the previous theorem, E is measurable. If $\lambda(E_k) = \infty$ for some k, then $\lambda(E_n) = \infty$ for all $n \geq k$ and $\lambda(E) = \infty$, thus proving the result. Hence we assume that $\lambda(E_n) < \infty$ for all n.

Set $E_o = \emptyset$, and for $n \in \mathbb{N}$, let $A_n = E_n \setminus E_{n-1}$. Then each A_n is measurable, the collection $\{A_n\}$ is pairwise disjoint, and

$$\bigcup_{n=1}^{\infty} A_n = E.$$

Thus by Theorem 10.4.5(b),

$$\lambda(E) = \sum_{n=1}^{\infty} \lambda(A_n) = \lim_{N \to \infty} \sum_{n=1}^{N} \lambda(E_n \setminus E_{n-1}).$$

But $E_n = (E_n \setminus E_{n-1}) \cup E_{n-1}$. Since the sets $E_n \setminus E_{n-1}$ and E_{n-1} are disjoint,

$$\lambda(E_n) = \lambda(E_n \setminus E_{n-1}) + \lambda(E_{n-1}).$$

Therefore

$$\sum_{n=1}^{N} \lambda(E_n \setminus E_{n-1}) = \sum_{n=1}^{N}[\lambda(E_n) - \lambda(E_{n-1})] = \lambda(E_N) - \lambda(E_o) = \lambda(E_N),$$

from which the result now follows.

(b) Let $E = \bigcap_{n=1}^{\infty} E_n$. Again by the previous theorem E, is measurable. Since

$$E_1 = E \bigcup \bigcup_{n=1}^{\infty} (E_n \setminus E_{n+1}),$$

which is a union of pairwise disjoint measurable sets, by Theorem 10.4.5(b)

$$\lambda(E_1) = \lambda(E) + \sum_{n=1}^{\infty} \lambda(E_n \setminus E_{n+1})$$

$$= \lambda(E) + \lim_{N \to \infty} \sum_{n=1}^{N} [\lambda(E_n) - \lambda(E_{n+1})]$$

$$= \lambda(E) + \lambda(E_1) - \lim_{N \to \infty} \lambda(E_{N+1}).$$

Since $\lambda(E_1) < \infty$, $\lambda(E) = \lim_{N \to \infty} \lambda(E_{N+1})$. □

The Sigma-Algebra of Measurable Sets

Let $\mathcal{M}$ denote the collection of measurable subsets of $\mathbb{R}$. By Theorem 10.3.6, every interval I is in $\mathcal{M}$ with $\lambda(I) = m(I)$. Since every open set U can be expressed as a finite or countable union of pairwise disjoint open intervals, by Theorem 10.4.5 every open set U is measurable with

$$\lambda(U) = m(U).$$

Since the complement of every measurable set is measurable, $\mathcal{M}$ also contains all the closed subsets of $\mathbb{R}$. In particular, every compact set K is measurable with $\lambda(K) = m(K)$, where $m(K)$ is as defined in Definition 10.2.10. These by no means exhaust the measurable sets. Any set obtained from a countable union or intersection of open sets or closed sets, or of sets obtained in this manner, is again measurable.

The collection $\mathcal{M}$ is very large. To illustrate just how large we use the fact that the Cantor set P has measure zero. Thus any subset of the Cantor set has outer measure zero, and as a consequence of Theorem 10.3.5 is measurable. By Property 6 of the Cantor set (Section 2.5), P has the same cardinality (Definition 1.7.1) as the set of all sequences of 0's and 1's. By Miscellaneous Exercise 5 of Chapter 1, the set of all sequences of 0's and 1's is equivalent to $[0, 1]$, and this set has the same cardinality as all of $\mathbb{R}$. Thus the set of all subsets of P is equivalent (not equal) to the set of all subsets of $\mathbb{R}$. As a consequence, in the terminology of equivalence of sets, $\mathcal{M}$ has the same cardinality as the set of all subsets of $\mathbb{R}$. However, nonmeasurable subsets of $\mathbb{R}$ do exist. The construction of such a set will be outlined in the miscellaneous exercises.

We conclude this section by summarizing some of the properties of $\mathcal{M}$.

THEOREM 10.4.7

(a) *If $E \in \mathcal{M}$, then $E^c \in \mathcal{M}$.*

(b) *$\emptyset, \mathbb{R} \in \mathcal{M}$.*

(c) *If $E_n \in \mathcal{M}$, $n = 1, 2, \ldots$, then*

$$\bigcup_{n=1}^{\infty} E_n \in \mathcal{M} \qquad and \qquad \bigcap_{n=1}^{\infty} E_n \in \mathcal{M}.$$

(d) *Every interval $I \in \mathcal{M}$ with $\lambda(I) = m(I)$.*

(e) *If $E \in \mathcal{M}$, then $E + x \in \mathcal{M}$ for all $x \in \mathbb{R}$ with*

$$\lambda(E + x) = \lambda(E).$$

Proof. The result (a) is Corollary 10.4.3, whereas (b) follows from Theorem 10.3.6 and Corollary 10.4.3. The statement (c) is Theorem 10.4.5, whereas (d) is Theorem 10.3.6. The proof of (e) follows from Exercise 1(b) of the previous section. $\square$

Any collection $\mathcal{A}$ of subsets of a set X satisfying (a), (b), and (c) of the previous theorem is called a **sigma-algebra** (σ-algebra) of sets. The σ denotes that the collection $\mathcal{A}$ is closed under countable unions.

Remark. An alternate approach to the theory of measure is due to Constantin Carathéodory (1873–1950) in which a subset E of $\mathbb{R}$ is said to be measurable if

$$\lambda^*(E \cap T) + \lambda^*(E^c \cap T) = \lambda^*(T) \qquad (5)$$

for every subset T of $\mathbb{R}$. Since $T = (E \cap T) \cup (E^c \cap T)$, by Theorem 10.3.9 one always has $\lambda^*(T) \le \lambda^*(E \cap T) + \lambda^*(E^c \cap T)$. Thus E satisfies (5) if and only if

$$\lambda^*(E \cap T) + \lambda^*(E^c \cap T) \le \lambda^*(T).$$

The advantage to this approach is that it does not require the concept of inner measure, and it includes both unbounded and bounded sets simultaneously.

If a subset E of $\mathbb{R}$ satisfies (5), taking $T = [a, b]$, $a, b \in \mathbb{R}$ gives

$$\lambda^*(E \cap [a, b]) + \lambda^*(E^c \cap [a, b]) = \lambda^*([a, b]) = b - a.$$

Thus E satisfies Theorem 10.4.2 and hence is measurable. In Exercise 6, the reader will be asked to prove that if E is measurable as defined in the text, then E satisfies (5) for every subset T of $\mathbb{R}$.

Exercises 10.4

1. Find a sequence $\{E_n\}_{n=1}^{\infty}$ of measurable sets with $E_1 \supset E_2 \supset \cdots$ such that
    $$\lambda\left(\bigcap_{n=1}^{\infty} E_n\right) \neq \lim_{n\to\infty} \lambda(E_n).$$

2. *If E is a measurable subset of $\mathbb{R}$, prove that given $\epsilon > 0$, there exists an open set $U \supset E$ and a closed set $F \subset E$ such that $\lambda(U \setminus E) < \epsilon$ and $\lambda(E \setminus F) < \epsilon$.

3. Let E be a bounded subset of $\mathbb{R}$. Prove that E is measurable if and only if given $\epsilon > 0$ there exists an open set U and a compact set K with $K \subset E \subset U$ such that $\lambda(U \setminus K) < \epsilon$.

4. *If E_1, E_2 are measurable subsets of $[0, 1]$, and if $\lambda(E_1) = 1$, prove that $\lambda(E_1 \cap E_2) = \lambda(E_2)$.

5. Suppose E_1 is a nonmeasurable subset of $[0, 1]$. Set $E = E_1 \cup (1, \infty)$. Prove that E is nonmeasurable but that
    $$\lambda_*(E) = \lambda^*(E) = \infty.$$

6. *(**Carathéodory**) Prove that a subset E of $\mathbb{R}$ is measurable if and only if $\lambda^*(E \cap T) + \lambda^*(E^c \cap T) = \lambda^*(T)$ for every subset T of $\mathbb{R}$.

10.5   Measurable Functions

In our discussion of Lebesgue's approach to integration, we defined the sets

$$E_j = \left\{ x \in [a, b] : (j-1)\frac{\beta}{n} \leq f(x) < j\frac{\beta}{n} \right\},$$

where $f : [a, b] \to [0, \beta)$ is a bounded real-valued function. As we saw in Section 10.1, in order to define the integral of f by partitioning the range, it is necessary that the sets E_j, $j = 1, ..., n$, be measurable. Thus we make the following definition.

DEFINITION 10.5.1 *Let f be a real-valued function defined on $[a, b]$. The function f is said to be **measurable** if for every $s \in \mathbb{R}$, the set*

$$\{x \in [a, b] : f(x) > s\}$$

*is measurable. More generally, if E is a measurable subset of $\mathbb{R}$, a function $f : E \to \mathbb{R}$ is **measurable** if*

$$\{x \in E : f(x) > s\}$$

is measurable for every $s \in \mathbb{R}$.

Since $f^{-1}((s, \infty)) = \{x : f(x) > s\}$, f is measurable if and only if $f^{-1}((s, \infty))$ is a measurable set for every $s \in \mathbb{R}$. We illustrate the idea of a measurable function with the following examples.

EXAMPLES 10.5.2 (a) Let A be a measurable subset of $\mathbb{R}$ and let χ_A denote the characteristic function of A. Then

$$\{x : \chi_A(x) > s\} = \begin{cases} \mathbb{R}, & s \le 0, \\ A, & 0 < s \le 1, \\ \emptyset, & s > 1. \end{cases}$$

Since each of the sets $\emptyset$, A, and $\mathbb{R}$ are measurable, χ_A is a measurable function on $\mathbb{R}$.

(b) Let $f : [0, 1] \to \mathbb{R}$ be defined by

$$f(x) = \begin{cases} 0, & x \in \mathbb{Q} \cap [0, 1], \\ x, & x \in [0, 1] \setminus \mathbb{Q}. \end{cases}$$

Then

$$\{x \in [0, 1] : f(x) > s\} = \begin{cases} [0, 1], & \text{if } s < 0, \\ \mathbb{Q}^c \cap (s, 1), & \text{if } 0 \le s < 1, \\ \emptyset, & \text{if } s \ge 1. \end{cases}$$

Again, since each of the sets is a measurable subset of $\mathbb{R}$, f is measurable. $\quad\square$

Properties of Measurable Functions

We now consider some properties of measurable functions. Our first result provides several equivalent conditions for measurability.

THEOREM 10.5.3 *Let f be a real-valued function defined on a measurable set E. Then f is measurable if and only if any of the following hold:*

(a) $\{x : f(x) > s\}$ *is measurable for every $s \in \mathbb{R}$.*
(b) $\{x : f(x) \ge s\}$ *is measurable for every $s \in \mathbb{R}$.*
(c) $\{x : f(x) < s\}$ *is measurable for every $s \in \mathbb{R}$.*
(d) $\{x : f(x) \le s\}$ *is measurable for every $s \in \mathbb{R}$.*

Proof. The set of (d) is the complement of the set in (a). Thus by Corollary 10.4.3, one is measurable if and only if the other is. Similarly for the sets of (b) and (c). Thus it suffices to prove that (a) is equivalent to (b).

Suppose (a) holds. For each $n \in \mathbb{N}$, let

$$E_n = \left\{x : f(x) > s - \frac{1}{n}\right\}$$

By (a), E_n is measurable for all $n \in \mathbb{N}$. But

$$\{x : f(x) \geq s\} = \bigcap_{n=1}^{\infty} E_n,$$

which is measurable by Theorem 10.4.5. Conversely, since

$$\{x : f(x) > s\} = \bigcup_{n=1}^{\infty} \left\{ x : f(x) \geq s + \frac{1}{n} \right\},$$

if (b) holds, then by Theorem 10.4.5, (a) also holds. $\quad \square$

THEOREM 10.5.4 *Suppose f, g are measurable real-valued functions defined on a measurable set E. Then*

(a) *$f + c$ and cf are measurable for every $c \in \mathbb{R}$.*

(b) *$f + g$ is measurable.*

(c) *fg is measurable.*

(d) *$1/g$ is measurable provided $g(x) \neq 0$ for all $x \in E$.*

Proof. The proof of (a) is straightforward and is omitted. The proof of (a) also follows from (b) and (c) upon showing that constant functions are measurable.

(b) Let $s \in \mathbb{R}$. Then $f(x) + g(x) > s$ if and only if $f(x) > s - g(x)$. If $x \in E$ is such that $f(x) > s - g(x)$, then there exists $r \in \mathbb{Q}$ such that

$$f(x) > r > s - g(x).$$

Let $\{r_n\}_{n=1}^{\infty}$ be an enumeration of $\mathbb{Q}$. Then

$$\{x : f(x) + g(x) > s\} = \bigcup_{n=1}^{\infty} \left(\{x : f(x) > r_n\} \bigcap \{x : r_n > s - g(x)\} \right).$$

Since f and g are measurable functions,

$$\{x : f(x) > r_n\} \quad \text{and} \quad \{x : r_n > s - g(x)\}$$

are measurable sets for every $n \in \mathbb{N}$. Thus their intersection and the resulting union is also measurable. Therefore $f + g$ is measurable.

(c) To prove (c) we first show that f^2 is measurable. If $s < 0$, then

$$\{x \in E : f^2(x) > s\} = E,$$

which is measurable. Assume $s \geq 0$. Then

$$\{x : f^2(x) > s\} = \{x : f(x) > \sqrt{s}\} \bigcup \{x : f(x) < -\sqrt{s}\}.$$

But each of these two sets are measurable. Thus their union is measurable. Since

$$fg = \frac{1}{4} \left[(f + g)^2 - (f - g)^2 \right],$$

the function fg is measurable.

(d) The proof of (d) is left as an exercise (Exercise 5). $\quad \square$

THEOREM 10.5.5 *Every continuous real-valued function on $[a,b]$ is measurable.*

Proof. Exercise 7. □

A Property Holding Almost Everywhere

A very important concept in the study of measure theory involves the idea of a *property* being true for all x except for a set of measure zero. This idea was previously encountered in the statement of Lebesgue's theorem in Chapter 6; namely, a bounded real-valued function f on $[a,b]$ is Riemann integrable if and only if $\{x : f$ is not continuous at $x\}$ has measure zero. An equivalent formulation is that f is continuous except on a set of measure zero. In this section, we will encounter several other properties that are assumed to hold except on sets of measure zero.

DEFINITION 10.5.6 *A property P is said to hold* **almost everywhere** *(abbreviated* **a.e.***) if the set of points where P does not hold has measure zero, i.e.,*

$$\lambda(\{x : \ P \ does \ not \ hold \ \}) = 0.$$

Remark. The assertion that a set is of measure zero includes the assertion that it is measurable. This however is not necessary. If instead we only require that $\lambda^*(\{x : \ P$ does not hold $\}) = 0$, then by Theorem 10.3.5 the set $\{x : \ P$ does not hold $\}$ is in fact measurable.

We will illustrate the concept of a property holding almost everywhere by means of the following examples.

EXAMPLES 10.5.7 (a) Suppose f and g are real-valued functions defined on $[a,b]$. The functions f and g are said to be **equal almost everywhere**, denoted $f = g$ a.e., if

$$\{x \in [a,b] : f(x) \neq g(x)\}$$

has measure zero. For example, if $g(x) = 1$ for all $x \in [0,1]$ and

$$f(x) = \begin{cases} 1, & x \in [0,1] \setminus \mathbb{Q}, \\ 0, & x \in [0,1] \cap \mathbb{Q}. \end{cases}$$

Then $\{x \in [0,1] : f(x) \neq g(x)\} = [0,1] \cap \mathbb{Q}$ which has measure zero. Therefore $f = g$ a.e.

(b) In Theorem 10.5.4 we proved that if g is a real valued measurable function on $[a,b]$ with $g(x) \neq 0$ for all $x \in [a,b]$, then $1/g$ is also measurable on $[a,b]$. Suppose we replace the hypothesis $g(x) \neq 0$ for all $x \in [a,b]$ with $g \neq 0$ a.e.; that is, the set

$$E = \{x \in [a,b] : g(x) = 0\}$$

has measure zero. If we now define f by

$$f(x) = \begin{cases} g(x), & x \in [a,b] \setminus E, \\ 1, & x \in E, \end{cases}$$

then $f(x) \neq 0$ for all $x \in [a,b]$ and $f(x) = g(x)$ except for $x \in E$, which has measure zero. Thus $f = g$ a.e. on $[a,b]$. As a consequence of our next theorem, the function f will also be measurable on $[a,b]$.

(c) A real-valued function f on $[a,b]$ is **continuous almost everywhere** if

$$\{x \in [a,b] : f \text{ is not continuous at } x\}$$

has measure zero. As in Example 6.1.14, consider the function f on $[0,1]$ defined by

$$f(x) = \begin{cases} 1, & x = 0, \\ 0, & \text{if } x \text{ is irrational} \\ \frac{1}{n}, & \text{if } x = \frac{m}{n} \text{ in lowest terms }, x \neq 0. \end{cases}$$

As was shown in Example 4.2.2(g), the function f is continuous at every irrational number in $[0,1]$, and discontinuous at every rational number in $[0,1]$. Therefore,

$$\lambda(\{x \in [0,1] : f \text{ is not continuous at } x\}) = \lambda(\mathbb{Q} \cap [0,1]) = 0.$$

Thus f is continuous a.e. on $[0,1]$.

(d) Let f and f_n, $n = 1,2,\dots$ be a sequence of real-valued functions defined on $[a,b]$. The sequence $\{f_n\}$ is said to **converge almost everywhere** to f, denoted $f_n \to f$ a.e., if

$$\{x \in [a,b] : \{f_n(x)\} \text{ does not converge to } f(x)\}$$

has measure zero. To illustrate this, consider the sequence $\{f_n\}$ defined in Example 8.1.2(c) as follows: Let $\{x_k\}$ be an enumeration of $\mathbb{Q} \cap [0,1]$. For each $n \in \mathbb{N}$, define f_n on $[0,1]$ by

$$f_n(x) = \begin{cases} 0, & x = x_k, 1 \leq k \leq n, \\ 1, & \text{otherwise.} \end{cases}$$

Then

$$\lim_{n \to \infty} f_n(x) = f(x) = \begin{cases} 0, & \text{if } x \text{ is rational,} \\ 1, & \text{if } x \text{ is irrational.} \end{cases}$$

Thus $\{x \in [0,1] : \{f_n(x)\} \text{ does not converge to } 1\} = \mathbb{Q} \cap [0,1]$, which has measure zero. Hence $f_n \to 1$ a.e. on $[0,1]$. □

One of the key results needed in the sequel is the following:

THEOREM 10.5.8 *Suppose f and g are real-valued functions defined on a measurable set A. If f is measurable and $g = f$ a.e., then g is measurable on A.*

Proof. Let $E = \{x \in A : g(x) \neq f(x)\}$. Then $\lambda(E) = 0$. Let $B = A \setminus E$. Since E is measurable, so is B. Also on B, $g(x) = f(x)$. Fix $s \in \mathbb{R}$. Then

$$\{x \in A : g(x) > s\} = \{x \in B : g(x) > s\} \bigcup \{x \in E : g(x) > s\}$$
$$= \{x \in B : f(x) > s\} \bigcup \{x \in E : g(x) > s\}.$$

Since $E_1 = \{x \in E : g(x) > s\}$ is a subset of E and $\lambda(E) = 0$, the set E_1 is measurable. Also, since f is measurable, $\{x \in B : f(x) > s\}$ is measurable. Therefore $\{x : g(x) > s\}$ is measurable, and thus g is measurable. $\square$

THEOREM 10.5.9 *Let $\{f_n\}_{n=1}^{\infty}$ be a sequence of real-valued measurable functions defined on a measurable set A such that $\{f_n(x)\}_{n=1}^{\infty}$ is bounded for every $x \in A$. Let*

$$\varphi(x) = \sup\{f_n(x) : n \in \mathbb{N}\} \quad and \quad \psi(x) = \inf\{f_n(x) : n \in \mathbb{N}\}.$$

Then φ and ψ are measurable on A.

Proof. The result follows by Theorem 10.4.5, and the fact that for every $s \in \mathbb{R}$,

$$\{x : \varphi(x) > s\} = \bigcup_{n=1}^{\infty} \{x : f_n(x) > s\} \quad and$$

$$\{x : \psi(x) < s\} = \bigcup_{n=1}^{\infty} \{x : f_n(x) < s\}. \quad \square$$

COROLLARY 10.5.10 *Let $\{f_n\}_{n=1}^{\infty}$ be a sequence of real-valued measurable functions defined on a measurable set A, and let f be a real-valued function on A. If $f_n \to f$ a.e. on A, then f is measurable on A.*

Proof. Let $E = \{x : \{f_n(x)\} \text{ does not converge to } f(x)\}$. By hypothesis $\lambda(E) = 0$. Set

$$g_n(x) = \begin{cases} f_n(x), & x \in A \setminus E, \\ 0, & x \in E. \end{cases}$$

Then $g_n = f_n$ a.e. and thus is measurable. Also, $\lim_{n \to \infty} g_n(x) = g(x)$ exists for all $x \in A$. But

$$g(x) = \lim_{n \to \infty} g_n(x) = \varlimsup_{n \to \infty} g_n(x) = \inf_n \sup\{f_k(x) : k \geq n\}.$$

By the previous theorem each of the functions

$$F_n(x) = \sup\{f_k(x) : k \geq n\} \qquad \text{and} \qquad g(x) = \inf\{F_n(x) : n \in \mathbb{N}\}$$

are measurable on A. Finally, since $f = g$ a.e., f itself is measurable. $\square$

Suppose $\{f_n\}$ is a sequence of measurable functions on $[a, b]$ such that $f_n \to f$ a.e. Then by definition there exists a subset E of $[a, b]$ such that $\lambda([a, b] \setminus E) = 0$, and $\lim_{n \to \infty} f_n(x) = f(x)$ for all $x \in E$. Exercise 15 provides a significant strengthening of this result. There you will be asked to prove that given $\epsilon > 0$, there exists a measurable set $E \subset [a, b]$, such that $\lambda([a, b] \setminus E) < \epsilon$, and $\{f_n\}$ converges uniformly to f on E. This result is known as **Egorov's theorem**.

Exercises 10.5

1. *Let f be defined on $[0, 1]$ by

    $$f(x) = \begin{cases} 0, & x = 0, \\ \dfrac{1}{x}, & 0 < x < 1, \\ 2, & x = 1. \end{cases}$$

    Prove directly that f is measurable.

2. Let f be a real-valued function on a measurable set A with finite range, i.e., Range $f = \{\alpha_1, ..., \alpha_n\}$, $\alpha_j \in \mathbb{R}$. Prove that f is measurable if and only if $f^{-1}(\alpha_j)$ is measurable for all $j = 1, ..., n$.

3. Let A be a measurable subset of $\mathbb{R}$, and let $f : A \to \mathbb{R}$ be measurable. Prove that for each $n \in \mathbb{N}$, the function f_n defined by

    $$f_n(x) = \begin{cases} f(x), & \text{if } |f(x)| \leq n, \\ n, & \text{if } |f(x)| > n, \end{cases}$$

    is measurable on A.

4. If f is measurable on $[a, b]$, prove that $f + c$ and cf are measurable on $[a, b]$ for every $c \in \mathbb{R}$.

5. *If g is measurable on $[a, b]$ with $g(x) \neq 0$ for all $x \in [a, b]$, prove that $1/g$ is measurable on $[a, b]$.

6. Let f be a real-valued function on $[a, b]$. Prove that f is measurable if and only if $f^{-1}(U)$ is a measurable subset of $[a, b]$ for every open subset U of $\mathbb{R}$.

7. *Prove that every continuous real-valued function f on $[a, b]$ is measurable.

8. Let A be a measurable subset of $\mathbb{R}$. If $g : A \to \mathbb{R}$ is measurable and $f : \mathbb{R} \to \mathbb{R}$ is continuous, prove that $f \circ g$ is measurable.

9. If f is monotone on $[a, b]$ and $g : \mathbb{R} \to [a, b]$ is measurable, prove that $f \circ g$ is measurable.

10. Let E be a measurable subset of $\mathbb{R}$, and let f be a measurable function on E. Define the functions f^+ and f^- on E as follows:
$f^+(x) = \max\{f(x), 0\}, \quad f^-(x) = \max\{-f(x), 0\}.$

*a. Prove directly that f^+ and f^- are nonnegative measurable functions on E with $f(x) = f^+(x) - f^-(x)$.

b. Prove that $|f(x)| = f^+(x) + f^-(x)$ and that $|f|$ is measurable.

*c. If f is a real-valued function on $[a, b]$ such that $|f|$ is measurable on $[a, b]$, is f measurable on $[a, b]$?

d. If $f(x) = \frac{1}{2} + \cos x$, $x \in [0, 2\pi]$, find $f^+(x)$ and $f^-(x)$.

11. Let A be a measurable subset of $\mathbb{R}$ and let $\{f_n\}$ be a sequence of measurable functions on A. Let $E = \{x \in A : \{f_n(x)\}_{n=1}^\infty$ converges $\}$. Prove that E is measurable.

12. *Let f be a bounded measurable function on $[0, 1]$ and let $\{f_n\}$ be defined on $[0, 1]$ by $f_n(x) = x^n f(x)$. Prove that each f_n is measurable and that $\{f_n\}$ converges to 0 almost everywhere on $[0, 1]$.

13. Let $\{x_k\}$ be an enumeration of the rational numbers in $[0, 1]$. For each $n \in \mathbb{N}$ let
$$f_n(x) = \begin{cases} 1 & \text{if } x = x_k, 1 \le k \le n, \\ 0 & \text{otherwise.} \end{cases}$$
Show that f_n is measurable for each $n \in \mathbb{N}$, and that $\{f_n\}$ converges to 0 almost everywhere on $[0, 1]$.

14. *If f is differentiable on $[a, b]$, prove that f' is measurable on $[a, b]$.

15. **Egorov's Theorem:** Let $\{f_n\}$ be a sequence of measurable functions on $[a.b]$ such that $f_n \to f$ a.e. on $[a, b]$. Given $\epsilon > 0$, prove that there exists a measurable subset E of $[a, b]$ such that $\lambda([a, b] \setminus E) < \epsilon$ and $\{f_n\}$ converges uniformly to f on E.

16. Construct a sequence $\{f_n\}$ of measurable functions on $[0, 1]$ such that $\{f_n(x)\}$ converges for each $x \in [0, 1]$ but that $\{f_n\}$ does not converges uniformly on any measurable set $E \subset [0, 1]$ with $\lambda([0, 1] \setminus E) = 0$.

17. Let f be a measurable function on $[a, b]$. Prove that the function
$$\lambda(\{x \in [a, b] : f(x) > t\}), t > 0,$$
is nonincreasing and right continuous.

18. If $\{f_n\}$ is a nondecreasing sequence of measurable functions on $[a, b]$ and $f = \lim_{n \to \infty} f_n$, then for all $t > 0$,
$$\lim_{n \to \infty} \lambda(\{x \in [a, b] : |f_n(x)| > t\}) = \lambda(\{x \in [a, b] : |f(x)| > t\}).$$

10.6   Lebesgue Integral of a Bounded Function

There are many different approaches to the development of the Lebesgue integral. One is the method outlined in the first section of this chapter which is pursued further in Exercise 1. The drawback to this approach is that it a priori assumes that f is a measurable function. The approach that we will follow is patterned on the Darboux approach to the Riemann integral.

DEFINITION 10.6.1 *Let E be a measurable subset of $\mathbb{R}$. A **measurable partition** of E is a finite collection $\mathcal{P} = \{E_1, ..., E_n\}$ of pairwise disjoint measurable subsets of E such that*

$$\bigcup_{k=1}^{n} E_k = E.$$

Suppose $\mathcal{P} = \{x_o, x_1, ..., x_n\}$ is a point partition of $[a, b]$ as considered in Chapter 6. Set $E_1 = [x_o, x_1]$, and for $k = 2, ..., n$, set $E_k = (x_{k-1}, x_k]$. Then the collection $\mathcal{P} = \{E_1, ..., E_n\}$ is a measurable partition of $[a, b]$. A measurable partition of $[a, b]$ however need not consist of intervals. For example if $E_1 = [a, b] \cap \mathbb{Q}$ and $E_2 = [a, b] \setminus E_1$, then $\{E_1, E_2\}$ is a measurable partition of $[a, b]$.

As for the Riemann integral, if f is a bounded real-valued function on $[a, b]$, and $\mathcal{P} = \{E_1, ..., E_n\}$ is a measurable partition of $[a, b]$, we define the **lower** and **upper Lebesgue sums** of f with respect to $\mathcal{P}$, denoted $\mathcal{L}_L(\mathcal{P}, f)$ and $\mathcal{U}_L(\mathcal{P}, f)$ respectively, by

$$\mathcal{L}_L(\mathcal{P}, f) = \sum_{k=1}^{n} m_k \lambda(E_k) \quad \text{and} \quad \mathcal{U}_L(\mathcal{P}, f) = \sum_{k=1}^{n} M_k \lambda(E_k),$$

where $m_k = \inf\{f(x) : x \in E_k\}$ and $M_k = \sup\{f(x) : x \in E_k\}$. Clearly $\mathcal{L}_L(\mathcal{P}, f) \leq \mathcal{U}_L(\mathcal{P}, f)$ for every measurable partition $\mathcal{P}$ of $[a, b]$. As for the Riemann integral, we could now define the upper and lower Lebesgue integrals of f, and then define a function to be Lebesgue integrable if and only if these two quantities are equal. The following theorem however shows that this is unnecessary.

THEOREM 10.6.2 *Let f be a bounded real-valued function on $[a, b]$. Then*

$$\sup_{\mathcal{P}} \mathcal{L}_L(\mathcal{P}, f) = \inf_{\mathcal{Q}} \mathcal{U}_L(\mathcal{Q}, f),$$

where the infimum and supremum are taken over all measurable partitions $\mathcal{Q}$ and $\mathcal{P}$ of $[a, b]$, if and only if f is measurable on $[a, b]$.

Remark. Although the previous theorem is stated for a closed interval $[a, b]$, the result is also true for f defined an any measurable subset A of $\mathbb{R}$ with $\lambda(A) < \infty$.

As a consequence of the previous theorem, which we will shortly prove, we make the following definition.

DEFINITION 10.6.3 *If* f *is a bounded real-valued measurable function on* $[a,b]$, *the* **Lebesgue integral** *of* f *over* $[a,b]$, *denoted* $\int_{[a,b]} f \, d\lambda$ *(or* $\int_a^b f \, d\lambda$*), is defined by*

$$\int_{[a,b]} f \, d\lambda = \sup_{\mathcal{P}} \mathcal{L}_L(\mathcal{P}, f),$$

where the supremum is taken over all measurable partitions of $[a,b]$. *If* A *is a measurable subset of* $[a,b]$, *the Lebesgue integral of* f *over* A, *denoted* $\int_A f \, d\lambda$, *is defined by*

$$\int_A f \, d\lambda = \int_{[a,b]} f \chi_A \, d\lambda.$$

Remarks. (a) In defining $\int_A f \, d\lambda$ it was implicitly assumed that f is defined on all of $[a,b]$. If f is only defined on the measurable set A, $A \subset [a,b]$, then $f\chi_A$ can still be defined on all of $[a,b]$ in the obvious manner; namely,

$$(f\chi_A)(x) = \begin{cases} f(x), & x \in A, \\ 0, & x \notin A. \end{cases}$$

Alternately, if f is a bounded measurable function defined on a measurable subset A of $[a,b]$, we could define

$$\int_A f \, d\lambda = \sup \mathcal{L}_L(\mathcal{Q}, f),$$

where the supremum is taken over all measurable partitions $\mathcal{Q}$ of A. However, if $\mathcal{Q} = \{E_1, ..., E_n\}$ is any measurable partition of A, then

$$\mathcal{P} = \{E_1, ..., E_n, [a,b] \setminus A\}$$

is a measurable partition of $[a,b]$ with

$$\mathcal{L}_L(\mathcal{Q}, f) = \mathcal{L}_L(\mathcal{P}, f\chi_A).$$

Conversely, if $\mathcal{P} = \{E_1, ..., E_n\}$ is any measurable partition of $[a,b]$, then $\mathcal{Q} = \{E_1 \cap A, ..., E_n \cap A\}$ is a measurable partition of A for which $\mathcal{L}_L(Q, f) = \mathcal{L}_L(\mathcal{P}, f\chi_A)$. Thus the two definitions for $\int_A f \, d\lambda$ give the same value.

(b) To distinguish between the Lebesgue and Riemann integral of a bounded real-valued function f on $[a,b]$, the Riemann integral of f, if it exists, will be denoted by

$$\int_a^b f(x)dx.$$

If in the Lebesgue integral we wish to emphasize the variable x, we will write $\int_a^b f(x) \, d\lambda(x)$ to denote the Lebesgue integral of f. The two different notations should cause no confusion. In fact, in Corollary 10.6.8 we will prove that every

Riemann integrable function on $[a, b]$ is also Lebesgue integrable, and the two integrals are equal.

Prior to proving Theorem 10.6.2, we first need the analogue of Theorem 6.1.4.

DEFINITION 10.6.4 *Let E be a measurable set and let $\mathcal{P}$ be a measurable partition of E. A measurable partition $\mathcal{Q}$ of E is a* **refinement** *of $\mathcal{P}$ if every set in $\mathcal{Q}$ is a subset of some set in $\mathcal{P}$.*

A useful fact about refinements is the following: If $\mathcal{P} = \{E_1, ..., E_n\}$ and $\mathcal{Q} = \{A_1, ..., A_m\}$ are measurable partitions of E, then the collection

$$\{E_i \cap A_j\}_{i=1, j=1}^{n, m}$$

is a measurable partition of E that is a refinement of both $\mathcal{P}$ and $\mathcal{Q}$.

LEMMA 10.6.5 *If $\mathcal{P}, \mathcal{Q}$ are measurable partitions of $[a, b]$ such that $\mathcal{Q}$ is a refinement of $\mathcal{P}$, then*

$$\mathcal{L}_L(\mathcal{P}, f) \leq \mathcal{L}_L(\mathcal{Q}, f) \leq \mathcal{U}_L(\mathcal{Q}, f) \leq \mathcal{U}_L(\mathcal{P}, f).$$

As a consequence,

$$\sup_{\mathcal{P}} \mathcal{L}_L(\mathcal{P}, f) \leq \inf_{\mathcal{Q}} \mathcal{U}_L(\mathcal{Q}, f).$$

Proof. The proof of the lemma is almost verbatim the proof of Lemma 6.1.3 and Theorem 6.1.4, and thus is left to the exercises (Exercise 2). ☐

EXAMPLE 10.6.6 In this example, we calculate the Lebesgue integral of what is commonly called a simple function on $[a, b]$. A **simple function** on $[a, b]$ is a measurable real-valued function on $[a, b]$ that assumes only a finite number of values.

Suppose s is a simple function on $[a, b]$ with Range $s = \{\alpha_1, ..., \alpha_n\}$, where $\alpha_i \neq \alpha_j$ whenever $i \neq j$. For each $i = 1, ..., n$, set

$$E_i = \{x \in [a, b] : s(x) = \alpha_i\} = s^{-1}(\{\alpha_i\}).$$

Since s is measurable, each E_i is a measurable set, and

$$s(x) = \sum_{i=1}^{n} \alpha_i \chi_{E_i}(x). \tag{6}$$

Furthermore, since $\alpha_i \neq \alpha_j$ if $i \neq j$, the sets E_i, $i = 1, ..., n$, are pairwise disjoint with $\bigcup_{i=1}^{n} E_i = [a, b]$. Equation (6) is called the **canonical representation** of s. If all the sets E_i are intervals, then s is a step function on $[a, b]$. ☐

We will now show that every simple function s is Lebesgue integrable on $[a, b]$ and compute the Lebesgue integral of s.

LEMMA 10.6.7 *If s is a simple function on $[a, b]$ with canonical representation*

$$s = \sum_{i=1}^{n} \alpha_i \chi_{E_i},$$

where $\{E_i\}_{i=1}^{n}$ are pairwise disjoint measurable subsets of $[a, b]$ with $\bigcup_{i=1}^{n} E_i = [a, b]$, then s is Lebesgue integrable on $[a, b]$ with

$$\int_{[a,b]} s \, d\lambda = \sum_{j=1}^{n} \alpha_j \lambda(E_j).$$

Proof. To show that s is Lebesgue integrable on $[a, b]$ we will prove that

$$\sup_{\mathcal{Q}} \mathcal{L}_L(\mathcal{Q}, s) = \inf_{\mathcal{Q}} \mathcal{U}_L(\mathcal{Q}, f),$$

where the supremum and infimum are taken over all measurable partitions $\mathcal{Q}$ of $[a, b]$. Since $\mathcal{P} = \{E_1, ..., E_n\}$ is a measurable partition of $[a, b]$ and $s(x) = \alpha_j$ for all $x \in E_j$,

$$\mathcal{L}_L(\mathcal{P}, f) = \mathcal{U}_L(\mathcal{P}, f) = \sum_{j=1}^{n} \alpha_j \lambda(E_j).$$

But then

$$\sum_{j=1}^{n} \alpha_j \lambda(E_j) \le \sup_{\mathcal{Q}} \mathcal{L}_L(\mathcal{Q}, f) \le \inf_{\mathcal{Q}} \mathcal{U}_L(\mathcal{Q}, f) \le \sum_{j=1}^{n} \alpha_j \lambda(E_j).$$

Therefore s is Lebesgue integrable on $[a, b]$ with $\int_{[a,b]} s \, d\lambda = \sum_{i=1}^{n} \alpha_i \lambda(E_i)$. $\square$

Remark. Suppose f is a bounded real-valued measurable function on $[a, b]$. If $\mathcal{P} = \{E_1, ..., E_n\}$ is a measurable partition of $[a, b]$, set

$$\varphi(x) = \sum_{k=1}^{n} m_k \chi_{E_k}(x) \quad \text{and} \quad \psi(x) = \sum_{k=1}^{n} M_k \chi_{E_k}(x), \qquad (7)$$

where $m_k = \inf\{f(x) : x \in E_k\}$ and $M_k = \sup\{f(x) : x \in E_k\}$. Then φ and ψ are simple functions on $[a, b]$ with $\varphi(x) \le f(x) \le \psi(x)$ for all $x \in [a, b]$. Furthermore, by Lemma 10.6.7

$$\int_{[a,b]} \varphi \, d\lambda = \sum_{k=1}^{n} m_k \lambda(E_k) = \mathcal{L}_L(\mathcal{P}, f), \quad \text{and}$$

$$\int_{[a,b]} \psi \, d\lambda = \sum_{k=1}^{n} M_k \lambda(E_k) = \mathcal{U}_L(\mathcal{P}, f).$$

Thus if f is a bounded real-valued measurable function on $[a, b]$,

$$\int_{[a,b]} f \, d\lambda = \sup_{\varphi \le f} \int_{[a,b]} \varphi \, d\lambda,$$

where the supremum is taken over all simple functions φ on $[a, b]$ satisfying $\varphi(x) \le f(x)$ for all $x \in [a, b]$.

Proof of Theorem 10.6.2. Suppose f is a measurable function on $[a, b]$ with $m \le f(x) < M$ for all $x \in [a, b]$. Let $\beta = M - n$, and for $n \in \mathbb{N}$, partition $[m, M)$ into n subintervals of length β/n. For each $k = 1, ..., n$, set

$$E_k = \left\{ x \in [a, b] : m + (k-1)\frac{\beta}{n} \le f(x) < m + k\frac{\beta}{n} \right\}.$$

Then $\mathcal{P}_n = \{E_1, ..., E_n\}$ is a measurable partition of $[a, b]$. Also, if $m_k = \inf\{f(x) : x \in E_k\}$ and $M_k = \sup\{f(x) : x \in E_k\}$, then

$$m_k \ge m + (k-1)\frac{\beta}{n} \quad \text{and} \quad M_k \le m + k\frac{\beta}{n}.$$

Therefore

$$0 \le \inf_{\mathcal{P}} \mathcal{U}_L(\mathcal{P}, f) - \sup_{\mathcal{Q}} \mathcal{L}_L(\mathcal{Q}, f) \le \mathcal{U}_L(\mathcal{P}_n, f) - \mathcal{L}_L(\mathcal{P}_n, f)$$

$$= \sum_{k=1}^{n} (M_k - m_k)\lambda(E_k)$$

$$\le \sum_{k=1}^{n} \left[(m + k\tfrac{\beta}{n}) - (m - (k-1)\tfrac{\beta}{n}) \right] \lambda(E_k)$$

$$= \frac{\beta}{n} \sum_{k=1}^{n} \lambda(E_k) = \frac{\beta}{n}(b - a).$$

Since $n \in \mathbb{N}$ is arbitrary, by letting $n \to \infty$ we obtain

$$\sup_{\mathcal{Q}} \mathcal{L}_L(\mathcal{Q}, f) = \inf_{\mathcal{P}} \mathcal{U}_L(\mathcal{P}, f). \tag{8}$$

Conversely, suppose (8) holds. By taking a common refinement if necessary, for each $n \in \mathbb{N}$, there exists a measurable partition $\mathcal{P}_n$ of $[a, b]$ such that

$$\mathcal{U}_L(\mathcal{P}_n, f) < \inf_{\mathcal{P}} \mathcal{U}_L(\mathcal{P}, f) + \frac{1}{2n}, \quad \text{and}$$

$$\mathcal{L}_L(\mathcal{P}_n, f) > \sup_{\mathcal{Q}} \mathcal{L}_L(\mathcal{Q}, f) - \frac{1}{2n}.$$

Since equality holds in (8),

$$\mathcal{U}_L(\mathcal{P}_n, f) - \mathcal{L}_L(\mathcal{P}_n, f) < \frac{1}{n}.$$

For the partition $\mathcal{P}_n$, let φ_n and ψ_n be simple functions on $[a, b]$ as defined by equation (7), satisfying $\varphi_n(x) \leq f(x) \leq \psi_n(x)$ for all $x \in [a, b]$, and

$$\int_{[a,b]} \varphi_n d\lambda = \mathcal{L}_L(\mathcal{P}_n, f), \qquad \int_{[a,b]} \psi_n d\lambda = \mathcal{U}_L(\mathcal{P}_n, f).$$

Define φ and ψ on $[a, b]$ by $\varphi(x) = \sup_n \varphi_n(x)$ and $\psi(x) = \inf_n \psi_n(x)$. By Theorem 10.5.9 the functions φ and ψ are measurable functions on $[a, b]$, with

$$\varphi(x) \leq f(x) \leq \psi(x)$$

for all $x \in [a, b]$.

To complete the proof we will show that $\varphi = \psi$ a.e. on $[a, b]$. Then as a consequence of Theorem 10.5.8, the function f will be measurable on $[a, b]$. Let

$$E = \{x \in [a, b] : \varphi(x) < \psi(x)\},$$

and for each $k \in \mathbb{N}$, let

$$E_k = \left\{ x \in [a, b] : \varphi(x) < \psi(x) - \frac{1}{k} \right\}.$$

Then $E = \bigcup_{k=1}^{\infty} E_k$. If $x \in E_k$, then $\varphi_n(x) < \psi_n(x) - \frac{1}{k}$ for all $n \in \mathbb{N}$. For $n, k \in \mathbb{N}$, let

$$A_{n,k} = \left\{ x : \varphi_n(x) < \psi_n(x) - \frac{1}{k} \right\}.$$

If $x \in A_{n,k}$, then $\psi_n(x) - \varphi_n(x) > 1/k$. Consider the simple function

$$s_n(x) = (\psi_n(x) - \varphi_n(x)) \chi_{A_{n,k}}(x).$$

Suppose the measurable partition $\mathcal{P}_n$ is given by $\{B_1, ..., B_N\}$. Then

$$s_n(x) = \sum_{j=1}^{N} (M_j - m_j) \chi_{B_j \cap A_{n,k}}(x),$$

where M_j and m_j denote the supremum and infimum of f respectively over B_j. The collection

$$\mathcal{Q} = \{B_j \cap A_{n,k}\}_{j=1}^{N} \cup \{[a, b] \setminus A_{n,k}\}$$

is a measurable partition of $[a, b]$. If $m_j^* = \inf\{s_n(x) : x \in B_j \cap A_{n,k}\}$, then $m_j^* > 1/k$ for all $j = 1, ..., N$. Also, since $s_n(x) = 0$ for all $x \in [a, b] \setminus A_{n,k}$,

$$\mathcal{L}_L(\mathcal{Q}, s_n) = \sum_{j=1}^{N} m_j^* \lambda(B_j \cap A_{n,k}) > \frac{1}{k} \sum_{j=1}^{N} \lambda(B_j \cap A_{n,k}) = \frac{\lambda(A_{n,k})}{k}.$$

On the other hand,

$$\mathcal{U}_L(\mathcal{Q}, s_n) = \sum_{j=1}^{N} (M_j - m_j)\lambda(B_j \cap A_{n,k})$$

$$\leq \sum_{j=1}^{N} (M_j - m_j)\lambda(B_j) = \mathcal{U}_L(\mathcal{P}_n, f) - \mathcal{L}_L(\mathcal{P}_n, f) < \frac{1}{n}.$$

Combining the above two inequalities gives $\lambda(A_{n,k}) < k/n$ for all $k, n \in \mathbb{N}$. Since $E_k \subset A_{n,k}$ for all n, for each k,

$$\lambda(E_k) < \frac{k}{n}$$

for all $n \in \mathbb{N}$. Therefore $\lambda(E_k) = 0$. Finally since

$$\lambda(E) \leq \sum_{k=1}^{\infty} \lambda(E_k),$$

we have $\lambda(E) = 0$. Thus $\varphi = \psi$ a.e. on $[a, b]$ which proves the result. $\square$

Comparison with the Riemann Integral

The definition of the Lebesgue integral is very similar to that of the Riemann integral, except that in the Lebesgue theory we use measurable partitions rather than point partitions. If $\mathcal{P} = \{x_0, x_1, ..., x_n\}$ is a partition of $[a, b]$, then

$$\mathcal{P}^* = \{[x_0, x_1]\} \cup \{(x_{k-1}, x_k]\}_{k=2}^{n}$$

is a measurable partition of $[a, b]$. Furthermore, if f is a bounded real-valued function on $[a, b]$, then

$$\mathcal{L}(\mathcal{P}, f) = \mathcal{L}_L(\mathcal{P}^*, f) \qquad \text{and} \qquad \mathcal{U}(\mathcal{P}, f) = \mathcal{U}_L(\mathcal{P}^*, f).$$

Therefore, the lower Riemann integral of f satisfies

$$\underline{\int_a^b} f = \sup\{\mathcal{L}(\mathcal{P}, f) : \mathcal{P} \text{ is a partition of } [a, b]\}$$

$$\leq \sup\{\mathcal{L}_L(\mathcal{Q}, f) : \mathcal{Q} \text{ is a measurable partition of } [a, b]\}.$$

Similarly, for the upper Riemann integral of f we have

$$\overline{\int_a^b} f \geq \inf\{\mathcal{U}_L(\mathcal{Q}, f) : \mathcal{Q} \text{ is a measurable partition of } [a, b]\}.$$

If f is Riemann integrable on $[a, b]$, then the upper and lower Riemann integrals of f are equal, and thus

$$\int_a^b f(x)\, dx \leq \sup_{\mathcal{Q}} \mathcal{L}_L(\mathcal{Q}, f) \leq \inf_{\mathcal{Q}} \mathcal{U}_L(\mathcal{Q}, f) \leq \int_a^b f(x)\, dx,$$

where the supremum and infimum are taken over all measurable partitions $\mathcal{Q}$ of $[a, b]$. As a consequence of Theorem 10.6.2, this proves the following result.

COROLLARY 10.6.8 *If f is Riemann integrable on $[a, b]$, then f is Lebesgue integrable on $[a, b]$, and*

$$\int_{[a,b]} f \, d\lambda = \int_a^b f(x) \, dx.$$

The converse however is false! This is illustrated by the following example.

EXAMPLE 10.6.9 Let $E = [0, 1] \setminus \mathbb{Q}$, and set

$$f(x) = \chi_E(x) = \begin{cases} 1, & \text{when } x \text{ is irrational,} \\ 0, & \text{when } x \text{ is rational.} \end{cases}$$

By Example 6.1.6(a) the function f is not Riemann integrable. On the other hand, since f is a simple function, f is Lebesgue integrable, and by Lemma 10.6.7,

$$\int_{[0,1]} f \, d\lambda = \lambda(E) = 1. \quad \square$$

Properties of the Lebesgue Integral for Bounded Functions

The following theorem summarizes some basic properties of the Lebesgue integral for bounded functions.

THEOREM 10.6.10 *Suppose f, g are bounded real-valued measurable functions on $[a, b]$. Then*

(a) *for all $\alpha, \beta \in \mathbb{R}$,* $\displaystyle\int_{[a,b]} (\alpha f + \beta g) \, d\lambda = \alpha \int_{[a,b]} f \, d\lambda + \beta \int_{[a,b]} g \, d\lambda.$

(b) *If A_1, A_2 are disjoint measurable subsets of $[a, b]$, then*

$$\int_{A_1 \cup A_2} f \, d\lambda = \int_{A_1} f \, d\lambda + \int_{A_2} f \, d\lambda.$$

(c) *If $f \geq g$ a.e. on $[a, b]$, then* $\displaystyle\int_{[a,b]} f \, d\lambda \geq \int_{[a,b]} g \, d\lambda.$

(d) *If $f = g$ a.e. on $[a, b]$, then* $\displaystyle\int_{[a,b]} f \, d\lambda = \int_{[a,b]} g \, d\lambda.$

(e) $\displaystyle\left| \int_{[a,b]} f \, d\lambda \right| \leq \int_{[a,b]} |f| \, d\lambda.$

Proof. Since the proof of (a) is similar to the proof of the corresponding result for the Riemann integral we leave it as an exercise (Exercise 4). For the proof of (b), by definition

$$\int_{A_1 \cup A_2} f \, d\lambda = \int_{[a,b]} f \chi_{A_1 \cup A_2} \, d\lambda.$$

Since $A_1 \cap A_2 = \emptyset$, $f\chi_{A_1 \cup A_2} = f\chi_{A_1} + f\chi_{A_2}$, and the result now follows by (a).

(c) Consider the function $h(x) = f(x) - g(x)$. By hypothesis $h \geq 0$ a.e. on $[a, b]$. Let

$$E_1 = \{x : h(x) \geq 0\} \qquad \text{and} \qquad E_2 = [a, b] \setminus E_1.$$

Consider the measurable partition $\mathcal{P} = \{E_1, E_2\}$ of $[a, b]$. Then

$$\int_{[a,b]} h \, d\lambda \geq \mathcal{L}_L(\mathcal{P}, f) = m_1 \lambda(E_1) + m_2 \lambda(E_2).$$

Since $h(x) \geq 0$ for all $x \in E_1$, $m_1 = \inf\{h(x) : x \in E_1\} \geq 0$. On the other hand, since $h \geq 0$ a.e., $\lambda(E_2) = 0$. Therefore $\int_a^b h \, d\lambda \geq 0$. The result now follows by (a).

The result (d) is an immediate consequence of (c), and (e) is left for the exercises. The measurability of $|f|$ follows from Exercise 8 or 10 of the previous section. $\square$

Bounded Convergence Theorem

One of the main advantages of the Lebesgue theory of integration involves the interchange of limits. If $\{f_n\}$ is a sequence of Riemann integrable functions on $[a, b]$ such that $f_n(x)$ converges to a function $f(x)$ for all $x \in [a, b]$, then there is no guarantee that f is Riemann integrable on $[a, b]$. An example of such a sequence was given in Example 8.1.2(c). For the Lebesgue integral however we have the following very useful result.

THEOREM 10.6.11 (Bounded Convergence Theorem) *Suppose $\{f_n\}$ is a sequence of real-valued measurable functions on $[a, b]$ for which there exists a positive constant M such that $|f_n(x)| \leq M$ for all $n \in \mathbb{N}$, and all $x \in [a, b]$. If*

$$\lim_{n \to \infty} f_n(x) = f(x) \quad \text{a.e. on } [a, b],$$

then f is Lebesgue integrable on $[a, b]$ and

$$\int_{[a,b]} f \, d\lambda = \lim_{n \to \infty} \int_{[a,b]} f_n \, d\lambda.$$

Remark. Although we state and prove the bounded convergence theorem for a closed and bounded interval $[a, b]$, the conclusion is still valid if the sequence

$\{f_n\}$ is defined on a bounded measurable set A. The necessary modifications to the proof are left to the exercises.

Proof. Since $f_n \to f$ a.e., f is measurable by Corollary 10.5.10, and thus Lebesgue integrable. Let

$$E = \{x \in [a,b] : f_n(x) \text{ does not converge to } f(x)\}.$$

Define the functions g and g_n, $n \in \mathbb{N}$, on $[a,b]$ as follows:

$$g_n(x) = \begin{cases} f_n(x), & x \in [a,b] \setminus E, \\ 0, & x \in E, \end{cases} \quad \text{and} \quad g(x) = \begin{cases} f(x), & x \in [a,b] \setminus E, \\ 0, & x \in E. \end{cases}$$

Since $\lambda(E) = 0$, $g_n = f_n$ a.e. and $g = f$ a.e. Therefore

$$\int_a^b g_n \, d\lambda = \int_a^b f_n \, d\lambda \quad \text{and} \quad \int_a^b g \, d\lambda = \int_a^b f \, d\lambda.$$

Furthermore, $g_n(x) \to g(x)$ for all $x \in [a,b]$. Let $\epsilon > 0$ be given. For $m \in \mathbb{N}$, set

$$E_m = \{x \in [a,b] : |g(x) - g_n(x)| < \epsilon \text{ for all } n \geq m\}.$$

Then $E_1 \subset E_2 \subset \cdots$ with $\bigcup_{n=1}^\infty E_m = [a,b]$. Therefore

$$\bigcap_{m=1}^\infty E_m^c = \emptyset.$$

Here $E_m^c = [a,b] \setminus E_m$. Thus by Theorem 10.4.6 $\lim_{m \to \infty} \lambda(E_m^c) = 0$. Choose $m \in \mathbb{N}$ such that $\lambda(E_m^c) < \epsilon$. Then $|g(x) - g_n(x)| < \epsilon$ for all $n \geq m$ and all $x \in E_m$. Therefore

$$\left| \int_a^b f \, d\lambda - \int_a^b f_n \, d\lambda \right| = \left| \int_a^b g \, d\lambda - \int_a^b g_n d\lambda \right| \leq \int_{[a,b]} |g - g_n| \, d\lambda$$

$$= \int_{E_m} |g - g_n| \, d\lambda + \int_{E_m^c} |g - g_n| \, d\lambda$$

$$< \epsilon \lambda(E_m) + 2M \lambda(E_m^c) < \epsilon[b - a + 2M].$$

Since $\epsilon > 0$ was arbitrary, we have $\lim_{n \to \infty} \int_{[a,b]} f_n d\lambda = \int_{[a,b]} f \, d\lambda$. $\square$

Combining the bounded convergence theorem with Corollary 10.6.8, we obtain the bounded convergence theorem for Riemann integrable functions previously stated in Chapter 8. The theorem does require the additional hypothesis that the limit function f is Riemann integrable.

THEOREM 8.4.3 Let f and f_n, $n \in \mathbb{N}$, be Riemann integrable functions on $[a,b]$ with $\lim_{n \to \infty} f_n(x) = f(x)$ for all $x \in [a,b]$. Suppose there exists a positive constant M such that $|f_n(x)| \leq M$ for all $x \in [a,b]$ and all $n \in \mathbb{N}$. Then

$$\lim_{n \to \infty} \int_a^b f_n(x) \, dx = \int_a^b f(x) \, dx.$$

EXAMPLES 10.6.12 (a) In the first example we show that the conclusion of the bounded convergence theorem is false if the sequence $\{f_n\}$ is not bounded; that is, there does not exist a finite constant M such that $|f_n(x)| \le M$ for all $n \in \mathbb{N}$, and all $x \in [a, b]$. For each $n \in \mathbb{N}$, define f_n on $[0, 1]$ by

$$f_n(x) = \begin{cases} n, & 0 < x \le \frac{1}{n}, \\ 0, & \text{otherwise.} \end{cases}$$

Then $\{f_n\}_{n=1}^{\infty}$ is a sequence of measurable functions on $[0, 1]$ that is not bounded but which satisfies

$$\lim_{n \to \infty} f_n(x) = f(x) = 0 \qquad \text{for all } x \in [0, 1].$$

However, $\int_0^1 f_n \, d\lambda = n\lambda((0, 1/n]) = 1$. Thus

$$\lim_{n \to \infty} \int_{[0,1]} f_n \, d\lambda = 1 \ne 0 = \int_{[0,1]} f \, d\lambda.$$

(b) As our second example we consider the sequence $\{f_n\}$ of Example 9.2.3. For each $n \in \mathbb{N}$, write $n = 2^k + j$, where $k = 0, 1, 2...$, and $0 \le j < 2^k$. Define f_n on $[0, 1]$ by

$$f_n(x) = \begin{cases} 1, & \frac{j}{2^k} \le x \le \frac{j+1}{2^k}, \\ 0, & \text{otherwise.} \end{cases}$$

The first few of these are as follows: $f_1 = \chi_{[0,1]}$, $f_2 = \chi_{[0,\frac{1}{2}]}$, $f_3 = \chi_{[\frac{1}{2},1]}$, $f_4 = \chi_{[0,\frac{1}{4}]}, ...$ For each $n \in \mathbb{N}$, $f_n \in \mathcal{R}[0, 1]$ with

$$\int_0^1 f_n(x) \, dx = \frac{1}{2^k}.$$

Thus $\lim\limits_{n \to \infty} \int_0^1 f_n(x) \, dx = 0$. On the other hand, if $x \in [0, 1]$, then the sequence $\{f_n(x)\}$ contains an infinite number of 0's and 1's, and thus does not converge. $\square$

Exercises 10.6.

1. *Let f be a bounded real-valued measurable function on $[a, b]$ with $m \le f(x) < M$ for all $x \in [a, b]$. Set $\beta = M - m$. For $n \in \mathbb{N}$ and $j = 1, ..., n$, let $E_j = \{x \in [a, b] : m + (j - 1)\frac{\beta}{n} \le f(x) < m + j\frac{\beta}{n}\}$.

    The **Lebesgue sums** for f are defined by

    $S_n(f) = \sum\limits_{j=1}^{n} (m + (j - 1)\frac{\beta}{n})\lambda(E_j)$. Prove that

    $$\lim_{n \to \infty} S_n(f) = \int_{[a,b]} f \, d\lambda.$$

2. Prove Lemma 10.6.5.

3. *Let f be a bounded measurable function on $[a,b]$. If A is a measurable subset of $[a,b]$ with $\lambda(A) = 0$, prove that $\int_A f\, d\lambda = 0$.

4. **a.** Prove Theorem 10.6.10(a).

   b. Prove Theorem 10.6.10(e).

5. *Let f be a nonnegative bounded measurable function on $[a,b]$. If E, F are measurable subsets of $[a,b]$ with $E \subset F$, prove that $\int\limits_E f\, d\lambda \le \int\limits_F f\, d\lambda$.

6. Let f be a bounded measurable function on $[a,b]$. For each $c > 0$, prove that

$$\lambda(\{x \in [a,b] : |f(x)| > c\}) \le \frac{1}{c}\int_{[a,b]} |f|\, d\lambda.$$

7. *Let f be a nonnegative bounded measurable function on $[a,b]$ satisfying $\int_{[a,b]} f\, d\lambda = 0$. Use the previous exercise to prove that $f = 0$ a.e. on $[a,b]$.

8. **(Fundamental Theorem of Calculus for the Lebesgue Integral)** If f is differentiable on $[a,b]$ and f' is bounded on $[a,b]$, then f' is Lebesgue integrable, and

$$\int_{[a,b]} f'\, d\lambda = f(b) - f(a).$$

9. Prove that $\sum\limits_{n=0}^{\infty} \int\limits_0^{\pi/2} (1 - \sqrt{\cos x})^n \sin x\, dx$ converges to a finite limit, and find that limit.

10. Let f be a bounded measurable function on $[a,b]$ such that $|f| < 1$ a.e. on $[a,b]$. Prove that $\lim\limits_{n\to\infty} \int\limits_{[a,b]} f^n d\lambda = 0$.

11. Let $\{f_n\}$ be a sequence of nonnegative measurable functions on $[a,b]$ satisfying $f_n(x) \le M$ for all $x \in [a,b]$ and $n \in \mathbb{N}$. If $\{f_n\}$ converges to f a.e. on $[a,b]$, prove that

$$\lim_{n\to\infty} \int_{[a,b]} f_n e^{-f_n}\, d\lambda = \int_{[a,b]} f e^{-f}\, d\lambda.$$

12. *If f is a bounded real-valued measurable function on $[a,b]$, prove that there exists a sequence $\{s_n\}$ of simple functions on $[a,b]$ such that $\lim\limits_{n\to\infty} s_n(x) = f(x)$ uniformly on $[a,b]$.

13. Modify the proof of the bounded convergence theorem where the interval $[a,b]$ is replaced by a bounded measurable set A.

14. Use Egorov's theorem (Exercise 15, Section 10.5) to provide an alternate proof of the bounded convergence theorem.

15. Let f be a bounded measurable function on $[a,b]$.

    a. Given $\epsilon > 0$, prove that there exists a simple function φ on $[a,b]$ such that $\int\limits_{[a,b]} |f - \varphi|\, d\lambda < \epsilon$.

    b. If φ is a simple function on $[a,b]$ and $\epsilon > 0$, prove that there exists a step function h on $[a,b]$ such that $\varphi(x) = h(x)$ except on a set of measure less than ϵ.

c. Given $\epsilon > 0$, prove that there exists a step function h on $[a,b]$ such that $\int\limits_{[a,b]} |f - h|\, d\lambda < \epsilon$.

16. Let $S_n(x) = \frac{1}{2}A_0 + \sum\limits_{k=1}^{n} A_k \cos kx + B_k \sin kx$. If $|S_n(x)| \le M$ for all $x \in [-\pi, \pi]$ and $n \in \mathbb{N}$ and $f(x) = \lim\limits_{n\to\infty} S_n(x)$ exists a.e. on $[-\pi, \pi]$, prove that f is measurable and that the A_k and B_k are the Fourier coefficients of f.

10.7   The General Lebesgue Integral

In this section, we extend the definition of the Lebesgue integral to include both the case where the function f is unbounded, and also where the domain of integration is unbounded. We will then prove the well known results of Fatou and Lebesgue on the interchange of limits and integration. We first consider the extension of the Lebesgue integral to nonnegative measurable functions.

The Lebesgue Integral of a Nonnegative Measurable Function

Suppose first that A is a bounded measurable subset of $\mathbb{R}$, and that f is a nonnegative measurable function defined on A. For each $n \in \mathbb{N}$, consider the function f_n defined on A by

$$f_n(x) = \min\{f(x), n\} = \begin{cases} f(x), & \text{if } f(x) \le n, \\ n, & \text{if } f(x) > n. \end{cases}$$

Then $\{f_n\}$ is a sequence of nonnegative bounded measurable functions defined on A, with $\lim\limits_{n\to\infty} f_n(x) = f(x)$ for all $x \in A$. Furthermore, if $m > n$, then

$$f_n(x) \le f_m(x) \le f(x)$$

for all $x \in A$, and thus the sequence

$$\left\{ \int_A f_n\, d\lambda \right\}_{n=1}^{\infty}$$

is monotone increasing, and therefore converges either to a real number, or diverges to ∞. This leads us to make the following definition:

DEFINITION 10.7.1

(a) *Let f be a nonnegative measurable function defined on a bounded measurable subset A of $\mathbb{R}$. The* **Lebesgue integral** *of f over A, denoted $\int_A f\,d\lambda$, is defined by*

$$\int_A f\,d\lambda = \lim_{n\to\infty} \int_A f_n\,d\lambda = \sup_n \int_A \min\{f,n\}\,d\lambda.$$

(b) *If A is an unbounded measurable subset of $\mathbb{R}$ and f is a nonnegative measurable function on A, the* **Lebesgue integral** *of f over A, denoted $\int_A f\,d\lambda$, is defined by*

$$\int_A f\,d\lambda = \lim_{n\to\infty} \int_{A\cap[-n,n]} f\,d\lambda.$$

In part (b) of the definition, the sequence

$$\left\{ \int_{A\cap[-n,n]} f\,d\lambda \right\}_{n\in\mathbb{N}}$$

is also monotone increasing, and thus converges either to a nonnegative real number, or diverges to ∞. In the definition we do not exclude the possible value of ∞ for the integral of f. If the integral however is finite, we make the following definition.

DEFINITION 10.7.2 *A nonnegative measurable function f defined on a measurable subset A of $\mathbb{R}$ is said to be* **Lebesgue integrable** *on A if*

$$\int_A f\,d\lambda < \infty.$$

Remark. If A is either a finite or infinite interval with endpoints $a, b \in \mathbb{R}\cup\{-\infty,\infty\}$, then the integral of a nonnegative measurable function f on A is also denoted by $\int_a^b f\,d\lambda$.

EXAMPLES 10.7.3 (a) For our first example we consider the function $f(x) = 1/\sqrt{x}$ defined on $(0,1)$. Then for each $n \in \mathbb{N}$,

$$f_n(x) = \min\{f(x),n\} = \begin{cases} n, & 0 < x < \frac{1}{n^2}, \\ \frac{1}{\sqrt{x}}, & \frac{1}{n^2} < x < 1. \end{cases}$$

Therefore

$$\int_0^1 f_n\,d\lambda = \int_0^{1/n^2} n\,d\lambda + \int_{1/n^2}^1 \frac{1}{\sqrt{x}}\,dx = \frac{1}{n} + \left(2 - \frac{2}{n}\right) = 2 - \frac{1}{n}.$$

As a consequence

$$\int_{(0,1)} f \, d\lambda = \lim_{n \to \infty} \int_{(0,1)} f_n \, d\lambda = \lim_{n \to \infty} \left(2 - \tfrac{1}{n}\right) = 2.$$

The answer in this example corresponds to the improper Riemann integral of the function f. This will always be the case for nonnegative functions for which the improper Riemann integral exists (Exercise 18).

(b) Let $g(x) = 1/x$, $0 < x \le 1$. For the function g,

$$\min\{g(x), n\} = \begin{cases} n, & 0 < x < \tfrac{1}{n}, \\ \tfrac{1}{x}, & \tfrac{1}{n} \le x < 1. \end{cases}$$

Therefore

$$\int_0^1 \min\{g, n\} \, d\lambda = \int_0^{1/n} n \, d\lambda + \int_{1/n}^1 \frac{1}{x} \, dx = 1 + \ln n.$$

Thus

$$\int_0^1 g \, d\lambda = \lim_{n \to \infty} (1 + \ln n) = \infty.$$

Since the Lebesgue integral of g is infinite, g is not integrable on $(0, 1]$.

(c) As our final example, consider $f(x) = x^{-2}$ defined on $A = (1, \infty)$. In this example, for $n \ge 2$, $A \cap [-n, n] = (1, n]$, and

$$\int_1^n f \, d\lambda = \int_1^n x^{-2} dx = 1 - \frac{1}{n}.$$

Therefore

$$\int_1^\infty f \, d\lambda = \lim_{n \to \infty} 1 - \frac{1}{n} = 1.$$

Thus f is integrable on $(1, \infty)$. $\square$

The following theorem summarizes some of the basic properties of the Lebesgue integral of nonnegative measurable functions. Integrability of the functions f and g are not required.

THEOREM 10.7.4 *Let f, g be nonnegative measurable functions defined on a measurable set A. Then*

(a) $\displaystyle\int_A (f + g) \, d\lambda = \int_A f \, d\lambda + \int_A g \, d\lambda$ *and* $\displaystyle\int_A cf \, d\lambda = c \int_A f \, d\lambda$ *for all $c > 0$.*

(b) *If A_1, A_2 are disjoint measurable subsets of A, then*

$$\int_{A_1 \cup A_2} f \, d\lambda = \int_{A_1} f \, d\lambda + \int_{A_2} f \, d\lambda.$$

(c) *If $f \le g$ a.e. on A, then $\displaystyle\int_A f \, d\lambda \le \int_A g \, d\lambda$, with equality if $f = g$ a.e. on A*

Proof. We will indicate the method of proof by proving part of (a). The remaining proofs are left to the exercises (Exercise 2). Suppose first that the set A is bounded. Let $h = f + g$. Since

$$\min\{f(x) + g(x), n\} \leq \min\{f(x), n\} + \min\{g(x), n\} \leq \min\{f(x) + g(x), 2n\},$$

we have $h_n \leq f_n + g_n \leq h_{2n}$ for all $n \in \mathbb{N}$. As a consequence

$$\int_A h_n \, d\lambda \leq \int_A f_n d\lambda + \int_A g_n d\lambda \leq \int_A h_{2n} d\lambda.$$

Suppose f, g are integrable on A. Then

$$\lim_{n \to \infty} \left(\int_A f_n d\lambda + \int_A g_n d\lambda \right) = \lim_{n \to \infty} \int_A f_n d\lambda + \lim_{n \to \infty} \int_A g_n d\lambda = \int_A f \, d\lambda + \int_A g \, d\lambda.$$

Therefore, since

$$\lim_{n \to \infty} \int_A h_n \, d\lambda = \lim_{n \to \infty} \int_A h_{2n} d\lambda = \int_A (f + g) \, d\lambda,$$

the result follows from the above by letting $n \to \infty$. If one, or both of the sequences $\{\int_A f_n d\lambda\}$, $\{\int_A g_n \, d\lambda\}$ diverges to ∞, then so does their sum. In this case, we obtain $\int_A (f + g) d\lambda = \infty$. If A is unbounded, then by the above, for each $n \in \mathbb{N}$,

$$\int_{A \cap [-n,n]} (f + g) \, d\lambda = \int_{A \cap [-n,n]} f \, d\lambda + \int_{A \cap [-n,n]} g \, d\lambda,$$

and the result again follows by letting $n \to \infty$. □

As a consequence of the previous theorem if f and g are nonnegative integrable functions on the measurable set A, then so is $f + g$ and cf for every $c > 0$. Furthermore, if $f \leq g$ a.e. and g is integrable, then so is f.

Fatou's Lemma

Our first major convergence theorem for integrals of nonnegative measurable functions is the following result of Fatou.

THEOREM 10.7.5 (Fatou's Lemma) *If $\{f_n\}$ is a sequence of nonnegative measurable functions on a measurable set A, and $\lim_{n \to \infty} f_n(x) = f(x)$ a.e. on A, then*

$$\int_A f \, d\lambda \leq \varliminf_{n \to \infty} \int_A f_n \, d\lambda.$$

Proof. Suppose first that the set A is bounded. For each $k \in \mathbb{N}$, let

$$h_n(x) = \min\{f_n(x), k\} \quad \text{and} \quad h(x) = \min\{f(x), k\}.$$

Then for each $k \in \mathbb{N}$, the sequence $\{h_n\}$ converges a.e. to h on A. Since $|h_n(x)| \le k$ for all $x \in A$, by the bounded convergence theorem

$$\int_A \min\{f, k\} \, d\lambda = \lim_{n \to \infty} \int_A \min\{f_n, k\} \, d\lambda \le \lim_{n \to \infty} \int_A f_n \, d\lambda.$$

Since the above holds for each $k \in \mathbb{N}$,

$$\int_A f \, d\lambda = \lim_{k \to \infty} \int_A \min\{f, k\} \, d\lambda \le \lim_{n \to \infty} \int_A f_n \, d\lambda.$$

If A is unbounded, then by the above, for each $k \in \mathbb{N}$

$$\int_{A \cap [-k,k]} f \, d\lambda \le \lim_{n \to \infty} \int_{A \cap [-k,k]} f_n \, d\lambda \le \lim_{n \to \infty} \int_A f_n \, d\lambda.$$

Letting $k \to \infty$ will give the desired result. $\square$

EXAMPLE 10.7.6 In this example, we show that equality need not hold in Fatou's lemma. Consider the sequence $\{f_n\}$ on $[0, 1]$ of Example 10.6.12(a). For each $n \in \mathbb{N}$, $f_n(x) = n$ if $0 < x \le \frac{1}{n}$, and $f_n(x) = 0$ elsewhere. This sequence satisfies

$$0 = \int_{[0,1]} (\lim_{n \to \infty} f_n) \, d\lambda < 1 = \lim_{n \to \infty} \int_{[0.1]} f_n \, d\lambda. \quad \square$$

Remark. Fatou's lemma is often used to prove that the limit function f of a convergent sequence of nonnegative Lebesgue integrable functions is Lebesgue integrable. For if $\underline{\lim} \int_A f_n d\lambda < \infty$ and if $f_n \to f$ a.e. on A with $f_n \ge 0$ a.e. for all n, then by Fatou's lemma $\int_A f \, d\lambda < \infty$. Thus f is integrable on A.

The General Lebesgue Integral

We now turn our attention to the case where f is an arbitrary real-valued measurable function defined on a measurable subset A of $\mathbb{R}$. As in Exercise 10 of Section 10.5, we define the functions f^+ and f^- on A as follows:

$$f^+(x) = \max\{f(x), 0\}, \quad f^-(x) = \max\{-f(x), 0\}.$$

If $f(x) > 0$, then $f^+(x) = f(x)$ and $f^-(x) = 0$. On the other hand, if $f(x) < 0$, then $f^+(x) = 0$ and $f^-(x) = -f(x)$. If f is measurable on A, then f^+ and f^- are nonnegative measurable functions on A with

$$f(x) = f^+(x) - f^-(x) \quad \text{and} \quad |f(x)| = f^+(x) + f^-(x)$$

for all $x \in A$.

Our natural inclination is to define the integral of f over A by

$$\int_A f \, d\lambda = \int_A f^+ \, d\lambda - \int_A f^- \, d\lambda.$$

The only problem with this definition is that it is possible that $\int_A f^+ d\lambda = \int_A f^- d\lambda = \infty$ giving the undefined $\infty - \infty$ in the above. However, if we assume that both f^+ and f^- are integrable on A, then the above definition makes sense. Furthermore, if f is measurable, and f^+ and f^- are both integrable on A, then $|f|$ is also integrable on A. Conversely, if f is measurable and $|f|$ is integrable on A, then since $f^+ \leq |f|$ and $f^- \leq |f|$, by Theorem 10.7.4(c) both f^+ and f^- are integrable on A. Therefore we make the following definition.

DEFINITION 10.7.7 *Let f be a measurable real-valued function defined on a measurable subset A of $\mathbb{R}$. The function f is said to be* **Lebesgue integrable** *on A if $|f|$ is Lebesgue integrable on A. The set of Lebesgue integrable functions on A is denoted by $\mathcal{L}(A)$. For $f \in \mathcal{L}(A)$, the* **Lebesgue integral** *of f on A is defined by*

$$\int_A f \, d\lambda = \int_A f^+ \, d\lambda - \int_A f^- \, d\lambda.$$

Remark. The set $\mathcal{L}(A)$ of Lebesgue integrable functions on A is often also denoted by $\mathcal{L}^1(A)$.

The definition of the general Lebesgue integral is consistent with our definition of the Lebesgue integral of a bounded function on $[a, b]$. If f is a bounded real-valued measurable function on $[a, b]$ then so are the functions f^+ and f^-. Let $E_1 = \{x : f(x) \geq 0\}$ and $E_2 = \{x : f(x) < 0\}$. Then E_1 and E_2 are disjoint measurable subsets of $[a, b]$ with $E_1 \cup E_2 = [a, b]$. Furthermore $f\chi_{E_1} = f^+$ everywhere and $f\chi_{E_2} = -f^-$ a.e.. By Theorem 10.6.10(b)

$$\int_a^b f \, d\lambda = \int_{E_1} f \, d\lambda + \int_{E_2} f \, d\lambda$$

$$= \int_a^b f\chi_{E_1} \, d\lambda + \int_a^b f\chi_{E_2} \, d\lambda = \int_a^b f^+ \, d\lambda - \int_a^b f^- \, d\lambda.$$

Remark. For a nonnegative measurable function, the Lebesgue integral and the improper Riemann integral are the same, provided of course that the latter exists (Exercise 18). This however is false for functions that are <u>not</u> nonnegative. For example, consider the function $f(x) = (\sin x)/x$, $x \in [\pi, \infty)$, of Example 6.4.4(b). By Exercise 7 of Section 6.4, the improper integral of f exists on $[\pi, \infty)$. However, as was shown in Example 6.4.4,

$$\int_\pi^\infty |f| \, d\lambda = \lim_{n \to \infty} \int_\pi^{(n+1)\pi} \frac{|\sin x|}{x} \, dx = \infty.$$

Thus f is not Lebesgue integrable on $[\pi, \infty)$. Another such example for a finite interval is given in Exercise 23. The crucial fact to remember is that a measurable function f is Lebesgue integrable on a measurable set A if and only if $\int_A |f| \, d\lambda < \infty$.

Our first result is the following extension of Theorem 10.7.4 to the class of integrable functions.

THEOREM 10.7.8 *Suppose f and g are Lebesgue integrable functions on the measurable set A. Then*

(a) *$f + g$ and cf, $c \in \mathbb{R}$, are integrable on A with*

$$\int_A (f + g)\, d\lambda = \int_A f\, d\lambda + \int_A g\, d\lambda \quad and \quad \int_A cf\, d\lambda = c \int_A f\, d\lambda.$$

(b) *If $f \leq g$ a.e. on A, then $\int_A f\, d\lambda \leq \int_A g\, d\lambda$, with equality if $f = g$ a.e.*

(c) *If A_1 and A_2 are disjoint measurable subsets of A, then*

$$\int_{A_1 \cup A_2} f\, d\lambda = \int_{A_1} f\, d\lambda + \int_{A_2} f\, d\lambda.$$

Proof. The proof that cf is integrable and that $\int_A cf\, d\lambda = c \int_A f\, d\lambda$ follows immediately from the definition. Before proving the result about the sum, we first note that the definition of the integral of f on A is independent of the decomposition $f = f^+ - f^-$. Suppose that $f = f_1 - f_2$ where f_1 and f_2 are nonnegative integrable functions on A. Then

$$f^+ + f_2 = f^- + f_1,$$

and thus by Theorem 10.7.4,

$$\int_A f^+\, d\lambda + \int_A f_2\, d\lambda = \int_A f^-\, d\lambda + \int_A f_1\, d\lambda.$$

Since all the integrals are finite,

$$\int_A f\, d\lambda = \int_A f^+\, d\lambda - \int_A f^-\, d\lambda = \int_A f_1\, d\lambda - \int_A f_2\, d\lambda.$$

If f and g are integrable on A, then by definition so are the functions $f^+ + g^+$ and $f^- + g^-$. Since $f + g = (f^+ + g^+) - (f^- + g^-)$, by the above

$$\int_A (f + g)\, d\lambda = \int_A (f^+ + g^+)\, d\lambda - \int_A (f^- + g^-)\, d\lambda,$$

which by Theorem 10.7.4

$$= \int_A f^+\, d\lambda - \int_A f^-\, d\lambda + \int_A g^+\, d\lambda - \int_A g^-\, d\lambda$$

$$= \int_A f\, d\lambda + \int_A g\, d\lambda.$$

(b) If $f \leq g$ a.e., then $g - f \geq 0$ a.e. Therefore by part (a),

$$0 \leq \int_A (g - f)\, d\lambda = \int_A g\, d\lambda - \int_A f\, d\lambda,$$

from which the result follows. (c) The proof of (c) also follows from (a) and the fact that since $A_1 \cap A_2 = \emptyset$, $f\chi_{A_1 \cup A_2} = f\chi_{A_1} + f\chi_{A_2}$. $\quad\square$

Lebesgue's Dominated Convergence Theorem

Our second major convergence result is the following theorem of Lebesgue.

THEOREM 10.7.9 (Lebesgue's Dominated Convergence Theorem)
Let $\{f_n\}$ be a sequence of measurable functions defined on a measurable set A such that $\lim\limits_{n\to\infty} f_n(x) = f(x)$ exists a.e. on A. Suppose there exists a nonnegative integrable function g on A such that $|f_n(x)| \le g(x)$ a.e. on A. Then f is integrable on A and

$$\int_A f\, d\lambda = \lim_{n\to\infty} \int_A f_n\, d\lambda.$$

Proof. Since g is integrable on A, and $|f_n| \le g$ a.e. on A, by Theorem 10.7.4 each f_n is also integrable on A. Also, by Corollary 10.5.10, the function f is measurable on A. Furthermore, by Fatou's lemma,

$$\int_A |f|\, d\lambda \le \underline{\lim} \int_A |f_n|\, d\lambda \le \int_A g\, d\lambda < \infty.$$

Thus f is also integrable on A.

By redefining all the f_n, $n \in \mathbb{N}$, on a set of measure zero if necessary, we can without loss of generality assume that $|f_n(x)| \le g(x)$ for all $x \in A$. Consider the sequence $\{g + f_n\}_{n\in\mathbb{N}}$ of nonnegative measurable functions on A. By Fatou's lemma,

$$\int_A (g+f)\, d\lambda = \int_A \lim_{n\to\infty}(g+f_n)\, d\lambda \le \underline{\lim_{n\to\infty}} \int_A (g+f_n)\, d\lambda$$
$$= \int_A g\, d\lambda + \underline{\lim_{n\to\infty}} \int_A f_n d\lambda.$$

Therefore,

$$\int_A f\, d\lambda \le \underline{\lim_{n\to\infty}} \int_A f_n d\lambda.$$

Similarly, by applying Fatou's lemma to the sequence $\{g - f_n\}_{n\in\mathbb{N}}$, which is again a sequence of nonnegative functions on A,

$$\int_A (g-f)\, d\lambda \le \underline{\lim_{n\to\infty}} \int_A (g-f_n)\, d\lambda = \int_A g\, d\lambda + \underline{\lim_{n\to\infty}} \int_A (-f_n)d\lambda.$$

But $\underline{\lim\limits_{n\to\infty}} \int_A (-f_n)d\lambda = -\overline{\lim\limits_{n\to\infty}} \int_A f_n d\lambda$. Therefore,

$$\int_A f\, d\lambda \ge \overline{\lim_{n\to\infty}} \int_A f_n d\lambda.$$

Combining the two inequalities gives the desired result. $\square$

Remark. The hypothesis that there exists an integrable function g satisfying $|f_n| \le g$ a.e. is required in the proof in order to subtract $\int g\, d\lambda$ in the above inequalities. This is not possible if $\int g\, d\lambda = \infty$. As the following example shows, if such a function g does not exist, then the conclusion may be false.

EXAMPLE 10.7.10 As in Example 10.6.12(a) consider the sequence $\{f_n\}$ on $[0, 1]$ defined by

$$f_n(x) = \begin{cases} n, & 0 < x \le \frac{1}{n}, \\ 0, & \text{elsewhere.} \end{cases}$$

For the sequence $\{f_n\}$ we have $\lim_{n \to \infty} f_n(x) = 0$ for all $x \in [0, 1]$ but $\int_{[0,1]} f_n d\lambda = 1$ for all n. We now show that any measurable function g satisfying $g(x) \ge f_n(x)$ for all $x \in [0, 1]$ and $n \in \mathbb{N}$ satisfies $\int_{[0,1]} g \, d\lambda = \infty$. Since $g(x) \ge f_n(x)$ for all $x \in [0, 1]$ we have $g(x) \ge n$ for all $x \in (\frac{1}{n+1}, \frac{1}{n}]$. Since the collection $\{(\frac{1}{k+1}, \frac{1}{k}]\}_{k=1}^{n}$ of intervals is pairwise disjoint, by Theorem 10.7.4

$$\int_{[0,1]} g \, d\lambda \ge \sum_{k=1}^{n} \int_{(\frac{1}{k+1}, \frac{1}{k}]} g \, d\lambda \ge \sum_{k=1}^{n} k \, \lambda((\tfrac{1}{k+1}, \tfrac{1}{k}]) = \sum_{k=1}^{n} \frac{1}{k+1}.$$

Since the series $\sum 1/(k+1)$ diverges, we have $\int_{[0,1]} g \, d\lambda = \infty$. $\square$

Exercises 10.7

1. Let f be a nonnegative measurable function on a measurable set A. If $\int_A f \, d\lambda = 0$, prove that $f = 0$ a.e. on A.

2. *a. Prove Theorem 10.7.4(b).

   b. Prove Theorem 10.7.4(c).

3. Let A be a measurable subset of $\mathbb{R}$.

   a. If f is integrable on A and g is bounded and measurable on A, prove that fg is integrable on A.

   b. If f and g are integrable on A, is the function fg integrable on A?

4. *Let $f_p(x) = x^{-p}$, $x \in (0, 1)$. Prove that f_p is integrable on $(0, 1)$ for all p, $0 < p < 1$, and that

$$\int_{(0,1)} f_p \, d\lambda = \frac{1}{1-p}.$$

5. Define f on $[1, \infty)$ by

$$f(x) = \begin{cases} \sqrt{n} & \text{if } x \in [n, n + 1/n^2), \ n = 1, 2, \dots, \\ 0 & \text{otherwise.} \end{cases}$$

   Show that $f \in \mathcal{L}([1, \infty))$ but $f^2 \notin \mathcal{L}([1, \infty))$.

6. Let P denote the Cantor set of Section 2.5. Define f on $[0, 1]$ as follows: $f(x) = 0$ for every $x \in P$, and $f(x) = k$ for each x in the open interval of length $1/3^k$ on $[0, 1] \setminus P$. Prove that f is integrable on $[0, 1]$ and that

$$\int_0^1 f \, d\lambda = 3.$$

7. *Let f be a nonnegative integrable function on $[a, b]$. For each $n \in \mathbb{N}$, let $E_n = \{x : n \le f(x) < n + 1\}$. Prove that $\sum_{n=1}^{\infty} n\lambda(E_n) < \infty$.

8. Evaluate each of the following limits:

   a. $\lim_{n \to \infty} \int_{[0,1]} (1 - e^{-x^2/n}) x^{-1/2} \, dx$.   **b.** $\lim_{n \to \infty} \int_{[0,n]} (1 - x/n)^n e^{x/2} \, dx$.

9. *Let f be an integrable function on $[a, b]$. Given $\epsilon > 0$, prove that there exists a bounded measurable function g on $[a, b]$ such that $\int_a^b |f - g| \, d\lambda < \epsilon$.

10. *Suppose f is integrable on a measurable set A with $\lambda(A) = \infty$. Given $\epsilon > 0$, prove that there exists a bounded measurable set $E \subset A$ such that

$$\int_{A \setminus E} |f| \, d\lambda < \epsilon.$$

11. Show by example that Fatou's lemma is false if the functions f_n, $n \in \mathbb{N}$, are not nonnegative.

12. Show by example that the bounded convergence theorem is false for a measurable set A with $\lambda(A) = \infty$.

13. *Let f be a Lebesgue integrable function on a measurable set A. Prove that given $\epsilon > 0$, there exists a $\delta > 0$ such that

$$\int_E |f| \, d\lambda < \epsilon$$

for all measurable subsets E of A with $\lambda(E) < \delta$.

14. Let f be an integrable function on (a, b), where $-\infty \le a < b \le \infty$. Define F on (a, b) by

$$F(x) = \int_a^x f \, d\lambda, \qquad x \in (a, b).$$

   a. Prove that F is continuous on (a, b).

   b. If f is continuous at $x_o \in (a, b)$, prove that F is differentiable at x_o with $F'(x_o) = f(x_o)$.

15. *If $f \in \mathcal{L}([0, 1])$, show that for every $t \in \mathbb{R}$ the function $x \to \sin(tf(x))$ is in $\mathcal{L}([0, 1])$, and that $g(t) = \int_{[0,1]} \sin(tf(x)) \, d\lambda(x)$ is a differentiable function of $t \in \mathbb{R}$. Find $g'(t)$.

16. *If $f \in \mathcal{L}(\mathbb{R})$, prove that $\lim_{n \to \infty} \int_{\mathbb{R}} f(x) \sin nx \, d\lambda = 0$.

17. Let $f \in \mathcal{L}(\mathbb{R})$.

   a. Prove that $\int_{\mathbb{R}} f(x + t) \, d\lambda = \int_{\mathbb{R}} f(x) \, d\lambda$.

   b. Prove that $\lim_{t \to 0} \int_{\mathbb{R}} |f(x + t) - f(x)| \, d\lambda = 0$.

18. Let f be a nonnegative measurable function on $(a, b]$ satisfying $f \in \mathcal{R}[c, b]$ for every c, $a < c < b$. Prove that

$$\int_a^b f \, d\lambda = \lim_{c \to a^+} \int_c^b f(x) \, dx.$$

19. *(**Monotone Convergence Theorem**) Let $\{f_n\}$ be a monotone increasing sequence of nonnegative measurable functions on a measurable set A. Prove that

$$\int_A (\lim_{n\to\infty} f_n)\, d\lambda = \lim_{n\to\infty} \int_A f_n d\lambda.$$

20. Show that the monotone convergence theorem is false for decreasing sequences of measurable functions.

21. *a. Let f be a nonnegative measurable function on a measurable set A, and let $\{A_n\}$ be a sequence of pairwise disjoint measurable subsets of A with $\bigcup_n A_n = A$. Prove that

$$\int_A f\, d\lambda = \sum_{n=1}^{\infty} \int_{A_n} f\, d\lambda.$$

   b. Prove that the conclusion of part (a) is still valid for arbitrary $f \in \mathcal{L}(A)$.

22. Let $\{f_n\}$ be a sequence of measurable functions on $[a,b]$ satisfying $|f_n(x)| \leq g(x)$ a.e., where g is integrable on $[a,b]$. If $\lim_{n\to\infty} f_n(x) = f(x)$ exists a.e. on $[a,b]$, and h is any bounded measurable function on $[a,b]$, prove that

$$\int_a^b fh\, d\lambda = \lim_{n\to\infty} \int_a^b f_n h\, d\lambda.$$

23. *As in Exercise 4, Section 6.4, let f be defined on $(0,1)$ by

$$f(x) = \frac{d}{dx}\left(x^2 \sin \frac{1}{x^2}\right).$$

   Prove that f is not Lebesgue integrable on $(0,1)$.

24. Let f be a nonnegative measurable function on $[a,b]$. For each $t \geq 0$, let

$$m_f(t) = \lambda(\{x \in [a,b] : f(x) > t\}).$$

   a. Prove that $m_f(t)$ is monotone decreasing on $[0,\infty)$.

   b. Prove that $\int\limits_a^b f\, d\lambda = \int\limits_0^\infty m_f(t)\, dt.$

10.8   Square Integrable Functions

In analogy with the space ℓ^2 of square summable sequences, we define the space $\mathcal{L}^2$ of **square integrable functions** as follows.

DEFINITION 10.8.1 *Let A be a measurable subset of $\mathbb{R}$. We denote by $\mathcal{L}^2(A)$ the set of all measurable functions f on A for which $|f|^2$ is integrable on A. For $f \in \mathcal{L}^2(A)$, set*

$$\|f\|_2 = \left(\int_A |f|^2 d\lambda\right)^{1/2}.$$

The quantity $\|f\|_2$ is called the 2-**norm** or **norm** of f. Clearly, $\|f\|_2 \geq 0$, and from the definition it follows that if $f \in \mathcal{L}^2(A)$ and $c \in \mathbb{R}$, then $cf \in \mathcal{L}^2(A)$ with $\|cf\|_2 = |c|\|f\|_2$. We will shortly prove that if $f, g \in \mathcal{L}^2(A)$, then $f + g \in \mathcal{L}^2(A)$ with $\|f + g\|_2 \leq \|f\|_2 + \|g\|_2$. Thus $\mathcal{L}^2(A)$ is a vector space over $\mathbb{R}$.

If $f \in \mathcal{L}^2(A)$ satisfies $\|f\|_2 = 0$, then by Exercise 1, Section 10.7, $|f|^2 = 0$ a.e., and thus $f = 0$ a.e. on A. This does not mean that $f(x) = 0$ for all $x \in A$; only that $f = 0$ except on a set of measure zero. Thus $\|\ \|_2$ satisfies all the properties of a norm except for $\|f\|_2 = 0$ if and only if $f = 0$. To get around this difficulty we will consider any two functions f and g in $\mathcal{L}^2(A)$ for which $f = g$ a.e. as representing the same function. Formally, we define two measurable functions f and g on A to be **equivalent** if $f = g$ a.e.. In this way it is possible to define $\mathcal{L}^2(A)$ as the set of equivalence classes of square integrable functions on A. Rather than proceeding in this formal fashion, we will take the customary approach of simply saying that two functions in $\mathcal{L}^2$ are **equal** if and only if they are equal almost everywhere. With this definition $\mathcal{L}^2(A)$ is a normed linear space.

EXAMPLES 10.8.2

(a) For our first example let $f(x) = 1/\sqrt{x}$, $x \in (0,1)$. By Exercise 4 of Section 10.7, f is integrable on $(0,1)$ with $\int_0^1 f\, d\lambda = 2$. Since $f^2(x) = 1/x$, by Example 10.7.3(a), f^2 is not integrable on $(0,1)$ and thus $f \notin \mathcal{L}^2((0,1))$. On the other hand, if $g(x) = x^{-1/3}$, then $g^2(x) = x^{-2/3}$, which by Exercise 4 of the previous section is integrable. Thus $g \in \mathcal{L}^2((0,1))$ with

$$\|g\|_2^2 = \int_0^1 x^{-2/3}\, dx = 3.$$

(b) Consider the function $f(x) = 1/x$ for $x \in [1, \infty)$. For any $n \in \mathbb{N}$, $n \geq 2$,

$$\int_1^n |f|^2 d\lambda = \int_1^n x^{-2} dx = \left[1 - \frac{1}{n}\right].$$

Thus

$$\|f\|_2^2 = \int_1^\infty |f|^2 d\lambda = \lim_{n \to \infty} \int_1^n |f|^2\, d\lambda = 1$$

Therefore $f \in \mathcal{L}^2([1, \infty))$ with $\|f\|_2 = 1$. It is easily shown that $f \notin \mathcal{L}([1, \infty))$.
□

Cauchy-Schwarz Inequality

Our first result will be the analogue of the Cauchy-Schwarz inequality for ℓ^2. The following inequality is sometimes also referred to as Hölder's inequality.

THEOREM 10.8.3 (Cauchy-Schwarz Inequality) *Let A be a measurable subset of $\mathbb{R}$. If $f, g \in \mathcal{L}^2(A)$, then fg is integrable on A with*

$$\int_A |fg| \, d\lambda \leq \|f\|_2 \|g\|_2.$$

Proof. By Theorem 10.5.4, the product fg is measurable, and for any $x \in A$, we have

$$|f(x)g(x)| \leq \frac{1}{2}(|f(x)|^2 + |g(x)|^2).$$

Since by hypothesis $f, g \in L^2(A)$, the function $|fg|$ is integrable on A, and thus fg is integrable on A. As in the proof of Theorem 7.4.3, for $\gamma \in \mathbb{R}$

$$0 \leq \int_A (|f| - \gamma|g|)^2 = \|f\|_2^2 + \gamma^2 \|g\|_2^2 - 2\gamma \int_A |fg| \, d\lambda. \tag{10}$$

If $\|g\|_2 = 0$, then by Exercise 1 of Section 10.7, $g = 0$ a.e. on A. As a consequence, $\int_A |fg| \, d\lambda = 0$, and the conclusion holds. If $\|g\|_2 \neq 0$, set $\gamma = (\int_A |fg| \, d\lambda)/\|g\|_2^2$. With γ as defined, (10) becomes

$$0 \leq \|f\|_2^2 - \frac{(\int_A |fg| \, d\lambda)^2}{\|g\|_2^2},$$

and thus $(\int_A |fg| \, d\lambda)^2 \leq \|f\|_2^2 \|g\|_2^2$, which proves the result. $\square$

Our next result is Minkowski's inequality for the space $\mathcal{L}^2$. Since the proof of this is identical to the proof of Theorem 7.4.5, we leave the details to the exercises.

THEOREM 10.8.4 (Minkowski's Inequality) *Let A be a measurable subset of $\mathbb{R}$. If $f, g \in \mathcal{L}^2(A)$, then $f + g \in \mathcal{L}^2(A)$ with*

$$\|f + g\|_2 \leq \|f\|_2 + \|g\|_2.$$

Proof. Exercise 4. $\square$

The Normed Linear Space $\mathcal{L}^2([a, b])$

If A is a measurable subset of $\mathbb{R}$, the norm $\| \ \|_2$ on $\mathcal{L}^2(A)$ satisfies the following properties:

(a) $\|f\|_2 \geq 0$ for any $f \in \mathcal{L}^2(A)$.
(b) $\|f\|_2 = 0$ if and only if $f = 0$ a.e. on A.
(c) $\|cf\|_2 = |c| \|f\|_2$ for all $f \in \mathcal{L}^2(A)$ and $c \in \mathbb{R}$.
(d) $\|f + g\|_2 \leq \|f\|_2 + \|g\|_2$ for all $f, g \in \mathcal{L}^2(A)$.

Properties (a) and (c) follow from the definition, and (d) is Minkowski's inequality. With the convention that two functions f and g are **equal** if and only if $f = g$ a.e., $\mathcal{L}^2(A)$ is a normed linear space.

By Definition 8.3.9, a sequence $\{f_n\}$ in $\mathcal{L}^2$ converges to a function f in $\mathcal{L}^2$ if and only if $\lim_{n\to\infty} \|f - f_n\|_2 = 0$. Example 10.6.12(b) shows that convergence in $\mathcal{L}^2$ does not imply pointwise convergence of the sequence. As in Chapter 9, convergence in $\mathcal{L}^2$ is usually called **convergence in the mean** (or **norm convergence**). We now prove that $\mathcal{L}^2([a,b])$ is a complete normed linear space.

THEOREM 10.8.5 *The normed linear space* $(\mathcal{L}^2([a,b]), \| \ \|_2)$ *is complete.*

Remark. Although we state and prove the result for a closed and bounded interval, the same method of proof will work for $\mathcal{L}^2(A)$ where A is any measurable subset of $\mathbb{R}$.

Before proving the theorem, we first state and prove the following lemma.

LEMMA 10.8.6 *Let A be a measurable subset of $\mathbb{R}$. Suppose $\{f_n\}$ is a monotone increasing sequence of nonnegative measurable functions on A satisfying*

$$\int_A f_n d\lambda \le C, \qquad \text{for all } n \in \mathbb{N},$$

and for some finite constant C. Then $f(x) = \lim_{n\to\infty} f_n(x)$ is finite a.e. on A.

Proof. Since $\{f_n(x)\}$ is monotone increasing for each $x \in A$, the sequence either converges to a real number or diverges to ∞. Let $f(x) = \lim_{n\to\infty} f_n(x)$, and let

$$E = \{x \in A : f(x) = \infty\}.$$

We will prove that $\lambda(E) = 0$. For each $k \in \mathbb{N}$, let

$$E_k = \{x \in A : f(x) > k\}.$$

Then $E_k \supset E_{k+1}$ for all $k \in \mathbb{N}$ with $\bigcap_{k\in\mathbb{N}} E_k = E$. For fixed $k \in \mathbb{N}$, set

$$A_{n,k} = \{x \in A : f_n(x) > k\}, \qquad n \in \mathbb{N}.$$

Then $A_{n,k} \subset A_{n+1,k}$ with $\bigcup_{n\in\mathbb{N}} A_{n,k} = E_k$. Thus by Theorem 10.4.6(a),

$$\lambda(E_k) = \lim_{n\to\infty} \lambda(A_{n,k}).$$

But

$$\lambda(A_{n,k}) = \int_{A_{n,k}} 1 \, d\lambda \le \frac{1}{k} \int_{A_{n,k}} f_n d\lambda \le \frac{1}{k} \int_A f_n d\lambda \le \frac{C}{k}.$$

Therefore $\lambda(E_k) \le C/k$ for all $k \in \mathbb{N}$. Since $\lambda(E_1) < \infty$, by Theorem 10.4.6(b),

$$\lambda(E) = \lim_{k\to\infty} \lambda(E_k) = 0.$$

Therefore f is finite a.e. on A. $\square$

Proof of Theorem 10.8.5. Let $\{f_n\}$ be a Cauchy sequence in $\mathcal{L}^2$. Then given $\epsilon > 0$, there exists $n_o \in \mathbb{N}$ such that $\|f_n - f_m\|_2 < \epsilon$ for all $n, m \geq n_o$. For each $k \in \mathbb{N}$, let n_k be the smallest integer such that $\|f_m - f_n\|_2 < 1/2^k$ for all $m, n \geq n_k$. Then $n_1 \leq n_2 \leq \cdots \leq n_k \leq \cdots$, and

$$\|f_{n_{k+1}} - f_{n_k}\|_2 < \frac{1}{2^k}.$$

For each $k \in \mathbb{N}$, set

$$g_k = |f_{n_1}| + |f_{n_2} - f_{n_1}| + \cdots + |f_{n_{k+1}} - f_{n_k}|.$$

By Minkowski's inequality,

$$\int_{[a,b]} g_k^2 \, d\lambda = \|g_k\|_2^2 \leq \left(\|f_{n_1}\|_2 + \sum_{j=1}^{k} \|f_{n_{j+1}} - f_{n_j}\|_2 \right)^2$$

$$\leq \left(\|f_{n_1}\|_2 + \sum_{j=1}^{k} \frac{1}{2^k} \right)^2 \leq (\|f_{n_1}\|_2 + 1)^2.$$

Thus the sequence $\{g_k^2\}$ satisfies the hypothesis of Lemma 10.8.6. Therefore $\lim_{k \to \infty} g_k^2$ is finite a.e. on $[a, b]$. But then $\lim_{k \to \infty} g_k$ is also finite a.e. on $[a, b]$. As a consequence the series

$$|f_{n_1}(x)| + \sum_{j=1}^{\infty} |f_{n_{j+1}}(x) - f_{n_j}(x)|$$

converges a.e. on $[a, b]$, and therefore so does the series

$$f_{n_1}(x) + \sum_{j=1}^{\infty} (f_{n_{j+1}}(x) - f_{n_j}(x)).$$

But the kth partial sum of this series is $f_{n_{k+1}}(x)$. Therefore the sequence $\{f_{n_k}\}_{k \in \mathbb{N}}$ converges a.e. on $[a, b]$. Let E denote the set of $x \in [a, b]$ for which this sequence converges. Then $\lambda([a, b] \setminus E) = 0$. Define

$$f(x) = \begin{cases} \lim_{k \to \infty} f_{n_k}(x), & x \in E, \\ 0, & \text{otherwise.} \end{cases}$$

Then $\{f_{n_k}\}$ converges to f a.e. on $[a, b]$.

It remains to be shown that $f \in \mathcal{L}^2$, and that $\{f_n\}$ converges to f in $\mathcal{L}^2$. Since

$$|f_{n_{k+1}}| \leq |f_{n_1}| + \sum_{j=1}^{k} |f_{n_{j+1}} - f_{n_j}| = g_k,$$

by Fatou's lemma

$$\int_{[a,b]} |f|^2 d\lambda \le \lim_{k\to\infty} \int_{[a,b]} g_k^2 < \infty.$$

Thus $f \in \mathcal{L}^2$. Finally, since

$$f(x) - f_{n_k}(x) = \lim_{j\to\infty} (f_{n_{j+1}}(x) - f_{n_k}(x)) \quad \text{a.e.},$$

by Fatou's lemma again,

$$\int_{[a,b]} |f - f_{n_k}|^2 d\lambda \le \lim_{j\to\infty} \int_{[a,b]} |f_{n_{j+1}} - f_{n_k}|^2 < \left(\frac{1}{2^k}\right)^2.$$

Therefore $\|f - f_{n_k}\|_2 < 1/2^k$ for all $k \in \mathbb{N}$. Thus the subsequence $\{f_{n_k}\}_{k\in\mathbb{N}}$ converges to f in the norm of $\mathcal{L}^2$. Finally by the triangle inequality,

$$\|f - f_n\|_2 \le \|f - f_{n_k}\|_2 + \|f_{n_k} - f_n\|_2 < \frac{1}{2^k} + \|f_{n_k} - f_n\|_2.$$

From this it now follows that the original sequence $\{f_n\}$ also converges to f in the norm of $\mathcal{L}^2$. $\square$

Fourier Series

We close this chapter by making a few observations about Fourier series and the space $\mathcal{L}^2([-\pi, \pi])$. As in Definition 9.3.1, if f is Lebesgue integrable on $[-\pi, \pi]$, the **Fourier coefficients** of f with respect to the orthogonal system $\{1, \cos nx, \sin nx\}_{n=1}^{\infty}$ are given by

$$a_n = \frac{1}{\pi} \int_{-\pi}^{\pi} f(x) \cos nx \, dx, \quad n = 0, 1, 2, ...,$$

$$b_n = \frac{1}{\pi} \int_{-\pi}^{\pi} f(x) \sin nx \, dx, \quad n = 1, 2, ...$$

Since f is Lebesgue integrable, the functions $f(x) \sin nx$ and $f(x) \cos nx$ are measurable on $[-\pi, \pi]$, and in absolute value less than or equal to $|f(x)|$. Thus the functions $f(x) \cos nx$ and $f(x) \sin nx$ are all integrable on $[-\pi, \pi]$.

The same method of proof used in proving Bessel's inequality for Riemann integrable functions proves the following (Exercise 8): If $f \in \mathcal{L}^2([-\pi, \pi])$ and $\{a_k\}$ and $\{b_k\}$ are the Fourier coefficients of f, then

$$\frac{1}{2}a_0^2 + \sum_{k=1}^{\infty} a_k^2 + b_k^2 \le \frac{1}{\pi} \int_{-\pi}^{\pi} f^2 \, d\lambda. \qquad \textbf{(Bessel's Inequality)}$$

Thus the sequences $\{a_k\}_{k=0}^{\infty}$ and $\{b_k\}_{k=1}^{\infty}$ are square summable. We now use completeness of the space $\mathcal{L}^2([-\pi, \pi])$ to prove the converse.

THEOREM 10.8.7 *If* $\{a_k\}_{k=0}^{\infty}$ *and* $\{b_k\}_{k=1}^{\infty}$ *are any sequences of real numbers satisfying*

$$\frac{1}{2}a_0^2 + \sum_{k=1}^{\infty} a_k^2 + b_k^2 < \infty,$$

then there exists $f \in \mathcal{L}^2([-\pi, \pi])$ *whose Fourier coefficients are precisely* $\{a_k\}$ *and* $\{b_k\}$.

Proof. For each $n \in \mathbb{N}$, set

$$S_n(x) = \frac{1}{2}a_0 + \sum_{k=1}^{n} a_k \cos kx + b_k \sin kx.$$

Since each S_n is continuous, S_n is square integrable on $[-\pi, \pi]$. If $m < n$, then

$$\|S_n - S_m\|_2^2 = \int_{-\pi}^{\pi} \left[\sum_{k=m+1}^{n} a_k \cos kx + b_k \sin kx \right]^2 dx$$

which by orthogonality

$$= \pi \sum_{k=m+1}^{n} (a_k^2 + b_k^2).$$

Since the series converges, the sequence $\{S_n\}$ is a Cauchy sequence in $\mathcal{L}^2([-\pi, \pi])$. Thus by Theorem 10.8.5, there exists a function $f \in \mathcal{L}^2([-\pi, \pi])$ such that S_n converges to f in $\mathcal{L}^2$; i.e., $\lim\limits_{n \to \infty} \|f - S_n\|_2 = 0$. If $n > m$, then

$$\int_{-\pi}^{\pi} S_n(x) \cos mx \, dx = \pi a_m \qquad \text{and} \qquad \int_{-\pi}^{\pi} S_n(x) \sin mx \, dx = \pi b_m.$$

Therefore

$$\left| \frac{1}{\pi} \int_{-\pi}^{\pi} f(x) \cos mx \, dx - a_m \right| = \left| \frac{1}{\pi} \int_{-\pi}^{\pi} (f(x) - S_n(x)) \cos mx \, dx \right|$$

which by the Cauchy-Schwarz inequality

$$\leq \frac{1}{\pi} \|f - S_n\|_2 \| \cos mx \|_2 = \frac{1}{\sqrt{\pi}} \|f - S_n\|_2.$$

Since this holds for all $n > m$, letting $n \to \infty$ gives

$$a_m = \frac{1}{\pi} \int_{-\pi}^{\pi} f(x) \cos mx \, dx.$$

A similar argument proves that the b_m are the sine coefficients of f. $\square$

Is every Trigonometric Series a Fourier Series?

In Section 9.3 we showed that the series $\sum_{n=2}^{\infty} \dfrac{\sin nx}{\ln n}$, even though it converges for all $x \in \mathbb{R}$, is not the Fourier series of a Riemann integrable function on $[-\pi, \pi]$. Since $\sum (\ln n)^{-2} = \infty$, by Bessel's inequality it is also not the Fourier series of a square integrable function. This however does not rule out the possibility that it is the Fourier series of a Lebesgue integrable function. The following interesting classical result is very useful in providing an answer to this question. Since the proof of the theorem is beyond the level of this text, we state the result without proof.[2]

THEOREM 10.8.8

(a) *If* $b_n > 0$ *for all* n *and* $\displaystyle\sum_{n=1}^{\infty} \dfrac{b_n}{n} = \infty$, *then*

$$\sum_{n=1}^{\infty} b_n \sin nx$$

is not the Fourier series of a Lebesgue integrable function.

(b) *If* $\{a_n\}$ *is a sequence of nonnegative real numbers with* $\lim_{n \to \infty} a_n = 0$, *satisfying* $a_n \leq \frac{1}{2}(a_{n-1} + a_{n+1})$, *then the series*

$$\sum_{n=1}^{\infty} a_n \cos nx$$

is the Fourier series of a nonnegative Lebesgue integrable function on $[-\pi, \pi]$.

Since the sequence $\{1/(\ln n)\}$ satisfies hypothesis (a), the series $\sum_{n=2}^{\infty} \dfrac{\sin nx}{\ln n}$ is not the Fourier series of any integrable function on $[-\pi, \pi]$. However, it is interesting to note that the sequence $\{1/(\ln n)\}$ also satisfies hypothesis (b) (Exercise 13), and thus the series

$$\sum_{n=2}^{\infty} \dfrac{\cos nx}{\ln n}$$

is the Fourier series of a nonnegative Lebesgue integrable function.

---

[2] A proof of the result can be found in Chapter V of the text by Zygmund.

Exercises 10.8

1. *For $x \in (0,1)$ let $f_p(x) = x^{-p}$, $p > 0$. Determine all values of p such that $f_p \in L^2((0,1))$.

2. For each $n \in \mathbb{N}$, let $I_n = (n, n + 1/n^2)$. For a given sequence $\{c_n\}$ of real numbers, define f on $[1,\infty)$ by $f(x) = \sum\limits_{n=1}^{\infty} c_n \chi_{I_n}(x)$. Show that $f \in L^2([1,\infty))$ if and only if $\sum\limits_{n=1}^{\infty} \frac{c_n^2}{n^2} < \infty$.

3. Find an example of a real-valued function f on $(0,\infty)$ such that f^2 is integrable on $(0,\infty)$, but $|f|^p$ is not integrable on $(0,\infty)$ for any p, $0 < p < \infty$, $p \neq 2$.

   (Hint: Consider the function $g(x) = \dfrac{1}{x(1 + |\ln x|)^2}$.)

4. *Prove Theorem 10.8.4.

5. Let A be a measurable subset of $\mathbb{R}$ with $\lambda(A) < \infty$. If $f \in L^2(A)$, prove that $f \in \mathcal{L}(A)$ with
$$\int_A |f|\, d\lambda \leq \|f\|_2 (\lambda(A))^{1/2}.$$

6. Let $\{f_n\}$ be a sequence in $\mathcal{L}^2([a,b])$. Suppose $\{f_n\}$ converges in $\mathcal{L}^2$ to $f \in \mathcal{L}^2$ and $\{f_n\}$ converges a.e. to some measurable function g. Prove that $f = g$ a.e. on $[a,b]$.

7. *Let $\{f_n\}$ be a sequence in $\mathcal{L}^2(A)$ that converges in $\mathcal{L}^2$ to a function $f \in \mathcal{L}^2(A)$. If $g \in \mathcal{L}^2(A)$, prove that
$$\lim_{n\to\infty} \int_A f_n g\, d\lambda = \int_A f g\, d\lambda.$$

8. If $f \in \mathcal{L}^2([-\pi,\pi])$ and $\{a_k\}$ and $\{b_k\}$ are the Fourier coefficients of f, prove that
$$\frac{1}{2}a_0^2 + \sum_{k=1}^{\infty} a_k^2 + b_k^2 \leq \frac{1}{\pi}\int_{-\pi}^{\pi} f^2\, d\lambda.$$

9. Which of the following trigonometric series are Fourier series of an $\mathcal{L}^2$ function?

   *a. $\sum\limits_{n=1}^{\infty} \dfrac{\cos nx}{n}$.   b. $\sum\limits_{n=1}^{\infty} \dfrac{\sin nx}{\sqrt{n}}$.   c. $\sum\limits_{n=2}^{\infty} \dfrac{\cos nx}{\sqrt{n}\ln n}$.

10. Suppose E is a measurable subset of $(-\pi,\pi)$ with $\lambda(E) > 0$. Prove that for each $\delta > 0$, there exist at most finitely many integers n such that $\sin nx \geq \delta$ for all $x \in E$.

11. Let $f \in \mathcal{L}^2([a,b])$. Prove that given $\epsilon > 0$, there exists a continuous function g on $[a,b]$ such that $\|f - g\|_2 < \epsilon$. (Hint: First prove that there exists a simple function having the desired properties, and then use Exercise 15, Section 10.6, and Lemma 9.4.8.)

12. For $f \in \mathcal{L}^2([-\pi,\pi])$, let $\{a_k\}$ and $\{b_k\}$ be the Fourier coefficients of f.

    a. Prove Parseval's equality:

$$\frac{1}{2}a_0^2 + \sum_{k=1}^{\infty} a_k^2 + b_k^2 = \frac{1}{\pi}\int_{-\pi}^{\pi} f^2\, d\lambda.$$

b. If $a_k = b_k = 0$ for all k, prove that $f = 0$ a.e.

13. *Show that the sequence $\{1/(\ln n)\}$ satisfies the hypothesis of Theorem 10.8.8(b).

14. Let $\{\phi_k\}_{k=1}^{\infty}$ be an orthonormal family in $\mathcal{L}^2([a,b])$. If $\{c_k\}_{k=1}^{\infty} \in \ell^2$, prove that there exists a function $f \in \mathcal{L}^2([a,b])$ such that $c_k = \langle f, \phi_k \rangle$.

Furthermore, for this f, $\lim_{n\to\infty} \|s_n - f\|_2 = 0$, where $s_n = \sum_{k=1}^{n} c_k \phi_k$.

Notes

Lebesgue's development of the theory of measure and integration was one of the great mathematical achievements of the twentieth century. His proof that every bounded measurable function is Lebesgue integrable was based on the new idea of partitioning the range of a function, rather than its domain. Lebesgue's theory of integration also permitted him to provide necessary and sufficient conditions for Riemann integrability of a bounded function f.

In addition to the fact that the Lebesgue integral enlarged the family of integrable functions, the power of the Lebesgue integral results from the ease with which it handles the interchange of limits and integration. For the Riemann integral, uniform convergence of the sequence $\{f_n\}$ is required. Otherwise, the limit function may not be Riemann integrable. On the other hand, if $\{f_n\}$ is a sequence of measurable functions on $[a,b]$, then its pointwise limit f is also measurable. Hence if f is also bounded, then f is integrable. The bounded convergence theorem is notable for its simplicity of hypotheses and proof. It only requires that $\{f_n\}$ be uniformly bounded and converge a.e. on $[a,b]$. This is sufficient to ensure that

$$\lim_{n\to\infty} \int_{[a,b]} f_n\, d\lambda = \int_{[a,b]} \left(\lim_{n\to\infty} f_n\right) d\lambda.$$

With the additional hypothesis that the pointwise limit f is Riemann integrable, the bounded convergence theorem is also applicable to a sequence $\{f_n\}$ of Riemann integrable functions.

The bounded convergence theorem, or the dominated convergence theorem, are also the tools required to prove the fundamental theorem of calculus for the Lebesgue integral.

Theorem A If f is differentiable and f' is bounded on $[a,b]$, then f' is Lebesgue integrable, and $\int_a^b f'\, d\lambda = f(b) - f(a)$.

The proof of this result was requested in Exercise 8 of Section 10.6 . It follows simply by applying the bounded convergence theorem to the sequence $\{g_n\}$ defined by $g_n(x) = n[f(x + \frac{1}{n}) - f(x)]$. Since f is differentiable, the sequence $\{g_n\}$ converges pointwise to f. Also, by the mean value theorem the sequences $\{g_n\}$ is uniformly bounded on $[a,b]$. This then establishes the analogue of Theorem 6.3.2 for the

Lebesgue integral. If instead of bounded, one assumes that f' is Lebesgue integrable, then the result also follows by Lebesgue's dominated convergence theorem.

The Riemann theory of integration allows us to prove that if $f \in \mathcal{R}[a,b]$, and F is defined by $F(x) = \int_a^x f(t)dt$, then $F'(x) = f(x)$ at any $x \in [a,b]$ at which f is continuous. Since $f \in \mathcal{R}[a,b]$ if and only if f is continuous a.e. on $[a,b]$, we have that $F'(x) = f(x)$ a.e. on $[a,b]$. Although not proved in the text, this result is still valid for the Lebesgue integral.

Theorem B Let f be Lebesgue integrable on $[a,b]$, and define $F(x) = \int_a^x f\, d\lambda$. Then $F'(x) = f(x)$ a.e. $[a,b]$.

By writing $f = f^+ - f^-$, it suffices to assume that $f \geq 0$, and thus F is monotone increasing on $[a,b]$. It is a fact independent of integration that every monotone function is differentiable a.e. on $[a,b]$.[3] A slight generalization of Theorem A then gives that

$$\int_a^x F'\, d\lambda = F(x) - F(a) = \int_a^x f\, d\lambda,$$

or that $\int_a^x [F' - f]\, d\lambda = 0$ for all $x \in [a,b]$. As a consequence of Miscellaneous Exercise 5 we have $F' = f$ a.e. Newton and Leibniz realized the inverse relationship of differentiation and integration. The above two versions of the fundamental theorem of calculus provide a rigorous formulation of this inverse relationship for a large class of functions.

The Lebesgue theory of integration also provides the proper setting for the study of Fourier series. The bounded convergence theorem was used by Lebesgue to prove uniqueness of a Fourier series. If

$$f(x) = \frac{1}{2}A_0 + \sum_{k=1}^{\infty}(A_k \cos kx + B_k \sin kx),$$

for all $x \in [-\pi, \pi]$, then f, being the pointwise limit of a sequence of continuous functions, is automatically measurable on $[-\pi, \pi]$. If f is also bounded, then f is integrable. If the sequence $\{S_n\}$ of partial sums of the trigonometric series is uniformly bounded, then by the bounded convergence theorem, the trigonometric series is the Fourier series of f. In his 1903 paper, "Sur les series trigonometric," [4] Lebesgue showed that uniform boundedness of the partial sums may be removed; that boundedness of the function f itself was sufficient. This result was extended in 1912 by de la Vallée-Poussin[5] to the case were the function f is integrable on $[-\pi, \pi]$. The reader is referred to the article by Alan Gluchoff for an overview of how trigonometric series has influenced the various theories of integration.

---

[3] See page 208 of the text by Natanson.

[4] *Annales Scientifiques de l'École Normale Supérieure*, (3)**20** (1903), 453–485.

[5] "Sur l'unicité du développement trigonométrique," *Bull de l'Acad. Royale de Belgique* (1912), 702–718; see also Chapter 9 of the text by Zygmund.

Micellaneous Exercises

1. Let A be a measurable subset of $\mathbb{R}$. For $f \in \mathcal{L}(A)$ set

$$\|f\|_1 = \int_A |f| \, d\lambda.$$

Prove that $(\mathcal{L}(A), \| \ \|_1)$ is a complete normed linear space.

2. Let A be a measurable subset of $\mathbb{R}$, and let $f, f_n, n = 1, 2, \ldots$ be measurable functions on A. The sequence $\{f_n\}$ is said to **converge in measure** to f if for every $\delta > 0$,

$\lim_{n \to \infty} \lambda(\{x \in A : |f_n(x) - f(x)| \geq \delta\}) = 0.$

a. If $\{f_n\}$ is a sequence in $\mathcal{L}(A)$, and $\{f_n\}$ converges in the norm of $\mathcal{L}(A)$ to $f \in \mathcal{L}(A)$, prove that $\{f_n\}$ converges in measure to f.

b. Find a sequence $\{f_n\}$ of measurable functions on a measurable set A that converges to a function f in measure, but does not converge to f in norm.

c. If $\lambda(A)$ is finite and $\{f_n\}$ is a sequence of measurable functions that converges in measure to a measurable function f, prove that there exists a subsequence $\{f_{n_k}\}$ of $\{f_n\}$ such that $f_{n_k} \to f$ a.e. on A.

d. Show that the sequence $\{f_n\}$ of Example 10.6.12(b) converges to 0 in measure.

e. Find a sequence $\{f_n\}$ of measurable functions on a measurable set A such that $f_n \to 0$ everywhere on A but $\{f_n\}$ does not converge to 0 in measure.

3. Let $\{\varphi_n\}$ be a sequence of orthogonal functions on $[a, b]$ having the property that the only continuous real-valued function f satisfying $\int_a^b f \varphi_n d\lambda = 0$ for all $n \in \mathbb{N}$, is the zero function. Prove that the system $\{\varphi_n\}$ is complete. (Hint: First use the hypothesis to prove that if $f \in \mathcal{L}^2([a, b])$ satisfies $\int_a^b f \varphi_n \, d\lambda = 0$ for all $n \in \mathbb{N}$ then $f = 0$ a.e. Next use completeness of $\mathcal{L}^2$ to prove that Parseval's equality holds for every $f \in \mathcal{L}^2([a, b])$.)

4. (**Construction of a Nonmeasurable Set**) For each $x \in [-\frac{1}{2}, \frac{1}{2}]$, define the set $K(x)$ by

$K(x) = \{y \in [-\frac{1}{2}, \frac{1}{2}] : y - x \in \mathbb{Q}\}.$

a. Prove that for any $x, y \in [-\frac{1}{2}, \frac{1}{2}]$, either $K(x) \cap K(y) = \emptyset$ or $K(x) = K(y)$. (Note: $K(x) = K(y)$ does not imply that $x = y$; it only implies that $x - y$ is rational).

Consider the family $\mathcal{F} = \{K(x) : x \in [-\frac{1}{2}, \frac{1}{2}]\}$ of disjoint subsets of $[-\frac{1}{2}, \frac{1}{2}]$. Choose one point from each distinct set in this family and let A denote the set of points selected. The ability to choose such a point from each of the disjoint sets requires an axiom from set theory known as the

axiom of choice. Further information about this very important axiom can be found in the text by Halmos.

Let r_k, $k = 0, 1, 2, \ldots$, be an enumeration of the rationals in $[-1, 1]$, with $r_0 = 0$, and for each $k = 0, 1, 2..$, set $A_k = A + r_k$.

b. Show that the collection $\{A_k\}$ is pairwise disjoint with $[-\frac{1}{2}, \frac{1}{2}] \subset \bigcup_{k=0}^{\infty} A_k \subset [-\frac{3}{2}, \frac{3}{2}]$.

c. Use the above to show that $\lambda_*(A) = 0$ and $\lambda^*(A) > 0$, thus proving that A is nonmeasurable.

5. Suppose f is Lebesgue integrable on $[a, b]$. If $\int_a^x f \, d\lambda = 0$ for every $x \in [a, b]$, prove that $f = 0$ a.e..

Supplemental Reading

Botts, T., "Probability theory and the Lebesgue integral," *Math. Mag.* **42** (1969), 105–111.

Burkill, H., "The periods of a periodic function," *Math Mag.* **47** (1974), 206–210.

Darst, R. B., "Some Cantor sets and Cantor functions," *Math. Mag.* **45** (1972), 2–7.

Dressler, R. E. and Stromberg, K. R., "The Tonelli integral," *Amer. Math. Monthly* **81** (1974), 67–68.

Gluchoff, A. D., "Trigonometric series and theories of integration," *Math. Mag.* **67** (1994), 3–20.

Katznelson, Y. and Stromberg, K., "Everywhere differentiable, nowhere monotone function," *Amer. Math. Monthly* **81** (1974), 349–354.

Koliha, J. J., "A fundamental theorem of calculus for Lebesgue integration," *Amer. Math. Monthly* **113** (2006), 551–555.

Kraft, R. L., "What's the difference between Cantor sets," *Amer. Math. Monthly* **101** (1994), 640–650.

Maligranda, L., "A simple proof of the Hölder and Minkowski inequality," *Amer. Math. Monthly* **102** (1995), 256–259.

Mc Shane, E. J., "A unified theory of integration," *Amer. Math. Monthly* **80** (1973), 349–359.

Priestly, W. M., "Sets thick and thin," *Amer. Math. Monthly* **83** (1976), 648–650.

Thompson, B. S., "Monotone convergence theorem for the Riemann integral," *Amer. Math. Monthly* **117** (2010), 547–550.

Varberg, D. E., "On absolutely continuous functions," *Amer. Math. Monthly* **72** (1965), 831–841.

Xiang, J. X., "A note on the Cauchy-Schwarz inequality," *Amer. Math. Monthly* **120** (2013), 456–459.

Wade, W. R., "The bounded convergence theorem," *Amer. Math. Monthly* **81** (1974), 387–389.

Bibliography

Berg, P. W. & McGregor, J. L., *Elementary Partial Differential Equations*, Holden-Day, Oakland, CA, 1966.

Hewitt, E. & Stromberg, K., *Real and Abstract Analysis*, Springer-Verlag, New York, 1965.

Katz, Victor J., *A History of Mathematics*, Harper Collins, New York, 1993.

Natanson, I. P., *Theory of Functions of a Real Variable*, vol. I, Frederick Ungar Publ. Co., New York, 1964

Rudin, W., *Principles of Mathematical Analysis*, McGraw-Hill, Inc., New York, 1976.

Titchmarch, E. C., *The Theory of Functions*, Oxford University Press, 1939.

Weinberger, H. F., *Partial Differential Equations*, John Wiley & Sons, New York, 1965.

Zygmund, A., *Trigonometric Series*, vol. I & II, Cambridge University Press, London, 1968.

Hints and Solutions

Chapter 1

Exercises 1.1   page 5

2. (a) $A \cap B = B$, $A \cap \mathbb{Z} = \{-1, 0, 1, 2, 3, 4, 5\}$, $B \cap C = \{x : 2 \leq x \leq 3\}$. **(b)** $A \times B = \{(x, y) : -1 \leq x \leq 5, 0 \leq y \leq 3\}$. **4. (a)** Suppose $x \in A \cap (B \cap C)$. Then $x \in A$ and $x \in B \cap C$. Since $x \in B \cap C$, $x \in B$ and $x \in C$. Thus $x \in A \cap B$ and $x \in C$. Therefore $x \in (A \cap B) \cap C$. This proves $A \cap (B \cap C) \subset (A \cap B) \cap C$. The reverse containment is proved similarly. **7.(a)** Let $x \in A \cup (B \cap C)$. Then $x \in A$ or $x \in B \cap C$. If $x \in A$, then $x \in A \cup B$ and $x \in A \cup C$. Therefore $x \in (A \cup B) \cap (A \cup C)$. If $x \in B \cap C$, then $x \in B$ and $x \in C$. Hence $x \in A \cup B$ and $x \in A \cup C$, i.e., $x \in (A \cup B) \cap (A \cup C)$. Thus $A \cup (B \cap C) \subset (A \cup B) \cap (A \cup C)$. The reverse containment is proved similarly. **(c)** Let $x \in C \setminus (A \cap B)$. Then $x \in C$ and $x \notin A \cap B$. Since $x \notin A \cap B$ we have $x \notin A$ or $x \notin B$. If $x \notin A$ then $x \in C \setminus A$. Likewise, if $x \notin B$, then $x \in C \setminus B$. In either case, $x \in (C \setminus A) \cup (C \setminus B)$. Therefore $C \setminus (A \cap B) \subset (C \setminus A) \cup (C \setminus B)$. The reverse containment is proved similarly.
8. If $A = \{1, 2, 3\}$ then $\mathcal{P}(A) = \{\emptyset, \{1\}, \{2\}, \{3\}, \{1, 2\}, \{1, 3\}, \{2, 3\}, A\}$. **11.** Let $(x, y) \in A \times (B_1 \cup B_2)$. Then $x \in A$ and $y \in B_1 \cup B_2$. Thus $x \in A$ and $y \in B_1$ or $y \in B_2$. If $y \in B_1$ then $(x, y) \in A \times B_1$. If $y \in B_2$, then $(x, y) \in A \times B_2$. In either case, $(x, y) \in (A \times B_1) \cup (A \times B_2)$. The reverse containment is proved similarly.

Exercises 1.2   page 14

1. (b) No. The ordered pairs $(1, -1)$ and $(1, 3)$ contradict the definition of function. **(d)** In terms of ordered pairs $k = \{(-1, 1), (0, 3), (1, 5), (4, 7)\}$ and thus is a function from A into B. **2. (a)** No! The ordered pairs $(0, 1)$ and $(0, -1)$ are both elements of A. This however contradicts the definition of function. **3. (a)** $f(\{1, 2, 3, 4\}) = \{1, 3, 5, 7\}$ and $f^{-1}(\{1, 2, 3, 4\}) = \{1, 2\}$. **4. (a)** $f(A) = \{y : 0 \leq y \leq 9\}$; $f^{-1}(A) = \{x : x^3 + 1 \in A\} = \{x : -\sqrt[3]{3} \leq x \leq 1\}$. **(c)** $f^{-1}(x) = \sqrt[3]{x - 1}$, $x \in \mathbb{R}$.
5. (b) For $k \in \mathbb{N}$, $(f \circ g)(k) = 2k + 3$. Therefore $(f \circ g)(\mathbb{N}) = \{2k + 3 : k = 1, 2, 3, \dots\}$ which is the set of odd integers greater or equal to 5. **6. (b)** Range $f = \mathbb{R}$, f is one-to-one, and $x = f^{-1}(y) = \frac{1}{3}(y + 2)$. **(e)** Range $f = \mathbb{R}$, f is not one-to-one: If $y_1 \neq y_2$, then $(x, y_1) \neq (x, y_2)$ for any x, yet $f(x, y_1) = f(x, y_2)$. **(f)** Range $f = \{y : \frac{1}{2} \leq y \leq 1\}$. f is not one-to-one. If $0 < x \leq 1$, then $f(-x) = f(x)$. **7. (a)** Range $f = \{(x, y) \in \mathbb{R} \times \mathbb{R} : x^2 + y^2 = 1\}$. **(b)** $f^{-1}((1, 0)) = 0$, $f^{-1}((0, -1)) = \frac{3\pi}{2}$. **11.** Assume f is not one-to-one and show that this leads to a contradiction.

Exercises 1.3   page 20

1. (b) For $n = 1$, $1 = 1^2$. Assume the result is true for $n = k$. Then for $n = k + 1$, $1 + 3 + 5 + \cdots + (2k - 1) + (2(k + 1) - 1) = k^2 + (2k + 1) = (k + 1)^2$. **(d)** When $n = 1$, $1^3 = [\frac{1}{2} \cdot 1 \cdot 2]^2$. Assume true for $n = k$. Then for $n = k + 1$, $1^3 + 2^3 + \cdots + k^3 + (k + 1)^3 =$

533

$[\frac{1}{2}k(k+1)]^2 + (k+1)^3 = \frac{1}{4}(k+1)^2(k^2+4k+4) = [\frac{1}{2}(k+1)(k+2)]^2$. **(f)** When $n=1$, $x^2 - y^2 = (x-y)(x+y)$ and equality holds. For $n = k+1$ write $x^{k+2} - y^{k+2} = x^{k+2} - xy^{k+1} + xy^{k+1} - y^{k+2} = x(x^{k+1} - y^{k+1}) + (x-y)y^{k+1}$, and now apply the induction hypothesis. **2. (a)** The result is true for $n=1$. Assume that for $k \in \mathbb{N}$, $2^k > k$. Then by the induction hypothesis, $2^{k+1} = 2^k \cdot 2 > k \cdot 2 = k+k \geq k+1$. **(c)** For $n=4$, $4! = 24 > 16 = 2^4$. Thus the inequality is true when $n=4$. Assume that $n! > 2^n$ for some $n \geq 4$. Then $(n+1)! = (n+1)n! > (n+1)2^n > 2 \cdot 2^n = 2^{n+1}$. Thus by the modified principle of mathematical induction the inequality holds for all $n \in \mathbb{N}$, $n \geq 4$. **(d)** True for $n=3$. Assume true for $k \geq 3$. Then for $n = k+1$, $1^3 + 2^2 + \cdots + k^3 + (k+1)^3 < \frac{1}{2}k^4 + (k+1)^3 = \frac{1}{2}[k^4 + 2k^3 + 6k^2 + 6k + 2]$. But for $k \geq 2$, $6k + 2 < 4k + 1 + 2k^3$ from which the result now follows. **4.** For $n \in \mathbb{N}$ let $P(n)$ be the statement $f(n) = 3 \cdot 2^n + (-1)^n$. Then $P(n)$ is true for $n = 1, 2$. For $k \geq 3$, assume that $P(j)$ is true for all $j \in \mathbb{N}$, $j < k$. Use the fact that $f(k) = 2f(k-2) + f(k-1)$ and the induction hypothesis to show that $P(k)$ is true. Thus by the second principle of mathematical induction the result holds for all $n \in \mathbb{N}$. **5. (b)** $f(n) = n^2$. **(d)** $f(n) = 0$ if n is even and $f(n) = (-1)^{(n-1)/2}/n!$ if n is odd. **(f)** $f(2k+1) = \frac{1}{k+2}a_1$ and $f(2k) = \frac{3}{2k+3}a_2$. These can be proved by induction on k. **7.** For each $n \in \mathbb{N}$ let $S_n = r + r^2 + \cdots + r^n$. Then $S_n - rS_n = r - r^{n+1}$, from which the result follows. **8.** Hint: Let $A = \frac{1}{n}(a_1 + \cdots + a_n)$ and write $a_{n+1} = xA$ for some $x \geq 0$. Use the induction hypothesis to prove that $(a_1 \cdots a_n \cdot a_{n+1})^{1/(n+1)} \leq x^{1/(n+1)}A$. Now use Bernoulli's inequality to prove that $x^{1/(n+1)} \leq (n+x)/(n+1)$. From this it now follows that
$$x^{1/(n+1)}A \leq \frac{n+x}{n+1}A = \frac{1}{n+1}(a_1 + \cdots + a_n + a_{n+1}).$$

Exercises 1.4. page 28

4. (a) Consider $(a-b)^2$. **5. (a)** $\inf A = 0$, $\sup A = 1$. **(c)** $\inf C = 0$, $\sup C = \infty$. **(f)** $\inf F = 1$, $\sup E = 3$. **(h)** $\inf H = -2$, $\sup H = 2$ **8.** Apply Theorem 1.4.4. **14. (b)** Since A and B are non-empty and bounded above, $\alpha = \sup A$ and $\beta = \sup B$ both exist in $\mathbb{R}$. Since $\alpha = \sup A$ we have $a \leq \alpha$ for all $a \in A$. Similarly $b \leq \beta$ for all $b \in B$. Therefore $a + b \leq \alpha + \beta$ for all $a \in A$, $b \in B$. Thus $\alpha + \beta$ is an upper bound for $A + B$, and thus $\gamma = \sup(A+B) \leq \alpha + \beta$. To prove the reverse inequality, we first note that since γ is an upper bound for $A + B$, $a + b \leq \gamma$ for all $a \in A$, $b \in B$. Let $b \in B$ be arbitrary, but fixed. Then $a \leq \gamma - b$ for all $a \in A$. Thus $\gamma - b$ is an upper bound for A and hence $\alpha \leq \gamma - b$. Since this holds for all $b \in B$, we also have that $b \leq \gamma - \alpha$ for all $b \in B$. Thus $\beta \leq \gamma - \alpha$; i.e., $\alpha + \beta \leq \gamma$. **15. (a)** Let $\alpha = \sup\{f(x) : x \in X\}$, $\beta = \sup\{g(x) : x \in X\}$. since the range of f and g are bounded, α and β are finite with $f(x) + g(x) \leq \alpha + \beta$ for every $x \in X$. Therefore $\alpha + \beta$ is an upper bound for $\{f(x) + g(x) : x \in X\}$. Thus $\sup\{f(x) + g(x) : x \in X\} \leq \alpha + \beta$. **(d)** Let $\alpha = \sup\{g(x) : x \in X\}$. Hence $g(x) \leq \alpha$ for all $x \in X$. Thus by hypothesis $f(x) \leq \alpha$ for all $x \in X$. Therefore α is an upper bound for $\{f(x) : x \in X\}$. As a consequence $\sup\{f(x) : x \in x\} \leq \alpha$. **16. (a)** $F(x) = 3x + 2$, $\sup\{F(x) : x \in [0,1]\} = 5$. **(c)** Range $f = [0,5]$. Therefore $\sup\{f(x,y) : (x,y) \in X \times Y\} = 5$. **20. (a)** $F(x) = 3x + 2$ and $H(y) = 2y$. Thus $\sup\{H(y) : y \in [0,1]\} = 2$ and $\inf\{F(x) : x \in [0,1]\} = 2$.

Exercises 1.5. page 32

1. First prove that if p and q are positive integers then there exists $n \in \mathbb{N}$ such that $np > q$. If $p > q$ then $n = 1$ works. If $p \leq q$, consider $(q+1)p$. **4.** Suppose r_1, r_2 are rational with $r_1 < r_2$. Then $r_2 - r_1 > 0$ and $r = r_1 + \frac{1}{2}(r_2 - r_1)$ is rational with $r_1 < r < r_2$. **6. (a)** Use the fact $\sqrt{2}/2$ is irrational. **(b)** Use Theorem 1.5.2 and (a).   **8.** Since $x < y$ and $u > 0$, we have $x/u < y/u$. Now apply Theorem 1.5.2 and (a).

Exercises 1.6.   page 36

1. (a) $.0202020\ldots$ **2. (c)** $.0101 = \frac{1}{2^2} + \frac{1}{2^4} = \frac{3}{8}$.   **(d)** $.010101\cdots = \frac{1}{2^2} + \frac{1}{2^4} +$

$\frac{1}{2^6}\cdots = \frac{1}{2^2} \sum_{n=0}^{\infty} (\frac{1}{4})^n = \frac{1}{4}\left[\frac{1}{1-\frac{1}{4}}\right] = \frac{1}{3}$.   **(f)** $.001001\cdots = \frac{1}{2^3} + \frac{1}{2^6} + \frac{1}{2^9} + \cdots =$

$\frac{1}{2^3}\sum_{n=0}^{\infty} (\frac{1}{8})^n = \frac{1}{8}\left[\frac{1}{1-\frac{1}{8}}\right] = \frac{1}{7}$.   **3. (a)** $.0022 = \frac{0}{3} + \frac{0}{3^2} + \frac{2}{3^3} + \frac{2}{3^4} = \frac{2}{27} + \frac{2}{81} = \frac{8}{81}$.

(d) $.101010\cdots = \frac{1}{3} + \frac{0}{3^2} + \frac{1}{3^3} + \frac{0}{3^4} + \frac{1}{3^5} + \cdots = \frac{1}{3}\sum_{n=0}^{\infty} (\frac{1}{9})^n = \frac{1}{3}\left[\frac{1}{1-\frac{1}{9}}\right] = \frac{3}{8}$.   **(f)**

$.121212\cdots = \sum_{k=0}^{\infty} \frac{1}{3^{2k+1}} + \sum_{k=1}^{\infty} \frac{2}{3^{2k}} = \frac{1}{3}\sum_{n=0}^{\infty} (\frac{1}{9})^n + \frac{2}{9}\sum_{n=0}^{\infty} (\frac{1}{9})^n = \frac{1}{3}\cdot\frac{9}{8} + \frac{2}{9}\cdot\frac{9}{8} = \frac{5}{8}$.

4. $.010101\cdots$.   **5.** Finite binary expansion is $.0011$ whereas the infinite binary expansion is $.0010111\cdots$.

Exercises 1.7.   page 45

1. (c) Let $f : \mathbb{N} \to \mathbb{O}$ be defined by $f(n) = 2n-1$.   **4. (a)** $g(x) = a+x(b-a)$ is a one-to-one mapping of $(0,1)$ onto (a,b).   **6. (a)** Since $A \sim X$, there exists a one-to-one function h from A onto X. Similarly, there exists a one-to-one function g from B onto Y. To prove the result, show that $F : A \times B \to X \times Y$ defined by $F(a,b) = (h(a), g(b))$ is one-to-one and onto.   **8. (a)** $\bigcup A_n = \mathbb{R}$, $\bigcap A_n = \{x : -1 < x < 1\}$.
(c) $\bigcup A_n = (-1,2)$, $\bigcap A_n = [0,1]$.   **(e)** $\bigcup A_n = (0,1)$, $\bigcap A_n = \{\frac{1}{2}\}$.   **12. (a)** Let $y \in f(\bigcup_\alpha E_\alpha)$. Then $y = f(x)$ for some $x \in \bigcup_\alpha E_\alpha$. But then $x \in E_\alpha$ for some α. Therefore $y = f(x) \in f(E_\alpha)$. Thus $f(\bigcup_\alpha E_\alpha) \subset \bigcup_\alpha f(E_\alpha)$. The reverse containment follows similarly.   **13. (b)** Since the set of rational number $\mathbb{Q}$ is countable and the set of real numbers $\mathbb{R}$ is uncountable, by part (a) the set of irrational numbers, namely $\mathbb{R} \setminus \mathbb{Q}$, is uncountable.   **15. (a)** For $n \in \mathbb{N}$ let P_n denote the set of all polynomials in x of degree less than or equal to n with rational coefficients, and let $\mathbb{Q}^{n+1} = \mathbb{Q} \times \cdots \times \mathbb{Q}$ ($n+1$ times). By repeated application of Exercise 6 (b), $\mathbb{Q}^{n+1}$ is countable. Define $f : \mathbb{Q}^{n+1} \to P_n$ by $f(a_0, a_1, \ldots, a_n) = a_n x^n + \cdots + a_1 x + a_0$. Since f maps $\mathbb{Q}^{n+1}$ onto P_n and P_n is infinite, P_n is countable.   **(b)** By part (a) P_n is countable. Thus by Theorem 1.7.15, $\bigcup_n P_n$, the set of all polynomials with rational coefficients, is countable.   **18. (a)** Consider the function on $(0,1)$ that for each $n \in \mathbb{N}$, $n \geq 2$, maps $\frac{1}{n}$ to $\frac{1}{n-1}$, and is the identity mapping elsewhere.   **19.** For a polynomial $p(x) = a_n x^n + \cdots + a_1 x + a_0$, consider the height h of the polynomial defined by $h = n + |a_0| + |a_1| + \cdots |a_n|$. Prove that there are only a finite number of polynomials with integer coefficients of a given height h, and therefore only a finite number of algebraic numbers arising from polynomials of a given height h.   **22.** If f is a function from $A \to \mathcal{P}(A)$, show that f is not onto by considering the set $\{x \in A : x \notin f(x)\}$.   **23.** For $a, b \in [0,1]$ with decimal expansion $a = .a_1 a_2 \ldots$ and $b = .b_1 b_2 \ldots$, consider the function $f : [0,1] \times [0,1] \to [0,1]$ by $f(a,b) = .a_1 b_1 a_2 b_2 \ldots$.

Chapter 2

Exercises 2.1.   page 56

2. (b) We first note that $|x| = |x - y + y| \leq |x - y| + |y|$. Therefore $|x| - |y| \leq |x - y|$. Interchanging x and y gives $|y| - |x| \leq |y - x| = |x - y|$. Now use the definition of $||x| - |y||$.   **5. (a)** $-3 \leq x \leq 13/3$.   **(c)** $-1 < x < 2$.   **7. (c)** This is a metric.

The only nontrivial part is the triangle inequality. This follows from the following. Since the ln function is increasing on $(0, \infty)$, for a, b positive we have $\ln(1 + a + b) \leq \ln((1 + a)(1 + b)) = \ln(1 + a) + \ln(1 + b)$.

11. (a)(i) Since the points are collinear, the distance is just the usual euclidean distance, i.e. $d((\frac{1}{2}, \frac{1}{4}), (-\frac{1}{2}, -\frac{1}{4})) = \frac{1}{2}\sqrt{5}$. **(ii)** The points are not collinear. So $d((\frac{1}{2}, \frac{1}{2}), (0, 1)) = \frac{1}{2}\sqrt{2} + 1$. **12. (b)** For $x \in [0, 1]$, $|x - x^2| = x - x^2$, which has a maximum at $x = \frac{1}{2}$. Therefore $d(f, g) = \frac{1}{4}$. **14.** Again the only non-trivial part is the triangle inequality and it follows from the following: for a, b positive,

$$\frac{a + b}{1 + a + b} = \frac{a}{1 + a + b} + \frac{b}{1 + a + b} \leq \frac{a}{1 + a} + \frac{b}{1 + b}.$$

Exercises 2.2   page 67

2. Let $p \in O = \bigcup_\alpha O_\alpha$ be arbitrary. Then $p \in O_{\alpha_o}$ for some $\alpha_o \in A$. Since O_{α_o} is open, there exists $\epsilon > 0$ such that $N_\epsilon(p) \subset O_{\alpha_o}$. But then $N_\epsilon(p) \subset O$; i.e., p is an interior point of O. **3. (a)** By Corollary 2.2.16 a finite set has no limit points. Now apply Theorem 2.2.14. This can also be proved directly by showing that the complement of a finite set is the finite union of open intervals.
5. (b) Since $\sqrt{x_1^2 + x_2^2} \leq |x_1| + |x_2|$ we obtain $N_\epsilon^1(\mathbf{p}) \subset N_\epsilon^2(\mathbf{p})$. Likewise, since $\max\{|x_1|, |x_2|\} \leq \sqrt{x_1^2 + x_2^2}$ it follows that $N_\epsilon^2(\mathbf{p}) \subset N_\epsilon^\infty(\mathbf{p})$. The last containment follows since $|x_1| + |x_2| \leq 2\max\{|x_1|, |x_2|\}$. **8. (a)** For $E = (0, 1) \cup \{2\}$, Int $E = (0, 1)$, $E' = [0, 1]$, isolated points $= \{2\}$, $\overline{E} = [0, 1] \cup \{2\}$. **(d)** For $E = \{\frac{1}{n} : n \in \mathbb{N}\}$, Int$(E) = \emptyset$, $E' = \{0\}$, isolated points $= E$, $\overline{E} = E \cup \{0\}$. **9. (a)** $\{a, b\}$.
13. (a) Closed in X. **(c)** Neither. **16.** Since $\alpha \notin A$, for every $\epsilon > 0$ there exists an $a \in A$ such that $\alpha - \epsilon < a < \alpha$. Therefore α is a limit point of A. **17. (a)** Let $p \in \text{Int}(E)$ be arbitrary. Since p is an interior point of E, there exists an $\epsilon > 0$ such that $N_\epsilon(p) \subset E$. To show that $N_\epsilon(p) \subset \text{Int}(E)$ it remains to be shown that every $q \in N_\epsilon(p)$ is an interior point of E. **19. (a)** Since $A \cup B$ is a subset of $\overline{A} \cup \overline{B}$ (which is closed), by Theorem 2.2.18(c) $\overline{A \cup B} \subset \overline{A} \cup \overline{B}$. The reverse containment follows analogously. **22.** Let $\{r_n\}_{n=1}^\infty$ be an enumeration of $\mathbb{Q}$, and $\{\epsilon_n\}_{n=1}^\infty$ an enumeration of the positive rational numbers. Take $\mathcal{I} = \{N_{\epsilon_j}(r_n) : j, n \in \mathbb{N}\}$.
23. Suppose $U \subset \mathbb{R}$ is open and suppose $E \subset U$ is open in U. Let $p \in E$ be arbitrary. Use the fact that E is open in U and that U is open to show that there exists an $\epsilon > 0$ such that $N_\epsilon(p) \subset E$. Thus p is an interior point of E. Since $p \in E$ was arbitrary, E is open in $\mathbb{R}$. The converse is obvious. **25. (a)** Let $U = (0, 1)$ and $V = (\frac{3}{2}, \frac{5}{2})$.
26. Hint: Use the fact that $\overline{A} = A \cup A'$, and if U open satisfies $U \cap A' \neq \emptyset$ then $U \cap A \neq \emptyset$. **27.** Let A be a non-empty subset of $\mathbb{R}$. If A contains at least two points and is not an interval, then there exist $r, s \in A$ with $r < s$ and $t \in \mathbb{R}$ with $r < t < s$, but $t \notin A$. The open sets $U = (-\infty, t)$ and $V = (t, \infty)$ will prove that A is not connected. Therefore, every connected set is an interval. Conversely, suppose A is an interval and A is not connected. Then there exist disjoint open sets U and V with $A \cap U \neq \emptyset$, $A \cap V \neq \emptyset$, and $A \subset U \cup V$. Suppose $a \in A \cap U$ and $b \in A \cap V$. By Theorem 2.2.20 applied to U and V there exist disjoint open intervals I and J such that $a \in I$ and $b \in J$. Suppose $a < b$ and $J = (t, s)$. Show that $t \notin U \cup V$, but $t \in A$. This contradiction proves that A is connected.

Exercises 2.3.   page 73

1. (b) Use the fact that 0 is a limit point of A. **3. (a)** Let $\mathcal{U} = \{U_\alpha\}_{\alpha \in A}$ be an open cover of $A \cup B$. Then $\mathcal{U}$ is also an open cover of A and B, respectively. Now use the compactness of A and B to obtain a finite subcover of $A \cup B$. **4.** Since K is compact, by Theorem 2.3.5 K is closed. Now show that K is bounded. Let

$\alpha = \sup K$. This exists since K is non-empty and bounded. Now use the fact that K is closed to show that $\alpha \in K$.

Exercises 2.4.   page 76

1. Take $K_n = [0, n]$.   **3.** Suppose $I_n = [a_n, b_n]$. Since $I_n \supset I_m$ for all $m \geq n$, we have $a_n \leq a_m \leq b_m \leq b_n$ for all $m \geq n$. Let $a = \sup\{a_n\}$ and $b = \inf\{b_n\}$. Now show that $\cap I_n = [a, b]$.

Exercises 2.5.   page 79

4. If $x \in P$ with $x = .a_1 a_2 \cdots$, $a_n \in \{0, 2\}$, set $b_n = \frac{1}{2} a_n$. Consider the function $x \to .b_1 b_2 \cdots$.

Chapter 3

Exercises 3.1.   page 87

1. (a) Let $a_n = (3n + 5)/(2n + 7)$. Then $|a_n - \frac{3}{2}| = 4/(4n + 7) < 1/n$. Given $\epsilon > 0$, choose $n_o \in \mathbb{N}$ such that $n_o \geq 1/\epsilon$. Then for all $n \geq n_o$, $|a_n - \frac{3}{2}| < \epsilon$. Therefore $\lim a_n = \frac{3}{2}$.   **(c)** Set $a_n = (n^2 + 1)/2n^2$. Then $|a_n - \frac{1}{2}| = \frac{1}{2n^2} < \frac{1}{2n}$. Given $\epsilon > 0$, choose $n_o \in \mathbb{N}$ such that $n_o \geq 1/2\epsilon$. Then for all $n \geq n_o$, $|a_n - \frac{1}{2}| < \epsilon$. Thus $\lim a_n = \frac{1}{2}$.   **(f)** First show that $\sqrt{n+1} - \sqrt{n} = 1/(\sqrt{n+1} + \sqrt{n}) < 1/(2\sqrt{n})$. Now given $\epsilon > 0$ choose n_o such that $1/(2\sqrt{n}) < \epsilon$ for all $n \geq n_o$.   **2. (a)** If n is even then $n(1 + (-1)^n) = 2n$. Thus the sequence is unbounded and diverges in $\mathbb{R}$. **(c)** When n is even, i.e., $n = 2k$, then $\sin \frac{n\pi}{2} = \sin k\pi = 0$. On the other hand, when n is odd, i.e. $n = 2k+1$, then $\sin \frac{n\pi}{2} = \sin(2k+1)\frac{\pi}{2} = (-1)^k$. Thus $\sin \frac{n\pi}{2}$ assumes the values $-1, 0$, and 1 for infinitely many values of n. Therefore the sequence diverges. **4. (a)** Write $b = 1 + a$ where $a > 0$. By Example 1.3.2(b), $b^n \geq 1 + na$. Now use the previous exercise.   **5.** First show that $|a_n^2 - a^2| \leq (|a_n| + |a|)|a_n - a|$. Now use the fact that since $\{a_n\}$ converges, there exists a positive constant M such that $|a_n| \leq M$ for all $n \in \mathbb{N}$.   **6.** Consider $a = 0$ and $a > 0$ separately. For $a > 0$, $\sqrt{a_n} - \sqrt{a} = (a_n - a)/(\sqrt{a_n} + \sqrt{a})$.   **8.** Take $\epsilon = a/2$. Then for this ϵ, there exists $n_o \in \mathbb{N}$ such that $|a_n - a| < a/2$ for all $n \geq n_o$. From this it now follows that $a_n > a/2$ for all $n \geq n_o$.

Exercises 3.2.   page 93

5. (b) If $p > 1$, let $x_n = \sqrt[n]{p} - 1$. Apply the inequality of Example 1.3.2(b) to $(1 + x_n)^n$.   **6. (a)** 3/5.   **(c)** −1.   **(e)** 0.   **(g)** $a/2$.   **7. (a)** converges to 1. **(c)** converges to 0.   **(e)** converges to 2/3.   **8.** Use the fact that $|\cos x| \leq 1$. **10. (a)** Suppose $\lim a_{n+1}/a_n = L < 1$. Choose $\epsilon > 0$ such that $L + \epsilon < 1$. For this ϵ there exists $n_o \in \mathbb{N}$ such that $a_{n+1}/a_n < L + \epsilon$ for all $n \geq n_o$. From this one obtains that for $n > n_o$, $0 < a_n \leq (L + \epsilon)^{n - n_o} a_{n_o} = M(L + \epsilon)^n$, where $M = a_{n_o}/(L + \epsilon)^{n_o}$. Since $(L + \epsilon) < 1$, by Theorem 3.2.6(e), $\lim_{n \to \infty} (L + \epsilon)^n = 0$. The result now follows by Theorem 3.2.4.   **11. (a)** With $a_n = n^2 a^n$, $0 < a < 1$, $L = \lim_{n \to \infty} \frac{a_{n+1}}{a_n} = a \lim_{n \to \infty} (1 + \frac{1}{n})^2 = a$. Thus since $L = a < 1$, the sequence converges. **12.** Set $x_n = (a_n - 1)/(a_n + 1)$ and solve for a_n **13. (b)** Verify the result for $n = 1$. Assume the result holds for $n = k$. For $n = k + 1$, $(1 + a)^{(k+1)} = (1 + a)(1 + a)^k$ which by the induction hypothesis

$$= (1 + a) \sum_{j=0}^{k} \binom{k}{j} a^j = \sum_{j=0}^{k} \binom{k}{j} a^j + \sum_{j=1}^{k+1} \binom{k}{j-1} a^j.$$

Using part (a) show that the above is equal to $\sum_{j=0}^{k+1} \binom{k+1}{j} a^j$.   **14. (a)** By con-

sidering the sequence $\{a_k - a\}$ show first that one can assume $a = 0$.

Exercises 3.3.   page 101

1. Take $I_n = [n, \infty)$.   **2. (a)** Set $a_n = \sqrt{n^2 + 1}/n = \sqrt{1 + 1/n^2}$. Since $1/(n+1)^2 <$
$1/n^2$ we have $a_{n+1} < a_n$ and $\lim a_n = 1$.   **4.** Use mathematical induction to show
that $a_n > 1$ for all $n \in \mathbb{N}$. From the inequality $2ab \leq a^2 + b^2$, $a, b \geq 0$, we have
$2a_n \leq a_n^2 + 1$ or $a_{n+1} = 2 - 1/a_n \leq a_n$. Therefore $\{a_n\}$ is monotone decreasing.
Finally, if $a = \lim a_n$, then $a = \lim_{n \to \infty} a_{n+1} = \lim_{n \to \infty} (2 - 1/a_n) = 2 - 1/a$. Therefore
$a = 1$.   **6. (a)** Use induction to show that $x_n > \sqrt{\alpha}$ for all n. The inequality
$ab \leq \frac{1}{2}(a^2 + b^2)$, $a, b \geq 0$ should prove useful. Use the fact that $x_n > \sqrt{\alpha}$ to prove
that $\{x_n\}$ is monotone decreasing. Consider $x_{n+1} - x_n$ and simplify.   **8. (c)** $\{a_n\}$
is monotone increasing with $a_n \leq 3$ for all n. If $a = \lim a_n$ then $a^2 = 2a + 3$. Thus
$a = 3$.   **(e)** $\{a_n\}$ is monotone decreasing with $a_n > 2$ for all n and $\lim a_n = 2$.
10. (a) e^2 **(c)** $e^{3/2}$.   **11.** To show that $\{s_n\}$ is unbounded show that $s_{2^n} > 1 + \frac{n}{2}$.
Hint: First show it for $n = 1, 2$, and 3, then use mathematical induction to prove the
result for all $n \in \mathbb{N}$.   **13.** Hint: For $k \geq 2$, $\frac{1}{k^2} \leq \frac{1}{k(k-1)} = \frac{1}{k-1} - \frac{1}{k}$.   **15. (a)** Write
$a = 1 + b$ with $b > 0$. Now use the binomial theorem to show that $a^n/n \geq cn$ for
some positive constant c and n sufficiently large.   **(c)** $n + \frac{(-1)^n}{n} \geq n - \frac{1}{n} \geq (n-1)$.
16. (d) The sequence is not monotone: If $x_n = n + (-1)^n \sqrt{n}$, then $x_{2n+1} < x_{2n}$.
21. This problem is somewhat tricky. It is not sufficient to just choose a monotone
increasing sequence in the set; one also has to guarantee that the sequence converges
to the least upper bound of the set. Let E be a non-empty subset of $\mathbb{R}$ that is bounded
above. Let $\mathcal{U}$ denote the set of upper bounds of E. Since $E \neq \emptyset$, we can choose an
element $x_1 \in E$. Also, since E is bounded above, $\mathcal{U} \neq \emptyset$. Choose $\beta_1 \in \mathcal{U}$. Let
$\alpha_1 = \frac{1}{2}(x_1 + \beta_1)$, and consider the two intervals $[x_1, \alpha_1]$ and $(\alpha_1, \beta_1]$. Since $x_1 \in E$,
one, or both of these intervals have non-empty intersection with E. If $(\alpha_1, \beta_1] \cap E \neq \emptyset$,
choose $x_2 \in E$ such that $\alpha_1 < x_2 \leq b_1$. In this case set $\beta_2 = \beta_1$. If $(\alpha_1, \beta_1] \cap E = \emptyset$,
choose $x_2 \in E$ such that $x_1 \leq x_2 \leq \alpha_1$, and set $\beta_2 = \alpha_1$. In this case $\beta_2 \in \mathcal{U}$.
Proceeding inductively construct two monotone sequences $\{x_n\}$ and $\{\beta_n\}$ such that
(a) $\{x_n\} \subset E$ with $x_n \leq x_{n+1}$ for all n, (b) $\{\beta_n\} \subset \mathcal{U}$ with $\beta_n \geq \beta_{n+1}$ for all n,
and (c) $0 \leq \beta_n - x_n \leq 2^{-n+1}(b_1 - x_1)$. Assuming that every bounded monotone
sequence converges, let $\beta = \lim \beta_n$. By (c) we also have $\beta = \lim x_n$. It only remains
to be shown that $\beta = \sup E$.   **22.** Suppose $A = \{x_n : n \in \mathbb{N}\}$ is a countable subset
of $[0, 1]$. To show that $A \subsetneq [0, 1]$ proceed as follows: At least one of the three closed
intervals $[0, \frac{1}{3}], [\frac{1}{3}, \frac{2}{3}], [\frac{2}{3}, 1]$ does not contain x_1. Call it I_1. Divide I_1 into three closed
intervals of length $1/3^2$. At least one of these, say I_2, does not contain x_2.

Exercises 3.4.   page 106

3. (a) $\{-1, 0, 1\}$.   **(c)** $\{1\}$.   **(e)** $\{1, -3\}$.   **4. (a)** e^2.   **10.** For convenience we
take $n = 2$. Let $\{\mathbf{p}_n\}$ be a bounded sequence in $\mathbb{R}^2$ where for each n, $\mathbf{p}_n = (a_n, b_n)$.
But then the sequences $\{a_n\}$ and $\{b_n\}$ are also bounded. Since $\{a_n\}$ is bounded, by
the Bolzano-Weierstrass theorem there exists a subsequence $\{a_{n_k}\}$ that converges,
to say a. Since the sequence $\{b_{n_k}\}$ is also bounded, it has a convergent subsequence,
say $\{b_{n_{k_j}}\}$ that converges to say b. But then the subsequence $\{\mathbf{p}_{n_{k_j}}\}$ converges to
(a, b).

Exercises 3.5. page 112

1. (a) $\{-\infty, \infty\}$. (c) $\{-1, 1\}$. (f) $\{-\frac{2}{3}, \frac{2}{3}\}$. 3. $0, 1$. 6. (a) Let $\alpha = \underline{\lim} a_n$ and $\gamma = \overline{\lim}(a_n + b_n)$. Since the sequences are all bounded, $\alpha, \gamma \in \mathbb{R}$. Let $\epsilon > 0$ be given. Then by Theorems 3.5.3 and 3.5.4 there exists $n_o \in \mathbb{N}$ such that $a_n > \alpha - \epsilon/2$ for all $n \geq n_o$ and $(a_n + b_n) < \gamma + \epsilon/2$ for all $n \geq n_o$. Therefore $b_n < \gamma + \epsilon/2 - a_n < \gamma - \alpha + \epsilon$. From this it now follows that $\overline{\lim} b_n \leq \gamma - \alpha + \epsilon$. Since $\epsilon > 0$ was arbitrary, we have $\overline{\lim} b_n \leq \gamma - \alpha$; i.e., $\underline{\lim} a_n + \overline{\lim} b_n \leq \overline{\lim}(a_n + b_n)$. The other inequality is proved similarly. 8. $\{\frac{1}{2}, 1\}$. Hint: Consider the subsequences $\{s_{2m}\}$ and $\{s_{2m+1}\}$.
10. By Theorem 3.5.7, there exists a subsequence $\{a_{n_k}\}$ of $\{a_n\}$ such that $a_{n_k} \to a$. Since $\{b_n\}$ converges to b, $a_{n_k} b_{n_k} \to ab$. Therefore ab is a subsequential limit of $\{a_n b_n\}$. Thus $\overline{\lim} a_n b_n \leq ab$. The reverse inequality follows similarly. The fact that $b \neq 0$ is crucial.

Exercises 3.6. page 118

2. (a) The sequence $\{(n+1)/n\}$ converges and thus is Cauchy. (d) The sequence converges to zero and thus is Cauchy. 4. (a) Consider $s_{2n} - s_n$. 10. For $n \geq 3$, $|a_{n+1} - a_n| < \frac{1}{4}|a_n - a_{n-1}|$. Therefore the sequence $\{a_n\}$ is contractive. If $a = \lim a_n$, then $0 < a < 1$ and is a solution of $a^2 + 2a - 1 = 0$. 13. (b) Since $(a_{n+1} - a_n) = (b-1)(a_n - a_{n-1})$, by induction $(a_{n+1} - a_n) = (b-1)^{n-1}(a_2 - a_1)$. Therefore,

$$a_{n+1} - a_1 = \sum_{k=1}^{n}(a_{k+1} - a_k) = (a_2 - a_1)\sum_{k=0}^{n-1}(b-1)^k = (a_2 - a_1)\frac{1 - (b-1)^n}{2 - b}.$$

Letting $n \to \infty$ gives $a = a_1 + \frac{1}{2-b}(a_2 - a_1)$.

Exercises 3.7. page 123

1. Let $s_n = \sum_{k=1}^{n} \frac{1}{k^2}$. Since $\frac{1}{k^2} \leq \frac{1}{k-1} - \frac{1}{k}$, $k \geq 2$, for $n \geq 2$, $s_n \leq 1 + \sum_{k=2}^{n}\left(\frac{1}{k-1} - \frac{1}{k}\right) = 2 - \frac{1}{n} \leq 2$. Therefore $\{s_n\}$ is bounded above and hence converges by Theorem 3.7.6.
5. Use the inequality $ab \leq \frac{1}{2}(a^2 + b^2)$, $a, b \geq 0$.

Chapter 4

Exercises 4.1. page 142

1. (a) If $f(x) = 2x - 7$, $L = -3$, then $|f(x) - L| = 2|x - 2|$. Given $\epsilon > 0$ take $\delta = \epsilon/2$. Then for all x with $|x - 2| < \delta$, $|f(x) - L| = 2|x - 2| < 2\delta = \epsilon$. (c) If $f(x) = x/(1+x)$, then $|f(x) - \frac{1}{2}| = \frac{|x-1|}{2|x+1|} < \frac{1}{2}|x-1|$ for all $x > 0$. Hence given $\epsilon > 0$, choose $\delta = \min\{2\epsilon, 1\}$. With this choice of δ, $x > 0$ and thus $|f(x) - \frac{1}{2}| < \frac{1}{2}\delta \leq \epsilon$.
(e) Let $f(x) = (x^3 + 1)/(x + 1)$. Since $x^3 + 1 = (x+1)(x^2 - x + 1)$, we have $f(x) = x^2 - x + 1$ for $x \neq -1$. Therefore $|f(x) - 3| = |x+1||x-2|$. But for $-2 < x < 0$ we have $|x - 2| < 4$. Therefore for all such x we have $|f(x) - 3| < 4|x - (-1)|$. Given $\epsilon > 0$ take $\delta = \min\{\frac{\epsilon}{4}, 1\}$. Then for $|x - (-1)| < \delta$ we have $|f(x) - 3| < \epsilon$.
2. (c) We first note that $x^3 - p^3 = (x - p)(x^2 + xp + p^2)$. For $|x - p| < 1$ we have $|x| < |p| + 1$. Therefore $|x^3 - p^3| < (3|p|^2 + 3|p| + 1)|x - p|$. Hence given $\epsilon > 0$ choose δ so that $0 < \delta \leq \min\{1, \epsilon/(3|p|^2 + 3|p| + 1)\}$. Then for $|x - p| < \delta$ we have $|x^3 - p^3| < \epsilon$. (e) Note that for $x > 0$, $\sqrt{x} - \sqrt{p} = (x - p)/(\sqrt{x} + \sqrt{p})$. Therefore $|\sqrt{x} - \sqrt{p}| < |x - p|/\sqrt{p}$. Given $\epsilon > 0$ choose δ so that $0 < \delta < \sqrt{p}\epsilon$. Then if $|x - p| < \delta$ we have $x > 0$ and $|\sqrt{x} - \sqrt{p}| < \epsilon$. 3. (a) The limit does not exist. For $x > 0$, $x/|x| = 1$, whereas for $x < 0$, $x/|x| = -1$. (c) The limit does not exist. Consider the sequence $\{1/n\pi\}$ which has limit 0 as $n \to \infty$. (e) Since

$(x+1)^2 - 1 = x^2 + 2x$ we have that for $x \neq 0$, $[(x+1)^2 - 1]/x = x + 2$. Thus the
limit as $x \to 0$ exists and equals 2.   **4.** $\lim\limits_{x \to -1} f(x) = -3$.   **5. (a)** By Figure 4.5,
for $0 < t < \pi/2$, $\sin t = $ length of $PQ <$ length of arc $PR = t$.   **7. (a)** Use the
inequality $||f(x)| - |L|| \leq |f(x) - L|$.   **(c)** Use induction on n and Theorem 4.1.6.
8. (a) 0.   **(c)** $\sqrt{\frac{4}{7}}$.   **(e)** $\frac{1}{4}$.   **(g)** 2. Note: $\sin 2x/x = 2(\sin 2x/2x)$.   **9.** Let $L = \lim\limits_{x \to p} f(x)$. By hypothesis $L > 0$. Take $\epsilon = \frac{1}{2}L$ and use the definition of limit.
12. Since g is bounded on E, there exists a positive constant M such that $|g(x)| \leq M$
for all $x \in E$. Thus $|f(x)g(x)| \leq M|f(x)|$ for all $x \in E$. Now use the fact that
$\lim\limits_{x \to p} f(x) = 0$.   **14. (a)** By Theorem 4.1.6 (a), $\lim\limits_{x \to p} g(x) = \lim\limits_{x \to p} (f(x) + g(x)) - \lim\limits_{x \to p} f(x)$, both of which are assumed to exist.   **16.** Suppose $\lim\limits_{x \to \infty} f(x) = L$. Let
$\epsilon > 0$ be given. By definition there exists $M > 0$ such that $|f(x) - L| < \epsilon$ for
all $x \in (a, \infty)$ with $x > M$. Let $\delta = 1/M$. Then for all $t \in (0, 1/a)$ with $t < \delta$,
$1/t \in (a, \infty)$ and $1/t > M$. Therefore $|g(t) - L| = |f(\frac{1}{t}) - L| < \epsilon$. The proof that
$\lim\limits_{t \to 0} g(t) = L$ implies $\lim\limits_{x \to \infty} f(x) = L$ is similar.   **17. (a)** 3/2.   **(c)** 2.   **(e)** 1/2.
(g) Limit does not exist. For all $x > \pi/3$, $\cos \frac{1}{x} > 1/2$. Thus $x \cos \frac{1}{x} > x/2$ and
$x \cos \frac{1}{x} \to \infty$ as $x \to \infty$.

Exercises 4.2.   page 155

1. (c) Since $1 - \cos x = 2 \sin^2(x/2)$, for $x \neq 0$, $g(x) = \frac{2}{x} \sin^2(x/2)$. Now use the
fact that $|\sin t| \leq |t|$.   **(d)** Since $\lim\limits_{x \to 2} k(x)$ does not exist, k is not continuous
at $x_o = 2$.   **2. (b)** The function f is not continuous at 1. Consider $f(p_n)$ where
$\{p_n\}$ is a sequence of irrational numbers with $p_n \to 1$.   **4.** If $p > 0$, $|f(x) - f(p)| =$
$|\sqrt{x} - \sqrt{p}| = |x - p|/(\sqrt{x} + \sqrt{p}) < \frac{1}{\sqrt{p}} |x - p|$. Let $\epsilon > 0$ be given. Set $\delta = \min\{p, \sqrt{p}\epsilon\}$.
Then $|x - p| < \delta$ implies that $|f(x) - f(p)| < \epsilon$. Therefore f is continuous at p. If
$p = 0$, set $\delta = \epsilon^2$. Alternately use Theorem 4.1.3 and Exercise 6 of Section 3.1.
5. (b) Verify that $\lim\limits_{x \to 0} f(x) = 0$. Hence if we set $f(0) = 0$, the function f is
continuous on $[0, 1]$.   **6. (a)** See Exercise 7(a) of Section 4.1.   **7.** Use Theorem
4.2.4. and the fact that x^n is continuous on $\mathbb{R}$ for all $n \in \mathbb{N}$.   **9. (a)** $\mathbb{R} \setminus \{-2, 0, 2\}$.
(c) $\mathbb{R}$.   **8. (a)** Use an identity for $\cos x - \cos y$ and Exercise 5(a) of Section 4.1.
12 (a) With the metric d_2, verify that
$$d_2(f(x_1, y_1), f(x_2, y_2)) = \sqrt{2[(x_1 - x_2) + (y_1 - y_2)]^2} \leq 2\sqrt{(x_1 - x_2)^2 + (y_1 - y_2)^2} = 2d((x_1, y_1), (x_2, y_2)).$$
Hence given $\epsilon > 0$, take $\delta < \frac{1}{2}\epsilon$.   **14. (a)** Hint: Use the fact that $\max\{f(x), g(x)\} = \frac{1}{2}(f(x) + g(x) + |f(x) - g(x)|)$.   **16.** If $p(x)$ is a polynomial of odd degree, show
that $\lim\limits_{x \to \infty} p(x)/p(-x) = -1$. Hence there exists an $r > 0$ such that $p(r)$ and
$p(-r)$ are of opposite sign. Now apply the Intermediate Value Theorem to $p(x)$ on
$[-r, r]$.   **17.** Consider $g(x) = f(x) - f(x - 1)$, $x \in [0, 1]$.   **19.** Let $p \in E$ be a limit
point of F. Use Theorem 3.1.4 and continuity of f to show that $p \in F$.   **22.** Take
$\epsilon = 1$. Then for this choice of ϵ there exists a $\delta > 0$ such that $|f(x) - f(p)| \leq 1$ for all
$x \in N_\delta(p) \cap E$. Show that this implies that $|f(x)| \leq (|f(p)| + 1)$ for all $x \in N_\delta(p) \cap E$.
25. First show that $f(K)$ is closed as follows. Let q be a limit point of $f(K)$. Then
there exists a sequence $\{p_n\}$ in K such that $f(p_n) \to q$. Now use Theorem 3.4.5 and
continuity of f to prove that $q \in f(K)$. Now assume $f(K)$ is not bounded. Then
there exists a sequence $\{p_n\}$ in K such that $|f(p_n)| \to \infty$. Obtain a contradiction as
above.   **27. (a)** Suppose $x \in (g \circ f)^{-1}(V)$. Then $g(f(x)) \in V$. Hence by definition
$f(x) \in g^{-1}(V)$. But then $x \in f^{-1}(g^{-1})(V)$. The reverse containment follows likewise.
29. By hypothesis, for each $x \in K$ there exists $\epsilon_x > 0$ and $M_x > 0$ such that

$|f(y)| \leq M_x$ for all $y \in N_{\epsilon_x}(x) \cap K$. The collection $\{N_{\epsilon_x}(x)\}_{x \in K}$ is an open cover of K. Now use compactness of K to show that there exists a positive constant M such that $|f(y)| \leq M$ for all $y \in K$.

Exercises 4.3.   page 161

2. (a) Suppose $f(x) = x^2$ is uniformly continuous on $[0, \infty)$. Then with $\epsilon = 1$, there exists a $\delta > 0$ such that $|f(x) - f(y)| < 1$ for all $x, y \in [0, \infty)$ satisfying $|x - y| < \delta$. Set $x_n = n$ and $y_n = n + \frac{1}{n}$. If $n_o \in \mathbb{N}$ is such that $n_o \delta > 1$, then $|y_n - x_n| = \frac{1}{n} < \delta$ for all $n \geq n_o$. But $|f(y_n) - f(x_n)| = 2 + \frac{1}{n^2} \geq 2$ for all n. This is a contradiction!
(c) Take $p_n = \frac{1}{(2n+1)\frac{\pi}{2}}$ and $q_n = \frac{1}{n\pi}$. Then $|h(p_n)| = 1$ and $h(q_n) = 0$ for all n. But $\{p_n\}$ and $\{q_n\}$ both converge to 0. Hence given any $\delta > 0$ there exists an integer n_o such that $|p_n - q_n| < \delta$ for all $n \geq n_o$ and $|h(p_n) - h(q_n)| = 1$. Hence h is not uniformly continuous on $(0, \infty)$.   **3. (a)** For all $x, y \in [0, \infty)$, $|f(x) - f(y)| = \left|\frac{x}{1+x} - \frac{y}{1+y}\right| = \frac{|x-y|}{(1+x)(1+y)} < |x - y|$. Thus given $\epsilon > 0$, the choice $\delta = \epsilon$ will work.   **(c)** We first note that $|h(x) - h(y)| = \frac{|y^2 - x^2|}{(x^2+1)(y^2+1)} < |y - x|\frac{|x|+|y|}{x^2+1}$. But for $|y - x| < 1$, $|y| < |x| + 1$. As a consequence $(|x| + |y|)/(x^2 + 1) < 2$. (Verify!) Therefore $|h(x) - h(y)| < 2|y - x|$. Hence given $\epsilon > 0$, choose δ so that $0 < \delta < \min\{1, \epsilon/2\}$.   **(f)** Set $g(x) = \sin x/x$, $x \in (0, 1]$ and $g(0) = 1$. Then g is continuous on $[0, 1]$, and thus by Theorem 4.3.4 uniformly continuous on $[0, 1]$. From this it now follows that f is uniformly continuous on $(0, 1)$.   **4. (a)** Show that $|f(x) - f(y)| = |\frac{1}{x} - \frac{1}{y}| \leq \frac{1}{a^2}|x - y|$ for all $x, y \in [a, \infty)$, $a > 0$.   **(c)** Using a trigonometric identity for $\sin A - \sin B$ we obtain

$$\left|\sin\frac{1}{x} - \sin\frac{1}{y}\right| = 2\left|\sin\frac{1}{2}\left(\frac{1}{x} - \frac{1}{y}\right)\cos\frac{1}{2}\left(\frac{1}{x} + \frac{1}{y}\right)\right| \leq 2\left|\sin\frac{1}{2}\left(\frac{y-x}{xy}\right)\right|.$$ Now us-

ing the inequality $|\sin h| \leq |h|$ we have
$$2\left|\sin\frac{1}{2}\left(\frac{y-x}{xy}\right)\right| \leq \frac{|y-x|}{xy} \leq \frac{1}{a^2}|y - x|$$ for all $x, y \in [a, \infty)$.
5. (a) $|\sqrt{x} - \sqrt{y}| = |x - y|/(\sqrt{x} + \sqrt{y}) \leq \frac{1}{2\sqrt{a}}|x - y|$ provided $x, y \in [a, \infty)$.
(c) Assume that it does and obtain a contradiction.   **7. (b)** Suppose $|f|$ and $|g|$ are bounded by M_1 and M_2 respectively. Then
$|f(x)g(x) - f(y)g(y)| \leq |f(x)||g(x) - g(y)| + |g(y)||f(x) - f(y)| \leq M_1|g(x) - g(y)| + M_2|f(x) - f(y)|$.
Now use the uniform continuity of f and g.   **10.** Suppose f is not bounded. Then there exists a sequence $\{x_n\}$ in E such that $|f(x_n)| \to \infty$. Since E is bounded, $\{x_n\}$ has a convergent subsequence $\{x_{n_k}\}$ in $\mathbb{R}$. Thus $\{x_{n_k}\}$ is Cauchy. Since f is uniformly continuous, $\{f(x_{n_k})\}$ is Cauchy. But then by Theorem 3.6.2 the sequence $\{f(x_{n_k})\}$ is bounded, which is a contradiction.   **12. (a)** By taking $\epsilon = 1$ show that there exists r_o such that $|f(x)| < |L| + 1$ for all $x \in [r_o, \infty)$. Now use the continuity of f on $[a, r_o]$ to conclude that f is bounded on $[a, \infty)$.   **13. (a)** Let $x_1 \in E$ be arbitrary. For $n \geq 1$ set $x_{n+1} = f(x_n)$. Show that the sequence $\{x_n\}$ is contractive.

Exercises 4.4.   page 175

2. (b) $\lim_{x \to 0^+} f(x) = \lim_{x \to 0^-} f(x) = 0$.   **(d)** For $0 < |x| < 1$, $[x^2 - 1] = -1$, Hence both the right and left limit at 0 exist and are equal to -1.   **(f)** $\lim_{x \to 0^+} f(x) = 1$.
Hint: For $\frac{1}{n+1} < x \leq \frac{1}{n}$, $[\frac{1}{x}] = n$.   **4.** Use the fact that $[x]$ is bounded near $x_o = 2$.   **6. (b)** $b = -10$.   **7. (b)** For $n \in \mathbb{N}$ set $x_n = n$ and $y_n = n + \frac{1}{n\pi}$, and show that $|f(x_n) - f(y_n)| = \sin 2$.   **10. (a)** Define F on $[a, b]$ by $F(x) = f(x)$

for $x \in (a, b]$ and $F(a) = f(a+)$, which is assumed to exist. Then F is continuous on $[a, b]$, and thus uniformly continuous by Theorem 4.3.4. Hence f is uniformly continuous on $(a, b]$.   **12.** If $n \in \mathbb{N}$, $g(x) = x^n$ is continuous and strictly increasing on $(0, \infty)$ with Range $g = (0, \infty)$. Therefore by Theorem 4.4.12, its inverse function $g^{-1}(x) = x^{1/n}$ is also continuous on $(0, \infty)$. From this it now follows that $f(x)$ is continuous on $(0, \infty)$.   **14. (a)** Suppose first that $U = (a, b) \subset I$. Then since f is strictly increasing and continuous on I, $f((a, b)) = (f(a), f(b))$, which is open. For an arbitrary open set $U \subset I$, write $U = \bigcup_n I_n$, where $\{I_n\}$ is a finite or countable collection of open intervals and use Theorems 1.7.14 and 2.2.9.   **16. (b)** f is strictly increasing on $[0, \frac{1}{2})$ and strictly decreasing on $[\frac{1}{2}, 1]$, and thus one-to-one on each of the intervals. **(c)** $f([0, \frac{1}{2})) = [0, 1)$ and $f([\frac{1}{2}, 1]) = [1, 2]$. Therefore $f([0, 1]) = [0, 2]$. **(d)** For $y \in [0, 1]$, $f^{-1}(y) = \frac{1}{2}y$, and for $y \in [1, 2]$, $f^{-1}(y) = \frac{1}{2}(3 - y)$. Therefore $f^{-1}(1-) = \frac{1}{2}$ and $f^{-1}(1+) = 1$. Thus f^{-1} is not continuous at $y_o = 1$.

Chapter 5

Exercises 5.1.   page 190

1. (a) For $f(x) = x^3$,
$$f'(x) = \lim_{h \to 0} \frac{f(x+h) - f(x)}{h} = \lim_{h \to 0} \frac{(x+h)^3 - x^3}{h} = \lim_{h \to 0} (3x^2 + 3xh + h^2) = 3x^2.$$
(c) For $h(x) = 1/x$, $x \neq 0$,
$$h'(x) = \lim_{h \to 0} \frac{\frac{1}{x+h} - \frac{1}{x}}{h} = \lim_{h \to 0} \frac{-1}{x(x+h)} = -\frac{1}{x^2}.$$
(e) For $f(x) = \frac{x}{x+1}$, $x \neq -1$, $\dfrac{f(x+h) - f(x)}{h} = \dfrac{1}{(x+1)(x+h+1)}.$
Thus $f'(x) = \lim_{h \to 0} \dfrac{1}{(x+1)(x+h+1)} = \dfrac{1}{(x+1)^2}.$
2. If $n \in \mathbb{N}$, by the binomial theorem
$$(x+h)^n - x^n = \sum_{k=1}^n \binom{n}{k} h^k x^{n-k} = nhx^{n-1} + h \sum_{k=2}^n \binom{n}{k} h^{k-1} x^{n-k}.$$
Dividing by h and taking the limit as $h \to 0$ proves the result for $n \in \mathbb{N}$. If $n \in \mathbb{Z}$ is negative, write $x^n = 1/x^m$, $m \in \mathbb{N}$, and use Theorem 5.1.5(c).   **3. (a)** Since $\cos x = \sin(x + \frac{\pi}{2})$, by the chain rule $\frac{d}{dx} \cos x = \cos(x + \frac{\pi}{2}) = -\sin x$. Alternately use the definition of the derivative.   **5. (a)** Yes.   **(c)** No.   **(e)** Yes.
7. (a) $f'(x)$ exists for all $x \in \mathbb{R} \setminus \mathbb{Z}$. For $x \in (k, k+1)$, $k \in \mathbb{Z}$, $f(x) = x[x] = kx$. Thus $f'(x) = k$.   **(c)** The function h is differentiable at all x where $\sin x \neq 0$. For $x \in (2k\pi, (2k+1)\pi)$, $k \in \mathbb{Z}$, $h'(x) = \cos x$. For $x \in ((2k-1)\pi, 2k\pi)$, $k \in \mathbb{Z}$, $h'(x) = -\cos x$.   **9. (b)** For $x \neq 0$, $g'(x) = 2x\sin\frac{1}{x} - \cos\frac{1}{x}$. Since $\lim_{x \to 0} \cos\frac{1}{x}$ does not exist, $\lim_{x \to 0} g'(x)$ does not exist.   **10. (a)** $2a + b = 6$.   **(b)** $a = 4, b = -2$. Justify why $b = -2$.   **12. (a)** $f'(x) = 2/(2x+1)$.   **(c)** $h'(x) = 3[L(x)]^2/x$.
14. (b) $f'_+(0) = \lim_{h \to 0+} h^{(b-1)} \sin\frac{1}{h}$ which exists and equals 0 if and only if $(b-1) > 0$; i.e., $b > 1$.   **15. (b)** No. Consider $f(x) = |x|$ at $x_o = 0$.

Exercises 5.2.   page 203

3. (a) Increasing on $\mathbb{R}$.   **(c)** Decreasing on $(-\infty, 0)$ and increasing on $(0, \infty)$, with an absolute minimum at $x = 0$.   **(e)** Increasing on $(-\infty, 2) \cup (2, \infty)$, decreasing on $(0, 2)$. The function has a local minimum at $x = 2$.   **5. (a)** Let $f(x) =$

$(1+x)^{\frac{1}{2}}, x > -1$. By the mean value theorem $f(x) - f(0) = f'(\zeta)$ where ζ is between 0 and x. If $x > 0$, then $f'(\zeta) < \frac{1}{2}$. On the other hand, if $x < 0$ and $x < \zeta < 0$, then $f'(\zeta) > \frac{1}{2}$. But then $f'(\zeta) < \frac{1}{2}x$. (c) Take $f(x) = x^\alpha, 0 < \alpha < 1$. Then $f(x) - f(a) = f'(\zeta)(x - a)$ where $a < \zeta < x$. But then $f'(\zeta) < \alpha a^{\alpha-1}$.
6. (a) Show that the function $f(x) = x^{1/n} - (x-1)^{1/n}$ is decreasing on the interval $1 \le x \le a/b$. **(b)** Set $f(x) = \alpha x - x^\alpha, x \ge 0$. Prove that $f(x) \ge f(1)$ to obtain $x^\alpha \le \alpha x + (1 - \alpha), x \ge 0$. Now take $x = a/b$. **7. (a)** If $f''(c) > 0$, then there exists a $\delta > 0$ such that $f'(x)/(x-c) > 0$ for all x, $|x - c| < \delta$. Therefore $f'(x) < 0$ on $(c - \delta, c)$ and $f'(x) > 0$ on $(c, c + \delta)$. Thus f has a local minimum at c.
8. Show that $f'(x) = 0$ for all $x \in (a, b)$. **9.** Since $P(2) = 0$ we can assume that $P(x) = a(x - 2)^2 + b(x - 2)$. Now use the fact that P must satisfy $P(1) = 1$ and $P'(1) = 2$ to determine a and b. **10.** $a = -2, b = 2, c = 1$. **12. (a)** See Example 4.1.10 (d). **(b)** By the result of (a), $f'(x) < 0$ for all $x \in (0, \frac{\pi}{2}]$. **14. (a)** Let $t_n \to c$. Since $f'(c)$ exists, $\lim\limits_{n \to \infty} (f(t_n) - f(c))/(t_n - c) = f'(c)$. Now apply the mean value theorem. **17.** Since $f'_+(a) = \lim\limits_{x \to a^+} (f(x) - f(a))/(x - a) > 0$, there exists a $\delta > 0$ such that $(f(x) - f(a))/(x - a) > 0$ for all x, $a < x < a + \delta$.
19. Hint: Consider $f'(x)$. **20. (b)** Check the values of $f'(x)$ at $p_n = 1/(n\pi), n \in \mathbb{N}$.
24. (a) For fixed $a > 0$ consider $f(x) = L(ax), x \in (0, \infty)$. **(c)** By (a) and (b), $L(b^n) = nL(b)$ for all $n \in \mathbb{Z}$ and $b \in (0, \infty)$. But then $L(b) = L((b^{1/n})^n) = nL(b^{1/n})$. From this it now follows the $L(b^r) = rL(b)$ for all $r \in \mathbb{Q}$. Now use the continuity of L to prove that $L(b^x) = xL(b)$ for all $x \in \mathbb{R}$ where $b^x = \sup\{b^r : r \in \mathbb{Q}, r \le x\}$. **25. (b)** Since $\tan(\arctan x) = x$, by Theorem 5.2.14 and the chain rule, $\frac{d}{dx}\tan(\arctan x) = (\sec^2(\arctan x))(\frac{d}{dx}\arctan x) = 1$. The result now follows from the identity $\sec^2(\arctan x) = 1 + x^2$. To prove this, consider the right triangle with sides of length 1, $|x|$, $1 + x^2$ respectively.

Exercises 5.3.   page 212

2. $\dfrac{f(x)}{g(x)} = \dfrac{f(x) - f(x_o)}{x - x_o} \Big/ \dfrac{g(x) - g(x_o)}{x - x_o}$. Now use Theorem 4.1.6(c) and the definition of the derivative. **4.** Use the fact that since $\lim\limits_{x \to a^+} f(x)$ exists, $f(x)$ is bounded on $(a, a + \delta)$ for some $\delta > 0$. **6. (a)** $\lim\limits_{x \to 1} \dfrac{x^5 + 2x - 3}{2x^3 - x^2 - 1} = \lim\limits_{x \to 1} \dfrac{5x^4 + 2}{6x^2 - 2x} = \dfrac{7}{4}$. **(c)** By l'Hopital's rule, $\lim\limits_{x \to \infty} \dfrac{\ln x}{x} = \lim\limits_{x \to \infty} \dfrac{1}{x} = 0$. **(e)** Make the substitution $x = 1/t$.
(g) 0. Use repeated applications of l'Hospitals rule until the exponent of $\ln x$ is less than or equal to zero. **(i)** 0. Use l'Hospitals rule twice on $(\sin x - x)/(x \sin x), x \ne 0$.
9. (a) $f'(0) = 0$. **(b)** $f''(0) = -\frac{1}{3}$.

Exercises 5.4.   page 219

1. Let $c_1 > 0$ be arbitrary. By Newton's method $c_{n+1} = (2c_n^3 + \alpha)/3c_n^2$.
2. (a) $f(0) = 1$ and $f(1) = -1$. Therefore f has a zero on the interval $[0, 1]$. With $c_1 = 0.5$, $c_2 = 0.33333333$, $f(c_2) = .037037037$, $c_3 = 0.34722222$, $f(c_3) = 0.000195587$, $c_4 = .34729635$, $f(c_4) = 0.000000015$.

Chapter 6

Exercises 6.1.   page 238

1.(a) Since f is increasing on $[-1, 0]$ and decreasing on $[0, 2]$, $m_1 = m_2 = 0$, $m_3 = -3$ and $M_1 = M_2 = 1$, $M_3 = 0$. Since $\Delta x_i = 1$, $\mathcal{L}(\mathcal{P}, f) = -3$ and $\mathcal{U}(\mathcal{P}, f) = 2$.

2. (a) $fx = \begin{cases} -1, & 0 \le x < 1, \\ 2, & 1 \le x \le 2. \end{cases}$ Let $\mathcal{P} = \{x_0, x_1, ..., x_n\}$ be any partition of $[0, 2]$

and let $k \in \{1, ..., n\}$ be such that $x_{k-1} < 1 \le x_k$. Then $\mathcal{L}(\mathcal{P}, f) = 4 - 3x_k$ and $\mathcal{U}(\mathcal{P}, f) = 4 - 3x_{k-1}$. Thus $\mathcal{U}(\mathcal{P}, f) - \mathcal{L}(\mathcal{P}, f) = 3(x_k - x_{k-1})$. By Theorem 6.1.7 it now follows that f is Riemann integrable on $[0, 2]$. Also, since $x_{k-1} < 1 \le x_k$, $\mathcal{L}(\mathcal{P}, f) \le 1 < \mathcal{U}(\mathcal{P}, f)$ for any partition $\mathcal{P}$. Hence $\int_0^2 f = 1$. Alternatively, consider the partition $\mathcal{P} = \{0, c, 1, 2\}$ where $0 < c < 1$ is arbitrary. For this partition, $\mathcal{L}(\mathcal{P}, f) = 1$ and $\mathcal{U}(\mathcal{P}, f) = 4 - 3c$. The results now follow as above.   **3. (a)** If $\mathcal{P} = \{x_o, x_1, ..., x_n\}$ is a partition of $[a, b]$, then $\inf\{f(t) : t \in [x_{i-1}, x_i]\} = \sup\{f(t) : t \in [x_{i-1}, x_i]\} = c$. Therefore $\mathcal{L}(\mathcal{P}, f) = \mathcal{U}(\mathcal{P}, f) = \sum_{i=1}^n c(x_i - x_{i-1}) = c(b - a)$.   **4. (a)** Since $f(x) = [3x]$ is monotone increasing on $[0, 1]$, f is Riemann integrable on $[0, 1]$. For $n \ge 4$ consider the partition $\mathcal{P}_n = \{0, \frac{1}{3} - \frac{1}{n}, \frac{1}{3}, \frac{2}{3} - \frac{1}{n}, \frac{2}{3}, 1 - \frac{1}{n}, 1\}$. For this partition $\mathcal{L}(\mathcal{P}_n, f) = 1$ and $\mathcal{U}(\mathcal{P}_n, f) = 1 + \frac{3}{n}$. From this it now follows that $\int_0^1 [3x]\, dx = 1$.

(c) Since f is increasing on $[0, 1]$ it is Riemann integrable. Take $\mathcal{P}_n = \{0, \frac{1}{n}, \frac{2}{n}, ..., 1\}$. Then $\mathcal{U}(\mathcal{P}_n, f) = \sum_{i=1}^n \left(3\frac{i}{n} + 1\right) \frac{1}{n} = \frac{3}{n^2} \sum_{i=1}^n i + \frac{1}{n} \sum_{i=1}^n 1 = \frac{3}{2} \frac{n(n+1)}{n^2} + 1$. Therefore $\int_0^1 (3x + 1) dx = \lim_{n \to \infty} \frac{3}{2}\left(1 + \frac{1}{n}\right) + 1 = \frac{5}{2}$.   **6.** Let $\mathcal{P} = \{x_0, x_1, ..., x_n\}$ be a partition of $[a, b]$. Since $f(x) \le g(x)$ for all $x \in [a, b]$, $\sup\{f(t) : t \in [x_{i-1}, x_i]\} \le \sup\{g(t) : t \in [x_{i-1}, x_i]\}$ for all $i = 1, ..., n$. As a consequence $\mathcal{U}(\mathcal{P}, f) \le \mathcal{U}(\mathcal{P}, g)$ for all partitions $\mathcal{P}$ of $[a, b]$. Taking the infimum over $\mathcal{P}$ gives $\overline{\int_a^b} f \le \overline{\int_a^b} g$. The result now follows from the fact that $f, g \in \mathcal{R}[a, b]$.

8. $\underline{\int_0^1} f = 0$, $\overline{\int_0^1} f = \frac{1}{2}$.   **10.** Since $a \ge 0$, f is increasing on $[a, b]$. Thus if $\mathcal{P} = \{x_0, x_1, ..., x_n\}$ is a partition of $[a, b]$, $m_i = x_{i-1}^2$ and $M_i = x_i^2$. Therefore $\mathcal{L}(\mathcal{P}, f) = \sum_{i=1}^n x_{i-1}^2 \Delta x_i$ and $\mathcal{U}(\mathcal{P}, f) = \sum_{i=1}^n x_i^2 \Delta x_i$. Now show that $x_{i-1}^2 \Delta x_i \le \frac{1}{3}(x_i^3 - x_{i-1}^3) \le x_i^2 \Delta x_i$. From this it will now follow that $\mathcal{L}(\mathcal{P}, f) \le \frac{1}{3}(b^3 - a^3) \le \mathcal{U}(\mathcal{P}, f)$. Since f is continuous, $f \in \mathcal{R}[a, b]$ with $\int_a^b f = \frac{1}{3}(b^3 - a^3)$.   **12. (a)** With $\mathcal{P}_n = \{\frac{k}{n} : k = 0, 1, ..., n\}$ and $f(x) = x$, $U(\mathcal{P}_n, f) = \sum_{k=1}^n \left(\frac{k}{n}\right) = \frac{1}{n^2} \sum_{k=1}^n k$. By Exercise 1(a), Section 1.3, $\sum_{k=1}^n k = \frac{1}{2}n(n+1)$. Therefore $\int_0^1 x dx = \lim_{n \to \infty} n(n+1)/(2n^2) = 1/2$.   **(c)** Take $\mathcal{P}_n = \{0, \frac{1}{n}, \frac{2}{n}, ..., 1\}$. Then $\mathcal{U}(\mathcal{P}_n, f) = \sum_{i=1}^n \left(\frac{i^3}{n^3}\right)\frac{1}{n} = \frac{1}{n^4} \sum_{i=1}^n i^3 = \frac{1}{n^4}\left[\frac{1}{2}n(n+1)\right]^2 = \frac{1}{4}\left[1 + \frac{1}{n}\right]^2$. Therefore, $\int_0^1 x^3 dx = \lim_{n \to \infty} \mathcal{U}(\mathcal{P}_n, f) = 1/4$.   **13.** Use Theorem 6.1.7.   **14. (a)** Let $\mathcal{P}_n = \{x_o, x_1, x_2, ..., x_n\}$ be a partition of $[a, b]$, and for each $i = 1, 2, ..., n$, let M_i and M_i^* denote the supremum of f and $|f|$ respectively on $[x_{i-1}, x_i]$. Likewise, let m_i and m_i^* denote the infimum of f and $|f|$ respectively on the same intervals. Use the

inequality $\|f(s)| - |f(t)\| \leq |f(s) - f(t)|$ to prove that $M_i^* - m_i^* \leq M_i - m_i$ from which the result follows. **15. (a)** Since f is bounded on $[a, b]$, $|f(x)| \leq M$ for all $x \in [a, b]$. Let $\mathcal{P} = \{x_0, x_1, ..., x_n\}$ be any partition of $[a, b]$. For each i let $M_i(f^2)$ and $m_i(f^2)$ denote the supremum and infimum of f^2 respectively over $[x_{i-1}, x_i]$, with an analogous definition for $M_i(f)$ and $m_i(f)$. Let $\epsilon > 0$ be given. Then for each i there exists $s_i, t_i \in [x_{i-1}, x_i]$ such that $M_i(f^2) < f^2(s_i) + \frac{1}{2}\epsilon$ and $m_i(f^2) > f^2(t_i) - \frac{1}{2}\epsilon$. Therefore
$$0 \leq M_i(f^2) - m_i(f^2) < f^2(s_i) - f^2(t_i) + \epsilon \leq |f(s_i) + f(t_i)||f(s_i) - f(t_i)| + \epsilon \leq 2M[M_i(f) - m_i(f)] + \epsilon.$$
Since this holds for any $\epsilon > 0$, $M_i(f^2) - m_i(f^2) \leq 2M[M_i(f) - m_i(f)]$. Now use Theorem 6.1.7. **16.** Assume that f is continuous on $[a, b]$ except perhaps at a or b, or both. Since f is bounded there exists $M > 0$ such that $|f(x)| \leq M$ for all $x \in [a, b]$. Let $\epsilon > 0$ be given. Choose y_1, y_2, $a < y_1 < y_2 < b$ so that $y_1 - a < \epsilon/8M$ and $b - y_2 < \epsilon/8M$. Then
$[\sup\{f(t) : t \in [a, y_1]\} - \inf\{f(t) : t \in [a, y_1]\}](y_1 - a) \leq 2M(y_1 - a) < \frac{1}{4}\epsilon$.
Similarly for the interval $[y_2, b]$. Since f is continuous on $[y_1, y_2]$ there exists a partition $\mathcal{P}$ of $[y_1, y_2]$ such that $\mathcal{U}(\mathcal{P}, f) - \mathcal{L}(\mathcal{P}, f) < \frac{1}{2}\epsilon$. Let $\mathcal{P}^* = \mathcal{P} \cup \{a, b\}$. Then $\mathcal{P}^*$ is a partition of $[a, b]$, and by the above $\mathcal{U}(\mathcal{P}^*, f) - \mathcal{L}(\mathcal{P}^*, f) < \epsilon$. Thus by Theorem 6.1.7, $f \in \mathcal{R}[a, b]$. Suppose f is continuous on $[a, b]$ except at a finite number of points $c_1, c_2, ..., c_n$ with $a \leq c_1 < c_2 < \cdots < c_n \leq b$. Apply the above to each of the intervals $[a, c_1], [c_1, c_2], ..., [c_n, b]$ to obtain a partition of $[a, b]$ for which Theorem 6.1.7 holds.

Exercises 6.2.   page 247

2. (a) Take $t_i = \frac{1}{3}(x_i^2 + x_i x_{i-1} + x_{i-1}^2)$.   **3. (a)** If $c_1 \neq a$, set $c_0 = a$. Similarly, set $c_{n+1} = b$ if $c_n \neq b$. First prove that $f \in \mathcal{R}[c_{i-1}, c_i]$, $i = 1, 2, ..., n+1$ with $\int_{c_{i-1}}^{c_i} f = 0$. Now use Theorem 6.2.3.   **(b)** Set $h = f - g$.   **5.** Since f is bounded, $|f(x)| \leq M$ for all $x \in [a, b]$ for some $M > 0$. Use equations (7), (8), and the fact that $f \in \mathcal{R}[c, b]$ for every $c \in (a, b)$ to prove that
$$0 \leq \overline{\int_a^b} f - \int_a^b f = \overline{\int_a^c} f - \int_a^c f \leq 2M(c - a).$$
Use this to show that $f \in \mathcal{R}[a, b]$.   **7. (a)** $\lim\limits_{n \to \infty} \dfrac{1}{n^3} \sum\limits_{k=1}^{n} k^2 = \int_0^1 x^2\, dx = \dfrac{1}{3}$.

(c) $\lim\limits_{n \to \infty} \sum\limits_{k=1}^{n} \dfrac{n}{n^2 + k^2} = \int_0^1 \dfrac{1}{1 + x^2} dx = \dfrac{\pi}{4}$.   **8.** For each fixed c, $0 < c < 1$, the series defining f becomes a finite sum on $[0, c]$. Evaluate $\int_0^c f$ and use Exercise 5.

Exercises 6.3.   page 255

2. Since f is bounded, $|f(x)| \leq M$ for all $x \in [a, b]$. If $x, y \in [a, b]$ with $x < y$, then
$$|F(y) - F(x)| = \left|\int_x^y f\right| \leq \int_x^y |f| \leq M|y - x|.$$
Thus F satisfies a Lipschitz condition on $[a, b]$ and hence is uniformly continuous.
3. (b) $F(x) = x$ for $0 \leq x \leq \frac{1}{2}$; $F(x) = \frac{3}{2} - 2x$ for $\frac{1}{2} < x \leq 1$.   **(d)** $F(x) = 0$ for $0 \leq x \leq \frac{1}{3}$; $F(x) = \frac{1}{2}(x^2 - \frac{1}{9})$ for $\frac{1}{3} < x \leq \frac{2}{3}$; $F(x) - x^2$ $\frac{5}{18}$ for $\frac{2}{3} < x \leq 1$.
6. (b) $F'(x) = \cos x^2$.   **(d)** $F'(x) = 2xf(x^2)$.   **7.** By the chain rule, $\frac{d}{dx}L(\frac{1}{x}) = -\frac{1}{x} = \frac{d}{dx}[-L(x)]$. Therefore $L(\frac{1}{x}) = -L(x) + C$ for some constant C. Taking $x = 1$ shows that $C = 0$.   **10.** Let m and M denote the minimum and maximum of f on $[a, b]$ respectively. Since $g(x) \geq 0$ for all x, $mg(x) \leq f(x)g(x) \leq Mg(x)$. If $\int_a^b g > 0$, then from the previous inequality one obtains

$$m \leq \int_a^b fg / \int_a^b g \leq M.$$

Now apply the intermediate value theorem to f. If $\int_a^b g = 0$, use Theorem 6.2.2 to prove that $\int_a^b fg = 0$ and thus the conclusion holds for any $c \in [a, b]$.   **12. (a)** With

$\varphi(x) = \ln x$, by Theorem 6.3.8 $\int_1^2 \frac{\ln x}{x} dx = \int_0^{\ln 2} x \, dx = \frac{1}{2}(\ln 2)^2$.   **(b)** Use integration

by parts.   **(e)** $2 \ln 2 - 1$. Use Exercise 9 with $\varphi(x) = \sqrt{x}$ and $f(t) = t/(1 + t)$.
13. (b) With the given change of variable, $dx = a \sec^2 t \, dt$ and $\sqrt{a^2 + x^2} = a \sec t$.

Therefore $\int_0^a \frac{1}{\sqrt{x^2 + a^2}} dx = \int_0^{\frac{\pi}{4}} \sec t \, dt$. To evaluate this last integral first establish

that $\sec t = \frac{\sin t}{\cos t} + \frac{\cos t}{1 + \sin t}$.   **14.** That $\left(\int_a^b |f(x)|^n dx \right)^{1/n} \leq M$ is straight forward.
For the reverse inequality, given ϵ, $0 < \epsilon < M$, using continuity of f show that there exists an interval $I \subset [a, b]$ such that $|f(x)| \geq M - \epsilon$ for all $x \in I$. Using this show

that $\left(\int_a^b |f(x)|^n dx \right)^{1/n} \geq (M - \epsilon)\ell(I)^{1/n}$. Explain why the result now follows.

15. First show that $F'_+(c) = \lim_{h \to 0+} \frac{1}{h} \int_c^{c+h} f(x) \, dx$. Since f is monotone increasing

$f(c^+)$ exists. Thus given $\epsilon > 0$, there exists $\delta > 0$ such that $f(c^+) \leq f(x) < f(c^+) + \epsilon$
for all $x, c < x < c + \delta$. Explain how the conclusion now follows.

Exercises 6.4   page 263

1. (a) Since $0 < p < 1$, $\int_0^1 x^{-p} dx = \lim_{c \to 0+} \int_c^1 x^{-p} dx = \lim_{c \to 0+} \frac{1}{1-p}[1 - c^{1-p}] = \frac{1}{1-p}$.
Note: If $p \leq 0$, then x^{-p} is continuous on $[0, 1]$, and if $p \geq 1$ then the improper integral diverges.   **(d)** Converges, with $\int_0^1 x \ln x \, dx = -\frac{1}{4}$.   **(f)** In this problem the

integrand is undefined at $x = 1$. For $0 < c < 1$, $\int_0^c \tan \frac{\pi}{2} x \, dx = -\frac{2}{\pi} \ln(\cos \frac{\pi}{2} c)$. Since

$\lim_{c \to 1+} \ln(\cos \frac{\pi}{2} c)$ does not exist, the improper integral diverges.   **2. (a)** Converges,

with $\int_0^\infty e^{-x} dx = \lim_{c \to \infty} \int_0^c e^{-x} dx = \lim_{c \to \infty} 1 - e^{-c} = 1$.   **(c)** The improper inte-

gral converges with $\lim_{c \to \infty} \int_1^c x^{-p} dx = \lim_{c \to \infty} \frac{1}{p-1} [1 - c^{1-p}] = \frac{1}{p-1}$.   **(e)** Diverges.

$\int_2^c \frac{1}{x \ln x} dx = \ln(\ln c) - \ln(\ln 2)$, which diverges to ∞ as $c \to \infty$.

(g) $\int_0^c \frac{x}{x^2 + 1} dx = \frac{1}{2} \ln(c^2 + 1)$. The improper integral diverges since $\lim_{c \to \infty} \ln(c^2 + 1)$

does not exist.   **3. (a)** For $p > -1$, the improper integral converges for all $q \in \mathbb{R}$,
and for $p < -1$, the improper integral diverges for all $q \in \mathbb{R}$. When $p = -1$,
the improper integral converges for all $q < -1$, and diverges for all $q \geq -1$.   **4.**
Since $\lim_{c \to 0+} \int_c^1 f(x) dx = \sin 1$ the improper integral converges. To show that the

improper integral of $|f|$ diverges, first show that $|f(x)| \geq \frac{1}{x} - 2x$ on each of the

intervals $\left[\frac{1}{\sqrt{(2n+\frac{1}{3})\pi}}, \frac{1}{\sqrt{(2n-\frac{1}{3})\pi}} \right]$. Use the above to find a partition $\mathcal{P}_N$ such that

$\mathcal{L}(\mathcal{P}_N, |f|) \geq C \sum_{k=1}^N 1/k$, for some positive constant C independent of N. From this

you can conclude that the improper integral of $|f|$ diverges.   **5.** Hint: Use the fact that $0 \leq |f(x)| - f(x) \leq 2|f(x)|$. **7. (a)** $\int_{\pi}^{c} \left| \dfrac{\cos x}{x^2} \right| dx \leq \int_{\pi}^{c} \dfrac{1}{x^2} dx = \dfrac{1}{\pi} - \dfrac{1}{c}$. Thus $\lim_{c \to \infty} \int_{\pi}^{c} |f| < \infty$. **(b)** By integration by parts, $\int_{\pi}^{c} \dfrac{\sin x}{x} dx = -\dfrac{1}{\pi} - \dfrac{\cos c}{c} - \int_{\pi}^{c} \dfrac{\cos x}{x^2} dx$.

By (a) and Exercise 5, $\lim_{c \to \infty} \int_{\pi}^{c} \dfrac{\cos x}{x^2} dx$ exists. Also, $\lim_{c \to \infty} \dfrac{\cos c}{c} = 0$. Therefore, $\lim_{c \to \infty} \int_{\pi}^{c} \dfrac{\sin x}{x} dx$ exists. **9. (a)** To show convergence of the improper integral consider the integrals of $t^{x-1} e^{-t}$ over the two intervals $(0, 1]$ and $[1, \infty)$. If $x \geq 1$, then $t^{x-1} e^{-t}$ is continuous on $[0, 1]$, and thus the integral over $[0, 1]$ clearly exists. If $0 < x < 1$, then since $e^{-t} \leq 1$ for $t \in (0, 1]$,

$$0 \leq \int_{c}^{1} t^{x-1} e^{-t} dt \leq \int_{c}^{1} t^{x-1} dt = \dfrac{1}{x}(1 - c^x).$$ Thus $\lim_{c \to 0+} \int_{c}^{1} t^{x-1} e^{-t} dt < \infty$. For $t \geq 1$, use l'Hospital's rule to show that there exists a $t_o \geq 1$ such that $t^{x-1} \leq e^{\frac{1}{2}t}$ for all $t \geq t_o$. Thus $t^{x-1} e^{-t} \leq e^{-t/2}$ for all $t \geq t_o$, and as a consequence $\lim_{R \to \infty} \int_{1}^{R} t^{x-1} e^{-t} dt < \infty$. Thus the improper integral defining $\Gamma(x)$ converges for all $x > 0$.

Exercises 6.5. page 278

1. $2f(0)$. **3. (a)** See Exercise 2, Section 6.3. **5. (a)** $\dfrac{\pi}{2} - 1$. Use Theorem 6.5.10. **(c)** $1^2 + 2^2 + 3^2$. For $x \in [0, 3]$, $[x] = I(x - 1) + I(x - 2) + I(x - 3)$. Now use formula (12). **(f)** $\int_{1}^{4} (x - [x]) dx^3 = \int_{1}^{4} (x - [x]) 3x^2 dx = \int_{1}^{2} 3x^3 dx + \int_{1}^{4} 3x^2 dx + \int_{2}^{3} 6x^2 dx + \int_{3}^{4} 9x^2 dx$ etc. **8. (a)** $\sum_{n=1}^{\infty} \dfrac{1}{n 2^n}$. **12. (a)** As in the solution of Exercise 15(a) of Section 6.1, if $\mathcal{P} = \{x_0, x_1, \ldots, x_n\}$ is a partition of $[a, b]$, $M_i(f^2) - m_i(f^2) \leq 2M[M_i(f) - m_i(f)]$ where $M > 0$ is such that $|f(x)| \leq M$ for all $x \in [a, b]$. From this it now follows that $0 \leq \mathcal{U}(\mathcal{P}, f^2, \alpha) - \mathcal{L}(\mathcal{P}, f^2, \alpha) \leq 2M[\mathcal{U}(\mathcal{P}, f, \alpha) - \mathcal{L}(\mathcal{P}, f, \alpha)]$. Now apply Theorem 6.5.5.

Exercises 6.6. page 290

2. (a) With $n = 4$, $h = .25$. Set $x_i = .25 i$, $y_i = f(x_i)$, $i = 0, 1, 2, 3, 4$. Then $y_0 = 1.00000$, $y_1 = 0.94118$, $y_2 = 0.80000$, $y_3 = 0.64000$, $y_4 = 0.50000$. Therefore, $T_4(f) = \dfrac{.25}{2}[y_0 + 2y_1 + 2y_2 + 2y_3 + y_4] = 0.782795$.

$S_4(f) = \dfrac{.25}{3}[y_0 + 4y_1 + 2y_2 + 4y_3 + y_4] = 0.785393$.

By computation $f''(t) = 2(3t^2 - 1)/(1 + t^2)^3$. Using the first derivative test, $f''(t)$ has a local maximum of $\frac{1}{2}$ at $t = 1$. Therefore $|f''(t)| \leq \frac{1}{2}$ for all $t \in [0, 1]$. Thus by equation (23) with $n = 4$,

$\left| \dfrac{\pi}{4} - T_4(f) \right| \leq \dfrac{1}{12} \left(\dfrac{1}{2} \right) \dfrac{1}{4^2} = 0.0026042$. Since $\pi/4 = 0.7853982$ (to seven decimal places), $|\pi/4 - T_4(f)| = 0.0026032$. **3. (b)** By computation, $f^{(4)}(x) = 3(4x^2 - 1)(1 + x^2)^{-\frac{7}{2}}$. By the first derivative test the function $f^{(4)}$ has a maximum on $[0, 2]$ at $x = \sqrt{3}/2$. Thus $|f^{(4)}(x)| \leq 6/\sqrt{(1 + \frac{3}{4})^7} < 1$. If we choose n (even) so that $|\int_{0}^{2} f - S_n(f)| < 10^{-5}$, then we will be guaranteed accuracy to four decimal places. By inequality (26) with $M = 1$, we need to choose n so that $2^5/180n^4 < 10^{-5}$, or $n^4 > 17{,}778$. The value $n = 12$ will work. This value of n will guarantee that $E_{12}(f) < 0.0000086$. Compare your answer with the exact answer of $\sqrt{5} + \frac{1}{2}\ln(2 + \sqrt{5})$.

Exercises 6.7.   page 296

2. Hint: Consider $f(x) = 1 - \chi_P(x)$, $0 \le x \le 1$, and use Exercise 21(b) of Section 6.1.   **4.** No. Consider $g = \chi_Q$ on $[0, 1]$ and let f be the zero function.

Chapter 7

Exercises 7.1.   page 313

2. (a) Diverges.   **(c)** Converges.   **(e)** Diverges.   **(g)** Converges by the ratio test.   **(i)** Converges.   **(k)** Diverges. First show that $\sqrt{k} \ge \ln k$ for $k \ge 4$. Thus since $\sum 1/k$ diverges, by the comparison test $\sum 1/(\ln k)^2$ diverges.   **(m)** Converges for $p > 1$; diverges for $0 < p \le 1$.   **(o)** Converges. Hint, rewrite $\frac{1}{k}\ln(1 + \frac{1}{k})$ as $\frac{1}{k^2}\ln(1 + \frac{1}{k})^k$ and use the comparison text.   **3. (a)** Converges to $1/(1 - \sin p)$ for all $p \in \mathbb{R}$ for which $|\sin p| < 1$; that is, for all $p \ne (2k + 1)\frac{\pi}{2}$, $k \in \mathbb{Z}$.
4. (b) Since $\sum a_k$ converges, $\lim a_k = 0$. Thus there exists $k_o \in \mathbb{N}$ such that $0 \le a_k \le 1$ for all $k \ge k_o$. But then $0 \le a_k^2 \le a_k$ for all $k \ge k_o$, and $\sum a_k^2$ converges by the comparison test.   **(d)** Take $a_k = 1/k^2$.   **(f)** Converges. Use the inequality $ab \le \frac{1}{2}(a^2 + b^2)$, $a, b \ge 0$.   **5.** The series diverges for all $q < 1$, $p \in \mathbb{R}$, and converges for all $q > 1$, $p \in \mathbb{R}$. If $q = 1$, the series diverges for $p \le 1$ and converges for $p > 1$.
6. Suppose $\lim(a_n/b_n) = L$, where $0 < L < \infty$. Take $\epsilon = \frac{1}{2}L$. For this ϵ, there exist $k_o \in \mathbb{N}$ such that $\frac{1}{2}L \le a_k/b_k \le \frac{3}{2}L$ for all $k \ge k_o$. The result now follows by the comparison test.   **8.** Since $a_n > 0$ for all n, we have $b_k \ge a_1/k$ for all k. Hence by the comparison test $\sum b_k$ diverges.   **11.** For a simple example take $a_k = (-1)^k$.
12. The proof uses the fact that $\lim_{k \to \infty} \left(\frac{k+1}{k}\right)^n = \lim_{k \to \infty} \sqrt[k]{k^n} = 1$ for all $n \in \mathbb{Z}$.
13. The given series is the sum of the two series $\sum_{k=1}^{\infty} \dfrac{1}{(2k-1)^2}$ and $\sum_{k=1}^{\infty} \dfrac{1}{(2k)^3}$, each of which converges.   **16.** Let $s_n = a_1 + a_2 + \cdots + a_n$ and $t_k = a_1 + 2a_1 + \cdots + 2^k a_{2^k}$. By writing $s_n = a_1 + (a_2 + a_3) + (a_4 + a_5 + a_6 + a_7) + \cdots$, show that if $n < 2^k$, then $s_n \le t_k$, and if $n > 2^k$, then $s_n \ge \frac{1}{2}t_k$. From these two inequalities it now follows that $\sum a_k < \infty$ if and only if $\sum 2^k a_{2^k} < \infty$.   **18. (a)** Diverges. If $a_k = 1/(k \ln k)$, then $2^k a_{2^k} = 1/(k \ln 2)$.   **19.** Use Example 5.2.7 to show that $c_k - c_{k+1} \ge 0$ for all k. Thus $\{c_k\}$ is monotone decreasing. Use the definition of $\ln k$ and the method of proof of the integral test to show that $c_k \ge 0$ for all k.   **21.** Write $a_{k+1}/a_k = 1 - x_k/k$ where $x_k = (q - p)(k/(q + k + 1))$.
22. (c) When $p = 2$, $a_k = \left(\dfrac{1 \cdot 3 \cdots (2k-1)}{2 \cdot 4 \cdots (2k)}\right)^2 = \prod_{j=1}^{k}\left(1 - \dfrac{1}{2j}\right)^2 \ge \left(1 - \dfrac{1}{2k}\right)^{2k}$. Now use the fact that $\lim_{h \to 0}(1 + h)^{1/h} = e$.   **23. (a)** Set $s_n = \sum_{k=1}^{n} a_k$, and let $s = \lim s_n$. Consider the series $\sum b_k$ where $b_1 = (\sqrt{s} - \sqrt{s - s_1})$ and for $k \ge 2$, $b_k = (\sqrt{s - s_{k-1}} - \sqrt{s - s_k})$.

Exercises 7.2.   page 320

1. If $\{b_n\}$ is monotone increasing to b, consider $\sum_{k=1}^{\infty}(b - b_k)a_k$.   **2.** Take $b_k = 1/k$ for k odd, and $b_k = 1/k^2$ for k even.   **4.** If $B_n = \sum_{k=1}^{n}\cos kt$, then

$(\sin \frac{1}{2}t) B_n = \frac{1}{2} \sum_{k=1}^{n} [\sin(k+\frac{1}{2})t - \sin(k-\frac{1}{2})t] = \frac{1}{2}[\sin(n+\frac{1}{2})t - \sin\frac{1}{2}t].$

5. (a) Converges. **(c)** Converges. **(d)** Diverges; $\lim_{k\to\infty} \dfrac{k^k}{(1+k)^k} = 1/e \neq 0.$

(f) Converges. **(h)** Converges for all $t \neq 2n\pi$, $n \in \mathbb{Z}$. If $t = 2n\pi$, then the series converges for $p > 1$ and diverges for $0 < p \leq 1$. **8.** Use the partial summation formula to prove that

$\sum_{k=1}^{n} ka_k = nA_n - \sum_{k=1}^{n-1} A_k$ where $A_k = \sum_{k=1}^{n} a_k.$

Now use Exercise 14 of Section 3.2.

Exercises 7.3. page 327

2. Use the inequality $|ab| \leq \frac{1}{2}(a^2 + b^2)$, $a, b \in \mathbb{R}$. **4.** Use the hypothesis on $|a_n|$ to show that $s_n = \sum_{k=1}^{n} |a_k| \leq b_1 - b_{n+1}$. **6. (a)** Converges conditionally. **(c)** Converges absolutely for $p > 1$ and conditionally for $0 < p \leq 1$. **(e)** Converges absolutely for $p > 1$, and conditionally for $0 < p \leq 1$. **(g)** Rewrite $a_k = \dfrac{k^k}{(k+1)^k}$ as $a_k = (1 + \frac{1}{k})^{-k}$. Since $\lim_{k\to\infty} a_k = e^{-1}$, the series $\sum (-1)^{k+1} a_k$ diverges. **(i)** By the comparison test (with $\sum 1/k^2$) the series converges absolutely. **9.** First note that $S_{3n} = \sum_{k=0}^{n-1} \left(\frac{1}{3k+1} + \frac{1}{3k+2} - \frac{1}{3k+3} \right)$. Now show that $S_{3n} \to \infty$ as $n \to \infty$.

11. By Theorem, 7.2.6 the series converges. To show that $\sum |\sin k|/k = \infty$, show that for any three consecutive integers, at least one satisfies $|\sin k| \geq \frac{1}{2}$.

Exercises 7.4. page 333

1. (a) $\|\{1/(\ln k)\}\|_2^2 = \sum_{k=2}^{\infty} 1/(\ln k)^2$, which diverges (Exercise 5, Section 7.1).

(c) $\|\{\ln k/\sqrt{k}\}\|_2^2 = \sum_{k=2}^{\infty} (\ln k)^2/k$ which diverges by the Comparison test.

2. (a) $|p| < 1$. **(c)** $p \geq \frac{1}{2}$. **3.** Since $\{1/k\} \in \ell^2$, the result follows by the Cauchy-Schwarz inequality. **9.** If we interpret the vectors $\mathbf{a}$ and $\mathbf{b}$ as forming two sides of a triangle, with the third side given by $\mathbf{b} - \mathbf{a}$, then by the law of cosines $\|\mathbf{b} - \mathbf{a}\|_2^2 = \|\mathbf{b}\|_2^2 + \|\mathbf{a}\|_2^2 - 2\|\mathbf{a}\|_2\|\mathbf{b}\|_2 \cos\theta$. Now apply Exercise 8(e). **11. (b)** Suppose $\{a_k\} \in \ell^1$. Since $\lim_{k\to\infty} a_k = 0$ there exists $k_o \in \mathbb{N}$ such that $|a_k| \leq 1$ for all $k \geq k_o$. But then $|a_k|^2 \leq |a_k|$ for all $k \geq k_o$. Hence $\{a_k\} \in \ell^2$.

Chapter 8

Exercises 8.1. page 343

2. (a) $\lim_{n\to\infty} \dfrac{nx}{1+nx} = \begin{cases} 0, & x = 0, \\ 1, & x > 0. \end{cases}$

(c) $\lim_{n\to\infty} (\cos x)^{2n} = \begin{cases} 0, & x \neq \pm(2k-1)\frac{\pi}{2}, k \in \mathbb{N}, \\ 1, & x = \pm(2k-1)\frac{\pi}{2}, k \in \mathbb{N}. \end{cases}$ **3. (a)** By the root test the

series converges for all $|x| < 2$; diverges for $|x| \geq 2$. **(c)** converges for all $x > 0$; diverges for $x \leq 0$. **5. (c)** $\int_0^1 f_n = \int_0^{1/n} n^2 x \, dx + \int_{1/n}^{2/n} (2n - n^2 x) \, dx = 1/2 + 1/2 =$ 1. **7. (a)** If $x = 0$, $f_n(0) = 0$ for all $n \in \mathbb{N}$. If $x > 0$, then $0 < f_n(x) < x/n$, from which the result follows. **(b)** For each $n \in \mathbb{N}$, $f_n(x)$ has a maximum of e^{-1} at

$x = n$. **8.** Use the fact that for $N, M \in \mathbb{N}$,

$$\sum_{n=1}^{N}\left(\sum_{m=1}^{M} a_{n,m}\right) = \sum_{m=1}^{M}\left(\sum_{n=1}^{N} a_{n,m}\right) \leq \sum_{m=1}^{M}\left(\sum_{n=1}^{\infty} a_{n,m}\right) \leq \sum_{m=1}^{\infty}\left(\sum_{n=1}^{\infty} a_{n,m}\right).$$

The above inequalities hold since $a_{n,m} \geq 0$ for all $n, m \in \mathbb{N}$. Now first let $M \to \infty$, and then $N \to \infty$, to obtain

$$\sum_{n=1}^{\infty}\left(\sum_{m=1}^{\infty} a_{n,m}\right) \leq \sum_{m=1}^{\infty}\left(\sum_{n=1}^{\infty} a_{n,m}\right).$$

The same argument also proves the reverse inequality.

Exercises 8.2.   page 350

2. (b) Suppose $\{f_n\}$ and $\{g_n\}$ converge uniformly to f and g respectively on E. Then $|f_n(x)g_n(x) - f(x)g(x)| \leq |g_n(x)||f_n(x) - f(x)| + |f(x)||g_n(x) - g(x)|$. By hypothesis $|g_n(x)| \leq N$ for all $x \in E$, $n \in \mathbb{N}$. Also, since $|f_n(x)| \leq M$ for all $x \in E$, $n \in \mathbb{N}$, $|f(x)| \leq M$ for all $x \in E$. Therefore
$|f_n(x)g_n(x) - f(x)g(x)| \leq N|f_n(x) - f(x)| + M|g_n(x) - g(x)|$.
Now use the definition of uniform convergence of $\{f_n\}$ and $\{g_n\}$ to show that given $\epsilon > 0$, there exists $n_o \in \mathbb{N}$ such that $|f_n(x)g_n(x) - f(x)g(x)| < \epsilon$ for all $x \in E$ and $n \geq n_o$.   **4.** Find $M_n = \max\{f_n(x) : x \in [0,1]\}$, and show that $M_n \to \infty$.
5. (a) For $x \in [0, a]$, $|f_n(x)| \leq a^n$. If $0 < a < 1$, then $\lim_{n\to\infty} a^n = 0$. Thus given $\epsilon > 0$ there exists $n_o \in \mathbb{N}$ so that $a^n < \epsilon$ for all $n \geq n_o$; that is $|f_n(x)| < \epsilon$ for all $x \in [0, a]$, $n \geq n_o$. Therefore $\{f_n\}$ converges uniformly to 0 on $[0, a]$ whenever $a < 1$.   **(b)** No. Obtain a contradiction to Theorem 8.2.5.   **8. (a)** $\dfrac{1}{k^2 + x^2} \leq \dfrac{1}{k^2}$

for all $x \in \mathbb{R}$. Since $\sum 1/k^2 < \infty$, the series $\sum \dfrac{1}{k^2 + x^2}$ converges uniformly by the Weierstrass M-test.   **(c)** For $x \geq 1$, $k^2 e^{-kx} \leq k^2(1/e)^k$. Since $1/e < 1$, the series $\sum k^2(1/e)^k$ converges.   **9. (a)** $\left|\dfrac{\sin 2kx}{(2k+1)^{3/2}}\right| \leq \dfrac{1}{(2k+1)^{3/2}} \leq C\dfrac{1}{k^{3/2}}$ for all

$x \in \mathbb{R}$. Since $\sum \dfrac{1}{k^{3/2}} < \infty$, the given series converges uniformly for all $x \in \mathbb{R}$ by the Weierstrass M-test.   **(d)** Since $|\sin h| \leq |h|$ we have $|\sin(x/k^p)| \leq 2/k^p$ for $|x| \leq 2$. Since $p > 1$, by the Weierstrass M-test the series converges uniformly and absolutely for $|x| \leq 2$.   **(e)** Hint: Let $S_n(x) = \sum_{k=0}^{n}\left(\dfrac{1}{kx+2} - \dfrac{1}{(k+1)x+2}\right)$.   **10. (a)** For

$x \geq a > 0$, $1 + k^2 x \geq ak^2$. Thus since $\sum 1/k^2 < \infty$, by the Weierstrass M-test the given series converges uniformly on $[a, \infty)$ for every $a > 0$. To show that it does not converge uniformly on $(0, \infty)$, consider $(S_{2n} - S_n)(1/n^2)$, where S_n is the nth partial sum of the series.   **12.** Since $|a_k x^k| \leq |a_k|$ for all $x \in [-1, 1]$ and $\sum |a_k|$ converges, the given series converges absolutely and uniformly by the Weierstrass M-test.

16. Set $A_n = \sum_{k=1}^{n} \sin kt$. By (1) of the proof of Theorem 7.2.6, $|A_n| \leq 1/|\sin \frac{1}{2}t|$. Thus $\{A_n\}$ is uniformly bounded on any closed interval that does not contain an integer multiple of 2π. The conclusion now follows by the Abel partial summation formula.   **18.** Suppose $|F_0(x)| \leq M$ for all $x \in [0,1]$. Show that $|F_n(x)| \leq M\dfrac{x^n}{n!}$ for all $x \in [0,1]$, $n \in \mathbb{N}$. Now use the Weierstrass M-test.

Exercises 8.3.   page 359

1. Show that $\displaystyle\sum_{k=0}^{\infty} x(1-x)^k = \begin{cases} 0, & x = 0, 1, \\ 1, & 0 < x < 1. \end{cases}$ Thus by Corollary 8.3.2 the convergence cannot be uniform on $[0, 1]$.   **4.** Since f is uniformly continuous on $\mathbb{R}$, given $\epsilon > 0$, there exists a $\delta > 0$ such that $|f(x) - f(y)| < \epsilon$ for all $x, y \in \mathbb{R}$, $|x - y| < \delta$. Choose $n_o \in \mathbb{N}$ such that $1/n_o < \delta$. Then for all $n \geq n_o$, $|f(x) - f_n(x)| < \epsilon$ for all $x \in \mathbb{R}$.   **6.** Let $\epsilon > 0$ be given. Since $\{f_n\}$ converges uniformly on D, there exists $n_o \in \mathbb{N}$ such that $|f_n(x) - f_m(x)| < \epsilon$ for all $x \in D$, $n, m \geq n_o$. Use continuity of the functions and the fact that D is dense in E to prove that $|f_n(y) - f_m(y)| \leq \epsilon$ for all $y \in E$, $n, m \geq n_o$.   **9.** Note, $\left(1 + \frac{x}{n}\right)^n = \left[\left(1 + \frac{x}{n}\right)^{\frac{n}{x}}\right]^x$. For $x > 0$, as in Example 3.3.5, the sequence $\left(1 + \frac{x}{n}\right)^n$ is an increasing sequence converging to e^x. Set $g_n(x) = e^x - f_n(x)$ and apply Dini's theorem on $[a, b]$ provided $a \geq 0$. This has to be modified if $a < 0$.   **11. (a)** We first note that $|(T\varphi)(x)| \leq \|\varphi\|_u \int_0^x dt = x\|\varphi\|_u$. It now follows that $|T(T\varphi)(x)| \leq \frac{1}{2}x^2\|\varphi\|_u$ for all $x \in [0, 1]$.

Exercises 8.4.   page 362

1. By the Weierstrass M-test and the hypothesis on $\{a_k\}$, the series $\sum a_k x^k$ converges uniformly on $[0, 1]$. Now apply Corollary 8.4.2.   **4.** Since $f \in \mathcal{R}[0, 1]$, f is bounded on $[0, 1]$, i.e., $|f(x)| \leq M$ for all $x \in [0, 1]$. Now apply the bounded convergence theorem to $g_n(x) = x^n f(x)$ which converges pointwise to $g(x) = 0$, $0 \leq x < 1$, and $f(1)$ when $x = 1$.   **6.** We first note that

$$\left|\int_0^1 f(x^n)dx - f(0)\right| \leq \int_0^1 |f(x^n) - f(0)|dx.$$

Now use the fact that $x^n \to 0$ uniformly on $[0, c]$ for any $c, 0 < c < 1$, and that $\int_c^1 |f(x^n) - f(0)|dx \leq M(1 - c)$ for some constant M.

7. For each $k \in \mathbb{N}$ the function $2^{-k}I(x - r_k)$ is Riemann integrable on $[0, 1]$ with $\int_0^1 2^{-k}I(x - r_k) = 2^{-k}(1 - r_k)$. By the Weierstrass M-test the series converges uniformly on $[0, 1]$. Thus $f \in \mathcal{R}[0, 1]$ with $\int_0^1 f = \sum_{k=1}^{\infty} 2^{-k}(1 - r_k)$.   **9.** By Theorem 6.2.1, $f_n g \in \mathcal{R}[a, b]$ for all $n \in \mathbb{N}$. Show that $\{f_n g\}$ converges uniformly to fg on $[a, b]$, and apply Theorem 8.4.1.   **10.** Since $|f_n(x)| \leq g(x)$ for all $x \in [0, \infty)$, $n \in \mathbb{N}$, the same is true for $|f(x)|$. By Exercise 5, Section 6.4, it now follows that the improper integrals of $f_n, n \in \mathbb{N}$, and f on $[0, \infty)$ converge. Since $\int_0^{\infty} g < \infty$, show that given $\epsilon > 0$, there exists $c \in \mathbb{R}$, $c > 0$, so that $\int_c^{\infty} g < \frac{1}{2}\epsilon$. Now show that

$$\left|\int_0^{\infty} f - \int_0^{\infty} f_n\right| \leq \int_0^c |f - f_n| + 2\int_c^{\infty} g.$$

Use the uniform convergence of $\{f_n\}$ to f on $[0, c]$ to finish the proof.   **11. (b)** To show that $(C[a, b], \|\ \|_1)$ is not complete it suffices to find a sequence $\{f_n\}$ of continuous functions that converges in the norm $\|\ \|_1$ to a Riemann integrable function f that is not continuous.

Exercises 8.5.   page 369

2. By the fundamental theorem of calculus, $f_n(x) = f_n(x_o) + \int_{x_o}^x f_n'(t)dt$ for all $x \in [a, b]$. If $\{f_n'\}$ converges uniformly to g on $[a, b]$, use Theorems 6.3.4 and 8.4.1 to prove that $\{f_n\}$ converges uniformly to a function f on $[a, b]$ with $f'(x) = g(x)$ for all $x \in [a, b]$.   **4.** Let $x \in (a, b)$ be arbitrary, and choose c, d such that $a < c < x < d < b$. Now apply Theorem 8.5.1 to the sequence $\{f_n\}$ on $[c, d]$, to obtain that f is differentiable at x with $f'(x) = \lim_{n \to \infty} f_n'(x)$.   **6. (a)** Use the comparison test

to show that the given series converges for all $x > 0$. Let $S(x) = \sum\limits_{k=1}^{\infty}(1 + kx)^{-2}$
and $S_n(x) = \sum\limits_{k=1}^{n}(1 + kx)^{-2}$. Then $S'_n(x) = -2\sum\limits_{k=1}^{n}k(1 + kx)^{-3}$. Use the Weierstrass
M-test and the comparison test to show that the sequences $\{S_n(x)\}$ and $\{S'_n(x)\}$
converge uniformly on $[a, \infty)$ for every $a > 0$. Thus by Theorem 8.5.1
$$S'(x) = \lim_{n\to\infty} S'_n(x) = -2\sum_{k=1}^{\infty}k(1 + kx)^{-3}$$
for all $x \in [a, \infty)$. Since this holds for every $a > 0$, the results holds for all $x \in (0, \infty)$.
(c) By the root test the given series converges for all x, $|x| < 1$. Let $S(x) = \sum_{k=0}^{\infty}x^k$
and $S_n(x) = \sum_{k=0}^{n}x^k$. Then $S'_n(x) = \sum_{k=0}^{n-1}(k + 1)x^k$. Again by the root test, the
series $\sum_{k=0}^{\infty}(k + 1)x^k$ converges pointwise for all x, $|x| < 1$, and uniformly for all
x, $|x| \leq a$ for any a, $0 < a < 1$. Hence by Theorem 8.5.1
$$S'(x) = \lim_{n\to\infty} S'_n(x) = \sum_{k=0}^{\infty}(k + 1)x^k$$
for all x, $|x| \leq a$. Since this holds for every a, $0 < a < 1$, the result holds for all
x, $|x| < 1$.

Exercises 8.6. page 377

2. Let $\mathcal{P} = \{x_o, x_1, ..., x_n\}$ be a partition of $[a, a + p]$. Set $y_j = x_j - a$. Then
$\mathcal{P}^* = \{y_o, y_1, ..., y_n\}$ is a partition of $[0, p]$. If $t \in [x_{j-1}, x_j]$, then $t = s + p$ for
some $s \in [y_{j-1}, y_j]$. Since f is periodic of period p, $f(t) = f(s + p) = f(s)$. Therefore, $\sup\{f(t) : t \in [x_{j-1}, x_j]\} = \sup\{f(s) : s \in [y_{j-1}, y_j]\}$, and as a consequence
$\mathcal{U}(\mathcal{P}, f) = \mathcal{U}(\mathcal{P}^*, f)$. From this it now follows that $\overline{\int_a^{a+p}} f = \overline{\int_0^p} f$. The proof for the
lower integral is similar. Thus $f \in \mathcal{R}[0, p]$ if and only if $f \in \mathcal{R}[a, a + p]$.
4. (a) $c_n = \frac{1}{2}(n + 1)$.   **6.** Set $A_n(\delta) = \sup\{Q_n(x) : x \in [-\delta, \delta]\}$. Then $0 < \delta_1 < \delta_2$
implies $A_n(\delta_1) \leq A_n(\delta_2)$. Suppose $\overline{\lim\limits_{n\to\infty}} A_n(\delta_1) < \infty$ for some $\delta_1 > 0$. Then there
exists a finite constant C and $n_o \in \mathbb{N}$ such that $A_n(\delta) \leq C$ for all $n \geq n_o$, $0 < \delta \leq \delta_1$.
Use this fact to obtain a contradiction to the hypothesis that $\{Q_n\}$ is an approximate
identity.

Exercises 8.7. page 396

1. (b) $R = 2$.   **(d)** $R = e$.   **(f)** $R = 2$.   **2. (b)** By the root test the series
converges absolutely for all x, $-2 < x < 2/5$, and diverges for all other $x \in \mathbb{R}$.
3. (a) $x/(1 - x)^2$.   **8. (a)** $\dfrac{1}{1 + x^2} = \dfrac{1}{1 - (-x^2)} = \sum\limits_{k=0}^{\infty}(-1)^k x^{2k}$, $|x| < 1$. Use the
previous exercise and the fact that $\arctan x = \int\limits_0^x (1 + t^2)^{-1}dt$ to find the Taylor
series expansion of $\arctan x$ at $c = 0$.   **(c)** Use Theorem 7.2.4.   **10.** Let $P(x) =$
$x^4 + 3x^2 - 2x + 5$ and use Taylor's theorem with $c = 1$.   **12. (b)** By Example
8.7.20(c), $\ln x = \ln(1 + (x - 1)) = \sum\limits_{k=1}^{\infty}\dfrac{(-1)^{k+1}}{k}(x - 1)^k$, which converges for all x, $0 <$
$x \leq 2$.   **(d)** By computation, $f^{(k)}(x) = \frac{1}{2} \cdot \frac{3}{2} \cdots (k - \frac{1}{2})(1 - x)^{-(k+\frac{1}{2})}$. Therefore
the Taylor series expansion of $(1 - x)^{-\frac{1}{2}}$ is given by $1 + \sum\limits_{k=1}^{\infty}\dfrac{\frac{1}{2} \cdot \frac{3}{2} \cdots (k - \frac{1}{2})}{k!}x^k$. For
$-1 < x \leq 0$, use Theorem 8.7.16 to show that $|R_n(x)| \leq \dfrac{1 \cdot 3 \cdots (n + \frac{1}{2})}{(n + 1)!}|x|^{n+1}$.
Use convergence of the series $\sum\limits_{k=1}^{\infty}\dfrac{1 \cdot 3 \cdots (n + \frac{1}{2})}{(n + 1)!}|x|^{n+1}$, $|x| < 1$, to conclude that

$\lim\limits_{n\to\infty} R_n(x) = 0$, $-1 < x \le 0$. If $0 < x < 1$, use Corollary 8.7.19 to show that

$$|R_n(x)| \le \frac{1\cdot 3\cdots(n+\frac{1}{2})}{n!}\ \frac{x}{(1-x)^{3/2}}\ \left(\frac{x-\zeta}{1-\zeta}\right)^n$$

for some ζ, $0 < \zeta < x$. Now use the method of Example 8.7.20(c) to show that $\lim\limits_{n\to\infty} R_n(x) = 0$ for all x, $0 < x < 1$. Thus the series converges to $(1-x)^{-\frac{1}{2}}$ for all x, $|x| < 1$. **(f)** Use the fact that $\arcsin x = \int\limits_0^x \frac{1}{\sqrt{1-t^2}}\,dt$, $|x| < 1$. **(h)** For p real,

$$(1+x)^p = 1 + px + \frac{p(p-1)}{2!}x^2 + \frac{p(p-1)(p-2)}{3!}x^3 + \cdots. \text{ If } p \text{ is a positive integer,}$$
then the expansion is finite.

Exercises 8.8. page 402

1. (a) $\Gamma(\frac{3}{2}) = \Gamma(\frac{1}{2}+1) = \frac{1}{2}\Gamma(\frac{1}{2}) = \frac{1}{2}\sqrt{\pi}$. **2.** Make the change of variable $t = -\ln s$.

3. (a) $\int_0^\infty e^{-t}t^{3/2}\,dt = \Gamma(\frac{5}{2}) = \frac{3}{4}\sqrt{\pi}$. **5. (a)** $\int\limits_0^{\pi/2}(\sin x)^{2n}\,dx = \frac{1}{2}\frac{\Gamma(n+\frac{1}{2})\Gamma(\frac{1}{2})}{\Gamma(n+1)}$.

Chapter 9

Exercises 9.1. page 417

1. $\int\limits_{-1}^1 \phi_1^2 = \int\limits_{-1}^1 1 = 2$ and $\int\limits_{-1}^1 \phi_2^2 = \int\limits_{-1}^1 x^2 = 2/3$. Therefore $c_1 = \frac{1}{2}\int\limits_{-1}^1 \sin\pi x\,dx = 0$ and $c_2 = \frac{3}{2}\int\limits_{-1}^1 x\sin\pi x\,dx = 3/\pi$. Thus by Theorem 9.1.4, $S_2(x) = (3x)/\pi$ gives the best approximation in the mean to $\sin\pi x$ on $[-1,1]$. **3. (a)** $a_1 = 1/2$, $a_2 = 1$, $a_3 = -1/6$. **(b)** $S_2(x) = \frac{2}{\pi} + \frac{60}{\pi^3}(\pi^2 - 12)(x^2 - x + \frac{1}{6})$.

5. (c) $b_n = \frac{2}{\pi}\int\limits_0^\pi x\sin nx\,dx = -\frac{2}{n}\cos n\pi = \frac{2}{n}(-1)^{n+1}$. Therefore

$$x \sim 2\sum_{k=1}^\infty \frac{(-1)^{k+1}}{k}\sin kx. \quad \textbf{6. (c)}\ x \sim \frac{\pi}{2} - \frac{4}{\pi}\sum_{k=0}^\infty \frac{1}{(2k+1)^2}\cos(2k+1)x.$$

12. (a) As in the proof of Theorem 7.4.3, for $\lambda \in \mathbb{R}$, $0 \le \|\mathbf{x} - \lambda\mathbf{y}\|^2 = \|\mathbf{x}\|^2 - 2\lambda\langle\mathbf{x},\mathbf{y}\rangle + \lambda^2\|\mathbf{y}\|^2$. If $\mathbf{y} \ne \mathbf{0}$, take $\lambda = \langle\mathbf{x},\mathbf{y}\rangle/\|\mathbf{y}\|^2$ to derive the inequality.

Exercises 9.2. page 422

4. (a) For the orthogonal system $\{\sin nx\}_{n=1}^\infty$ on $[0,\pi]$, $\int_0^\pi \sin^2 nx\,dx = \frac{1}{2}\pi$. Therefore $b_n = \frac{2}{\pi}\int_0^\pi f(x)\sin nx\,dx$. Thus Parseval's equality for the orthogonal system $\{\sin nx\}$ becomes $\sum\limits_{n=1}^\infty b_n^2 = \frac{2}{\pi}\int\limits_0^\pi f^2(x)\,dx$. **(b) (i)** $\frac{\pi^2}{8}$. **(ii)** $\frac{\pi^2}{6}$. **5.** Use Parseval's equality and the fact that $fg = \frac{1}{2}[(f+g)^2 - f^2 - g^2]$. **6.** Any function that is identically zero except at a finite number of points will satisfy $\int_a^b f(x)\phi_n(x)\,dx = 0$.

Exercises 9.3. page 431

1. (a) If f is even on $[-\pi,\pi]$, then $f(x)\sin nx$ is odd and $f(x)\cos nx$ is even. Thus $b_n = 0$ for all $n = 1, 2, \ldots$ and $a_n = \frac{2}{\pi}\int\limits_0^\pi f(x)\cos nx\,dx$, $n = 0, 1, 2\ldots$.

3. (a) $f(x) \sim \frac{2}{\pi}\sum\limits_{k=1}^\infty \frac{(1-(-1)^k)}{k}\sin kx = \frac{4}{\pi}\sum\limits_{k=0}^\infty \frac{1}{2k+1}\sin(2k+1)x.$

(c) $|x| \sim \dfrac{\pi}{2} - \dfrac{4}{\pi} \sum\limits_{k=0}^{\infty} \dfrac{1}{(2k+1)^2} \cos(2k+1)x.$ (e) $1 + x \sim 1 + 2 \sum\limits_{k=1}^{\infty} \dfrac{(-1)^{k+1}}{k} \sin kx.$

5. (a) $1,\ \dfrac{4}{\pi} \sum\limits_{k=0}^{\infty} \dfrac{1}{2k+1} \sin(2k+1)x.$ (f) $\dfrac{1}{2} - \dfrac{2}{\pi} \sum\limits_{k=0}^{\infty} \dfrac{(-1)^k}{2k+1} \cos(2k+1)x,$

$-\dfrac{2}{\pi} \sum\limits_{n=0}^{\infty} \dfrac{1}{n} \left[(-1)^n - \cos\tfrac{1}{2}n\pi \right] \sin nx.$

6. (c) Since h_e is even, the Fourier series of h_e is the cosine series of h. Therefore

$$a_0 = \dfrac{2c}{\pi} \int\limits_0^{\pi/2} x\,dx + \dfrac{2c}{\pi} \int\limits_{\pi/2}^{\pi} (\pi - x)\,dx = \dfrac{\pi c}{2}, \text{ and}$$

$$a_n = \dfrac{2c}{\pi} \int\limits_0^{\pi/2} x \cos nx\,dx + \dfrac{2c}{\pi} \int\limits_{\pi/2}^{\pi} (\pi - x) \cos nx\,dx = \dfrac{2c}{\pi n^2} \left[2 \cos\tfrac{n\pi}{2} + (-1)^{n+1} - 1 \right].$$

Thus $h_e(x) \sim \dfrac{c\pi}{4} - \dfrac{c}{\pi} \sum\limits_{k=1}^{\infty} \dfrac{1 - (-1)^k}{k^2} \cos 2kx.$ 7. (b) $f(x) \sim \dfrac{2}{\pi} -$

$\dfrac{4}{\pi} \sum\limits_{k=1}^{\infty} \dfrac{1}{(4k^2 - 1)} \cos 2kx.$ 8. By integration by parts, $b_k = -\dfrac{1}{\pi k^2} \int\limits_{-\pi}^{\pi} f''(x) \sin kx\,dx.$

Since $f'' \in \mathcal{R}[-\pi, \pi]$, it is bounded on $[-\pi, \pi]$; i.e., $|f''(x)| \leq M$. Therefore $|b_k| \leq 2M/k^2$. Similarly for a_k. Thus by the Weierstrass M-test, the Fourier series of f converges uniformly on $[-\pi, \pi]$.

Exercises 9.4.   page 442

2. $\dfrac{\pi^4}{90}$.   3. To show that $\{\sin nx\}_{n=1}^{\infty}$ is complete on $[0, \pi]$ it suffices to show that Parseval's equality holds for every $f \in \mathcal{R}[0, \pi]$; i.e., $\sum\limits_{n=1}^{\infty} b_n^2 = \dfrac{2}{\pi} \int\limits_0^{\pi} f^2(x)\,dx.$ To accomplish this, let f_o denote the odd extension of f to $[-\pi, \pi]$. Since f_o is odd, $a_n = 0$ for all $n = 0, 1, 2\ldots$, and $b_n = \dfrac{2}{\pi} \int_0^{\pi} f(x) \sin nx\,dx.$ Since the orthogonal system $\{1, \cos nx, \sin nx\}_{n=1}^{\infty}$ is complete on $[-\pi, \pi]$, $\sum\limits_{n=1}^{\infty} b_n^2 = \dfrac{1}{\pi} \int\limits_{-\pi}^{\pi} f_o^2(x)\,dx = \dfrac{2}{\pi} \int\limits_0^{\pi} f^2(x)\,dx.$

6. Let $S_n(t) = \tfrac{1}{2}a_0 = \sum\limits_{k=1}^{n} a_k \cos kt + b_k \sin kt.$ If $x \in [-\pi, \pi]$, then

$$\left| \int\limits_{-\pi}^{x} f(t)\,dt - \int\limits_{-\pi}^{x} S_n(t)\,dt \right| \leq \int\limits_{-\pi}^{x} |f(t) - S_n(t)|\,dt.$$

Thus by Exercise 5(a) and Theorem 9.4.7, $\lim\limits_{n \to \infty} \int\limits_{-\pi}^{x} S_n(t)\,dt = \int\limits_{-\pi}^{x} f(t)\,dt,$ with the convergence being uniform on $[-\pi, \pi]$. But $\int\limits_{-\pi}^{x} S_n(t)\,dt = \tfrac{1}{2}a_0(x + \pi) +$

$\sum\limits_{k=1}^{n} \left(\dfrac{a_k}{k} \sin kx - \dfrac{b_k}{k}(\cos kx - \cos k\pi) \right),$ from which the result follows.   10. Use Lemma 9.4.8 and the Weierstrass approximation theorem.

Exercises 9.5.   page 453

2. $\dfrac{\pi^2}{12}, \dfrac{\pi^2}{6}.$ 3. (a) $\dfrac{1}{2}.$ (b) $\dfrac{1}{2} - \dfrac{\pi}{4}.$ 6. (a) $\dfrac{(e^{a\pi} - 1)}{a\pi} + \dfrac{2a}{\pi} \sum\limits_{k=1}^{\infty} \dfrac{((-1)^n e^{a\pi} - 1)}{a^2 + k^2} \cos kx.$

(b) On $[-\pi, \pi]$, the series converges to $e^{|ax|}$, and thus to the 2π-periodic extension of $e^{|ax|}$ on all of $\mathbb{R}$.

Chapter 10

Exercises 10.2.   page 474

2. Since U is open and non-empty, there exists $x \in U$ and $r > 0$ such that $(x-r, x+r) \subset U$. Thus by Theorem 10.2.4, $m(U) \geq m((x-r, x+r)) = 2r > 0$.   **3.** Let $V_\epsilon = (-\epsilon, 1+\epsilon)$. Then V_ϵ is an open set containing P. Show that $m(V_\epsilon \setminus P) \geq 1$ for all $\epsilon > 0$ and thus $m(P) < 2\epsilon$ for every $\epsilon > 0$. Alternately, show that $m(P^c \cap [0,1]) = 1$ and use Theorem 10.2.15.   **6.** First show that there exist disjoint bounded open sets U_1, U_2 with $U_1 \supset K_1$ and $U_2 \supset K_2$. Then $m(K_1 \cup K_2) = m(U_1 \cup U_2) - m((U_1 \cup U_2) \setminus (K_1 \cup K_2))$. But $(U_1 \cup U_2) \setminus (K_1 \cup K_2) = (U_1 \setminus K_1) \cup (U_2 \setminus K_2)$. Now use Theorem 10.2.9.

Exercises 10.3.   page 480

1. (b) First show that if U is any open set, then $U + x$ is open and $m(U+x) = m(U)$. Use this and the definition to prove that $\lambda^*(E+x) = \lambda^*(E)$. If K is compact and U is a bounded open set containing K, show that $(U+x) \setminus (K+x) = (U \setminus K) + x$. Use this to show that $m(K+x) = m(K)$ and $\lambda_*(E+x) = \lambda_*(E)$.   **3.** Since $E_1 \cap E_2 \subset E_1$ and $\lambda^*(E_1) = 0$, $\lambda^*(E_1 \cap E_2) = 0$. Thus by Theorem 10.3.5, $E_1 \cap E_2$ is measurable. For $E_1 \cup E_2$ apply Theorem 10.3.9.   **6.** If $\lambda^*(E) < \infty$, then for each $k \in \mathbb{N}$ there exists an open set U_k with $U_k \supset E$ such that $m(U_k) < \lambda^*(E) + \frac{1}{k}$. Now use the fact that $E \subset \bigcap U_n \subset U_k$ for all $k \in \mathbb{N}$.   **8.** Set $E_k = E \cap [-k, k]$, $k \in \mathbb{N}$. Then $\{\lambda_*(E_k)\}$ is monotone increasing with $\lambda_*(E_k) \leq \lambda_*(E)$ for all $k \in \mathbb{N}$. Let $\alpha = \lim_{k \to \infty} \lambda_*(E_k)$. Suppose $\alpha < \lambda_*(E)$. Choose $\beta \in \mathbb{R}$ such that $\alpha < \beta < \lambda_*(E)$. By definition there exists a compact set K with $K \subset E$ such that $m(K) > \beta$. Use this to show that there exists $k_o \in \mathbb{N}$ such that $\lambda_*(E_k) > \beta$ for all $k \geq k_o$, which is a contradiction.

Exercises 10.4.   page 488

2. If E is bounded, the result follows from the definition of $\lambda_*(E)$ and $\lambda^*(E)$, and Theorem 10.4.5(b) (for a finite union). If E is unbounded, let $E_n = E \cap (-n, n)$. Given $\epsilon > 0$, choose U_n open such that $E \subset U_n$ and $\lambda(U_n \setminus E) < \epsilon/2^n$. Let $U = \bigcup U_n$. Show that $U \setminus E \subset \bigcup(U_n \setminus E_n)$. Now use Theorem 10.3.5 to show that $\lambda(U \setminus E) < \epsilon$. To obtain a closed set $F \subset E$ satisfying $\lambda(F \setminus F) < \epsilon$, apply the result for open sets to E^c.   **4.** First show that $\lambda(E_1 \cup E_2) = 1$; then use Theorem 10.4.1.   **6.** If E satisfies $\lambda^*(E \cup T) + \lambda^*(E^c \cup T) = \lambda^*(T)$ for every $T \subset \mathbb{R}$, then E satisfies Theorem 10.4.2 and thus is measurable. Conversely, suppose E is measurable and $T \subset \mathbb{R}$. If $\lambda^*(T) = \infty$, the result is true. Assume $\lambda^*(T) < \infty$. Let $\epsilon > 0$ be arbitrary. Then there exists an open set $U \supset T$ such that $\lambda(U) < \lambda^*(T) + \epsilon$. Since E and U are measurable, $E \cap U$ and $E^c \cap U$ are disjoint measurable sets with $(E \cap U) \cup (E^c \cap U) = U$. Furthermore, $E \cap U \supset E \cap T$ and $E^c \cap U \supset E^c \cap T$. Thus by Theorem 10.3.9
$$\lambda^*(T) \leq \lambda^*(E \cap T) + \lambda^*(E^c \cap T) \leq \lambda(E \cap U) + \lambda(E^c \cap U) = \lambda(U) < \lambda^*(T) + \epsilon.$$
Since the above holds for every $\epsilon > 0$, we have $\lambda^*(E \cap T) + \lambda^*(E^c \cap T) = \lambda^*(T)$.

Exercises 10.5.   page 494

1. $\{x : f(x) > c\} = \begin{cases} [0, 1], & \text{if } c < 0, \\ (0, 1], & \text{if } 0 \leq c < 1, \\ (0, \frac{1}{c}) \cup \{1\}, & \text{if } 1 \leq c < 2, \\ (0, \frac{1}{c}), & \text{if } 2 \leq c. \end{cases}$

5. If $c > 0$, then $\{x : 1/g(x) > c\} = \{x : g(x) > 0\} \cap \{x : g(x) < 1/c\}$. Since g is measurable, each of the sets $\{g(x) > 0\}$ and $\{g(x) < 1/c\}$ is measurable. Thus their intersection is measurable. The case $c < 0$ is treated similarly.   **7.** If f is continuous on $[a, b]$, then $f^{-1}((s, \infty))$ is open in $[a, b]$ for every $s \in \mathbb{R}$. Thus for a fixed s, $f^{-1}((s, \infty)) = U \cap [a, b]$ where U is open in $\mathbb{R}$. Since both U and $[a, b]$ are measurable, so is $f^{-1}((s, \infty))$, i.e., f is measurable.   **10. (a)** If $c \geq 0$, then $\{x : f^+(x) > c\} = \{x : f(x) > c\}$, and if $c < 0$, then $\{x : f^+(x) > c\} = E$. Since each of the sets $\{f(x) > c\}$ and E are measurable, f^+ is measurable.   **(c)** Not in general. If E is a non-measurable set, consider the function that is 1 on E and -1 on E^c.   **12.** Since both x^n and f are measurable, by Theorem 10.5.4 their product is measurable. Suppose $|f(x)| \leq M$ for all $x \in [0, 1]$, then $|f_n(x)| \leq M x^n$ from which the result now follows.   **14.** Since f is differentiable on $[a, b]$,

$$f'(x) = \lim_{n \to \infty} n[f(x + \tfrac{1}{n}) - f(x)]$$

for all $x \in [a, b]$. For each $n \in \mathbb{N}$, $g_n(x) = n[f(x + \tfrac{1}{n}) - f(x)]$ is measurable (Justify). Thus by Corollary 10.5.10, the function f' is measurable.   **15.** First show that given $\epsilon, \delta > 0$, there exists a measurable set $E \subset [a, b]$ and $n_o \in \mathbb{N}$ such that $\lambda([a, b] \setminus E) < \epsilon$ and $|f_n(x) - f(x)| < \delta$ for all $x \in E$ and $n \geq n_o$. To accomplish this, for each $k \in \mathbb{N}$ consider

$$A_k = \{x : |f_n(x) - f(x)| < \delta \text{ for all } n \geq k\}.$$

Now show that $\lim\limits_{k \to \infty} \lambda(A_k^c) = 0$. Here $A_k^c = [a, b] \setminus A_k$. Complete the proof of Egorov's theorem as follows: By the above, for each $k \in \mathbb{N}$, there exists a measurable set E_k and an integer n_k such that $\lambda(E_k^c) < \epsilon/2^k$ and $|f(x) - f_n(x)| < 1/k$ for all $x \in E_k$ and $n \geq n_k$. The set $E = \bigcap E_k$ will have the desired properties.

Exercises 10.6.   page 506

1. For each $n \in \mathbb{N}$, let $\varphi_n = \sum\limits_{j=1}^{n} (m + (j - 1)\tfrac{\beta}{n}) \chi_{E_j}$. Then φ_n is a simple function on $[a, b]$ with $\int_a^b \varphi_n d\lambda = S_n(f)$. Furthermore, for each $x \in [a, b]$, $0 \leq f(x) - \varphi_n(x) \leq \beta/n$. Therefore $\lim\limits_{n \to \infty} \varphi_n(x) = f(x)$ for all $x \in [a, b]$. Now apply the bounded convergence theorem.   **3.** Suppose $|f| \leq M$. Then $|\int_A f d\lambda| \leq \int_A |f| d\lambda \leq M \lambda(A)$, which proves the result.   **5.** By Theorem 10.6.10(b), $\int_F f d\lambda = \int_E f d\lambda + \int_{F \setminus E} f d\lambda \geq \int_E f d\lambda$.

7. For each $n \in \mathbb{N}$, let $E_n = \{x : f(x) > \tfrac{1}{n}\}$. Then $\bigcup E_n = \{x : f(x) > 0\}$. Use the previous exercise to show $\lambda(E_n) = 0$. Now use Theorem 10.4.5.   **12.** The function φ_n defined in the solution to Exercise 1 satisfies $|f(x) - \varphi_n(x)| < \beta/n$ for all $x \in [a, b]$. Thus $\{\varphi_n\}$ converges uniformly to f on $[a, b]$.   **15. (b)** Suppose first that $\varphi = \chi_A$ where A is a measurable subset of $[a, b]$. By Exercise 2, Section 10.4, there exists an open set $U \supset A$ such that $\lambda(U \setminus A) < \epsilon/2$. Use the set U to show that there exists a finite number of disjoint closed intervals $\{J_n\}_{n=1}^{N}$ such that

$$V = \bigcup\nolimits_{n=1}^{N} J_n \subset U \text{ and } \lambda(U \setminus V) < \epsilon/2. \text{ Let } h = \sum_{n=1}^{N} \chi_{J_n}. \text{ Then } h \text{ is a step function}$$

on $[a, b]$ and $\{x : h(x) \neq \varphi(x)\} \subset (U \setminus V) \cup (U \setminus A)$. If $\varphi = \sum\limits_{j=1}^{n} \alpha_j \chi_{A_j}$, where the A_j are disjoint measurable subsets of $[a, b]$, approximate each χ_{A_j} by a step function h_j which agrees with χ_{A_j} except on a set of measure less than ϵ/n.

Exercises 10.7.   page 516

2. (a) Assume first that A_1 and A_2 are bounded measurable sets. For each $n \in \mathbb{N}$, set $f_n = \min\{f, n\}$. By Theorem 10.7.4(b)

$$\int_{A_1 \cup A_2} f_n d\lambda = \int_{A_1} f_n d\lambda + \int_{A_2} f_n d\lambda.$$

Since each of the sequences $\{\int_{A_i} f_n\}_{n=1}^{\infty}$, $i = 1, 2$ are monotone increasing, they converge either to a finite number, or to ∞. In either case,
$$\int_{A_1 \cup A_2} f\, d\lambda = \lim_{n \to \infty} \int_{A_1 \cup A_2} f_n d\lambda = \lim_{n \to \infty} \int_{A_1} f_n d\lambda + \int_{A_2} f_n d\lambda = \int_{A_1} f\, d\lambda + \int_{A_2} f\, d\lambda.$$
If either A_1 or A_2 is unbounded, consider the integral of f over $(A_1 \cup A_2) \cap [-n, n]$, and use the above. **4.** For $f_p(x) = x^{-p}$, $x \in (0,1)$, $f_n(x) = \min\{f_p(x), n\} = \begin{cases} n, & 0 < x < n^{-1/p}, \\ x^{-p}, & n^{-1/p} \le x < 1. \end{cases}$ Therefore

$$\int_0^1 f_n d\lambda = \int_0^{n^{-1/p}} n\, dx + \int_{n^{-1/p}}^1 x^{-p} dx = \frac{1}{n^{(1-p)/p}} + \frac{1}{1-p}\left[1 - \frac{1}{n^{(1-p)/p}}\right] = $$
$$\frac{1}{1-p}\left[1 - \frac{p}{n^{(1-p)/p}}\right].$$

Since $(1-p) > 0$, $\lim_{n \to \infty} \int_0^1 f_n d\lambda = 1/(1-p)$. **7.** $0 \le \sum_{n=1}^{N} n\lambda(E_n) \le \sum_{n=1}^{N} \int_{E_n} f\, d\lambda$

$$= \int_{\cup_{n=1}^{N} E_n} f\, d\lambda \le \int_a^b f\, d\lambda < \infty.$$ **9.** Justify first why f can be assumed to be non-

negative and then use Definition 10.7.1. **10.** Let $A_n = A \cap [-n, n]$. By definition, $\int_A f\, d\lambda = \lim_{n \to \infty} \int_{A_n} f\, d\lambda$. Since f is integrable, given $\epsilon > 0$, there exists $n \in \mathbb{N}$ such that $0 \le \int_A f\, d\lambda - \int_{A_n} f\, d\lambda < \epsilon$. By Theorem 10.7.4(b), $\int_A f\, d\lambda - \int_{A_n} f\, d\lambda = \int_{A \setminus A_n} f\, d\lambda$. Thus $E = A_n$ is the desired set. **13.** Let $f_n = \min\{|f|, n\}$. Then $0 \le f_n \le |f|$. Since f is integrable, by Lebesgue's dominated convergence theorem $\lim_{n \to \infty} \int_A f_n\, d\lambda = \int_A |f|\, d\lambda$. Therefore, given $\epsilon > 0$, there exists $n \in \mathbb{N}$ such that $\int_A (|f| - f_n)\, d\lambda < \epsilon/2$. In particular, if E is any measurable subset of A, $\int_E |f|\, d\lambda < \int_E f_n d\lambda + \frac{1}{2}\epsilon$. But if $\lambda(E) < \infty$, $\int_E f_n d\lambda \le n\lambda(E)$. Hence choose $\delta > 0$ so that $n\delta < \epsilon/2$. **15.** For $n \in \mathbb{N}$ set $h_n(t, x) = n\left[\sin((t + \frac{1}{n})f(x)) - \sin(tf(x))\right]$. Show that $|h_n(t, x)| \le 2|f(x)|$ and apply Lebesgue's dominated convergence theorem. **16.** First prove the result for the characteristic function of an interval. Then use Exercise 9 above and Exercise 15 of Section 10.6. **19.** Since $\{f_n\}$ is monotone increasing on A, $f(x) = \lim_{n \to \infty} f_n(x)$ exists, either as a finite number or as ∞, for every $x \in A$. By Fatou's lemma, $\int_A f\, d\lambda \le \underline{\lim} \int_A f_n d\lambda$. On the other hand, since $f_n \le f$ for all $n \in \mathbb{N}$, $\overline{\lim} \int_A f_n d\lambda \le \int_A f\, d\lambda$. Combining the two inequalities proves the result. **21. (a)** Use the monotone convergence theorem. **23.** Hint: $|f(x)| \ge (1/x)|\cos 1/x^2| - 2x \ge x^{-1} - 2x$ on each of the intervals $((2n + \frac{1}{3})\pi)^{-\frac{1}{2}} \le x \le ((2n - \frac{1}{3})\pi)^{-\frac{1}{2}}$.

Exercises 10.8. page 526

1. $0 < p < \frac{1}{2}$. **4.** Since $|f + g|^2 \le 2(|f|^2 + |g|^2)$, the function $f + g \in \mathcal{L}^2(A)$. Assume $\|f + g\|_2 \neq 0$. Then
$$\|f + g\|_2^2 = \int_A |f + g|^2 \le \int_A |f + g||f| + \int_A |f + g||g|$$
which by the Cauchy Schwarz inequality $\le \|f + g\|_2 \|f\|_2 + \|f + g\|_2 \|g\|_2$. The result follows upon simplification. **7.** Use the Cauchy-Schwarz inequality. **9. (a)** By Bessel's inequality it is the Fourier series of an $\mathcal{L}^2([0, \pi])$ function. **13.** Let $f(x) = 1/\ln x$, $x \in (1, \infty)$. Since $f''(x) > 0$ for all $x \in (1, \infty)$, f is convex on $(1, \infty)$ (see Miscellaneous Exercise 3, Chapter 5). Therefore $f(\frac{1}{2}x + \frac{1}{2}y) \le \frac{1}{2}f(x) + \frac{1}{2}f(y)$ for all $x, y \in (1, \infty)$. Since $n = \frac{1}{2}(n - 1) + \frac{1}{2}(n + 1)$ the inequality follows.

Index

De Morgan's laws, 4, 42
decreasing, 167
Dedekind, Richard, 1
dense, 31, 65
denumerable, 38
derivative, 183
differentiable, 183
Dini's theorem, 355
Dini, Ulisse, 450
Dirac sequence, 372
Dirichlet kernel, 433
Dirichlet test, 316, 351
Dirichlet's theorem, 445
Dirichlet, Peter Lejeune, 315, 407, 454
discontinuity of first kind, 165
discontinuity of second kind, 165
disjoint, 3
distance, 333
distance from a point to a set, 158
distance function, 53
distribution function, 172
distributive, 21
distributive laws, 4, 42
diverge, 120
divergent improper integral, 258, 260
divergent sequence, 84
divergent series, 302
diverges to ∞, 100, 206
diverges to infinity, 100, 302
domain, 7

Egorov's theorem, 495
element of, 2
empty set, 2
enumerable, 38
enumeration, 39
equal, 3, 5
equal almost everywhere, 491
equivalence class, 125
equivalent functions, 519
equivalent sequences, 125
equivalent sets, 37
error, 283
error function, 388
Euclid, 334
euclidean distance, 53
euclidean length, 80
euclidean norm, 54
Euler's constant, 315

Euler's number e, 99
Euler, Leonhard, 99, 144, 220, 280, 297, 398
even extension, 430
even function, 247, 425
exponential function, 257

factorial, 19, 91
family, 2
Fatou's lemma, 511
Fejér kernel, 436, 437
Fejér's theorem, 438
Fejér, L., 438, 452
Fermat, Pierre de, 297
field, 21
finite expansion, 35
finite set, 38
first derivative test, 199
first order method, 283
fixed point, 162
Fourier coefficient, 415, 424, 523
Fourier cosine coefficient, 430
Fourier cosine series, 430
Fourier series, 415, 424
Fourier sine coefficient, 430
Fourier sine series, 430
Fourier, Joseph, 407, 454
fourth order method, 288
function, 6
fundamental theorem of calculus, 249, 251
fundamental theorem of calculus for the Lebegue integral, 507

Gödel, Kurt, 47
Galileo, 37
Gamma function, 263, 398
geometric series, 120
Gibbs phenomenon, 455
Gibbs, Josiah, 455
graph, 7
greatest integer, 166
greatest lower bound, 24
Greatest Lower Bound Property, 25
Gregory, James, 339

half-closed interval, 27
half-open interval, 27
harmonic series, 307

Printed in the United States
By Bookmasters